Lie and non-Lie Symmetries: Theory and Applications for Solving Nonlinear Models

Special Issue Editor

Roman M. Cherniha

MDPI • Basel • Beijing • Wuhan • Barcelona • Belgrade

Special Issue Editor
Roman M. Cherniha
National Academy of Sciences of Ukraine
Ukraine

Editorial Office
MDPI AG
St. Alban-Anlage 66
Basel, Switzerland

This edition is a reprint of the Special Issue published online in the open access journal *Symmetry* (ISSN 2073-8994) from 2015–2017 (available at: http://www.mdpi.com/journal/symmetry/special_issues/lie_theory).

For citation purposes, cite each article independently as indicated on the article page online and as indicated below:

Author 1, Author 2. Article title. *Journal Name*. **Year**. Article number/page range.

First Edition 2017

ISBN 978-3-03842-526-7 (Pbk)
ISBN 978-3-03842-527-4 (PDF)

Table of Contents

Chapter 3

Chapter 4

Chapter 5

Chapter 6

About the Special Issue Editor

Roman M. Cherniha graduated (with honours) in Mathematics from the Taras Shevchenko Kyiv State University (1981), completed his PhD (1987) and habilitation (2003) dissertations at the Institute of Mathematics, NAS of Ukraine. Since 2012, he has been Professor in Mathematics. During his early career, Roman gained substantial experience in the field of Applied Mathematics and Physics at Institute of Technical Heat Physics (Kyiv). Since 1992, he has held a permanent research position at the Institute of Mathematics. He spent a few years abroad working at the Universite Henri Poincare Nancy I (a temporary CNRS position) and the University of Nottingham (Marie Curie Research Fellow). Roman has a wide range of research interests including non-linear partial differential equations (especially reaction-diffusion equations): Lie and conditional symmetries, exact solutions and their properties; development of new methods for analytical solving non-linear PDEs; application of modern methods for analytical solving nonlinear boundary-value problems, arising in real-world application; analytical and numerical solving boundary-value problems with free boundaries; development of mathematical models describing the specific processes arising in physics, biology and medicine. Professor R. Cherniha also has substantial pedagogical experience, in particular, at the National University 'Kyiv Mohyla Academy', where he was professor in Mathematics for five years. He was the supervisor for five PhD students. Roman is the co-author of the monograph Nonlinear Reaction-Diffusion Systems—Conditional Symmetry, Exact Solutions and Their Applications in Biology (Springer, Lecture Notes in Mathematics, vol. 2196, 2017).

Preface to "Lie and non-Lie Symmetries: Theory and Applications for Solving Nonlinear Models"

Nowadays, the most powerful methods for construction of exact solutions to nonlinear partial differential equations (PDEs) are symmetry-based methods. These methods originated from the Lie method, which was created by the prominent Norwegian mathematician Sophus Lie in the second half of the 19th century. The method was essentially developed using modern mathematical language by L.V. Ovsiannikov, G. Bluman, N. Ibragimov, W.F. Ames and some other researchers in the 1960s and 1970s. Although the technique of the Lie method is well known, the method still attracts the attention of many researches and new results are published on a regular basis.

However, it is well known that the Lie method is not efficient for solving PDEs with a 'poor' Lie symmetry (i.e., their maximal algebra of invariance is trivial). Thus, other symmetry-based methods that use non-Lie symmetries (conditional symmetry, weak symmetry, nonlocal symmetry, generalized conditional symmetry etc.) have been developed over the last decades. The best known among them is the method of nonclassical symmetries proposed by G.Bluman and J.Cole in 1969. Notwithstanding this approach was suggested almost 50 years ago, it was successfully applied in searching for nonclassical symmetries (Q-conditional symmetries) of nonlinear equations only in the 1990s–2000s. Moreover, one may say that progress is still modest concerning the successful application of symmetry-based methods (except the Lie method) to systems of PDEs and multidimensional PDEs, especially to those arising in real-world applications.

Another hot topic is the application of symmetry-based methods for solving nonlinear boundary-value problems (BVPs). One may note that symmetry-based methods have not been widely used for solving boundary-value problems (initial value problems belong to such problems as particular cases), although the first rigorous definitions of Lie invariance for BVPs were formulated in the 1970s. The obvious reason for this lack of application is highlighted in the following observation: the relevant boundary and initial conditions are usually not invariant under any transformations, i.e. they do not admit any symmetry of the governing PDE of the problem in question.. However, it was shown very recently that there are new classes of nonlinear BVPs (including multidimensional BVPs), which possess non-trivial Lie and/or non-Lie symmetry. As a result, such BVPs can be reduced, simplified and exactly solved (additional restrictions are usually needed).

This book is a collection of the papers published in the journal Symmetry within two Special Issues: Lie Theory and Its Applications and Lie and Conditional Symmetries and Their Applications for Solving Nonlinear Models, for which I served as the Guest Editor in 2015–2017. The book consists of six chapters. Each chapter contains papers that are devoted to similar topics and/or methods.

The papers forming Chapter 1 are devoted to searching for Lie symmetries and their applications for nonlinear differential equations (including ODEs) that arise in the modelling of some real-world processes (especially in biology).

Chapter 2 consists of papers concerned with different aspects of non-classical (Q-conditional) and generalized conditional symmetries. In particular, the above-mentioned symmetries are constructed for nonlinear reaction-diffusion systems and wave equations with variable coefficients.

The papers belonging to Chapter 3 are devoted to the search for and application of Lie and Q-conditional symmetries for nonlinear BVPs (including multidimensional problems). Notably, new definitions of conditional invariance of a given BVP with a wide range of boundary conditions are proposed and applied to reduce and solve some nonlinear real-world models.

Chapter 4 consists of papers in which some Lie algebras of infinite or very high dimensionality are studied. It transpires that such Lie algebras are related to some physical processes (such as phase transitions) and generate Lie symmetries of physically interesting PDEs.

The papers forming Chapter 5 can be grouped into two subsets. The first three papers are devoted to the classical problem: How is symmetry of a given PDE related to its integrability? The last two papers discuss another well-known problem: How can conservation laws be identified using symmetry of the relevant equation?

Finally, Chapter 6 contains papers devoted to some theoretical aspects of Lie algebras, which are common in Lie symmetry analysis of differential equations.

Roman M. Cherniha
Special Issue Editor

Chapter 1:

symmetry

Article

Symmetries, Lagrangians and Conservation Laws of an Easter Island Population Model

M.C. Nucci * and G. Sanchini

Dipartimento di Matematica e Informatica, Università degli Studi di Perugia, 06123 Perugia, Italy
giampaolo.sanchini@istruzione.it
* Author to whom correspondence should be addressed; nucci@unipg.it

Academic Editor: Roman M. Cherniha
Received: 6 June 2015; Accepted: 19 August 2015; Published: 8 September 2015

Abstract: Basener and Ross (2005) proposed a mathematical model that describes the dynamics of growth and sudden decrease in the population of Easter Island. We have applied Lie group analysis to this system and found that it can be integrated by quadrature if the involved parameters satisfy certain relationships. We have also discerned hidden linearity. Moreover, we have determined a Jacobi last multiplier and, consequently, a Lagrangian for the general system and have found other cases independently and dependently on symmetry considerations in order to construct a corresponding variational problem, thus enabling us to find conservation laws by means of Noether's theorem. A comparison with the qualitative analysis given by Basener and Ross is provided.

Keywords: lie group analysis; Jacobi last multiplier; Lagrangians; Noether's theorem; Easter Island population

1. Introduction

Lie group analysis has been applied in a multitude of physics problems for more than a century, but rarely in biology, maybe because the ordinary differential equations studied in these fields are generally of first order, in contrast with those in physics, which are usually of second order. Yet, when Lie group analysis is successfully applied to biology models, then several instances of integrability, even linearity, are found, which lead to the general solution of the model [1–6].

In [7], a mathematical model that describes the dynamics of growth and sudden decrease in the population of Easter Island was proposed. The model is a nonlinear system of two first-order ordinary differential equations, *i.e.*,

$$\dot{w}_1 = cw_1\left(1 - \tfrac{w_1}{K}\right) - hw_2, \tag{1}$$

$$\dot{w}_2 = aw_2\left(1 - \tfrac{w_2}{w_1}\right), \tag{2}$$

where w_1 is the amount of resources, R in the original notation, w_2 that of population, P in the original notation, c is the growth rate of the resources, K the carrying capacity, h the harvesting constant and a the growth rate of the population. More details on the construction of this model can be found in [7], where numerical solutions of system (1)–(2) were given and a qualitative analysis of the general behavior of solutions both in finite and infinite time was presented.

In this paper, we apply Lie group analysis to an equivalent second-order equation that can be easily derived from system (1)–(2) and find that it can either be integrated by quadrature or reduced to a linear equation if the involved parameters satisfy certain relationships. We also determine a Jacobi last multiplier for system (1)–(2) in order to construct the corresponding Lagrangians (*i.e.*, variational problems) for both system (1)–(2), and the equivalent second-order equation [8]. We find other Jacobi last multipliers and, consequently, Lagrangians independently and dependently on symmetry considerations. Then, we apply Noether's theorem [9] in order to find conservation laws. Finally, we compare our results to the qualitative analysis in [7].

2. Jacobi Last Multiplier and Its Properties

In this section, we recall the definition and properties of the Jacobi last multiplier, its connection to Lie symmetries and to Lagrangians (namely, calculus of variations).

The method of the Jacobi last multiplier [10,11] (an English translation of [11] is available in [12]) provides a means to determine all of the solutions of the partial differential equation:

$$\mathcal{A}f = \sum_{i=1}^{n} a_i(x_1, \cdots, x_n)\frac{\partial f}{\partial x_i} = 0 \tag{3}$$

or its equivalent associated Lagrange's system:

$$\frac{\mathrm{dx_1}}{a_1} = \frac{\mathrm{dx_2}}{a_2} = \ldots = \frac{\mathrm{dx_n}}{a_n}. \tag{4}$$

In fact, if one knows the Jacobi last multiplier and all, but one of the solutions, namely $n-2$ solutions, then the last solution can be obtained by a quadrature. The Jacobi last multiplier M is given by:

$$\frac{\partial(f, \omega_1, \omega_2, \ldots, \omega_{n-1})}{\partial(x_1, x_2, \ldots, x_n)} = M\mathcal{A}f, \tag{5}$$

where

$$\frac{\partial(f, \omega_1, \omega_2, \ldots, \omega_{n-1})}{\partial(x_1, x_2, \ldots, x_n)} = \det\begin{bmatrix} \frac{\partial \mathrm{f}}{\partial \mathrm{x_1}} & \cdots & \frac{\partial \mathrm{f}}{\partial \mathrm{x_n}} \\ \frac{\partial \omega_1}{\partial \mathrm{x_1}} & & \frac{\partial \omega_1}{\partial \mathrm{x_n}} \\ \vdots & & \vdots \\ \frac{\partial \omega_{\mathrm{n}-1}}{\partial \mathrm{x_1}} & \cdots & \frac{\partial \omega_{\mathrm{n}-1}}{\partial \mathrm{x_n}} \end{bmatrix} = 0 \tag{6}$$

and $\omega_1, \ldots, \omega_{n-1}$ are $n-1$ solutions of (3) or, equivalently, first integrals of (4) independent of each other. This means that M is a function of the variables $(x_1, \ldots, x_n)$ and depends on the chosen $n-1$ solutions, in the sense that it varies as they vary. The essential properties of the Jacobi last multiplier are:

(a) If one selects a different set of $n-1$ independent solutions $\eta_1, \ldots, \eta_{n-1}$ of Equation (3), then the corresponding last multiplier N is linked to M by the relationship:

$$N = M\frac{\partial(\eta_1, \ldots, \eta_{n-1})}{\partial(\omega_1, \ldots, \omega_{n-1})}.$$

(b) Given a non-singular transformation of variables:

$$\tau : (x_1, x_2, \ldots, x_n) \longrightarrow (x'_1, x'_2, \ldots, x'_n),$$

then the last multiplier M' of $\mathcal{A}'F = 0$ is given by:

$$M' = M\frac{\partial(x_1, x_2, \ldots, x_n)}{\partial(x'_1, x'_2, \ldots, x'_n)},$$

where M obviously comes from the $n-1$ solutions of $\mathcal{A}F = 0$, which correspond to those chosen for $\mathcal{A}'F = 0$ through the inverse transformation τ^{-1}.

(c) One can prove that each multiplier M is a solution of the following linear partial differential equation:

$$\sum_{i=1}^{n} \frac{\partial(Ma_i)}{\partial x_i} = 0; \tag{7}$$

and *vice versa*, every solution M of this equation is a Jacobi last multiplier.

(d) If one knows two Jacobi last multipliers M_1 and M_2 of Equation (3), then their ratio is a solution ω of (3) or, equivalently, a first integral of (4). Naturally, the ratio may be quite trivial, namely a constant; *vice versa*, the product of a multiplier M_1 times any solution ω yields another last multiplier $M_2 = M_1\omega$.

Since the existence of a solution/first integral is consequent upon the existence of symmetry, an alternative formulation in terms of symmetries was provided by Lie [13,14]. A clear treatment of the formulation in terms of solutions/first integrals and symmetries is given by Bianchi [15]. If we know $n-1$ symmetries of (3)/4), say:

$$\Gamma_i = \sum_{j=1}^{n} \xi_{ij}(x_1, \cdots, x_n)\partial_{x_j}, i = 1, n-1, \tag{8}$$

a Jacobi last multiplier is given by $M = \Delta^{-1}$, provided that $\Delta \neq 0$, where:

$$\Delta = \det \begin{bmatrix} \mathrm{a}_1 & \cdots & \mathrm{a}_\mathrm{n} \\ \xi_{1,1} & & \xi_{1,\mathrm{n}} \\ \vdots & & \vdots \\ \xi_{\mathrm{n}-1,1} & \cdots & \xi_{\mathrm{n}-1,\mathrm{n}} \end{bmatrix}. \tag{9}$$

There is an obvious corollary to the results of the Jacobi mentioned above. In the case that there exists a constant multiplier, the determinant is a first integral. This result is potentially very useful in the search for first integrals of systems of ordinary differential equations. In particular, if each component of the vector field of the equation of motion is missing, the variable associated with that component, *i.e.*, $\partial a_i/\partial x_i = 0$, the last multiplier is a constant, and any other Jacobi last multiplier is a first integral.

Another property of the Jacobi last multiplier is its (almost forgotten) relationship with the Lagrangian, $L = L(t, x, \dot{x})$, for any second-order equation:

$$\ddot{x} = \phi(t, x, \dot{x}), \tag{10}$$

namely [11,16]:

$$M = \frac{\partial^2 L}{\partial \dot{x}^2}, \tag{11}$$

where $M = M(t, x, \dot{x})$ satisfies the following equation:

$$\frac{\mathrm{d}}{\mathrm{d}t}(\log M) + \frac{\partial \phi}{\partial \dot{x}} = 0. \tag{12}$$

Then, Equation (10) becomes the Euler–Lagrange equation:

$$-\frac{\mathrm{d}}{\mathrm{d}t}\left(\frac{\partial L}{\partial \dot{x}}\right) + \frac{\partial L}{\partial x} = 0. \tag{13}$$

The proof is given by taking the derivative of (13) by $\dot{x}$ and showing that this yields (12). If one knows a Jacobi last multiplier, then L can be obtained by a double integration, *i.e.*:

$$L = \int \left(\int M \, \mathrm{d}\dot{x} \right) \mathrm{d}\dot{x} + \ell_1(t,x)\dot{x} + \ell_2(t,x), \tag{14}$$

where ℓ_1 and ℓ_2 are functions of t and x, which have to satisfy a single partial differential equation related to (10) [17]. As was shown in [17], ℓ_1, ℓ_2 are related to the gauge function $F = F(t,x)$. In fact, we may assume:

$$\begin{aligned} \ell_1 &= \tfrac{\partial F}{\partial x} \\ \ell_2 &= \tfrac{\partial F}{\partial t} + \ell_3(t,x) \end{aligned} \tag{15}$$

where ℓ_3 has to satisfy the mentioned partial differential equation and F is obviously arbitrary.

In [18], it was shown that a system of two first-order ordinary differential equations:

$$\begin{aligned} \dot{u}_1 &= \phi_1(t,u_1,u_2) \\ \dot{u}_2 &= \phi_2(t,u_1,u_2) \end{aligned} \tag{16}$$

always admits a linear Lagrangian of the form:

$$L = U_1(t,u_1,u_2)\dot{u}_1 + U_2(t,u_1,u_2)\dot{u}_2 - V(t,u_1,u_2). \tag{17}$$

The key is a function W, such that:

$$W = -\frac{\partial U_1}{\partial u_2} = \frac{\partial U_2}{\partial u_1} \tag{18}$$

and

$$\frac{\mathrm{d}}{\mathrm{d}t}(\log W) + \frac{\partial \phi_1}{\partial u_1} + \frac{\partial \phi_2}{\partial u_2} = 0. \tag{19}$$

It is obvious that Equation (19) is Equation (7) of the Jacobi last multiplier for system (16), as it was point out in [8]. Therefore, once a Jacobi last multiplier $M(t,u_1,u_2)$ has been found, then a Lagrangian of system (16) can be obtained by two integrations, *i.e.*,

$$L = \left(\int M \, \mathrm{d}u_1 \right) \dot{u}_2 - \left(\int M \, \mathrm{d}u_2 \right) \dot{u}_1 + g(t,u_1,u_2) + \frac{\mathrm{d}}{\mathrm{d}t} G(t,u_1,u_2), \tag{20}$$

where $g(t,u_1,u_2)$ satisfies two linear differential equations of first order that can be always integrated and $G(t,u_1,u_2)$ is the arbitrary gauge function that should be taken into consideration in order to apply correctly Noether's theorem [9]. If a Noether symmetry:

$$\Gamma = \xi(t,u_1,u_2)\partial_t + \eta_1(t,u_1,u_2)\partial_{u_1} + \eta_2(t,u_1,u_2)\partial_{u_2} \tag{21}$$

exists for the Lagrangian L in (20), then a first integral of system (16) is:

$$-\xi L - \frac{\partial L}{\partial \dot{u}_1}(\eta_1 - \xi \dot{u}_1) - \frac{\partial L}{\partial \dot{u}_2}(\eta_2 - \xi \dot{u}_2) + G(t,u_1,u_2). \tag{22}$$

We underline that $\dot{u}_1$ and $\dot{u}_2$ always disappear from the expression of the first integral (22) thanks to the linearity of the Lagrangian (20) and formula (20).

3. Lie Symmetries of System (1)–(2)

It is well known to practitioners of Lie group analysis that a first-order system of ordinary differential equations admits an infinite-dimensional Lie symmetry algebra; e.g., see [19]. Lie's theory allows us to integrate system (1)–(2) by quadrature, if we find at least a two-dimensional Lie algebra. In order to find it, we derive w_1 from (2), *i.e.*,

$$w_1 = -\frac{aw_2^2}{\dot{w}_2 - aw_2}. \tag{23}$$

Consequently we obtain a second-order ordinary differential equation (ODE) in $u \equiv w_2$,

$$\ddot{u} = \frac{(2a-h)\dot{u}^2}{au} - (a+c-2h)\dot{u} - \frac{ac}{K}u^2 + a(c-h)u. \tag{24}$$

and search for its Lie symmetry algebra. An operator Γ:

$$\Gamma = V(t,u)\partial_t + G(t,u)\partial_u \tag{25}$$

is said to generate a Lie point symmetry group of an equation of second-order, e.g.,

$$\ddot{u} = F(t,u,\dot{u}), \tag{26}$$

if its second prolongation:

$$\underset{2}{\Gamma} = \Gamma + \left(\frac{\mathrm{d}G}{\mathrm{d}t} - \dot{u}\frac{\mathrm{d}V}{\mathrm{d}t}\right)\partial_{\dot{u}} + \left[\frac{\mathrm{d}}{\mathrm{d}t}\left(\frac{\mathrm{d}G}{\mathrm{d}t} - \dot{u}\frac{\mathrm{d}V}{\mathrm{d}t}\right) - \ddot{u}\frac{\mathrm{d}V}{\mathrm{d}t}\right]\partial_{\ddot{u}}$$

applied to (26), on its solutions, is identically equal to zero, *i.e.*:

$$\underset{2}{\Gamma}(26)\Big|_{(26)} = 0. \tag{27}$$

A trivial Lie point symmetry of system (1)–(2) and also of Equation (24) is translation in time, *i.e.*,

$$\partial_t. \tag{28}$$

Using *ad hoc* REDUCE interactive programs [20], we find that Equation (24) admits at least another symmetry in two cases. Of course, any of those symmetries corresponds to a symmetry of system (1)–(2), as we show below.

Case (A)

If the following relationship among the parameters h, a and c is satisfied (condition $3a - 2c = 0$ does not yield a second Lie symmetry of Equation (24)):

$$h = \frac{a(2a-c)}{3a-2c}, \tag{29}$$

then Equation (24) admits a two-dimensional Lie symmetry algebra generated by (if $a = c$, then (29) implies that $h = c$; this is a particular example of Case (A), and we discuss it in Remark 3 of Section 4):

$$\Gamma_1 = \partial_t, \Gamma_2 = \exp((a-c)t)(\partial_t - 2(a-c)u\partial_u). \tag{30}$$

This Lie symmetry algebra is non-Abelian and transitive, *i.e.*, of Type III in Lie's classification of two-dimensional algebras in the real plane [14] (this classification can also be found in Bianchi's book

[15] and in modern textbooks, e.g., [21]). Therefore, to integrate Equation (24), we have to transform it into its canonical form, as given by Lie himself, *i.e.*,

$$\frac{\mathrm{d}^2\widetilde{u}}{\mathrm{d}\widetilde{t}^2} = \frac{1}{\widetilde{t}}\mathfrak{F}\left(\frac{\mathrm{d}\widetilde{u}}{\mathrm{d}\widetilde{t}}\right), \tag{31}$$

where $\widetilde{t}, \widetilde{u}$ are the new independent and dependent variables, respectively, $\mathfrak{F}$ is an arbitrary function of $\frac{\mathrm{d}\widetilde{u}}{\mathrm{d}\widetilde{t}}$ and the two-dimensional canonical Lie algebra of Type III is generated by the following operators:

$$\partial_{\widetilde{t}}, \widetilde{t}\partial_{\widetilde{t}} + \widetilde{u}\partial_{\widetilde{u}}. \tag{32}$$

We determine:

$$\widetilde{t} = \frac{\exp(-(a-c)t)}{\sqrt{u}}, \widetilde{u} = -\frac{\exp(-(a-c)t)}{(a-c)^2} \tag{33}$$

and consequently, (24) becomes:

$$\frac{\mathrm{d}^2\widetilde{u}}{\mathrm{d}\widetilde{t}^2} = \frac{a}{2K(3a-2c)\widetilde{t}}\left(c\left(3a^3 - 2c^3 + 7ac^2 - 8a^2c\right)\left(\frac{\mathrm{d}\widetilde{u}}{\mathrm{d}\widetilde{t}}\right)^2 + 2K\right)\frac{\mathrm{d}\widetilde{u}}{\mathrm{d}\widetilde{t}} \tag{34}$$

which can be integrated by two quadratures to yield:

$$\widetilde{u} = \int\sqrt{\frac{2A_1K^2(3a-2c)}{\widetilde{t}^{-2aK^2} + A_1cK(20ac^3 - 9a^4 - 37a^2c^2 + 30a^3c - 4c^4)}}\,d\widetilde{t} + A_2, \tag{35}$$

with A_1, A_2 two arbitrary constants.

Case (B)

If the following relationship among the parameters a and c is satisfied:

$$a = 2c, \tag{36}$$

then Equation (24) admits a two-dimensional Lie symmetry algebra generated by:

$$\Gamma_1 = \partial_t, \Gamma_2 = \exp(-ct)(\partial_t + 2cu\partial_u). \tag{37}$$

This Lie symmetry algebra is also non-Abelian and transitive. Therefore, we can derive the corresponding canonical transformation, *i.e.*,

$$\widetilde{t} = \frac{\exp(ct)}{\sqrt{u}}, \widetilde{u} = -\frac{\exp(ct)}{c}, \tag{38}$$

and consequently, (24) becomes:

$$\frac{\mathrm{d}^2\widetilde{u}}{\mathrm{d}\widetilde{t}^2} = \frac{1}{cK}\left(K(c-h) - c^3\left(\frac{\mathrm{d}\widetilde{u}}{\mathrm{d}\widetilde{t}}\right)^2\right)\frac{\mathrm{d}\widetilde{u}}{\mathrm{d}\widetilde{t}}, \tag{39}$$

which can be integrated by two quadratures to yield:

$$\widetilde{u} = A_1\sqrt{K(h-c)}\int\frac{\exp(\widetilde{t})}{\sqrt{A_1^{2h/c}\exp(2h\widetilde{t}/c) - c^3A_1^2\exp(2\widetilde{t})}}\,d\widetilde{t} + A_2, \tag{40}$$

with A_1, A_2 two arbitrary constants.

Subcase (B.1)

If in addition to the condition (36), the following relationship among the parameters h and c is satisfied:

$$h = \frac{3}{2}c \tag{41}$$

then Equation (24), *i.e.*,

$$\ddot{u} = \frac{5}{4u}\dot{u}^2 - 2\frac{c^2}{K}u^2 - c^2 u, \tag{42}$$

admits a three-dimensional Lie symmetry algebra generated by:

$$\Gamma_1 = \partial_t, \Gamma_2 = \exp(-ct)(\partial_t + 2cu\partial_u)\Gamma_3 = \exp(ct)(\partial_t - 2cu\partial_u). \tag{43}$$

This algebra is a representation of $sl(2,\mathbb{R})$. It was shown in [22] that if we treat (42) as a first integral, namely if we solve the equation with respect to $1/K$ (this choice, instead of K, is just for convenience), *i.e.*:

$$\frac{1}{K} = -\frac{1}{2c^2u^2}\ddot{u} - \frac{1}{2u} + \frac{5}{8c^2u^3}\dot{u}^2, \tag{44}$$

then a linearizable third-order equation is obtained, *i.e.*,

$$\dddot{u} = \dot{u}\left(c^2 + \frac{9}{2u}\ddot{u} - \frac{15}{4u^2}\dot{u}^2\right). \tag{45}$$

In fact, this equation admits a seven-dimensional Lie symmetry algebra generated by the following operators:

$$X_1 = \exp(ct)(\partial_t - 2uc\partial_u),\ X_2 = \exp(-ct)(\partial_t + 2uc\partial_u),\ X_3 = \partial_t,$$
$$X_4 = u\partial_u,\ X_5 = u^{3/2}\exp(ct)\partial_u,\ X_6 = u^{3/2}\exp(-ct)\partial_u,\ X_7 = u^{3/2}\partial_u. \tag{46}$$

We remark that we could not solve Equation (42) with respect to c since it is not an essential constant: in fact, we could have eliminated it from system (1)–(2) by rescaling. Indeed, the third-order equation that one gets through c, namely:

$$\dddot{u} = \frac{\dot{u}}{2u^2(K+2u)}\left(18u^2\ddot{u} + 7Ku\ddot{u} - 5K\dot{u}^2 - 15u\dot{u}^2\right),$$

admits a two-dimensional Lie symmetry algebra, generated by $\partial_t,\ t\partial_t$, and consequently, it is not linearizable.

Following Lie [14], the linearizing transformation of Equation (45) is given by finding a two-dimensional Abelian intransitive subalgebra and putting it into the canonical form $\partial_{\widetilde{u}},\ \widetilde{t}\partial_{\widetilde{u}}$. Since a two-dimensional Abelian intransitive subalgebra is that generated by X_5 and X_6, then the point transformation:

$$\widetilde{t} = \exp(-2ct), \widetilde{u} = -2u^{-1/2}\exp(-ct) \tag{47}$$

takes (45) into the following linear equation:

$$\frac{\mathrm{d}^3\widetilde{u}}{\mathrm{d}\widetilde{t}^3} = -\frac{3}{2\widetilde{t}}\frac{\mathrm{d}^2\widetilde{u}}{\mathrm{d}\widetilde{t}^2}. \tag{48}$$

Its general solution simply is:

$$\widetilde{u} = A_3 + A_2\widetilde{t} + A_1\sqrt{\widetilde{t}}, \tag{49}$$

with $A_i (i = 1,2,3)$ arbitrary constants, that, replaced into (45), yields the general solution:

$$u = \frac{4}{\exp(2ct)\left(A_3 + A_2\exp(-2ct) + A_1\exp(-ct)\right)^2}. \tag{50}$$

Since the original Equation (42) is of second order, then one of the constants of integrations A_i should be superfluous. Indeed, if we replace (50) into (44), then the following condition is obtained:

$$\frac{1}{K} = \frac{1}{2}A_2A_3 - \frac{1}{8}A_1^2 \tag{51}$$

that yields:

$$A_3 = \frac{8 + A_1^2K}{4A_2K}. \tag{52}$$

Finally, the general solution of Equation (42) is:

$$u = \frac{64A_2^2K}{\exp(2ct)\left(8 + A_1^2K + 4A_2^2K\exp(-2ct) + 4A_1A_2K\exp(-ct)\right)^2}, \tag{53}$$

and consequently, by means of (23), the general solution of system (1)–(2), *i.e.*,

$$\dot{w}_1 = cw_1\left(1 - \frac{w_1}{K}\right) - \frac{3c}{2}w_2, \tag{54}$$

$$\dot{w}_2 = 2cw_2\left(1 - \frac{w_2}{w_1}\right), \tag{55}$$

is given by:

$$w_1 = \frac{32A_2^2K^2\exp(-2ct)}{\left(8+A_1^2K+4A_2^2K\exp(-2ct)+4A_1A_2K\exp(-ct)\right)\left(8+A_1^2K+2A_1A_2K\exp(-ct)\right)},$$

$$w_2 = \frac{64A_2^2K\exp(-2ct)}{\left(8+A_1^2K+4A_2^2K\exp(-2ct)+4A_1A_2K\exp(-ct)\right)^2}. \tag{56}$$

Since $\lim_{t\to+\infty} w_1 = 0, \lim_{t\to+\infty} w_2 = 0$, this solution corresponds to asymptotic death for all [7]. We present two particular instances of the general solution (56) in Figures 1 and 2.

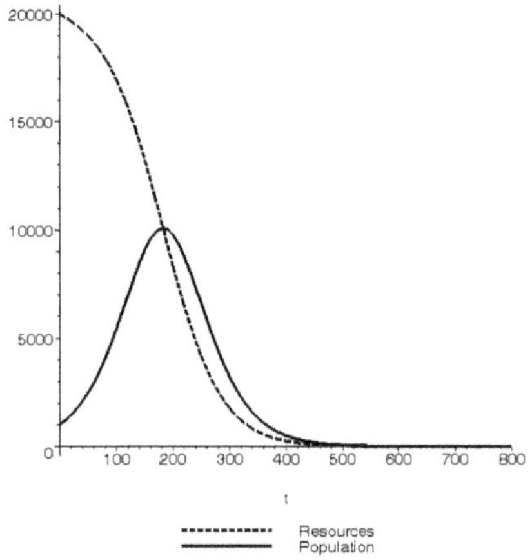

Figure 1. The amount of resources $R \equiv w_1$ and that of the population $P \equiv w_2$ for the values of the parameters $K = 20000, c = 0.01$, and for the initial conditions $P(0) = 1000$, $R(0) = 20000$.

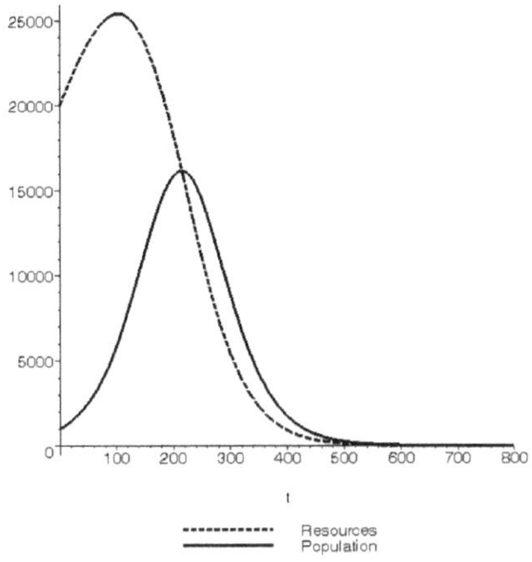

Figure 2. The amount of resources $R \equiv w_1$ and that of the population $P \equiv w_2$ for the values of the parameters $K = 40000, c = 0.01$, and for the initial conditions $P(0) = 1000$, $R(0) = 20000$.

It is interesting to remark that if we double the carrying capacity K leaving the other parameters and initial conditions unaltered, the amount of resources $R \equiv w_1$ actually grows along with the population $P \equiv w_2$ at least for a period of time (approximately a century in Figure 2) before decreasing dramatically.

4. Jacobi Last Multipliers, Lagrangians and First Integrals of System (1)–(2)

A Jacobi last multiplier for system (1)–(2) has to satisfy Equation (7), *i.e.*,

$$\frac{\mathrm{d}}{\mathrm{d}t}\left(\log M_{[w]}\right) + \frac{\partial a_1}{\partial w_1} + \frac{\partial a_2}{\partial w_2} = 0. \tag{57}$$

As in [8], we assume that $M_{[w]}$ has the following form:

$$M_{[w]} = w_1^{a_1} w_2^{a_2} \exp(a_0 t), \tag{58}$$

where $a_i, (i = 0, 1, 2)$ are constants to be determined. Replacing this $M_{[w]}$ into system (1)–(2) yields:

$$M_{[w]} = w_1^{-2} w_2^{2h/a-2} \exp((a + c - 2h)t). \tag{59}$$

Therefore, the following Lagrangian of system (1) and (2) is obtained by means of Equation (20):

$$\begin{aligned} L_{[w]} \quad &= \quad w_1^{-2} w_2^{2h/a-2} \exp((a + c - 2h)t)\left(\tfrac{a}{a-2h} w_2 \dot{w}_1 - w_1 \dot{w}_2 \right. \\ &\quad \left. + a w_2 \tfrac{2cw_1^2 + K(a-c-2h)w_1 + K(2h-a)w_2}{K(a-2h)}\right). \end{aligned} \tag{60}$$

It is obvious that if $a = 2h$, then the Lagrangian $L_{[w]}$ does not exist. Indeed, if:

$$a = 2h, \tag{61}$$

then the Jacobi Last Multiplier (59) becomes:

$$M^{\bigstar}_{[w]} = w_1^{-2} w_2^{-1} \exp(ct), \tag{62}$$

and consequently, the Lagrangian of system (1)–(2) is:

$$L^{\star}_{[w]} = w_1^{-2}w_2^{-1}\exp(ct)\big(-w_2\log(w_2)\dot{w}_1 - w_1\dot{w}_2 - \tfrac{2c}{K}\log(w_2)w_1^2w_2 -2hw_2^2 + c\log(w_2)w_1w_2 + 4hw_1w_2\big). \tag{63}$$

The Jacobi Last Multiplier (59) yields the following Jacobi last multiplier of Equation (24) by means of the application of Property (b):

$$M_{[u]} = M_{[w]}\frac{\partial(w_2,w_1)}{\partial(u,\dot{u})} = \begin{vmatrix} 1 & 0 \\ -au\frac{2\dot{u}-au}{(\dot{u}-au)^2} & \frac{au^2}{(\dot{u}-au)^2} \end{vmatrix} = \frac{u^{2h/a-4}}{a}\exp[(a+c-2h)t], \tag{64}$$

and consequently, the following Lagrangian is derived by means of (14):

$$L_{[u]} = u^{\frac{2h}{a}-4}\exp[(a+c-2h)t]\left(\frac{\dot{u}^2}{2a} + \frac{acu^3}{K(a-2h)} - \frac{a(c-h)u^2}{2(a-h)}\right). \tag{65}$$

If the following condition on the parameters is satisfied, *i.e.*,

$$h = \frac{a+c}{2}, \tag{66}$$

then the trivial Lie symmetry (28) admitted by system (1)–(2) and Equation (24) is also a Noether symmetry (Equation (24) does not admit another Lie symmetry). Consequently, Noether's theorem [9] yields the following first integral:

$$Int = \frac{u^{\frac{c}{a}-3}\left(K\dot{u}^2 + (2u-K)a^2u^2\right)}{2aK} \tag{67}$$

of Equation (24), *i.e.*,

$$\ddot{u} = \frac{-a^3Ku^2 + a^2cKu^2 - 2a^2cu^3 + 3aK\dot{u}^2 - cK\dot{u}^2}{2aKu}. \tag{68}$$

Replacing u with w_2 and $\dot{u}$ with $\dot{w}_2 = aw_2(1-w_2/w_1)$, namely the RHS of Equation (2), into Int yields the following first integral of system (1)–(2):

$$I = \frac{w_2^{c/a}}{w_1^2}\left(2w_1^2 - 2Kw_1 + Kw_2\right). \tag{69}$$

Remark 1: It was shown in [7], Proposition 6, that if $c > a$, then the first integral (69) yields periodic orbits. In particular, if $a = 1, c = 2$, it is easy to show that the general solution depends on elliptic functions.

It is obvious that if either $a = 2h$ or $a = h$, then the Lagrangian $L_{[u]}$ in (65) does not exist. In fact, if:

$$a = 2h, \tag{70}$$

then from $M_{[u]}$ in (64), *i.e.*,

$$M^{\star}_{[u]} = \frac{u^{-3}}{2h}\exp(ct), \tag{71}$$

the following Lagrangian is obtained by means of (14), *i.e.*,

$$L^{\star}_{[u]} = \frac{u^{-3}}{K}\exp(ct)\left(\frac{K}{4h}\dot{u}^2 - c\log(u)u^3 + K(h-c)u^2\right). \tag{72}$$

This Lagrangian does not admit any Noether point symmetry unless $c = h$, and consequently, the Lagrangian (72) admits the Lie symmetry $\exp(-ht)(\partial_t + 2hu\partial_u)$ as Noether symmetry and the following first integral is derived:

$$Int_0 = \frac{4h^2\log(u)u^3 + 4h^2Ku^2 - 8th^3u^3 - 4hKu\dot{u} + K\dot{u}^2}{4hKu^3}, \tag{73}$$

the gauge function being $F = h(K - 2htu)/(Ku)$.

Replacing u with w_2 and $\dot{u}$ with $\dot{w}_2 = aw_2(1 - w_2/w_1)$, namely the RHS of Equation (2), into Int_0 yields the following first integral of system (1)–(2):

$$I_0 = \log(w_2) - 2ht + K\frac{w_2}{w_1^2}. \tag{74}$$

Remark 2: The case $a = 2h, c = h$ is a particular example of Case (B), and consequently, Equation (24), *i.e.*,

$$\ddot{u} = \frac{3K\dot{u}^2 - 2Kuh\dot{u} - 4u^3h^2}{2Ku}, \tag{75}$$

admits the two Lie symmetries (37), *i.e.*,

$$\partial_t, (\partial_t + 2hu\partial_u)\exp(-ht). \tag{76}$$

Another Jacobi last multiplier can be obtained from the reciprocal of (9), *i.e.*,

$$\begin{aligned}\Delta &= \det\begin{bmatrix} 1 & \dot{u} & \frac{3K\dot{u}^2 - 2Kuh\dot{u} - 4u^3h^2}{2Ku} \\ 1 & 0 & 0 \\ \exp(-ht) & 2hu\exp(-ht) & h(3\dot{u} - 2hu)\exp(-ht) \end{bmatrix} \\ &= -\frac{4h^3u^3}{K}\exp(-ht).\end{aligned} \tag{77}$$

Surprisingly, we do not obtain another Jacobi last multiplier. Indeed, $\frac{1}{\Delta} = -\frac{K\exp(ht)}{4h^3u^3}$ is equal to $M^{\star}_{[u]} = \frac{\exp(ht)}{2hu^3}$ in (71), apart from an unessential multiplicative constant.

If

$$a = h, \tag{78}$$

then from $M_{[u]}$ in (64), *i.e.*,

$$M^{\star\star}_{[u]} = \frac{\exp((c - h)t)}{hu^2}, \tag{79}$$

the following Lagrangian is obtained by means of (14), *i.e.*,

$$L^{\star\star}_{[u]} = \frac{\exp((c - h)t)\left(K\dot{u}^2 - 2chu^3 + 2hK(c - h)\log(u)u^2\right)}{2hKu^2}. \tag{80}$$

This Lagrangian does not admit any Noether point symmetry unless either of the following further conditions are also satisfied, *i.e.*,

$$h = c \tag{81}$$

or

$$h = 2c. \tag{82}$$

If $h = c$, then the Lagrangian (80) admits the trivial Lie symmetry (28) as Noether symmetry, and consequently, the following first integral is derived:

$$Int_{1a} = \frac{K\dot{u}^2 + 2c^2u^3}{2cKu^2}. \tag{83}$$

Replacing u with w_2 and $\dot{u}$ with $\dot{w}_2 = aw_2(1 - w_2/w_1)$, namely the RHS of Equation (2), into Int_{1a} yields the following first integral of system (1)–(2):

$$I_{1a} = \frac{-2Kw_1w_2 + Kw_2^2 + 2w_1^2w_2}{w_1^2}. \tag{84}$$

Remark 3: The case $a = h = c$ is a particular example of Case (A), and consequently, Equation (24), *i.e.*,

$$\ddot{u} = \frac{K\dot{u}^2 - c^2u^3}{Ku}, \tag{85}$$

admits the two Lie symmetries (30), *i.e.*,

$$\partial_t, t\partial_t - 2u\partial_u. \tag{86}$$

Its general solution can be determined through the canonical variables, as explained in the previous section, although in implicit form, *i.e.*,

$$\frac{A_1 - 3(c^2+1)u}{3u\sqrt{u}} + \sqrt{\frac{K}{2A_1}}\log\left(\frac{K}{A_1u}\right) + \sqrt{\frac{2K}{A_1}}\log\left(A_1 + \sqrt{A_1(A_1 - c^2u)}\right) = t + A_2. \tag{87}$$

Another Jacobi last multiplier (hence, another Lagrangian) for Equation (85) can be obtained from the reciprocal of (9), *i.e.*,

$$\Delta = \det\begin{bmatrix} 1 & \dot{u} & \frac{K\dot{u}^2 - c^2u^3}{Ku} \\ 1 & 0 & 0 \\ t & -2u & -3\dot{u} \end{bmatrix} = \frac{K\dot{u}^2 + 2c^2u^3}{K}. \tag{88}$$

Because of Property (d), the ratio of the Jacobi last multiplier in (79), *i.e.*, $M_{[u]}^{\star\star} = \frac{1}{cu^2}$, with that found in Equation (88), *i.e.*, $\frac{1}{\Delta} = \frac{K}{K\dot{u}^2 + 2c^2u^3}$, yields a first integral that surprisingly is just Int_{1a}, apart from an unessential multiplicative constant.

If $h = 2c$, then the Lagrangian (80) admits the Lie symmetry $\exp(-ct)(\partial_t + 2cu\partial_u)$ as Noether symmetry, and the following first integral is derived:

$$Int_{1b} = \frac{4c^2Ku^2 + 4c^2u^3 - 4cKu\dot{u} + K\dot{u}^2}{4\exp(2ct)cKu^2}, \tag{89}$$

the gauge function being $F = c(1 - \log(u))\exp(-2ct)$.

Replacing u with w_2 and $\dot{u}$ with $\dot{w}_2 = aw_2(1 - w_2/w_1)$, namely the RHS of Equation (2), into Int_{1b} yields the following first integral of system (1)–(2):

$$I_{1b} = \exp(-2ct)w_2\frac{Kw_2 + w_1^2}{w_1^2}. \tag{90}$$

Remark 4: The case $a = h = 2c$ is another instance of Case (B), and consequently, Equation (24), *i.e.*,

$$\ddot{u} = \frac{K\dot{u}^2 + Kcu\dot{u} - 2Kc^2u^2 - 2c^2u^3}{Ku}, \tag{91}$$

admits the two Lie symmetries (37), *i.e.*,

$$\partial_t, (\partial_t + 2cu\partial_u)\exp(-ct), \tag{92}$$

and its general solution can be determined through the canonical variables, as explained in the previous section. Thus, the general solution of system (1)–(2) is that both w_1 and w_2 can go to infinity at the same finite time, *i.e.*, if:

$$t = \frac{1}{c}\log\left(cA_2 - \frac{c\sqrt{A_1K}}{4}\log\left(c^4A_1^4K^2\right)\right)$$

$$w_1 = \frac{4c(A_1K)^{3/2}\exp\left(ct+\frac{2(cA_2-\exp(ct))}{c\sqrt{A_1K}}\right)}{c^4A_1^4K^2-\exp\left(4\frac{cA_2-\exp(ct)}{c\sqrt{A_1K}}\right)}, \tag{93}$$

$$w_2 = \frac{4A_1K\exp\left(2ct+2\frac{cA_2-\exp(ct)}{c\sqrt{A_1K}}\right)}{\left(c^2A_1^2K+\exp\left(2\frac{cA_2-\exp(ct)}{c\sqrt{A_1K}}\right)\right)^2}. \tag{94}$$

Another Jacobi last multiplier for Equation (91) can be obtained from the reciprocal of Equation (9), *i.e.*,

$$\begin{aligned}\Delta &= \det\begin{bmatrix} 1 & \dot{u} & \frac{K\dot{u}^2+Kcu\dot{u}-2Kc^2u^2-2c^2u^3}{Ku} \\ 1 & 0 & 0 \\ \exp(-ct) & 2cu\exp(-ct) & c(3\dot{u}-2cu)\exp(-ct) \end{bmatrix} \\ &= -\tfrac{c}{K}\left(K(\dot{u}-2cu)^2 + 4c^2u^3\right)\exp(-ct). \end{aligned}\tag{95}$$

Because of Property (d), the ratio of the Jacobi last multiplier in (79), *i.e.*, $\frac{\exp(-ct)}{2cu^2}$, with that found in (95), *i.e.*, $\frac{1}{\Delta} = \frac{K\exp(ct)}{c\left(K(\dot{u}-2cu)^2+4c^2u^3\right)}$, yields a first integral that is just Int_{1a}, apart from an unessential multiplicative constant.

Case (A)

If condition (29) is satisfied, then the Lagrangian (65) admits the Lie symmetry Γ_2 in (30) as Noether symmetry, with gauge function $F = u^{\frac{2(c-a)}{3a-2c}}\exp[2a(a-c)t/(3a-2c)](3a^2-5ac+2c^2)/(2a)$, and consequently, Noether's theorem [9] yields the following first integral:

$$Int_A = u^{\frac{2(c-a)}{3a-2c}}\exp\left(\frac{2a(a-c)t}{3a-2c}\right)\left(K\left(\frac{\dot{u}}{u}-2c\right)^2 + 4aK\frac{\dot{u}}{u} + 2c(3a-2c)u + 4aK(a-2c)\right) \tag{96}$$

of Equation (24), *i.e.*,

$$\ddot{u} = \frac{(4a-3c)\dot{u}^2}{(3a-2c)u} + \frac{(a-c)(a-2c)\dot{u}}{3a-2c} - \frac{au}{K(3a-2c)}\left(2K(a-c)^2 + c(3a-2c)u\right). \tag{97}$$

Replacing u with w_2 and $\dot{u}$ with $\dot{w}_2 = aw_2(1-w_2/w_1)$, namely the RHS of Equation (2), into Int_A yields the following first integral of system (1)–(2):

$$I_A = w_2^{\frac{2(c-a)}{3a-2c}}\exp\left(\frac{2a(a-c)t}{3a-2c}\right)\left(K\left(a\frac{w_2}{w_1} - (3a-2c)\right)^2 + 2c(3a-2c)w_2\right). \tag{98}$$

Another Jacobi last multiplier, and consequently Lagrangian, can be obtained by means of the two Lie symmetries (30) through (9), *i.e.*,

$$M_{A[1,2]} = \exp((c-a)t)\frac{3a-2c}{a(a-c)}\left[k\dot{u}^2 + 4Ku(a-c)\dot{u} + 2u^2\left(2K(a-c)^2 + c(3a-2c)u\right)\right]^{-1}. \quad (99)$$

The corresponding Lagrangian admits one Lie symmetry as Noether symmetry, with gauge function $F = t/K$, *i.e.*, Γ_2 in (30), and consequently, the following first integral:

$$\begin{aligned} Int_{A[1,2]} &= \tfrac{1}{2(a-c)aK}((3a-2p)\log(4K(a-c)^2u^2 + 2(3a-2c)cu^3 \\ &+4K(a-c)u\dot{u} + K\dot{u}^2) + 2(a-c)at - 2(4a-3c)\log(u)). \end{aligned} \quad (100)$$

We remark that the two first integrals Int_A and $Int_{A[1,2]}$ are not functionally independent.

Case (B)

If condition (36) is satisfied, then the Lagrangian (65) admits the Lie symmetry Γ_2 in (37) as Noether symmetry, with gauge function $F = c^2u^{h/c}\exp[2(c-h)t]/((2c-h)u^2)$, and consequently, Noether's theorem [9] yields the following first integral ($h \neq c$):

$$\begin{aligned} Int_B &= \tfrac{u^{h/c}\exp[2(c-h)t]}{4cK(c-h)u^4}(4c^3Ku^2 - 4c^2hKu^2 - 4c^3u^3 - 4c^2Ku\dot{u} \\ &+4chKu\dot{u} + cK\dot{u}^2 - hK\dot{u}^2\Big) \end{aligned} \quad (101)$$

of Equation (24), *i.e.*,

$$\ddot{u} = \frac{4c^3Ku^2 - 4c^2hKu^2 - 4c^3u^3 - 6c^2Ku\dot{u} + 4Khcu\dot{u} + 4cK\dot{u}^2 - Kh\dot{u}^2}{2cKu}. \quad (102)$$

Replacing u with w_2 and $\dot{u}$ with $\dot{w}_2 = aw_2(1 - w_2/w_1)$, namely the RHS of Equation (2), into Int_B yields the following first integral of system (1)–(2):

$$I_B = w_2^{\frac{h-c}{c}}\exp[2(c-h)t]\left(K(c-h)\frac{w_2}{w_1^2} - c\right). \quad (103)$$

Another Jacobi last multiplier, and consequently Lagrangian, can be obtained by means of the two Lie symmetries (37) through (9), *i.e.*,

$$M_{B[1,2]} = \frac{K\exp(ct)}{K(c-h)\dot{u}^2 - 4cK(c-h)u\dot{u} + 4c^2u^2(Kc - Kh - cu)}. \quad (104)$$

The corresponding Lagrangian admits only the Lie symmetry Γ_2 in (37) as Noether symmetry, with gauge function $F = t$, and consequently, the following first integral:

$$\begin{aligned} Int_{B[1,2]} &= \tfrac{1}{2c(c-h)}(2c(c-h)t + (h-4c)\log(u) \\ &+\log\left(4Kc^2(c-h)^2u^2 - 4c^3u^3 - 4cK(c-h)u\dot{u} + K(c-h)\dot{u}^2\right)\Big). \end{aligned} \quad (105)$$

We remark that the two first integrals Int_B and $Int_{B[1,2]}$ are not functionally independent.

Case (B.1.)

If condition (41) is satisfied, then the Lagrangian (65) admits all three Lie symmetries in (43) as Noether symmetries, with gauge functions $F_2 = \frac{2c}{\sqrt{u}\exp(ct)}$, $F_3 = \frac{2c\exp(ct)}{\sqrt{u}}$, and consequently, Noether's theorem [9] yields the following three first integrals:

$$\begin{aligned} Int1_{B_1} &= \frac{8c^2u^3-4c^2Ku^2+K\dot{u}^2}{4cKu^2\sqrt{u}}, \\ Int2_{B_1} &= \frac{4c^2Ku^2+8c^2u^3-4cKu\dot{u}+K\dot{u}^2}{4cKu^2\sqrt{u}\exp(ct)}, \\ Int3_{B_1} &= \exp(ct)\frac{4c^2Ku^2+8c^2u^3+4cKu\dot{u}+K\dot{u}^2}{4cKu^2\sqrt{u}}, \end{aligned} \tag{106}$$

of Equation (24), *i.e.*,

$$\ddot{u} = \frac{5K\dot{u}^2 - 4c^2Ku^2 - 8c^2u^3}{4Ku}. \tag{107}$$

Three more Jacobi last multipliers, and consequently, three Lagrangians, can be obtained by means of the three Lie symmetries (43) through (9), *i.e.*,

$$\begin{aligned} M_{B_1[1,2]} &= -\frac{2K\exp(ct)}{c\left(4c^2Ku^2+8c^2u^3-4cKu\dot{u}+K\dot{u}^2\right)}, \\ M_{B_1[1,3]} &= \frac{2K}{c\exp(ct)\left(4c^2Ku^2+8c^2u^3+4cKu\dot{u}+K\dot{u}^2\right)}, \\ M_{B_1[2,3]} &= -\frac{K}{c\left(4c^2Ku^2-8c^2u^3-K\dot{u}^2\right)}. \end{aligned} \tag{108}$$

Since the ratio of two Jacobi last multipliers is a first integral, then we can obtain three first integrals of Equation (107), *i.e.*,

$$\begin{aligned} \frac{M_{B_1[1,2]}}{M_{B_1[1,3]}} &= -\exp(2ct)\frac{4c^2Ku^2+8c^2u^3+4cKu\dot{u}+K\dot{u}^2}{4c^2Ku^2+8c^2u^3-4cKu\dot{u}+K\dot{u}^2}, \\ \frac{M_{B_1[1,2]}}{M_{B_1[2,3]}} &= 2\exp(ct)\frac{4c^2Ku^2-8c^2u^3-K\dot{u}^2}{4c^2Ku^2+8c^2u^3-4cKu\dot{u}+K\dot{u}^2}, \\ \frac{M_{B_1[1,3]}}{M_{B_1[2,3]}} &= -2\exp(-ct)\frac{4c^2Ku^2-8c^2u^3-K\dot{u}^2}{4c^2Ku^2+8c^2u^3+4cKu\dot{u}+K\dot{u}^2}. \end{aligned} \tag{109}$$

The Lagrangian corresponding to $M_{B_1[1,2]}$ admits only the Lie symmetry Γ_2 in (43) as Noether symmetry, with gauge function $F = t$, and consequently, Noether's theorem [9] yields the following first integral:

$$Int_{B_1[1,2]} = \frac{1}{2c}\left(5\log(u) + 2ct - 2\log\left(4c^2Ku^2 + 8c^2u^3 - 4cKu\dot{u} + K\dot{u}^2\right)\right). \tag{110}$$

The Lagrangian corresponding to $M_{B_1[1,3]}$ admits only the Lie symmetry Γ_3 in (43) as Noether symmetry, with gauge function $F = t$, and consequently, Noether's theorem [9] yields the following first integral:

$$Int_{B_1[1,3]} = \frac{1}{2c}\left(-5\log(u) + 2ct + 2\log\left(4c^2Ku^2 + 8c^2u^3 + 4cKu\dot{u} + K\dot{u}^2\right)\right). \tag{111}$$

The Lagrangian corresponding to $M_{B_1[2,3]}$ admits only the Lie symmetry Γ_1 in (43) as Noether symmetry, and consequently, Noether's theorem [9] yields the following first integral:

$$Int_{B_1[2,3]} = \frac{1}{4c}\left(-5\log(u) + 2\log\left(4c^2Ku^2 - 8c^2u^3 - K\dot{u}^2\right)\right). \tag{112}$$

Of course, of all these nine first integrals, only two are functionally independent.

5. Discussion and Final Remarks

In this paper, we have determined different cases that fit the qualitative prediction (in [7], it was assumed $a = 1$) given in [7]. Assuming $a = 1$, then the solution (56) with $c = 1/2, h = 3/4$ corresponds (since $h > c$ and $h < 1$) to Proposition 2, namely the case of asymptotic death for all, with both the population and the resources going to zero at infinity. Furthermore, the ratio of population over resources goes asymptotically to 2, *i.e.*, $\frac{c-1}{h-1}$, as predicted in [7]. Indeed, solution (56) shows that resources decrease, while the population grows exponentially *for an extended period of time* followed by a *catastrophic elimination* [7], as can also be seen in Figure 1. However, for at least a period of time, resources may increase along with the population, as shown in Figure 2, if the carrying capacity K is large enough.

In [7], one first integral was derived under the condition $h = (c+1)/2$ (Proposition 6). It corresponds to our first integral I in (69) under the condition $h = (c+a)/2$. We have derived many other first integrals and also the general solution of system (1)–(2) in closed form, as well as the general solution of Equation (24) in implicit form:

the first integral I_0 in (74) if $a = 2h, c = h$;
the first integral I_{1a} in (84) and also the general solution (87) if $a = h = c$;
the first integral I_{1b} in (90) and also the general solution (94) if $a = h, h = 2c$;
the first integral I_A in (98) if $h = \frac{a(2a-c)}{3a-2c}$, that corresponds to Proposition 3 ($h = 2, c = 4/3$);
the first integral I_B in (103) if $a = 2c$, that corresponds to Proposition 1 ($h = 3/4, c = 1/2$).

Last, but not least, we were able to find a Lagrangian (60) for system (1)–(2) and any value of the involved parameters (if $a = 2h$ the Lagrangian is (63)).

Recently, archeological findings have pointed out that Polynesian rats may have greatly contributed to the tree destruction on Easter Island [23]. Consequently, in [24], rats have been included in the mathematical model, leading to three nonlinear first-order differential equations. The chaotic nature of that system has subsequently been determined in [25]. Can symmetries help since they have always been associated with non-chaotic systems? We recall that in [26], Lie symmetries yielded an exact transformation that turned a butterfly into a tornado (and *vice versa*), and the chaotic features of the Lorenz system [27] were thus tamed. Could the chaotic features of the rats-trees-islanders model be equally tamed? We hope to address this question in the near future.

Acknowledgments: M.C.N. acknowledges the support of the Italian Ministry of University and Scientific Research through PRIN 2010-2011, Prot. n. 2010JJ4KPA_004.

Author Contributions: The authors contributed equally to this work.

Conflicts of Interest: The authors declare no conflict of interest.

References

1. Edwards, M.; Nucci, M.C. Application of Lie group analysis to a core group model for sexually transmitted diseases. *J. Nonlinear Math. Phys.* **2006**, *13*, 211–230. [CrossRef]
2. Gradassi, A.; Nucci, M.C. Hidden linearity in systems for competition with evolution in ecology and finance. *J. Math. Anal. Appl.* **2007**, *333*, 274–294. [CrossRef]
3. Nucci, M.C. Using lie symmetries in epidemiology. *Electron. J. Diff. Eqns. Conf.* **2005**, *12*, 87–101.
4. Nucci, M.C.; Leach, P.G.L. An integrable S-I-S model. *J. Math. Anal. Appl.* **2004**, *290*, 506–518. [CrossRef]
5. Nucci, M.C.; Leach, P.G.L. Lie integrable cases of the simplified multistrain/two-stream model for tuberculosis and Dengue fever. *J. Math. Anal. Appl.* **2007**, *333*, 430–449. [CrossRef]
6. Torrisi, V.; Nucci, M.C. Application of Lie group analysis to a mathematical model which describes HIV transmission. In *The Geometrical Study of Differential Equations*; Leslie, J.A., Hobart, T.P., Eds.; American Mathematical Society: Providence, RI, USA, 2011; pp. 11–20.
7. Basener, B.; Ross, D.S. Booming and crashing populations and Easter Island. *SIAM J. Appl. Math.* **2005**, *65*, 684–701. [CrossRef]

8. Nucci, M.C.; Tamizhmani, K.M. Lagrangians for biological models. *J. Nonlinear Math. Phys.* **2012**, *19*, 1250021:1–1250021:23. [CrossRef]
9. Noether, E. Invariante Variationsprobleme. *Nachr. Ges. Wiss. Gött. Math. Phys. Kl.* **1918**, *1918*, 235–257.
10. Jacobi, C.G.J. Theoria novi multiplicatoris systemati æquationum differentialium vulgarium applicandi. *J. Reine Angew. Math.* **1844**, *27*, 199–268, **1845**, *29*, 213–279, 333–376. [CrossRef]
11. Jacobi, C.G.J. *Vorlesungen über Dynamik. Nebst Fünf Hinterlassenen Abhandlungen Desselben Herausgegeben von A. Clebsch*; Druck und Verlag von Georg Reimer: Berlin, Germany, 1884.
12. Jacobi, C.G.J. *Jacobi's Lectures on Dynamics*; Clebsch, A., Ed.; Industan Book Agency: New Delhi, India, 2009.
13. Lie, S. Veralgemeinerung und neue Verwerthung der Jacobischen Multiplicator-Theorie. *Christ. Forh.* **1874**, 255–274.
14. Lie, S. *Vorlesungen über Differentialgleichungen mit Bekannten Infinitesimalen Transformationen*; Teubner: Leipzig, Germany, 1912.
15. Bianchi, L. *Lezioni Sulla Teoria Dei Gruppi Continui Finiti di Trasformazioni*; Enrico Spoerri: Pisa, Italy, 1918.
16. Whittaker, E.T. *A Treatise on the Analytical Dynamics of Particles and Rigid Bodies*; Cambridge University Press: Cambridge, UK, 1999.
17. Nucci, M.C.; Leach, P.G.L. Lagrangians galore. *J. Math. Phys.* **2007**, *48*, 123510:1–123510:16. [CrossRef]
18. Trubatch, S.L.; Franco, A. Canonical procedures for population dynamics. *J. Theor. Biol.* **1974**, *48*, 299–324. [CrossRef]
19. Olver, P.J. *Applications of Lie Groups to Differential Equations*; Springer-Verlag: Berlin, Germany, 1986.
20. Nucci, M.C. Interactive REDUCE programs for calculating Lie point, non-classical, Lie-Bäcklund, and approximate symmetries of differential equations: Manual and floppy disk. In *CRC Handbook of Lie Group Analysis of Differential Equations. Vol. 3: New Trends in Theoretical Developments and Computational Methods*; Ibragimov, N.H., Ed.; CRC Press: Boca Raton, FL, USA, 1996; pp. 415–481.
21. Ibragimov, N.H. *Elementary Lie Group Analysis and Ordinary Differential Equations*; Wiley: Chichester, UK, 1999.
22. Leach, P.G.L. Equivalence classes of second-order ordinary differential equations with only a three-dimensional Lie algebra of point symmetries and linearisation. *J. Math. Anal. Appl.* **2003**, *284*, 31–48. [CrossRef]
23. Hunt, T. Rethinking the fall of Easter Island: New evidence points to an alternative explanation for a civilization's collapse. *Am. Sci.* **2006**, *94*, 412–419. [CrossRef]
24. Basener, W.; Brooks, B.P.; Radin, M.A.; Wiandt, T. Dynamics of a discrete population model for extinction and sustainability in ancient civilizations. *Nonlinear Dyn. Psychol. Life Sci.* **2008**, *12*, 29–53.
25. Sprott, J.C. Chaos in Easter Island ecology. *Nonlinear Dyn. Psychol. Life Sci.* **2011**, *15*, 445–454.
26. Nucci, M.C. Lorenz integrable system moves à la Poinsot. *J. Math. Phys.* **2003**, *44*, 4107–4118. [CrossRef]
27. Lorenz, E.N. Deterministic nonperiodic flow. *J. Atmos. Sci.* **1963**, *20*, 130–141. [CrossRef]

symmetry

Article

Noether Symmetries Quantization and Superintegrability of Biological Models

Maria Clara Nucci * and Giampaolo Sanchini

Dipartimento di Matematica e Informatica, Università degli Studi di Perugia, 06123 Perugia, Italy; giampaolo.sanchini@istruzione.it

* Correspondence: nucci@unipg.it

Academic Editor: Roman M. Cherniha

Received: 17 October 2016; Accepted: 12 December 2016; Published: 20 December 2016

Abstract: It is shown that quantization and superintegrability are not concepts that are inherent to classical Physics alone. Indeed, one may quantize and also detect superintegrability of biological models by means of Noether symmetries. We exemplify the method by using a mathematical model that was proposed by Basener and Ross (2005), and that describes the dynamics of growth and sudden decrease in the population of Easter Island.

Keywords: Lie symmetries; Jacobi last multiplier; Lagrangians; Noether symmetries; classical quantization; superintegrability; Easter Island population

---

1. Introduction

In a review to honor the 50th Anniversary Year of *the Journal of Theoretical Biology* [1], one reads:

> "It is frequently claimed that— like Newton's invention of calculus—biological theory will require 'new mathematics'.... There are, however, many areas of mathematics that have been neglected by theoretical biology that could prove to be of great value. Einstein's work on general relativity, for instance, made good use of mathematical ideas, in particular differential geometry that had previously been developed with completely different motivation. More likely than not, the formal structures have been set forth in some context, and await their discovery and subsequent development in representing biological theory."

Since many mathematical tools used in physics have also been used in biology with different success, we present a somewhat forgotten and neglected tool, a tool that in one of its outcomes, Noether symmetries, helped Einstein and Klein in their quarrel with Hilbert about the energy-momentum conservation of general relativity theory [2]. This tool is Lie continuous symmetries, which yield conservation laws, calculus of variation setting, and ultimately quantization.

The applications of Lie symmetries to various biological models have already been shown to either provide more accurate predictions [3] or implement [4–8] the usual techniques related to qualitative and numerical analysis, which are common tools for any mathematical biologist.

We would like to stir up some controversy with the purpose of making both mathematicians and biologists ponder over some missed opportunities [9].

In [10], a mathematical model that describes the dynamics of growth and sudden decrease in the population of Easter Island was proposed. The model is a nonlinear system of two first-order ordinary differential equations, i.e.,

$$\dot{w}_1 = cw_1\left(1-\frac{w_1}{K}\right) - hw_2, \tag{1}$$

$$\dot{w}_2 = aw_2\left(1-\frac{w_2}{w_1}\right), \tag{2}$$

where w_1 is the amount of resources, w_2 is the population, c is the growth rate of the resources, K the carrying capacity, h the harvesting constant, and a the growth rate of the population. More details on the construction of this model can be found in [10] where numerical solutions of the system (1)–(2) were given and a qualitative analysis of the general behavior of solutions both in finite and infinite time was presented.

In [11], we have applied Lie group analysis to this system and found that it can be integrated by quadrature if the involved parameters satisfy certain relationships. We have also discerned hidden linearity. Moreover, we have determined a Jacobi last multiplier and consequently a Lagrangian for the general system [12], and have found other cases, independent of and dependent on symmetry considerations in order to construct a corresponding variational problem, thus enabling us to find conservation laws by means of Noether's theorem. A comparison with the qualitative analysis given in [10] was provided.

In the present paper, we perform the quantization of the system (1)–(2); namely, we derive the Schrödinger equation by means of the method that preserves the Noether symmetries [13].

Moreover, in the wake of Volterra's last papers [14,15], the same mathematical model (1)–(2) is transformed into two second-order Lagrangian equations, and the superintegrability of this system is proven; namely three conservation laws, including the Hamiltonian, are found by means of Noether's theorem.

2. Quantizing with Noether Symmetries

It has been known for sixty-five years that quantization and nonlinear canonical transformations have no guarantee of consistency [16]. For a more recent perspective see [17] where an up-to-date account of the various approaches to tackle canonical transformation is also provided.

In [13,18], a procedure which obviates the constraint imposed by the conflict between consistent quantization and the invariance of the Hamiltonian description under nonlinear canonical transformation was proposed. It is based on the preservation of Noether symmetries when going from classical to quantum mechanics. The quantization of classical problems is achieved by constructing a suitable time-dependent Schrödinger equation.

This method was reformulated in [19] for problems that are linearizable by Lie point symmetries, and also successfully applied to various classical problems: second-order Riccati equation [20], dynamics of a charged particle in a uniform magnetic field and a non-isochronous Calogero's goldfish system [18], an equation related to a Calogero's goldfish equation [21], two nonlinear equations somewhat related to the Riemann problem [22], a Liénard I nonlinear oscillator [19], a family of Liénard II nonlinear oscillators [23], N planar rotors and an isochronous Calogero's goldfish system [24], the motion of a free particle and that of a harmonic oscillator on a double cone [25].

If a system of second-order equations is considered, i.e.,

$$\ddot{\mathbf{x}}(t) = \mathbf{F}(t, \mathbf{x}, \dot{\mathbf{x}}), \quad \mathbf{x} \in \boldsymbol{R}^N, \tag{3}$$

that comes from a variational principle with a Lagrangian of first-order, then the quantization method that was first proposed in [13] consists of the following steps:

Step I. Find the Lie symmetries of the Lagrange equations

$$\Upsilon = W(t, \mathbf{x})\partial_t + \sum_{k=1}^{N} W_k(t, \mathbf{x})\partial_{x_k}.$$

Step II. Among them, find the Noether symmetries

$$\Gamma = V(t, \mathbf{x})\partial_t + \sum_{k=1}^{N} V_k(t, \mathbf{x})\partial_{x_k}.$$

This may require searching for the Lagrangian yielding the maximum possible number of Noether symmetries [26–29].

Step III. Construct the Schrödinger equation, where we assume $\hbar = 1$ without loss of generality, admitting these Noether symmetries as Lie symmetries, namely

$$2i\psi_t + \sum_{k,j=1}^{N} f_{kj}(\mathbf{x})\psi_{x_j x_k} + \sum_{k=1}^{N} h_k(\mathbf{x})\psi_{x_k} + f_0(\mathbf{x})\psi = 0 \tag{4}$$

admitting the Lie symmetries

$$\Omega = V(t,\mathbf{x})\partial_t + \sum_{k=1}^{N} V_k(t,\mathbf{x})\partial_{x_k} + G(t,\mathbf{x},\psi)\partial_\psi,$$

without adding any other symmetries apart from the two symmetries that are present in any linear homogeneous partial differential equation, namely

$$\psi\partial_\psi, \qquad \alpha(t,\mathbf{x})\partial_\psi,$$

where $\alpha = \alpha(t,\mathbf{x})$ is any solution of the Schrödinger Equation (4).

In [11], we derived w_1 from Equation (2), i.e.,

$$w_1 = -\frac{au^2}{\dot{u} - au}. \tag{5}$$

and consequently obtained a second-order ordinary differential equation in $u \equiv w_2$,

$$\ddot{u} = \frac{(2a-h)\dot{u}^2}{au} - (a+c-2h)\dot{u} - \frac{ac}{K}u^2 + a(c-h)u. \tag{6}$$

Among other cases, we found that if the following relationships among the parameters a, h and c are satisfied

$$a = 2c, \quad h = \frac{3}{2}c \tag{7}$$

then Equation (6), i.e.,

$$\ddot{u} = \frac{5\dot{u}^2}{4u} - \frac{2c^2}{K}u^2 - c^2 u \tag{8}$$

admits a three-dimensional Lie symmetry algebra generated by

$$\Gamma_1 = \partial_t, \qquad \Gamma_2 = e^{-ct}\left(\partial_t + 2cu\partial_u\right) \qquad \Gamma_3 = e^{ct}\left(\partial_t - 2cu\partial_u\right). \tag{9}$$

In [11], we proved that these three operators generate a representation of the complete symmetry group of Equation (8), determined a straightforward Jacobi last multiplier, i.e., [11]

$$M = \frac{\sqrt{u}}{2cu^3}. \tag{10}$$

and consequently the following Lagrangian

$$L = \sqrt{u}\left(\frac{\dot{u}^2}{4cu^3} + \frac{c}{u} - \frac{2c}{K}\right), \tag{11}$$

that admits (9) as Noether symmetries. The three symmetries (9) are a representation of the complete symmetry group of Equation (8), namely a group that completely specifies a given differential equation through its algebraic representation [30]. Indeed, if we impose to the following general second-order ordinary differential equation

$$\ddot{u} = F(t, u, \dot{u}) \tag{12}$$

to admit the symmetry algebra with generators (9), then we obtain the equation

$$\ddot{u} = \frac{5\dot{u}^2}{4u} - Bu^2 - c^2 u, \tag{13}$$

namely a family of equations characterized by the parameter B, which can be replaced with $-2c^2/K$ without any loss of generality.

We remark that it is important to have a Lagrangian, otherwise the quantization of Equation (8) cannot be pursued [31].

The Hamiltonian corresponding to the Lagrangian (11) is

$$H = c\sqrt{u}\left(u^2 p^2 - \frac{1}{u} + \frac{2}{K}\right). \tag{14}$$

In the present paper, we quantize Equation (8), namely we derive the Schrödinger equation by preserving the Noether symmetries. We follow the three steps given above, i.e.,

Step I. We have found the three Lie symmetries, i.e., (9).
Step II. Those symmetries are also the Noether symmetries of the Lagrangian (11).
Step III. We consider the general equation

$$2i\Psi_t + f_{11}(u)\Psi_{uu} + h_1(u)\Psi_u + f_0(u)\Psi = 0 \tag{15}$$

and impose that it should admit the following three Lie symmetries

$$\Omega_1 = \Gamma_1 + G_1(t, u, \Psi)\partial_\Psi,\ \Omega_2 = \Gamma_2 + G_2(t, u, \Psi)\partial_\Psi,\ \Omega_3 = \Gamma_3 + G_3(t, u, \Psi)\partial_\Psi, \tag{16}$$

without adding any other symmetries apart from the two symmetries that are present in any linear homogeneous partial differential equation, namely

$$\Psi\partial_\Psi, \qquad \alpha(t, u)\partial_\Psi,$$

where $\alpha = \alpha(t, u)$ is any solution of the Schrödinger Equation (15).

Using ad hoc REDUCE interactive programs [32], we derive the following Schrödinger equation:

$$2i\Psi_t + u^2\sqrt{u}\Psi_{uu} - \sqrt{u}\left(\eta + 4\frac{c^2}{u}\right)\Psi = 0, \tag{17}$$

with η any arbitrary constant.

Indeed, Equation (17) admits a Lie symmetry algebra generated by the following operators:

$$\begin{aligned}
\Omega_1 &= \Gamma_1, \\
\Omega_2 &= \Gamma_2 + ce^{-ct}\left(\frac{3}{2} + 4i\frac{c}{\sqrt{u}}\right)\Psi\partial_\Psi \\
\Omega_3 &= \Gamma_3 + ce^{ct}\left(-\frac{3}{2} + 4i\frac{c}{\sqrt{u}}\right)\Psi\partial_\Psi.
\end{aligned} \tag{18}$$

The spectrum of this Schrödinger equation is continuous. This is not a surprise. In fact, in [11] we have derived that the general solution of Equation (8) is of exponential type, i.e.,

$$u = \frac{64A_2^2K}{\exp(2ct)\left(8 + A_1^2K + 4A_2^2K\exp(-2ct) + 4A_1A_2K\exp(-ct)\right)^2}, \tag{19}$$

with A_1, A_2 arbitrary constants.

However, if we replace t with $\tau = it$, then Equation (8) becomes:

$$\frac{\mathrm{d}^2u}{\mathrm{d}\tau^2} = \frac{5}{4u}\left(\frac{\mathrm{d}u}{\mathrm{d}\tau}\right)^2 + \frac{2c^2}{K}u^2 + c^2u. \tag{20}$$

This equation is one of the isochronous Liénard II equations [33], $\forall h = h(u)$, i.e.,

$$\frac{\mathrm{d}^2u}{\mathrm{d}\tau^2} + \frac{h''}{h'}\left(\frac{\mathrm{d}u}{\mathrm{d}\tau}\right)^2 + \omega^2\frac{h}{h'} + \frac{A}{h'h^3} = 0, \tag{21}$$

with the identification $c^2 = 4\omega^2$, and $h(u) = (AK/(2\omega^2u))^{1/4}$. The quantization of Equation (21) was given in [34]. In [23], the quantization that preserves the Noether symmetries was applied to all equations (21). Equation (20) gives rise to the following Schrödinger equation

$$2i\psi_\tau + \frac{\psi_{uu}}{(h')^2} - \frac{h''\psi_u}{(h')^3} + \left(\frac{A}{h^2} - \omega^2h^2\right)\psi = 0. \tag{22}$$

Its eigenfunctions are:

$$\psi_n = h^{\frac{k+1}{2}}e^{-i\frac{k+2}{2}\omega t - \frac{\omega}{2}h^2}L_n^{k/2}\left(\omega h^2\right), \tag{23}$$

with $k = \sqrt{1-4A}$, and $L_n^{k/2}$ the associated Laguerre polynomials, while the energy eigenvalues are:

$$E_n = 2\omega\left(n + \frac{1}{2} + \frac{k}{4}\right). \tag{24}$$

More details can be found in [23].

3. In the Wake of Volterra: A Superintegrable System

In 1939 Volterra wrote [15]:

> "I have been able to show that the equations of the struggle for existence depend on a question of Calculus of Variations (omissis). In order to obtain this result, I have replaced the notion of *population* by that of *quantity of life* [14]. In this manner I have also obtained some results by which dynamics is brought into relation to problems of the struggle for existence."

The quantity of life X and the population N of a species are connected by the relation

$$N = \frac{\mathrm{d}X}{\mathrm{d}t}. \tag{25}$$

It is immediately obvious that this idea of raising the order of each equation is totally different from that by Trubach and Franco [35], who provided a method for finding a linear Lagrangian for systems of first-order equations. Also, Volterra's method is different from that of deriving a single second-order equation from a system of two first-order equations: indeed, Volterra takes a system of first-order equations and transforms it into a system of second-order equations.

In [12], Volterra's examples were analyzed and their connection to the Jacobi last multiplier was shown.

Let us apply Volterra's method to the system (1)–(2).

We mimic Volterra by introducing what we call the quantity of natural life u_2, such that:

$$w_2 = \frac{du_2}{dt}. \tag{26}$$

Then, Equation (1) becomes a linear differential form in w_1 and u_2, and the substitution $w_1 = r_1 - hu_2$ simplifies it. Consequently, a Riccati equation is obtained, and the following linearizing transformation

$$r_1 = \frac{K}{c}\frac{du_1}{dt}, \tag{27}$$

yields

$$\ddot{u}_1 = \frac{c}{K^2}\left(K(2hu_2 + K)\dot{u}_1 - hc(hu_2 + K)u_2u_1\right), \tag{28}$$

while Equation (2) becomes:

$$\ddot{u}_2 = \frac{a\dot{u}_2}{K\dot{u}_1 - hcu_2u_1}\left(K\dot{u}_1 - hcu_2u_1 - cu_1\dot{u}_2\right). \tag{29}$$

Next, we check whether the system (28)–(29) admits a Lagrangian by applying Douglas's method [36], which consists of determining whether (at least) a Lagrangian exists for systems of two second-order ordinary differential equations that satisfy given conditions. We found that a Lagrangian (and only one) exists if $a = h = c$, namely if the system (28)–(29) becomes the following:

$$\ddot{u}_1 = \frac{c^2u_2}{K^2}(2K\dot{u}_1 - c^2u_1u_2) - \frac{c}{K}(c^2u_1u_2 - K\dot{u}_1), \tag{30}$$

$$\ddot{u}_2 = c\dot{u}_2 + \frac{c^2u_1\dot{u}_2^2}{c^2u_1u_2 - K\dot{u}_1}. \tag{31}$$

In [11], the case $a = h = c$ was discussed in Remark 3 of Section 4 where the general solution of w_2 in (1)–(2) was determined in implicit form, i.e.,

$$\frac{A_1 - 3(c^2+1)u}{3u\sqrt{u}} + \sqrt{\frac{K}{2A_1}}\log\left(\frac{K}{A_1u}\right) + \sqrt{\frac{2K}{A_1}}\log\left(A_1 + \sqrt{A_1(A_1 - c^2u)}\right) = t + A_2.$$

The unique Lagrangian is

$$L = 2K\log(c^2u_1u_2 - K\dot{u}_1)\dot{u}_2 + cu_2(cu_2 + 2K). \tag{32}$$

It admits three Noether symmetries generated by the following operators:

$$\Gamma_1 = \partial_t, \quad \Gamma_2 = c^2tu_1\partial_{u_1} + K\partial_{u_2}, \quad \Gamma_3 = u_1\partial_{u_1}. \tag{33}$$

The corresponding first integrals are:

$$\begin{aligned} I_1 &= -cu_2(cu_2 + 2K) - \frac{2K^2\dot{u}_1\dot{u}_2}{c^2u_1u_2 - K\dot{u}_1}, \\ I_2 &= 2K\log(c^2u_1u_2 - K\dot{u}_1) - 2c^2tu_2 - 2cKt - \frac{2c^2Ktu_1\dot{u}_2}{c^2u_1u_2 - K\dot{u}_1}, \\ I_3 &= u_2 + \frac{Ku_1\dot{u}_2}{c^2u_1u_2 - K\dot{u}_1}, \end{aligned} \tag{34}$$

with gauge functions $F_1 = 0$, $F_2 = 2cKt(cu_2 + K)$, and $F_3 = u_2$, respectively.

The first integral I_1 in (34) is the Hamiltonian. Indeed, if one introduces the momenta p_1, p_2 by means of the Legendre transformation applied to the Lagrangian (32), i.e.,

$$\begin{aligned} p_1 &\equiv \frac{\partial L}{\partial \dot{u}_1} = -\frac{2K^2\dot{u}_2}{c^2u_1u_2 - K\dot{u}_1}, \\ p_2 &\equiv \frac{\partial L}{\partial \dot{u}_2} = 2K\log(c^2u_1u_2 - K\dot{u}_1), \end{aligned} \tag{35}$$

then the corresponding Hamiltonian, i.e. $H \equiv p_1\dot{u}_1 + p_2\dot{u}_2 - L$, is

$$H = -cu_2(cu_2 + 2K) + \frac{1}{K}\left(c^2p_1u_1u_2 - p_1\exp\left(\frac{p_2}{2K}\right)\right), \tag{36}$$

that is I_1 in (34), if $\dot{u}_1, \dot{u}_2$ are replaced with p_1, p_2, as given in (35).

Introducing the momenta into I_2, and I_3 in (34) yields:

$$\begin{aligned} Int_2 &= p_2 - 2c^2tu_2 - 2cKt + \frac{c^2tp_1u_1}{K}, \\ Int_3 &= u_2 - \frac{p_1u_1}{2K}, \end{aligned} \tag{37}$$

respectively. Then the Hamiltonian system is superintegrable, since we have been able to determine three first integrals.

4. Discussion and Final Remarks

We have shown that it is possible to quantize and also detect superintegrability of biological models. The key is to find a Lagrangian and its admitted Noether symmetries. The Easter Island model (1)–(2) was used as an example. Both classical quantization and superintegrability could be achieved if the involved parameters satisfy certain relationships, namely $a = 2c, h = \frac{3}{2}c$, and $a = h = c$, respectively.

Classical quantization was achieved by applying the quantization method that preserves the Noether symmetries [13]. We have derived the Schrödinger equation corresponding to Equation (8) by preserving the three Noether symmetries (9). Also, we have shown that if we replace the independent variable t with $\tau = it$, then Equation (8) is transformed into Equation (20), which is one of the isochronous Liénard II equations [33]. Its corresponding Schrödinger equation was derived in [23,34]. The eigenfunctions and energy eigenvalues are given in (23) and (24), respectively.

About superintegrability, we have applied Volterra's idea [14] to system (1)–(2) by introducing the quantity of natural life u_2 as given in (26), and therefore increasing the order of the system (1)–(2). Consequently, a Riccati equation also appeared and we have applied the known transformation (27) that yields a second-order equation. Thus, we have obtained a system of two second-order equations (28) and (29). Then, we have applied Douglas's method and derived a unique Lagrangian if the involved parameters satisfy the relationships $a = h = c$. Finally, we have shown that the corresponding Hamiltonian (36) yields a superintegrable Hamiltonian system, since we have been able to determine three first integrals (34) by means of Noether's theorem.

We would like to conclude with the following statement in [37]:

> "Only rarely does one find mention, at post-graduate level, of any problem in connection with the process of actually solving such equations. The electronic computer may perhaps be partly to blame for this, since the impression prevails in many quarters that almost any differential equation problem can be merely put on the machine, so that finding an analytical solution is largely a waste of time. This, however, is only a small part of the truth, for at higher levels there are generally so many parameters or boundary conditions involved that numerical solutions, even if practicable, give no real idea of the properties of the equation. Moreover, any analyst of sensibility will feel that to fall back on numerical

techniques savours somewhat of breaking a door with a hammer when one could, with a little trouble, find the key".

Acknowledgments: M.C.N. acknowledges the support of University of Perugia through *Fondi Ricerca di base 2015 (years 2016–2017).*

Author Contributions: The authors contributed equally to this work.

Conflicts of Interest: The authors declare no conflict of interest.

References

1. Krakauer, D.C.; Collins, J.P.; Erwin, D.; Flack, J.C.; Fontana, W.; Laubichler, M.D.; Prohaska, S.J.; West, G.B.; Stadler, P.F. The challenges and scope of theoretical biology. *J. Theor. Biol.* **2011**, *276*, 269–276.
2. Rowe, D. Einstein meets Hilbert: At the Crossroads of Physics and Mathematics. *Phys. Perspect.* **2001**, *3*, 379–424.
3. Nucci, M.C. Using Lie symmetries in epidemiology. *Electron. J. Differ. Equ.* **2004**, *12*, 87–101.
4. Torrisi, V.; Nucci, M.C. Application of Lie group analysis to a mathematical model which describes HIV transmission. In *The Geometrical Study of Differential Equations*; Leslie, J.A., Hobart, T.P., Eds.; American Mathematical Society: Providence, RI, USA, 2001; pp. 11–20.
5. Nucci, M.C.; Leach, P.G.L. An integrable S-I-S model. *J. Math. Anal. Appl.* **2004**, *290*, 506–518.
6. Edwards, M.; Nucci, M.C. Application of Lie group analysis to a core group model for sexually transmitted diseases. *J. Nonlinear Math. Phys.* **2006**, *13*, 211–230.
7. Gradassi, A.; Nucci, M.C. Hidden linearity in systems for competition with evolution in ecology and finance. *J. Math. Anal. Appl.* **2007**, *333*, 274–294.
8. Nucci, M.C.; Leach, P.G.L. Lie integrable cases of the simplified multistrain/two-stream model for tuberculosis and Dengue fever. *J. Math. Anal. Appl.* **2007**, *333*, 430–449.
9. Freeman, J.D. Missed opportunities. *Bull. Am. Math. Soc.* **1972**, *78*, 635–652.
10. Basener, B.; Ross, D.S. Booming and crashing populations and Easter Island. *SIAM J. Appl. Math.* **2005**, *65*, 684–701.
11. Nucci, M.C.; Sanchini, G. Symmetries, Lagrangians and Conservation Laws of an Easter Island Population Model. *Symmetry* **2015**, *7*, 1613–1632.
12. Nucci, M.C.; Tamizhmani, K.M. Lagrangians for biological models. *J. Nonlinear Math. Phys.* **2012**, *19*, 1250021.
13. Nucci, M.C. Quantization of classical mechanics: Shall we Lie? *Theor. Math. Phys.* **2011**, *168*, 994–1001.
14. Volterra, V. Population growth, equilibria, and extinction under specified breeding conditions: A development and extension of the theory of the logisitc curve. *Hum. Biol.* **1938**, *10*, 1–11.
15. Volterra, V. Calculus of Variations and the Logistic Curve. *Hum. Biol.* **1939**, *11*, 173–178.
16. Van Hove, L. Sur certaines représentations unitaires d'un groupe infini de transformations. *Memoires Acad. R. Belg. Cl. Sci.* **1951**, *26*, 1–102.
17. Błaszak, M.; Domański, Z. Canonical transformations in quantum mechanics. *Ann. Phys.* **2013**, *331*, 70–96.
18. Nucci, M.C. Quantizing preserving Noether symmetries. *J. Nonlinear Math. Phys.* **2013**, *20*, 451–463.
19. Gubbiotti, G.; Nucci, M.C. Noether symmetries and the quantization of a Liénard-type nonlinear oscillator. *J. Nonlinear Math. Phys.* **2014**, *21*, 248–264.
20. Nucci, M.C. From Lagrangian to Quantum Mechanics with Symmetries. *J. Phys. Conf. Ser.* **2012**, *380*, 012008.
21. Nucci, M.C. Symmetries for thought. *Math. Notes Miskolc* **2013**, *14*, 461–474.
22. Nucci, M.C. Spectral realization of the Riemann zeros by quantizing $H = w(x)(p + \ell_p^2/p)$: The Lie-Noether symmetry approach. *J. Phys. Conf. Ser.* **2014**, *482*, 012032.
23. Gubbiotti, G.; Nucci, M.C. Quantization of quadratic Liénard-type equations by preserving Noether symmetries. *J. Math. Anal. Appl.* **2015**, *422*, 1235–1246.
24. Nucci, M.C. Ubiquitous symmetries. *Theor. Math. Phys.* **2016**, *188*, 1361–1370.
25. Gubbiotti, G.; Nucci, M.C. Quantization of the dynamics of a particle on a double cone by preserving Noether symmetries. *arXiv* **2016**, arXiv:1607.00543.
26. Nucci, M.C.; Leach, P.G.L. Lagrangians galore. *J. Math. Phys.* **2007**, *48*, 123510.
27. Nucci, M.C.; Leach, P.G.L. Jacobi last multiplier and Lagrangians for multidimensional linear systems. *J. Math. Phys.* **2008**, *49*, 073517.

28. Nucci, M.C.; Tamizhmani, K.M. Using an old method of Jacobi to derive Lagrangians: A nonlinear dynamical system with variable coefficients. *Nuovo Cimento B* **2010**, *125*, 255–269.
29. Nucci, M.C.; Tamizhmani, K.M. Lagrangians for dissipative nonlinear oscillators: The method of Jacobi Last Multiplier. *J. Nonlinear Math. Phys.* **2010**, *17*, 167–178.
30. Krause, J. On the complete symmetry group of the classical Kepler system. *J. Math. Phys.* **1994**, *35*, 5734–5748.
31. Hojman, S.A.; Shepley, L.C. No Lagrangian? No quantization! *J. Math. Phys.* **1991**, *32*, 142–146.
32. Nucci, M.C. Interactive REDUCE programs for calculating Lie point, non-classical, Lie-Bäcklund, and approximate symmetries of differential equations: Manual and floppy disk. In *CRC Handbook of Lie Group Analysis of Differential Equations. Vol. 3: New Trends in Theoretical Developments and Computational Methods*; Ibragimov, N.H., Ed.; CRC Press: Boca Raton, FL, USA, 1996; pp. 415–481.
33. Tiwari, A.K.; Pandey, S.N.; Senthilvelan, M.; Lakshmanan, M. Classification of Lie point symmetries for quadratic Liénard type equation $\ddot{x} + f(x)\dot{x}^2 + g(x) = 0$. *J. Math. Phys.* **2013**, *54*, 053506.
34. Choudhury, A.G.; Guha, P. Quantization of the Liénard II equation and Jacobi's last multiplier. *J. Phys. A Math. Theor.* **2013**, *46*, 165202.
35. Trubatch, S.L.; Franco, A. Canonical Procedures for Population Dynamics. *J. Theor. Biol.* **1974**, *48*, 299–324.
36. Douglas, J. Solution of the Inverse Problem of the Calculus of Variations. *Trans. Am. Math. Soc.* **1941**, *50*, 71–128.
37. Arscott, F.M. *Periodic Differential Equations*; Pergamon Press: Oxford, UK, 1964.

symmetry

MDPI

Article

Symmetry Analysis and Conservation Laws of the Zoomeron Equation

Tanki Motsepa [1], Chaudry Masood Khalique [1] and Maria Luz Gandarias [2,*]

1 International Institute for Symmetry Analysis and Mathematical Modelling, Department of Mathematical Sciences, North-West University, Mafikeng Campus, Private Bag X 2046, Mmabatho 2735, South Africa; ttmotsepa@gmail.com (T.M.); Masood.Khalique@nwu.ac.za (C.M.K.)

2 Departamento de Matemáticas, Universidad de Cádiz, P. O. Box 40, 11510 Puerto Real, Cádiz, Spain

* Correspondence: marialuz.gandarias@uca.es

Academic Editor: Roman M. Cherniha

Received: 27 October 2016; Accepted: 10 February 2017; Published: 21 February 2017

Abstract: In this work, we study the (2 + 1)-dimensional Zoomeron equation which is an extension of the famous (1 + 1)-dimensional Zoomeron equation that has been studied extensively in the literature. Using classical Lie point symmetries admitted by the equation, for the first time we develop an optimal system of one-dimensional subalgebras. Based on this optimal system, we obtain symmetry reductions and new group-invariant solutions. Again for the first time, we construct the conservation laws of the underlying equation using the multiplier method.

Keywords: (2 + 1)-dimensional Zoomeron equation; Lie point symmetries; optimal system; exact solutions; conservation laws; multiplier method

1. Introduction

Many physical phenomena of the real world are governed by nonlinear partial differential equations (NLPDEs). It is therefore absolutely necessary to analyse these equations from the point of view of their integrability and finding exact closed form solutions. Although this is not an easy task, many researchers have developed various methods to find exact solutions of NLPDEs. These methods include the sine-cosine method [1], the extended tanh method [2], the inverse scattering transform method [3], the Hirota's bilinear method [4], the multiple exp-function method [5], the simplest equation method [6,7], non-classical method [8], method of generalized conditional symmetries [9], and the Lie symmetry method [10,11].

This paper aims to study one NLPDE; namely, the (2 + 1)-dimensional Zoomeron equation [12]

$$\left(\frac{u_{xy}}{u}\right)_{tt}-\left(\frac{u_{xy}}{u}\right)_{xx}+2(u^2)_{tx}=0, \tag{1}$$

which has attracted some attention in recent years. Many authors have found closed-form solutions of this equation. For example, the $(G'/G)-$expansion method [12,13], the extended tanh method [14], the tanh-coth method [15], the sine-cosine method [16,17], and the modified simple equation method [18] have been used to find closed-form solutions of (1). The (2 + 1)-dimensional Zoomeron equation with power-law nonlinearity was studied in [19] from a Lie point symmetries point of view and symmetry reductions, and some solutions were obtained. Additionally, in [19], the authors have given a brief history of the (1 + 1)-dimensional Zoomeron equation. See also [20–22].

In this paper we first use the classical Lie point symmetries admitted by Equation (1) to find an optimal system of one-dimensional subalgebras. These are then used to perform symmetry reductions and determine new group-invariant solutions of (1). It should be noted that such approach was previously used for examination of a wide range of nonlinear PDEs [23–31]. Furthermore, we derive the conservation laws of (1) using the multiplier method [32,33].

The paper is organized as follows: in Section 2, we compute the Lie point symmetries of (1) and use them to construct the optimal system of one-dimensional subalgebras. These are then used to perform symmetry reductions and determine new group-invariant solutions of (1). In Section 3, we derive conservation laws of (1) by employing the multiplier method. Finally, concluding remarks are presented in Section 4.

2. Symmetry Reductions and Exact Solutions of (1) Based on Optimal System

In this section, firstly we use the Lie point symmetries admitted by (1) to construct an optimal system of one-dimensional subalgebras. Thereafter, we obtain symmetry reductions and group-invariant solutions based on the optimal system of one-dimensional subalgebras [23,24].

2.1. Lie Point Symmetries of (1)

The Lie point symmetries of the Zoomeron Equation (1) are given by [19]

$$X_1 = \frac{\partial}{\partial t},\ X_2 = \frac{\partial}{\partial x},\ X_3 = \frac{\partial}{\partial y},\ X_4 = t\frac{\partial}{\partial t} + x\frac{\partial}{\partial x} - y\frac{\partial}{\partial y},\ X_5 = 2y\frac{\partial}{\partial y} - u\frac{\partial}{\partial u},$$

which generate a five-dimensional Lie algebra L_5.

2.2. Optimal System of One-Dimensional Subalgebras

In this subsection, we use the Lie point symmetries of (1) to compute an optimal system of one-dimensional subalgebras. We employ the method given in [23,24], which takes a general element from the Lie algebra and reduces it to its simplest equivalent form by using the chosen adjoint transformations

$$\mathrm{Ad}(\exp(\varepsilon X_i))X_j = \sum_{n=0}^{\infty} \frac{\varepsilon^n}{n!}(\mathrm{ad}X_i)^n(X_j) = X_j - \varepsilon[X_i, X_j] + \frac{\varepsilon^2}{2!}[X_i,[X_i,X_j]] - \cdots,$$

where ε is a real number, and $[X_i, X_j]$ denotes the commutator defined by

$$[X_i, X_j] = X_iX_j - X_jX_i.$$

The table of commutators of the Lie point symmetries of Equation (1) and the adjoint representations of the symmetry group of (1) on its Lie algebra are given in Tables 1 and 2, respectively. Then, Tables 1 and 2 are used to construct the optimal system of one-dimensional subalgebras for Equation (1).

Using Tables 1 and 2, we can construct an optimal system of one-dimensional subalgebras, which is given by $\{X_3, X_4, X_5, X_1 + X_3, X_2 + X_3, X_1 + X_5, X_2 + X_5, X_4 + X_5, X_1 + X_2 + X_3, X_1 + X_2 + X_5\}$.

Table 1. Lie brackets for Equation (1).

[,]	X_1	X_2	X_3	X_4	X_5
X_1	0	0	0	X_1	0
X_2	0	0	0	X_2	0
X_3	0	0	0	$-X_3$	$2X_3$
X_4	$-X_1$	$-X_2$	X_3	0	0
X_5	0	0	$-2X_3$	0	0

Table 2. Adjoint representation of subalgebras.

Ad	X_1	X_2	X_3	X_4	X_5
X_1	X_1	X_2	X_3	$X_4-\varepsilon X_1$	X_5
X_2	X_1	X_2	X_3	$X_4-\varepsilon X_2$	X_5
X_3	X_1	X_2	X_3	$X_4+\varepsilon X_3$	$X_5-2\varepsilon X_3$
X_4	$e^{\varepsilon}X_1$	$e^{\varepsilon}X_2$	$e^{-\varepsilon}X_3$	X_4	X_5
X_5	X_1	X_2	$e^{2\varepsilon}X_3$	X_4	X_5

2.3. Symmetry Reductions

In this subsection, we use the optimal system of one-dimensional subalgebras computed in the previous subsection, and present symmetry reductions of (1) to two-dimensional partial differential equations.

For the first operator X_3 of the optimal system, we have the three invariants $s=t,\ r=x,\ f=u$, and using these invariants, (1) reduces to

$$(f^2)_{sr}=0.$$

Likewise for X_4, the invariants $s=ty,\ r=xy,\ f=u$ transforms (1) to

$$\left(\frac{f_{rr}}{f^2}-\frac{f_{ss}}{f^2}+\frac{2f_s^2}{f^3}-\frac{2f_r^2}{f^3}\right)(sf_s+rf_r)_r+\frac{2f_r}{f^2}(sf_s+rf_r)_{rr}-\frac{2f_s}{f^2}(sf_s+rf_r)_{sr}$$
$$+\frac{1}{f}(sf_s+rf_r)_{ssr}-\frac{1}{f}(sf_s+rf_r)_{rrr}+2\left(f^2\right)_{sr}=0.$$

The invariants $s=t,\ r=x,\ f=u\sqrt{y}$ of X_5 reduces (1) to

$$\left(\frac{f_r}{2f}\right)_{rr}-\left(\frac{f_r}{2f}\right)_{ss}+2\left(f^2\right)_{sr}=0.$$

Using the invariants $s=x,\ r=y-t,\ f=u$ of X_1+X_3, (1) reduces to

$$\left(\frac{f_{sr}}{f}\right)_{rr}-\left(\frac{f_{sr}}{f}\right)_{ss}-2\left(f^2\right)_{sr}=0.$$

Similarly, the invariants $s=t,\ r=y-x,\ f=u$ of X_2+X_3 reduces (1) to

$$\left(\frac{f_{rr}}{f}\right)_{rr}-\left(\frac{f_{rr}}{f}\right)_{ss}-2\left(f^2\right)_{sr}=0.$$

The symmetry X_1+X_5 has invariants $s=x,\ r=ye^{-2t},\ f=uy^{1/2}$, and these reduce (1) to

$$\frac{8r^2f_r^2f_{sr}}{f^3}+\frac{f_s^3}{rf^3}-\frac{2f_s^2f_{sr}}{f^3}-\frac{4rf_r^2f_s}{f^3}-\frac{4r^2f_{rr}f_{sr}}{f^2}-\frac{8r^2f_rf_{srr}}{f^2}+\frac{2f_rf_s}{f^2}+\frac{2rf_{rr}f_s}{f^2}-\frac{3f_{ss}f_s}{2rf^2}$$
$$+\frac{2f_sf_{ssr}}{f^2}-\frac{8rf_rf_{sr}}{f^2}+\frac{f_{ss}f_{sr}}{f^2}+\frac{2f_{sr}}{f}+\frac{10rf_{srr}}{f}+\frac{4r^2f_{srrr}}{f}+\frac{f_{sss}}{2rf}-\frac{f_{sssr}}{f}-4\left(f^2\right)_{sr}=0.$$

The invariants $s=t,\ r=ye^{-2x},\ f=uy^{1/2}$ of X_2+X_5 transform (1) to

$$\frac{16r^3 f_r^2 f_{rr}}{f^3} + \frac{8r^2 f_r^3}{f^3} - \frac{4r f_{rr} f_s^2}{f^3} - \frac{2f_r f_s^2}{f^3} - \frac{8r^3 f_{rr}^2}{f^2} - \frac{16r^3 f_r f_{rrr}}{f^2} - \frac{52r^2 f_r f_{rr}}{f^2}$$
$$+ \frac{4r f_s f_{srr}}{f^2} - \frac{12r f_r^2}{f^2} + \frac{2r f_{rr} f_{ss}}{f^2} + \frac{2f_s f_{sr}}{f^2} + \frac{f_r f_{ss}}{f^2} - \frac{2r f_{ssrr}}{f} + \frac{44r^2 f_{rrr}}{f} + \frac{44r f_{rr}}{f}$$
$$+ \frac{8r^3 f_{rrrr}}{f} + \frac{4f_r}{f} - \frac{f_{ssr}}{f} - 8f_{sr} - 8f_r f_s = 0.$$

Using the invariants $s = x/t,\ r = y/t,\ f = tu$ of $X_4 + X_5$, (1) reduces to

$$\frac{2r^2 f_r^2 f_{sr}}{f^3} - \frac{r^2 f_{rr} f_{sr}}{f^2} - \frac{2r^2 f_r f_{srr}}{f^2} + \frac{r^2 f_{srrr}}{f} + \frac{2s^2 f_s^2 f_{sr}}{f^3} - \frac{s^2 f_{ss} f_{sr}}{f^2} - \frac{2s^2 f_s f_{ssr}}{f^2}$$
$$+ \frac{s^2 f_{sssr}}{f} - \frac{2rs f_{sr}^2}{f^2} - 4rf f_{sr} - \frac{6r f_r f_{sr}}{f^2} + \frac{4rs f_r f_s f_{sr}}{f^3} - \frac{2rs f_s f_{srr}}{f^2} + \frac{6r f_{srr}}{f}$$
$$- \frac{2rs f_r f_{ssr}}{f^2} + \frac{2rs f_{ssrr}}{f} - 12 f_s f - \frac{6s f_s f_{sr}}{f^2} + \frac{6 f_{sr}}{f} - 4s f_{ss} f + \frac{6s f_{ssr}}{f}$$
$$- 4r f_r f_s - 4s f_s^2 + \left(\frac{f_{sr}}{f}\right)_{ss} = 0.$$

The operator $X_1 + X_2 + X_3$ has invariants $s = x - t,\ r = y - t,\ f = u$, and with the use of these invariants, (1) reduces to

$$\left(\frac{(f_s + f_r)_r}{f}\right)_{rr} - \left(\frac{(f_s + f_r)_r}{f}\right)_{ss} + 2\left(f^2\right)_{sr} = 0.$$

Finally, $X_1 + X_2 + X_5$ has invariants $s = x - t,\ r = ye^{-2t},\ f = ue^t$, and their use reduces (1) to

$$\frac{8r^2 f_r^2 f_{sr}}{f^3} + \frac{8r f_r f_s f_{sr}}{f^3} - \frac{4r^2 f_{rr} f_{sr}}{f^2} - \frac{8r^2 f_r f_{srr}}{f^2} - \frac{4r f_{sr}^2}{f^2} - \frac{12r f_r f_{sr}}{f^2}$$
$$- \frac{4r f_s f_{srr}}{f^2} - \frac{4r f_r f_{ssr}}{f^2} - \frac{4 f_s f_{sr}}{f^2} + \frac{4r f_{ssrr}}{f} + \frac{12r f_{srr}}{f} + \frac{4 f_{sr}}{f}$$
$$+ \frac{4 f_{ssr}}{f} + \frac{4r^2 f_{srrr}}{f} - 8 f_s f - 8r f_r f_s - 8r f f_{sr} - 4 f_s^2 - 4 f_{ss} f = 0.$$

2.4. Group-Invariant Solutions

We now obtain group-invariant solutions based on the optimal system of one-dimensional subalgebras. However, in this paper we are looking only at some interesting cases.

Case 1. $X_5 = 2y\partial/\partial y - u\partial/\partial u$

The associated Lagrange system to the operator X_4 yields three invariants

$$s = t, \quad r = x, \quad u = y^{-1/2} U(r, s),$$

which give group-invariant solution $u = y^{-1/2} U(s, r)$ and transforms (1) to

$$\left(\frac{U_r}{U}\right)_{ss} - \left(\frac{U_r}{U}\right)_{rr} - 4\left(U^2\right)_{rs} = 0. \qquad (2)$$

This equation has three Lie point symmetries, viz.,

$$\Gamma_1 = \frac{\partial}{\partial s}, \ \Gamma_2 = \frac{\partial}{\partial r}, \ \Gamma_3 = 2s\frac{\partial}{\partial s} + 2r\frac{\partial}{\partial r} - U\frac{\partial}{\partial U}.$$

The symmetry $\Gamma_1 - \nu\Gamma_2$ gives the two invariants $z = r + \nu s$ and $F = U$. Using these invariants, (2) transforms to the nonlinear third-order ordinary differential equation

$$\left(\frac{F'}{F}\right)'' + \frac{4\nu}{1-\nu^2}\left(F^2\right)'' = 0. \tag{3}$$

Integrating (3) twice with respect to z, we obtain

$$F'(z) + \frac{4\nu}{1-\nu^2}(F(z))^3 - k_1 z F(z) - k_2 F(z) = 0, \tag{4}$$

where k_1 and k_2 are constants of integration. The solutions of this equation are given by

$$F(z) = \pm\sqrt{\frac{\sqrt{k_1}(1-\nu^2)\exp\left\{(k_1 z + k_2)^2/k_1\right\}}{k_3\sqrt{k_1}(1-\nu^2)\exp\left\{k_2^2/k_1\right\} + 4\nu\sqrt{\pi}\,\mathrm{erfi}\left((k_1 z + k_2)/\sqrt{k_1}\right)}},$$

where k_3 is a constant of integration and $\mathrm{erfi}(z)$ is the imaginary error function [34]. Thus, solutions of (1) are

$$u(t,x,y) = \pm y^{-1/2}\sqrt{\frac{\sqrt{k_1}(1-\nu^2)\exp\left\{(k_1(x+\nu t) + k_2)^2/k_1\right\}}{k_3\sqrt{k_1}(1-\nu^2)\exp\left\{k_2^2/k_1\right\} + 4\nu\sqrt{\pi}\,\mathrm{erfi}\left((k_1(x+\nu t) + k_2)/\sqrt{k_1}\right)}}.$$

Case 2. $X_1 + X_5 = \partial/\partial t + 2y\partial/\partial y - u\partial/\partial u$

The associated Lagrange system to this operator yields the three invariants

$$s = x, \quad r = ye^{-2t}, \quad u = e^{-t}U,$$

which give group-invariant solution $u = e^{-t}U(s,r)$ and transforms (1) to

$$\begin{aligned} &U\left(U_{sr}\left(4r^2U_{rr} - U_{ss}\right) + 4rU_r\left(3U_{sr} + 2rU_{srr}\right) - 2U_sU_{ssr}\right) + 8U^4\left(rU_{sr} + U_s\right) + 8rU_rU_sU^3 \\ &+U^2\left(U_{sssr} - 4\left(U_{sr} + r\left(3U_{srr} + rU_{srrr}\right)\right)\right) + 2\left(U_s^2 - 4r^2U_r^2\right)U_{sr} = 0. \end{aligned} \tag{5}$$

The Lie point symmetries of the above equation are

$$\Gamma_1 = \frac{\partial}{\partial s}, \ \Gamma_2 = 2r\frac{\partial}{\partial r} - U\frac{\partial}{\partial U}.$$

The symmetry Γ_2 gives the two invariants $z = s$ and $U = r^{-1/2}F$, and using these invariants, (2) transforms to the nonlinear third-order ordinary differential equation

$$\left(\frac{F'}{F}\right)'' = 0. \tag{6}$$

Integrating (6) twice with respect to z, we obtain

$$F'(z) = k_1 z F(z) + k_2 F(z), \tag{7}$$

where k_1 and k_2 are constants of integration. The solution of this equation is given by

$$F(z) = k_3 \exp\left(\frac{k_1}{2}z^2 + k_2 z\right),$$

where k_3 is a constant of integration. Thus, a solution of (1) is

$$u(t,x,y) = k_3 y^{-1/2} \exp\left(\frac{k_1}{2}x^2 + k_2 x\right),$$

which is a steady-state solution.

Case 3. $X_1 + X_2 + X_3$

The associated Lagrange system to this symmetry operator gives three invariants, viz.,

$$s = x - t, \quad r = y - t, \quad U = u,$$

which give group-invariant solution $u = U(s,r)$ and reduces (1) to

$$\begin{aligned} &U^2\left(U_{srrr} + 2U_{ssrr}\right) - 4U^4\left(U_{sr} + U_{ss}\right) - 4U_s U^3\left(U_r + U_s\right) + 2U_r\left(U_r + 2U_s\right)U_{sr} \\ &-U\left(U_{rr}U_{sr} + 2\left(U_{sr}{}^2 + U_s U_{srr} + U_r\left(U_{srr} + U_{ssr}\right)\right)\right) = 0. \end{aligned} \tag{8}$$

The Lie point symmetries of the above equation are

$$\Gamma_1 = \frac{\partial}{\partial s}, \ \Gamma_2 = \frac{\partial}{\partial r}, \ \Gamma_3 = s\frac{\partial}{\partial s} + r\frac{\partial}{\partial r} - U\frac{\partial}{\partial U}.$$

The symmetry $\Gamma_1 - \nu\Gamma_2$ gives the two invariants $z = r + \nu s$ and $F = U$. Using these invariants, (8) transforms to the nonlinear fourth-order ordinary differential equation

$$\left(\frac{F''}{F}\right)'' - \frac{2(\nu+1)}{2\nu+1}\left(F^2\right)'' = 0. \tag{9}$$

Integrating (9) twice with respect to z, we obtain

$$F'' - \frac{2(\nu+1)}{2\nu+1}F^3 - k_1 z F - k_2 F = 0, \tag{10}$$

where k_1 and k_2 are constants of integration. This equation can not be integrated in the closed form. However, by taking $k_1 = 0$, one can obtain its solution in the closed form in the following manner. Multiplying (10) with $k_1 = 0$ by F' and integrating, we obtain

$$F'^2 = \frac{\nu+1}{2\nu+1}F^4 + k_2 F^2 + k_3, \tag{11}$$

where k_3 is a constant of integration. The solution of this equation is given by

$$F(z) = \sqrt{\frac{2k_3(2\nu+1)}{C}}\,\mathrm{sn}\left(\sqrt{\frac{C}{2(2\nu+1)}}\,z + k_4, 2\sqrt{\frac{-k_3(\nu+1)}{Ck_2 + 4k_3 + 4k_3\nu}}\right),$$

where k_4 is a constant of integration, $C = \sqrt{4k_2^2\nu^2 + 4k_2^2\nu + k_2^2 - 16k_3\nu^2 - 24k_3\nu - 8k_3 - 2k_2\nu - k_2} \neq 0$ and sn is the Jacobi elliptic sine function [35]. Thus, a solution of (1) is

$$u(t,x,y) = \sqrt{\frac{2k_3(2\nu+1)}{C}}\,\mathrm{sn}\left(\sqrt{\frac{C}{2(2\nu+1)}}\,(y+\nu x-(\nu+1)t)+k_4, 2\sqrt{\frac{-k_3(\nu+1)}{Ck_2+4k_3+4k_3\nu}}\right).$$

For $k_3 = 0$ we have the solution given by

$$u(t,x,y) = \frac{2k_2(2\nu+1)\exp[\sqrt{k_2}\{\pm(\nu x+y-(\nu+1)t)\}]}{2\nu+1-k_2(\nu+1)\exp[2\sqrt{k_2}\{\pm(\nu x+y-(\nu+1)t)\}]}$$

and when $C = 0$ we have

$$u(t,x,y) = \left\{\sqrt{\frac{\nu+1}{2\nu+1}}(\nu x+y-(\nu+1)t)\right\}^{-1}.$$

Likewise, one may obtain more group-invariant solutions using the other symmetry operators of the optimal system of one-dimensional subalgebras. For example, the symmetry operator $X_2 + X_3$ of the optimal system gives us the group-invariant solution (2.9) of [19] in terms of the Airy functions.

3. Conservation Laws of (1)

Conservation laws describe physical conserved quantities, such as mass, energy, momentum and angular momentum, electric charge, and other constants of motion [32]. They are very important in the study of differential equations. Conservation laws can be used in investigating the existence, uniqueness, and stability of the solutions of nonlinear partial differential equations. They have also been used in the development of numerical methods and in obtaining exact solutions for some partial differential equations.

A local conservation law for the (2 + 1)-dimensional Zoomeron Equation (1) is a continuity equation

$$D_tT + D_xX + D_yY = 0 \tag{12}$$

holding for all solutions of Equation (1), where the conserved density T and the spatial fluxes X and Y are functions of t, x, y, u. The results in [11] show that all non-trivial conservation laws arise from multipliers. Specifically, when we move off of the set of solutions of Equation (1), every non-trivial local conservation law (12) is equivalent to one that can be expressed in the characteristic form

$$D_t\tilde{T} + D_x\tilde{X} + D_y\tilde{Y} = \left(\left(\frac{u_{xy}}{u}\right)_{tt} - \left(\frac{u_{xy}}{u}\right)_{xx} + 2(u^2)_{tx}\right)Q \tag{13}$$

holding off of the set of solutions of Equation (1) where $Q(x,y,t,u\ldots)$ is the multiplier, and where $(\tilde{T},\tilde{X},\tilde{Y})$ differs from (T,X,Y) by a trivial conserved current. On the set of solutions $u(x,y,t)$ of Equation (1), the characteristic form (13) reduces to the conservation law (12).

In general, a function $Q(x,t,u\ldots)$ is a multiplier if it is non-singular on the set of solutions $u(x,y,t)$ of Equation (1), and if its product with Equation (1) is a divergence expression with respect to t, x, y. There is a one-to-one correspondence between non-trivial multipliers and non-trivial conservation laws in characteristic form.

The determining equation to obtain all multipliers is

$$\frac{\delta}{\delta u}\left(\left(\frac{u_{xy}}{u}\right)_{tt} - \left(\frac{u_{xy}}{u}\right)_{xx} + 2(u^2)_{tx}\right)Q = 0, \tag{14}$$

where $\delta/\delta u$ is the Euler–Lagrange operator given by

$$\frac{\delta}{\delta u} = \frac{\partial}{\partial u} + \sum_{s\geq 1}(-1)^s D_{i_1}\cdots D_{i_s}\frac{\partial}{\partial u_{i_1 i_2 \cdots i_s}}.$$

Equation (14) must hold off of the set of solutions of Equation (1). Once the multipliers are found, the corresponding non-trivial conservation laws are obtained by integrating the characteristic Equation (13) [11].

We will now find all multipliers $Q(x,y,t,u)$ and obtain corresponding non-trivial (new) conservation laws. The determining Equation (14) splits with respect to the variables $u_t, u_x, u_y, u_{tt}, u_{tx}, u_{ty}, u_{xy}, u_{yy}, u_{ttt}, u_{ttx}, u_{txy}, u_{tyy}, u_{xyy}, u_{tttx}, u_{txyy}$. This yields a linear determining system for $Q(x,y,t,u)$ which can be solved by the same algorithmic method used to solve the determining equation for infinitesimal symmetries. By applying this method, for Equation (1), we obtain the following linear determining equations for the multipliers:

$$Q_u\,(t,x,y,u) = 0, \tag{15}$$

$$Q_{ty}\,(t,x,y,u) = 0, \tag{16}$$

$$Q_{yyy}\,(t,x,y,u) = 0, \tag{17}$$

$$Q_{tt}\,(t,x,y,u) - Q_{yy}\,(t,x,y,u) = 0. \tag{18}$$

It is straightforward using Maple to set up and solve this determining system (15)–(18), and we get the four multipliers given by

$$Q_1 = \frac{1}{2}\left(t^2+y^2\right) f_1(x), \tag{19}$$

$$Q_2 = f_2(x)y, \tag{20}$$

$$Q_3 = f_3(x)t, \tag{21}$$

$$Q_4 = f_4(x). \tag{22}$$

For each solution Q, a corresponding conserved density and flux can be derived (up to local equivalence) by integration of the divergence identity (13) [11,36]. We obtain the following results.

Corresponding to these multipliers, we obtain four conservation laws. Thus, the multiplier (19) gives the conservation law with the following conserved vector:

$$\begin{aligned} T_1 &= f_1(x)\left\{\frac{1}{2}(t^2+y^2)\left(\frac{{u_t}^2 u_x}{u^3} - \frac{u_x u_{tt}}{u^2}\right) + \frac{t u_x u_t}{u^2} - 2yu^2\right\} \\ &\quad + f_1'(x)\left\{\frac{1}{2}(t^2+y^2)\left(\frac{u_{tt}}{u} - \frac{1}{2}\frac{{u_t}^2}{u^2}\right) - \frac{t u_t}{u}\right\}, \\ X_1 &= f_1(x)\left\{\frac{1}{2}(t^2+y^2)\left(\frac{2u_t u_{tt}}{u^2} - \frac{u_{ttt}}{u} - \frac{{u_t}^3}{u^3}\right) - \frac{1}{2}\frac{{u_t}^2 t}{u^2} + \frac{u_t}{u}\right\}, \\ Y_1 &= f_1(x)\left\{\frac{1}{2}(t^2+y^2)\left(4uu_t + \frac{u_{txy}}{u} - \frac{u_y u_{tx}}{u^2}\right) - \frac{y u_{tx}}{u}\right\}. \end{aligned}$$

Likewise, the multiplier (20) yields

$$\begin{aligned} T_2 &= f_2(x)y\left(4\,uu_y - \frac{u_x u_{tt}}{u^2} + \frac{{u_t}^2 u_x}{u^3}\right) + f_2'(x)y\left(\frac{u_{tt}}{u} - \frac{1}{2}\frac{{u_t}^2}{u^2}\right), \\ X_2 &= f_2(x)y\left(\frac{2u_t u_{tt}}{u^2} - \frac{u_{ttt}}{u} - \frac{{u_t}^3}{u^3}\right), \\ Y_2 &= f_2(x)\left(\frac{y u_{txy}}{u} - \frac{y u_y u_{tx}}{u^2} - \frac{u_{tx}}{u}\right) \end{aligned}$$

as conserved vector.

Similarly, the multiplier (21) results in the following conserved vector

$$
\begin{aligned}
T_3 &= f_3(x)\left(4\,tuu_y - \frac{tu_xu_{tt}}{u^2} + \frac{tu_t{}^2u_x}{u^3} + \frac{u_xu_t}{u^2}\right) + f_3'(x)\left(\frac{tu_{tt}}{u} - \frac{1}{2}\frac{tu_t{}^2}{u^2} - \frac{u_t}{u}\right),\\
X_3 &= f_3(x)\left(\frac{2tu_tu_{tt}}{u^2} - \frac{tu_t{}^3}{u^3} - \frac{1}{2}\frac{u_t{}^2}{u^2} - \frac{tu_{ttt}}{u}\right),\\
Y_3 &= f_3(x)\frac{\left(tuu_{txy} - 2\,u^4 - tu_yu_{tx}\right)}{u^2}.
\end{aligned}
$$

Lastly, the multiplier (22) gives the conserved vector whose components are

$$
\begin{aligned}
T_4 &= f_4(x)\left(4uu_y - \frac{u_xu_{tt}}{u^2} + \frac{u_t{}^2u_x}{u^3}\right) + f_4'(x)\left(\frac{u_{tt}}{u} - \frac{1}{2}\frac{u_t{}^2}{u^2}\right),\\
X_4 &= f_4(x)\left(\frac{2u_tu_{tt}}{u^2} - \frac{u_{ttt}}{u} - \frac{u_t{}^3}{u^3}\right),\\
Y_4 &= f_4(x)\left(\frac{u_{txy}}{u} - \frac{u_yu_{tx}}{u^2}\right).
\end{aligned}
$$

4. Concluding Remarks

In this paper, we studied the (2 + 1)-dimensional Zoomeron Equation (1). For the first time, the classical Lie point symmetries were used to construct an optimal system of one-dimensional subalgebras. This system was then used to obtain symmetry reductions and new group-invariant solutions of (1). Again for the first time, we derived the conservation laws for (1) by employing the multiplier method. We note that since we had arbitrary functions in the multipliers, we obtained infinitely many conservation laws for Equation (1).

Acknowledgments: Tanki Motsepa would like to thank the North-West University, Mafikeng Campus and DST-NRF CoE-MaSS of South Africa for their financial support. Maria Luz Gandarias thanks the support of DGICYT project MTM2009-11875 and Junta de Andalucía group FQM-201.

Author Contributions: Tanki Motsepa, Chaudry Masood Khalique and Maria Luz Gandarias worked together in the derivation of the mathematical results. All authors read and approved the final manuscript.

Conflicts of Interest: The authors declare no conflict of interest.

References

1. Wazwaz, M. The Tanh and Sine-Cosine Method for Compact and Noncompact Solutions of Nonlinear Klein Gordon Equation. *Appl. Math. Comput.* **2005**, *167*, 1179–1195.
2. Wazwaz, A.M. The extended tanh method for the Zakharo-Kuznetsov (ZK) equation, the modified ZK equation, and its generalized forms. *Commun. Nonlinear Sci. Numer. Simul.* **2008**, *13*, 1039–1047.
3. Ablowitz, M.J.; Clarkson, P.A. *Soliton, Nonlinear Evolution Equations and Inverse Scattering*; Cambridge University Press: Cambridge, UK, 1991.
4. Hirota, R. *The Direct Method in Soliton Theory*; Cambridge University Press: Cambridge, UK, 2004.
5. Ma, W.X.; Huang, T.; Zhang, Y. A multiple exp-function method for nonlinear differential equations and its applications. *Phys. Scr.* **2010**, *82*, 065003.
6. Kudryashov, N.A. Simplest equation method to look for exact solutions of nonlinear differential equations. *Chaos Solitons Fractals* **2005**, *24*, 1217–1231.
7. Kudryashov, N.A. One method for finding exact solutions of nonlinear differential equations. *Commun. Nonlinear Sci. Numer. Simul.* **2012**, *17*, 2248–2253.
8. Bluman, W.; Cole, J.D. The general similarity solutions of the heat equation. *J. Math. Mech.* **1969**, *18*, 1025–1042.

9. Fokas, A.S.; Liu, Q.M. Generalized conditional symmetries and exact solutions of nonintegrable equations. *Theor. Math. Phys.* **1994**, *99*, 263–277.
10. Olver, P.J. *Applications of Lie Groups to Differential Equations*; 2nd ed.; Springer: Berlin, Germany, 1993.
11. Bluman, G.W.; Cheviakov, A.F.; Anco, S.C. *Applications of Symmetry Methods to Partial Differential Equations*; Springer: New York, NY, USA, 2010.
12. Abazari, R. The solitary wave solutions of Zoomeron equation. *Appl. Math. Sci.* **2011**, *5*, 2943–2949.
13. Khan, K.; Akbar, M.A.; Salam, M.A.; Islam, M.H. A note on enhanced $(G'/G)-$expansion method in nonlinear physics. *Ain Shams Eng.* **2014**, *5*, 877–884.
14. Alquran, M.; Al-Khaled, K. Mathematical methods for a reliable treatment of the (2 + 1)-dimensional Zoomeron equation. *Math. Sci.* **2012**, *6*, 1–5.
15. Irshad, A.; Mohyud-Din, S.T. Solitary wave solutions for Zoomeron equation. *Walailak J. Sci. Technol.* **2013**, *10*, 201–208.
16. Qawasmeh, A. Soliton solutions of (2 + 1)-Zoomeron equation and Duffing equation and SRLW equation. *J. Math. Comput. Sci.* **2013**, *3*, 1475–1480.
17. Gao, H. Symbolic computation and new exact travelling solutions for the (2 + 1)-dimensional Zoomeron equation. *Int. J. Mod. Nonlinear Theory Appl.* **2014**, *3*, 23–28.
18. Khan, K.; Akbar, M.A. Traveling wave solutions of the (2 + 1)-dimensional Zoomeron equation and the Burgers equations via the MSE method and the Exp-function method. *Ain Shams Eng. J.* **2014**, *5*, 247–256.
19. Morris, R.M.; Leach, P.G.L. Symmetry reductions and solutions to the Zoomeron equation. *Phys. Scr.* **2015**, *90*, 015202.
20. Calogero, F.; Degasperis, A. Nonlinear Evolution Equations Solvable by the Inverse Spectral Transform Associated with the Matrix Schrödinger Equation; In *Solitons*; Bullough, R.K., Caudrey, P.J., Eds.; Springer: Berlin, Germany, 1980; pp. 301–324.
21. Calogero, F.; Degasperis, A. *Spectral Transform and Solitons II: Tools to Solve and Investigate Nonlinear Evolution Equations*; North-Holland: Amsterdam, The Netherlands, 1982.
22. Calogero, F.; Degasperis, A. New integrable PDEs of boomeronic type. *J. Phys. A Math. Gen.* **2006**, *39*, 8349–8376.
23. Patera, J.; Winternitz, P.; Zassenhaus, H. Continuous subgroups of the fundamental groups of physics: I. General method and the Poincare group. *J. Math. Phys.* **1975**, *16*, 1597–1615.
24. Patera, J.; Winternitz, P.; Zassenhaus, H. Continuous subgroups of the fundamental groups of physics: II. The similitude group. *J. Math. Phys.* **1975**, *16*, 1616–1624.
25. Boyer, C.P.; Sharp, R.T.; Winternitz, P. Symmetry breaking interactions for the time dependent Schrödinger equation. *J. Math. Phys.* **1976**, *17*, 1439–1451.
26. Cherniha, R. Galilei-invariant non-linear PDEs and their exact solutions. *J. Nonlinear Math. Phys.* **1995**, *2*, 374–383.
27. Cherniha, R. Lie Symmetries of nonlinear two-dimensional reaction-diffusion systems. *Rept. Math. Phys.* **2000**, *46*, 63–76.
28. Fedorchuk, V.; Fedorchuk, V. On Classification of symmetry reductions for the Eikonal equation. *Symmetry* **2016**, *8*, 51.
29. Fushchich, W.; Cherniga, R. Galilei-invariant nonlinear equations of Schrödinger-type and their exact solutions I. *Ukr. Math. J.* **1989**, *41*, 1161–1167
30. Cherniha, R. Exact Solutions of the Multidimensional Schrödinger Equation with Critical Nonlinearity. In *Group and Analytic Methods in Mathematical Physics, Proceedings of Institute of Mathematics*; NAS of Ukraine: Kiev, Ukraine, 2001; Volume 36, pp. 304–315.
31. Cherniha, R.; Kovalenko, S. Lie symmetries and reductions of multidimensional boundary value problems of the Stefan type. *J. Phys. A Math. Theor.* **2011**, *44*, 485202.
32. Anco, S.C.; Bluman, G. Direct construction method for conservation laws for partial differential equations Part II: General treatment. *Eur. J. Appl. Math.* **2002**, *13*, 567–585.
33. Bruzón, M.S.; Garrido, T.M.; de la Rosa, R. Conservation laws and exact solutions of a Generalized Benjamin–Bona–Mahony–Burgers equation. *Chaos Solitons Fractals* **2016**, *89*, 578–583.
34. Weisstein, E.W. "Erfi." From MathWorld–A Wolfram Web Resource. Available online: http://mathworld.wolfram.com/Erfi.html (accessed on 27 October 2016).

35. Abramowitz, M.; Stegun, I.A. *Handbook of Mathematical Functions with Formulas, Graphs, and Mathematical Tables*; 10th ed.; Dover Publications: New York, NY, USA, 1972.
36. Anco, S.C. Generalization of Noether's Theorem in Modern Form to Non-Variational Partial Differential Equations. In *Recent progress and Modern Challenges in Applied Mathematics, Modeling and Computational Science*; Springer: New York, NY, USA, 2016.

symmetry

MDPI

Article

An Application of Equivalence Transformations to Reaction Diffusion Equations

Mariano Torrisi †,* and Rita Tracinà †

Department of Mathematics and Computer Sciences, University of Catania, Catania I 95125, Italy; tracina@dmi.unict.it

* Author to whom correspondence should be addressed; torrisi@dmi.unict.it or m.torrisi12@gmail.it.

† These authors contributed equally to this work.

Academic Editor: Roman M. Cherniha

Received: 15 July 2015; Accepted: 15 October 2015; Published: 23 October 2015

Abstract: In this paper, we consider a quite general class of advection reaction diffusion systems. By using an equivalence generator, derived in a previous paper, the authors apply a projection theorem to determine some special forms of the constitutive functions that allow the extension by one of the two-dimensional principal Lie algebra. As an example, a special case is discussed at the end of the paper.

Keywords: equivalence transformations; groups of transformations; classical symmetries; biomathematical models

---

1. Introduction

In this paper, we focus our attention on the following family of 2×2 nonlinear advection reaction diffusion systems in $(1+1)$ independent variables:

$$\begin{cases} u_t = (f(u)u_x)_x + g(u, v, u_x), \\ v_t = h(u, v), \end{cases} \tag{1}$$

with $f(u), g(u, v, u_x), h(u, v)$ analytic functions. These systems, apart from their own mathematical interest, offer the possibility to be analyzed as possible biomathematical models for two interacting species u and v, where one of them, the species v, does not suffer diffusion. The dependence of the function g on the gradient u_x shows advection effects; in fact, the individuals of the species u could be influenced by external stimuli as wind velocity or water currents. Of course, the absence of the advective phenomena brings to the following system:

$$\begin{cases} u_t = (f(u)u_x)_x + g(u, v), \\ v_t = h(u, v), \end{cases} \tag{2}$$

that can describe the evolution of the *Aedes aegypti* mosquito population in a region where wind effects are negligible or the evolution of a *Proteus mirabilis* bacterial colony when the diffusion coefficient depends only on the species u, that is when the system (2) is a subclass of the following wider class:

$$\begin{cases} u_t = (f(u, v)u_x)_x + g(u, v), \\ v_t = h(u, v), \end{cases} \tag{3}$$

considered in [1–4].

One of the most important problems in modeling the phenomena of life sciences and natural sciences is to select "good" forms of arbitrary functions (constitutive equations) that fit well with the experimental data and possess mathematical properties that allow scientists to get some solutions or much news about them.

A powerful tool of investigation in this field is given from transformation groups, in particular from equivalence transformations and symmetries.

In the framework of the group analysis, the literature concerning the systems of the type (1) is scarce. There are no papers devoted to a complete Lie symmetry analysis of PDE systems with advection (convection) terms of the form (1). In [5], it is possible to find a complete solution of this problem for a class of diffusion systems with convection terms in both equations. Moreover, the paper [6] contains some description of Lie symmetries for a class of systems, which includes cases having a structure similar to system (3). However, it is possible to find some papers devoted to the complete Lie symmetry analysis of a single advection (convection) reaction diffusion equation (see, e.g., [7–9]).

Following, e.g., [10], an equivalence transformation for the system (1) is a non-degenerate change of the independent and dependent variables t, x, u, v into $\hat{t}, \hat{x}, \hat{u}, \hat{v}$:

$$\begin{cases} x = x(\hat{x},\hat{t},\hat{u},\hat{v}), \\ t = t(\hat{x},\hat{t},\hat{u},\hat{v}), \\ u = u(\hat{x},\hat{t},\hat{u},\hat{v}), \\ v = v(\hat{x},\hat{t},\hat{u},\hat{v}), \end{cases} \tag{4}$$

that transforms a system of the class (1) in another one of the same class. That is, an equivalence transformation brings the system of the form (1) in a system preserving the differential structure, but, in general, with:

$$\hat{f}(\hat{u}) \neq f(u), \hat{g}(\hat{u},\hat{v},\hat{u}_{\hat{x}}) \neq g(u,v,u_x), \hat{h}(\hat{u},\hat{v}) \neq h(u,v). \tag{5}$$

It maps a solution of a system in a solution of the equivalent system.

It could occur that the transformed equations show still the same structure, but the arbitrary functions are depending on additional variables. In this case, the equivalence is said *weak*.

Of course, in the case:

$$\hat{f}(\hat{u}) = f(u), \hat{g}(\hat{u},\hat{v},\hat{u}_{\hat{x}}) = g(u,v,u_x), \hat{h}(\hat{u},\hat{v}) = h(u,v), \tag{6}$$

an equivalence transformation becomes a symmetry (a transformation of variables that leaves invariant the transformed system).

A symmetry allows one to reduce the number of independent variables of an equation so that, for instance, a PDE in $1+1$ independent variables can become an ODE. Once solved this last one, going back to the original variables, we get a solution that is invariant with respect to the symmetries used for the reduction. It is worthwhile to note that a symmetry transforms invariant solutions into invariant solutions that are not essentially different (see Ovsiannikov [11]), but, having a different form, they could satisfy different suitable initial/boundary conditions.

The aim of this paper is an improvement of the results that we have shown in [12], bearing in mind some generalization of the special form assumed from the constitutive functions f, g and h already used in some previous papers about [4,12–14]. In this paper, we use the infinitesimal generator of equivalence transformations derived in [12] for the class (1) in order to obtain some extensions of the principal Lie algebra for the following subclass:

$$\begin{aligned} u_t &= (f(u)u_x)_x + u^r u_x + \Gamma_1(u) + \Gamma_2(v), \\ v_t &= h(u,v). \end{aligned} \tag{7}$$

Here, we assumed:

$$g(u, v, u_x) = u^r u_x + \Gamma_1(u) + \Gamma_2(v) \tag{8}$$

that is a generalization of that ones used in [12,13].

In the next section, after recalling, for the sake of completeness, some elements about equivalence transformations (for additional mathematical and methodical details, the interested reader can see [12, 15,16]), we write the equivalence generator derived in [12]. In Section 3, the principal Lie algebra and its extensions are discussed; moreover, a simple example that could be related to the biomathematical model of Aedes aegypti is considered. The conclusions are given in the last section.

2. On Equivalence Transformations and Their Calculation for the Class (1)

It is easy to ascertain that, often, in papers on differential equations, it is possible to find several examples of equivalence transformations and their applications. In general, in the past and now, the direct search for the most general equivalence transformations through the finite form of the transformation has been used. Quite often, this search is connected to considerable computational difficulties and does not always lead to the complete solution of the problem (e.g., [17,18]).

2.1. Elements on Equivalence Transformations

Following [11,15,16,19,20] (see also, e.g., [10,21–23]), we look for the infinitesimal generator of the equivalence transformations of the system (1) of the form:

$$Y = \xi^1\partial_x + \xi^2\partial_t + \eta^1\partial_u + \eta^2\partial_v + \mu^1\partial_f + \mu^2\partial_g + \mu^3\partial_h \tag{9}$$

where the infinitesimal components ξ^1, ξ^2, η^1, η^2 are sought depending on x, t, u, v, while the infinitesimal components μ^i $(i = 1,2,3)$ can also depend on u_t, u_x, v_t, v_x, f, g and h. Here, we are interested in obtaining the infinitesimal coordinates ξ^i, η^i and μ^j $(i = 1,2$ and $j = 1,2,3)$, by applying the Lie–Ovsiannikov infinitesimal criterion [11] by requiring the invariance, with respect to a suitable prolongations $Y^{(1)}$ and $Y^{(2)}$ of generator (9), of the following equations:

$$u_t - (f u_x)_x - g = 0, \tag{10}$$

$$v_t - h = 0, \tag{11}$$

without requiring the invariance of the so-called *auxiliary conditions* [15,16,24,25]:

$$f_t = f_x = f_v = f_{u_x} = f_{u_t} = f_{v_x} = f_{v_t} = g_t = g_x = g_{u_t} = g_{v_t} = g_{v_x} = 0, \tag{12}$$

$$h_t = h_x = h_{u_x} = h_{u_t} = h_{v_x} = h_{v_t} = 0, \tag{13}$$

that characterize the functional dependence of f, g and h.

In this way, we obtain the weak equivalence transformations [15,16].

The main difference with respect to the classical one is that the infinitesimal operators of weak equivalence transformations can generate transformations that do not preserve the functional dependence of the arbitrary elements.

With respect to the application in biomathematical models, equivalence and weak equivalence transformations were applied not only to study tumor models [26,27], but also the population dynamics in [1,3,4].

2.2. Calculation of Weak Equivalence Transformations

We need the following prolongations $Y^{(1)}$ and $Y^{(2)}$:

$$Y^{(1)} = Y + \zeta_1^1\partial_{u_x} + \zeta_2^1\partial_{u_t} + \zeta_1^2\partial_{v_x} + \zeta_2^2\partial_{v_t} + \omega_u^1\partial_{f_u}, \tag{14}$$

$$Y^{(2)} = Y^{(1)} + \zeta^1_{xx}\partial_{u_{xx}}, \tag{15}$$

with (see [12] for more details),

$$\zeta^1_1 = D_x\eta^1 - u_x D_x\xi^1 - u_t D_x\xi^2, \tag{16}$$

$$\zeta^1_2 = D_t\eta^1 - u_x D_t\xi^1 - u_t D_t\xi^2, \tag{17}$$

$$\zeta^2_1 = D_x\eta^2 - v_x D_x\xi^1 - v_t D_x\xi^2, \tag{18}$$

$$\zeta^2_2 = D_t\eta^2 - v_x D_t\xi^1 - v_t D_t\xi^2, \tag{19}$$

$$\zeta^1_{11} = D_x\zeta^1_1 - u_{xx} D_x\xi^1 - u_{tx} D_x\xi^2, \tag{20}$$

$$\omega^1_u = \widetilde{D}_u\mu^1 - f_u\widetilde{D}_u\eta^1, \tag{21}$$

where D_x and D_t are, respectively, the total derivatives with respect to x and t, while in our case, the operator $\widetilde{D}_u$ is defined as:

$$\widetilde{D}_u = \partial_u + f_u\partial_f + g_u\partial_g + h_u\partial_h. \tag{22}$$

The invariant conditions read:

$$\zeta^1_2 - 2\zeta^1_1 u_x f_u - u_x^2\omega^1_u - u_{xx}\mu^1 - f\zeta^1_{11} - \mu^2 = 0, \tag{23}$$

$$\zeta^2_2 - \mu^3 = 0, \tag{24}$$

both under the constraints (10) and (11).

Following the usual techniques, we derive the following infinitesimal components for the weak equivalence generators:

$$\xi^1 = \alpha(x), \xi^2 = \beta(t), \eta^1 = \delta(t,u), \eta^2 = \lambda(x,t,v), \tag{25}$$

$$\mu^1 = (2\alpha' - \beta')f, \mu^2 = \delta_t + (\delta_u - \beta')g + \left(\alpha'' u_x - \delta_{uu}u_x^2\right)f, \mu^3 = (\lambda_v - \beta')h + \lambda_t, \tag{26}$$

where $\alpha(x), \beta(t), \delta(t,u), \lambda(x,t,v)$ are arbitrary real functions of their arguments. The corresponding infinitesimal generator is:

$$\begin{aligned} Y &= \alpha(x)\partial_x + \beta(t)\partial_t + \delta(t,u)\partial_u + \lambda(x,t,v)\partial_v + (2\alpha' - \beta')f\partial_f \\ &+ (\delta_t + (\delta_u - \beta')g + (\alpha'' u_x - \delta_{uu}u_x^2)f)\partial_g + ((\lambda_v - \beta')h + \lambda_t)\partial_h. \end{aligned} \tag{27}$$

3. Symmetries for a Subclass of Advection Reaction Diffusion Systems

In this section, we apply the projection theorem, introduced in [28] and successively generalized in [15,16,24], in order to carry out a symmetry classification for the following subclass of system (1):

$$\begin{aligned} u_t &= (f(u)u_x)_x + u^r u_x + \Gamma_1(u) + \Gamma_2(v), \\ v_t &= h(u,v), \end{aligned} \tag{28}$$

with $r \neq 0$, ${\Gamma_1}' \neq 0$ and ${\Gamma_2}' \neq 0$.

For the system (28), we can affirm the following:

Theorem 1. *The projection of the infinitesimal weak equivalence generator Y for the system* (1) *on the space (x,t,u,v):*

$$X = \alpha(x)\partial_x + \beta(t)\partial_t + \delta(t,u)\partial_u + \lambda(x,t,v)\partial_v \tag{29}$$

is an infinitesimal symmetry generator of a system of the class (28) *if and only if the constitutive equations, specifying the forms of f, g and h, are invariant with respect to Y.*

Applying the previous theorem, in order to obtain the determining system for the subclass (28), we request the invariance with respect to Y of the following constitutive equations:

$$\begin{aligned} f &= f(u), \\ g &= u^r u_x + \Gamma_1(u) + \Gamma_2(v), \\ h &= h(u,v), \end{aligned} \tag{30}$$

that is

$$\begin{aligned} Y(f - f(u)) &= 0, \\ Y(g - u^r u_x - \Gamma_1(u) - \Gamma_2(v)) &= 0, \\ Y(h - h(u,v)) &= 0, \end{aligned} \tag{31}$$

under the constraints (30). Then, taking into account the form (27) of generator Y, we have the following determining equations:

$$(2\alpha' - \beta')f - \delta f_u = 0, \tag{32}$$

$$\begin{aligned} \delta_t + \ & (\delta_u - \beta')(u^r u_x + \Gamma_1 + \Gamma_2) - (\delta_{uu} u_x^2 - \alpha'' u_x) f - (\delta_u - \alpha') u_x u^r - \\ & \delta(r u^{r-1} u_x + \Gamma_1{}') - \lambda \Gamma_2{}' = 0, \end{aligned} \tag{33}$$

$$(\lambda_v - \beta')h + \lambda_t - \delta h_u - \lambda h_v = 0. \tag{34}$$

We recall here that the principal Lie algebra $L_{\mathcal{P}}$ [10,19] is the Lie algebra that leaves invariant the system (28) for any form of the functions $f(u), \Gamma_1(u), \Gamma_2(v)$ and $h(u,v)$. Then, the principal Lie algebra is the generator (29) where the functions α, β, δ and λ are solutions of the system (32)–(34) for arbitrary functions $f(u), \Gamma_1(u), \Gamma_2(v)$ and $h(u,v)$. Consequently, it is a simple matter to ascertain the following:

Corollary 2. *The projection* (29) *of infinitesimal weak equivalence generator Y for the system* (1) *on the space (x,t,u,v) is the infinitesimal symmetry generator corresponding to the principal Lie algebra of the class* (28) *if and only if $\eta^i = 0$, $\mu^j = 0$, $i = 1, 2$, $j = 1, 2, 3$.*

Then, the principal Lie algebra $L_{\mathcal{P}}$ is spanned by the following generators corresponding respectively to translations in time and in the space:

$$X_1 = \partial_t, \ X_2 = \partial_x. \tag{35}$$

3.1. On the Extensions of the $L_{\mathcal{P}}$

Here, we analyze some particular cases of the extension of the principal algebra for the class (28). That is, we look for a family of particular functions $f(u), \Gamma_1(u), \Gamma_2(v)$ and $h(u,v)$, such that the solution of system (32)–(34) is different from $\delta = \lambda = 0$ and α and β constants, which corresponds to the generators (35).

From Equation (32), deriving with respect to x, we get:

$$2\alpha'' f = 0, \tag{36}$$

that is

$$\alpha(x) = a_1 x + a_0, \tag{37}$$

with a_0 and a_1 arbitrary constants. Consequently, from Equation (33), deriving with respect to x, we get:

$$-\lambda_x \Gamma_2{}' = 0, \tag{38}$$

and taking into account that $\Gamma_2{}' \neq 0$, we obtain:

$$\lambda = \lambda(t,v). \tag{39}$$

Equation (33) becomes:

$$\delta_t + (\delta_u - \beta')(\Gamma_1 + \Gamma_2) - \delta\Gamma_1' - \lambda\Gamma_2' + u_x\left((a_1 - \beta')u^r - \delta r u^{r-1}\right) + u_x^2(-\delta_{uu} f) = 0. \tag{40}$$

As any function does not depend on u_x, from Equation (40), we derive:

$$\delta_{uu} f = 0, \tag{41}$$

$$(a_1 - \beta')u^r - \delta r u^{r-1} = 0, \tag{42}$$

$$\delta_t + (\delta_u - \beta')(\Gamma_1 + \Gamma_2) - \delta\Gamma_1' - \lambda\Gamma_2' = 0. \tag{43}$$

Then, from Equation (41):

$$\delta(t, u) = uA_1(t) + A_2(t), \tag{44}$$

with $A_1(t)$ and $A_2(t)$ arbitrary functions of t. After these partial results, for the sake of clarity, we rewrite the determining system:

$$(2a_1 - \beta')f - (uA_1 + A_2)f_u = 0, \tag{45}$$

$$u(a_1 - \beta' - rA_1) - rA_2 = 0, \tag{46}$$

$$(uA_1' + A_2') + (A_1 - \beta')(\Gamma_1 + \Gamma_2) - (uA_1 + A_2)\Gamma_1' - \lambda\Gamma_2' = 0, \tag{47}$$

$$(\lambda_v - \beta')h + \lambda_t - (uA_1 + A_2)h_u - \lambda h_v = 0. \tag{48}$$

From Equation (46), taking into account that any function does not depend on u, we get:

$$A_1 = \tfrac{a_1 - \beta'}{r}, A_2 = 0, \tag{49}$$

then the other equations become:

$$(2a_1 - \beta')f - u\tfrac{a_1 - \beta'}{r} f_u = 0, \tag{50}$$

$$-\tfrac{\beta''}{r}u + \tfrac{a_1 - (1+r)\beta'}{r}(\Gamma_1 + \Gamma_2) - u\tfrac{a_1 - \beta'}{r}\Gamma_1' - \lambda\Gamma_2' = 0, \tag{51}$$

$$(\lambda_v - \beta')h + \lambda_t - u\tfrac{a_1 - \beta'}{r}h_u - \lambda h_v = 0. \tag{52}$$

We observe that from Equation (50), if f is arbitrary, it follows $\beta = b_0$, $a_1 = 0$, from Equation (51) $\lambda = 0$, while the Equation (52) is satisfied. Therefore, for f arbitrary, we do not obtain the extension of the principal Lie algebra. Then, in order to look for extensions of the principal algebra, we observe that from Equation (50), the form of function f must have the following structure:

1. $f = f_0 u^r$.
2. $f = f_0 u^s$ and $s \neq r$.

We study these cases separately.

1. $f = f_0 u^r$

   In this case, from Equation (50), we have $a_1 = 0$. Moreover, by differentiating Equation (51) with respect to u, we have:

$$\beta'' + \beta'(r\Gamma_1' - u\Gamma_1'') = 0. \tag{53}$$

   We observe that if Γ_1 is arbitrary, we have $\beta = b_0$, while from Equation (51), we have $\lambda = 0$, and Equation (52) is satisfied; however, we do not obtain the extension of the principal Lie algebra.

Then, in order to have extensions of the principal algebra, the following conditions must be satisfied:

$$u\Gamma_1'' - r\Gamma_1' = \gamma_0, \tag{54}$$

$$\beta'' = \gamma_0 \beta'. \tag{55}$$

We distinguish two cases: $r \neq -1$ and $r = -1$.

(a) If $r \neq -1$, from Equation (54), we get:

$$\Gamma_1(u) = \frac{c_1 u^{1+r}}{1+r} - \frac{\gamma_0 u}{r} + c_2. \tag{56}$$

Consequently, from Equation (51), we obtain:

$$\lambda(t, v) = -\frac{(c_2+\Gamma_2)(r+1)\beta'}{r\Gamma_2'}, \tag{57}$$

while Equation (52) becomes:

$$\frac{\beta'}{r\Gamma_2'^2} J_1 = 0 \tag{58}$$

with:

$$J_1 \equiv h(1+r)(c_2+\Gamma_2)\Gamma_2'' - (h(1+2r) - uh_u)\Gamma_2'^2 + (1+r)(h_v - \gamma_0)(c_2+\Gamma_2)\Gamma_2'. \tag{59}$$

We observe that if $\beta' = 0$, then $\lambda = 0$, and we do not obtain the extension of the principal Lie algebra. Consequently, in order to have extensions of the principal algebra, the functions Γ_2 and h must satisfy the equation $J_1 = 0$. In this case, we have two possible generators depending on γ_0.

i. If $\gamma_0 \neq 0$, as from Equation (55), we have:

$$\beta(t) = b_0 + b_1 e^{\gamma_0 t}, \tag{60}$$

the additional generator is:

$$X_3 = e^{\gamma_0 t}\partial_t - \frac{\gamma_0 e^{\gamma_0 t}}{r} u\partial_u - \frac{(c_2+\Gamma_2)(r+1)\gamma_0 e^{\gamma_0 t}}{r\Gamma_2'}\partial_v. \tag{61}$$

ii. If $\gamma_0 = 0$, as from Equation (55), we have:

$$\beta(t) = b_0 + b_1 t, \tag{62}$$

the additional generator is:

$$X_3 = t\partial_t - \frac{u}{r}\partial_u - \frac{(c_2+\Gamma_2)(r+1)}{r\Gamma_2'}\partial_v. \tag{63}$$

(b) If $r = -1$, from Equation (54), we get:

$$\Gamma_1(u) = c_1 \ln(u) + \gamma_0 u + c_2. \tag{64}$$

Consequently, from Equation (51), we obtain:

$$\lambda(t, v) = -\frac{c_1\beta'}{\Gamma_2'}, \tag{65}$$

while Equation (52) becomes:

$$\frac{\beta'}{\Gamma_2'^2} J_2 = 0 \tag{66}$$

with:

$$J_2 \equiv hc_1\Gamma_2'' - (h + uh_u)\Gamma_2'^2 + c_1(h_v - \gamma_0)\Gamma_2'. \tag{67}$$

We observe that if $\beta' = 0$, then $\lambda = 0$, and we do not obtain extension of the principal Lie algebra. Consequently, in order to have extensions of the principal algebra, the functions Γ_2 and h must satisfy the equation $J_2 = 0$. In this case, we have two possible generators depending on γ_0.

i. If $\gamma_0 \neq 0$, as from Equation (55), we have:

$$\beta(t) = b_0 + b_1 e^{\gamma_0 t}, \tag{68}$$

the additional generator is:

$$X_3 = e^{\gamma_0 t}\partial_t + \gamma_0 e^{\gamma_0 t} u\partial_u - \frac{c_1\gamma_0 e^{\gamma_0 t}}{\Gamma_2'}\partial_v. \tag{69}$$

ii. If $\gamma_0 = 0$, as from Equation (55), we have:

$$\beta(t) = b_0 + b_1 t, \tag{70}$$

the additional generator is:

$$X_3 = t\partial_t - \frac{u}{r}\partial_u - \frac{c_1}{\Gamma_2'}\partial_v. \tag{71}$$

2. $f = f_0 u^s$ and $s \neq r$

In this case, from Equation (50), we have:

$$\beta(t) = \frac{a_1(2r-s)}{r-s}t + b_0, \tag{72}$$

and Equation (51) becomes:

$$a_1 u\Gamma_1' - a_1(1 + 2r - s)(\Gamma_1 + \Gamma_2) - (r - s)\lambda\Gamma_2' = 0. \tag{73}$$

Moreover, by differentiating with respect to u, we get:

$$a_1\left(u\Gamma_1'' + (s - 2r)\Gamma_1'\right) = 0. \tag{74}$$

We observe that if Γ_1 is arbitrary, then we have $a_1 = 0$, while from Equation (51) $\lambda = 0$ and Equation (52) is satisfied, but we do not obtain the extension of the principal Lie algebra. Then, in order to have extensions of the principal algebra, the following condition must be satisfied:

$$u\Gamma_1'' + (s - 2r)\Gamma_1' = 0. \tag{75}$$

We distinguish the following two cases.

(a) If $s \neq 2r + 1$, from Equation (75), we get:

$$\Gamma_1(u) = c_1 + c_2 u^{1+2r-s}. \tag{76}$$

From Equation (51):

$$\lambda(t, v) = \frac{a_1(s - 2r - 1)(c_1 + \Gamma_2)}{(r - s)\Gamma_2'}, \tag{77}$$

while Equation (52) becomes:

$$\frac{a_1}{(r - s)\Gamma_2'^2} J_3 = 0 \tag{78}$$

with:

$$J_3 \equiv (1+2r-s)(c_1+\Gamma_2)(h\Gamma_2'' + h_v\Gamma_2') + (uh_u - h(1+4r-2s))\Gamma_2'^2. \tag{79}$$

We observe that if the functions Γ_2 and h do not satisfy the equation $J_3 = 0$, we do not obtain the extension of the principal Lie algebra. Then, in order to have extensions of the principal algebra, the functions Γ_2 and h must satisfy the equation $J_3 = 0$. In this case, we obtain the following additional generator:

$$X_3 = x\partial_x + \tfrac{2r-s}{r-s}t\partial_t + \tfrac{1}{s-r}u\partial_u + \tfrac{(s-2r-1)(c_1+\Gamma_2)}{(r-s)\Gamma_2'}\partial_v. \tag{80}$$

(b) If $s = 2r+1$, from Equation (75), we get:

$$\Gamma_1(u) = c_1 \ln(u) + c_2, \tag{81}$$

and from Equation (72):

$$\beta(t) = \tfrac{a_1}{r+1}t + b_0. \tag{82}$$

Consequently, from Equation (51), we obtain:

$$\lambda(t,v) = -\tfrac{c_1 a_1}{(r+1)\Gamma_2'}, \tag{83}$$

while Equation (52) becomes:

$$\tfrac{a_1}{(r+1)\Gamma_2'^2}J_4 = 0 \tag{84}$$

with:

$$J_4 \equiv hc_1\Gamma_2'' - (h+uh_u)\Gamma_2'^2 + c_1h_v\Gamma_2'. \tag{85}$$

We observe that if the functions Γ_2 and h do not satisfy the equation $J_4 = 0$, we do not obtain the extension of the principal Lie algebra. Then, in order to have extensions of the principal algebra, the functions Γ_2 and h must satisfy the equation $J_4 = 0$. In this case, we obtain the following additional generator:

$$X_3 = x\partial_x + \tfrac{1}{r+1}t\partial_t + \tfrac{1}{r+1}u\partial_u - \tfrac{c_1}{(r+1)\Gamma_2'}\partial_v. \tag{86}$$

Summarizing, we obtained six subclasses of class (28), which admit a three-dimensional Lie algebra.

1. $f = f_0u^r$ with $r \neq -1$, $\Gamma_1(u) = \frac{c_1u^{1+r}}{1+r} - \frac{\gamma_0 u}{r} + c_2$ with $\gamma_0 \neq 0$, the functions h and Γ_2 linked from the following relation:

$$h(1+r)(c_2+\Gamma_2)\Gamma_2'' - (h(1+2r) - uh_u)\Gamma_2'^2 + (1+r)(h_v - \gamma_0)(c_2+\Gamma_2)\Gamma_2' = 0. \tag{87}$$

2. $f = f_0u^r$ with $r \neq -1$, $\Gamma_1(u) = \frac{c_1u^{1+r}}{1+r} + c_2$ and the functions h and Γ_2 linked from the following relation:

$$h(1+r)(c_2+\Gamma_2)\Gamma_2'' - (h(1+2r) - uh_u)\Gamma_2'^2 + (1+r)(h_v)(c_2+\Gamma_2)\Gamma_2' = 0. \tag{88}$$

3. $f = \frac{f_0}{u}$, $\Gamma_1(u) = c_1\ln(u) + \gamma_0 u + c_2$ with $\gamma_0 \neq 0$ and the functions h and Γ_2 linked from the following relation:

$$hc_1\Gamma_2'' - (h+uh_u)\Gamma_2'^2 + c_1(h_v - \gamma_0)\Gamma_2' = 0. \tag{89}$$

4. $f = \frac{f_0}{u}, \Gamma_1(u) = c_1 \ln(u) + c_2$ and the functions h and Γ_2 linked from the following relation:

$$hc_1\Gamma_2'' - (h + uh_u)\Gamma_2'^2 + c_1(h_v)\Gamma_2' = 0. \tag{90}$$

5. $f = f_0 u^s$ with $s \neq r,\ 2r+1, \Gamma_1(u) = c_1 + c_2 u^{1+2r-s}$ and the functions h and Γ_2 linked from the following relation:

$$(1 + 2r - s)(c_1 + \Gamma_2)(h\Gamma_2'' + h_v\Gamma_2') + (uh_u - h(1 + 4r - 2s))\Gamma_2'^2 = 0. \tag{91}$$

6. $f = f_0 u^{2r+1}, \Gamma_1(u) = c_1 \ln(u) + c_2$ and the functions h and Γ_2 linked from the following relation:

$$hc_1\Gamma_2'' - (h + uh_u)\Gamma_2'^2 + c_1 h_v\Gamma_2' = 0. \tag{92}$$

3.2. A Special Case

In agreement with some news about the biological compatibility of the form of g derived from some previous papers (see, e.g., [29,30] and references insides), in this subsection, we show an example of the application of the previous results.

By selecting the case 1a from the obtained cases and assuming $r = 1$ and $c_2 = 0$ in (56), we consider f, Γ_1, of the following form:

$$f = f_0 u, \Gamma_1(u) = \gamma_1 u^2 - \gamma_0 u, \tag{93}$$

with f_0, γ_0, γ_1, arbitrary constants. Moreover, we assume:

$$\Gamma_2(v) = \gamma_2 v + \gamma_3, \tag{94}$$

with γ_2,γ_3, arbitrary constants.

In this case, in order to have an extension on the principal algebra, the function $h(u,v)$ must satisfy the equation $J_1 = 0$, that is:

$$(uh_u - 3h)\gamma_2^2 + 2(h_v - \gamma_0)(\gamma_2 v + \gamma_3)\gamma_3 = 0. \tag{95}$$

Solutions of this equation are functions $h(u,v)$ of the form:

$$h(u,v) = u^3 H(\sigma) - 2\frac{\gamma_0}{\gamma_2}(\gamma_2 v + \gamma_3), \tag{96}$$

where H is an arbitrary function of $\sigma = \frac{\gamma_2 v + \gamma_3}{\gamma_2 u^2}$. By assuming $H(\sigma) = \sigma$ in agreement with [13,29,30], we get:

$$h(u,v) = \frac{\gamma_2 v + \gamma_3}{\gamma_2}(u - 2\gamma_0). \tag{97}$$

The system (28) becomes:

$$\begin{cases} u_t = f_0 u_x^2 + f_0 u u_{xx} + u u_x + \gamma_1 u^2 - \gamma_0 u + \gamma_2 v + \gamma_3, \\ v_t = \frac{\gamma_2 v + \gamma_3}{\gamma_2}(u - 2\gamma_0). \end{cases} \tag{98}$$

While the third generator is obtained by specializing generator (61) and has the form:

$$X_3 = e^{\gamma_0 t}\partial_t - \gamma_0 e^{\gamma_0 t} u\partial_u - \frac{2\gamma_0}{\gamma_2}(\gamma_2 v + \gamma_3)e^{\gamma_0 t}\partial_v. \tag{99}$$

By considering the generator $kX_2 + X_3$, we get:

$$u(t,x) = U(z)e^{-\gamma_0 t}, \; v(t,x) = \frac{V(z)e^{-2\gamma_0 t} - \gamma_3}{\gamma 2}, \tag{100}$$

with $z = \frac{\gamma_0 x}{k} + e^{-\gamma_0 t}$, while the functions $U(z)$, $V(z)$ are solutions of the reduced system:

$$U'' U f_0 \gamma_0^2 + U'^2 f_0 \gamma_0^2 + k\gamma_0(U+k)U' + k^2 V + \gamma_1 k^2 U^2 = 0, \tag{101}$$

$$\gamma_0 V' + UV = 0. \tag{102}$$

This reduced system, as well as other cases of biological specializations, will be studied in later research.

Of course, the systems studied here cannot be considered, *strictu sensu*, as mathematical models. In fact, their constitutive parameters need to be characterized carefully from the biological point view. However, having in mind some previous models concerned with *Aedes aegypti* [13,14,29–31], we try to stress some structural features of the system (98). To this aim, we rewrite system (98) as:

$$\begin{cases} u_t = (f_0 u u_x)_x + u u_x + \gamma_1 u\left(u - \frac{\gamma_0}{\gamma_1}\right) + \gamma_2\left(v + \frac{\gamma_3}{\gamma_2}\right) \\ v_t = (u - 2\gamma_0)\left(v + \frac{\gamma_3}{\gamma_2}\right) \end{cases} \tag{103}$$

It is easy to ascertain a weak interaction of the equation for the aquatic population on the equation for the winged population. Moreover, in this last one appears a growth for population u having a logistic structure. By identifying γ_1 as a positive rate of maturation of the aquatic forms in winged female mosquitoes and γ_0 as the positive winged population mortality, it is possible to find a threshold value $u_{trs} = \frac{\gamma_0}{\gamma_1}$. Finally, the aquatic population evolution equation shows a growth rate $u - 2\gamma_0$ ruled by the density of mosquitoes and their mortality.

Remark 1. It is a simple matter to ascertain that the system (103) admits as the special solution:

$$u = \frac{\gamma_0}{\gamma_1}, v = -\frac{\gamma_3}{\gamma_2}. \tag{104}$$

Moreover, it is possible to get other solutions by assuming $v = -\frac{\gamma_3}{\gamma_2}$, while u is obtained as a solution of equation:

$$u_t = (f_0 u u_x)_x + u u_x + \gamma_1 u\left(u - \frac{\gamma_0}{\gamma_1}\right). \tag{105}$$

For the interested reader, it could be worthwhile noticing that Equation (105) is a particular case of equation considered in [8,9]. Moreover, we can get the results obtained in [8,9] (see Table 1, Case 8 of both papers) by projection. Indeed, the Lie symmetry generator (99) projected in the space $\left(t, x, u, v = -\frac{\gamma_3}{\gamma_2}\right)$ becomes the Lie symmetry T_1 of [8,9].

4. Conclusions

In this paper, we have considered a class of advection reaction diffusion systems of interest in biomathematics. After having recalled a weak equivalence generator, obtained in a previous work [12], we find some particular cases of the nonlinear system (28) admitting three-dimensional Lie algebras by using a specialization of a projection theorem [24,28]. In this subclass, the constitutive equation characterizing g is assigned as:

$$g = u^r u_x + \Gamma_1(u) + \Gamma_2(v)$$

that generalizes [12]:

$$g = \rho u^r u_x^s + \Gamma_1 u^a + \Gamma_2 v^b, \tag{106}$$

where the constants $\rho(\neq 0)$, Γ_1, Γ_2, r, s, a and b are constitutive parameters. We derive the principal Lie algebra and the functions admitting at least an extension by one. These results are summarized at the end of Section 3.1. A special case is considered in Section 3.2.

Acknowledgments: The authors wish to thank the reviewers for their interesting observations and the editor for his very useful comments. Mariano Torrisi was supported from Gruppo Nazionale per la Fisica Matematica of Istituto Nazionale di Alta Matematica.

Author Contributions: The authors contributed equally to this work.

Conflicts of Interest: The authors declare no conflict of interest.

References

1. Cardile, V.; Torrisi, M.; Tracinà, R. On a reaction-diffusion system arising in the development of Bacterial Colonies. In Proceedings of the 10th International Conference in Modern Group Analysis, Larnaca, Cyprus, 24–31 October 2004; Volume 32, p. 38.
2. Medvedev, G.S.; Kaper, T.J.; Kopell, N. A reaction diffusion system with periodic front Dynamics. *SIAM J. Appl. Math.* **2000**, *60*, 1601–1638. [CrossRef]
3. Torrisi, M.; Tracinà, R. On a class of reaction diffusion systems: Equivalence transformations and symmetries. In *Asymptotic Methods in Nonlinear Wave Phenomena*; Ruggeri, T., Sammartino, M., Eds.; World Science Publishing Co. Pte. Ltd.: Singapore, 2007; pp. 207–216.
4. Torrisi, M.; Tracinà, R. Exact solutions of a reaction-diffusion system for *Proteus Mirabilis* bacterial colonies. *Nonlinear Anal. Real World Appl.* **2011**, *12*, 1865–1874. [CrossRef]
5. Cherniha, R.; Serov, M. Nonlinear systems of the burgers-type equations: Lie and Q-conditional symmetries, ansätze and solutions. *J. Math. Anal. Appl.* **2003**, *282*, 305–328. [CrossRef]
6. Cherniha, R.; Wilhelmsson, H. Symmetry and exact solution of heat-mass transfer equations in thermonuclear plasma. *Ukr. Math. J.* **1996**, *48*, 1434–1449. [CrossRef]
7. Cherniha, R.; Serov, M. Lie and non-Lie symmetries of nonlinear diffusion equations with convection term. *Symmetry Nonlinear Math. Phys.* **1997**, *2*, 444–449.
8. Cherniha, R.; Serov, M. Symmetries, ansätze and exact solutions of nonlinear second-order evolution equations with convection terms. *Eur. J. Appl. Math.* **1998**, *9*, 527–542. [CrossRef]
9. Cherniha, R.; Serov, M. Symmetries, ansätze and exact solutions of nonlinear second-order evolution equations with convection terms, II. *Eur. J. Appl. Math.* **2006**, *17*, 597–605. [CrossRef]
10. Ibragimov, N.H.; Torrisi, M.; Valenti, A. Preliminary group classification of equation $v_{tt} = f(x, v_x)v_{xx} + g(x, v_x)$. *J. Math. Phys.* **1991**, *32*, 2988–2995. [CrossRef]
11. Ovsiannikov, L.V. *Group Analysis of Differential Equations*; Academic Press: New York, NY, USA, 1982.
12. Freire, I.L.; Torrisi, M. Weak equivalence transformations for a class of models in biomathematics. *Abstr. Appl. Anal.* **2014**. [CrossRef]
13. Freire, I.L.; Torrisi, M. Symmetry methods in mathematical modeling *Aedes aegypti* dispersal dynamics. *Nonlinear Anal. Real World Appl.* **2013**, *14*, 1300–1307. [CrossRef]
14. Freire, I.L.; Torrisi, M. Similarity solutions for systems arising from an Aedes aegypti model. *Commun. Nonlinear Sci. Numer. Simul.* **2014**, *19*, 872–879. [CrossRef]
15. Romano, V.; Torrisi, M. Application of weak equivalence transformations to a group analysis of a drift-diffusion model. *J. Phys. A Math. Gen.* **1999**, *32*, 7953–7963. [CrossRef]
16. Torrisi, M.; Tracinà, R. Equivalence transformations and symmetries for a heat conduction model. *Int. J. Non-Linear Mech.* **1998**, *33*, 473–487. [CrossRef]
17. Gazeau, J.P.; Winternitz, P. Symmetries of variable-coefficient Korteweg-de Vries equations. *J. Math. Phys.* **1992**, *33*, 4087–4102. [CrossRef]
18. Winternitz, P.; Gazeau, J.P. Allowed transformations and symmetry classes of variable coefficient Korteweg-de Vries equations. *Phys. Lett. A* **1992**, *167*, 246–250. [CrossRef]
19. Akhatov, I.S.H.; Gazizov, R.K.; Ibragimov, N.H. Nonlocal symmetries. Heuristic approach. *J. Sov. Math.* **1991**, *55*, 1401–1450. [CrossRef]
20. Lisle, I.G. Equivalence Transformation for Classes of Differential Equations. Ph.D. Thesis, University of British Columbia, Vancouver, BC, Canada, 1992.

21. Khalique, C.M.; Mahomed, F.M.; Ntsime, B.P. Group classification of the generalized Emden-Fowler-type equation. *Nonlinear Anal. Real World Appl.* **2009**, *10*, 3387–3395. [CrossRef]
22. Ibragimov, N.H. *CRC Handbook of Lie Group Analysis of Differential Equations*; CRC Press: Boca Raton, FL, USA, 1996.
23. Molati, M.; Khalique, C.M. Lie group classification of a generalized Lane–Emden Type system in two dimensions. *J. Appl. Math.* **2012**. [CrossRef]
24. Torrisi, M.; Tracinà, R. Equivalence transformations for systems of first order quasilinear partial differential equations. In *Modern Group Analysis VI: Developments in Theory, Computation and Application*; New Age International(P) Ltd.: New Delhi, India, 1996; pp. 115–135.
25. Torrisi, M.; Tracinà, R.; Valenti, A. Group analysis approach for a non linear differential system arising in diffusion phenomena. *J. Math. Phys.* **1996**, *37*, 4758–4767. [CrossRef]
26. Gambino, G.; Greco, A.M.; Lombardo, M.C. A group analysis via weak equivalence transformations for a model of tumour encapsulation. *J. Phys. A* **2004**, *37*, 3835–3846. [CrossRef]
27. Ibragimov, N.H.; Säfström, N. The equivalence group and invariant solutions of a tumour growth model. *Commun. Nonlinear Sci. Num. Simul.* **2004**, *9*, 61–69. [CrossRef]
28. Ibragimov, N.H.; Torrisi, M. A simple method for group analysis and its application to a model of detonation. *J. Math. Phys.* **1992**, *33*, 3931–3937. [CrossRef]
29. Maidana, N.A.; Yang, H.M. Describing the geographic spread of dengue disease by traveling waves. *Math. Biosci.* **2008**, *215*, 64–77. [CrossRef] [PubMed]
30. Takahashi, L.T.; Maidana, N.A.; Ferreira, W.C., Jr.; Pulino, P.; Yang, H.M. Mathematical models for the *Aedes aegypti* dispersal dynamics: Traveling waves by wing and wind. *Bull. Math. Biol.* **2005**, *67*, 509–528. [CrossRef] [PubMed]
31. Bacani, F.; Freire, I.L.; Maidana, N.A.; Torrisi, M. Modelagem para a dinâmica populacional do *Aedes aegypti* via simetrias de Lie. *Proc. Ser. Braz. Soc. Appl. Comput. Math.* **2015**, *3*. [CrossRef]

Chapter 2:

symmetry

MDPI

Article

Nonclassical Symmetries of a Nonlinear Diffusion–Convection/Wave Equation and Equivalents Systems

Daniel J. Arrigo, Brandon P. Ashley, Seth J. Bloomberg and Thomas W. Deatherage

Department of Mathematics, University of Central Arkansas, Conway, AR 72035, USA;
brandonpashley@live.com (B.A.); sbloomberg1@cub.uca.edu (S.B.); tdeatherage1@cub.uca.edu (T.D.)
* Correspondence: darrigo@uca.edu; Tel.: +501-450-5668

Academic Editor: Roman M. Cherniha
Received: 07 August 2016; Accepted: 22 November 2016; Published: 26 November 2016

Abstract: It is generally known that classical point and potential Lie symmetries of differential equations (the latter calculated as point symmetries of an equivalent system) can be different. We question whether this is true when the symmetries are extended to nonclassical symmetries. In this paper, we consider two classes of nonlinear partial differential equations; the first one is a diffusion–convection equation, the second one a wave, where we will show that the majority of the nonclassical point symmetries are included in the nonclassical potential symmetries. We highlight a special case were the opposite is true.

Keywords: nonclassical symmetry; nonclassical potential symmetry; diffusion equation; wave equation

1. Introduction

Symmetry analysis plays a fundamental role in the construction of exact solutions to nonlinear partial differential equations. Based on the original work of Lie [1] on continuous groups, symmetry analysis provides a unified explanation for the seemingly diverse and ad hoc integration methods used to solve ordinary differential equations. At the present time, there is extensive literature on the subject, and we refer the reader to the books by Arrigo [2], Bluman and Kumei [3], and Olver [4].

A particular class of equation that has benefited from this type of analysis is the nonlinear diffusion equation

$$u_t = (K(u)u_x)_x \tag{1}$$

From a symmetry point of view, this equation was first considered by Ovsjannikov [5] (see also [3] and [6]), where it was found that (1) admits nontrivial symmetries for a variety of different diffusivities. In particular, power law diffusion, where

$$u_t = (u^m u_x)_x \tag{2}$$

admits the symmetry generator

$$\Gamma = T\frac{\partial}{\partial t} + X\frac{\partial}{\partial x} + U\frac{\partial}{\partial u} \tag{3}$$

where T, X, and U are

$$\begin{aligned} T &= c_1 + c_2 t \\ X &= c_3 + c_4 x \\ U &= \frac{1}{m}(2c_4 - c_2)\, u \end{aligned} \tag{4}$$

(where c_i are arbitrary constants) for general powers m ($m \neq 0$), and in the special case $m = -4/3$, where (2) admits an additional symmetry with generator

$$\Gamma = x^2 \frac{\partial}{\partial x} - 3xu \frac{\partial}{\partial u}$$

In 1988, Bluman, Reid, and Kumei [7] considered the equivalent system

$$v_t = K(u)\, u_x, \quad v_x = u \tag{5}$$

and found that this system possesses a rather rich symmetry structure and identified new forms of $K(u)$ that admitted new nontrivial symmetries. Of particular interest are again power law diffusivities $K(u) = u^m$, where (5) admits the symmetry generator

$$\Gamma = T\frac{\partial}{\partial t} + X\frac{\partial}{\partial x} + U\frac{\partial}{\partial u} + V\frac{\partial}{\partial v} \tag{6}$$

where $T, X, U,$ and V are given by, in the case of $m \neq -2$,

$$\begin{aligned} T &= c_1 + c_2 t \\ X &= c_3 + \frac{c_2 + mc_4}{m+2} x \\ U &= \frac{2c_4 - c_2}{m+2} u \\ V &= c_5 + c_4 v \end{aligned} \tag{7}$$

and in the case of $m = -2$,

$$\begin{aligned} T &= c_1 + 2c_2 t + 4c_3 t^2 \\ X &= \left(c_6 - 2c_3 t - c_5 v - c_3 v^2\right) x + F(t, v) \\ U &= \left(c_2 - c_6 + 6c_3 t + c_5 v + c_3 v^2\right) u + \left(c_5 x + 2c_3 x v - F_v\right) u^2 \\ V &= c_4 + 2c_5 t + \left(c_2 + 4c_3 t\right) v \end{aligned} \tag{8}$$

where F satisfies $F_t = F_{vv}$. Clearly, the powers $m = -4/3$ and $m = -2$ show themselves as special, and—as this example demonstrates—the symmetries of equations and equivalent systems can be different. A natural question to ask is whether this holds true for nonclassical symmetries; that the nonclassical symmetries of a particular equation and a equivalent system (nonclassical potential symmetries) are different.

The nonclassical method, first introduced by Bluman and Cole [8] (see, for example, [2] or [3]), seeks invariance of a given partial differential equation (PDE) augmented with the invariant surface condition. As the determining equations for these nonclassical symmetries are nonlinear, there seemed to be little hope for this new method; however, with the development of computer algebra systems, the nineties saw a huge explosion of interest as several authors took interest in the nonclassical method and continues today to be an active area of interest (e.g., [9–23] and references within).

Of particular interest here is the paper by Bluman and Yan [24]. They consider two algorithms that extend the nonclassical method to potential systems and potential equations. They consider the nonlinear diffusion Equation (1), an equivalent potential system (Algorithm 1)

$$v_x = u, \quad v_t = K(u) u_x \tag{9}$$

and potential equation (Algorithm 2)

$$v_t = K\left(v_x\right) v_{xx} \tag{10}$$

In the case where $K(u) = \dfrac{1}{u^2+u}$, they were able to show that (10) admits nonclassical symmetries that the original Equation (1) does not. So, there is some evidence that the nonclassical symmetries of a PDE and a potential equation/equivalent system can be different (see also [25] and references within). Although we will not address this question in general here, we will use Algorithm 1 to consider a large class of nonlinear diffusion–convection and wave equations to show that—in the majority of cases—the nonclassical potential system symmetries contain the nonclassical symmetries of the original equation. We also highlight a special case where the opposite is true.

2. Nonclassical Symmetries

In this section, we consider the nonclassical symmetries of the following nonlinear partial differential equations

$$(i) \quad u_t = (F(u)u_x + G(u))_x \tag{11a}$$

$$(ii) \quad u_{tt} = (F(u)u_x + G(u))_x \tag{11b}$$

These equations are of considerable interest because of their applications. For example, (11a), sometimes known as Richard's equation, has been used to model the one-dimensional, nonhysteretic infiltration in uniform nonswelling soil (Broadbridge and White [26]) and to model two phase filtration under gravity (Rogers, Stallybrass, and Clement [27]). Furthermore, (11b)—sometimes known as the nonlinear telegraph equation—has been used to model the telegraphy of a two-conductor transmission line (Katayev [28]) and the motion of a hyperelastic homogeneous rod whose cross-sectional area varies exponentially along the rod (Jeffery [29]).

In what follows, we omit the cases where (11) are linear or linearizable via a point transformation, as it is known that all solutions of linear PDEs can be obtained via classical Lie symmetries [30]. Each equation will be considered separately.

2.1. Nonlinear Diffusion–Convection Equation

We first consider the nonclassical symmetries of (11a). These are calculated by appending to (11a) the invariant surface condition

$$Tu_t + Xu_x = U \tag{12}$$

As usual, if $T \neq 0$, we set $T = 1$ in (12) without loss of generality. This gives rise to the following determining equations for the infinitesimals $X(t, x, u)$ and $U(t, x, u)$:

$$FX_{uu} - F_u X_u = 0 \tag{13a}$$

$$F^2 U_{uu} - 2F^2 X_{xu} + FF_u U_u + (2FG_u + 2FX)\, X_u + UFF_{uu} - UF_u^2 = 0 \tag{13b}$$

$$FU_t - F^2 U_{xx} - FG_u U_x + 2FUX_x - U^2 F_u = 0 \tag{13c}$$

$$\begin{aligned} 2F^2 U_{xu} + FX_t - F^2 X_{xx} + 2FF_u U_x - 2FUX_u + F\,(2X + G_u)\, X_x \\ + (FG_{uu} - F_u G_u - XF_u)\, U = 0 \end{aligned} \tag{13d}$$

A variation of these determining equations are given in Cherniha and Serov [31], and in the case of $G = 0$, appear in Arrigo and Hill [32]. To calculate the nonclassical potential symmetries, we calculate the nonclassical symmetries for the associated system

$$v_t = F(u)\, u_x + G(u), \quad v_x = u \tag{14}$$

augmented with the two associated invariant surface conditions

$$Tu_t + Xu_x = U, \quad Tv_t + Xv_x = V \tag{15}$$

again noting that we will set $T = 1$, as we are assuming that $T \neq 0$. Our approach to obtaining the determining equations is through compatibility. Several authors have shown that this is equivalent to the nonclassical method (see [33–36]). Solving (14) and (15) for the first order derivatives u_t, u_x, v_t, and v_x gives

$$u_t = \frac{XG + UF - XV + uX^2}{F}, \quad u_x = \frac{V - uX - G}{F}, \quad v_t = V - uX, \quad v_x = u \tag{16}$$

Requiring compatibility by eliminating partial derivatives by cross-differentiation gives

$$FV_x + (V - uX - G)\,V_u + uFV_v - uFX_x + u\,(uX + G - V)\,X_u - u^2FX_v - FU = 0 \tag{17a}$$

$$\begin{aligned} FV_t + FXV_x + FUV_u + FVV_v + (FX_x - 2(G + uX)X_u - FU_u - UF_u)\,V \\ +X_uV^2 - F^2U_x + F(uX + G)U_u - uF^2U_v - uFX_t - F(2uX + G)X_x \\ + \left((uX + G)^2 - uFU\right)X_u - uF\,(uX + G)\,X_v + (uX + G)UF_u - FUG_u - FXU = 0 \end{aligned} \tag{17b}$$

In the case of $G = 0$, these determining equations are equivalent to those that appear in Bluman and Shtelen [37]. It is interesting to note that at first appearance, (17) seems to be underdetermined—two equations for the three unknowns X, U and V. However, if we let $V = FW + uX + G$, where $W = W(t, x, u, v)$, then (16) becomes

$$u_t = U - XW, \quad u_x = W, \quad v_t = FW + G, \quad v_x = u \tag{18}$$

and compatibility of (18) again, by cross-differentiation gives rise to the determining equations

$$W_t + XW_x + UW_u + (FW + G + uX)W_v + X_uW^2 + (X_x + uX_v - U_u)W - U_x - uU_v = 0 \tag{19a}$$

$$FW_x + W(FW + G)_u + uFW_v + XW - U = 0 \tag{19b}$$

To show that the nonclassical symmetries of (11a) are included in the nonclassical symmetries of (14) is to show that V exists satisfying (17) if X and U satisfy (13). As we have defined V in terms of W, it suffices to have X, U, and W functions of t, x, and u only. Doing so and requiring that (19) be compatible via cross-differentiation gives rise to

$$\begin{aligned} F\,(FX_{uu} - F_uX_u)\,W^3 \\ - \left(F^2U_{uu} - 2F^2X_{xu} + FF_uU_u + (2FG_u + 2FX)\,X_u + UFF_{uu} - UF_u^2\right)W^2 \\ - \Big(2F^2U_{xu} + FX_t - F^2X_{xx} + 2FF_uU_x - 2FUX_u \\ +F\,(2X + G_u)\,X_x + (FG_{uu} - F_uG_u - XF_u)\,U\,\Big)\,W \\ +FU_t - F^2U_{xx} - FG_uU_x + 2FUX_x - U^2F_u \quad = 0 \end{aligned} \tag{20}$$

By virtue of (13), this is identically satisfied given that a W exists satisfying (19), which in turn gives that a V exists satisfying (17), thus proving our claim.

2.2. Nonlinear Wave Equation

We now consider the nonclassical symmetries of (11b). Again, we set $T = 1$. For this particular class of equations, it is necessary to consider two cases: (i) $X^2 \neq F$ and (ii) $X^2 = F$. Each will be considered separately.

Case (i) $X^2 \neq F$

In this case, we have the following determining equations for the infinitesimals $X(t, x, u)$ and $U(t, x, u)$:

$$\left(X^2 - F\right) X_{uu} + F_u X_u - 2XX_u^2 = 0 \tag{21a}$$

$$\begin{aligned}&\left(X^2 - F\right)(2XUX_{uu} + 2FX_{xu} + 2XX_{tu} - (F_u + 2XX_u)U_u - 2G_uX_u - F_{uu}U)\\&+\left(X^2 - F\right)^2 U_{uu} - 2\left(2X^2X_t + 2FXX_x - 2XUF_u + 4X^2UX_u\right) X_u\\&+2XF_u(X_t + XX_x) - F_u^2 U = 0\end{aligned} \tag{21b}$$

$$\begin{aligned}&\left(X^2 - F\right)\left(2XUU_{uu} + 2XU_{tu} + 2FU_{xu} + U^2X_{uu} + 2UX_{tu} - FX_{xx} + X_{tt}\right)\\&-\left(X^2 - F\right)(2X_uU_t - 2F_uU_x + 2X_tU_u + G_uX_x - G_{uu}) - 2X\left(X_t^2 - FX_x^2 + U^2X_u^2\right)\\&-2(2FX_t + 2FXX_x + 2X^2UX_u - XUF_u)U_u + U(F_uU - 2G_uX)X_u\\&-(4XUX_u + 2G_uX - F_uU)X_t - (F_uXU + 2FG_u)X_x + F_uG_uU = 0\end{aligned} \tag{21c}$$

$$\begin{aligned}&\left(X^2 - F\right)\left(U^2U_{uu} + 2UU_{tu} + U_{tt} - FU_{xx} - F_uU_x\right)\\&-(U_t + UU_x)(2XX_t + 2FX_x + 2XUX_u - F_uU)\\&+(2FX_t + 2FX_x + 2FUX_u - F_uXU)U_x = 0\end{aligned} \tag{21d}$$

In the case of $G = 0$, these determining equations appear in Näslund [38]. To calculate the nonclassical potential symmetries, we calculate the nonclassical symmetries for the associated system

$$v_t = F(u)\,u_x + G(u), \quad v_x = u_t \tag{22}$$

augmented with the two associated invariant surface conditions

$$Tu_t + Xu_x = U, \quad Tv_t + Xv_x = V \tag{23}$$

with $T = 1$. This gives rise to two determining equations that have 43 and 44 terms, respectively. As we did in the previous section, we can simplify these determining equations. Solving (22) and (23) for u_t, u_x, v_t, and v_x gives

$$u_t = \frac{XG + UF - XV}{F - X^2}, \quad u_x = \frac{V - XU + G}{F - X^2}, \quad v_t = \frac{V - uX}{F - X^2}, \quad v_x = \frac{u}{F - X^2} \tag{24}$$

Letting $V = (F - X^2)\,W + XU + G$, where $W = W(t, x, u, v)$ gives (24) as

$$u_t = U - XW, \quad u_x = W, \quad v_t = FW + G, \quad v_x = U - XW \tag{25}$$

Requiring compatibility through cross-differentiation gives rise to the following determining equations:

$$\begin{aligned}W_t + XW_x + UW_u + \left(XU + \left(F - X^2\right)W + G\right)W_v + (X_u - XX_v)\,W^2&\\+(X_x + UX_v - U_u + XU_v)\,W - U_x - UU_v = 0&\end{aligned} \tag{26a}$$

$$\begin{aligned}XW_t + FW_x + \left(XU + (F - W^2)W\right)W_u + (FU + GX)\,W_v - U_t - UU_u - FU_v&\\+(F_u - XX_u + FX_v)\,W^2 + (X_t + UX_u + GX_v + XU_u - FU_v + G_u)\,W = 0&\end{aligned} \tag{26b}$$

To show that the nonclassical symmetries of (11b) are included in the nonclassical symmetries of (22) is to show that W exists satisfying (26) if X and U satisfy (21). Eliminating derivatives of W in (26) through cross-differentiation shows that (26) is compatible, provided that

$$\left(X^2 - F\right) AW^3 - BW^2 + CW - D = 0, \tag{27}$$

where $A, B, C,$ and D are precisely the expressions given in (21a)–(21d), thus showing that (27) is identically satisfied, again proving our claim.

Case (ii) $X^2 = F$

For this special case, we will show the opposite is true. The nonclassical symmetries of the system are contained within the nonclassical symmetries of the single equation. For the system (22), we find determining equations give rise to $V = XU + G$, and that U satisfies

$$U_t - XU_x + UU_u + (G - XU)\, U_v = 0 \tag{28a}$$

$$U_u - XU_v + \frac{X_u}{2X}U + \frac{G_u}{2X} = 0 \tag{28b}$$

Compatibility of (28) by eliminating U_t gives rise to the third equation

$$X_u U_x - G_u U_v + \left(\frac{2XX_{uu} - 3X_u^2}{4X^2}\right) U^2 + \left(\frac{XG_{uu} - 2X_u G_u}{2X^2}\right) U - \frac{G_u^2}{4X^2} = 0 \tag{29}$$

Further compatibility between (28a) and (29) by eliminating all derivatives of U gives rise to

$$\left(2XG_u X_{uu} - 2XX_u G_{uu} + G_u X_u^2\right) U^2 + 2X_u G_u^2 U + G_u^3 = 0 \tag{30}$$

If either $U_t \neq 0$, $U_x \neq 0$, or $U_v \neq 0$, then from (30) $G_u = 0$ and (22) is linearizable via a hodograph transformation. Thus, the only case to consider is when $U = U(u)$. In this case, (28) can be solved, giving

$$U = -c_1, \quad G = c_1 X + c2 \tag{31}$$

where c_1 and c_2 are arbitrary constants, and $X(u)$ is arbitrary.

We now turn our attention to the single Equation (11b). In the special case where $F = X^2$, we are restricted in the number of differential consequences of our invariant surface condition to be combined with our original PDE. Differential consequences of (12) (with $T = 1$ and $X_t = X_x = 0$) are

$$u_{tt} + Xu_{tx} + X_u u_t u_x = U_t + U_u u_t \tag{32a}$$

$$u_{tx} + Xu_{xx} + X_u u_x^2 = U_x + U_u u_x \tag{32b}$$

In the case where $X^2 \neq F$, we can solve the original PDE (11b) along with differential consequence of the invariant surface condition (12) for u_{tt}, u_{tx}, and u_{xx}. In this special case where $X^2 = F$, we can only solve for two second order derivatives of u. If we solve (11b) and (32b) for u_{tt} and u_{tx}, the second determining equation in (32) becomes

$$-U_t - U_u u_t + (XU_u + G_u)\, u_x + X_u u_t u_x + XX_u u_x^2 = 0 \tag{33}$$

and using the invariant surface condition (12), we obtain

$$(2XU_u + UX' + G')\, u_x - U_t + XU_x - UU_u = 0 \tag{34}$$

From (34) we see two cases emerge. If $2XU_u + UX_u + G_u = 0$, then $U_t - XU_x + UU_u = 0$, and comparing with (28) shows they are identical if $U_v = 0$. However, our analysis there showed the only solution is (31), and so the two results coincide. If $2XU_u + UX_u + G_u \neq 0$, then we obtain the single determining equation

$$\begin{aligned}(U_t - XU_x + UU_u)\left(2XU_{tu} + 2X^2U_{xu} + 2XUU_{uu} + 2XX_{uu}U_x + 2X_uUU_u + X_{uu}U^2 + G_{uu}U\right)\\ - (2XU_u + X_uU + G_u)\left(U_{tt} - X^2U_{xx} + 2UU_{tu} + U^2U_{uu} - (2X_uU + G_u)U_x\right) = 0 \quad (35)\end{aligned}$$

We make no effort to solve (35) in general; however, if $U = U(u)$, then (35) can be solved giving

$$U = -\frac{G(u) + c_1}{X(u) + c_2} \quad (36)$$

where c_1 and c_2 are arbitrary constants showing that the nonclassical symmetries of the single Equation (11b) contain the nonclassical symmetries of the equivalent system (22), and are in fact more general.

3. $T = 0$

In applying the nonclassical method in the previous section, we assumed that $T \neq 0$, letting us set $T = 1$ without loss of generality. We now consider the case when $T = 0$. Without loss of generality, we can set $X = 1$. Again, we will consider the nonlinear diffusion–convection and wave equations separately.

3.1. Nonlinear Diffusion Equation

In the case of the nonlinear heat Equation (11a), the nonclassical method gives rise to the following single equation for U:

$$U_t - FU_{xx} - 2FUU_{xu} - FU^2U_{uu} - (3F_uU + G_u)\,U_x - 2F_uU^2U_u - F_{uu}U^3 - G_{uu}U^2 = 0 \quad (37)$$

Applying the nonclassical method to the system (14) gives the single equation

$$u_t - F\,(U_x + UU_u + uU_v) - F_uU^2 - G_uU = 0 \quad (38)$$

At this point, we set the coefficients of the derivatives to zero. This gives

$$u_t = 0, \quad F\,(U_x + UU_u + uU_v) + F_uU^2 + G_uU = 0 \quad (39)$$

showing that the only solutions to (14) are of the form $u = f(x)$. This was also noted in Bluman and Yan [24] in the case of $G = 0$. However, we could continue to refine the nonclassical method and solve (38) for u_t and impose compatibility with $u_x = U$. This would give

$$\begin{aligned}U_t - FU_{xx} - 2FUU_{xu} - 2uFU_{xv} - FU^2U_{uu} - 2uFUU_{uv} - u^2FU_{vv}\\ -(3F_uU + G_u)U_x - 2F_uU^2U_u + (G - uG_u - 3uF_uU)U_v - F_{uu}U^3 - G_{uu}U^2 = 0 \quad (40)\end{aligned}$$

Setting $U_v = 0$ in (40) recovers (37), showing that a refinement in nonclassical method applied to the system (14) includes the nonclassical symmetries of the original Equation (11a).

3.2. Nonlinear Wave Equation

In the case of the nonlinear wave Equation (11b), the determining equations are:

$$\begin{aligned} U_{uu} &= 0 \\ U_{tu} &= 0 \\ U_{tt} - F\left(U_{xx} + 2UU_{xu} + U^2U_{uu}\right) - 2F_uU^2U_u & \\ - \left(3F_uU + G_u\right)U_x - F_{uu}U^3 - G_{uu}U^2 &= 0 \end{aligned} \tag{41}$$

For the system (22), they are

$$\begin{aligned} V_t + VV_u + (FU+G)V_v - FU_x - FUU_u - FVU_v - F_uU^2 - G_uU &= 0 \\ V_x + UV_u + VV_v - U_t - VU_u - (FU+G)U_v &= 0 \end{aligned} \tag{42}$$

Setting $U_v = 0$ and requiring that (42) be compatible gives

$$\begin{aligned} U_{uu}V^2 + 2U_{tu}V + U_{tt} - F\left(U_{xx} + 2UU_{xu} + U^2U_{uu}\right) - 2F_uU^2U_u & \\ - \left(3F_uU + G_u\right)U_x - F_{uu}U^3 - G_{uu}U^2 &= 0 \end{aligned} \tag{43}$$

which by virtue of (41) shows that this is identically satisfied, proving that nonclassical symmetries of (11a) are included to those of (22).

4. Conclusions

In this paper, we have considered the symmetries of a nonlinear diffusion–convection and wave equation and equivalent systems. It is well known that classical Lie symmetries of differential equations and equivalent systems can be different. We question whether this is true if we extend the symmetries to include nonclassical symmetries. We have shown that in the majority of cases, the nonclassical symmetries of equivalent systems (sometimes termed potential symmetries) contain the nonclassical symmetries of the single equation counterpart. However, we have found a special case where the opposite is true, for the nonlinear wave equation when $X^2 = F$, where we have found that the nonclassical symmetries of the single equation contain the nonclassical symmetries of a system equivalent. A natural question is whether this is true for more general equations

$$\begin{aligned} (i) \quad & u_t = \left(F(t,x,u)u_x + G(t,x,u)\right)_x \\ (ii) \quad & u_{tt} = \left(F(t,x,u)u_x + G(t,x,u)\right)_x \end{aligned}$$

There seems to be some indication that this is true, but further study is needed.

Acknowledgments: The authors B.A., S.B. and T.D. gratefully acknowledge the University of Central Arkansas for support through a graduate teaching assistantship.

Author Contributions: The authors contributed equally to this work.

Conflicts of Interest: The authors declare no conflict of interest.

References

1. Lie, S. Klassifikation und Integration von gewohnlichen Differentialgleichen zwischen x, y die eine Gruppe von Transformationen gestatten. *Math. Ann.* **1888**, *32*, 213–281.
2. Arrigo, D.J. *Symmetries Analysis of Differential Equations—An Introduction*; Wiley: New York, NY, USA, 2015.
3. Bluman, G.; Kumei, S. *Symmetries and Differential Equations*; Springer: New York, NY, USA, 1989.
4. Olver, P.J. *Applications of Lie Groups to Differential Equations*, 2nd ed.; Springer: New York, NY, USA, 1993.
5. Ovsiannikov, L.V. Group properties of nonlinear heat equation. *Dokl. AN SSSR* **1959**, *125*, 492–495.

6. Ovsiannikov, L.V. *Group Analysis of Differential Equations*; Academic Press: New York, NY, USA, 1982.
7. Bluman, G.W.; Reid, G.J.; Kumei, S. New classes of symmetries for partial differential equations. *J. Math. Phys.* **1988**, *29*, 806–881.
8. Bluman, G.W.; Cole, J.D. The general similarity solution of the heat equation. *J. Math. Phys.* **1969**, *18*, 1025–1042.
9. Bradshaw-Hajek, B.H.; Edwards, M.P.; Broadbridge, P.; Williams, G.H. Nonclassical symmetry solutions for reaction diffusion equations with explicit spatial dependence. *Nonliner Anal.* **2007**, *67*, 2541–2552.
10. Cherniha, R. New Q-conditional symmetries and exact solutions of some reaction- diffusion-convection equations arising in mathematical biology. *J. Math. Anal. Appl.* **2007**, *326*, 783–799.
11. Popovych, R.O.; Vaneeva, O.O.; Ivanova N.M. Potential nonclassical symmetries and solutions of fast diffusion equation. *Phys. Lett. A* **2007**, *362*, 166–173.
12. Bruzon, M.S.; Gandarias, M.L. Applying a new algorithm to derive nonclassical symmetries. *Commun. Nonlinear Sci. Numer. Simul.* **2008**, *13*, 517–523.
13. Arrigo, D.J.; Ekrut, D.A.; Fliss, J.R.; Le, L. Nonclassical symmetries of a class of Burgers' systems. *J. Math. Anal. Appl.* **2010**, *371*, 813–820.
14. Bluman, G.W.; Tian, S.F.; Yang, Z. Nonclassical analysis of the nonlinear Kompaneets equation. *J. Eng. Math.* **2014**, *84*, 87–97.
15. Cherniha, R.; Davydovych, V. Conditional symmetries and exact solutions of nonlinear reaction-diffusion systems with nonconstant diffusivities, Commun. *Nonlinear Sci. Numer. Simulat.* **2012**, *17*, 3177–3188.
16. Hashemi, M.S.; Nucci, M.C. Nonclassical symmetries for a class of reaction-diffusion equations: The method of heir-equations. *J. Non. Math Phys.* **2012**, *20*, 44–60.
17. Huang, D.J.; Zhou, S. Group-theoretical analysis of variable coefficient nonlinear telegraph equations. *Acta Appl. Math.* **2012**, *117*, 135–183.
18. Vaneeva, O.O.; Popovych, R.O.; Sophocleous, C. Extended group analysis of variable coefficient reaction diffusion equations with exponential nonlinearities. *J. Math. Anal. Appl.* **2012**, *396*, 225–242.
19. Broadbridge, P.; Bradshaw-Hajek, B.H.; Triadis, D. Exact non-classical symmetry solutions of Arrhenius reaction-diffusion. *Proc. R. Soc. Lond.* **2015**, *471*, doi:10.1098/rspa.2015.0580.
20. Louw, K.; Moitsheki, R.J. Group-invariant solutions for the generalised fisher type equation. *Nat. Sci.* **2015**, *7*, 613–624.
21. Pliukhin, O. Q-conditional symmetries and exact solutions of nonlinear reaction-diffusion systems. *Symmetry* **2015**, *7*, 1841–1855.
22. Yun, Y.; Temuer, C. Classical and nonclassical symmetry classifications of nonlinear wave equation with dissipation. *Appl. Math. Mech. Eng. Ed.* **2015**, *36*, 365–378.
23. Broadbridge, P.; Bradshaw-Hajek, B.H. Exact solutions for logistic reaction-diffusion equations in biology. *ZAMP* **2016**, arXiv:1602.07370.
24. Bluman, G.W.; Yan, Y.S. Nonclassical potential solutions of partial differential equations. *Eur. J. Appl. Math.* **2005**, *16*, 239–261.
25. Gandarias, M.L.; Bruzon, M.S. Solutions through nonclassical potential symmetries for a generalized inhomogeneous nonlinear diffusion equation. *Math. Meth. Appl. Sci.* **2008**, *31*, 753–767.
26. Broadbridge, P.; White, I. Constant rate rainfall infiltration: A versatile nonlinear model 1. Analytic solution. *Water Res. Res.* **1988**, *24*, 145–154.
27. Rogers, C.; Stallybrass, M.P.; Clements, D.L. On two phase filtration under gravity and with boundary infiltration: Application of a Ba äcklund transformation. *J. Nonliner Anal. Meth. Appl.* **1983**, *7*, 785–799.
28. Katayev, I.G. *Electromagnetic Shock Waves*; Iliffe: London, UK, 1966.
29. Jeffery, A. Acceleration wave propagation in hyperelastic rods of variable cross-section. *Wave Motion* **1982**, *4*, 173–180.
30. Broadbridge, P.; Arrigo, D.J. All solutions of standard symmetric linear partial differential equations have classical Lie symmetry. *J. Math. Anal. Appl.* **1999**, *234*, 109–122.
31. Cherniha, R.; Serov, M. Symmetries, ansätze and exact solutions of nonlinear second-order evolution equations with convection terms. *Eur. J. Appl. Math.* **1998**, *72*, 21–39.
32. Arrigo, D.J.; Hill, J.M. Nonclassical symmetries for nonlinear diffusion and absorption. *Stud. Appl. Math.* **1995**, *72*, 21–39.

33. Arrigo, D.J.; Beckham, J.R. Nonclassical symmetries of evolutionary partial differential equations and compatibility. *J. Math. Anal. Appl.* **2004**, *289*, 55–65.
34. Niu, X.H.; Pan, Z.L. Nonclassical symmetries of a class of nonlinear partial differential equations with arbitrary order and compatibility. *J. Math. Anal. Appl.* **2005**, *311*, 479–488.
35. Wan, W.T.; Chen, Y. A note on nonclassical symmetries of a class of nonlinear partial differential equations and compatibility. *Commun. Theor. Phys.* **2009**, *52*, 398–402.
36. El-Sabbagh, M.F.; Ali, A.T. Nonclassical symmetries for nonlinear partial differential equations via compatibility. *Commun. Theor. Phys.* **2011**, *56*, 611–616.
37. Bluman, G.W.; Shtelen, V. Developments in similarity methods related to pioneering work of Julian Cole. In *Mathematics Is for Solving Problems*; Cook, S.L.P., Roytburd,V., Tulin, M., Eds.; SIAM: Philadelphia, PA, USA, 1996; pp. 105–118.
38. Näslund, R.N. *On Conditional Q-Symmetries of Some Quasi-Linear Hyperbolic Wave Equations*; Reprint Department of Mathematics, Lulea University of Technology: Lulea, Sweden, 2003.

symmetry MDPI

Article

Q-Conditional Symmetries and Exact Solutions of Nonlinear Reaction–Diffusion Systems

Oleksii Pliukhin

Department of Mathematics, Poltava National Technical Yuriy Kondratyuk University, 24, Pershotravnevyi Prospekt, 36601 Poltava, Ukraine; pliukhin@gmail.com; Tel.: +38-095-823-29-84

Academic Editor: Roman M. Cherniha
Received: 28 June 2015; Accepted: 1 October 2015; Published: 16 October 2015

Abstract: A wide range of reaction–diffusion systems with constant diffusivities that are invariant under Q-conditional operators is found. Using the symmetries obtained, the reductions of the corresponding systems to the systems of ODEs are conducted in order to find exact solutions. In particular, the solutions of some reaction–diffusion systems of the Lotka–Volterra type in an explicit form and satisfying Dirichlet boundary conditions are obtained. An biological interpretation is presented in order to show that two different types of interaction between biological species can be described.

Keywords: Q-conditional symmetry; reaction–diffusion systems; exact solution; Lotka–Volterra system

1. Introduction

In 1952, Alan Turing published his prominent paper [1]. In this paper he proposed the Turing hypothesis of pattern formation. He used reaction–diffusion equations of the form

$$\begin{aligned} \lambda_1 u_t &= (D_1(u)u_x)_x + F(u,v), \\ \lambda_2 v_t &= (D_2(v)v_x)_x + G(u,v) \end{aligned} \tag{1}$$

which are central to the field of pattern formation.

In system (1), F and G are arbitrary smooth functions, $u = u(t,x)$ and $v = v(t,x)$ are unknown functions of the variables t and x, while the subscripts t and x denote differentiation with respect to this variable. Nonlinear system (1) generalizes many well-known nonlinear second-order models used to describe various processes in physics [2], biology [3–5] and ecology [6].

Here we concentrate ourselves on the most important subclass of RD systems with the form of (1), namely that with constant coefficients of diffusivity

$$\begin{aligned} \lambda_1 u_t &= u_{xx} + F(u,v), \\ \lambda_2 v_t &= v_{xx} + G(u,v) \end{aligned} \tag{2}$$

System (2) has been intensely studied using different mathematical methods (see, e.g., [3,4,7] and papers cited therein). All possible Lie symmetries of system (2) have been found, in [8–11]. In particular, Q-conditional symmetries of (2) were found in [12]. Reference [13] also contains some results related with system (2).

System (1) is a natural generalization of the well-known RD equation

$$u_t = [D(u)u_x]_x + F(u) \tag{3}$$

There are many papers devoted to the construction of Q-conditional symmetries for this equation [14–21], starting from the pioneering work in [22]. There is also a non-trivial generalization of these results for the case of the reaction–diffusion–convection equation ([21] and papers cited therein).

In contrast to (3), there are not many results for searching Q-conditional symmetries of system (2). Construction of the Q-conditional symmetries (non-classical symmetries) of such systems is a very difficult task. Only a few papers have been devoted to the search of such symmetries. In [23] the Q-conditional symmetries of the system

$$\begin{aligned} u_t &= \left(u^k u_x\right)_x + F(u,v), \\ v_t &= \left(v^l v_x\right)_x + G(u,v), l^2 + k^2 \neq 0, \end{aligned}$$

have been obtained; in [24] the Q-conditional symmetries of the Lotka–Volterra system

$$\begin{aligned} \lambda_1 u_t &= u_{xx} + u(a_1 + b_1 u + c_1 v), \\ \lambda_2 v_t &= v_{xx} + v(a_2 + b_2 u + c_2 v) \end{aligned} \tag{4}$$

were obtained.

The paper is organized as follows. In Section 2 three theorems are presented which contain the main result for Q-conditional symmetries of system (2). In Section 3, ansätze for all systems and solutions for one of the systems are derived. In Section 4, the solutions for a generalization of the Lotka–Volterra system are obtained and analyzed. Some graphs of the exact solutions are also presented. Finally, we present some conclusions.

2. Main Result

Let us consider the reaction–diffusion system with constant diffusivities: (2). We want to find Q-conditional operators of the form

$$Q = \partial_t + \xi(t,x,u,v)\partial_x + \eta^1(t,x,u,v)\partial_u + \eta^2(t,x,u,v)\partial_v \tag{5}$$

under which system (2) is invariant.

The most general form of the Q-conditional operators is

$$Q = \xi^0(t,x,u,v)\partial_t + \xi^1(t,x,u,v)\partial_x + \eta^1(t,x,u,v)\partial_u + \eta^2(t,x,u,v)\partial_v.$$

In the case $\xi^0(t,x,u,v) \neq 0$, this operator can be reduced to that with $\xi^0(t,x,u,v) = 1$ [25]. So we investigate operator (5).

We write down system (2) in the following form:

$$\begin{aligned} u_{xx} &= \lambda_1 u_t + C^1(u,v), \lambda_1 \neq 0, \\ v_{xx} &= \lambda_2 v_t + C^2(u,v), \lambda_2 \neq 0 \end{aligned} \tag{6}$$

where $C^1(u,v) = -F(u,v), C^2(u,v) = -G(u,v)$.

The determining equations for finding coefficients of operator (5) and functions $C^1(u,v)$, $C^2(u,v)$ from system (6) have the form

$$
\begin{array}{l}
1)\xi_{uu} = \xi_{vv} = \xi_{uv} = 0, \\
2)\eta^1_{vv} = 0, \\
3)\eta^2_{uu} = 0, \\
4)2\lambda_1\xi\xi_u + \eta^1_{uu} - 2\xi_{xu} = 0, \\
5)2\lambda_2\xi\xi_v + \eta^2_{vv} - 2\xi_{xv} = 0, \\
6)(\lambda_1+\lambda_2)\xi\xi_v + 2\eta^1_{uv} - 2\xi_{xv} = 0, \\
7)(\lambda_1+\lambda_2)\xi\xi_u + 2\eta^2_{uv} - 2\xi_{xu} = 0, \\
8)(\lambda_1-\lambda_2)\xi\eta^1_v + 2\eta^1_{xv} - 2\xi_v C^1 - 2\lambda_1\xi_v\eta^1 = 0, \\
9)(\lambda_2-\lambda_1)\xi\eta^2_u + 2\eta^2_{xu} - 2\xi_u C^2 - 2\lambda_2\xi_u\eta^2 = 0, \\
10)\lambda_1(2\xi_u\eta^1 - \xi_t - \xi_v\eta^2 - 2\xi\xi_x) + \lambda_2\xi_v\eta^2 + 3\xi_u C^1 + \xi_v C^2 - 2\eta^1_{xu} + \xi_{xx} = 0, \\
11)\lambda_2(2\xi_v\eta^2 - \xi_t - \xi_u\eta^1 - 2\xi\xi_x) + \lambda_1\xi_u\eta^1 + 3\xi_v C^2 + \xi_u C^1 - 2\eta^2_{xv} + \xi_{xx} = 0, \\
12)\lambda_1(\eta^1_t + \eta^2\eta^1_v + 2\xi_x\eta^1) - \lambda_2\eta^2\eta^1_v + \eta^1 C^1_u + \eta^2 C^1_v - \eta^1_u C^1 + 2\xi_x C^1 - \eta^1_v C^2 - \eta^1_{xx} = 0, \\
13)\lambda_2(\eta^2_t + \eta^1\eta^2_u + 2\xi_x\eta^2) - \lambda_1\eta^1\eta^2_u + \eta^1 C^2_u + \eta^2 C^2_v - \eta^2_u C^1 + 2\xi_x C^2 - \eta^2_v C^2 - \eta^2_{xx} = 0.
\end{array}
\quad (7)
$$

System (7) is an over-determined system of partial differential equations and there are no any general method for solving of such systems [26,27]. Thus, we were not able to find the general solution of system (7), hence we have solved it with conditions

$$\xi = \xi(u,v), \eta^i = \eta^i(u,v), i = 1,2 \quad (8)$$

Solving Equations 1)–3) of system (7), we obtain

$$\xi = au + bv + c, \eta^1 = p^1(u)v + q^1(u), \eta^2 = p^2(v)u + q^2(v) \quad (9)$$

where a, b, c are arbitrary constants, p^1, p^2, q^1, q^2 are arbitrary smooth functions. Substituting (9) into 6), 7) from (7) and splitting the obtained equations with respect to the powers of u and v, we arrive at the system

$$
\begin{array}{c}
a^2(\lambda_1+\lambda_2) = 0, b^2(\lambda_1+\lambda_2) = 0, \\
(\lambda_1+\lambda_2)a(bv+c) + 2p^2_v = 0, (\lambda_1+\lambda_2)b(au+c) + 2p^1_u = 0
\end{array}
\quad (10)
$$

Obviously, that solutions of first pair of equations of (10) will be $\lambda_2 = -\lambda_1$, or $a = b = 0$.

Let us consider the case $\lambda_2 = -\lambda_1$ (the case $a = b = 0$ will be considered later). In this case we obtain $p^1 = const = \alpha_1, p^2 = const = \alpha_2$. Substituting (9) $p^1 = \alpha_1, p^2 = \alpha_2$ into Equations 4) and 5) of system (7) and splitting with respect to the powers of u and v, we arrive at

$$2\lambda_1 ab = 0, q^1_{uu} + 2\lambda_1 a(au+c) = 0, q^2_{vv} - 2\lambda_1 b(bv+c) = 0 \quad (11)$$

Since $\lambda_1 \neq 0$, we conclude that $ab = 0$. Consider the case $a = 0$, $b \neq 0$ (the case $b = 0$, $a \neq 0$ is symmetrical). From Equation 8), we obtain $\alpha_2 = 0$. Substituting (9) with the specified coefficients, namely

$$\xi = bv + c, \eta^1 = \alpha_1 v + q^1, \eta^2 = q^2,$$

into Equations 10), 11) of system (7), we arrive at

$$b\left(C^2 - 2\lambda_1 q^2\right) = 0, b\left(3C^2 - 2\lambda_1 q^2\right) = 0 \quad (12)$$

Substituting $q^2 = 0$, obtained from (12), into the third equation of system (11), we obtain $\lambda_1 b(bv+c) = 0$, that is $\lambda_1 b = 0$, but that contradicts the above restrictions.

Thus, in the case $\lambda_2 = -\lambda_1$ we do not obtain any Q-conditional operator of the form (5).

Consider the case $a = b = 0$. In this case, from Equations $4), 5), 6)$ and $7)$ of system (7), we obtain

$$p^i = const = \alpha_i, i = 1,2, q^1 = \beta_1 u + \gamma_1, q^2 = \beta_2 v + \gamma_2,$$

where $\beta_i, \gamma_i, i = 1,2$ are the arbitrary constants. Thus, expressions (9) take the form

$$\xi = c, \eta^1 = \alpha_1 v + \beta_1 u + \gamma_1, \eta^2 = \alpha_2 u + \beta_2 v + \gamma_2 \quad (13)$$

Substituting (13) into Equations $8)$ and $9)$ of system (7), we arrive at

$$c\alpha_1(\lambda_2 - \lambda_1) = 0, c\alpha_2(\lambda_2 - \lambda_1) = 0 \quad (14)$$

Solving the system of algebraic Equations (14), we obtain three solutions $\lambda_2 = \lambda_1, \alpha_1 = \alpha_2 = 0$ and $c = 0$, therefore we obtain three cases. Let us consider all these cases.

Theorem 1. *In the cases $\lambda_2 = \lambda_1$ or $\eta^1_v = \eta^2_u = 0$ with conditions (8), the system of determining equations for finding of the Q-conditional operators of the form (5) for system (6) coincide with the system of determining equations for finding Lie operators.*

Proof. Substituting (13), with $\lambda_2 = \lambda_1$, into system (7) we find that Equations $1) - 11)$ are transformed into identities, and Equations $12)$ and $13)$ take the form

$$\eta^1 C^1_u + \eta^2 C^1_v - \eta^1_u C^1 - \eta^1_v C^2 = 0, \eta^1 C^2_u + \eta^2 C^2_v - \eta^2_u C^1 - \eta^2_v C^2 = 0 \quad (15)$$

In [11] the determining equations for finding of Lie symmetries with condition $\lambda_2 = \lambda_1$ are written down in explicit form. Substituting conditions (8) into these equations, we see that the result is completely identical to Equations (15).

Substituting (13), with $\alpha_1 = \alpha_2 = 0$ into system (7), we see that Equations $1) - 11)$ also transform into identities, and equations $12)$ and $13)$ take the form

$$\eta^1 C^1_u + \eta^2 C^1_v - \eta^1_u C^1 = 0, \eta^1 C^2_u + \eta^2 C^2_v - \eta^2_v C^2 = 0. \quad (16)$$

Comparing equations (16) with equations for finding of Lie symmetries of system (6) with conditions (8) from [9], we see that they are completely identical. □

Thus, in the following we assume that $\lambda_1 \neq \lambda_2, \alpha_1^2 + \alpha_2^2 \neq 0$.

Let us consider the case $c = 0$, which is on the one hand the most interesting and on the other the most difficult. In this case, (13) takes the form

$$\xi = 0, \eta^1 = \alpha_1 v + \beta_1 u + \gamma_1, \eta^2 = \alpha_2 u + \beta_2 v + \gamma_2 \quad (17)$$

Equations $1) - 11)$ satisfy expressions (17) and Equations $12), 13)$ take the form

$$\begin{aligned} &(\alpha_1 v + \beta_1 u + \gamma_1)C^1_u + (\alpha_2 u + \beta_2 v + \gamma_2)C^1_v - \beta_1 C^1 - \alpha_1 C^2 \\ &\quad = \alpha_1(\lambda_2 - \lambda_1)(\alpha_2 u + \beta_2 v + \gamma_2), \\ &(\alpha_1 v + \beta_1 u + \gamma_1)C^2_u + (\alpha_2 u + \beta_2 v + \gamma_2)C^2_v - \alpha_2 C^1 - \beta_2 C^2 \\ &\quad = \alpha_2(\lambda_1 - \lambda_2)(\alpha_1 v + \beta_1 u + \gamma_1) \end{aligned} \quad (18)$$

Thus, we can formulate the following theorem.

Theorem 2. *The nonlinear reaction–diffusion system (6) is Q-conditionally invariant under operator (5) with coefficients (17) if and only if the nonlinearities C^1, C^2 are the solutions of linear system (18).*

To find the general solution of system (18), one need to analyze two cases $\alpha_2 = 0$ and $\alpha_2 \neq 0$. In the case $\alpha_2 = 0$, system (18) takes the form

$$\alpha_1(\lambda_2 \qquad + \\ (\alpha_1 v + \beta_1 u + \gamma_1)C^2_u + (\beta_2 v + \gamma_2)C^2_v = \beta_2 C^2 \tag{19}$$

Since $\alpha_1 \neq 0$, renaming $C^1 \to \alpha_1 C^1$, u
and $\gamma_1 \to \alpha_1\gamma_1$, and taking into account that with any coefficients β_1, β_2 we can remove the parameter γ_1 using linear substitutions of u, v, system (19) reduces to the form

$$\begin{gathered} (v + \beta_1 u)C^1_u + (\beta_2 v + \gamma_2)C^1_v = \beta_1 C^1 + C^2 + (\lambda_2 - \lambda_1)(\beta_2 v + \gamma_2), \\ (v + \beta_1 u)C^2_u + (\beta_2 v + \gamma_2)C^2_v = \beta_2 C^2 \end{gathered} \tag{20}$$

One notes a particular solution of system (20), of the form

$$C^1_{part} = \frac{1}{2}(\lambda_2 - \lambda_1)(v + \beta_1 u), C^2_{part} = \frac{1}{2}(\lambda_1 - \lambda_2)(\beta_2 v + \gamma_2) \tag{21}$$

Now to construct the general solution of (20), we need to solve the corresponding homogeneous system, that is

$$\begin{gathered} (v + \beta_1 u)C^1_u + (\beta_2 v + \gamma_2)C^1_v = \beta_1 C^1 + C^2, \\ (v + \beta_1 u)C^2_u + (\beta_2 v + \gamma_2)C^2_v = \beta_2 C^2 \end{gathered} \tag{22}$$

As a result, the following statement was proved.

Theorem 3. *Reaction–diffusion system (6) is Q-conditionally invariant under operator (5) with conditions (8), and* $\eta^2_{\bar u} = 0$, *if and only if the system and corresponding operator have one of the seven following forms (moreover* $\lambda_2 \neq \lambda_1$*):*

$$\begin{gathered} u_{xx} = \lambda_1 u_t + g(\omega) v \ln(v) + h(\omega) v + \tfrac{1}{2}(\lambda_2 - \lambda_1)(v + \beta_1 u), \\ v_{xx} = \lambda_2 v_t + \beta_1 g(\omega) v + \tfrac{1}{2}(\lambda_1 - \lambda_2)\beta_1 v, \\ Q = \partial_t + (v + \beta_1 u)\partial_u + \beta_1 v \partial_v, \\ \omega = v^{-1} \exp\left(\tfrac{\beta_1 u}{v}\right), \beta_1 \neq 0 \end{gathered} \tag{23}$$

$$\begin{gathered} u_{xx} = \lambda_1 u_t + h(\omega) v^{\frac{\beta_1}{\beta_2}} + g(\omega) v + \tfrac{1}{2}(\lambda_2 - \lambda_1)(v + \beta_1 u), \\ v_{xx} = \lambda_2 v_t + (\beta_2 - \beta_1) g(\omega) v + \tfrac{1}{2}(\lambda_1 - \lambda_2)\beta_2 v, \\ Q = \partial_t + (v + \beta_1 u)\partial_u + \beta_2 v \partial_v, \\ \omega = v^{-\frac{\beta_1}{\beta_2}}((\beta_1 - \beta_2)u + v), \beta_1\beta_2(\beta_1 - \beta_2) \neq 0 \end{gathered} \tag{24}$$

$$\begin{gathered} u_{xx} = \lambda_1 u_t + u g(v) + h(v), \\ v_{xx} = \lambda_2 v_t + v g(v), \\ Q = \partial_t + v \partial_u \end{gathered} \tag{25}$$

$$\begin{gathered} u_{xx} = \lambda_1 u_t + g(\omega) v + h(\omega) + \tfrac{1}{2}(\lambda_2 - \lambda_1) v, \\ v_{xx} = \lambda_2 v_t + g(\omega) + \tfrac{1}{2}(\lambda_1 - \lambda_2), \\ Q = \partial_t + v \partial_u + \partial_v, \\ \omega = 2u - v^2 \end{gathered} \tag{26}$$

$$\begin{gathered} u_{xx} = \lambda_1 u_t + g(\omega) v + h(\omega) + \tfrac{1}{2}(\lambda_2 - \lambda_1) v, \\ v_{xx} = \lambda_2 v_t + \beta_2 g(\omega)(v + \tau_2) + \tfrac{1}{2}\beta_2(\lambda_1 - \lambda_2)(v + \tau_2), \\ Q = \partial_t + v \partial_u + \beta_2(v + \tau_2)\partial_v, \\ \omega = \beta_2 u - v + \tau_2 \ln(v + \tau_2), \beta_2 \neq 0 \end{gathered} \tag{27}$$

$$\begin{aligned} u_{xx} &= \lambda_1 u_t + (v+u)h(v) - g(v), \\ v_{xx} &= \lambda_2 v_t + g(v), \\ Q &= \partial_t + \beta_1(v+u)\partial_u, \beta_1 \neq 0 \end{aligned} \tag{28}$$

$$\begin{aligned} u_{xx} &= \lambda_1 u_t + h(\omega)\exp\left(\tfrac{\beta_1}{\gamma_2}v\right) - g(\omega) + \tfrac{1}{2}(\lambda_2-\lambda_1)(v+\beta_1 u), \\ v_{xx} &= \lambda_2 v_t + \beta_1 g(\omega) + \tfrac{1}{2}\gamma_2(\lambda_1-\lambda_2), \\ Q &= \partial_t + (v+\beta_1 u)\partial_u + \gamma_2\partial_v, \\ \omega &= \exp\left(-\tfrac{\beta_1}{\gamma_2}v\right)(\beta_1^2 u + \beta_1 v + \gamma_2), \beta_1\gamma_2 \neq 0 \end{aligned} \tag{29}$$

Proof. To prove this theorem, it is necessary and sufficient to construct the general solution of system (22) for all possible ratios between parameters $\beta_1, \beta_2, \gamma_2$. To do this we need to investigate the following seven cases:

1. $\beta_1\beta_2 \neq 0, \beta_1 = \beta_2$;
2. $\beta_1\beta_2 \neq 0, \beta_1 \neq \beta_2$;
3. $\beta_2 = 0, \beta_1 = 0, \gamma_2 = 0$;
4. $\beta_2 = 0, \beta_1 = 0, \gamma_2 \neq 0$;
5. $\beta_2 \neq 0, \beta_1 = 0$.
6. $\beta_2 = 0, \beta_1 \neq 0, \gamma_2 = 0$;
7. $\beta_2 = 0, \beta_1 \neq 0, \gamma_2 \neq 0$.

These cases take into account all possibilities that arise when we solve system (22). Let us consider these cases.

Case 1. Solving the second equation of (22), we get $C^2 = \beta_1 v g(\omega)$ and $\omega = v^{-1}\exp\left(\frac{\beta_1 u}{v}\right)$. So the first equation of (22) reduces to an ODE for finding of the function C^1:

$$C^1_v - \frac{C^1}{v} = g(\omega).$$

Solving it, we get that $C^1 = g(\omega)v\ln(v) + vh(\omega)$. Taking into account the expressions for C^1, C^2, ω, obtained above, C^1_{part}, C^2_{part} from Formulas (21) and restrictions (obtained above), finally we arrive at the reaction–diffusion system and the Q-conditional operator listed in (23) of Theorem 3.

Cases 2–7. Considering similarly these cases and using simple renamings, we arrive at systems and operators (24)–(29) of Theorem 3. □

In the case $\alpha_2 \neq 0$ we should also assume that $\alpha_1 \neq 0$, otherwise we obtain the case $\alpha_2 = 0$ up to renaming. We seek a solution of system (18) of the form

$$C^1 = r_1 u + r_2 v + r_3, C^2 = s_1 u + s_2 v + s_3 \tag{30}$$

Substituting (30) into (18), we obtain the system of algebraic equations

$$\begin{aligned} \alpha_2 r_2 - \alpha_1 s_1 + \alpha_1\alpha_2(\lambda_1-\lambda_2) &= 0, \\ \alpha_1 r_1 + (\beta_2-\beta_1)r_2 - s_2\alpha_1 + \alpha_1\beta_2(\lambda_1-\lambda_2) &= 0, \\ \gamma_1 r_1 + \gamma_2 r_2 - \beta_1 r_3 - \alpha_1 s_3 + \alpha_1\gamma_2(\lambda_1-\lambda_2) &= 0, \\ \alpha_2 r_1 + (\beta_2-\beta_1)s_1 - s_2\alpha_2 + \alpha_2\beta_1(\lambda_1-\lambda_2) &= 0, \\ \alpha_2 r_3 - \gamma_1 s_1 - \gamma_2 s_2 + \beta_2 s_3 + \alpha_2\gamma_1(\lambda_1-\lambda_2) &= 0 \end{aligned} \tag{31}$$

Solving system (31), we arrive at two possibilities depending on $\Delta = \alpha_1\alpha_2 - \beta_1\beta_2$:

I) $\Delta = 0$,

$$\begin{aligned} r_1 &= \tfrac{\beta_1}{\alpha_2}s_1 + \beta_1(\lambda_2-\lambda_1), r_2 = \tfrac{\alpha_1}{\alpha_2}s_1 + \alpha_1(\lambda_2-\lambda_1), \\ r_3 &= \tfrac{\alpha_2\gamma_1+\beta_2\gamma_2}{\alpha_2^2}s_1 - \tfrac{\beta_2}{\alpha_2}s_3 + \gamma_1(\lambda_2-\lambda_1), s_2 = \tfrac{\beta_2}{\alpha_2}s_1. \end{aligned}$$

II) $\Delta \neq 0$,

$$r_1 = \frac{(\beta_1-\beta_2)}{\alpha_2}s_1 + s_2 + \beta_1(\lambda_2 - \lambda_1), r_2 = \frac{\alpha_1}{\alpha_2}s_1 + \alpha_1(\lambda_2 - \lambda_1),$$
$$r_3 = \frac{(\Delta+\beta_2^2)\gamma_1 - \alpha_1\beta_2\gamma_2}{\alpha_2\Delta}s_1 + \frac{\alpha_1\gamma_2 - \beta_2\gamma_1}{\Delta}s_2 + \gamma_1(\lambda_2 - \lambda_1), s_3 = \frac{\alpha_1\gamma_2 - \beta_2\gamma_1}{\Delta}s_1 + \frac{\alpha_2\gamma_1 - \beta_1\gamma_2}{\Delta}s_2.$$

In Case I) $s_1 = s_3 = 0$, we obtain the solution of system (18)

$$C^1_{part} = (\lambda_2 - \lambda_1)\left(\frac{\beta_1}{\alpha_2}(\alpha_2 u + \beta_2 v) + \gamma_1\right), C^2_{part} = 0 \tag{32}$$

In Case II) $s_1 = s_2 = 0$ we obtain the solution of system (18)

$$C^1_{part} = (\lambda_2 - \lambda_1)(\alpha_1 v + \beta_1 u + \gamma_1), C^2_{part} = 0 \tag{33}$$

Furthermore, we must solve the homogeneous system

$$\begin{aligned} (\alpha_1 v + \beta_1 u + \gamma_1)C^1_u + (\alpha_2 u + \beta_2 v + \gamma_2)C^1_v &= \beta_1 C^1 + \alpha_1 C^2, \\ (\alpha_1 v + \beta_1 u + \gamma_1)C^2_u + (\alpha_2 u + \beta_2 v + \gamma_2)C^2_v &= \alpha_2 C^1 + \beta_2 C^2 \end{aligned} \tag{34}$$

Let us consider Case I). Using the condition $\Delta = 0$ for system (34), we get

$$\begin{aligned} \left(\tfrac{\beta_1}{\alpha_2}(\alpha_2 u + \beta_2 v) + \gamma_1\right)C^1_u + (\alpha_2 u + \beta_2 v + \gamma_2)C^1_v &= \tfrac{\beta_1}{\alpha_2}(\alpha_2 C^1 + \beta_2 C^2), \\ \left(\tfrac{\beta_1}{\alpha_2}(\alpha_2 u + \beta_2 v) + \gamma_1\right)C^2_u + (\alpha_2 u + \beta_2 v + \gamma_2)C^2_v &= \alpha_2 C^1 + \beta_2 C^2 \end{aligned} \tag{35}$$

Multiplying the second equation of (35) by $-\frac{\beta_1}{\alpha_2}$, adding to the first and renaming $u \to u - \frac{\gamma_2}{\alpha_2}, \gamma_1 \to \gamma_1 + \frac{\beta_1\gamma_2}{\alpha_2}$, we arrive at

$$\left(\frac{\beta_1}{\alpha_2}(\alpha_2 u + \beta_2 v) + \gamma_1\right)\left(C^1 - \frac{\beta_1}{\alpha_2}C_2\right)_u + (\alpha_2 u + \beta_2 v)\left(C^1 - \frac{\beta_1}{\alpha_2}C_2\right)_v = 0 \tag{36}$$

Using the substitution

$$C^1 = S(u,v) + \frac{\beta_1}{\alpha_2}C_2 \tag{37}$$

we obtain the equation

$$\left(\frac{\beta_1}{\alpha_2}(\alpha_2 u + \beta_2 v) + \gamma_1\right)S_u + (\alpha_2 u + \beta_2 v)S_v = 0 \tag{38}$$

Solving Equation (38), we arrive at three subcases:
1) $\beta_2 = -\beta_1, \gamma_1 = 0, S = S(\omega), \omega = \alpha_2 u - \beta_1 v$;
2) $\beta_2 = -\beta_1, \gamma_1 \neq 0, S = S(\omega), \omega = (\alpha_2 u - \beta_1 v)^2 - 2\alpha_2\gamma_1 v$;
3) $\beta_2 \neq -\beta_1, S = S(\omega), \omega = \alpha_2 u - \beta_1 v - \frac{\alpha_2\gamma_1}{\beta_1+\beta_2}\ln\left|\alpha_2 u + \beta_2 v + \frac{\alpha_2\gamma_1}{\beta_1+\beta_2}\right|$.
Substituting (37) together with the function S from subcase 1) into the second equation of (35), we obtain

$$(\alpha_2 u - \beta_1 v)\left(\alpha_2 {C_v}^2 + \beta_1 {C_u}^2\right) = \alpha_2^2 f(\omega) \tag{39}$$

Solving (39), using (37), (32) and renaming $u \to \beta_1 u, v \to \alpha_2 v$ we obtain the system

$$\begin{aligned} u_{xx} &= \lambda_1 u_t + f(\omega)u + \beta_1(g(\omega) - \lambda_1)(u - v), \\ v_{xx} &= \lambda_2 v_t + f(\omega)v + \beta_1(g(\omega) - \lambda_2)(u - v), \omega = u - v, \end{aligned}$$

Q-conditionally invariant under the operator

$$Q = \partial_t + \beta_1(u - v)(\partial_u + \partial_v).$$

Similarly, for subcase 2), we arrive at the system

$$\begin{aligned} u_{xx} &= \lambda_1 u_t + (f(\omega) - \lambda_1)(\beta_1(u - v) + \gamma_1) + g(\omega), \\ v_{xx} &= \lambda_2 v_t + \beta_1(f(\omega) - \lambda_2)(u - v) + g(\omega), \omega = (u - v)^2 - 2\tfrac{\gamma_1}{\beta_1} v, \gamma_1 \neq 0 \end{aligned}$$

and the operator

$$Q = \partial_t + (\beta_1(u - v) + \gamma_1)\partial_u + \beta_1(u - v)\partial_v.$$

In the subcase 3), we obtain the system

$$\begin{aligned} u_{xx} &= \lambda_1 u_t + k(f(\omega) - \lambda_1) + \beta_1(g(\omega) - \lambda_1)(u + v), \\ v_{xx} &= \lambda_2 v_t - k(f(\omega) - \lambda_2) + \beta_2(g(\omega) - \lambda_2)(u + v), \\ \omega &= \beta_2 u - \beta_1 v - k \ln|u + v|, k = \tfrac{\gamma_1}{\beta_1 + \beta_2}, \end{aligned}$$

and the operator

$$Q = \partial_t + (\beta_1(u + v) + k)\partial_u + (\beta_2(u + v) - k)\partial_v.$$

Examination of Case II) is highly nontrivial and will be reported in another paper.

3. Ansätze and Exact Solutions of the Reaction–Diffusion System

Using standard procedures, we obtain ansätze for all operators of Theorem 3. Substituting these anzätze in the corresponding reaction–diffusion systems, we obtain the reduction systems of equations. All anzätze and reduction systems are presented in Table 1.

Table 1. Ansätze and reduction systems of Theorem 3.

No.	Ansätze	Systems of ODEs
(23)	$u = (t + \psi)e^{\beta_1 t + \varphi}$	$\varphi'' + (\varphi')^2 - \beta_1\left(g\left(e^{\beta_1\psi - \varphi}\right) + \frac{\lambda_1 + \lambda_2}{2}\right) = 0$
	$v = e^{\beta_1 t + \varphi}$	$\psi'' + 2\varphi'\psi' + g\left(e^{\beta_1\psi - \varphi}\right)(\beta_1\psi - \varphi) - h\left(e^{\beta_1\psi - \varphi}\right) - \frac{\lambda_1 + \lambda_2}{2} = 0$
(24)	$u = \psi\left(e^{\beta_2 t + \varphi}\right)^{\frac{\beta_1}{\beta_2}} - \frac{e^{\beta_2 t + \varphi}}{\beta_1 - \beta_2}$	$\varphi'' + (\varphi')^2 + g((\beta_1 - \beta_2)\psi)(\beta_1 - \beta_2) - \beta_2\frac{\lambda_1 + \lambda_2}{2} = 0$
	$v = e^{\beta_2 t + \varphi}$	$\psi'' + \frac{2\beta_1}{\beta_2}\varphi'\psi' + \frac{\beta_1}{\beta_2}\left(\frac{\beta_1}{\beta_2} - 1\right)\psi\left((\varphi')^2 - \beta_2 g((\beta_1 - \beta_2)\psi)\right) - h((\beta_1 - \beta_2)\psi) = 0$
(25)	$u = \varphi t + \psi$	$\varphi'' - \varphi g(\varphi) = 0$
	$v = \varphi$	$\psi'' - g(\varphi)\psi - h(\varphi) - \lambda_1\varphi = 0$
(26)	$u = \frac{1}{2}t^2 + \varphi t + \psi$	$\varphi'' - g(2\psi - \varphi^2) - \frac{\lambda_1 + \lambda_2}{2} = 0$
	$v = t + \varphi$	$\psi'' - g(2\psi - \varphi^2)\varphi - h(2\psi - \varphi^2) - \frac{\lambda_1 + \lambda_2}{2}\varphi = 0$
(27)	$u = \frac{1}{\beta_2}e^{\beta_2 t + \varphi} - \tau_2 t + \psi$	$\varphi'' + (\varphi')^2 - \beta_2 g(\tau_2(\varphi + 1) + \beta_2\psi) - \frac{\lambda_1 + \lambda_2}{2}\beta_2 = 0$
	$v = e^{\beta_2 t + \varphi} - \tau_2$	$\psi'' - h(\tau_2(\varphi + 1) + \beta_2\psi) + \tau_2 g(\tau_2(\varphi + 1) + \beta_2\psi) + \frac{\lambda_1 + \lambda_2}{2}\tau_2 = 0$
(28)	$u = \psi e^{\beta_1 t} - \varphi$	$\varphi'' - g(\varphi) = 0$
	$v = \varphi$	$\psi'' - (h(\varphi) + \beta_1\lambda_1)\psi = 0$
(29)	$u = \psi e^{\beta_1 t} - \frac{\gamma_2 t}{\beta_1} - \frac{\varphi}{\beta_1} - \frac{\gamma_2}{\beta_1^2}$	$\varphi'' - g\left(\beta_1^2\psi e^{-\frac{\beta_1\varphi}{\gamma_2}}\right)\beta_1 - \frac{(\lambda_1 + \lambda_2)}{2}\gamma_2 = 0$
	$v = \gamma_2 t + \varphi$	$\psi'' - e^{\frac{\beta_1\varphi}{\gamma_2}} h\left(\beta_1^2\psi e^{-\frac{\beta_1\varphi}{\gamma_2}}\right) - \beta_1\frac{\lambda_1 + \lambda_2}{2}\psi = 0$

It is impossible to find the general solution of the systems from Table 1 for arbitrary functions g and h. However, if we correctly specify these functions we can find the solutions of these systems.

System (27) is the most interesting one from the point of view of applicability. Let us consider system (27) with $\tau_2 = 0, g(\omega) = b_1\omega + b_0 - \frac{\lambda_1 + \lambda_2}{2}, h(\omega) = a_2\omega^2 + a_1\omega + a_0$. In this case, the reduction system has the form

$$\varphi'' + (\varphi')^2 - \beta_2(b_1\beta_2\psi + b_0) = 0 \tag{40}$$

$$\psi'' - a_2\beta_2^2\psi^2 - a_1\beta_2\psi - a_0 = 0 \tag{41}$$

The solution of Equation (41) has the form

$$\psi = k = const \tag{42}$$

Substituting (42) into (40), we arrive at

$$\varphi'' + (\varphi')^2 = A, A = \beta_2(b_1\beta_2 k + b_0) \tag{43}$$

The solutions of Equation (43) depend on the parameter A. Solving Equation (43) we get three different solutions (up to transformations $x \to x + C_1, C_1 = const$)

$$\varphi = \begin{cases} \ln|Cx|, A = 0; \\ \ln|C\cosh\left(\sqrt{A}x\right)|, A > 0; \\ \ln|C\cos\left(\sqrt{-A}x\right)|, A < 0. \end{cases}$$

Substituting φ and (42) into corresponding ansatz from Table 1, and renaming $C \to \beta_2 C$, we arrive at the exact solutions

$$u = Ce^{\beta_2 t}x + k, v = C\beta_2 e^{\beta_2 t}x,$$
$$u = Ce^{\beta_2 t}\cosh\left(\sqrt{A}x\right) + k, v = C\beta_2 e^{\beta_2 t}\cosh\left(\sqrt{A}x\right)$$
$$u = Ce^{\beta_2 t}\cos\left(\sqrt{-A}x\right) + k, v = C\beta_2 e^{\beta_2 t}\cos\left(\sqrt{-A}x\right) \tag{44}$$

of the reaction–diffusion system

$$\begin{aligned} u_{xx} &= \lambda_1 u_t + (v - u\beta_2)(-a_1 + a_2 v - a_2\beta_2 u) + (b_0 - \lambda_1^* - b_1 v + b_1\beta_2 u)v + a_0, \\ v_{xx} &= \lambda_2 v_t + \beta_2(b_0 - \lambda_2^* - b_1 v + b_1\beta_2 u)v \end{aligned} \tag{45}$$

where k is the solution of the equation $a_2\beta_2^2 k^2 + a_1\beta_2 k + a_0 = 0$.

4. Solutions and Their Properties of Some Generalization of the Lotka–Volterra System

Let us consider in detail the case $A < 0$. Renaming $\beta_2 = -\frac{B_2}{C_2}, b_1 = -\frac{C_2^2}{B_2}, b_0 = \frac{A_2C_2+B_2\lambda_2}{B_2}, a_0 = -e_0, a_1 = \frac{A_1C_2}{B_2}, a_2 = -\frac{B_1C_2^2}{B_2^2}$, we obtain the exact solution

$$u = Ce^{-\frac{B_2}{C_2}t}\cos\left(\sqrt{-A}x\right) + k, v = -\frac{B_2C}{C_2}e^{-\frac{B_2}{C_2}t}\cos\left(\sqrt{-A}x\right) \tag{46}$$

where $A = -\frac{kB_2C_2+A_2C_2+B_2\lambda_2}{C_2}$, and k is the solution of $B_1k^2 + A_1k + e_0 = 0$, of the reaction–diffusion system

$$\begin{aligned} \lambda_1 u_t &= u_{xx} + u(A_1 + B_1 u + C_1 v) + e_2 v^2 + e_1 v + e_0, \\ \lambda_2 v_t &= v_{xx} + v(A_2 + B_2 u + C_2 v) \end{aligned} \tag{47}$$

where $C_1 = \left(\frac{2B_1}{B_2} - 1\right)C_2, e_2 = \frac{(B_1-B_2)C_2^2}{B_2^2}, e_1 = \frac{(A_1-A_2)C_2}{B_2} + \lambda_1^* - \lambda_2^*$.

System (47) is the generalized Lotka–Volterra system. With $e_2 = e_1 = e_0 = 0$ system (47) becomes the classical Lotka–Volterra system

$$\begin{aligned} \lambda_1 u_t &= u_{xx} + u(A_1 + B_1 u + C_1 v), \\ \lambda_2 v_t &= v_{xx} + v(A_2 + B_2 u + C_2 v) \end{aligned} \tag{48}$$

Note that exact solutions of the form (46) for the classical Lotka–Volterra system (48) have been found in [24].

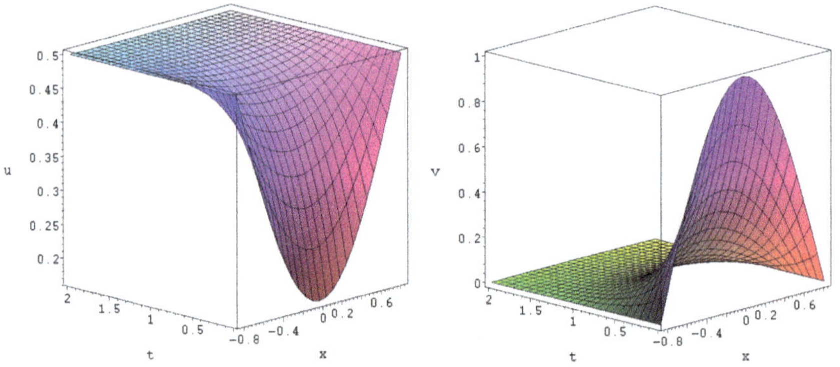

Figure 1. Exact solution to (51).

System (48) can be obtained from system (47) with $e_0 = 0, B_1 = B_2, A_1 = \frac{A_2C_2+(\lambda_2^*-\lambda_1^*)B_2}{C_2}$. Also, the coefficients of (46) and (48) must satisfy the equation $k\left(k + \frac{A_2}{B_2} + \frac{\lambda_2^*-\lambda_1^*}{C_2}\right) = 0$.

It is well known [3] that three main kinds of interactions between two biological species are simulated by system (48):

(i) predator u–prey v interaction,

(ii) competition of the species,

(iii) mutualism or symbiosis.

It turns out that solution (46) can describe the predator-prey interaction on the space interval $[-l, l]$, (here $l = \frac{\pi}{2\sqrt{-A}}$) provided that

$$B_2 < 0, C_2 < 0, C < 0, k > |C| \tag{49}$$

One can easily check that solution (46) is non-negative, bounded in the domain $\Omega = \{(t, x) \in (0, +\infty) \times (-l, l)\}$ and satisfies the given Dirichlet boundary conditions, *i.e.*,

$$u|_{x=-l} = k, v|_{x=-l} = 0, u|_{x=l} = k, \ v|_{x=l} = 0 \tag{50}$$

Choosing the coefficients $\lambda_1 = 2, \lambda_2 = 1, A_2 = 1, B_2 = -1, C_2 = -\frac{1}{3}, B_1 = 0, C = -\frac{1}{3}, e_0 = 1$, gives that $A_1 = -2, C_1 = \frac{1}{3}, k = \frac{1}{2}$. Thus, from solution (46) we obtain the solution

$$u = \frac{1}{2} - \frac{1}{3}e^{-3t}\cos\left(\sqrt{\frac{7}{2}}x\right), v = e^{-3t}\cos\left(\sqrt{\frac{7}{2}}x\right) \tag{51}$$

of the system

$$\begin{aligned} 2u_t &= u_{xx} + u\left(-2 + \tfrac{v}{3}\right) + \tfrac{v^2}{9} + 1, \\ v_t &= v_{xx} + v\left(1 - u - \tfrac{v}{3}\right) \end{aligned} \tag{52}$$

which can describe predator u–prey v interaction, as its coefficients satisfy the conditions for this type of the interaction [3]. System (52) is some generalization of the Lotka–Volterra system (48) with additional nonlinearity $\frac{v^2}{9} + 1$ in the first equation.

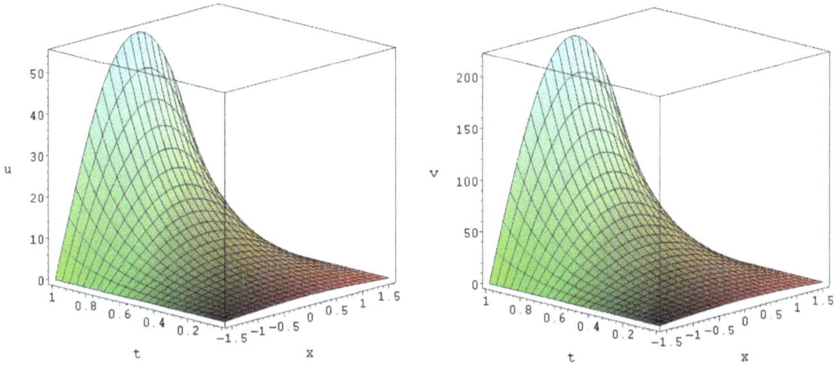

Figure 2. Exact solution of (55).

Solution (51) satisfies Dirichlet boundary conditions (50) with $l = \frac{\pi}{\sqrt{14}}, k = \frac{1}{2}$.

As an example, we present solution (51) in Figure 1. This solution can describe the predator u–prey v interaction between the species u and v when population of predator u becomes $\frac{1}{2}$ and prey eventually dies, *i.e.*, $(u, v) \to \left(\frac{1}{2}, 0\right)$ as $t \to +\infty$.

Choosing coefficients $\lambda_1 = \frac{3}{2}, \lambda_2 = \frac{5}{4}, A_2 = 1, B_2 = -4, C_2 = -\frac{1}{2}, B_1 = 1, C = -\frac{1}{4}, e_0 = 0$, we get $A_1 = -1, C_1 = \frac{3}{4}, k = 1$. Renaming $t \to \frac{1}{4}t$, from solution (46) we obtain the solution

$$u = 1 - \frac{1}{4}e^{-2t}\cos\left(\sqrt{7}x\right), v = 2e^{-2t}\cos\left(\sqrt{7}x\right) \tag{53}$$

of the system

$$\begin{aligned} 6u_t &= u_{xx} + u(-1 + u + \tfrac{3}{4}v) + \tfrac{5}{64}v^2, \\ 5v_t &= v_{xx} + v(1 - 4u - \tfrac{v}{2}) \end{aligned} \tag{54}$$

which can also describe the predator u–prey v interaction, as its coefficients satisfy conditions for this type of interaction [3]. System (52) is some generalization of Lotka–Volterra system (48) with additional nonlinearity $\frac{5v^2}{64}$ in the first equation.

Solution (53) satisfies Dirichlet boundary conditions (50) with $l = \frac{\pi}{2\sqrt{7}}$ and $k = 1$. This solution can describe the predator u–prey v interaction between the species u and v when population of predator u becomes 1 and prey eventually die, *i.e.*, $(u, v) \to (1, 0)$ as $t \to +\infty$.

If we consider system (48) with solution (46), then we obtain the solution that can describe competition of the species. Such a solution is presented in [24].

Also, system (47) can describe mutualism—or symbiosis—of two species. Choosing the coefficients $\lambda_1 = 2, \lambda_2 = 1, A_2 = 5, B_2 = 2, C_2 = -\frac{1}{2}, B_1 = -\frac{1}{10}, k = 0, C = 1, e_0 = 0$, we obtain $A_1 = 9, C_1 = \frac{11}{20}$. So, from solution (46) we obtain the solution

$$u = e^{4t}\cos(x), v = 4e^{4t}\cos(x) \tag{55}$$

of the system

$$\begin{aligned} 2u_t &= u_{xx} + u\left(9 - \tfrac{u}{10} + \tfrac{11v}{20}\right) - \tfrac{21v^2}{160}, \\ v_t &= v_{xx} + v(5 - \tfrac{v}{2} + 2u) \end{aligned} \tag{56}$$

which is a generalization of Lotka–Volterra system (48) with additional nonlinearity $-\frac{21}{160}v^2$ in the first equation.

Solution (55) satisfies Dirichlet boundary conditions (50) with $l = \frac{\pi}{2}, k = 0$. As an example, we present solution (55) in Figure 2. Solution (55) can describe the type of the interaction between the species u and v when both populations grow unboundedly, *i.e.*, $(u, v) \to (+\infty, +\infty)$ if $t \to +\infty$.

5. Conclusions

In this paper, the nonlinear RD system (2) was examined in order to find the Q-conditional operators under which this system is invariant and to construct exact solutions. Because the system of differential (7) is too complicated, we were unable (and believe it is not possible) to find all the solutions of the determining system (7) and thence to find all possible Q-conditional operators. We have found the Q–conditional operators with restrictions (8) (in the case $\eta^2_u = 0$ we have found all possible systems and operators, in the case $\eta^2_u \neq 0$ we have presented some examples) with respect to which the reaction–diffusion system of equations with constant diffusion (2) is invariant. All these operators are given in Theorem 3 of Section 2. In Section 3 the ansätze for all Q-conditional operators of Theorem 3 and the reduction systems are constructed. Section 4 contains the solutions of some generalization of the Lotka–Volterra system. These solutions are analyzed in order to present of biological interpretation. Some graphs of obtained solutions are also presented. It is shown that the obtained solutions satisfy Dirichlet boundary conditions, which are typical for biological interpretation.

Acknowledgments: The author is grateful to the referees for the useful comments.

Conflicts of Interest: The author declares no conflict of interest.

References

1. Turing, A.M. The chemical basis of morphogenesis. *Philos. Trans. Royal Soc. Lond. B Biol. Sci.* **1952**, *237*, 37–72. [CrossRef]
2. Ames, W.F. *Nonlinear Partial Differential Equations in Engineering*; Academic Press: New York, NY, USA; London, UK, 1965; p. xii+511.
3. Murray, J.D. *Mathematical Biology I: An Introduction*, 3rd ed.; Interdisciplinary Applied Mathematics; Springer-Verlag: New York, NY, USA, 2002; Volume 17, p. xxiv+551.
4. Murray, J.D. *Mathematical biology II: Spatial Models and Biomedical Applications*, 3rd ed.; Interdisciplinary Applied Mathematics; Springer-Verlag: New York, NY, USA, 2003; Volume 18, p. xxvi+811.
5. Britton, N.F. *Essential Mathematical Biology*; Springer Undergraduate Mathematics Series; Springer-Verlag London, Ltd.: London, UK, 2003; p. xvi+335.
6. Okubo, A.; Levin, S.A. *Diffusion and Ecological Problems: Modern Perspectives*, 2nd ed.; Interdisciplinary Applied Mathematics; Springer-Verlag: New York, NY, USA, 2001; Volume 14, p. xx+467.
7. Henry, D. *Geometric Theory of Semilinear Parabolic Equations*; Lecture Notes in Mathematics; Springer-Verlag: Berlin, Germany; New York, NY, USA, 1981; Volume 840, p. iv+348.
8. Cherniha, R. Lie symmetries of nonlinear two-dimensional reaction–diffusion systems. *Rep. Math. Phys.* **2000**, *46*, 63–76. [CrossRef]
9. Cherniha, R.; King, J.R. Lie symmetries of nonlinear multidimensional reaction–diffusion systems. I. *J. Phys. A* **2000**, *33*, 267–282. [CrossRef]
10. Cherniha, R.; King, J.R. Addendum: "Lie symmetries of nonlinear multidimensional reaction–diffusion systems. I". *J. Phys. A* **2000**, *33*, 7839–7841. [CrossRef]
11. Cherniha, R.; King, J.R. Lie symmetries of nonlinear multidimensional reaction–diffusion systems. II. *J. Phys. A* **2003**, *36*, 405–425. [CrossRef]
12. Barannik, T.A. Conditional symmetry and exact solutions of a multidimensional diffusion equation. *Ukr. Math. J.* **2002**, *54*, 1416–1420.
13. Barannyk, T. Symmetry and exact solutions for systems of nonlinear reaction–diffusion equations. Available online: http://eqworld.ipmnet.ru/en/solutions/interesting/barannyk.pdf (accessed on 13 October 2015).
14. Serov, N.I. Conditional invariance and exact solutions of a nonlinear heat equation. *Ukr. Math. J.* **1990**, *42*, 1370–1376.

15. Fushchych, W.; Shtelen, W.; Serov, M. *Symmetry Analysis and Exact Solutions of Equations of Nonlinear Mathematical Physics*; Kluwer: Dordrecht, The Netherland, 1993.
16. Nucci, M.C. Symmetries of linear, *C*-integrable, *S*-integrable, and nonintegrable equations. In *Nonlinear Evolution Equations and Dynamical Systems (Baia Verde, 1991)*; World Sci. Publ.: River Edge, NJ, USA, 1992; pp. 374–381.
17. Clarkson, P.A.; Mansfield, E.L. Symmetry reductions and exact solutions of a class of nonlinear heat equations. *Physica D* **1994**, *70*, 250–288. [CrossRef]
18. Arrigo, D.J.; Hill, J.M.; Broadbridge, P. Nonclassical symmetry reductions of the linear diffusion equation with a nonlinear source. *IMA J. Appl. Math.* **1994**, *52*, 1–24. [CrossRef]
19. Arrigo, D.J.; Hill, J.M. Nonclassical symmetries for nonlinear diffusion and absorption. *Stud. Appl. Math.* **1995**, *94*, 21–39. [CrossRef]
20. Pucci, E.; Saccomandi, G. Evolution equations, invariant surface conditions and functional separation of variables. *Physica D* **2000**, *139*, 28–47. [CrossRef]
21. Bluman, G.W.; Cheviakov, A.F.; Anco, S.C. *Applications of Symmetry Methods to Partial Differential Equations*; Applied Mathematical Sciences; Springer: New York, NY, USA, 2010; Volume 168.
22. Bluman, G.W.; Cole, J.D. The general similarity solution of the heat equation. *J. Math. Mech.* **1968/69**, *18*, 1025–1042.
23. Cherniha, R.; Pliukhin, O. New conditional symmetries and exact solutions of reaction–diffusion systems with power diffusivities. *J. Phys. A* **2008**, *41*, 185208:1–185208:14. [CrossRef]
24. Cherniha, R.; Davydovych, V. Conditional symmetries and exact solutions of the diffusive Lotka–Volterra system. *Math. Comput. Model.* **2011**, *54*, 1238–1251. [CrossRef]
25. Cherniha, R. Conditional symmetries for systems of PDEs: New definitions and their application for reaction–diffusion systems. *J. Phys. A* **2010**, *43*, 405207:1–405207:13. [CrossRef]
26. Sidorov, A.F.; Shapeev, V.P.; Yanenko, N.N. *Metod Differentsialnykh Svyazei i Ego Prilozheniya v Gazovoi Dinamike*; "Nauka" Sibirsk. Otdel.: Novosibirsk, Russia, 1984; p. 272. (In Russian)
27. Carini, M.; Fusco, D.; Manganaro, N. Wave-like solutions for a class of parabolic models. *Nonlinear Dynam.* **2003**, *32*, 211–222. [CrossRef]

symmetry

MDPI

Article

Reduction Operators and Exact Solutions of Variable Coefficient Nonlinear Wave Equations with Power Nonlinearities

Dingjiang Huang *, Yan Zhu and Qinmin Yang

Department of Mathematics, East China University of Science and Technology, Shanghai 200237, China; zy0205s@163.com (Y.Z.); qmyang@ecust.edu.cn(Q.Y.)
* Correspondence: djhuang@ecust.edu.cn; Tel.: +86-21-6425-3147

Academic Editor: Roman M. Cherniha
Received: 13 October 2016; Accepted: 14 December 2016; Published: 22 December 2016

Abstract: Reduction operators, i.e., the operators of nonclassical (or conditional) symmetry of a class of variable coefficient nonlinear wave equations with power nonlinearities, are investigated within the framework of a singular reduction operator. A classification of regular reduction operators is performed with respect to generalized extended equivalence groups. Exact solutions of some nonlinear wave models, which are invariant under certain reduction operators, are also constructed.

Keywords: symmetry analysis; reduction operators; equivalence group; nonlinear wave equation; exact solutions

1. Introduction

The investigation of geometric and algebraic structures, as well as exact solutions of nonlinear wave equations plays an important role in the study of nonlinear physical phenomena. Generally, when the equation under consideration is an exact solvable or integrable model (e.g., KdVequation), we can use many efficient methods, such as inverse scattering transform, Bäcklund and Darboux transform, Painleve analysis, nonlocal symmetry reductions, and so on (see, for example, References [1–5] and the references cited therein), to construct its exact solutions. While, if the equation is a non-integrable model, we usually turn to use the so-called Lie group analysis and its extension [6,7] to investigate it.

In this paper, we study reduction operators, i.e., the operators of nonclassical (or conditional) symmetry associated with a class of non-integrable variable coefficient nonlinear wave equations with power nonlinearities of the form:

$$f(x)u_{tt} = (g(x)u^n u_x)_x + h(x)u^m, \tag{1}$$

where $f = f(x), g = g(x)$ and $h = h(x)$ are three arbitrary functions, $fg \neq 0$, n and m are arbitrary constants, t is the time coordinate and x is the one-space coordinate. The linear case is excluded from consideration because it was well investigated. We also assume the variable wave speed coefficient u^n to be nonlinear, i.e., $n \neq 0$. The case $n = 0$ is quite singular and will be investigated separately.

Many specific nonlinear wave models derived from mechanics and engineering, such as the one-dimensional gas flow, longitudinal wave propagation on a moving thread line and electromagnetic transmission line, and so on, can be reduced to Equation (1) (see [8] pp. 50–52 and [9]). Since the 1970s, Lie symmetries and invariant solutions of various kinds of quasi-linear wave equations in two independent variables that intersect Class (1) have been investigated [7,10–24] because of the importance of the wave equation for various applications. Recently, we have presented a complete Lie symmetry and conservation law classification of Class (1) [25,26]. Classical Lie symmetry reduction and invariant solutions of some variable coefficients wave models, which are singled out from the classification results, are also investigated [25].

In general, using classical Lie symmetries' reduction of partial differential equations can provide part of the exact solutions of these equations [6,27,28]. In order to find more other types of exact solutions, one should generalize Lie's original reduction. The first approach to such a generalization was present by Bluman and Cole in 1969 [29] (see also [7]) in which they introduced a wider class of infinitesimal generators than Lie symmetries. Later, such infinitesimal generators, were named nonclassical symmetries [8,30–33] or Q-conditional symmetries [34–41] and were also extended by many authors to some concepts, such as conditional symmetries [34] (the notion 'conditional symmetry' used in this well-known monograph does not coincide with the notion 'non-classical', but is an essential generalization of non-classical symmetry (see Chapter 5.7 therein)), weak symmetry [33], differential constraints, etc. [42–44]. Recently, Boyko, Kunzinger and Popovych [45,46] named it as 'reduction operators' and present a novel framework, namely singular reduction operators or singular reduction modules, for finding an optimal way of obtaining the determining equation of conditional symmetries. As an application, they have investigated the properties of singular reduction operators for a number of (1 + 1)-dimensional evolution equations and a specific wave equations by using this new framework [46]. However, for more general nonlinear wave Equation (1), there exist no general results. In this paper, we employ Popovych's singular reduction operators theory to investigate the properties of nonclassical symmetries of Class (1). We propose a complete classification of regular reduction operators for (1) with respect to generalized extended equivalence groups and construct some non-Lie exact solutions for the nonlinear wave model, which are invariant under certain reduction operators. It should be noted that the terminology 'non-classical symmetry' and 'Q-conditional symmetry' is still used very widely in the literature in contrast to 'reduction operators'; however, we prefer to use this relatively new terminology.

The rest of paper is organized as follows. In Section 2, singular reduction operators and, in particular, regular reduction operators classification for the class under consideration are investigated. Section 3 contains the nonclassical symmetry reduction of some nonlinear wave models. New non-Lie exact solutions of the models are constructed by means of the reduction. Conclusions and a discussion are given in Section 4.

2. Nonclassical Symmetries

Nonclassical symmetries of Class (1) are performed in the framework of the singular reduction operator [46]. All necessary objects (singular and regular reduction operator, etc.) can be found there [46]. Before we proceed with the investigation, we can first simplify the class (1). Using the transformation:

$$\tilde{t} = t, \quad \tilde{x} = \int \frac{dx}{g(x)}, \quad \tilde{u} = u \tag{2}$$

from Theorem 1 in [25], we can reduce Equation (1) to $\tilde{f}(\tilde{x})\tilde{u}_{\tilde{t}\tilde{t}} = (\tilde{u}^n\tilde{u}_{\tilde{x}})_{\tilde{x}} + \tilde{h}(\tilde{x})\tilde{u}^m$, where $\tilde{f}(\tilde{x}) = g(x)f(x), \tilde{g}(\tilde{x}) = 1$ and $\tilde{h}(\tilde{x}) = g(x)h(x)$. Thus, without loss of generality, we can restrict ourselves to investigation of the equation:

$$f(x)u_{tt} = (u^n u_x)_x + h(x)u^m. \tag{3}$$

For convenience, we can further rewrite it as the form:

$$L[u] := f(x)u_{tt} - (u^n u_x)_x - h(x)u^m = 0. \tag{4}$$

All results on symmetries and solutions of Class (3) or (4) can be extended to Class (1) with transformations (2).

According to the algorithm in [29], we seek a reduction operator of Class (4) in the form:

$$Q = \tau(t,x,u)\partial_t + \xi(t,x,u)\partial_x + \eta(t,x,u)\partial_u, \quad (\tau,\xi) \neq (0,0), \tag{5}$$

which is a first-order differential operator on the space $\mathbb{R}^2 \times \mathbb{R}^1$ with coordinates t, x and u, where the coefficients τ and ξ do not simultaneously vanish. This operator allows one to construct an ansatz reducing the original Equation (4) to an ordinary differential equation. The conditional invariance criterion [32,36,41] for Equation (4) to be invariant with respect to the operator (5) reads as:

$$\mathrm{pr}^{(2)}\, Q(L[u])\Big|_{\mathcal{L}\cap\mathcal{Q}_{(2)}} = 0, \tag{6}$$

where $\mathrm{pr}^{(2)}\, Q$ is the usual second-order prolongation [6,28] of the operator (5), $\mathcal{L}$ is the manifold in the second-order jet space $J^{(2)}$ determined by the wave equation $L[u] = 0$ and $\mathcal{Q}_{(2)} \subset J^{(2)}$ is the the first prolongation of the invariant surface condition:

$$Q[u] := \tau u_t + \xi u_x - \eta = 0. \tag{7}$$

The system $\mathcal{Q}_{(2)}$ consists of (7) and the equations obtained by t- and x-differentiation of (7).

Below, according to the singular reduction operator theory [46], we first partition the set of reduction operators of Class (4) into two subsets, i.e., the singular reduction operator and the regular one. Then, we utilize the two kinds of operators to derive determining equations (overdetermined system of nonlinear PDEs with respect to the coefficients of the reduction operator (5)) from the conditional invariance criterion (6) separately. Solving the two systems, we can obtain the final reduction operators. In particular, we will present an exhausted classification of the regular operators of class (4) by solving the corresponding determining equations. In general, every Lie symmetry operator is also a reduction operator. Therefore, in this paper, we will concentrate on the regular reduction operators, which are inequivalent to Lie symmetry operators, called nontrivial.

2.1. Singular Reduction Operators

Using the procedure given by [46], we can obtain the following proposition.

Proposition 1. *A necessary and sufficient condition for a vector field $Q = \tau(t,x,u)\partial_t + \xi(t,x,u)\partial_x + \eta(t,x,u)\partial_u$, which is singular for the differential function $L = f(x)u_{tt} - (u^n u_x)_x - h(x)u^m$ from Equation (4), is that $\xi^2 f(x) = \tau^2 u^n$.*

Proof. Suppose that $\tau \neq 0$, and using the characteristic equation $\tau u_t + \xi u_x - \eta = 0$, we can obtain:

$$\begin{aligned} u_t &= \frac{\eta}{\tau} - \frac{\xi}{\tau}u_x, \\ u_{tt} &= (\frac{\eta}{\tau})_t - (\frac{\xi}{\tau})_t u_x + [(\frac{\eta}{\tau})_u - (\frac{\xi}{\tau})_u u_x](\frac{\eta}{\tau} - \frac{\xi}{\tau}u_x) - (\frac{\xi}{\tau})[(\frac{\eta}{\tau})_x + (\frac{\eta}{\tau})_u u_x - (\frac{\xi}{\tau})_x u_x \\ &\quad - (\frac{\xi}{\tau})_u u_x^2 - (\frac{\xi}{\tau})u_{xx}]. \end{aligned}$$

Now, we substitute the second equation of the above system into L; we can get a differential function:

$$\begin{aligned} \tilde{L} = {} & [f(x)(\frac{\xi}{\tau})^2 - u^n]u_{xx} + f(x)\Big\{(\frac{\eta}{\tau})_t - (\frac{\xi}{\tau})_t u_x + [(\frac{\eta}{\tau})_u - (\frac{\xi}{\tau})_u u_x](\frac{\eta}{\tau} - \frac{\xi}{\tau}u_x) \\ & - (\frac{\xi}{\tau})[(\frac{\eta}{\tau})_x + (\frac{\eta}{\tau})_u u_x - (\frac{\xi}{\tau})_x u_x - (\frac{\xi}{\tau})_u u_x^2]\Big\} - nu^{n-1}u_x^2 - hu^m u_x. \end{aligned}$$

With the aid of Definition 4 of the singular vector field in [46], we derive that a necessary and sufficient condition for ord $\tilde{L} < 2$ is that $f(x)(\frac{\xi}{\tau})^2 - u^n = 0$. □

Therefore, for any f, h, n and m with $fu^n > 0$, there exist exactly two sets of singular vector fields in the reduced form for the differential function $L = f(x)u_{tt} - (u^n u_x)_x - h(x)u^m$ from Equation (4), i.e., $S = \{\partial_t + \sqrt{u^n/f}\partial_x + \hat{\eta}\partial_u\}$ and $S^* = \{\partial_t - \sqrt{u^n/f}\partial_x + \hat{\eta}\partial_u\}$, where $\hat{\eta} = \frac{\eta}{\tau}$. Any singular vector field of L is equivalent to one of the above fields. Because the singular sets are mapped to each other by alternating the sign of x, we can just choose one of them for consideration.

Proposition 2. *For any variable coefficient nonlinear wave equations in the form of* (4)*, there exists exactly one set of singular vector fields in the reduced form, i.e.,* $S = \{\partial_t + \sqrt{u^n/f}\partial_x + \hat{\eta}\partial_u\}$ *for the differential function* $L = f(x)u_{tt} - (u^n u_x)_x - h(x)u^m$ *from Equation* (4).

Thus, for an equation from Class (4) and the operator $\partial_t + \sqrt{u^n/f}\partial_x + \eta\partial_u$, considering its conditional invariance criterion leads to:

Theorem 1. *Every singular reduction operator of an equation from Class* (4) *is equivalent to an operator of the form:*

$$Q = \partial_t + \sqrt{u^n/f(x)}\partial_x + \eta(t,x,u)\partial_u,$$

where the real-valued function $\eta(t,x,u)$ *satisfies the system of determining equations:*

$$\begin{array}{l}(-1/2nhu^{m-1} - 2f\eta_{tu} - 2f\eta\eta_{uu})\sqrt{u^n/f} + (3/4f_x^2/f - 1/2f_{xx})(u^n/f)^{3/2} \\ +(-1/4n^2\eta^2u^{n-2} - n\eta\eta_u u^{n-1} + 1/2n\eta^2u^{n-2})(u^n/f)^{-1/2} - n\eta_x u^{n-1} \\ -1/2n\eta u^{n-1}f_x/f - 2\eta_{xu}u^n = 0, \\ -\eta_{xx}u^n - n\eta\eta_x u^{n-1}(u^n/f)^{-1/2} + h\eta_u u^m + 2f\eta\eta_{tu} + f\eta^2\eta_{uu} \\ -mh\eta u^{m-1} + f\eta_{tt} + (hf_x/f - h_x)u^m\sqrt{u^n/f} = 0.\end{array} \qquad (8)$$

Remark 1. *Equation* (8) *is highly nonlinear, and it is difficult to solve; hence, we do not discuss it here and will try to investigate it in a future publication.*

2.2. Regular Reduction Operators

The above results for singular reduction operators of the class of nonlinear wave Equation (4) show that the regular case of the natural partition of the corresponding sets of reduction operators for the equation under consideration is singled out by the conditions $\xi \neq \pm\sqrt{u^n/f}\tau$. We factorize it with respect to the equivalence relation of vector fields, then we can derive the defining conditions of the regular subset of reduction operator, that is $\tau = 1, \xi \neq \pm\sqrt{u^n/f}$. Hence, we have:

Proposition 3. *For any variable coefficient nonlinear wave equations in the form of* (4)*, there exists exactly one set of regular vector fields in the reduced form, i.e.,* $S = \{\partial_t + \hat{\xi}\partial_x + \hat{\eta}\partial_u\}$ *with* $\hat{\xi} \neq \pm\sqrt{u^n/f}$ *for the differential function* $L = f(x)u_{tt} - (u^n u_x)_x - h(x)u^m$ *from Equation* (4).

Taking into accountant the conditional invariance criterion for an equation from Class (4) and the operator $\partial_t + \xi(t,x,u)\partial_x + \eta(t,x,u)\partial_u$ with $\xi(t,x,u) \neq \pm\sqrt{u^n/f}$, we can obtain the following determining equations for the coefficients ξ and η:

$$\begin{array}{l}\xi_u = 0, \quad 2f\xi_t - n\eta u^{n-1} + (2\xi_x + \xi f_x/f)u^n = 0, \\ (2n\xi_x - n\eta_u + n\xi f_x/f)u^{n-1} + (n\eta - n^2\eta)u^{n-2} - \eta_{uu}u^n + f\xi^2\eta_{uu} = 0, \\ 2f\xi_t\xi_x - 2f\xi_t\eta_u - 2n\eta_x u^{n-1} - f\xi_{tt} - 2f\xi\eta\eta_{uu} - 2f\xi\eta_{tu} + (\xi_{xx} - 2\eta_{xu})u^n = 0, \\ (\xi hf_x/f - \xi h_x + h\eta_u)u^m + f\eta^2\eta_{uu} + 2f\eta\eta_{tu} - 2f\xi_t\eta_x - \eta_{xx}u^n + f\eta_{tt} - mh\eta u^{m-1} = 0.\end{array} \qquad (9)$$

From the first two equations of System (9), we have:

$$\xi = \xi(t,x), \quad \eta = \frac{2}{n}f\xi_t u^{1-n} + \frac{1}{n}(2\xi_x + \xi\frac{f_x}{f})u.$$

Substituting the above formulas into the last three equations of System (9), we have the following assertion.

Theorem 2. *Every regular reduction operator of an equation from Class* (4) *is equivalent to an operator of the form:*

$$Q = \partial_t + \xi(t,x)\partial_x + \eta(t,x,u)\partial_u \quad \textit{with} \quad \eta(t,x,u) = \frac{2}{n} f\xi_t u^{1-n} + \frac{1}{n}(2\xi_x + \xi\frac{f_x}{f})u, \tag{10}$$

where the real-valued function $\xi(t,x)$ *satisfies the overdetermined system of partial differential equations:*

$$\begin{aligned}
&2(1-n)f\xi_t = 0, \quad 2(1-n)f^2\xi^2\xi_t = 0, \quad 8(1-n)f^3\xi\xi_t^2 = 0,\\
&4(1-n)[(f\xi_t)^2 - f^2\xi\xi_t(2\xi_x + \xi\tfrac{f_x}{f}) + f^2\xi\xi_{tt}] = 0,\\
&\xi_{xx} - 2(1+\tfrac{1}{n})(2\xi_x + \xi\tfrac{f_x}{f})_x = 0,\\
&2f\xi_t\xi_x - \tfrac{2}{n}f\xi_t(2\xi_x + \xi\tfrac{f_x}{f}) - 4(f\xi_t)_x - f\xi_{tt} - \tfrac{2}{n}f\xi(2\xi_x + \xi\tfrac{f_x}{f})_t - \tfrac{4}{n}(1-n)(f\xi_t)_x = 0,\\
&(\xi h\tfrac{f_x}{f} - \xi h_x + \tfrac{1}{n}h(1-m)(2\xi_x + \xi\tfrac{f_x}{f}))u^m + \tfrac{2}{n}(1-n-m)fh\xi_t u^{m-n}\\
&-\tfrac{8}{n^2}(1-n)f^4(\xi_t)^3u^{1-3n} + \tfrac{8}{n^2}(1-n)[f^2\xi_t(f\xi_t)_t - f^3\xi_t^2(2\xi_x + \xi\tfrac{f_x}{f})]u^{1-2n}\\
&+[\tfrac{2}{n}f(f\xi_t)_{tt} - \tfrac{4}{n}f\xi_t(f\xi_t)_x + \tfrac{4}{n^2}f^2\xi_t(2\xi_x + \xi\tfrac{f_x}{f})_t - \tfrac{2}{n^2}(1-n)f^2\xi_t(2\xi_x + \xi\tfrac{f_x}{f})^2\\
&+\tfrac{4}{n^2}(1-n)f(f\xi_t)_t(2\xi_x + \xi\tfrac{f_x}{f})]u^{1-n} + [\tfrac{2}{n^2}f(2\xi_x + \xi\tfrac{f_x}{f})(2\xi_x + \xi\tfrac{f_x}{f})_t\\
&+\tfrac{1}{n}f(2\xi_x + \xi\tfrac{f_x}{f})_{tt} - \tfrac{2}{n}f\xi_t(2\xi_x + \xi\tfrac{f_x}{f})_x - \tfrac{2}{n}(f\xi_t)_{xx}]u - \tfrac{1}{n}(2\xi_x + \xi\tfrac{f_x}{f})_{xx}u^{n+1} = 0.
\end{aligned} \tag{11}$$

Solving the above system with respect to the coefficient functions ξ, f and h under the equivalence group $G_1^{\sim}$ of the class (4), which consists of the transformations (see Theorem 3 and 4 in [25] for more details): for $n \neq -1$:

$$\begin{aligned}
&\tilde{t} = \epsilon_1 t + \epsilon_2, \quad \tilde{x} = \tfrac{\epsilon_3 x + \epsilon_4}{\epsilon_5 x + \epsilon_6} =: X(x), \quad \tilde{u} = \epsilon_7 X_x^{\frac{1}{2n+2}} u,\\
&\tilde{f} = \epsilon_1^2\epsilon_7^n X_x^{-\frac{3n+4}{2n+2}} f, \quad \tilde{h} = \epsilon_7^{-m+n+1} X_x^{-\frac{m+3n+3}{2n+2}} h, \quad \tilde{n} = n, \quad \tilde{m} = m,
\end{aligned}$$

where ϵ_j $(j = 1,\ldots,7)$ are arbitrary constants, $\epsilon_1\epsilon_7 \neq 0, \epsilon_3\epsilon_6 - \epsilon_4\epsilon_5 = \pm 1$, and for $n = -1$:

$$\begin{aligned}
&\tilde{t} = \epsilon_1 t + \epsilon_2, \quad \tilde{x} = \epsilon_3 x + \epsilon_4, \quad \tilde{u} = \epsilon_5 e^{\epsilon_6 x} u,\\
&\tilde{f} = \epsilon_1^2\epsilon_3^{-2}\epsilon_5^{-1} e^{-\epsilon_6 x} f, \quad \tilde{h} = \epsilon_3^{-2}\epsilon_5^{-m} e^{-m\epsilon_6 x} h, \quad \tilde{n} = n, \quad \tilde{m} = m,
\end{aligned}$$

where ϵ_j $(j = 1,\ldots,6)$ are arbitrary constants, $\epsilon_1\epsilon_3\epsilon_5 \neq 0$; we can get a classification of the regular reduction operator for the class (4). It is easy to know that some of the regular reduction operator are equivalent to Lie symmetry operators, which have been given in [25], while some of them are nontrivial. Below, we give a detailed investigations for these cases.

In fact, the first three equations of System (11) imply that there are two cases that should be considered: $n \neq 1$ or not. (It should be noted that $\xi = 0$ should be exclude from the consideration because it leads to $\eta = 0$).

Case 1: $n \neq 1$. In this case, we have $\xi_t = 0$, and System (11) can be reduced to:

$$\begin{aligned}
&(3n+4)\xi_{xx} + 2(n+1)(\xi\tfrac{f_x}{f})_x = 0,\\
&(\xi h\tfrac{f_x}{f} - \xi h_x + \tfrac{1}{n}(1-m)h(2\xi_x + \xi\tfrac{f_x}{f}))u^m - \tfrac{1}{n}(2\xi_x + \xi\tfrac{f_x}{f})_{xx}u^{n+1} = 0.
\end{aligned} \tag{12}$$

Thus, there are two cases that should be considered: $m \neq n+1$ or not.

Case 1.1: For $m \neq n+1$, from the second Equation of (12), we obtain:

$$\xi h \frac{f_x}{f} - \xi h_x + \frac{1}{n}(1-m)h(2\xi_x + \xi\frac{f_x}{f}) = 0, \quad (2\xi_x + \xi\frac{f_x}{f})_{xx} = 0. \tag{13}$$

Because $\xi_t = 0$ and f is a function of x, the first Equation of (12) suggests that $(3n+4)\xi_x + 2(n+1)\xi\frac{f_x}{f}$ is independent of the variables x and t. Thus, there exists a constant r, such that $(3n+4)\xi_x + 2(n+1)\xi\frac{f_x}{f} = nr$. The second Equation of (13) suggests that there exist two constants a and b, such that $2\xi_x + \xi\frac{f_x}{f} = nax + nb$. By solving the last two equations, we obtain:

$$\xi_x = 2(n+1)(ax+b) - r, \xi\frac{f_x}{f} = 2r - (3n+4)(ax+b),$$

which together with the first Equation of (13) imply:

$$\begin{aligned} &\xi = a(n+1)x^2 + [2b(n+1) - r]x + s, \\ &f(x) = \exp\big(\int \frac{2r - (3n+4)(ax+b)}{a(n+1)x^2 + [2b(n+1)-r]x + s}\,dx\big), \\ &h(x) = \exp\big(\int \frac{2r - (m+3n+3)(ax+b)}{a(n+1)x^2 + [2b(n+1)-r]x + s}\,dx\big), \end{aligned} \tag{14}$$

where a, b, r, s are arbitrary constants. Thus, the corresponding regular reduction operator has the form:

$$Q = \partial_t + [a(n+1)x^2 + (2b(n+1) - r)x + s]\partial_x + (ax+b)u\partial_u,$$

which is equivalent to the Lie symmetry operator.

Case 1.2: $m = n+1$. In this case, System (12) can be rewritten as:

$$(3n+4)\xi_{xx} + 2(n+1)(\xi\tfrac{f_x}{f})_x = 0, \quad \xi h_x + 2h\xi_x + \tfrac{1}{n}(2\xi_x + \xi\tfrac{f_x}{f})_{xx} = 0. \tag{15}$$

Integrating these two equations with respect to functions $f(x)$ and $h(x)$, we can obtain:

$$f(x) = |\,\xi|^{-\frac{3n+4}{2n+2}} \exp\big(r\int \frac{1}{\xi}\,dx\big), \quad h(x) = \frac{\xi_x^2 - 2\xi\xi_{xx} - p}{4(n+1)\xi^2}.$$

where p, r are arbitrary constants, ξ is an arbitrary smooth function and $n \neq -1$. In addition, $\eta = \frac{1}{n}(2\xi_x + \xi\frac{f_x}{f})u = (\frac{r}{n} + \frac{\xi_x}{2n+2})u$. Thus, we have a nontrivial regular reduction operator:

$$Q = \partial_t + \xi(x)\partial_x + [(\frac{r}{n} + \frac{\xi_x}{2n+2})u]\partial_u, \quad n \neq -1. \tag{16}$$

It should be noted that the reduction operator for $n = -1$ is also equivalent to the Lie symmetry operator.

Case 2: $n = 1$. In this case, we have $\eta = 2f\xi_t + (2\xi_x + \xi\frac{f_x}{f})u$. Thus, System (11) can be reduced to:

$$\begin{aligned} &(7\xi_x + 4\xi\tfrac{f_x}{f})_x = 0, \quad 2\big[\xi_x + 2(\xi+1)\tfrac{f_x}{f}\big]\xi_t + 4(\xi+1)\xi_{tx} + \xi_{tt} = 0, \\ &\big[2f(2\xi_x + \xi\tfrac{f_x}{f})(2\xi_x + \xi\tfrac{f_x}{f})_t + f(2\xi_x + \xi\tfrac{f_x}{f})_{tt} - 2f\xi_t(2\xi_x + \xi\tfrac{f_x}{f})_x - 2(f\xi_t)_{xx}\big]u \\ &-(2\xi_x + \xi\tfrac{f_x}{f})_{xx}u^2 + 2f^2(2\xi_t\xi_{tx} + \xi_{ttt}) + \big[\xi h\tfrac{f_x}{f} - \xi h_x + (1-m)h(2\xi_x + \xi\tfrac{f_x}{f})\big]u^m \\ &-2mhf\xi_t u^{m-1} = 0. \end{aligned} \tag{17}$$

Comparing different powers of u, we conclude that five cases should be considered, that is (i) Case 2.1: $m = 0$; (ii) case 2.2: $m = 1$; (iii) Case 2.3: $m = 2$; (iv) Case 2.4: $m = 3$ and (v) Case 2.5:

$m \neq 0, 1, 2, 3$. Below, we mainly give a detailed discussion for Cases 2.2 and 2.3, and the other cases can be considered in a similar way.

Case 2.2: When $m = 1$, the third Equation of (17) implies:

$$\begin{aligned}&\xi h\tfrac{f_x}{f} - \xi h_x + f(2\xi_x + \xi\tfrac{f_x}{f})_{tt} + 2f(2\xi_x + \xi\tfrac{f_x}{f})(2\xi_x + \xi\tfrac{f_x}{f})_t\\&-2f\xi_t(2\xi_x + \xi\tfrac{f_x}{f})_x - 2(f\xi_t)_{xx} = 0,\\&2f\xi_t\xi_{tx} + f\xi_{ttt} - h\xi_t = 0, \quad (2\xi_x + \xi\tfrac{f_x}{f})_{xx} = 0.\end{aligned} \tag{18}$$

From the last equation of System (18), we can know that there exist two functions $a(t)$ and $b(t)$, such that $2\xi_x + \xi\frac{f_x}{f} = a(t)x + b(t)$. On the other hand, the first Equation of (17) implies that there exists a function $c(t)$, such that $7\xi_x + 4\xi\frac{f_x}{f} = c(t)$. Solving the last two equations gives:

$$\xi_x = 4a(t)x + 4b(t) - c(t), \quad \xi\frac{f_x}{f} = -7a(t)x - 7b(t) + 2c(t),$$

from which we can get:

$$\xi = 2a(t)x^2 + 4b(t)x - c(t)x + d(t), \quad f(x) = \exp\Big(\int \frac{-7a(t)x - 7b(t) + 2c(t)}{2a(t)x^2 + 4b(t)x - c(t)x + d(t)}\,\mathrm{d}x\Big), \tag{19}$$

where $d(t)$ is an arbitrary function. Since $\frac{f_x}{f}$ is independent of t, we see that:

$$\Big[\frac{-7a(t)x - 7b(t) + 2c(t)}{2a(t)x^2 + 4b(t)x - c(t)x + d(t)}\Big]_t = 0,$$

which leads to:

$$\begin{cases}14\big[a(t)b'(t) - a'(t)b(t)\big] + 3\big[a'(t)c(t) - a(t)c'(t)\big] = 0,\\ \big[b(t)c'(t) - b'(t)c(t)\big] + 7\big[a(t)d'(t) - a'(t)d(t)\big] = 0,\\ 2\big[c'(t)d(t) - c(t)d'(t)\big] + 7\big[b(t)d'(t) - b'(t)d(t)\big] = 0.\end{cases} \tag{20}$$

Now, we multiply both sides of the second Equation of (17) by ξ and substitute (19) into it, then simplify the equation and compare the coefficient of $x^i (i = 0, 1, \ldots, 5)$ to obtain:

$$\begin{cases}a'(t) = 0, \quad a^2[-4b'(t) + c'(t)] = 0,\\ a[8c(t)c'(t) + 20ad'(t) + c''(t) - 4b''(t) - 32c(t)b'(t) + 112b(t)b'(t) - 28b(t)c'(t)] = 0,\\ -4c''(t)b(t) + 32ac(t)d'(t) - 80ab'(t) + 20ac'(t) - 4b''(t)c(t) - 120ab(t)d'(t)\\ +c''(t)c(t) + 2d''(t)a + 2c^2(t)c'(t) - 12b(t)c(t)c'(t) + 16b(t)^2c'(t) - 8c^2(t)b'(t)\\ +48b(t)c(t)b'(t) + 16b''(t)b(t) - 64b^2(t)b'(t) - 16ad(t)b'(t) + 4ad(t)c'(t) = 0,\\ [-20ad'(t) + 48b(t)b'(t) + 2c(t)c'(t) - 12b(t)c'(t) - 8c(t)b'(t) + 4b''(t) - c''(t)]d(t)\\ -80b^2(t)d'(t) + 4d''(t)b(t) - d''(t)c(t) - 4c(t)c'(t) + 44b(t)c(t)d'(t) - 28ad'(t)\\ -48b(t)b'(t) - 6c^2(t)d'(t) + 12b(t)c'(t) + 16c(t)b'(t) = 0,\\ [16b'(t) - 4c'(t)]d^2(t) + [-20b(t)d'(t) - 4c'(t) + 6c(t)d'(t) + d''(t) + 16b'(t)]d(t)\\ +8c(t)d'(t) - 28b(t)d'(t) = 0.\end{cases} \tag{21}$$

Note that ξ is assumed not to be identical with zero; thus, we find that Systems (20) and (21) can be reduced to:

$$a'(t) = 0, \quad b'(t) = 0, \quad c'(t) = 0, \quad d'(t) = 0 \tag{22}$$

or:

$$a=0,\quad c(t)=4b(t),\quad d(t)=qb(t),\quad qb''(t)+4qb(t)b'(t)+4b'(t)=0 \tag{23}$$

or:

$$a=0,\quad 2c(t)=7b(t),\quad b''(t)=-3b(t)b'(t),\quad b(t)d'(t)+2b'(t)(d(t)+1)+d''(t)=0 \tag{24}$$

or:

$$a=0,\quad c(t)=3b(t),\quad d=qb(t),\quad b''(t)+2b(t)b'(t)=0, \tag{25}$$

where q is an arbitrary constant.

Case 2.2a: If System (22) is satisfied, then $\xi_t=0$; the second equation of (18) is an identity. The expression (19) can be rewritten as:

$$\xi=2ax^2+4bx-cx+d,\quad f(x)=\exp\Big(\int\frac{-7ax-7b+2c}{2ax^2+4bx-cx+d}\,\mathrm{d}x\Big), \tag{26}$$

where a,b,c and d are arbitrary constants. The first Equation of (18) is reduced to $h_x/h=f_x/f$, which leads to $h(x)=\epsilon f(x)$ $(\epsilon=\pm1)$ $\mod G_1^{\sim}$. In addition, $\eta=2f\xi_t+(2\xi_x+\xi\frac{f_x}{f})u=(ax+b)u$. Thus, we have:

$$\begin{aligned}&\xi=2ax^2+4bx-cx+d,\quad \eta=(ax+b)u,\\ &f(x)=\exp\Big(\int\frac{-7ax-7b+2c}{2ax^2+4bx-cx+d}\,\mathrm{d}x\Big),\quad h(x)=\epsilon f(x),\end{aligned} \tag{27}$$

where a,b,c,d are arbitrary constants and $\epsilon=\pm1$. Thus, the corresponding regular reduction operator has the form:

$$Q=\partial_t+(2ax^2+4bx-cx+d)\partial_x+(ax+b)u\partial_u,$$

which is equivalent to the Lie symmetry operator.

Case 2.2b: If System (23) is satisfied, then the expression (19) can be rewritten as:

$$\xi=qb(t),\quad f(x)=\exp\big(\frac{x}{q}\big). \tag{28}$$

Hence, $\xi_x=0, k=b(t)$. If $b'(t)=0$, then $a(t),b(t),c(t),d(t)$ satisfy System (22), and the solution is included in Case 2.2a. We suppose that $b'(t)\neq0$. From the second Equation of (18), we see that $h=f\xi_{ttt}/\xi_t$. Substitute it into the first Equation of (18) to get $fk_{tt}+2fkk_t-2f_{xx}\xi_t=0$. Further, it can be reduced to $qb''(t)+2qb(t)b'(t)-2b'(t)=0$. Combine it with the fourth Equation of (23) to get $b(t)=-3/q$, which is a contradiction to the hypothesis $b'(t)\neq0$.

Case 2.2c: If System (24) is satisfied, then the expression (19) can be rewritten as:

$$\xi=\tfrac{1}{2}b(t)x+d(t),\quad f(x)=1\mod G_1^{\sim}. \tag{29}$$

If $b'(t)=d'(t)=0$, then $a(t),b(t),c(t),d(t)$ satisfy both Systems (22) and (24), and the solution is included in Case 2.2a. We suppose that $b'^2(t)+d'^2(t)\neq0$. Substitute (29) into the first Equation of (18) to obtain:

$$h_x=\frac{2[b''(t)+2b(t)b''(t)]}{b(t)x+2d(t)}. \tag{30}$$

Substitute (29) into the second Equation of (18) to obtain:

$$h(x)=b'(t)+\frac{b'''(t)x+2d'''(t)}{b'(t)x+2d'(t)}. \tag{31}$$

Substituting it into (30) yields:

$$\frac{b'''(t)d'(t)-b'(t)d'''(t)}{[b'(t)x+2d'(t)]^2}=\frac{b''(t)+2b(t)b'(t)}{b(t)x+2d(t)}.$$

Compare the coefficient of x^2 to obtain $b'^2(t)[b''(t)+2b(t)b'(t)]=0$. Substitute the third Equation of (24) into it to obtain $b(t)b'^3(t)=0$; hence, $b'(t)=0$. Thus, the fourth Equation of (24) can be reduced to $bd'(t)+d''(t)=0$. Solving this linear ordinary differential equation gives $d(t)=\gamma_1 e^{-bt}+\gamma_0$, where γ_1 and γ_0 are two arbitrary constants. Therefore, the expressions (29) and (31) can be rewritten as:

$$\xi=\tfrac{1}{2}bx+\gamma_1 e^{-bt}+\gamma_0,\quad f(x)=1,\quad h(x)=b^2 \mod G_1^{\sim}.$$

System (18) is verified to be true. In addition, $\eta=2f\xi_t+(2\xi_x+\xi\frac{f_x}{f})u=bu-2\gamma_1 be^{-bt}$. Therefore, we have:

$$\xi=\tfrac{1}{2}bx+\gamma_1 e^{-bt}+\gamma_0,\quad \eta=bu-2\gamma_1 be^{-bt},\quad f(x)=1,\quad h(x)=b^2, \tag{32}$$

where b,γ_1,γ_0 are arbitrary constants. Thus, we have a nontrivial regular reduction operator:

$$Q=\partial_t+(\frac{1}{2}bx+\gamma_1 e^{-bt}+\gamma_0)\partial_x+(bu-2\gamma_1 be^{-bt})\partial_u. \tag{33}$$

Case 2.2d: If System (25) is satisfied, then the expression (19) can be rewritten as:

$$\xi=b(t)(x+q),\quad f(x)=\frac{1}{x+q} \mod G_1^{\sim}. \tag{34}$$

Substitute it into the first Equation of (18) to obtain $b(t)(x+q)[h+(x+q)h_x]=[b''(t)+2b(t)b'(t)]$. Substitute the fourth Equation of (25) into it to get $b(t)(x+q)[h+(x+q)h_x]=0$. It follows that $h(x)=r/(x+q)$, where r is a nonzero constant. Substitute it and (34) into the second Equation of (18) to obtain $2b'^2(t)+b'''(t)-rb'(t)=0$. From the fourth Equation of (25), we find $b'''(t)=4b^2(t)b'(t)-2b'^2(t)$. Substitute it into the preceding equation to get $b'(t)[4b^2(t)-r]=0$, which leads to $b'(t)=0$. Then, $a(t),b(t),c(t),d(t)$ satisfy System (22), and the solution is included in Case 2.2a.

Case 2.3: When $m=2$, System (17) implies:

$$\begin{aligned}
&(7\xi_x+4\xi\tfrac{f_x}{f})_x=0,\quad 2(\xi_x+2\xi\tfrac{f_x}{f})\xi_t+4\xi_t\tfrac{f_x}{f}+4\xi\xi_{tx}+4\xi_{tx}+\xi_{tt}=0,\\
&2\xi_t\xi_{tx}+\xi_{ttt}=0,\quad \xi h_x+2h\xi_x+(2\xi_x+\xi\tfrac{f_x}{f})_{xx}=0,\\
&2f(2\xi_x+\xi\tfrac{f_x}{f})(2\xi_x+\xi\tfrac{f_x}{f})_t+f(2\xi_x+\xi\tfrac{f_x}{f})_{tt}-4hf\xi_t-2f\xi_t(2\xi_x+\xi\tfrac{f_x}{f})_x-2(f\xi_t)_{xx}=0.
\end{aligned} \tag{35}$$

From the first and the fourth equation of System (35), we can get:

$$f(x)=|\xi|^{-7/4}\exp(\alpha(t)\int\frac{dx}{\xi}),\quad h(x)=\frac{\xi_x^2-2\xi\xi_{xx}+q}{8\xi^2},$$

where $\alpha(t)$ is an arbitrary function and q is a constant. Substituting these expressions into the rest of the equations of System (35), we can see that $\xi(t,x)$ and $\alpha(t)$ satisfy the overdetermined system of partial differential equations:

$$\begin{aligned} &2\xi_t\xi_{tx} + \xi_{ttt} = 0, \quad \xi_{tt} - 3\xi_{tx} - 5\xi_t\xi_x + 4\xi\xi_{tx} + 4\alpha\xi_t + 4\alpha_t = 0, \\ &2\xi^2(\tfrac{1}{4}\xi_x + \alpha)_{tt} + 2\xi^2[(\tfrac{1}{4}\xi_x + \alpha)^2]_t - \xi^2\xi_t\xi_{xx} - \xi_t(\xi_x^2 - 2\xi\xi_{xx} + q) \\ &-4[(\alpha_t - \tfrac{3}{4}\xi_{tx})(\alpha - \tfrac{7}{4}\xi_x)\xi - \tfrac{3}{4}\xi^2\xi_{txx}] = 0. \end{aligned} \tag{36}$$

In addition, we have:

$$\eta = 2f\xi_t + (2\xi_x + \xi\frac{f_x}{f})u = 2\xi_t|\xi|^{-7/4}\exp(\alpha(t)\int\frac{\mathrm{d}x}{\xi}) + [\frac{1}{4}\xi_x + \alpha(t)]u.$$

Thus, we have a nontrivial regular reduction operator:

$$Q = \partial_t + \xi(t,x)\partial_x + \{2\xi_t|\xi|^{-7/4}\exp(\alpha(t)\int\frac{\mathrm{d}x}{\xi}) + [\frac{1}{4}\xi_x + \alpha(t)]u\}\partial_u, \tag{37}$$

where $\xi(t,x)$ and $\alpha(t)$ satisfy the overdetermined system of partial differential equations (36).

In particular, if $\xi_t = 0$, from System (35), we can obtain:

$$\xi = \xi(x), \quad \eta = \tfrac{1}{4}(\xi_x + a)u, \quad f(x) = |\xi|^{-7/4}\exp(\frac{a}{4}\int\frac{\mathrm{d}x}{\xi}), \quad h(x) = \frac{\xi_x^2 - 2\xi\xi_{xx} + q}{8\xi^2}. \tag{38}$$

where a, q are arbitrary constants. Thus, we have a nontrivial regular reduction operator:

$$Q = \partial_t + \xi(x)\partial_x + (\frac{1}{4}\xi_x + a)u\partial_u, \tag{39}$$

which is equivalent to Operator (16) with $n = 1$. Therefore, this special case can be included in Case 1.2, and we can impose an additional constraint $\xi_t \neq 0$ on the regular reduction operator (37).

Remark 2. *It should be noted that we can also give detailed analysis for Cases 2.1, 2.4 and 2.5 in a way similar to Cases 2.2 and 2.3. However, we find that all of the regular reduction operators resulting from these Cases 2.1, 2.4 and 2.5 are trivial, which are equivalent to Lie symmetry operators. Thus, we omit these results in order to avoid tediousness.*

From the above discussion, we can arrive at the following two theorems.

Theorem 3. *A complete list of $G_1^{\sim}$-inequivalent Equation* (3) *having a nontrivial regular reduction operator is exhausted by the ones given in Table 1.*

Table 1. Results of regular reduction operator classification of Class (3).

N	n	m	$f(x)$	$h(x)$	Regular Reduction Operator Q
1	$\neq -1$	$n+1$	$\lvert\xi\rvert^{-\frac{3n+4}{2n+2}}\exp\left(r\int\frac{1}{\xi}\,\mathrm{d}x\right)$	$\dfrac{\xi_x^2 - 2\xi\xi_{xx} - p}{4(n+1)\xi^2}$	$\partial_t + \xi(x)\partial_x + (\frac{r}{n} + \frac{\xi_x}{2n+2})u\partial_u$
2	1	1	1	b^2	$\partial_t + (\frac{1}{2}bx + \gamma_1 e^{-bt} + \gamma_0)\partial_x + (bu - 2\gamma_1 b e^{-bt})\partial_u$
3	1	2	$\lvert\xi\rvert^{-\frac{7}{4}}\exp(\alpha(t)\int\frac{\mathrm{d}x}{\xi})$	$\frac{\xi_x^2 - 2\xi\xi_{xx} + q}{8\xi^2}$	$\partial_t + \xi(t,x)\partial_x + \{2\xi_t\lvert\xi\rvert^{-\frac{7}{4}}\exp(\alpha(t)\int\frac{\mathrm{d}x}{\xi})$ $+[\frac{1}{4}\xi_x + \alpha(t)]u\}\partial_u$

Here $r, p, b, \gamma_1, \gamma_0$ are arbitrary constants; $b \neq 0$, $\xi(x)$ in Case 2.1 is an arbitrary functions of the variables x, $\xi(t,x)$; and $\alpha(t)$ in Case 2.3 satisfy the overdetermined system of partial differential Equation (36) and $\xi_t \neq 0$.

Theorem 4. *Any reduction operator of an equations from Class* (3) *having the form* (10) *with $\xi_t = 0$, $\xi_{xxx} = 0$ is equivalent to a Lie symmetry operator of this equation.*

3. Exact Solutions

In this section, we construct the nonclassical reduction and exact solutions for the classification models in Table 1 by using the corresponding regular reduction operator. Lie reduction and exact solutions of the equation from Class (3) have been investigated in some literature works; see, for example, References [9,15,16,18,21,24,25]. We choose Case 1 in Table 1 as an example to implement the reduction; Case 3 can be considered in a similar way. While for Case 2, the nonclassical reduction and exact solution of the corresponding equation $u_{tt} = (uu_x)_x + b^2u$ have been investigated by Fushchich et al. in [34,47]; thus, we do not consider it here.

For the first case in Table 1, the corresponding equation is:

$$[|\xi|^{-\frac{3n+4}{2n+2}}\exp\left(r\int\frac{1}{\xi}\mathrm{d}x\right)]u_{tt}-(u^nu_x)_x-\frac{\xi_x^2-2\xi\xi_{xx}-p}{4(n+1)\xi^2}u^{n+1}=0, \tag{40}$$

which admits the regular reduction operator:

$$Q=\partial_t+\xi(x)\partial_x+\left(\frac{r}{n}+\frac{\xi_x}{2n+2}\right)u\partial_u.$$

An ansatz constructed by this operator has the form:

$$u(t,x)=\varphi(\omega)|\xi|^{\frac{1}{2n+2}}\exp\left(\frac{r}{n}\int\frac{1}{\xi}\mathrm{d}x\right),\quad\text{where}\quad\omega=t-\int\frac{1}{\xi}\mathrm{d}x.$$

Substituting this ansatz into Equation (40) leads to the reduced ODE:

$$\begin{aligned}&[(4r^2-p)n^2+4(2n+1)r^2]\varphi^{n+1}(\omega)+4n(n+1)[n\varphi''(\omega)-2(n+1)r\varphi'(\omega)]\varphi^n(\omega)\\&+4n^3(n+1)\varphi'^2(\omega)\varphi^{n-1}(\omega)-4n^2(n+1)\varphi''(\omega)=0.\end{aligned} \tag{41}$$

Because there are higher nonlinear terms, we were not able to completely solve the above equation. Thus, we try to solve this equation under different additional constraints imposed on p and r.

We first rewrite Equation (41) as:

$$\begin{aligned}&4n^2(n+1)[\varphi'(\omega)\varphi^n(\omega)]'-4n^2(n+1)\varphi''(\omega)-8n(n+1)^2r\varphi'(\omega)\varphi^n(\omega)\\&+[4(n+1)^2r^2-pn^2]\varphi^{n+1}(\omega)=0.\end{aligned} \tag{42}$$

If we take $p=4(1+\frac{1}{n})^2r^2$, then the general solution of (42) can be written in the implicit form:

$$\int\frac{n(\varphi^n-1)}{2r\varphi^{n+1}+c_1}\mathrm{d}\varphi=\omega+c_2. \tag{43}$$

Up to the similarity of solutions of Equation (3), the constant c_2 is inessential and can be set to equal zero by a translation of ω, which is always induced by a translation of t.

If we further set $n=1$, the general solution (43) can be rewritten in the following implicit forms, that is if $\frac{c_1}{2r}>0$, then we have:

$$\frac{1}{4r}\ln|2r\varphi^2+c_1|-\frac{\arctan(\sqrt{\frac{2r}{c_1}}\varphi)}{\sqrt{2rc_1}}+C=\omega, \tag{44}$$

while if $\frac{c_1}{2r}<0$, then we have:

$$\frac{1}{4r}\ln|2r\varphi^2+c_1|-\frac{\operatorname{arctanh}(\sqrt{-\frac{2r}{c_1}}\varphi)}{\sqrt{-2rc_1}}+C=\omega; \tag{45}$$

and if $c_1 = 0$ and $r \neq 0$, then we have:

$$\frac{1}{2r}(\ln|\varphi| + \varphi^{-1}) + C = \omega, \tag{46}$$

while if $c_1 \neq 0$ and $r = 0$, then we have:

$$\frac{1}{c_1}(\frac{1}{2}\varphi^2 - \varphi) + C = \omega, \tag{47}$$

where C is an arbitrary constant.

Thus, we obtain the following solution:

$$u(t,x) = \varphi(\omega)|\xi|^{\frac{1}{4}} \exp\left(r\int \frac{1}{\xi}\,\mathrm{d}x\right), \qquad \omega = t - \int \frac{1}{\xi}\,\mathrm{d}x$$

for the equation:

$$[|\xi|^{-\frac{7}{4}} \exp\left(r\int \frac{1}{\xi}\,\mathrm{d}x\right)]u_{tt} - (uu_x)_x - \frac{\xi_x^2 - 2\xi\xi_{xx} - 8r^2}{8\xi^2}u^2 = 0,$$

where φ satisfies Equations (44)–(47), ξ is an arbitrary function and r is a non-zero constant.

If we further set $r = 0$, the general solution (43) can be rewritten in the implicit form:

$$\frac{1}{n+1}\varphi^{n+1}(\omega) - \varphi(\omega) = c_1\omega + c_2. \tag{48}$$

Thus, we obtain the following solution:

$$u(t,x) = \varphi(\omega)|\xi|^{\frac{1}{2n+2}}, \qquad \omega = t - \int \frac{1}{\xi}\,\mathrm{d}x$$

for the equation:

$$|\xi|^{-\frac{3n+4}{2n+2}}u_{tt} - (u^n u_x)_x - \frac{\xi_x^2 - 2\xi\xi_{xx}}{4(n+1)\xi^2}u^{n+1} = 0,$$

where φ satisfies Equation (48) and ξ is an arbitrary function. In particular, when $n = 1$, Equation (48) is reduced to Equation (47); thus, we have:

$$\varphi(\omega) = 1 \pm \sqrt{1 + 2(c_1\omega - C)}.$$

Thus, we obtain an explicit solution:

$$u(t,x) = [1 \pm \sqrt{1 + 2(c_1\omega - C)}]|\xi|^{\frac{1}{4}}, \qquad \omega = t - \int \frac{1}{\xi}\,\mathrm{d}x$$

for the equation:

$$|\xi|^{-\frac{7}{4}}u_{tt} - (uu_x)_x - \frac{\xi_x^2 - 2\xi\xi_{xx}}{8\xi^2}u^2 = 0.$$

If we take different functions for ξ, then we can obtain a series of solutions for the corresponding equations. In order to avoid tediousness, we do not make a further discussion here.

4. Conclusions and Discussion

In this paper, we have given a detailed investigation of the reduction operators of the variable coefficient nonlinear wave equations (1) (equivalently to (3)) by using the singular reduction operator theory. A classification of regular reduction operators is performed with respect to generalized extended equivalence groups. The main results on classification for Equation (3) are collected in Table 1, where we list three inequivalent cases with the corresponding regular reduction operators. The nonclassical symmetry reduction of a class of nonlinear wave Model (40), which singled out the classification models, is also performed. These are utilized to construct some non-Lie exact solutions, which are invariant under certain conditional symmetries for the corresponding model.

The nonclassical symmetry analysis of the class of hyperbolic type nonlinear wave Equation (1) presented in this paper is preliminary. In fact, it is easy to know that the well-known wave equation $u_{tt} = uu_{xx}$ possesses much wider Q-conditional (nonclassical) symmetry, which includes 12 nontrivial reduction operators (see Chapter 7 in [34] for details). Therefore, for Equation (1), we may find more reduction operators besides the ones given in Table 1. One possible way is that we can try to present a particular solution of (8) leading to a Q-conditional symmetry. However, Equation (8) is highly nonlinear, and it is difficult to solve; hence, we do not discuss it in this paper. Another way is to consider the special case $\tau = 0$, which is also excluded from our above discussion. Further studies along these lines may lead to more reduction operators. Furthermore, other different properties of Equation (1), including nonclassical potential symmetries, nonclassical potential exact solutions and physical application, will also be investigated in subsequent publications.

Acknowledgments: This work was partially supported by the National Natural Science Foundation of China (Grant No. 11501204), the Natural Science Foundation of Shanghai (Grant No. 15ZR1408300) and the Shanghai Key Laboratory of Intelligent Information Processing (Grant No. IIPL-2014-001).

Author Contributions: All the authors Dingjiang Huang, Yan Zhu and Qinmin Yang make contributions to Section 2.2. Dingjiang Huang make contribution to Sections 1, 2.1, 3 and 4.

Conflicts of Interest: The authors declare no conflict of interest.

References

1. Ablowitz, M.J.; Clarkson, P.A. *Solitons, Nonlinear Evolution Equations and Inverse Scattering*; Cambridge University Press: Cambridge, UK, 1991.
2. Rogers, C.; Schief, W.K. *Bäcklund and Darboux Transformations, Geometry and Morden Applications in Soliton Theory*; Cambridge University Press: Cambridge, UK, 2002.
3. Roy-Chowdhury, A.K. *Painleve Analysis and Its Applications*; Chapman and Hall/CRC Press: Boca Raton, FL, USA, 1999.
4. Ablowitz, M.J.; Musslimani, Z.H. Integrable nonlocal nonlinear Schrodinger equation. *Phys. Rev. Lett.* **2013**, *110*, 064105.
5. Ablowitz, M.J.; Musslimani, Z.H. Inverse scattering transform for the integrable nonlocal nonlinear Schrodinger equations. *Nonlinearity* **2016**, *29*, 915–946.
6. Olver, P.J. *Application of Lie Groups to Differential Equations*; Springer: New York, NY, USA, 1986.
7. Bluman, G.W.; Kumei, S. *Symmetries and Differential Equations*; Springer: Berlin, Germany, 1989.
8. Ames, W.F. *Nonlinear Partial Differential Equations in Engineering*; Academic: New York, NY, USA, 1972; Volume II, pp. 50–52.
9. Ames, W.F.; Adams, E.; Lohner, R.G. Group properties of $u_{tt} = [f(u)u_x]_x$. *Int. J. Non-Linear Mech.* **1981**, *16*, 439–447.
10. Ibragimov, N.H. (Ed.) *Lie Group Analysis of Differential Equations—Symmetries, Exact Solutions and Conservation Laws*; CRC Press: Boca Raton, FL, USA, 1994; Volume 1.
11. Arrigo, D.J. Group properties of $u_{xx} - u_y^m u_{yy} = f(u)$. *Int. J. Non-Linear Mech.* **1991**, *26*, 619–629.
12. Bluman, G.W.; Cheviakov, A.F. Nonlocally related systems, linearization and nonlocal symmetries for the nonlinear wave equation. *J. Math. Anal. Appl.* **2007**, *333*, 93–111.
13. Bluman, G.W.; Temuerchaolu; Sahadevan, R. Local and nonlocal symmetries for nonlinear telegraph equation. *J. Math. Phys.* **2005**, *46*, 023505.
14. Chikwendu, S.C. Non-linear wave propagation solutions by Fourier transform perturbation. *Int. J. Non-Linear Mech.* **1981**, *16* 117–128.
15. Donato, A. Similarity analysis and nonlinear wave propagation. *Int. J. Non-Linear Mech.* **1987**, *22* 307–314.

16. Gandarias, M.L.; Torrisi, M.; Valenti, A. Symmetry classification and optimal systems of a non-linear wave equation. *Int. J. Non-Linear Mech.* **2004**, *39*, 389–398.
17. Huang, D.J.; Ivanova, N.M. Group analysis and exact solutions of a class of variable coefficient nonlinear telegraph equations. *J. Math. Phys.* **2007**, *48*, 073507.
18. Huang, D.J.; Zhou, S.G. Group properties of generalized quasi-linear wave equations. *J. Math. Anal. Appl.* **2010**, *366*, 460–472.
19. Huang, D.J.; Zhou, S.G. Group-theoretical analysis of variable coefficient nonlinear telegraph equations. *Acta Appl. Math.* **2012**, *117*, 135–183.
20. Ibragimov, N.H.; Torrisi, M.; Valenti, A. Preliminary group classification of equations $v_{tt} = f(x, v_x)v_{xx} + g(x, v_x)$. *J. Math. Phys.* **1991**, *32*, 2988–2995.
21. Oron, A.; Rosenau, P. Some symmetries of the nonlinear heat and wave equations. *Phys. Lett. A* **1986**, *118*, 172–176.
22. Pucci, E. Group analysis of the equation $u_{tt} + \lambda u_{xx} = g(u, u_x)$. *Riv. Mat. Univ. Parma* **1987**, *12*, 71–87.
23. Pucci, E.; Salvatori, M.C. Group properties of a class of semilinear hyperbolic equations. *Int. J. Non-Linear Mech.* **1986**, *21*, 147–155.
24. Torrisi, M.; Valenti, A. Group properties and invariant solutions for infinitesimal transformations of a nonlinear wave equation. *Int. J. Non-Linear Mech.* **1985**, *20*, 135–144.
25. Huang, D.J.; Yang, Q.M.; Zhou, S.G. Lie symmetry classification and equivalence transformation of variable coefficient nonlinear wave equations with power nonlinearities. *Chin. J. Contemp. Math.* **2012**, *33*, 205–214.
26. Huang, D.J.; Yang, Q.M.; Zhou, S.G. Conservation law classification of variable coefficient nonlinear wave equation with power Nonlinearity. *Chin. Phys. B* **2011**, *20*, 070202.
27. Huang, D.J.; Ivanova, N.M. Algorithmic framework for group analysis of differential equations and its application to generalized Zakharov-Kuznetsov equations. *J. Differ. Equ.* **2016**, *260*, 2354–2382.
28. Ovsiannikov, L.V. *Group Analysis of Differential Equations*; Academic Press: New York, NY, USA, 1982.
29. Bluman, G.W.; Cole, J.D. The general similarity solution of the heat equation. *J. Math. Mech.* **1969**, *18*, 1025–1042.
30. Levi, D.; Winternitz, P. Non-classical symmetry reduction: Example of the Boussinesq equation. *J. Phys. A Math. Gen.* **1989**, *22*, 2915–2924.
31. Arrigo, D.J.; Broadbridge, P.; Hill, J.M. Nonclassical symmetry solutions and the methods of Bluman-Cole and Clarkson-Kruskal. *J. Math. Phys.* **1993**, *34*, 4692–4703.
32. Zhdanov, R.Z.; Tsyfra, I.M.; Popovych, R.O. A precise definition of reduction of partial differential equations. *J. Math. Anal. Appl.* **1999**, *238*, 101–123.
33. Olver, P.J.; Rosenau, P. Group-invariant solutions of differential equations. *SIAM J. Appl. Math.* **1987**, *47*, 263–278.
34. Fushchych, W.I.; Shtelen, W.M.; Serov, N.I. *Symmetry Analysis and Exact Solutions of Nonlinear Equations of Mathematical Physics*; Kluwer: Dordrecht, The Netherlands, 1993.
35. Fushchych, W.I.; Shtelen, W.M.; Serov, M.I.; Popovych, R.O. Q-conditional symmetry of the linear heat equation. *Proc. Acad. Sci. Ukraine* **1992**, *12*, 28–33.
36. Fushchych, W.I.; Zhdanov, R.Z. Conditional symmetry and reduction of partial differential equations. *Ukr. Math. J.* **1992**, *44*, 970–982.
37. Cherniha, R. Conditional symmetries for systems of PDEs: New definitions and their application for reaction-diffusion systems. *J. Phys. A Math. Theor.* **2010**, *43*, 405207.
38. Cherniha, R.; Pliukhin, O. New conditional symmetries and exact solutions of reaction-diffusion systems with power diffusivities. *J. Phys. A Math. Theor.* **2008**, *41*, 185208.
39. Cherniha, R.; Pliukhin, O. New conditional symmetries and exact solutions of reaction-diffusion-convection equations with exponential nonlinearities. *J. Math. Anal. Appl.* **2013**, *403*, 23–37.
40. Cherniha, R. New Q-conditional symmetries and exact solutions of some reaction-diffusion-convection equations arising in mathematical biology. *J. Math. Anal. Appl.* **2007**, *326*, 783–799.
41. Kunzinger, M.; Popovych, R.O. Is a nonclassical symmetry a symmetry. In Proceedings of the 4th Workshop "Group Analysis of Differential Equations and Integrability", Protaras, Cyprus, 26–30 October 2008.
42. Olver, P.J. Direct reduction and differential constraints. *Proc. R. Soc. Lond. A* **1994**, *444*, 509–523.
43. Olver, P.J.; Rosenau, P. The construction of special solutions to partial differential equations. *Phys. Lett. A* **1986**, *114*, 107–112.

44. Pucci, E.; Saccomandi, G. On the weak symmetry groups of partial differential equations. *J. Math. Anal. Appl.* **1992**, *163*, 588–598.
45. Boyko, V.M.; Kunzinger, M.; Popovych, R.O. Singular reduction modules of differential equations. *J. Math. Phys.* **2016**, *57*, 101503.
46. Kunzinger, M.; Popovych R.O. Singular reduction operators in two dimensions. *J. Phys. A* **2008**, *41*, 505201.
47. Fushchich, V.I.; Serov, N.I.; Repeta, V.K. Conditional symmetry, reduction and exact solutions of the nonlinear wave equation. *Dokl. Akad. Nauk Ukr. SSR* **1991**, *161*, 29–34. (In Russian)

symmetry

MDPI

Article

The Method of Linear Determining Equations to Evolution System and Application for Reaction-Diffusion System with Power Diffusivities

Lina Ji [1,2]

[1] Department of Mathematics, Shanghai University, Shanghai 200444, China; jilina@henau.edu.cn; Tel.: +86-21-6613-6865

[2] Department of Information and Computational Science, He'nan Agricultural University, Zhengzhou 450002, China

Academic Editor: Roman M. Cherniha

Received: 6 September 2016; Accepted: 13 December 2016; Published: 20 December 2016

Abstract: The method of linear determining equations is constructed to study conditional Lie–Bäcklund symmetry and the differential constraint of a two-component second-order evolution system, which generalize the determining equations used in the search for classical Lie symmetry. As an application of the approach, the two-component reaction-diffusion system with power diffusivities is considered. The conditional Lie–Bäcklund symmetries and differential constraints admitted by the reaction-diffusion system are identified. Consequently, the reductions of the resulting system are established due to the compatibility of the corresponding invariant surface conditions and the original system.

Keywords: linear determining equation; conditional Lie–Bäcklund symmetry; differential constraint; evolution system; reaction-diffusion system

1. Introduction

The method of differential constraint (DC) is pretty old, dating back at least to the time of Lagrange. Lagrange used DC to find the total integral of a first-order nonlinear equation. Darboux applied DC to integrate the partial differential equation (PDE) of second-order. Yanenko proposed the key idea of DC in [1]. The survey of this method was presented by Sidorvo, Shapeev and Yanenko in [2], where the method of DC was successfully introduced into practice on gas dynamics.

The general formulation of the method of DC requires that the original system of PDEs

$$\tilde{F}^{(1)} = 0,\ \tilde{F}^{(2)} = 0,\ \cdots,\tilde{F}^{(m)} = 0 \tag{1}$$

be enlarged by appending additional equations

$$h_1 = 0,\ h_2 = 0,\ \cdots,h_p = 0 \tag{2}$$

such that the over-determined system (1), (2) is compatible. The differential equations in (2) are called DCs. The requirements for the compatibility of system (1), (2) are so general that the method of DC does not allow us to find all the forms of DCs for the system of PDEs in question. A number of different names for the parent notions of DC (2) leads to many methods for finding exact particular solutions of PDEs can be unified within the general framework of the method of DC.

The "side condition" is proposed to unify different methods for constructing particular solutions of PDEs by Olver and Rosenau in [3], where it is stated that appending of a suitable "side condition" is responsible for different kinds of methods for obtaining explicit solutions, including Lie's classical method [4,5], Bluman and Cole's nonclassical method [6] and Ovsiannikov's partial invariance method [7]. "The invariant surface condition" is used as a unifying theme for finding special solutions to PDEs by Pucci and Saccomandi in [8], where it is shown that "the invariant surface condition" and "its general integral" are the key to understanding the link between the so-called direct method [9], separation method [10,11], nonclassical symmetry [6] and weak symmetry [12,13]. The "additional generating condition" first raised by Cherniha [14,15] is exactly the linear case of DC, which is very effective to study reductions of variant forms of diffusion equations and diffusion systems [15–17]. The method of invariant subspace initially presented by Galaktionov and his collaborators [18] can also be understood within the framework of DC due to certain linear DC. The DCs that are responsible for Clarkson and Kruskal's first-order direct reduction [9] and Galaktionov's higher-order direct reduction [11] are discussed by Olver in [19]. The equivalence relationship between weak symmetry and DC is studied by Olver and Rosenau in [13].

The method of conditional Lie–Bäcklund symmetry (CLBS) provides an appropriate symmetry background for the method of DC. The base of symmetry reduction for CLBS is the fact that the corresponding invariant surface condition is formally compatible with the governing system, which is extensively discussed in [20,21], where it is shown that the problem of discussing the DC of the evolution system is equivalent to studying the CLBS of this system.

CLBS for the scalar evolution equation is introduced by Zhdanov [22], and another term for CLBS is used by Fokas and Liu [23,24]. A family of physically important exact solutions including the multi-shock solution and multi-soliton solution is constructed for a large class of non-integrable evolution equations by using the method of CLBS [23–26]. The CLBS for the evolution system is studied by Sergyeyev in [27] and independently by Qu et al. in [28].

The procedure for determining whether or not a given DC is compatible with the original equations is straightforward. However, for a given system of differential equations, one can never know in full detail the entire range of possible DCs since the associated determining equations are an over-determined nonlinear system. Nevertheless, as is known, even finding particular DCs can lead to new explicit solutions of the considered system. In practice, the principal direction of such research is to content oneself with finding DCs in some classes, and these classes must be chosen using additional considerations. From the symmetry point of view, CLBSs related to sign-invariants [29–33], separation of variables [34] and invariant subspaces [35–37] are proved to be very effective to study the classifications and reductions of second-order nonlinear diffusion equations. These particular subclasses related to sign-invariants [29–33] and invariant subspaces [35–37] are also extended to consider CLBSs of nonlinear diffusion systems in [21] and [28,38,39].

The purpose of this paper is to construct a practical way for finding the general form of DCs

$$\begin{cases} \eta_1 = u_n + g\left(t, x, u, u_1, u_2, \cdots, u_{n-1}\right) = 0, \\ \eta_2 = v_n + h\left(t, x, v, u_1, v_2, \cdots, v_{n-1}\right) = 0 \end{cases} \quad (3)$$

compatible with a two-component second-order evolution system

$$\begin{cases} u_t = F(t, x, u, v, u_1, v_1, u_2, v_2), \\ v_t = G(t, x, u, v, u_1, v_1, u_2, v_2), \end{cases} \quad (4)$$

which is equivalent to presenting an effective method to find the general form of CLBS with the characteristics

$$\begin{cases} \eta_1 = u_n + g\left(t, x, u, u_1, u_2, \cdots, u_{n-1}\right), \\ \eta_2 = v_n + h\left(t, x, v, u_1, v_2, \cdots, v_{n-1}\right) \end{cases} \quad (5)$$

admitted by evolution system (4). It is noted that $u_k = \partial^k u/\partial x^k$ and $v_k = \partial^k v/\partial x^k$ with $k = 1, 2, \cdots, n$ in (3)–(5) and hereafter.

The constructive method of the additional generating condition is presented by Cherniha in [14], where exact solutions of the variant form of the system (4) are derived by appending an additional condition in the form of a linear system of ordinary differential equations to the original system. Here, we will present the linear determining equations to identify DC (3) and CLBS (5) in the general form of second-order evolution system (4), which is exactly the extension of the results for the scalar evolution equation in [40–42].

The method of linear determining equations is proposed for finding the general form of DC

$$\eta = u_n + g(t, x, u, u_1, u_2, \cdots, u_{n-1}) = 0 \tag{6}$$

to evolution equation

$$u_t = F(t, x, u, u_1, u_2, \cdots, u_N) \tag{7}$$

by Kaptsov in [40]. The linear determining equation

$$D_t\eta = \sum_{i=0}^{N}\sum_{k=0}^{i} b_{ik} D_x^{i-k}(F_{u_{N-k}}) D_x^{N-i}(\eta) \tag{8}$$

presented there generalizes the classical determining equations within the framework of Lie's classical symmetry. It is clear that it is workable to find the DC with the general form (6) of evolution Equation (7) by solving linear determining equation (8) about η.

The principal direction of the research on applying the method to second-order nonlinear diffusion equations [40–42] gains an appreciation of its usefulness. The two-component reaction-diffusion (RD) system with power law diffusivities

$$\begin{cases} u_t = (u^k u_x)_x + P(u, v), \\ v_t = (v^l v_x)_x + Q(u, v) \end{cases} \tag{9}$$

will be considered here to demonstrate the applicability of this method for a two-component second-order evolution system.

The RD system (9) generalizes many well-known nonlinear second-order models and is used to describe various processes in physics, chemistry and biology. A complete description of Lie symmetries of the system is presented in [16]. The conditional symmetries for (9) are studied in [43–46]. The second-order CLBS (DC) admitted by the system (9) is discussed in [21]. Once the symmetries of the considered system (9) have been identified, one can algorithmically implement the reduction procedure and thereby determine all solutions that are invariant under the resulting symmetries. In [16,21,43–46], a wide range of exact solutions has been established due to various symmetry reductions therein.

The structure of this paper is organized as follows. The necessary definitions and notations about CLBS and DC of evolution system are displayed in Section 2. In Section 3, the linear determining equations to second-order evolution system (4) are constructed. The DCs (3) and CLBSs (5) of the system (9) are identified by solving the linear determining equation for the RD system (9) in Section 4. The exact solutions of the resulting RD system (9) are constructed due to the compatibility of the DC (3) and the governing system (9) in Section 5. The last section is devoted to the final discussions and conclusions.

2. Preliminaries

Let us review some theoretical elements of the tools about CLBS method and DC method of evolution system. Set

$$
\begin{aligned}
V = \sum_{i=1}^{m} \Big[& h_i\left(t, x, u^{(1)}, u^{(2)}, \cdots, u^{(m)}, u_1^{(1)}, u_1^{(2)}, \cdots, u_1^{(m)}, \cdots\right) \frac{\partial}{\partial u^{(i)}} \\
& + D_x h_i\left(t, x, u^{(1)}, u^{(2)}, \cdots, u^{(m)}, u_1^{(1)}, u_1^{(2)}, \cdots, u_1^{(m)}, \cdots\right) \frac{\partial}{\partial u_1^{(i)}} \\
& + D_t h_i\left(t, x, u^{(1)}, u^{(2)}, \cdots, u^{(m)}, u_1^{(1)}, u_1^{(2)}, \cdots, u_1^{(m)}, \cdots\right) \frac{\partial}{\partial u_t^{(i)}} \\
& + D_x^2 h_i\left(t, x, u^{(1)}, u^{(2)}, \cdots, u^{(m)}, u_1^{(1)}, u_1^{(2)}, \cdots, u_1^{(m)}, \cdots\right) \frac{\partial}{\partial u_2^{(i)}} \\
& + \cdots \Big]
\end{aligned} \tag{10}
$$

to be a certain smooth Lie–Bäcklund vector field (LBVF) and

$$u_t^{(i)} = F^{(i)}\left(t, x, u^{(1)}, u^{(2)}, \cdots, u^{(m)}, u_1^{(1)}, u_1^{(2)}, \cdots, u_1^{(m)}, \cdots\right), \ i = 1, 2, \cdots, m \tag{11}$$

to be a nonlinear evolution system, where $u_k^{(i)} = \partial^k u^{(i)} / \partial x^k$ with $i = 1, 2, \cdots, m$ and $k = 1, 2, \cdots$.

Definition 1. *[4,5] The evolutionary vector field (10) is said to be a Lie–Bäcklund symmetry of the evolution system (11) if*

$$V\left(u_t^{(i)} - F^{(i)}\right) |_S = 0, \ i = 1,\ 2,\ \cdots,\ m,$$

where S denotes the set of all differential consequences of the system (11).

Definition 2. *[27,28] The evolutionary vector field (10) is said to be a CLBS of (11) if*

$$V\left(u_t^{(i)} - F^{(i)}\right) |_{S \cap H_x} = 0, \ i = 1,\ 2,\ \cdots,\ m, \tag{12}$$

where H_x denotes the set of all differential consequences of the invariant surface condition

$$h_i\left(t, x, u^{(1)}, u^{(2)}, \cdots, u^{(m)}, u_1^{(1)}, u_1^{(2)}, \cdots, u_1^{(m)}, \cdots\right) = 0 \ (i = 1,\ 2,\ \cdots,\ m) \tag{13}$$

with respect to x.

A direct computation will yield that the conditional invariant criterion (12) can be reduced to [27,28]

$$D_t h_i |_{S \cap H_x} = 0, \ i = 1, 2, \cdots, m. \tag{14}$$

The fact that LBVF (10) is a CLBS of system (11) leads to the compatibility of the invariant surface condition (13) and the governing system (11).

Definition 3. *[47] The differential constraints (13) and the evolution system (11) satisfy the compatibility condition if*

$$D_t h_i |_{S_x \cap H_x} = 0, \ i = 1, 2, \cdots, m, \tag{15}$$

where S_x denotes the set of all differential consequences of the system (11) with respect to x.

The compatibility condition (15) is nothing but the conditional invariance criterion (14).

3. Linear Determining Equations for the DC (3) and CLBS (5) of Two-Component Second-Order Evolution System (4)

In this section, we discuss the method of linear determining equations to construct the DC (3) and CLBS (5) for second-order evolution system (4). The compatibility condition (15) can be reformulated as nonlinear equations. We now prove this result for DC (3) of the system (4), which is a natural generalization of what was the case for scalar evolution equation (7) in [40,41]. Let E_x be the union of all differential consequences of the second-order evolution system (4) with respect to x and M_x be the union of all differential consequences of DC (3) with respect to x.

Theorem 1. *The DC (3) with $n \geq 4$ is compatible with two-component second-order evolution system (4) if and only if η_1 and η_2 satisfy the following equations*

$$\begin{aligned} D_t\eta_1|_{E_x} =& F_{u_2}D_x^2\eta_1 + F_{v_2}D_x^2\eta_2 + (F_{u_1} + nD_xF_{u_2})D_x\eta_1 \\ &+ \left[F_{v_1} + nD_xF_{v_2} + \left(\eta_{1u_{n-1}} - \eta_{2v_{n-1}}\right)F_{v_2}\right]D_x\eta_2 \\ &+ \left[F_u + nD_xF_{u_1} + \frac{n(n-1)}{2}D_x^2F_{u_2} - \eta_{1u_{n-1}}D_xF_{u_2}\right. \\ &\left. - \left(2D_x\eta_{1u_{n-1}} - \eta_1\eta_{1u_{n-1}u_{n-1}}\right)F_{u_2}\right]\eta_1 + \left[F_v + nD_xF_{v_1}\right. \\ &+ \frac{n(n-1)}{2}D_x^2F_{v_2} + \left(\eta_{1u_{n-1}} - \eta_{2v_{n-1}}\right)\left(F_{v_1} + nD_xF_{v_2}\right) \\ &- \eta_{1u_{n-1}}D_xF_{v_2} + \left(\eta_{1u_{n-2}} - \eta_{2v_{n-2}}\right)F_{v_2} - \eta_{2v_{n-1}}(\eta_{1u_{n-1}} \\ &\left. - \eta_{2v_{n-1}})F_{v_2} - \left(2D_x\eta_{2v_{n-1}} - \eta_2\eta_{2v_{n-1}v_{n-1}}\right)F_{v_2}\right]\eta_2 \end{aligned} \tag{16}$$

and

$$\begin{aligned} D_t\eta_2|_{E_x} =& G_{v_2}D_x^2\eta_2 + G_{u_2}D_x^2\eta_1 + (G_{v_1} + nD_xG_{v_2})D_x\eta_2 \\ &+ \left[G_{u_1} + nD_xG_{u_2} + \left(\eta_{2v_{n-1}} - \eta_{1u_{n-1}}\right)G_{u_2}\right]D_x\eta_1 \\ &+ \left[G_v + nD_xG_{v_1} + \frac{n(n-1)}{2}D_x^2G_{v_2} - \eta_{2v_{n-1}}D_xG_{v_2}\right. \\ &\left. - \left(2D_x\eta_{2v_{n-1}} - \eta_2\eta_{2v_{n-1}v_{n-1}}\right)G_{v_2}\right]\eta_2 + \left[G_u + nD_xG_{u_1}\right. \\ &+ \frac{n(n-1)}{2}D_x^2G_{u_2} + \left(\eta_{2v_{n-1}} - \eta_{1u_{n-1}}\right)\left(G_{u_1} + nD_xG_{u_2}\right) \\ &- \eta_{2v_{n-1}}D_xG_{u_2} + \left(\eta_{2v_{n-2}} - \eta_{1u_{n-2}}\right)G_{u_2} - \eta_{1u_{n-1}}(\eta_{2v_{n-1}} \\ &\left. - \eta_{1u_{n-1}})G_{u_2} - \left(2D_x\eta_{1u_{n-1}} - \eta_1\eta_{1u_{n-1}u_{n-1}}\right)G_{u_2}\right]\eta_1. \end{aligned} \tag{17}$$

Proof. Assume that η_1 and η_2 satisfy (16) and (17). It is easy to see that all terms on the right-hand side of (16) and (17) vanish on M_x. Hence

$$D_t\eta_1|_{E_x \cap M_x} = 0 \tag{18}$$

and

$$D_t\eta_2|_{E_x \cap M_x} = 0,$$

that is, the DC (3) is compatible with the second-order evolution system (4). We now prove the converse result. Let $\alpha \simeq \beta$ indicate that there are no terms containing $u_n, v_n, u_{n+1}, v_{n+1}, u_{n+2}$ and v_{n+2} in the difference $\alpha - \beta$. Then, we can derive that

$$D_t\eta_1|_{E_x} \simeq D_x^n F + \eta_{1u_{n-1}} D_x^{n-1} F + \eta_{1u_{n-2}} D_x^{n-2} F. \tag{19}$$

Since

$$\begin{aligned} D_x^{n-2}F &\simeq u_n F_{u_2} + v_n F_{v_2}, \\ D_x^{n-1}F &\simeq u_n \left[F_{u_1} + (n-1)D_x F_{u_2}\right] + u_{n+1}F_{u_2} \\ &\quad + v_n \left[F_{v_1} + (n-1)D_x F_{v_2}\right] + v_{n+1}F_{v_2}, \\ D_x^n F &\simeq u_n \left[F_u + nD_x F_{u_1} + \frac{n(n-1)}{2} D_x^2 F_{u_2}\right] \\ &\quad + u_{n+1}(F_{u_1} + nD_x F_{u_2}) + u_{n+2}F_{u_2} \\ &\quad + v_n \left[F_v + nD_x F_{v_1} + \frac{n(n-1)}{2} D_x^2 F_{v_2}\right] \\ &\quad + v_{n+1}(F_{v_1} + nD_x F_{v_2}) + v_{n+2}F_{v_2} \end{aligned} \tag{20}$$

holds naturally for $n \geq 4$, (19) can be written as

$$\begin{aligned} D_t\eta_1|_{E_x} \simeq & u_{n+2}F_{u_2} + u_{n+1}\left(F_{u_1} + nD_x F_{u_2} + \eta_{1u_{n-1}}F_{u_2}\right) + u_n\Big\{F_u + nD_x F_{u_1} \\ & + \frac{n(n-1)}{2} D_x^2 F_{u_2} + \eta_{1u_{n-1}}\left[F_{u_1} + (n-1)D_x F_{u_2}\right] + \eta_{1u_{n-2}}F_{u_2}\Big\} \\ & + v_{n+2}F_{v_2} + v_{n+1}\left(F_{v_1} + nD_x F_{v_2} + \eta_{1u_{n-1}}F_{v_2}\right) + v_n\Big\{F_v + nD_x F_{v_1} \\ & + \frac{n(n-1)}{2} D_x^2 F_{v_2} + \eta_{1u_{n-1}}\left[F_{v_1} + (n-1)D_x F_{v_2}\right] + \eta_{1u_{n-2}}F_{v_2}\Big\}. \end{aligned}$$

It is easy to see that

$$\begin{aligned} & D_x\eta_1 \simeq u_{n+1} + u_n\eta_{1u_{n-1}}, \\ & D_x^2\eta_1 \simeq u_{n+2} + u_{n+1}\eta_{1u_{n-1}} + u_n\left(\eta_{1u_{n-2}} + 2D_x\eta_{1u_{n-1}} - u_n\eta_{1u_{n-1}u_{n-1}}\right), \\ & D_x\eta_2 \simeq v_{n+1} + v_n\eta_{2v_{n-1}}, \\ & D_x^2\eta_2 \simeq v_{n+2} + v_{n+1}\eta_{2v_{n-1}} + v_n\left(\eta_{2v_{n-2}} + 2D_x\eta_{2v_{n-1}} - v_n\eta_{2v_{n-1}v_{n-1}}\right). \end{aligned}$$

Consequently, a direct calculation will yield

$$\begin{aligned} D_t\eta_1|_{E_x} & - F_{u_2}D_x^2\eta_1 - F_{v_2}D_x^2\eta_2 - (F_{u_1} + nD_x F_{u_2})D_x\eta_1 \\ & - \left[F_{v_1} + nD_x F_{v_2} + \left(\eta_{1u_{n-1}} - \eta_{2v_{n-1}}\right)F_{v_2}\right]D_x\eta_2 \\ & - \Big[F_u + nD_x F_{u_1} + \frac{n(n-1)}{2} D_x^2 F_{u_2} - \eta_{1u_{n-1}}D_x F_{u_2} \\ & - \left(2D_x\eta_{1u_{n-1}} - \eta_1\eta_{1u_{n-1}u_{n-1}}\right)F_{u_2}\Big]\eta_1 - \Big[F_v + nD_x F_{v_1} \\ & + \left(\eta_{1u_{n-1}} - \eta_{2v_{n-1}}\right)(F_{v_1} + nD_x F_{v_2}) - \eta_{1u_{n-1}}D_x F_{v_2} \\ & + \left(\eta_{1u_{n-2}} - \eta_{2v_{n-2}}\right)F_{v_2} - \eta_{2v_{n-1}}\left(\eta_{1u_{n-1}} - \eta_{2v_{n-1}}\right)F_{v_2} \\ & - \left(2D_x\eta_{2v_{n-1}} - \eta_2\eta_{2v_{n-1}v_{n-1}}\right)F_{v_2}\Big]\eta_2 \\ \simeq & \, 0. \end{aligned} \tag{21}$$

Equation (18) holds since DC (3) is compatible with the system (4). Let γ denote the left-hand side of (21); it is easy to see that

$$\gamma|_{E_x \cap M_x} = 0,$$

which is equivalent to

$$\gamma|_{M_x} = 0 \tag{22}$$

since γ is independent of $u_t, v_t, u_{tx}, v_{tx}, \cdots$. As shown above, γ depends only on $u_{n-1}, v_{n-1}, u_{n-2}, v_{n-2} \cdots$. On the other hand, η_1 depends on u_n, and η_2 depends on v_n. Hence, (22) holds only for $\gamma = 0$, which yields nonlinear determining equation (16). In analogy with the discussion above, we can derive another nonlinear determining equation (17) if DC (3) is compatible with the system (4). □

In fact, the problem of solving nonlinear determining equations (16) and (17) is a very difficult, if not an impossible, problem. A practical way to identify DC (3) of the system (4) is to keep the linear part of (16) and (17). A general form of the corresponding linear determining equations will finally lead to the following definition.

Definition 4. *The linear determining equations for DCs (3) of the two-component second-order evolution system (4) are the linear equations*

$$\begin{aligned} D_t\eta_1|_{E_x} =& F_{u_2}D_x^2\eta_1 + \left(\tilde{b}_{11}F_{u_1} + \tilde{b}_{12}D_xF_{u_2}\right)D_x\eta_1 \\ &+ \left(\tilde{b}_{13}F_u + \tilde{b}_{14}D_xF_{u_1} + \tilde{b}_{15}D_x^2F_{u_2}\right)\eta_1 \\ &+ F_{v_2}D_x^2\eta_2 + \left(\tilde{b}_{16}F_{v_1} + \tilde{b}_{17}D_xF_{v_2}\right)D_x\eta_2 \\ &+ \left(\tilde{b}_{18}F_v + \tilde{b}_{19}D_xF_{v_1} + \tilde{b}_{10}D_x^2F_{v_2}\right)\eta_2 \end{aligned} \tag{23}$$

and

$$\begin{aligned} D_t\eta_2|_{E_x} =& G_{v_2}D_x^2\eta_2 + \left(\tilde{b}_{21}G_{v_1} + \tilde{b}_{22}D_xG_{v_2}\right)D_x\eta_2 \\ &+ \left(\tilde{b}_{23}G_v + \tilde{b}_{24}D_xG_{v_1} + \tilde{b}_{25}D_x^2G_{v_2}\right)\eta_2 \\ &+ G_{u_2}D_x^2\eta_1 + \left(\tilde{b}_{26}G_{u_1} + \tilde{b}_{27}D_xG_{u_2}\right)D_x\eta_1 \\ &+ \left(\tilde{b}_{28}G_u + \tilde{b}_{29}D_xG_{u_1} + \tilde{b}_{20}D_x^2G_{u_2}\right)\eta_1. \end{aligned} \tag{24}$$

Linear determining equations (23) and (24) are the sufficient condition to justify whether DC (3) is compatible with the second-order evolution system (4). This family of linear determining equations is also effective to construct CLBS (5) of evolution system (4).

4. DCs (3) and CLBSs (5) of RD system (9)

Substituting $F = u^ku_2 + ku^{k-1}u_1^2 + P(u,v)$ and $G = v^lv_2 + lv^{l-1}v_1^2 + Q(u,v)$ into linear determining equations (23) and (24), we can derive the sufficient condition to identify DCs (3) and CLBSs (5) of RD system (9)

$$\begin{aligned} D_t\eta_1|_{E_x} =& u^kD_x^2\eta_1 + k\left(2\tilde{b}_{11} + \tilde{b}_{12}\right)u^{k-1}u_1D_x\eta_1 + \Big[k\left(\tilde{b}_{13} + 2\tilde{b}_{14} + \tilde{b}_{15}\right)u^{k-1}u_2 \\ &+ k(k-1)\left(\tilde{b}_{13} + 2\tilde{b}_{14} + \tilde{b}_{15}\right)u^{k-2}u_1^2 + \tilde{b}_{13}P_u\Big]\eta_1 + \tilde{b}_{18}P_v\eta_2 \end{aligned} \tag{25}$$

and

$$D_t\eta_2|_{E_x} = v^l D_x^2\eta_2 + l\left(2\tilde{b}_{21} + \tilde{b}_{22}\right) v^{l-1} v_1 D_x\eta_2 + \Big[l\left(\tilde{b}_{23} + 2\tilde{b}_{24} + \tilde{b}_{25}\right) v^{l-1} v_2 + l(l-1)\left(\tilde{b}_{23} + 2\tilde{b}_{24} + \tilde{b}_{25}\right) v^{l-2} v_1^2 + \tilde{b}_{23} Q_v \Big] \eta_2 + \tilde{b}_{28} Q_u \eta_1. \tag{26}$$

Here, we use the general form of (25) and (26)

$$D_t\eta_1|_{E_x} = u^k D_x^2\eta_1 + b_{11} u^{k-1} u_1 D_x\eta_1 + \left(b_{12} u^{k-1} u_2 + b_{13} u^{k-2} u_1^2 + b_{14} P_u\right) \eta_1 + b_{15} P_v \eta_2 \tag{27}$$

and

$$D_t\eta_2|_{E_x} = v^l D_x^2\eta_2 + b_{21} v^{l-1} v_1 D_x\eta_2 + \left(b_{22} v^{l-1} v_2 + b_{23} v^{l-2} v_1^2 + b_{24} Q_v\right) \eta_2 + b_{25} Q_u \eta_1 \tag{28}$$

to construct DCs (3) and CLBSs (5) of the RD system (9).

It would be quite enlightening to give the order estimate for DCs (3) and CLBSs (5) admitted by the considered system (4). However, this is another problem, which we leave to future research. Here, we restrict our consideration to $2 \le n \le 5$.

Firstly, we consider the case of $n = 3$. A direct computation will give

$$\begin{aligned} D_t\eta_1|_{E_x} =& u^k u_5 + \left(5ku^{k-1}u_1 + u^k g_{u_2}\right) u_4 + \Big[4ku^{k-1}u_1 g_{u_2} + u^k g_{u_1} + P_u \\ &+ 10ku^{k-1}u_2 + 10k(k-1)u^{k-2}u_1^2\Big] u_3 + P_v v_3 + \Big[u_1^2 P_{uu} + v_1^2 P_{vv} \\ &+ 2u_1 v_1 P_{uv} + u_2 P_u + v_2 P_v + 3ku^{k-1}u_2^2 + 6k(k-1)u^{k-2}u_2 u_1^2 \\ &+ k(k-1)(k-2)u^{k-3}u_1^4\Big] g_{u_2} + \big[u_1 P_u + v_1 P_v + 3ku^{k-1}u_2 u_1 \\ &+ k(k-1)u^{k-2}u_1^3\big] g_{u_1} + \left(u^k u_2 + ku^{k-1}u_1^2 + P\right) g_u + u_1^3 P_{uuu} \\ &+ v_1^3 P_{vvv} + 3u_1^2 v_1 P_{uuv} + 3u_1 v_1^2 P_{uvv} + 3u_1 u_2 P_{uu} + 3v_1 v_2 P_{vv} \\ &+ 3(u_1 v_2 + u_2 v_1) P_{uv} + k(k-1)(k-2)(k-3)u^{k-4}u_1^5 \\ &+ 10k(k-1)(k-2)u^{k-3}u_2 u_1^3 + 15k(k-1)u^{k-2}u_2^2 u_1 + g_t \end{aligned}$$

and

$$\begin{aligned} &u^k D_x^2\eta_1 + b_{11}u^{k-1}u_1 D_x\eta_1 + \left(b_{12}u^{k-1}u_2 + b_{13}u^{k-2}u_1^2 + b_{14}P_u\right)\eta_1 + b_{15}P_v\eta_2 \\ =& u^k u_5 + \left(b_{11}u^{k-1}u_1 + u^k g_{u_2}\right) u_4 + u^k g_{u_2u_2} u_3^2 + \big(2u^k u_2 g_{u_1u_2} + 2u^k u_1 g_{uu_2} \\ &+ 2u^k g_{xu_2} + b_{11}u^{k-1}u_1 g_{u_2} + u^k g_{u_1} + b_{14}P_u + b_{12}u^{k-1}u_2 + b_{13}u^{k-2}u_1^2\big) u_3 \\ &+ b_{15}P_v v_3 + u^k u_2^2 g_{u_1u_1} + u^k u_1^2 g_{uu} + 2u^k u_1 u_2 g_{uu_1} + 2u^k u_1 g_{xu} + u^k g_{xx} \\ &+ 2u^k u_2 g_{xu_1} + b_{11}u^{k-1}u_1 u_2 g_{u_1} + b_{11}u^{k-1}u_1^2 g_u + u^k u_2 g_u + b_{11}u^{k-1}u_1 g_x \\ &+ \left(b_{12}u^{k-1}u_2 + b_{13}u^{k-2}u_1^2 + b_{14}P_u\right) g + b_{15}P_v h. \end{aligned}$$

Since the left-hand side and right-hand side of (27) are both polynomials about u_5, u_4, u_3, v_3, equating the coefficients of similar terms will give $b_{11} = 5k$, $b_{15} = 1$ and

$$\begin{aligned}
& g_{u_2u_2} = 0, \\
& 2u^k u_2 g_{u_1u_2} + 2u^k u_1 g_{uu_2} + 2u^k g_{xu_2} + (b_{11} - 4k)u^{k-1}u_1 g_{u_2} + (b_{14} - 1)P_u \\
& + (b_{12} - 10k)u^{k-1}u_2 + [b_{13} - 10k(k-1)]u^{k-2}u_1^2 = 0, \\
& u^k u_2^2 g_{u_1u_1} + u^k u_1^2 g_{uu} + 2u^k u_1 u_2 g_{uu_1} + 2u^k u_1 g_{xu} + u^k g_{xx} + 2u^k u_2 g_{xu_1} \\
& - \Big[u_1^2 P_{uu} + v_1^2 P_{vv} + 2u_1 v_1 P_{uv} + u_2 P_u + v_2 P_v + 6k(k-1)u^{k-2}u_2 u_1^2 \\
& + 3ku^{k-1}u_2^2 + k(k-1)(k-2)u^{k-3}u_1^4\Big] g_{u_2} - \big[(3k - b_{11})u^{k-1}u_2 u_1 \\
& + k(k-1)u^{k-2}u_1^3 + u_1 P_u + v_1 P_v\big] g_{u_1} - \big[(k - b_{11})u^{k-1}u_1^2 + P\big] g_u \\
& + b_{11}u^{k-1}u_1 g_x - g_t + \left(b_{12}u^{k-1}u_2 + b_{13}u^{k-2}u_1^2 + b_{14}P_u\right) g + b_{15}P_v h \\
& - \Big[u_1^3 P_{uuu} + v_1^3 P_{vvv} + 3u_1^2 v_1 P_{uuv} + 3u_1 v_1^2 P_{uvv} + 3(u_1 v_2 + u_2 v_1)P_{uv} \\
& + 3u_1 u_2 P_{uu} + 3v_1 v_2 P_{vv} + k(k-1)(k-2)(k-3)u^{k-4}u_1^5 \\
& + 10k(k-1)(k-2)u^{k-3}u_2 u_1^3 + 15k(k-1)u^{k-2}u_2^2 u_1\Big] = 0.
\end{aligned} \tag{29}$$

Similar discussion about (28) will yield $b_{21} = 5l$, $b_{25} = 1$ and

$$\begin{aligned}
& h_{v_2v_2} = 0, \\
& 2v^l v_2 h_{v_1v_2} + 2v^l v_1 h_{vv_2} + 2v^l h_{xv_2} + (b_{21} - 4l)v^{l-1}v_1 h_{v_2} + (b_{24} - 1)Q_v \\
& + (b_{22} - 10l)v^{l-1}v_2 + [b_{23} - 10l(l-1)]v^{l-2}v_1^2 = 0, \\
& v^l v_2^2 h_{v_1v_1} + v^l v_1^2 h_{vv} + 2v^l v_1 v_2 h_{vv_1} + 2v^l v_1 h_{xv} + v^l h_{xx} + 2v^l v_2 h_{xv_1} \\
& - \Big[v_1^2 Q_{vv} + u_1^2 Q_{uu} + 2u_1 v_1 Q_{uv} + u_2 Q_u + v_2 Q_v + 6l(l-1)v^{l-2}v_2 v_1^2 \\
& + 3lv^{l-1}v_2^2 + l(l-1)(l-2)v^{l-3}v_1^4\Big] h_{v_2} - \big[(3l - b_{21})v^{l-1}v_2 v_1 \\
& + l(l-1)v^{l-2}v_1^3 + u_1 Q_u + v_1 Q_v\big] h_{v_1} - \big[(l - b_{21})v^{l-1}v_1^2 + Q\big] h_v \\
& + b_{21}v^{l-1}v_1 h_x - h_t + \left(b_{22}v^{l-1}v_2 + b_{23}v^{l-2}v_1^2 + b_{24}Q_v\right) h + b_{25}Q_u g \\
& - \Big[u_1^3 Q_{uuu} + v_1^3 Q_{vvv} + 3u_1^2 v_1 Q_{uuv} + 3u_1 v_1^2 Q_{uvv} + 3(u_1 v_2 + u_2 v_1)Q_{uv} \\
& + 3u_1 u_2 Q_{uu} + 3v_1 v_2 Q_{vv} + l(l-1)(l-2)(l-3)v^{l-4}v_1^5 \\
& + 10l(l-1)(l-2)v^{l-3}v_2 v_1^3 + 15l(l-1)u^{l-2}v_2^2 v_1\Big] = 0.
\end{aligned} \tag{30}$$

It is easy to know that g and h can be represented as

$$g(t, x, u, u_1, u_2) = g_1(t, x, u, u_1)u_2 + g_2(t, x, u, u_1)$$

and

$$h(t, x, v, v_1, v_2) = h_1(t, x, v, v_1)v_2 + h_2(t, x, v, v_1).$$

Substituting g into the second one of (29), we will derive that

$$\begin{aligned}&\left[2ug_{1u_1} + (b_{12} - 10k)\right] u^{k-1}u_2 + \left[(2ug_{1u} + kg_1)u_1 + 2ug_{1x}\right] u^{k-1}\\&+ \left[b_{13} - 10k(k-1)\right] u^{k-2}u_1^2 + (b_{14} - 1)P_u = 0.\end{aligned}$$

The vanishing of the coefficient of u_2 will yield

$$g_1(t,x,u,u_1) = \frac{10k - b_{12}}{2u}u_1 + g_3(t,x,u).$$

As a consequence, (29) can be simplified as

$$\left(b_{13} - 5k^2 - \frac{1}{2}kb_{12} + b_{12}\right) u^{k-2}u_1^2 + (2ug_{3u} + kg_3)\, u^{k-1}u_1 + 2u^k g_{3x} + (b_{14} - 1)P_u = 0,$$

which is a polynomial about u_1. Thus, $b_{13} = 5k^2 + \frac{1}{2}kb_{12} - b_{12}$ and $g_3(t,x,u) = g_4(t,x)u^{-\frac{k}{2}}$ can be derived by equating the coefficients of this polynomial to be zero. Subsequently, (29) finally becomes

$$2u^{\frac{k}{2}}g_{4x} + (b_{14} - 1)P_u = 0.$$

Since $P(u,v)$ must depend on v, we will arrive at $g_4(t,x) = g_5(t)$ and $b_{14} = 1$ or $P(u,v) = P_1(v)$ from the above equality.

A similar computational procedure for the first one and second one of (30) will give

$$\begin{aligned}&h(t,x,,v,v_1,v_2) = h_1(t,x,v,v_1)v_2 + h_2(t,x,v,v_1),\\&h_1(t,x,v,v_1) = \frac{10l - b_{22}}{2v}v_1 + h_3(t,x,v),\\&h_3(t,x,v) = h_4(t,x)v^{-\frac{l}{2}},\ h_4(t,x) = h_5(t),\\&b_{23} = 5l^2 + \frac{1}{2}lb_{22} - b_{22}\end{aligned}$$

and

$$b_{24} = 1 \text{ or } Q(u,v) = Q_1(u).$$

We will consider four different cases, including

$$\begin{aligned}&(i)\ b_{14} = 1,\ b_{24} = 1;\\&(ii)\ b_{14} = 1,\ Q(u,v) = Q_1(u);\\&(iii)\ P(u,v) = P_1(v),\ b_{24} = 1;\\&(iv)\ P(u,v) = P_1(v),\ Q(u,v) = Q_1(u)\end{aligned}$$

to further study. Further research about the last ones of (29) and (30) will finally identify DCs (3) and CLBSs (5) of RD system (9). The comprehensive computational procedure is omitted here, and the obtained results are listed in Table 1. The procedure to identify DCs (3) and CLBSs (5) of RD system (4) for $n = 2$, $n = 4$ and $n = 5$ is almost the same as that for the case of $n = 3$. We just list the obtained results in Table 1. It is noted that the results for $n = 2$ are all presented in [21], so we will not list these cases in Table 1.

Table 1. conditional Lie–Bäcklund symmetry (CLBS) (5) of reaction-diffusion (RD) System (9).

No.	RD System (9)	CLBS (5)
1	$\begin{cases} u_t = \left(u^{-\frac{3}{2}}u_x\right)_x - \frac{s}{r}b_1u + a_1u^{\frac{5}{2}} + b_1u^{\frac{5}{2}}v^{-\frac{3}{2}}, \\ v_t = \left(v^{-\frac{3}{2}}v_x\right)_x - \frac{r}{s}b_2v + a_2v^{\frac{5}{2}} + b_2v^{\frac{5}{2}}u^{-\frac{3}{2}} \end{cases}$	$\begin{cases} \eta_1 = u_3 - \frac{15}{2u}u_1u_2 + \frac{35}{4u^2}u_1^3 + ru^{\frac{5}{2}}, \\ \eta_2 = v_3 - \frac{15}{2v}v_1v_2 + \frac{35}{4v^2}v_1^3 + sv^{\frac{5}{2}} \end{cases}$
2	$\begin{cases} u_t = \left(u^{-\frac{4}{3}}u_x\right)_x + a_1u + b_1u^{\frac{5}{3}} - \frac{3s}{4}u^{-\frac{1}{3}} + c_1u^{\frac{5}{3}}v^{-\frac{2}{3}}, \\ v_t = \left(v^{-\frac{4}{3}}v_x\right)_x + a_2v + b_2v^{\frac{5}{3}} - \frac{3s}{4}v^{-\frac{1}{3}} + c_2v^{\frac{5}{3}}u^{-\frac{2}{3}} \end{cases}$	$\begin{cases} \eta_1 = u_3 - \frac{5}{u}u_1u_2 + \frac{40}{9u^2}u_1^3 + su_1, \\ \eta_2 = v_3 - \frac{5}{v}v_1v_2 + \frac{40}{9v^2}v_1^3 + sv_1 \end{cases}$
3	$\begin{cases} u_t = \left(u^{-\frac{4}{3}}u_x\right)_x + a_1u + b_1u^{\frac{5}{3}} - \frac{3s}{4}u^{-\frac{1}{3}} + c_1u^{\frac{5}{3}}v^l, \\ v_t = \left(v^lv_x\right)_x + a_2v + b_2v^{1-l} + \frac{(l+1)s}{l^2}v^{1+l} + c_2v^{1-l}u^{-\frac{2}{3}} \end{cases}$	$\begin{cases} \eta_1 = u_3 - \frac{5}{u}u_1u_2 + \frac{40}{9u^2}u_1^3 + su_1, \\ \eta_2 = v_3 + \frac{3(l-1)}{v}v_1v_2 + \frac{(l-1)(l-2)}{v^2}v_1^3 + sv_1 \end{cases}$
4	$\begin{cases} u_t = \left(u^ku_x\right)_x + a_1u + b_1u^{1-k} + \frac{(k+1)s}{k^2}u^{1+k} + c_1u^{1-k}v^l, \\ v_t = \left(v^lv_x\right)_x + a_2v + b_2v^{1-l} + \frac{(l+1)s}{l^2}v^{1+l} + c_2v^{1-l}u^k \end{cases}$	$\begin{cases} \eta_1 = u_3 + \frac{3(k-1)}{u}u_1u_2 + \frac{(k-1)(k-2)}{u^2}u_1^3 + su_1, \\ \eta_2 = v_3 + \frac{3(l-1)}{v}v_1v_2 + \frac{(l-1)(l-2)}{v^2}v_1^3 + sv_1 \end{cases}$
5	$\begin{cases} u_t = \left(u^{-\frac{3}{2}}u_x\right)_x + a_1u + b_1u^{\frac{5}{2}} + c_1u^{\frac{5}{2}}v^{-\frac{3}{2}}, \\ v_t = \left(v^{-\frac{3}{2}}v_x\right)_x + a_2v + b_2v^{\frac{5}{2}} + c_2v^{\frac{5}{2}}u^{-\frac{3}{2}} \end{cases}$	$\begin{cases} \eta_1 = u_4 - \frac{10}{u}u_1u_3 - \frac{15}{2u}u_2^2 + \frac{105}{2u^2}u_1^2u_2 - \frac{315}{8u^3}u_1^4, \\ \eta_2 = v_4 - \frac{10}{v}v_1v_3 - \frac{15}{2v}v_2^2 + \frac{105}{2v^2}v_1^2v_2 - \frac{315}{8v^3}v_1^4 \end{cases}$
6	$\begin{cases} u_t = \left(u^{-\frac{4}{3}}u_x\right)_x + a_1u + b_1u^{\frac{7}{3}} - \frac{3s}{20}u^{-\frac{1}{3}} + c_1u^{\frac{7}{3}}v^{-\frac{4}{3}}, \\ v_t = \left(v^{-\frac{4}{3}}v_x\right)_x + a_2v + b_2v^{\frac{7}{3}} - \frac{3s}{20}v^{-\frac{1}{3}} + c_2v^{\frac{7}{3}}u^{-\frac{4}{3}} \end{cases}$	$\begin{cases} \eta_1 = u_5 - \frac{35}{3u}u_1u_4 + \left(-\frac{70}{3u}u_2 + \frac{700}{9u^2}u_1^2 + s\right)u_3 \\ \quad + \frac{350}{3u^2}u_1u_2^2 - \left(\frac{7s}{u}u_1 + \frac{9100}{27u^3}u_1^3\right)u_2 + \frac{14560}{81u^4}u_1^5 \\ \quad + \frac{70s}{9u^2}u_1^3 + \frac{4s^2}{25}u_1, \\ \eta_2 = v_5 - \frac{35}{3v}v_1v_4 + \left(-\frac{70}{3v}v_2 + \frac{700}{9v^2}v_1^2 + s\right)v_3 \\ \quad + \frac{350}{3v^2}v_1v_2^2 - \left(\frac{7s}{v}v_1 + \frac{9100}{27v^3}v_1^3\right)v_2 + \frac{14560}{81v^4}v_1^5 \\ \quad + \frac{70s}{9v^2}v_1^3 + \frac{4s^2}{25}v_1, \end{cases}$

5. Reductions of RD System (9)

The compatibility of the RD system (9) and the invariant surface condition (DC) (3) is the basic reduction idea of CLBS. Therefore, the evolution system (9) and the admitted DC (3) share a common manifold of solutions. We first solve the DC (3) to identify the form of u and v and then substitute the obtained results into (9) to finally determine the solutions. Here, we will construct the reductions of the resulting systems (9) in Table 1.

Example 1. *RD system*

$$\begin{cases} u_t = \left(u^{-\frac{3}{2}}u_x\right)_x - \frac{s}{r}b_1u + a_1u^{\frac{5}{2}} + b_1u^{\frac{5}{2}}v^{-\frac{3}{2}}, \\ v_t = \left(v^{-\frac{3}{2}}v_x\right)_x - \frac{r}{s}b_2v + a_2v^{\frac{5}{2}} + b_2v^{\frac{5}{2}}u^{-\frac{3}{2}} \end{cases}$$

admits CLBS

$$\begin{cases} \eta_1 = u_3 - \frac{15}{2u}u_1u_2 + \frac{35}{4u^2}u_1^3 + ru^{\frac{5}{2}}, \\ \eta_2 = v_3 - \frac{15}{2v}v_1v_2 + \frac{35}{4v^2}v_1^3 + sv^{\frac{5}{2}}. \end{cases}$$

The solutions of this system are listed as

$$\begin{cases} u(x,t) = \left[\frac{r}{4}x^3 + C_1^{(1)}(t)x^2 + C_2^{(1)}(t)x + C_3^{(1)}(t)\right]^{-\frac{2}{3}}, \\ v(x,t) = \left[\frac{s}{4}x^3 + C_1^{(2)}(t)x^2 + C_2^{(2)}(t)x + C_3^{(2)}(t)\right]^{-\frac{2}{3}}, \end{cases}$$

where $C_1^{(1)}(t), C_2^{(1)}(t), C_3^{(1)}(t), C_1^{(2)}(t), C_2^{(2)}(t)$ and $C_3^{(2)}(t)$ satisfy the six-dimensional dynamical system

$$\begin{cases} C_1^{(1)\prime} = -\frac{2}{3}C_1^{(1)^2} + \frac{3s}{2r}b_1C_1^{(1)} + \frac{r}{2}C_2^{(1)} - \frac{3}{2}b_1C_1^{(2)}, \\ C_2^{(1)\prime} = -\frac{2}{3}C_1^{(1)}C_2^{(1)} + \frac{3s}{2r}b_1C_2^{(1)} + \frac{3r}{2}C_3^{(1)} - \frac{3}{2}b_1C_2^{(2)}, \\ C_3^{(1)\prime} = 2C_1^{(1)}C_3^{(1)} - \frac{2}{3}C_2^{(1)^2} + \frac{3s}{2r}b_1C_3^{(1)} - \frac{3}{2}b_1C_3^{(2)} - \frac{3}{2}a_1, \\ C_1^{(2)\prime} = -\frac{2}{3}C_1^{(2)^2} + \frac{3r}{2s}b_2C_1^{(2)} + \frac{s}{2}C_2^{(2)} - \frac{3}{2}b_2C_1^{(1)}, \\ C_2^{(2)\prime} = -\frac{2}{3}C_1^{(2)}C_2^{(2)} + \frac{3r}{2s}b_2C_2^{(2)} + \frac{3s}{2}C_3^{(2)} - \frac{3}{2}b_2C_2^{(1)}, \\ C_3^{(2)\prime} = 2C_1^{(2)}C_3^{(2)} - \frac{2}{3}C_2^{(2)^2} + \frac{3r}{2s}b_2C_3^{(2)} - \frac{3}{2}b_2C_3^{(1)} - \frac{3}{2}a_2. \end{cases}$$

Example 2. *RD system*

$$\begin{cases} u_t = \left(u^{-\frac{4}{3}}u_x\right)_x + a_1u + b_1u^{\frac{5}{3}} - \frac{3s}{4}u^{-\frac{1}{3}} + c_1u^{\frac{5}{3}}v^{-\frac{2}{3}}, \\ v_t = \left(v^{-\frac{4}{3}}v_x\right)_x + a_2v + b_2v^{\frac{5}{3}} - \frac{3s}{4}v^{-\frac{1}{3}} + c_2v^{\frac{5}{3}}u^{-\frac{2}{3}} \end{cases}$$

admits CLBS

$$\begin{cases} \eta_1 = u_3 - \frac{5}{u}u_1u_2 + \frac{40}{9u^2}u_1^3 + su_1, \\ \eta_2 = v_3 - \frac{5}{v}v_1v_2 + \frac{40}{9v^2}v_1^3 + sv_1. \end{cases}$$

The solutions of this system are given as below.

- *For $s > 0$,*

$$\begin{cases} u(x,t) = \left[C_1^{(1)}(t) + C_2^{(1)}(t)\sin\left(\sqrt{s}x\right) + C_3^{(1)}(t)\cos\left(\sqrt{s}x\right)\right]^{-\frac{3}{2}}, \\ v(x,t) = \left[C_1^{(2)}(t) + C_2^{(2)}(t)\sin\left(\sqrt{s}x\right) + C_3^{(2)}(t)\cos\left(\sqrt{s}x\right)\right]^{-\frac{3}{2}}, \end{cases}$$

where $C_1^{(1)}(t), C_2^{(1)}(t), C_3^{(1)}(t), C_1^{(2)}(t), C_2^{(2)}(t)$ *and* $C_3^{(2)}(t)$ *satisfy the six-dimensional dynamical system*

$$
\begin{cases}
C_1^{(1)'} = \frac{s}{2}C_1^{(1)}\left({C_1^{(1)}}^2 - {C_2^{(1)}}^2 - {C_3^{(1)}}^2\right) - \frac{2}{3}a_1C_1^{(1)} - \frac{2}{3}c_1C_1^{(2)} - \frac{2}{3}b_1, \\
C_2^{(1)'} = \frac{s}{2}C_2^{(1)}\left({C_1^{(1)}}^2 - {C_2^{(1)}}^2 - {C_3^{(1)}}^2\right) - \frac{2}{3}a_1C_2^{(1)} - \frac{2}{3}c_1C_2^{(2)}, \\
C_3^{(1)'} = \frac{s}{2}C_3^{(1)}\left({C_1^{(1)}}^2 - {C_2^{(1)}}^2 - {C_3^{(1)}}^2\right) - \frac{2}{3}a_1C_3^{(1)} - \frac{2}{3}c_1C_3^{(2)}, \\
C_1^{(2)'} = \frac{s}{2}C_1^{(2)}\left({C_1^{(2)}}^2 - {C_2^{(2)}}^2 - {C_3^{(2)}}^2\right) - \frac{2}{3}a_2C_1^{(2)} - \frac{2}{3}c_2C_1^{(1)} - \frac{2}{3}b_2, \\
C_2^{(2)'} = \frac{s}{2}C_2^{(2)}\left({C_1^{(2)}}^2 - {C_2^{(2)}}^2 - {C_3^{(2)}}^2\right) - \frac{2}{3}a_2C_2^{(2)} - \frac{2}{3}c_2C_2^{(1)}, \\
C_3^{(2)'} = \frac{s}{2}C_3^{(2)}\left({C_1^{(2)}}^2 - {C_2^{(2)}}^2 - {C_3^{(2)}}^2\right) - \frac{2}{3}a_2C_3^{(2)} - \frac{2}{3}c_2C_3^{(1)}.
\end{cases}
$$

- *For* $s = 0$,

$$
\begin{cases}
u(x,t) = \left[C_1^{(1)}(t)x^2 + C_2^{(1)}(t)x + C_3^{(1)}(t)\right]^{-\frac{3}{2}}, \\
v(x,t) = \left[C_1^{(2)}(t)x^2 + C_2^{(2)}(t)x + C_3^{(2)}(t)\right]^{-\frac{3}{2}},
\end{cases}
$$

where $C_1^{(1)}(t), C_2^{(1)}(t), C_3^{(1)}(t), C_1^{(2)}(t), C_2^{(2)}(t)$ *and* $C_3^{(2)}(t)$ *satisfy the six-dimensional dynamical system*

$$
\begin{cases}
C_1^{(1)'} = 2{C_1^{(1)}}^2C_3^{(1)} - \frac{1}{2}C_1^{(1)}{C_2^{(1)}}^2 - \frac{2}{3}a_1C_1^{(1)} - \frac{2}{3}c_1C_1^{(2)}, \\
C_2^{(1)'} = 2C_1^{(1)}C_2^{(1)}C_3^{(1)} - \frac{1}{2}{C_2^{(1)}}^3 - \frac{2}{3}a_1C_2^{(1)} - \frac{2}{3}c_1C_2^{(2)}, \\
C_3^{(1)'} = 2C_1^{(1)}{C_3^{(1)}}^2 - \frac{1}{2}{C_2^{(1)}}^2C_3^{(1)} - \frac{2}{3}a_1C_3^{(1)} - \frac{2}{3}c_1C_3^{(2)} - \frac{2}{3}b_1, \\
C_1^{(2)'} = 2{C_1^{(2)}}^2C_3^{(2)} - \frac{1}{2}C_1^{(2)}{C_2^{(2)}}^2 - \frac{2}{3}a_2C_1^{(2)} - \frac{2}{3}c_2C_1^{(1)}, \\
C_2^{(2)'} = 2C_1^{(2)}C_2^{(2)}C_3^{(2)} - \frac{1}{2}{C_2^{(2)}}^3 - \frac{2}{3}a_2C_2^{(2)} - \frac{2}{3}c_2C_2^{(1)}, \\
C_3^{(2)'} = 2C_1^{(2)}{C_3^{(2)}}^2 - \frac{1}{2}{C_2^{(2)}}^2C_3^{(2)} - \frac{2}{3}a_2C_3^{(2)} - \frac{2}{3}c_2C_3^{(1)} - \frac{2}{3}b_2.
\end{cases}
$$

- *For* $s < 0$,

$$
\begin{cases}
u(x,t) = \left[C_1^{(1)}(t) + C_2^{(1)}(t)\sinh\left(\sqrt{-s}x\right) + C_3^{(1)}(t)\cosh\left(\sqrt{-s}x\right)\right]^{-\frac{3}{2}}, \\
v(x,t) = \left[C_1^{(2)}(t) + C_2^{(2)}(t)\sinh\left(\sqrt{-s}x\right) + C_3^{(2)}(t)\cosh\left(\sqrt{-s}x\right)\right]^{-\frac{3}{2}},
\end{cases}
$$

where $C_1^{(1)}(t), C_2^{(1)}(t), C_3^{(1)}(t), C_1^{(2)}(t), C_2^{(2)}(t)$ *and* $C_3^{(2)}(t)$ *satisfy the six-dimensional dynamical system*

$$
\begin{cases}
C_1^{(1)'} = \frac{s}{2}C_1^{(1)}\left({C_1^{(1)}}^2 + {C_2^{(1)}}^2 - {C_3^{(1)}}^2\right) - \frac{2}{3}a_1C_1^{(1)} - \frac{2}{3}c_1C_1^{(2)} - \frac{2}{3}b_1, \\
C_2^{(1)'} = \frac{s}{2}C_2^{(1)}\left({C_1^{(1)}}^2 + {C_2^{(1)}}^2 - {C_3^{(1)}}^2\right) - \frac{2}{3}a_1C_2^{(1)} - \frac{2}{3}c_1C_2^{(2)}, \\
C_3^{(1)'} = \frac{s}{2}C_3^{(1)}\left({C_1^{(1)}}^2 + {C_2^{(1)}}^2 - {C_3^{(1)}}^2\right) - \frac{2}{3}a_1C_3^{(1)} - \frac{2}{3}c_1C_3^{(2)}, \\
C_1^{(2)'} = \frac{s}{2}C_1^{(2)}\left({C_1^{(2)}}^2 + {C_2^{(2)}}^2 - {C_3^{(2)}}^2\right) - \frac{2}{3}a_2C_1^{(2)} - \frac{2}{3}c_2C_1^{(1)} - \frac{2}{3}b_2, \\
C_2^{(2)'} = \frac{s}{2}C_2^{(2)}\left({C_1^{(2)}}^2 + {C_2^{(2)}}^2 - {C_3^{(2)}}^2\right) - \frac{2}{3}a_2C_2^{(2)} - \frac{2}{3}c_2C_2^{(1)}, \\
C_3^{(2)'} = \frac{s}{2}C_3^{(2)}\left({C_1^{(2)}}^2 + {C_2^{(2)}}^2 - {C_3^{(2)}}^2\right) - \frac{2}{3}a_2C_3^{(2)} - \frac{2}{3}c_2C_3^{(1)}.
\end{cases}
$$

Example 3. *RD system*

$$\begin{cases} u_t = \left(u^{-\frac{4}{3}}u_x\right)_x + a_1 u + b_1 u^{\frac{5}{3}} - \frac{3s}{4}u^{-\frac{1}{3}} + c_1 u^{\frac{5}{3}} v^l, \\ v_t = \left(v^l v_x\right)_x + a_2 v + b_2 v^{1-l} + \frac{(l+1)s}{l^2} v^{1+l} + c_2 v^{1-l} u^{-\frac{2}{3}} \end{cases}$$

admits CLBS

$$\begin{cases} \eta_1 = u_3 - \frac{5}{u}u_1 u_2 + \frac{40}{9u^2}u_1^3 + s u_1, \\ \eta_2 = v_3 + \frac{3(l-1)}{v} v_1 v_2 + \frac{(l-1)(l-2)}{v^2} v_1^3 + s v_1. \end{cases}$$

The solutions of this system are given as below.

- *For* $s > 0$,

$$\begin{cases} u(x,t) = \left[C_1^{(1)}(t) + C_2^{(1)}(t)\sin\left(\sqrt{s}x\right) + C_3^{(1)}(t)\cos\left(\sqrt{s}x\right)\right]^{-\frac{3}{2}}, \\ v(x,t) = \left[C_1^{(2)}(t) + C_2^{(2)}(t)\sin\left(\sqrt{s}x\right) + C_3^{(2)}(t)\cos\left(\sqrt{s}x\right)\right]^{\frac{1}{l}}, \end{cases}$$

where $C_1^{(1)}(t), C_2^{(1)}(t), C_3^{(1)}(t), C_1^{(2)}(t), C_2^{(2)}(t)$ *and* $C_3^{(2)}(t)$ *satisfy the six-dimensional dynamical system*

$$\begin{cases} {C_1^{(1)}}' = \frac{s}{2}C_1^{(1)}\left({C_1^{(1)}}^2 - {C_2^{(1)}}^2 - {C_3^{(1)}}^2\right) - \frac{2}{3}a_1 C_1^{(1)} - \frac{2}{3}c_1 C_1^{(2)} - \frac{2}{3}b_1, \\ {C_2^{(1)}}' = \frac{s}{2}C_2^{(1)}\left({C_1^{(1)}}^2 - {C_2^{(1)}}^2 - {C_3^{(1)}}^2\right) - \frac{2}{3}a_1 C_2^{(1)} - \frac{2}{3}c_1 C_2^{(2)}, \\ {C_3^{(1)}}' = \frac{s}{2}C_3^{(1)}\left({C_1^{(1)}}^2 - {C_2^{(1)}}^2 - {C_3^{(1)}}^2\right) - \frac{2}{3}a_1 C_3^{(1)} - \frac{2}{3}c_1 C_3^{(2)}, \\ {C_1^{(2)}}' = \frac{(l+1)s}{l}{C_1^{(2)}}^2 + \frac{s}{l}\left({C_2^{(2)}}^2 + {C_3^{(2)}}^2\right) + l a_2 C_1^{(2)} + l c_2 C_1^{(1)} + l b_2, \\ {C_2^{(2)}}' = \frac{(l+2)s}{l}C_1^{(2)} C_2^{(2)} + l a_2 C_2^{(2)} + l c_2 C_2^{(1)}, \\ {C_3^{(2)}}' = \frac{(l+2)s}{l}C_1^{(2)} C_3^{(2)} + l a_2 C_3^{(2)} + l c_2 C_3^{(1)}. \end{cases}$$

- *For* $s = 0$,

$$\begin{cases} u(x,t) = \left[C_1^{(1)}(t)x^2 + C_2^{(1)}(t)x + C_3^{(1)}(t)\right]^{-\frac{3}{2}}, \\ v(x,t) = \left[C_1^{(2)}(t)x^2 + C_2^{(2)}(t)x + C_3^{(2)}(t)\right]^{\frac{1}{l}}, \end{cases}$$

where $C_1^{(1)}(t), C_2^{(1)}(t), C_3^{(1)}(t), C_1^{(2)}(t), C_2^{(2)}(t)$ *and* $C_3^{(2)}(t)$ *satisfy the six-dimensional dynamical system*

$$\begin{cases} {C_1^{(1)}}' = 2{C_1^{(1)}}^2 C_3^{(1)} - \frac{1}{2}C_1^{(1)}{C_2^{(1)}}^2 - \frac{2}{3}a_1 C_1^{(1)} - \frac{2}{3}c_1 C_1^{(2)}, \\ {C_2^{(1)}}' = 2C_1^{(1)}C_2^{(1)}C_3^{(1)} - \frac{1}{2}{C_2^{(1)}}^3 - \frac{2}{3}a_1 C_2^{(1)} - \frac{2}{3}c_1 C_2^{(2)}, \\ {C_3^{(1)}}' = 2C_1^{(1)}{C_3^{(1)}}^2 - \frac{1}{2}{C_2^{(1)}}^2 C_3^{(1)} - \frac{2}{3}a_1 C_3^{(1)} - \frac{2}{3}c_1 C_3^{(2)} - \frac{2}{3}b_1, \\ {C_1^{(2)}}' = \frac{2(l+2)}{l}{C_1^{(2)}}^2 + l a_2 C_1^{(2)} + l c_2 C_1^{(1)}, \\ {C_2^{(2)}}' = \frac{2(l+2)}{l}C_1^{(2)}C_2^{(2)} + l a_2 C_2^{(2)} + l c_2 C_2^{(1)}, \\ {C_3^{(2)}}' = \frac{1}{l}{C_2^{(2)}}^2 + 2C_1^{(2)}C_3^{(2)} + l a_2 C_3^{(2)} + l c_2 C_3^{(1)} + l b_2. \end{cases}$$

- *For* $s < 0$,

$$\begin{cases} u(x,t) = \left[C_1^{(1)}(t) + C_2^{(1)}(t)\sinh\left(\sqrt{-s}x\right) + C_3^{(1)}(t)\cosh\left(\sqrt{-s}x\right)\right]^{\frac{3}{2}}, \\ v(x,t) = \left[C_1^{(2)}(t) + C_2^{(2)}(t)\sinh\left(\sqrt{-s}x\right) + C_3^{(2)}(t)\cosh\left(\sqrt{-s}x\right)\right]^{\frac{1}{l}}, \end{cases}$$

where $C_1^{(1)}(t), C_2^{(1)}(t), C_3^{(1)}(t), C_1^{(2)}(t), C_2^{(2)}(t)$ *and* $C_3^{(2)}(t)$ *satisfy the six-dimensional dynamical system*

$$\begin{cases} C_1^{(1)'} = \frac{s}{2}C_1^{(1)}\left(C_1^{(1)^2} + C_2^{(1)^2} - C_3^{(1)^2}\right) - \frac{2}{3}a_1C_1^{(1)} - \frac{2}{3}c_1C_1^{(2)} - \frac{2}{3}b_1, \\ C_2^{(1)'} = \frac{s}{2}C_2^{(1)}\left(C_1^{(1)^2} + C_2^{(1)^2} - C_3^{(1)^2}\right) - \frac{2}{3}a_1C_2^{(1)} - \frac{2}{3}c_1C_2^{(2)}, \\ C_3^{(1)'} = \frac{s}{2}C_3^{(1)}\left(C_1^{(1)^2} + C_2^{(1)^2} - C_3^{(1)^2}\right) - \frac{2}{3}a_1C_3^{(1)} - \frac{2}{3}c_1C_3^{(2)}, \\ C_1^{(2)'} = \frac{(l+1)s}{l}C_1^{(2)^2} + \frac{s}{l}\left(C_3^{(2)^2} - C_2^{(2)^2}\right) + la_2C_1^{(2)} + lc_2C_1^{(1)} + lb_2, \\ C_2^{(2)'} = \frac{(l+2)s}{l}C_1^{(2)}C_2^{(2)} + la_2C_2^{(2)} + lc_2C_2^{(1)}, \\ C_3^{(2)'} = \frac{(l+2)s}{l}C_1^{(2)}C_3^{(2)} + la_2C_3^{(2)} + lc_2C_3^{(1)}. \end{cases}$$

Example 4. *RD system*

$$\begin{cases} u_t = \left(u^k u_x\right)_x + a_1u + b_1u^{1-k} + \frac{(k+1)s}{k^2}u^{1+k} + c_1u^{1-k}v^l, \\ v_t = \left(v^l v_x\right)_x + a_2v + b_2v^{1-l} + \frac{(l+1)s}{l^2}v^{1+l} + c_2v^{1-l}u^k \end{cases}$$

admits CLBS

$$\begin{cases} \eta_1 = u_3 + \frac{3(k-1)}{u}u_1u_2 + \frac{(k-1)(k-2)}{u^2}u_1^3 + su_1, \\ \eta_2 = v_3 + \frac{3(l-1)}{v}v_1v_2 + \frac{(l-1)(l-2)}{v^2}v_1^3 + sv_1. \end{cases}$$

The solutions of this system are given as below.

- *For* $s > 0$,

$$\begin{cases} u(x,t) = \left[C_1^{(1)}(t) + C_2^{(1)}(t)\sin\left(\sqrt{s}x\right) + C_3^{(1)}(t)\cos\left(\sqrt{s}x\right)\right]^{\frac{1}{k}}, \\ v(x,t) = \left[C_1^{(2)}(t) + C_2^{(2)}(t)\sin\left(\sqrt{s}x\right) + C_3^{(2)}(t)\cos\left(\sqrt{s}x\right)\right]^{\frac{1}{l}}, \end{cases}$$

where $C_1^{(1)}(t), C_2^{(1)}(t), C_3^{(1)}(t), C_1^{(2)}(t), C_2^{(2)}(t)$ *and* $C_3^{(2)}(t)$ *satisfy the six-dimensional dynamical system*

$$\begin{cases} C_1^{(1)'} = \frac{(k+1)s}{k}C_1^{(1)^2} + \frac{s}{k}\left(C_2^{(1)^2} + C_3^{(1)^2}\right) + ka_1C_1^{(1)} + kc_1C_1^{(2)} + kb_1, \\ C_2^{(1)'} = \frac{(k+2)s}{k}C_1^{(1)}C_2^{(1)} + ka_1C_2^{(1)} + kc_1C_2^{(2)}, \\ C_3^{(1)'} = \frac{(k+2)s}{k}C_1^{(1)}C_3^{(1)} + ka_1C_3^{(1)} + kc_1C_3^{(2)}, \\ C_1^{(2)'} = \frac{(l+1)s}{l}C_1^{(2)^2} + \frac{s}{l}\left(C_2^{(2)^2} + C_3^{(2)^2}\right) + la_2C_1^{(2)} + lc_2C_1^{(1)} + lb_2, \\ C_2^{(2)'} = \frac{(l+2)s}{l}C_1^{(2)}C_2^{(2)} + la_2C_2^{(2)} + lc_2C_2^{(1)}, \\ C_3^{(2)'} = \frac{(l+2)s}{l}C_1^{(2)}C_3^{(2)} + la_2C_3^{(2)} + lc_2C_3^{(1)}. \end{cases}$$

- *For* $s = 0$,

$$\begin{cases} u(x,t) = \left[C_1^{(1)}(t)x^2 + C_2^{(1)}(t)x + C_3^{(1)}(t)\right]^{\frac{1}{k}}, \\ v(x,t) = \left[C_1^{(2)}(t)x^2 + C_2^{(2)}(t)x + C_3^{(2)}(t)\right]^{\frac{1}{l}}, \end{cases}$$

where $C_1^{(1)}(t), C_2^{(1)}(t), C_3^{(1)}(t), C_1^{(2)}(t), C_2^{(2)}(t)$ *and* $C_3^{(2)}(t)$ *satisfy the six-dimensional dynamical system*

$$\left\{\begin{array}{l} C_1^{(1)\prime} = \frac{2(k+2)}{k} C_1^{(1)^2} + ka_1 C_1^{(1)} + kc_1 C_1^{(2)}, \\ C_2^{(1)\prime} = \frac{2(k+2)}{k} C_1^{(1)} C_2^{(1)} + ka_1 C_2^{(1)} + kc_1 C_2^{(2)}, \\ C_3^{(1)\prime} = \frac{1}{k} C_2^{(1)^2} + 2C_1^{(1)} C_3^{(1)} + ka_1 C_3^{(1)} + kc_1 C_3^{(2)} + kb_1, \\ C_1^{(2)\prime} = \frac{2(l+2)}{l} C_1^{(2)^2} + la_2 C_1^{(2)} + lc_2 C_1^{(1)}, \\ C_2^{(2)\prime} = \frac{2(l+2)}{l} C_1^{(2)} C_2^{(2)} + la_2 C_2^{(2)} + lc_2 C_2^{(1)}, \\ C_3^{(2)\prime} = \frac{1}{l} C_2^{(2)^2} + 2C_1^{(2)} C_3^{(2)} + la_2 C_3^{(2)} + lc_2 C_3^{(1)} + lb_2. \end{array}\right.$$

- *For* $s < 0$,

$$\left\{\begin{array}{l} u(x,t) = \left[C_1^{(1)}(t) + C_2^{(1)}(t)\sinh\left(\sqrt{-s}x\right) + C_3^{(1)}(t)\cosh\left(\sqrt{-s}x\right)\right]^{\frac{1}{k}}, \\ v(x,t) = \left[C_1^{(2)}(t) + C_2^{(2)}(t)\sinh\left(\sqrt{-s}x\right) + C_3^{(2)}(t)\cosh\left(\sqrt{-s}x\right)\right]^{\frac{1}{l}}, \end{array}\right.$$

where $C_1^{(1)}(t), C_2^{(1)}(t), C_3^{(1)}(t), C_1^{(2)}(t), C_2^{(2)}(t)$ *and* $C_3^{(2)}(t)$ *satisfy the six-dimensional dynamical system*

$$\left\{\begin{array}{l} C_1^{(1)\prime} = \frac{(k+1)s}{k} C_1^{(1)^2} + \frac{s}{k}\left(C_3^{(1)^2} - C_2^{(1)^2}\right) + ka_1 C_1^{(1)} + kc_1 C_1^{(2)} + kb_1, \\ C_2^{(1)\prime} = \frac{(k+2)s}{k} C_1^{(1)} C_2^{(1)} + ka_1 C_2^{(1)} + kc_1 C_2^{(2)}, \\ C_3^{(1)\prime} = \frac{(k+2)s}{k} C_1^{(1)} C_3^{(1)} + ka_1 C_3^{(1)} + kc_1 C_3^{(2)}, \\ C_1^{(2)\prime} = \frac{(l+1)s}{l} C_1^{(2)^2} + \frac{s}{l}\left(C_3^{(2)^2} - C_2^{(2)^2}\right) + la_2 C_1^{(2)} + lc_2 C_1^{(1)} + lb_2, \\ C_2^{(2)\prime} = \frac{(l+2)s}{l} C_1^{(2)} C_2^{(2)} + la_2 C_2^{(2)} + lc_2 C_2^{(1)}, \\ C_3^{(2)\prime} = \frac{(l+2)s}{l} C_1^{(2)} C_3^{(2)} + la_2 C_3^{(2)} + lc_2 C_3^{(1)}. \end{array}\right.$$

Example 5. *RD system*

$$\left\{\begin{array}{l} u_t = \left(u^{-\frac{3}{2}} u_x\right)_x + a_1 u + b_1 u^{\frac{5}{2}} + c_1 u^{\frac{5}{2}} v^{-\frac{3}{2}}, \\ v_t = \left(v^{-\frac{3}{2}} v_x\right)_x + a_2 v + b_2 v^{\frac{5}{2}} + c_2 u^{-\frac{3}{2}} v^{\frac{5}{2}} \end{array}\right.$$

admits CLBS

$$\left\{\begin{array}{l} \eta_1 = u_4 - \frac{10}{u} u_1 u_3 - \frac{15}{2u} u_2^2 + \frac{105}{2u^2} u_1^2 u_2 - \frac{315}{8u^3} u_1^4, \\ \eta_2 = v_4 - \frac{10}{v} v_1 v_3 - \frac{15}{2v} v_2^2 + \frac{105}{2v^2} v_1^2 v_2 - \frac{315}{8v^3} v_1^4. \end{array}\right.$$

The solutions of this system are given by

$$\left\{\begin{array}{l} u(x,t) = \left[C_1^{(1)}(t)x^3 + C_2^{(1)}(t)x^2 + C_3^{(1)}(t)x + C_4^{(1)}(t)\right]^{-\frac{2}{3}}, \\ v(x,t) = \left[C_1^{(2)}(t)x^3 + C_2^{(2)}(t)x^2 + C_3^{(2)}(t)x + C_4^{(2)}(t)\right]^{-\frac{2}{3}}, \end{array}\right.$$

where $C_1^{(1)}(t), C_2^{(1)}(t), C_3^{(1)}(t), C_4^{(1)}(t), C_1^{(2)}(t), C_2^{(2)}(t), C_3^{(2)}(t)$ *and* $C_4^{(2)}(t)$ *satisfy the eight-dimensional dynamical system*

$$\begin{cases} C_1^{(1)\prime} = -\frac{3}{2}a_1C_1^{(1)} - \frac{3}{2}c_1C_1^{(2)}, \\ C_2^{(1)\prime} = -\frac{2}{3}C_2^{(1)2} + 2C_1^{(1)}C_3^{(1)} - \frac{3}{2}a_1C_2^{(1)} - \frac{3}{2}c_1C_2^{(2)}, \\ C_3^{(1)\prime} = -\frac{2}{3}C_2^{(1)}C_3^{(1)} + 6C_1^{(1)}C_4^{(1)} - \frac{3}{2}a_1C_3^{(1)} - \frac{3}{2}c_1C_3^{(2)}, \\ C_4^{(1)\prime} = -\frac{2}{3}C_3^{(1)2} + 2C_2^{(1)}C_4^{(1)} - \frac{3}{2}a_1C_4^{(1)} - \frac{3}{2}c_1C_4^{(2)} - \frac{3}{2}b_1, \\ C_1^{(2)\prime} = -\frac{3}{2}a_2C_1^{(2)} - \frac{3}{2}c_2C_1^{(1)}, \\ C_2^{(2)\prime} = -\frac{2}{3}C_2^{(2)2} + 2C_1^{(2)}C_3^{(2)} - \frac{3}{2}a_2C_2^{(2)} - \frac{3}{2}c_2C_2^{(1)}, \\ C_3^{(2)\prime} = -\frac{2}{3}C_2^{(2)}C_3^{(2)} + 6C_1^{(2)}C_4^{(2)} - \frac{3}{2}a_2C_3^{(2)} - \frac{3}{2}c_2C_3^{(1)}, \\ C_4^{(2)\prime} = -\frac{2}{3}C_3^{(2)2} + 2C_2^{(2)}C_4^{(2)} - \frac{3}{2}a_2C_4^{(2)} - \frac{3}{2}c_2C_4^{(1)} - \frac{3}{2}b_2. \end{cases}$$

Example 6. *RD system*

$$\begin{cases} u_t = \left(u^{-\frac{4}{3}}u_x\right)_x + a_1u + b_1u^{\frac{7}{3}} - \frac{3s}{20}u^{-\frac{1}{3}} + c_1u^{\frac{7}{3}}v^{-\frac{4}{3}}, \\ v_t = \left(v^{-\frac{4}{3}}v_x\right)_x + a_2v + b_2v^{\frac{7}{3}} - \frac{3s}{20}v^{-\frac{1}{3}} + c_2v^{\frac{7}{3}}u^{-\frac{4}{3}} \end{cases}$$

admits CLBS

$$\begin{cases} \eta_1 = u_5 - \frac{35}{3u}u_1u_4 + \left(-\frac{70}{3u}u_2 + \frac{700}{9u^2}u_1^2 + s\right)u_3 + \frac{350}{3u^2}u_1u_2^2 \\ \quad - \left(\frac{7s}{u}u_1 + \frac{9100}{27u^3}u_1^3\right)u_2 + \frac{14560}{81u^4}u_1^5 + \frac{70s}{9u^2}u_1^3 + \frac{4s^2}{25}u_1, \\ \eta_2 = v_5 - \frac{35}{3v}v_1v_4 + \left(-\frac{70}{3v}v_2 + \frac{700}{9v^2}v_1^2 + s\right)v_3 + \frac{350}{3v^2}v_1v_2^2 \\ \quad - \left(\frac{7s}{v}v_1 + \frac{9100}{27v^3}v_1^3\right)v_2 + \frac{14560}{81v^4}v_1^5 + \frac{70s}{9v^2}v_1^3 + \frac{4s^2}{25}v_1. \end{cases}$$

The solutions of this system are given as below.

- *For* $s > 0$,

$$\begin{cases} u(x,t) = \left[C_1^{(1)}(t) + C_2^{(1)}(t)\sin\left(\frac{\sqrt{5s}}{5}x\right) + C_3^{(1)}(t)\cos\left(\frac{\sqrt{5s}}{5}x\right)\right. \\ \qquad \left. +C_4^{(1)}(t)\sin\left(\frac{2\sqrt{5s}}{5}x\right) + C_5^{(1)}(t)\cos\left(\frac{2\sqrt{5s}}{5}x\right)\right]^{-\frac{3}{4}}, \\ v(x,t) = \left[C_1^{(2)}(t) + C_2^{(2)}(t)\sin\left(\frac{\sqrt{5s}}{5}x\right) + C_3^{(2)}(t)\cos\left(\frac{\sqrt{5s}}{5}x\right)\right. \\ \qquad \left. +C_4^{(2)}(t)\sin\left(\frac{2\sqrt{5s}}{5}x\right) + C_5^{(2)}(t)\cos\left(\frac{2\sqrt{5s}}{5}x\right)\right]^{-\frac{3}{4}}, \end{cases}$$

where $C_1^{(1)}(t), C_2^{(1)}(t), C_3^{(1)}(t), C_4^{(1)}(t), C_5^{(1)}(t), C_1^{(2)}(t), C_2^{(2)}(t), C_3^{(2)}(t), C_4^{(2)}(t)$ *and* $C_5^{(2)}(t)$ *satisfy the ten-dimensional dynamical system*

$$\begin{cases} C_1^{(1)'} = \frac{s}{5}C_1^{(1)^2} - \frac{3s}{40}C_2^{(1)^2} - \frac{3s}{40}C_3^{(1)^2} - \frac{3s}{5}C_4^{(1)^2} - \frac{3s}{5}C_5^{(1)^2} - \frac{4}{3}a_1C_1^{(1)} - \frac{4}{3}c_1C_1^{(2)} - \frac{4}{3}b_1, \\ C_2^{(1)'} = \frac{3s}{5}C_2^{(1)}C_5^{(1)} - \frac{3s}{5}C_3^{(1)}C_4^{(1)} + \frac{s}{5}C_1^{(1)}C_2^{(1)} - \frac{4}{3}a_1C_2^{(1)} - \frac{4}{3}c_1C_2^{(2)}, \\ C_3^{(1)'} = -\frac{3s}{5}C_2^{(1)}C_4^{(1)} - \frac{3s}{5}C_3^{(1)}C_5^{(1)} + \frac{s}{5}C_1^{(1)}C_3^{(1)} - \frac{4}{3}a_1C_3^{(1)} - \frac{4}{3}c_1C_3^{(2)}, \\ C_4^{(1)'} = \frac{3s}{20}C_2^{(1)}C_3^{(1)} - \frac{2s}{5}C_1^{(1)}C_4^{(1)} - \frac{4}{3}a_1C_4^{(1)} - \frac{4}{3}c_1C_4^{(2)}, \\ C_5^{(1)'} = -\frac{3s}{40}C_2^{(1)^2} + \frac{3s}{40}C_3^{(1)^2} - \frac{2s}{5}C_1^{(1)}C_5^{(1)} - \frac{4}{3}a_1C_5^{(1)} - \frac{4}{3}c_1C_5^{(2)}, \\ C_1^{(2)'} = \frac{s}{5}C_1^{(2)^2} - \frac{3s}{40}C_2^{(2)^2} - \frac{3s}{40}C_3^{(2)^2} - \frac{3s}{5}C_4^{(2)^2} - \frac{3s}{5}C_5^{(2)^2} - \frac{4}{3}a_2C_1^{(2)} - \frac{4}{3}c_2C_1^{(1)} - \frac{4}{3}b_2, \\ C_2^{(2)'} = \frac{3s}{5}C_2^{(2)}C_5^{(2)} - \frac{3s}{5}C_3^{(2)}C_4^{(2)} + \frac{s}{5}C_1^{(2)}C_2^{(2)} - \frac{4}{3}a_2C_2^{(2)} - \frac{4}{3}c_2C_2^{(1)}, \\ C_3^{(2)'} = -\frac{3s}{5}C_2^{(2)}C_4^{(2)} - \frac{3s}{5}C_3^{(2)}C_5^{(2)} + \frac{s}{5}C_1^{(2)}C_3^{(2)} - \frac{4}{3}a_2C_3^{(2)} - \frac{4}{3}c_2C_3^{(1)}, \\ C_4^{(2)'} = \frac{3s}{20}C_2^{(2)}C_3^{(2)} - \frac{2s}{5}C_1^{(2)}C_4^{(2)} - \frac{4}{3}a_2C_4^{(2)} - \frac{4}{3}c_2C_4^{(1)}, \\ C_5^{(2)'} = -\frac{3s}{40}C_2^{(2)^2} + \frac{3s}{40}C_3^{(2)^2} - \frac{2s}{5}C_1^{(2)}C_5^{(2)} - \frac{4}{3}a_2C_5^{(2)} - \frac{4}{3}c_2C_5^{(1)}. \end{cases}$$

- *For $s = 0$,*

$$\begin{cases} u(x,t) = \left[C_1^{(1)}(t)x^4 + C_2^{(1)}(t)x^3 + C_3^{(1)}(t)x^2 + C_4^{(1)}(t)x + C_5^{(1)}(t)\right]^{-\frac{3}{4}}, \\ v(x,t) = \left[C_1^{(2)}(t)x^4 + C_2^{(2)}(t)x^3 + C_3^{(2)}(t)x^2 + C_4^{(2)}(t)x + C_5^{(2)}(t)\right]^{-\frac{3}{4}}, \end{cases}$$

where $C_1^{(1)}(t), C_2^{(1)}(t), C_3^{(1)}(t), C_4^{(1)}(t), C_5^{(1)}(t), C_1^{(2)}(t), C_2^{(2)}(t), C_3^{(2)}(t), C_4^{(2)}(t)$ and $C_5^{(2)}(t)$ satisfy the ten-dimensional dynamical system

$$\begin{cases} C_1^{(1)'} = 2C_1^{(1)}C_3^{(1)} - \frac{3}{4}C_2^{(1)^2} - \frac{4}{3}a_1C_1^{(1)} - \frac{4}{3}c_1C_1^{(2)}, \\ C_2^{(1)'} = -C_2^{(1)}C_3^{(1)} + 6C_1^{(1)}C_4^{(1)} - \frac{4}{3}a_1C_2^{(1)} - \frac{4}{3}c_1C_2^{(2)}, \\ C_3^{(1)'} = \frac{3}{2}C_2^{(1)}C_4^{(1)} + 12C_1^{(1)}C_5^{(1)} - C_3^{(1)^2} - \frac{4}{3}a_1C_3^{(1)} - \frac{4}{3}c_1C_3^{(2)}, \\ C_4^{(1)'} = 6C_2^{(1)}C_5^{(1)} - C_3^{(1)}C_4^{(1)} - \frac{4}{3}a_1C_4^{(1)} - \frac{4}{3}c_1C_4^{(2)}, \\ C_5^{(1)'} = 2C_3^{(1)}C_5^{(1)} - \frac{3}{4}C_4^{(1)^2} - \frac{4}{3}a_1C_5^{(1)} - \frac{4}{3}c_1C_5^{(2)} - \frac{4}{3}b_1, \\ C_1^{(2)'} = 2C_1^{(2)}C_3^{(2)} - \frac{3}{4}C_2^{(2)^2} - \frac{4}{3}a_2C_1^{(2)} - \frac{4}{3}c_2C_1^{(1)}, \\ C_2^{(2)'} = -C_2^{(2)}C_3^{(2)} + 6C_1^{(2)}C_4^{(2)} - \frac{4}{3}a_2C_2^{(2)} - \frac{4}{3}c_2C_2^{(1)}, \\ C_3^{(2)'} = \frac{3}{2}C_2^{(2)}C_4^{(2)} + 12C_1^{(2)}C_5^{(2)} - C_3^{(2)^2} - \frac{4}{3}a_2C_3^{(2)} - \frac{4}{3}c_2C_3^{(1)}, \\ C_4^{(2)'} = 6C_2^{(2)}C_5^{(2)} - C_3^{(2)}C_4^{(2)} - \frac{4}{3}a_2C_4^{(2)} - \frac{4}{3}c_2C_4^{(1)}, \\ C_5^{(2)'} = 2C_3^{(2)}C_5^{(2)} - \frac{3}{4}C_4^{(2)^2} - \frac{4}{3}a_2C_5^{(2)} - \frac{4}{3}c_2C_5^{(1)} - \frac{4}{3}b_2, \end{cases}$$

- *For $s < 0$,*

$$\begin{cases} u(x,t) = \Big[C_1^{(1)}(t) + C_2^{(1)}(t)\sinh\left(\frac{\sqrt{-5s}}{5}x\right) + C_3^{(1)}(t)\cosh\left(\frac{\sqrt{-5s}}{5}x\right) \\ \qquad + C_4^{(1)}(t)\sinh\left(\frac{2\sqrt{-5s}}{5}x\right) + C_5^{(1)}(t)\cosh\left(\frac{2\sqrt{-5s}}{5}x\right)\Big]^{-\frac{3}{4}}, \\ v(x,t) = \Big[C_1^{(2)}(t) + C_2^{(2)}(t)\sinh\left(\frac{\sqrt{-5s}}{5}x\right) + C_3^{(2)}(t)\cosh\left(\frac{\sqrt{-5s}}{5}x\right) \\ \qquad + C_4^{(2)}(t)\sinh\left(\frac{2\sqrt{-5s}}{5}x\right) + C_5^{(2)}(t)\cosh\left(\frac{2\sqrt{-5s}}{5}x\right)\Big]^{-\frac{3}{4}}, \end{cases}$$

where $C_1^{(1)}(t), C_2^{(1)}(t), C_3^{(1)}(t), C_4^{(1)}(t), C_5^{(1)}(t), C_1^{(2)}(t), C_2^{(2)}(t), C_3^{(2)}(t), C_4^{(2)}(t)$ *and* $C_5^{(2)}(t)$ *satisfy the ten-dimensional dynamical system*

$$\left\{\begin{array}{l}
C_1^{(1)\prime} = \frac{s}{5}C_1^{(1)^2} + \frac{3s}{40}C_2^{(1)^2} - \frac{3s}{40}C_3^{(1)^2} + \frac{3s}{5}C_4^{(1)^2} - \frac{3s}{5}C_5^{(1)^2} - \frac{4}{3}a_1C_1^{(1)} - \frac{4}{3}c_1C_1^{(2)} - \frac{4}{3}b_1, \\
C_2^{(1)\prime} = \frac{3s}{5}C_2^{(1)}C_5^{(1)} - \frac{3s}{5}C_3^{(1)}C_4^{(1)} + \frac{s}{5}C_1^{(1)}C_2^{(1)} - \frac{4}{3}a_1C_2^{(1)} - \frac{4}{3}c_1C_2^{(2)}, \\
C_3^{(1)\prime} = \frac{3s}{5}C_2^{(1)}C_4^{(1)} - \frac{3s}{5}C_3^{(1)}C_5^{(1)} + \frac{s}{5}C_1^{(1)}C_3^{(1)} - \frac{4}{3}a_1C_3^{(1)} - \frac{4}{3}c_1C_3^{(2)}, \\
C_4^{(1)\prime} = \frac{3s}{20}C_2^{(1)}C_3^{(1)} - \frac{2s}{5}C_1^{(1)}C_4^{(1)} - \frac{4}{3}a_1C_4^{(1)} - \frac{4}{3}c_1C_4^{(2)}, \\
C_5^{(1)\prime} = \frac{3s}{40}C_2^{(1)^2} + \frac{3s}{40}C_3^{(1)^2} - \frac{2s}{5}C_1^{(1)}C_5^{(1)} - \frac{4}{3}a_1C_5^{(1)} - \frac{4}{3}c_1C_5^{(2)}, \\
C_1^{(2)\prime} = \frac{s}{5}C_1^{(2)^2} + \frac{3s}{40}C_2^{(2)^2} - \frac{3s}{40}C_3^{(2)^2} + \frac{3s}{5}C_4^{(2)^2} - \frac{3s}{5}C_5^{(2)^2} - \frac{4}{3}a_2C_1^{(2)} - \frac{4}{3}c_2C_1^{(1)} - \frac{4}{3}b_2, \\
C_2^{(2)\prime} = \frac{3s}{5}C_2^{(2)}C_5^{(2)} - \frac{3s}{5}C_3^{(2)}C_4^{(2)} + \frac{s}{5}C_1^{(2)}C_2^{(2)} - \frac{4}{3}a_2C_2^{(2)} - \frac{4}{3}c_2C_2^{(1)}, \\
C_3^{(2)\prime} = \frac{3s}{5}C_2^{(2)}C_4^{(2)} - \frac{3s}{5}C_3^{(2)}C_5^{(2)} + \frac{s}{5}C_1^{(2)}C_3^{(2)} - \frac{4}{3}a_2C_3^{(2)} - \frac{4}{3}c_2C_3^{(1)}, \\
C_4^{(2)\prime} = \frac{3s}{20}C_2^{(2)}C_3^{(2)} - \frac{2s}{5}C_1^{(2)}C_4^{(2)} - \frac{4}{3}a_2C_4^{(2)} - \frac{4}{3}c_2C_4^{(1)}, \\
C_5^{(2)\prime} = \frac{3s}{40}C_2^{(2)^2} + \frac{3s}{40}C_3^{(2)^2} - \frac{2s}{5}C_1^{(2)}C_5^{(2)} - \frac{4}{3}a_2C_5^{(2)} - \frac{4}{3}c_2C_5^{(1)}.
\end{array}\right.$$

6. Conclusions

The method of linear determining equations to construct DC (3) and CLBS (5) of two-component second-order evolution system (4) is provided. The linear determining equations (23) and (24) generalize the classical determining equations within the framework of Lie's operator. The general form of CLBS (5) and DC (3) admitted by the system (4) can be identified by solving the resulting linear determining equations.

As an application of this approach, the general form of DC (3) and CLBS (5) with $n = 3, 4, 5$ of RD system (9) is established in this paper. The reductions of the resulting equations are also constructed due to the compatibility of the admitted DC (3) and the governing system (9). These reductions cannot be obtained within the framework of Lie's classical symmetry method and conditional symmetry method.

All examples except Example 4 in Section 5 involve the power diffusivities with the exponent either $-4/3$ or $-2/3$. Exact solutions of the nonlinear diffusion equations $u_t = (u^{-4/3}u_x)_x$ and $u_t = (u^{-2/3}u_x)_x$ are firstly studied by using local and non-local symmetries by King [48]. The polynomial solutions like the ones in the examples of Section 5 for scalar nonlinear diffusion equations are also constructed by King [49,50]. Moreover, a range of more complicated exact solutions for scalar nonlinear diffusion equations are derived by Cherniha [15] due to the method of the additional generating condition. In addition, the results of Examples 4, 5 and the case of $s = 0$ for Example 6 in Section 5 have been given by Cherniha and King [16] by using the method of the additional generating condition. All of the reductions of the obtained RD system (9) constructed in Section 5, involving either a polynomial, trigonometric or hyperbolic function, are used in [14] for the first time within the framework of the method of the additional generating condition.

The method of linear determining equations can be extended to consider DCs and CLBSs of other types of evolution systems, including a multi-component diffusion system and a high-order evolution system. The discussion about the linear determining equation for evolution system (4) to identify CLBS and DC with η_1 and η_2 possessing different orders is another interesting problem. All of these problems will be involved in our future research.

Acknowledgments: This work was supported by NSFC (Grant No. U1204104 and No. 11501175) and the National Science Foundation for Post-doctoral Scientists of China (Grant No. 2014M561454).

Conflicts of Interest: The author declares no conflict of interest.

References

1. Yanenko, N.N. Theory of consistency and methods of integrating systems of nonlinear partial differential equations. In *Proceedings of the 4th All-Union Mathematical Congress*; Nauka: Leningrad, USSR, 1964; pp. 247–259. (In Russian)
2. Sidorov, A.F.; Shapeev, V.P.; Yanenko, N.N. *Method of Differential Constraints and Its Application to Gas Dynamics*; Nauka: Novosibirsk, Russia, 1984.
3. Olver, P.J.; Rosenau, P. The construction of special solutions to partial differential equations. *Phys. Lett. A* **1986**, *144*, 107–112.
4. Bluman, G.W.; Kumei, S. *Symmetries and Differential Equations*; Springer: New York, NY, USA, 1989.
5. Olver, P.J. *Applications of Lie Groups to Differential Equations*, 2nd ed.; Springer: New York, NY, USA, 1993.
6. Bluman, G.W.; Cole, J.D. The general similarity solution of the heat equation. *J. Math. Mech.* **1969**, *18*, 1025–1042.
7. Ovsiannikov, L.V. *Group Analysis of Differential Equations*; Academic Press: New York, NY, USA, 1982.
8. Pucci, E.; Saccomandi, G. Evolution equations, invariant surface conditions and functional separation of variables. *Phys. D* **2000**, *139*, 28–47.
9. Clarkson, P.; Kruskal, M. New similarity reductions of the Boussinesq equation. *J. Math. Phys.* **1989**, *30*, 2201–2213.
10. Miller, W., Jr. Mechanism for variable separation in partial differential equations and their relationship to group theory. In *Symmetries and Nonlinear Phenomena*; Levi, D., Winternitz, P., Eds.; World Scientific: London, UK, 1989.
11. Galaktionov, V.A. On new exact blow-up solutions for nonlinear heat conduction equations with source and applications. *Differ. Integral Equ.* **1990**, *3*, 863–874.
12. Olver, P.J.; Vorob'ev, E.M. Nonclassical and conditional symmetries. In *CRC Handbook of Lie Group Analysis*; Ibragiminov, N.H., Ed.; CRC Press: Boca Raton, FL, USA, 1994; Volume 3.
13. Olver, P.J.; Rosenau, P. Group invariant solutions of differential equations. *SIAM J. Appl. Math.* **1987**, *47*, 263–278.
14. Cherniha, R. A constructive method for construction of new exact solutions of nonlinear evolution equations. *Rep. Math. Phys.* **1996**, *38*, 301–312.
15. Cherniha, R. New non-Lie ansatze and exact solutions of nonlinear reaction-diffusion-convection equations. *J. Phys. A Math. Gen.* **1998**, *31*, 8179–8198.
16. Cherniha, R.; King, J.R. Non-linear reaction-diffusion systems with variable diffusivities: Lie symmetries, ansatze and exact solutions. *J. Math. Anal. Appl.* **2005**, *308*, 11–35.
17. Cherniha, R.; Myroniuk, L. New exact solutions of a nonlinear cross-diffusion system. *J. Phys. A Math. Theor.* **2008**, *41*, 395204.
18. Galaktionov, V.A.; Svirshchevskii, S.R. *Exact Solutions and Invariant Subapaces of Nonlinear Partial Differential Equations in Mechanics and Physics*; Chapman and Hall: London, UK, 2007.
19. Olver, P.J. Direct reduction and differential constraints. *Proc. Math. Phys. Sci.* **1994**, *444*, 509–523.
20. Kunzinger, M.; Popovych, R.O. Generalized conditional symmetries of evolution equaitons. *J. Math. Anal. Appl.* **2011**, *379*, 444–460.
21. Wang, J.P.; Ji, L.N. Conditional Lie–Bäcklund symmetry, second-order differential constraint and direct reduction of diffusion systems. *J. Math. Anal. Appl.* **2015**, *427*, 1101–1118.
22. Zhdanov, R.Z. Conditional Lie–Bäcklund symmetry and reduction of evolution equation. *J. Phys. A Math. Gen.* **1995**, *28*, 3841–3850.
23. Fokas, A.S.; Liu, Q.M. Nonlinear interaction of traveling waves of nonintegrable equations. *Phys. Rev. Lett.* **1994**, *72*, 3293–3296.
24. Fokas, A.S.; Liu, Q.M. Generalized conditional symmetries and exact solutions of non-integrable equations. *Theor. Math. Phys.* **1994**, *99*, 571–582.
25. Liu, Q.M.; Fokas, A.S. Exact interaction of solitary waves for certain non-integrable equations. *J. Math. Phys.* **1996**, *37*, 324–345.
26. Liu, Q.M. Exact solutions to nonlinear equations with quadratic nonlinearity. *J. Phys. A Math. Gen.* **2000**, *34*, 5083–5088.

27. Sergyeyev, A. Constructing conditionally integrable evolution systems in (1 + 1)-dimensions: A generalization of invariant modules approach. *J. Phys. A Math. Gen.* **2002**, *35*, 7563–7660.
28. Ji, L.N.; Qu, C.Z.; Shen, S.F. Conditional Lie–Bäcklund symmetry of evolution system and application for reaction-diffusion system. *Stud. Appl. Math.* **2014**, *133*, 118–149.
29. Qu, C.Z. Exact solutions to nonlinear diffusion equations obtained by generalized conditional symmetry. *IMA J. Appl. Math.* **1999**, *62*, 283–302.
30. Qu, C.Z. Group classification and generalized conditional symmetry reduction of the nonlinear diffusion-convection equation with a nonlinear source. *Stud. Appl. Math.* **1997**, *99*, 107–136.
31. Qu, C.Z.; Ji, L.N.; Wang, L.Z. Conditional Lie–Bäcklund symmetries and sign-invarints to quasi-linear diffusion equations. *Stud. Appl. Math.* **2007**, *119*, 355–391.
32. Ji, L.N.; Qu, C.Z. Conditional Lie–Bäcklund symmetries and solutions to (n + 1)-dimensional nonlinear diffusion equations. *J. Math. Phys.* **2007**, *484*, 103509.
33. Ji, L.N.; Qu, C.Z.; Ye, Y.J. Solutions and symmetry reductions of the n-dimensional nonlinear convection-diffusion equations. *IMA J. Appl. Math.* **2010**, *75*, 17–55.
34. Qu, C.Z.; Zhang, S.L.; Liu, R.C. Separation of vairables and exact solutions to quasilinear diffusion equations with nonlinear source. *Phys. D* **2000**, *144*, 97–123.
35. Ji, L.N.; Qu, C.Z. Conditional Lie–Bäcklund symmetries and invariant subspaces to nonlinear diffusion equations. *IMA J. Appl. Math.* **2011**, *76*, 610–632.
36. Ji, L.N.; Qu, C.Z. Conditional Lie–Bäcklund symmetries and invariant subspaces to nonlinear diffusion equaitons with convection and source. *Stud. Appl. Math.* **2013**, *131*, 266–301.
37. Qu, C.Z.; Ji, L.N. Invariant subspaces and condtioanl Lie–Bäcklund symmetries of inhomogeneous nonlinear diffusion equations. *Sci. China Math.* **2013**, *56*, 2187–2203.
38. Qu, C.Z.; Zhu, C.R. Classification of coupled systems with two-component nonlinear diffusion equations by the invariant subspace method. *J. Phys. A Math. Theor.* **2009**, *42*, 475201, doi:10.1088/1751-8113/42/47/475201.
39. Zhu, C.R.; Qu, C.Z. Maximal dimension of invariant subspaces admitted by nonlinear vector differential operators. *J. Math. Phys.* **2011**, *52*, 043507, doi: 10.1063/1.3574534.
40. Kaptsov, O.V. Linear determining equations for differential constraints. *Sb. Math.* **1998**, *189*, 1839–1854.
41. Kaptsov, O.V.; Verevkin, I.V. Differential constraints and exact solutions of nonlinear diffusion equations. *J. Phys. A Math. Gen.* **2003**, *36*, 1401–1414.
42. Ji, L.N. Conditional Lie–Bäcklund symmetries and differential constraints for inhomogeneous nonlinear diffusion equations due to linear determining equations. *J. Math. Anal. Appl.* **2016**, *440*, 286–299.
43. Cherniha, R.; Pliukhin, O. New conditional symmetries and exact solutions of reaction-diffusion systems with power diffusivities. *J. Phys. A Math. Theor.* **2008**, *41*, 185208, doi:10.1088/1751-8113/41/18/185208.
44. Cherniha, R. Conditional symmetries for systems of PDEs: New definitions and their applications for reaction-diffusion systems. *J. Phys. A Math. Theor.* **2010**, *43*, 405207, doi:10.1088/1751-8113/43/40/405207.
45. Cherniha, R.; Davydovych, V. Conditional symmetries and exact solutions of nonlinear reaction-diffusion systems with non-constant diffusibities. *Commun. Nonlinear Sci. Numer. Simul.* **2012**, *17*, 3177–3188.
46. Cherniha, R.; Davydovych, V. Nonlinear reaction-diffusion systems with a non-constant diffusivities: Conditional symmetries in no-go case. *Appl. Math. Comput.* **2015**, *268*, 23–34.
47. Andreev, V.K.; Kaptsov, O.V.; Pukhnachev, V.V.; Rodionov, A.A. *Applications of Group-Theoretic Methods in Hydrodynamics*; Kluwer: Dordrecht, The Netherlands, 1998.
48. King, J.R. Exact results for the nonlinear diffusion equations $u_t = (u^{-4/3}u_x)_x$ and $u_t = (u^{-2/3}u_x)_x$. *J. Phys. A Math. Gen.* **1991**, *24*, 5721–5745.
49. King, J.R. Exact polynomial solutions to some nonlinear diffusion equations. *Phys. D* **1993**, *64*, 35–65.
50. King, J.R. Exact multidimensional solutions to some nonlinear diffusion equations. *Q. J. Mech. Appl. Math.* **1993**, *46*, 419–436.

symmetry

MDPI

Article

Invariant Subspaces of the Two-Dimensional Nonlinear Evolution Equations

Chunrong Zhu [1] and Changzheng Qu [2,*]

[1] College of Mathematics and Computer Science, Anhui Normal University, Wuhu 241000, Anhui, China; zcr2009@mail.ahnu.edu.cn

[2] Center for Nonlinear Studies, Ningbo University, Ningbo 315211, Zhejiang, China

* Correspondence: quchangzheng@nbu.edu.cn; Tel.: +86-574-87609976

Academic Editor: Roman M. Cherniha

Received: 1 September 2016; Accepted: 7 November 2016; Published: 15 November 2016

Abstract: In this paper, we develop the symmetry-related methods to study invariant subspaces of the two-dimensional nonlinear differential operators. The conditional Lie–Bäcklund symmetry and Lie point symmetry methods are used to construct invariant subspaces of two-dimensional differential operators. We first apply the multiple conditional Lie–Bäcklund symmetries to derive invariant subspaces of the two-dimensional operators. As an application, the invariant subspaces for a class of two-dimensional nonlinear quadratic operators are provided. Furthermore, the invariant subspace method in one-dimensional space combined with the Lie symmetry reduction method and the change of variables is used to obtain invariant subspaces of the two-dimensional nonlinear operators.

Keywords: symmetry group; invariant subspace; conditional Lie–Bäcklund symmetry; finite-dimensional dynamical system; nonlinear differential operator

MSC: 37K35; 37K25; 53A55

1. Introduction

The invariant subspace method is an effective one to perform reductions of nonlinear partial differential equations (PDEs) to finite-dimensional dynamical systems. In [1], Galaktionov and Svirshchevskii provide a systematic account of this approach and its various applications for a large variety of nonlinear PDEs. They also addressed some fundamental and open questions on the invariant subspaces of nonlinear PDEs. Many interesting results were obtained in this book. In [2–20], the extensions of the invariant subspace method and various applications to other nonlinear PDEs were also discussed. It is noticed that a large number of exact solutions, such as N-solitons of integrable equations, similarity solutions of nonlinear evolution equations and the generalized functional separable solutions to nonlinear PDEs, can be recovered by the invariant subspace methods [1,21–31]. In the one-dimensional space case, the invariant subspace method can be implemented by the conditional Lie–Bäcklund symmetry introduced independently by Zhdanov [32] and Fokas-Liu [33]. A key point for the invariant subspace approach is the estimate of maximal dimension of the invariant subspaces [1,5,6,15,16]. It was shown in [1,5] that for k-th order one-dimensional nonlinear operator of the form:

$$F[u] = F(x, u, u_x, \cdots, u^{(k)})$$

where $u^{(k)} = \partial^k u/\partial x^k$, the dimension of their invariant subspaces is bounded by $2k+1$. Such an estimate can be extended to the k-th order m-component nonlinear vector operators:

$$\vec{F}[\vec{u}] = \vec{F}(x, \vec{u}, \vec{u}_x, \cdots, \vec{u}^{(k)}). \quad (1)$$

In [15], we proved that the maximal dimension of the invariant subspaces for operator (1) is bounded by $2mk+1$. This enables us to determine the maximal dimension preliminarily of the invariant subspaces of the nonlinear evolution equations. In contrast with the one-dimensional space case, only very limited results on the invariant subspaces of multi-dimensional PDEs were obtained. These results were obtained mostly by the ansatz-based method, and there are no systematic approaches to obtain these results. As mentioned in [1], the general problem of finding invariant subspaces for a wide class of nonlinear differential operators in the multi-dimensional case is not completely solved. A open question still remains: what is the maximal dimension of the two-dimensional k-th order scalar nonlinear operators of the form:

$$F[u] = F(x, y, u, u_x, u_y, u_{xx}, u_{xy}, u_{yy}, \cdots, u^{(k)}),$$

where $u^{(k)} = \partial^{r+s}u/\partial x^r \partial y^s$, $r+s=k$ denotes all k-th order derivatives with respect to x and y?

It is of great interest to develop the invariant subspace method to study the multi-dimensional nonlinear evolution equations. Indeed, there are a number of examples whose exact solutions can be derived from the invariant subspace method; please refer to [1,2] for more examples on invariant subspaces of the $2+1$-dimensional nonlinear evolution equations. For instance, it is discovered that the operators:

$$J[u] = u\Delta_2 u - |\nabla u|^2, \quad (x,y) \in \mathbb{R}^2$$

and:

$$Q[u] = u\Delta_2^2 u - (\Delta_2 u)^2 + 2 \bigtriangledown u \bigtriangledown \Delta_2 u, \quad (x,y) \in \mathbb{R}^2$$

with $\Delta_2 = \partial_x^2 + \partial_y^2$ admit the following invariant subspaces:

$$\begin{aligned} W_6 &= \mathcal{L}\{1, x, y, x^2, y^2, xy\}, \\ W_6 &= \mathcal{L}\{1, \cosh x, \cos y, \cosh(2x), \cos(2y), \cosh x \cos y\}, \\ W_{91} &= \mathcal{L}\{1, x, y, x^2+y^2, xy, xr^2, yr^2, r^4\}, \quad r^2 = x^2+y^2, \\ W_{92} &= \mathcal{L}\{1, \cosh(2x), \sinh(2x), \cos(2y), \sin(2y), \cosh x \cos y, \sinh x \cos y, \cosh x \sin y, \sinh x \sin y\}. \end{aligned}$$

It was proven in [1] that the quadratic operator defined in $\mathbb{R}^N$:

$$K[u] = \alpha(\Delta_n u)^2 + \beta u \Delta_n u + \gamma |\nabla u|^2, \quad x \in \mathbb{R}^N$$

admits the invariant subspaces:

$$\begin{aligned} W_2^r &= \mathcal{L}\{1, |x|^2, \}, \\ W_{N+1}^q &= \mathcal{L}\{1, x_1^2, x_2^2, \cdots, x_N^2\}, \\ W_n^q &= \mathcal{L}\{1, x_i x_j, 1 \le i, j \le N\}, \quad n = \frac{N(N+1)}{2} + 1, \\ W_N^{lin} &= \mathcal{L}\{x_1, x_2, \cdots, x_N\} \end{aligned}$$

and the direct sum of subspaces:

$$W_N^{lin} \bigoplus W_n^q.$$

The purpose of this paper is to develop symmetry-related method to study invariant subspaces of nonlinear evolution equations in the two- or multi-dimensional case. The outline of this paper is as follows. In Section 2, we first give two direct extensions of the concept of invariant subspace in

$\mathbb{R}^2$. Then, the algorithm of this approach will be shown by looking for the invariant subspaces of the operator:

$$A[u] \equiv \alpha(\Delta_2 u)^2 + \gamma u \Delta_2 u + \delta|\nabla u|^2 + \varepsilon u^2 \quad \text{in} \quad \mathbb{R}^2,$$

where α, γ, δ and ε are constants, and $\alpha^2 + \gamma^2 + \delta^2 + \varepsilon^2 \neq 0$. In Section 3, the general description of the changes of variables for the two-dimensional invariant subspace method is given, which can be regarded as an extension to the invariant subspace method in the one-dimensional case. Since the two-dimensional nonlinear evolution equations can be reduced to one-dimensional equations by the Lie symmetry method, this fact combined with the invariant subspace method in the one-dimensional case will be used to obtain invariant subspaces of the corresponding two-dimensional nonlinear operators, which will be discussed in Section 4. As an example, we obtain many new invariant subspaces admitted by a quadratic differential operator $J[u]$. Section 5 is the concluding remarks on this work.

2. Direct Extensions of Invariant Subspaces

2.1. Direct Extensions in $\mathbb{R}^2$

Let us first give a brief account of the invariant subspace method as presented in [1]. Consider the general evolution equation:

$$u_t = F(x, u, u_x, u_{xx}, \cdots, u^{(k)}) \equiv F[u], \quad x \in \mathbb{R} \tag{2}$$

where F is a k-th-order ordinary differential operator with respect to the variable x and $F(\cdot)$ is a given sufficiently smooth function of the indicated variables. Let $\{f_i(x),\ i = 1, \cdots, n\}$ be a finite set of $n \geqslant 1$ linearly independent functions, and W_n^x denotes their linear span $W_n^x = \mathcal{L}\{f_1(x), \cdots, f_n(x)\}$. The subspace W_n^x is said to be invariant under the given operator F, if $F[W_n^x] \subseteq W_n^x$, and then operator F is said to preserve or admit W_n^x, which means:

$$F[\sum_{i=1}^{n} C_i f_i(x)] = \sum_{i=1}^{n} \Psi_i(C_1, \cdots, C_n) f_i(x)$$

for any $C(t) = (C_1(t), \cdots, C_n(t)) \in \mathbb{R}^n$, where Ψ_i are the expansion coefficients of $F[u] \in W_n^x$ in the basis $\{f_i\}$. It follows that if the linear subspace W_n^x is invariant with respect to F, then Equation (2) has solutions of the form:

$$u(t,x) = \sum_{i=1}^{n} C_i(t) f_i(x),$$

where $C_i(t)$ satisfy the n-dimensional dynamical system:

$$C_i' = \Psi_i(C_1, \cdots, C_n), \quad i = 1, \cdots, n.$$

Moreover, assume that the invariant subspace W_n^x is defined as the space of solutions of the linear n-th-order ODE:

$$L_x[v] \equiv \frac{d^n v}{dx^n} + a_{n-1}(x)\frac{d^{n-1} v}{dx^{n-1}} + \cdots + a_1(x)\frac{dv}{dx} + a_0(x)v = 0. \tag{3}$$

If the operator $F[u]$ admits the invariant subspace W_n^x, then the invariant condition with respect to F takes the form:

$$L_x[F[u]]|_{[H]} \equiv 0, \tag{4}$$

where $[H]$ denotes the equation $L_x[u] = 0$ and its differential consequences with respect to x. The invariant condition leads to the following theorem on the maximal dimension of an invariant subspace preserved by the operator F.

Theorem 1. *[1] If a linear subspace W_n^x determined by the space of solutions of linear Equation (3) is invariant under a nonlinear differential operator F of order k, then:*

$$n \leqslant 2k + 1.$$

It is inferred from Equation (4) and the invariant criteria for conditional Lie–Bäcklund symmetry [32,33] that Equation (2) admits the conditional Lie–Bäcklund symmetry:

$$\sigma = L_x[u].$$

To look for the exact solutions of the form:

$$u(t,x,y) = \sum_{i,j} C_{ij}(t) f_i(x) g_j(y) \tag{5}$$

of the two-dimensional nonlinear evolution equations:

$$u_t = F[u] \equiv F(x, y, u, u_x, u_y, u_{xx}, u_{xy}, u_{yy}, \cdots, u^{(k)}), \tag{6}$$

we now introduce the linear subspace:

$$\begin{aligned} W_{nm}^{xy} &= \mathcal{L}\{f_1(x)g_1(y), \cdots, f_n(x)g_1(y), \cdots, f_1(x)g_m(y), \cdots, f_n(x)g_m(y)\} \\ &\equiv \{\sum_{i,j} C_{ij} f_i(x) g_j(y),\ \forall (C_{11}, \cdots, C_{1m}, \cdots, C_{n1}, \cdots, C_{nm}) \in \mathbb{R}^{nm}\} \end{aligned}$$

as an extension to W_n^x. Assume that $F[u] = F(x, y, u, u_x, u_y, u_{xx}, u_{xy}, u_{yy}, \cdots, u^{(k)})$ is a k-th-order differential operator with respect to the variables x and y, and $\{g_j(y),\ j = 1, \cdots, m\}$ is a finite set of $m \geqslant 1$ linearly independent functions of variable y. It is easy to see that the space $\{f_i(x)g_j(y),\ i = 1, \cdots, n,\ j = 1, \cdots, m\}$ is also a set of linearly independent functions. Let W_m^y denote the linear span of the set $\{g_j(y),\ j = 1, \cdots, m\}$, i.e., $W_m^y = \mathcal{L}\{g_1(y), \cdots, g_m(y)\}$. Similarly, the space W_m^y is defined as the space of solutions of the linear m-th-order ODE:

$$L_y[w] \equiv \frac{d^m w}{dy^m} + b_{m-1}(y)\frac{d^{m-1}w}{dy^{m-1}} + \cdots + b_1(y)\frac{dw}{dy} + b_0(y)w = 0. \tag{7}$$

If $u \in W_{nm}^{xy}$, then there exists a vector $(C_{11}(t), \cdots, C_{1m}(t), \cdots, C_{n1}(t), \cdots, C_{nm}(t)) \in \mathbb{R}^{nm}$, such that:

$$u = \sum_{i,j} C_{ij}(t) f_i(x) g_j(y). \tag{8}$$

We rewrite u as:

$$u = \sum_{i=1}^{n} (\sum_{j=1}^{m} (C_{ij}(t) g_j(y))) f_i(x) = \sum_{j=1}^{m} (\sum_{i=1}^{n} (C_{ij}(t) f_i(x))) g_j(y),$$

which means that:

$$L_x[u] = 0, \text{ and } L_y[u] = 0. \tag{9}$$

On the other hand, if the function $u = u(t,x,y)$ satisfies the condition (9), then u has the form (8). Indeed, $L_x[u] = 0$ means that there exists a vector function $(C_1(t,y), \cdots, C_n(t,y))$, such that:

$$u = \sum_{i=1}^{n} C_i(t,y) f_i(x),$$

while $L_y[u] = 0$ means that:

$$L_y[u] = L_y[\sum_{i=1}^{n} C_i(t,y) f_i(x)] = \sum_{i=1}^{n} f_i(x) L_y[C_i(t,y)] = 0.$$

Since $f_i(x)$ $(i = 1, \cdots, n)$ are linearly independent, the above equation leads to:

$$L_y[C_i(t,y)] = 0, \; i = 1, \cdots, n.$$

Hence, there exists a set of vectors $(C_{i1}(t), \cdots, C_{im}(t)) \in \mathbb{R}^m$, such that:

$$C_i(t,y) = \sum_{j=1}^{m} C_{ij}(t) g_j(y), \; i = 1, \cdots, n.$$

As above, we are able to obtain the invariance condition of the subspace W_{nm}^{xy} with respect to F, i.e., $F[W_{nm}^{xy}] \subseteq W_{nm}^{xy}$, which takes the form:

$$L_x[F[u]]|_{[H_x] \cap [H_y]} \equiv 0, \text{ and } L_y[F[u]]|_{[H_x] \cap [H_y]} \equiv 0, \tag{10}$$

where $[H_x] \cap [H_y]$ denotes $L_x[u] = 0$, $L_y[u] = 0$, and their differential consequences with respect to x and y. If $F[u]$ admits the invariant subspace W_{nm}^{xy}, then Equation (6) has solutions (5) and can be reduced to an nm-dimensional dynamic system.

We next consider a special case of the function (5). If $1 \in W_n^x \cap W_m^y$, then $a_0(x) = 0$ in (3) and $b_0(y) = 0$ in (7). Without loss of generality, we assume $f_1(x) = 1$ and $g_1(y) = 1$. Note that the function of the form:

$$u(t,x,y) = C_1(t) + \sum_{i=2}^{n} C_i(t) f_i(x) + \sum_{j=2}^{m} B_j(t) g_j(y) \tag{11}$$

is a special case of (5), which is a separable function with respect to spacial variables x and y. We denote:

$$\begin{aligned} W_{n+m-1}^{xy} &= \mathcal{L}\{1, f_2(x), \cdots, f_n(x), g_2(y), \cdots, g_m(y)\} \\ &\equiv \left\{ C_1(t) + \sum_{i=2}^{n} C_i(t) f_i(x) + \sum_{j=2}^{m} B_j(t) g_j(y) \right\}, \end{aligned}$$

which is a linear span of the set $\{1, f_i(x), g_j(y), i = 2, \cdots, n, j = 2, \cdots, m\}$. Clearly, if $u \in W_{n+m-1}^{xy}$, then:

$$L_x[u] = 0, \;\; L_y[u] = 0, \;\; \text{and } u_{xy} = 0. \tag{12}$$

On the other hand, if $u_{xy} = 0$, then the function u has the form:

$$u = f(t,x) + g(t,y).$$

From $L_x[u] = 0$ (notice that $a_0(x) = 0$), we obtain:

$$L_x[f(t,x) + g(t,y)] = L_x[f(t,x)] = 0,$$

which means that there exists a vector $(A_1(t), C_2(t), \cdots, C_n(t))$, such that:

$$f(t,x) = A_1(t) + \sum_{i=2}^{n} C_i(t) f_i(x).$$

Similarly, $L_y[u] = 0$ leads to:

$$g(t,y) = B_1(t) + \sum_{j=2}^{m} B_j(t) g_j(y),$$

where $B_j (j = 1, \cdots, m)$ are functions of t. We denote $C_1 = A_1 + B_1$. Hence, $u \in W^{xy}_{n+m-1}$ if and only if u satisfies the condition (12). Then, we can obtain the invariance condition of the subspace W^{xy}_{n+m-1} with respect to F, i.e., $F[W^{xy}_{n+m-1}] \subseteq W^{xy}_{n+m-1}$, which takes the form:

$$L_x[F[u]]|_{[H]} \equiv 0, \quad L_y[F[u]]|_{[H]} \equiv 0, \quad \text{and } (F[u])_{xy}|_{[H]} \equiv 0. \tag{13}$$

where $[H]$ denotes the set $\{L_x[u] = 0\} \cap \{L_y[u] = 0\} \cap \{u_{xy} = 0\}$, and their differential consequences with respect to x and y. In this case, Equation (6) has the solution of the form (11) and can be reduced to an $(n + m - 1)$-dimensional dynamic system.

Assume that the k-th-order differential operator $F[u]$, including the term $\partial^k u / \partial x^k$, admits the invariant subspace W^{xy}_{nm} (or W^{xy}_{n+m-1}), and note that the operator $F[u]$ can also be regarded as a differential operator only with respect to x; the first identity in the condition (10) (or (13)) leads to the estimate $n \leqslant 2k + 1$. The same estimate is also true for m.

Remark 1. *It is noted that the W^{xy}_{mn} and W^{xy}_{n+m-1} demonstrate two special forms of invariant subspaces of the operator $F[u]$. The general form can be introduced as below, which will be used in the following sections.*

Let $\{f_i(x,y),\ i = 1, \cdots, n\}$ be a finite set of $n \geqslant 1$ linearly independent functions, and W_n denote their linear span $W_n = \mathcal{L}\{f_1(x,y), \cdots, f_n(x,y)\}$. The subspace W_n is said to be invariant under the given operator $F[u]$, if $F[W_n] \subseteq W_n$, and then, operator $F[u]$ is said to preserve or admit W_n.

2.2. *Invariant Subspaces of a Quadratic Operator in $\mathbb{R}^2$*

Consider the quadratic operator:

$$A[u] \equiv \alpha(\Delta_2 u)^2 + \gamma u \Delta_2 u + \delta |\nabla u|^2 + \varepsilon u^2.$$

We will look for the invariant subspaces W^{xy}_{n+m-1} and W^{xy}_{nm} of $A[u]$. Note that the operator $A[u]$ is symmetric with respect to x and y; we assume that $n = m$. The cases of $n = 2,\ 3,\ 4,\ 5$ will be considered respectively. In the rest of this paper, the following notations will be used:

$$u_{r0} = \frac{\partial^r u}{\partial x^r}, \quad u_{0s} = \frac{\partial^s u}{\partial y^s}, \quad u_{rs} = \frac{\partial^{r+s} u}{\partial x^r \partial y^s}, \quad r, s = 1, 2, \cdots.$$

2.2.1. The Space W^{xy}_{n+n-1}

We first consider the case of $n = 3$. In this case, we look for the invariant subspaces W^{xy}_{3+3-1} of the operator $A[u]$, which are determined by the following ODEs:

$$L_x^3[v] \equiv \frac{d^3 v}{dx^3} + a_2 \frac{d^2 v}{dx^2} + a_1 \frac{dv}{dx} = 0, \quad L_y^3[w] \equiv \frac{d^3 w}{dy^3} + b_2 \frac{d^2 w}{dy^2} + b_1 \frac{dw}{dy} = 0. \tag{14}$$

Here and hereafter, a_i, b_i are constants. The invariant conditions take the form:

$$G_1 = L_x^3[A[u]]|_{[H]} \equiv 0, \;\; G_2 = L_y^3[A[u]]|_{[H]} \equiv 0, \;\; \text{and } G_3 = (A[u])_{xy}|_{[H]} \equiv 0, \tag{15}$$

where $[H]$ denotes the set $\{L_x^3[u] = 0\} \cap \{L_y^3[u] = 0\} \cap \{u_{xy} = 0\}$ and their differential consequences with respect to x and y.

Substituting $A[u]$ into (15), we obtain:

$$\begin{aligned}
G_1 =& (-4a_2\delta - 4a_2^3\alpha - 3a_2\gamma + 6a_1\alpha a_2)u_{20}^2 \\
&+ (6\varepsilon - 6a_1\gamma - 8a_2^2\alpha a_1 + a_2^2\gamma + 6a_1^2\alpha - 6a_1\delta)u_{10}u_{20} + (2a_2\varepsilon + a_2\gamma a_1 - 4a_2\alpha a_1^2)u_{10}^2, \\
G_2 =& (-4b_2\delta - 4b_2^3\alpha - 3b_2\gamma + 6b_1\alpha b_2)u_{02}^2 \\
&+ (6\varepsilon - 6b_1\gamma - 8b_2^2\alpha b_1 + b_2^2\gamma + 6b_1^2\alpha - 6b_1\delta)u_{01}u_{02} + (2b_2\varepsilon - 4b_2\alpha b_1^2 + b_2\gamma b_1)u_{01}^2, \\
G_3 =& 2\alpha b_2 a_2 u_{02}u_{20} + (2\alpha b_1 a_2 - \gamma a_2)u_{01}u_{20} + (-\gamma b_2 + 2\alpha b_2 a_1)u_{10}u_{02} \\
&+ (-\gamma b_1 - \gamma a_1 + 2\alpha b_1 a_1 + 2\varepsilon)u_{01}u_{10}.
\end{aligned}$$

In view of the coefficients in G_i $(i = 1, 2, 3)$, we deduce a system of $a_i, b_i, \alpha, \gamma, \delta$ and ε, which includes ten equations. Solving the resulting system, we arrive at the following results.

Proposition 1. *Assume that the subspaces W_{3+3-1}^{xy} are determined by the system (14). Then, the quadratic operators $A[u]$ in $\mathbb{R}^2$ preserving the invariant subspaces W_{3+3-1}^{xy} determined by $u_{xy} = 0$ and the following constraints are presented as below, where $\alpha, \gamma, \delta, \varepsilon, a_i, b_i (i = 1, 2)$ are arbitrary constants.*

(1) $A[u] = \gamma[u\Delta_2 u - |\nabla u|^2]$, *with:*

$$L_x^3[v] = \frac{d^3v}{dx^3} - b_1\frac{dv}{dx} = 0, \quad L_y^3[w] = \frac{d^3w}{dy^3} + b_1\frac{dw}{dy} = 0;$$

(2) $A[u] = \alpha[(\Delta_2 u)^2 - b_2^2|\nabla u|^2]$, *with:*

$$L_x^3[v] = \frac{d^3v}{dx^3} = 0, \quad L_y^3[w] = \frac{d^3w}{dy^3} + b_2\frac{d^2w}{dy^2} = 0;$$

(3) $A[u] = \alpha[(\Delta_2 u)^2 - \frac{8}{9}b_2^2 u\Delta_2 u + \frac{16}{81}b_2^4 u^2]$, *with:*

$$L_x^3[v] = \frac{d^3v}{dx^3} - \frac{4}{9}b_2^2\frac{dv}{dx} = 0, \quad L_y^3[w] = \frac{d^3w}{dy^3} + b_2\frac{d^2w}{dy^2} + \frac{2}{9}b_2^2\frac{dw}{dy} = 0;$$

(4) $A[u] = \gamma[(a_1 + b_1)(\Delta_2 u)^2 + 4a_1 b_1 u\Delta_2 u + (a_1 - b_1)^2|\nabla u|^2 + a_1 b_1(a_1 + b_1)u^2]$, *with:*

$$L_x^3[v] = \frac{d^3v}{dx^3} + a_1\frac{dv}{dx} = 0, \quad L_y^3[w] = \frac{d^3w}{dy^3} + b_1\frac{dw}{dy} = 0;$$

(5) $A[u] = \alpha[(\Delta_2 u)^2 + b_1|\nabla u|^2]$, *with:*

$$L_x^3[v] = \frac{d^3v}{dx^3} = 0, \quad L_y^3[w] = \frac{d^3w}{dy^3} + b_1\frac{dw}{dy} = 0;$$

(6) $A[u] = \alpha(\Delta_2 u)^2 + \gamma u\Delta_2 u + (\gamma b_1 - \alpha b_1^2)u^2$, *with:*

$$L_x^3[v] = \frac{d^3v}{dx^3} + b_1\frac{dv}{dx} = 0, \, L_y^3[w] = \frac{d^3w}{dy^3} + b_1\frac{dw}{dy} = 0;$$

(7) $A[u] = \alpha(\Delta_2 u)^2 + \gamma u \Delta_2 u + \delta|\nabla u|^2$, *with:*

$$L_x^3[v] = \frac{d^3 v}{dx^3} = 0, \quad L_y^3[w] = \frac{d^3 w}{dy^3} = 0;$$

Solving the systems (14) yields the corresponding invariant subspaces. Here, we just present the invariant subspaces in the fourth case. The invariant subspaces for the other cases can be obtained in a similar manner. In the fourth case, we get the following invariant subspaces:

$$W_{3+3-1}^{xy} = \begin{cases} \mathcal{L}\{1, \cos(\sqrt{a_1}x), \sin(\sqrt{a_1}x), \cos(\sqrt{b_1}y), \sin(\sqrt{b_1}y)\}, & a_1 > 0, b_1 > 0, \\ \mathcal{L}\{1, \cos(\sqrt{a_1}x), \sin(\sqrt{a_1}x), \cosh(\sqrt{-b_1}y), \sinh(\sqrt{-b_1}y)\}, & a_1 > 0, b_1 < 0, \\ \mathcal{L}\{1, \cos(\sqrt{a_1}x), \sin(\sqrt{a_1}x), y, y^2)\}, & a_1 > 0, b_1 = 0, \\ \mathcal{L}\{1, \cosh(\sqrt{-a_1}x), \sinh(\sqrt{-a_1}x), \cosh(\sqrt{-b_1}y), \sinh(-\sqrt{-b_1}y)\}, & a_1 < 0, b_1 < 0, \\ \mathcal{L}\{1, \cosh(\sqrt{-a_1}x), \sinh(\sqrt{-a_1}x), y, y^2)\}, & a_1 < 0, b_1 = 0, \\ \mathcal{L}\{1, x, x^2, y, y^2\}, & a_1 = 0, b_1 = 0. \end{cases}$$

In the case of $n = 2$, we assume that the subspace W_{2+2-1}^{xy} is determined by the system:

$$L_x^2[v] \equiv \frac{d^2 v}{dx^2} + a_1 \frac{dv}{dx} = 0, \quad L_y^2[w] \equiv \frac{d^2 w}{dy^2} + b_1 \frac{dw}{dy} = 0. \tag{16}$$

By the similar calculation, we obtain the following results.

Proposition 2. *Any operators $A[u]$ that admit the subspaces W_{2+2-1}^{xy} determined by the system* (16) *are presented as follows:*

(1) $A[u] = \gamma[(a_1^2 + b_1^2)(\Delta_2 u)^2 - 4a_1^2 b_1^2 u \Delta_2 u - (a_1^2 - b_1^2)^2 |\nabla u|^2 + a_1^2 b_1^2 (a_1^2 + b_1^2) u^2]$, *with:*

$$L_x^2[v] = \frac{d^2 v}{dx^2} + a_1 \frac{dv}{dx} = 0, \quad L_y^2[w] = \frac{d^2 w}{dy^2} + b_1 \frac{dw}{dy} = 0;$$

(2) $A[u] = \alpha[(\Delta_2 u)^2 - b_1^2 |\nabla u|^2]$, *with:*

$$L_x^2[v] = \frac{d^2 v}{dx^2} = 0, \quad L_y^2[w] = \frac{d^2 w}{dy^2} + b_1 \frac{dw}{dy} = 0;$$

(3) $A[u] = \alpha(\Delta_2 u)^2 + \gamma u \Delta_2 u - (\alpha b_1^2 + \gamma) b_1^2 u^2$, *with:*

$$L_x^2[v] = \frac{d^2 v}{dx^2} + b_1 \frac{dv}{dx} = 0, \quad L_y^2[w] = \frac{d^2 w}{dy^2} + b_1 \frac{dw}{dy} = 0;$$

(4) $A[u] = \alpha(\Delta_2 u)^2 + \gamma u \Delta_2 u - (\alpha b_1^2 + \gamma) b_1^2 u^2$, *with:*

$$L_x^2[v] = \frac{d^2 v}{dx^2} - b_1 \frac{dv}{dx} = 0, \quad L_y^2[w] = \frac{d^2 w}{dy^2} + b_1 \frac{dw}{dy} = 0;$$

(5) $A[u] = \alpha(\Delta_2 u)^2 + \gamma u \Delta_2 u + \delta|\nabla u|^2$, *with:*

$$L_x^2[v] = \frac{d^2 v}{dx^2} = 0, \quad L_y^2[w] = \frac{d^2 w}{dy^2} = 0;$$

In the case of $n = 4$, we consider the invariant subspaces W^{xy}_{4+4-1} admitted by the operator $A[u]$, which are determined by the following ODEs:

$$\begin{aligned} L_x^4[v] &\equiv \frac{d^4v}{dx^4} + a_3\frac{d^3v}{dx^3} + a_2\frac{d^2v}{dx^2} + a_1\frac{dv}{dx} = 0, \\ L_y^4[w] &\equiv \frac{d^4w}{dy^4} + b_3\frac{d^3w}{dy^3} + b_2\frac{d^2w}{dy^2} + b_1\frac{dw}{dy} = 0. \end{aligned} \tag{17}$$

By the similar calculation as that in the case of $n = 3$, the invariant condition:

$$(A[u])_{xy}|_{[H]} = 2\alpha u_{03}u_{30} + \gamma u_{10}u_{03} + \gamma u_{01}u_{30} + 2\varepsilon u_{10}u_{01} \equiv 0$$

leads to $\alpha = \gamma = \varepsilon = 0$, where $[H]$ denotes the set $\{L_x^4[u] = 0\} \cap \{L_y^4[u] = 0\} \cap \{u_{xy} = 0\}$, and their differential consequences with respect to x and y. The invariant condition:

$$L_x^4[A[u]]|_{[H]} \equiv 0, \;\; L_y^4[A[u]]|_{[H]} \equiv 0$$

yields $\delta = 0$, which shows that there are no operators $A[u]$ preserving the invariant subspaces determined by (17). Similarly, we are able to show that there are no operators $A[u]$ to preserve the subspace W^{xy}_{5+5-1} defined by the following ODEs:

$$\begin{aligned} L_x^5[v] &\equiv \frac{d^5v}{dx^5} + a_4\frac{d^4v}{dx^4} + a_3\frac{d^3v}{dx^3} + a_2\frac{d^2v}{dx^2} + a_1\frac{dv}{dx} = 0, \\ L_y^5[w] &\equiv \frac{d^5w}{dy^5} + b_4\frac{d^4w}{dy^4} + b_3\frac{d^3w}{dy^3} + b_2\frac{d^2w}{dy^2} + b_1\frac{dw}{dy} = 0. \end{aligned}$$

2.2.2. The Space W^{xy}_{nn}

From the invariant condition (10), a similar calculation as above leads to the following results.

Proposition 3. *There are no operators $A[u]$ admitting the invariant subspaces W^{xy}_{nn} determined by the system:*

$$\begin{aligned} L_x^n[v] &\equiv \frac{d^nv}{dx^n} + a_{n-1}\frac{d^{n-1}v}{dx^{n-1}} + \cdots + a_1\frac{dv}{dx} + a_0v = 0, \\ L_y^n[w] &\equiv \frac{d^nw}{dy^n} + b_{n-1}\frac{d^{n-1}w}{dy^{n-1}} + \cdots + b_1\frac{dw}{dy} + b_0w = 0, \end{aligned} \tag{18}$$

for $n = 3, 4, 5$. The operators $A[u]$, which preserve the invariant subspaces W^{xy}_{22} determined by the system (18) *for $n = 2$, are given as follows:*

(1) $A[u] = \alpha(\Delta_2 u)^2 + \gamma u\Delta_2 u - (a_0 + b_0)[\alpha(a_0 + b_0) - \gamma]u^2$, *with:*

$$L_x^2[v] = \frac{d^2v}{dx^2} + a_0v = 0, \;\; L_y^2[w] = \frac{d^2v}{dy^2} + b_0v = 0;$$

(2) $A[u] = \alpha[(\Delta_2 u)^2 - b_1^2(2u\Delta_2 u + |\nabla u|^2) + 2b_1^4u^2]$, *with:*

$$L_x^2[v] = \frac{d^2v}{dx^2} + b_1\frac{dv}{dx} = 0, \;\; L_y^2[w] = \frac{d^2v}{dy^2} + b_1\frac{dw}{dy} = 0;$$

(3) $A[u] = \alpha[(\Delta_2 u)^2 - b_1^2(2u\Delta_2 u + |\nabla u|^2) + 2b_1^4u^2]$, *with:*

$$L_x^2[v] = \frac{d^2v}{dx^2} - b_1\frac{dv}{dx} = 0, \;\; L_y^2[w] = \frac{d^2v}{dy^2} + b_1\frac{dw}{dy} = 0.$$

The invariant spaces of the following two nonlinear equations can be constructed in a similar manner.

Example 1. *Consider the Jacobian:*

$$J(u,\Delta u) = u_x\Delta_2 u_y - u_y\Delta_2 u_x \equiv u_x(u_{xxy}+u_{yyy}) - u_y(u_{xxx}+u_{xyy})$$

which is the nonlinear term in two-dimensional Rossby waves equation [34]:

$$\Delta u_t + J(u,\Delta u) + \beta u_x = 0.$$

It preserves the following invariant subspaces:

(1) W^{xy}_{2+2-1}, determined by the system:

$$L_x^2[v] = \frac{d^2v}{dx^2} + a_1\frac{dv}{dx} = 0, \quad L_y^2[w] = \frac{d^2w}{dy^2} + b_1\frac{dw}{dy} = 0, \quad \text{with } a_1b_1(a_1^2 - b_1^2) = 0;$$

(2) W^{xy}_{3+3-1}, determined by any of the following systems:

$$L_x^3[v] = \frac{d^3v}{dx^3} + a_1\frac{dv}{dx} = 0, \quad L_y^3[w] = \frac{d^3w}{dy^3} + a_1\frac{dw}{dy} = 0;$$
$$L_x^3[v] = \frac{d^3v}{dx^3} - b_2^2\frac{dv}{dx} = 0, \quad L_y^3[w] = \frac{d^3w}{dy^3} + b_2\frac{d^2w}{dy^2} = 0;$$
$$L_x^3[v] = \frac{d^3v}{dx^3} \pm a_2\frac{d^2v}{dx^2} = 0, \quad L_y^3[w] = \frac{d^3w}{dy^3} + a_2\frac{d^2w}{dy^2} = 0;$$

(3) W^{xy}_{4+4-1}, determined by the system:

$$L_x^4[v] = \frac{d^4v}{dx^4} + a_2\frac{d^2v}{dx^2} = 0, \quad L_y^4[w] = \frac{d^4w}{dy^4} + a_2\frac{d^2w}{dy^2} = 0;$$

(4) W^{xy}_{22}, determined by any of the following systems:

$$L_x^2[v] = \frac{d^2v}{dx^2} = 0, \quad L_y^2[w] = \frac{d^2w}{dy^2} + b_1\frac{dw}{dy} = 0;$$
$$L_x^2[v] = \frac{d^2v}{dx^2} + a_0v = 0, \quad L_y^2[w] = \frac{d^2w}{dy^2} + b_0w = 0.$$

Example 2. *The invariant subspaces* $W_3 = \mathcal{L}\{1, x^2, y^2\}$ *and* $W_6 = \mathcal{L}\{1, x^2, y^2, x^2y^2, x^4, y^4\}$ *admitted by Monge–Ampère operator* $M[u] = u_{xx}u_{yy} - u_{xy}^2$ *were given in [1]. Here, we are looking for more invariant subspaces of this operator. Indeed, it still admits the following invariant subspaces:*

(1) W^{xy}_{2+2-1}*, determined by the system:*

$$L_x^2[v] = \frac{d^2v}{dx^2} + a_1\frac{dv}{dx} = 0, \quad L_y^2[w] = \frac{d^2w}{dy^2} + b_1\frac{dw}{dy} = 0, \quad \textit{with } a_1b_1 = 0;$$

(2) W^{xy}_{3+3-1}*, determined by any of the following systems:*

$$L_x^3[v] = \frac{d^3v}{dx^3} = 0, \quad L_y^3[w] = \frac{d^3w}{dy^3} + b_2\frac{d^2w}{dy^2} + b_1\frac{dw}{dy} = 0;$$

(3) W_{22}^{xy}, *determined by any of the following systems:*

$$L_x^2[v] = \frac{d^2v}{dx^2} = 0, \quad L_y^2[w] = \frac{d^2w}{dy^2} + b_1\frac{dw}{dy} + b_0 w = 0, \quad \textit{with } b_0 b_1 = 0;$$

$$L_x^2[v] = \frac{d^2v}{dx^2} + a_0 v = 0, \quad L_y^2[w] = \frac{d^2w}{dy^2} + b_0 w = 0;$$

(4) $W_{33}^{xy} = \mathcal{L}\{1, x, x^2, y, y^2, xy, x^2y, xy^2, x^2y^2\}$, *determined by the system:*

$$L_x^3[v] = \frac{d^3v}{dx^3} = 0, \quad L_y^3[w] = \frac{d^3w}{dy^3} = 0.$$

3. Invariant Subspaces under the General Change of Variables

In King's papers [2,12], the formal solution of two-dimensional nonlinear diffusion equations:

$$C_1(t) + C_2(t)x + C_3(t)y + C_4(t)x^2 + C_5(t)xy + C_6(t)y^2 \tag{19}$$

was proposed as a non-group-invariant exact solution, which belongs to the subspace $W_6 = \mathcal{L}\{1, x, y, x^2, xy, y^2\}$. The solution:

$$\begin{aligned} U = &C_1(t) + C_2(t)x + C_3(t)x^2 + C_4(t)y + C_5(t)y^2 + C_6(t)xy \\ &+ C_7(t)x(x^2+y^2) + C_8(t)y(x^2+y^2) + C_9(t)(x^2+y^2)^2 \end{aligned} \tag{20}$$

of the equation:

$$U_t = U\Delta_2 U - |\nabla U|^2 \equiv J[U], \quad (x, y) \in \mathbb{R}^2, \tag{21}$$

was presented as a generalization of solution (19). The derivation was based on the change of variables. King [2] discovered that Equation (21) was invariant under the following change of variables:

$$U^{(1)} = (x^2+y^2)^{-2}U, \ x^{(1)} = \frac{x}{x^2+y^2}, \ y^{(1)} = \frac{y}{x^2+y^2}, \ t^{(1)} = t, \tag{22}$$

which means that:

$$U_t = J[U] \ \longrightarrow \ U_{t^{(1)}}^{(1)} = J[U^{(1)}], \ \text{i.e.,} \ U_t = (x^2+y^2)^2 J[U^{(1)}].$$

Hence, $J[U] = (x^2+y^2)^2 J[U^{(1)}]$. On the other hand, since the operator $J[U^{(1)}]$ preserves the invariant subspace:

$$\begin{aligned} W_6^{(1)} &= \mathcal{L}\{1, x^{(1)}, (x^{(1)})^2, y^{(1)}, (y^{(1)})^2, x^{(1)}y^{(1)}\} \\ &\equiv \mathcal{L}\left\{1, \frac{x}{x^2+y^2}, \frac{x^2}{(x^2+y^2)^2}, \frac{y}{x^2+y^2}, \frac{y^2}{(x^2+y^2)^2}, \frac{xy}{(x^2+y^2)^2}\right\}, \end{aligned}$$

then the operator $J[U]$ preserves the corresponding subspace:

$$\widehat{W}_6 = \mathcal{L}\{x^2, y^2, xy, x(x^2+y^2), y(x^2+y^2), (x^2+y^2)^2\}.$$

In [1], Galaktionov and Svirshchevskii used the Lie symmetry of Equation (21) to give the invariant transformations of variables as (21). Then, they applied the invariant transformations and invariant subspaces of the corresponding one-dimensional equation of (21), i.e., $U_t = UU_{xx} - U_x^2$, to obtain the invariant subspaces W_{91} and W_{92}. In general, we have the following result.

Proposition 4. *Given a two-dimensional nonlinear differential operator $F[u]$ with respect to the variables x and y, if the nonlinear evolution Equation (6) is invariant under the transformation:*

$$u^{(1)} = r(x,y)u, \;\; x^{(1)} = p(x,y), \;\; y^{(1)} = q(x,y), \;\; t^{(1)} = t, \tag{23}$$

and operator $F[u]$ admits the linear space $W_n = \mathcal{L}\{f_1(x,y), \cdots, f_n(x,y)\}$, then $F[u]$ also admits the linear space:

$$\widehat{W}_n = \mathcal{L}\{f_1(p(x,y),q(x,y))/r(x,y), \cdots, f_n(p(x,y),q(x,y))/r(x,y)\}.$$

Proof: Equation (6) is invariant under the transformation (23), which means $u^{(1)}_{t^{(1)}} = F[u^{(1)}]$. On the other hand, $u^{(1)}_{t^{(1)}} = r(x,y)u_t$. Hence, $F[u^{(1)}] = r(x,y)F[u]$. Assume that:

$$u^{(1)} = \sum_{i=1}^{n} C_i f_i(x^{(1)}, y^{(1)}),$$

where $C_i (i = 1, \cdots, n)$ are arbitrary functions of t. Correspondingly,

$$u = \frac{1}{r(x,y)} \sum_{i=1}^{n} C_i f_i(x^{(1)}, y^{(1)}).$$

$F[u^{(1)}]$ admits the subspace $W_n^{(1)} = \mathcal{L}\{f_1(x^{(1)}, y^{(1)}), \cdots, f_n(x^{(1)}, y^{(1)})\}$, which means that there exist functions $\Psi_i (i = 1, \cdots, n)$, such that:

$$F[u^{(1)}] = F[\sum_{i=1}^{n} C_i f_i(x^{(1)}, y^{(1)})] = \sum_{i=1}^{n} \Psi_i(C_1, \cdots, C_n) f_i(x^{(1)}, y^{(1)}),$$

i.e.,

$$r(x,y)F[u] = r(x,y)F[\frac{1}{r(x,y)} \sum_{i=1}^{n} C_i f_i(x^{(1)}, y^{(1)})] = \sum_{i=1}^{n} \Psi_i(C_1, \cdots, C_n) f_i(x^{(1)}, y^{(1)}).$$

Then, $F[u]$ admits the subspace:

$$\widehat{W}_n = \mathcal{L}\{f_1(p(x,y),q(x,y))/r(x,y), \cdots, f_n(p(x,y),q(x,y))/r(x,y)\}.$$

This completes the proof of the proposition. □

Example 3. *In Proposition 1, we find that the operator $J[U]$ admits the invariant subspaces:*

$$W_{51} = \mathcal{L}\{1, \cos(b_1 x), \sin(b_1 x), \cosh(b_1 y), \sinh(b_1 y)\} \text{ and.}$$

Hence, by the changes of variables (22), the following subspace:

$$\begin{aligned} \widehat{W}^{xy}_{3+3-1} = \mathcal{L}\Big\{ & (x^2+y^2)^2, (x^2+y^2)^2 \cos(b_1 \frac{x}{x^2+y^2}), (x^2+y^2)^2 \sin(b_1 \frac{x}{x^2+y^2}), \\ & (x^2+y^2)^2 \cosh(b_1 \frac{y}{x^2+y^2}), (x^2+y^2)^2 \sinh(b_1 \frac{y}{x^2+y^2}) \Big\} \end{aligned}$$

is invariant under $J[U]$.

Note that the transformation (22) is a special one, under which Equation (21) is invariant. We can introduce a general transformation. As for the one-dimensional case [1]; two two-dimensional operators $F[u]$ and $\widetilde{F}[\widetilde{u}]$ are said to be equivalent, if there exists the change of variables:

$$u = r(x,y)\widetilde{u}, \;\; \widetilde{x} = p(x,y), \;\; \widetilde{y} = q(x,y)$$

such that:

$$\widetilde{F}[\widetilde{u}] = F[u]/r(x,y).$$

It implies that if the operator $F[u]$ preserves the invariant subspace $W_n = \mathcal{L}\{f_1(x,y),\cdots,f_n(x,y)\}$, then the equivalent operator $\widetilde{F}[\widetilde{u}]$ preserves the invariant subspace $\widetilde{W}_n = \mathcal{L}\{\widetilde{f}_1(\widetilde{x},\widetilde{y}),\cdots,\widetilde{f}_n(\widetilde{x},\widetilde{y})\}$, where $\widetilde{f}_i(\widetilde{x},\widetilde{y}) = f_i(x(\widetilde{x},\widetilde{y}),y(\widetilde{x},\widetilde{y}))/r(x(\widetilde{x},\widetilde{y}),y(\widetilde{x},\widetilde{y}))(i=1,\cdots,n)$.

4. Invariant Subspace in $\mathbb{R}$ and Lie's Classical Symmetries

The Lie theory of the symmetry group plays an important role for differential equations, which is a useful method to explore various properties and obtain exact solutions of nonlinear PDEs. The approach and its several extensions are illustrated in the books [35,36] and the papers [32,33,37,38]. One of the multiple applications of the Lie symmetry method is the similarity reduction of PDEs to ones with fewer variables. As usual, if an n-dimensional PDE admits one symmetry, then it can be reduced to an $n-1$-dimensional PDE equation and even to a ODE. It has been known that the invariant subspaces of one-dimensional differential operator were used to construct solutions of multi-dimensional nonlinear evolution equations of the radially symmetry form, which are one-dimensional evolution equations. For the two-dimensional case, the radially-symmetric solution can be regarded as the rotational-invariant solution. Accordingly, more invariant subspaces of two-dimensional operators can be obtained by combining the Lie symmetry method with the invariant subspaces of one-dimensional operators.

Example 4. *Consider the invariant subspaces preserved by the quadratic operator $J[U]$. The equation:*

$$u_t = \nabla \times (u^{-1}\nabla u) = (u^{-1}u_x)_x + (u^{-1}u_y)_y$$

can be changed into Equation (21) by the transformation $u = 1/U$. Indeed, for $u > 0$, the above equation can be rewritten as:

$$u_t = \triangle \ln u, \tag{24}$$

which is a well-known equation for describing the Ricci flow in a two-dimensional space [39]. Lie's classical symmetries of Equation (24) were computed in [40–45]. Indeed, Equation (24) admits the Lie group of symmetry with infinitesimal generator:

$$X = \xi\partial_x + \eta\partial_y + \tau\partial_t + \phi\partial_u,$$

where $\tau = k_1 + k_2 t, \xi = \xi(x,y), \eta = \eta(x,y)$ and ξ, η and ϕ satisfy the following constraints:

$$\phi = (2k_2 - 2\xi_x)u, \quad \xi_x - \eta_y = 0, \quad \eta_x + \xi_y = 0. \tag{25}$$

Clearly, the function $\xi = \xi(x,y)$ satisfies the two-dimensional Laplace equation:

$$\xi_{xx} + \xi_{yy} = 0.$$

Solving Equation (25), we obtain the following infinitesimal generators admitted by Equation (24):

$$\begin{aligned}
&X_1 = \partial_x + \partial_y, \quad X_2 = y\partial_x - x\partial_y, \quad X_3 = x\partial_x + y\partial_y - 2u\partial_u\\
&X_4 = xy\partial_x + \frac{1}{2}(y^2 - x^2)\partial_y - 2yu\partial_u, \quad X_5 = \frac{1}{2}(x^2 - y^2)\partial_x + xy\partial_y - 2xu\partial_u,\\
&X_6 = \sinh(ax)\sin(ay)\partial_x - \cosh(ax)\cos(ay)\partial_y - 2a\cosh(ax)\sin(ay)u\partial_u,\\
&X_7 = \sinh(ax)\cos(ay)\partial_x + \cosh(ax)\sin(ay)\partial_y - 2a\cosh(ax)\cos(ay)u\partial_u,\\
&X_8 = \sinh(ay)\sin(ax)\partial_x + \cosh(ay)\cos(ax)\partial_y - 2a\sinh(ay)\cos(ax)u\partial_u,\\
&X_9 = \sinh(ay)\cos(ax)\partial_x - \cosh(ay)\sin(ax)\partial_y + 2a\sinh(ay)\sin(ax)u\partial_u, \quad \text{etc.}
\end{aligned}$$

Here, a is a non-zero arbitrary constant. On the other hand, the corresponding infinitesimal generators admitted by the Equation (21) can be obtained by the transformation $u = 1/U$, i.e.,

$$u \longrightarrow \frac{1}{U}, \ \partial_u \longrightarrow -U^2\partial_U,$$

which reduce Equation (21) to one-dimensional equations. We denote them by $\widetilde{X}_i$ $(i = 1, \cdots, 9)$.

(1) $\widetilde{X}_1$. For $\widetilde{X}_1$, its invariants are $\widetilde{U} = U$ and $z = x + y$. The corresponding invariant solutions of (21) are $U = \widetilde{U}(z,t)$, where $\widetilde{U}(z,t)$ satisfies:

$$\widetilde{U}_t = 2(\widetilde{U}\widetilde{U}_{zz} - \widetilde{U}_z^2) \equiv \widetilde{J}^1[\widetilde{U}].$$

(2) $\widetilde{X}_2$. For $\widetilde{X}_2$, its invariants are $\widetilde{U} = U$ and $z = x^2 + y^2$. The corresponding invariant solutions of (21) are $v = \widetilde{U}(z,t)$, where $\widetilde{U}(z,t)$ satisfies:

$$\widetilde{U}_t = 4z\widetilde{U}\widetilde{U}_{zz} - 4z\widetilde{U}_z^2 + 4\widetilde{U}\widetilde{U}_z \equiv \widetilde{J}^2[\widetilde{U}].$$

(3) $\widetilde{X}_3$. For $\widetilde{X}_3$, its invariants are $\widetilde{U} = Ux^{-2}$ and $z = y/x$. The corresponding invariant solutions of (21) are $U = x^2\widetilde{U}(z,t)$, where $\widetilde{U}(z,t)$ satisfies:

$$\widetilde{U}_t = (1+z^2)\widetilde{U}\widetilde{U}_{zz} - (1+z^2)\widetilde{U}_z^2 + 2z\widetilde{U}\widetilde{U}_z - 2\widetilde{U}^2 \equiv \widetilde{J}^3[\widetilde{U}].$$

(4) $\widetilde{X}_4$. For $\widetilde{X}_4$, its invariants are $\widetilde{U} = vx^{-2}$ and $z = x + y^2/x$. The corresponding invariant solutions of (21) are $U = x^2\widetilde{U}(z,t)$, where $\widetilde{U}(z,t)$ satisfies:

$$\widetilde{U}_t = z^2\widetilde{U}\widetilde{U}_{zz} - z^2\widetilde{U}_z^2 + 2z\widetilde{U}\widetilde{U}_z - 2\widetilde{U}^2 \equiv \widetilde{J}^4[\widetilde{U}].$$

(5) $\widetilde{X}_5$. For $\widetilde{X}_5$, its invariants are $\widetilde{U} = y^{-2}U$ and $z = y + x^2/y$. The invariant solutions of (21) are $U = y^2\widetilde{U}(z,t)$, where $\widetilde{U}(z,t)$ satisfies $\widetilde{U}_t = \widetilde{J}^4[\widetilde{U}]$.
(6) $\widetilde{X}_6$. For $\widetilde{X}_6$, its invariants are $\widetilde{U} = \sinh^{-2}(ax)U$ and $z = \cos(ay)/\sinh(ax)$. The invariant solutions of (21) are $U = \sinh^2(ax)\widetilde{U}(z,t)$, where $\widetilde{U}(z,t)$ satisfies $\widetilde{U}_t = a^2\widetilde{J}^3[\widetilde{U}]$.
(7) $\widetilde{X}_7$. For $\widetilde{X}_7$, its invariants are $\widetilde{U} = \sinh^{-2}(ax)U$ and $z = \sin(ay)/\sinh(ax)$. The invariant solutions of (21) are $U = \sinh^2(ax)\widetilde{v}(z,t)$, where $\widetilde{U}(z,t)$ satisfies $\widetilde{U}_t = a^2\widetilde{J}^3[\widetilde{U}]$.
(8) $\widetilde{X}_8$. For $\widetilde{X}_8$, its invariants are $\widetilde{U} = \cosh^{-2}(ay)U$ and $z = \sin(ax)/\cosh(ay)$. The invariant solutions of (21) are $U = \cosh^2(ay)\widetilde{U}(z,t)$, where $\widetilde{U}(z,t)$ satisfies:

$$\widetilde{U}_t = a^2(1-z^2)\widetilde{U}\widetilde{U}_{zz} + a^2(z^2-1)\widetilde{U}_z^2 - 2a^2z\widetilde{U}\widetilde{U}_z + 2a^2\widetilde{U}^2 \equiv \widetilde{J}^5[\widetilde{U}].$$

(9) $\widetilde{X}_9$. For $\widetilde{X}_9$, its invariants are $\widetilde{U} = U\cosh^{-2}(ay)$ and $z = \cos(ax)/\cosh(ay)$. The invariant solutions of (21) are $U = \cosh^2(ay)\widetilde{U}(z,t)$, where $\widetilde{U}(z,t)$ satisfies $\widetilde{U}_t = \widetilde{J}^5[\widetilde{U}]$.

Using the invariant subspace method for the one-dimensional case, we find that the nonlinear operators $\widetilde{J}^i[\widetilde{U}] (i = 1, \cdots, 5)$ only admit two- and three-dimensional subspaces determined by spaces of solutions of linear ODEs as:

$$\frac{d^n w}{dz^n} + b_{n-1}(z)\frac{d^{n-1}w}{dz^{n-1}} + \cdots + b_0(z)w = 0.$$

We concentrate on the three-dimensional invariant subspaces, which are listed as below:

(1) The operator $\widetilde{J}^1[\widetilde{U}]$ admits the invariant subspaces:

$$\widetilde{W}_3 = \begin{cases} \mathcal{L}\{1, z, z^2\}, & b = 0, \\ \mathcal{L}\{1, \cos(cz), \sin(cz)\}, & b = c^2, \\ \mathcal{L}\{1, \exp(cz), \exp(-cz)\}, & b = -c^2, \end{cases}$$

determined by the spaces of solutions of the ODE:

$$\frac{d^3w}{dz^3} + b\frac{dw}{dz} = 0.$$

(2) The operator $\widetilde{J}^2[\widetilde{U}]$ admits the invariant subspaces:

$$W_3 = \begin{cases} \mathcal{L}\{z, z\ln z, z(\ln z)^2\}, & b = -1, \\ \mathcal{L}\{z, z^{1-c}, z^{1+c}\}, & b = -1 + c^2, \\ \mathcal{L}\{z, z\sin(c\ln z), z\cos(c\ln z)\}, & b = -1 - c^2, \end{cases}$$

determined by the spaces of solutions of the ODE:

$$\frac{d^3w}{dz^3} + \frac{b}{z^2}\frac{dw}{dz} - \frac{b}{z^3}w = 0.$$

(3) The operator $\widetilde{J}^3[\widetilde{U}]$ admits the invariant subspaces:

$$\widetilde{W}_3 = \begin{cases} \mathcal{L}\{(1+z^2), (1+z^2)\arctan z, (1+z^2)(\arctan z)^2\}, & b = -4, \\ \mathcal{L}\{(1+z^2), (1+z^2)\sin(c\arctan z), (1+z^2)\cos(c\arctan z)\}, & b = -4 + c^2, \\ \mathcal{L}\{(1+z^2), (1+z^2)\exp(c\arctan z), (1+z^2)\exp(-c\arctan z)\}, & b = -4 - c^2, \\ \mathcal{L}\{1, z, z^2\}, & b = 0, \end{cases}$$

determined by the spaces of solutions of the ODE:

$$\frac{d^3w}{dz^3} + \frac{b}{(1+z^2)^2}\frac{dw}{dz} - \frac{2bz}{(1+z^2)^3}w = 0.$$

(4) The operator $\widetilde{J}^4[\widetilde{U}]$ admits the invariant subspaces:

$$\widetilde{W}_3 = \begin{cases} \mathcal{L}\{z^2, z^2\exp(-\frac{c}{z}), z^2\exp(\frac{c}{z})\}, & b = 2c^2, \\ \mathcal{L}\{z^2, z^2\sin(\frac{c}{z}), z^2\cos(\frac{c}{z})\}, & b = -2c^2, \\ \mathcal{L}\{1, z, z^2\}, & b = 0, \end{cases}$$

determined by the spaces of solutions of the ODE:

$$\frac{d^3w}{dz^3} - \frac{b}{2z^4}\frac{dw}{dz} + \frac{b}{z^5}w = 0.$$

(5) The operator $\widetilde{J}^5[\widetilde{U}]$ admits the invariant subspaces:

$$\widetilde{W}_3 = \begin{cases} \mathcal{L}\{(z^2-1), (z^2-1)\ln(\frac{z+1}{z-1}), (z^2-1)(\ln(\frac{z+1}{z-1}))^2\}, & b = 8, \\ \mathcal{L}\{(z^2-1), (z^2-1)\exp(\text{carctanhz}), (z^2-1)\exp(-\text{carctanhz})\}, & b = -8 + 8c^2, \\ \mathcal{L}\{(z^2-1), (z^2-1)\sin(\text{carctanhz}), (z^2-1)\cos(\text{carctanhz})\}, & b = -8 - 8c^2, \\ \mathcal{L}\{1, z, z^2\}, & b = 0, \end{cases}$$

determined by the spaces of solutions of the ODE:

$$\frac{d^3w}{dz^3} - \frac{b}{2(z^2-1)^2}\frac{dw}{dz} + \frac{bz}{(z^2-1)^3}w = 0,$$

where and hereafter b is an arbitrary constant, and c is a non-zero arbitrary constant.

Then, we can obtain the corresponding invariant subspaces preserved by $J[U]$, which are presented as below:

$$
\begin{aligned}
W_3 &= \mathcal{L}\{1, x+y, (x+y)^2\},\\
W_3 &= \mathcal{L}\{1, \cos(c(x+y)), \sin(c(x+y))\},\\
W_3 &= \mathcal{L}\{1, \cosh(c(x+y)), \sinh(c(x+y))\},\\
W_3 &= \mathcal{L}\{x^2+y^2, (x^2+y^2)\ln(x^2+y^2), (x^2+y^2)(\ln(x^2+y^2))^2\},\\
W_3 &= \mathcal{L}\{x^2+y^2, (x^2+y^2)^{1-c}, (x^2+y^2)^{1+c}\},\\
W_3 &= \mathcal{L}\{x^2+y^2, (x^2+y^2)\sin(c\ln(x^2+y^2)), (x^2+y^2)\cos(c\ln(x^2+y^2))\},\\
W_3 &= \mathcal{L}\{x^2+y^2, (x^2+y^2)\arctan(\frac{y}{x}), (x^2+y^2)(\arctan(\frac{y}{x}))^2\},\\
W_3 &= \mathcal{L}\{x^2+y^2, (x^2+y^2)\sin(c\arctan(\frac{y}{x})), (x^2+y^2)\cos(c\arctan(\frac{y}{x}))\},\\
W_3 &= \mathcal{L}\{x^2+y^2, (x^2+y^2)\cosh(c\arctan(\frac{y}{x})), (x^2+y^2)\sinh(c\arctan(\frac{y}{x}))\},\\
W_3 &= \mathcal{L}\{x^2, xy, y^2\},\\
W_3 &= \mathcal{L}\{(x^2+y^2)^2, (x^2+y^2)^2\cosh(\frac{cx}{x^2+y^2}), (x^2+y^2)^2\sinh(\frac{cx}{x^2+y^2})\},\\
W_3 &= \mathcal{L}\{(x^2+y^2)^2, (x^2+y^2)^2\sin(\frac{cx}{x^2+y^2}), (x^2+y^2)^2\cos(\frac{cx}{x^2+y^2})\},\\
W_3 &= \mathcal{L}\{x^2, x(x^2+y^2), (x^2+y^2)^2\},\\
W_3 &= \mathcal{L}\{(\cos^2 ay+\sinh^2 ax), (\cos^2 ay+\sinh^2 ax)\arctan\frac{\cos ay}{\sinh ax},\\
&\qquad (\cos^2 ay+\sinh^2 ax)(\arctan\frac{\cos ay}{\sinh ax})^2\},\\
W_3 &= \mathcal{L}\{(\cos^2 ay+\sinh^2 ax), (\cos^2 ay+\sinh^2 ax)\sin(c\arctan\frac{\cos ay}{\sinh ax}),\\
&\qquad (\cos^2 ay+\sinh^2 ax)\cos(c\arctan\frac{\cos ay}{\sinh ax})\},\\
W_3 &= \mathcal{L}\{(\cos^2 ay+\sinh^2 ax), (\cos^2 ay+\sinh^2 ax)\cosh(c\arctan\frac{\cos ay}{\sinh ax}),\\
&\qquad (\cos^2 ay+\sinh^2 ax)\sinh(c\arctan\frac{\cos ay}{\sinh ax})\},\\
W_3 &= \mathcal{L}\{\sinh^2 ax, \cos ay\sinh ax, \cos^2 ay\},\\
W_3 &= \mathcal{L}\{(\sin^2 ax-\cosh^2 ay), (\sin^2 ax-\cosh^2 ay)\ln\frac{\sin ax+\cosh ay}{\sin ax-\cosh ay},\\
&\qquad (\sin^2 ax-\cosh^2 ay)(\ln\frac{\sin ax+\cosh ay}{\sin ax-\cosh ay})^2\},
\end{aligned}
$$

$$
\begin{aligned}
W_3 &= \mathcal{L}\{(\sin^2 ax-\cosh^2 ay), (\sin^2 ax-\cosh^2 ay)\sin(c\operatorname{arctanh}\frac{\sin ax}{\cosh ay})\\
&\qquad (\sin^2 ax-\cosh^2 ay)\cos(c\operatorname{arctanh}\frac{\sin ax}{\cosh ay})\},\\
W_3 &= \mathcal{L}\{(\sin^2 ax-\cosh^2 ay), (\sin^2 ax-\cosh^2 ay)\cosh(c\operatorname{arctanh}\frac{\sin ax}{\cosh ay}),\\
&\qquad (\sin^2 ax-\cosh^2 ay)\sinh(c\operatorname{arctanh}\frac{\sin ax}{\cosh ay})\}.
\end{aligned}
$$

Since the operator $J[U]$ is symmetric with respect to the variables x and y, the following invariant subspaces can also be obtained from the above invariant subspaces:

$$W_3 = \mathcal{L}\{\cos^2(ax), \sinh(ay)\cos(ax), \sinh^2(ay)\},$$
$$W_3 = \mathcal{L}\{\sin^2(ay), \cosh(ax)\sin(ay), \cosh^2(ax)\}.$$

Example 5. *Consider the two-dimensional porous medium equation:*

$$u_t = (u^p u_x)_x + (u^p u_y)_y, \quad p \neq 0, -1, \tag{26}$$

which can be changed into the equation:

$$U_t = U(U_{xx} + U_{yy}) + \frac{1}{p}(U_x^2 + U_y^2) \equiv J_p[U] \tag{27}$$

by the transformation $U = u^p$. *Equation* (27) *admits the scaling invariance with the infinitesimal generator:*

$$\widetilde{X}_3 = x\partial_x + y\partial_y + 2U\partial_U, \tag{28}$$

which possesses the invariants:

$$\widetilde{U} = \frac{1}{x^2}U(z,t), \quad z = \frac{y}{x}, \quad \widetilde{t} = t.$$

Under the Lie symmetry $\widetilde{X}_3$, this equation is reduced to:

$$\widetilde{U}_t = (1+z^2)\widetilde{U}\widetilde{U}_{zz} + \frac{1}{p}(1+z^2)\widetilde{U}_z^2 - \frac{2(p+2)}{p}z\widetilde{U}\widetilde{U}_z + \frac{2(p+2)}{p}\widetilde{U}^2 \equiv \widetilde{J}_p[\widetilde{U}].$$

The operator $\widetilde{J}_p[\widetilde{U}]$ admits invariant subspace $W_3 = \mathcal{L}\{1, z, z^2\}$ determined by ODE $d^3w/dz^3 = 0$. Hence, the operator $J_p[U]$ admits the invariant subspaces $W_3 = \mathcal{L}\{x^2, xy, y^2\}$. On the other hand, for $p = -4/3$, the operator $\widetilde{J}_p[\widetilde{U}]$ admits another invariant subspace $W_3 = \mathcal{L}\{1, z^2+1, \sqrt{z^2+1}\}$ determined by the ODE:

$$\frac{d^3w}{dz^3} - \frac{3}{z(z^2+1)}\frac{d^2w}{dz^2} + \frac{3}{z^2(1+z^2)}\frac{dw}{dz} = 0.$$

Therefore, the corresponding invariant subspace admitted by the operator $J_{-\frac{4}{3}}[U]$ is:

$$W_3 = \mathcal{L}\{x^2, x^2+y^2, x\sqrt{x^2+y^2}\} \equiv \mathcal{L}\{x^2, y^2, x\sqrt{x^2+y^2}\}.$$

Accordingly, some invariant subspaces of $J[U]$ can be obtained from the invariant subspace $\mathcal{L}\{1, z, z^2\}$ admitted by the operator $\widetilde{J}'[\widetilde{U}]$, which is the polynomial subspace. The polynomial subspaces of nonlinear operators are studied in many papers, which were used to construct exact solutions of nonlinear evolution equations, including porous medium equations, thin film equations and Euler equations [1–3,12–14,28,29,46–49]. Using the Lie symmetry method, we may obtain polynomial invariant subspaces of some two-dimensional nonlinear operators. Note that in Examples 4 and 5, the invariant subspace $W_3 = \mathcal{L}\{x^2, xy, y^2\}$ can be obtained from the one-dimensional invariant subspace $\widetilde{W}_3 = \mathcal{L}\{1, z, z^2\}$ and the Lie group of symmetry (28). The subspace $\mathcal{L}\{1, z, z^2\}$ is determined by the space of solutions of linear ODE $\widetilde{U}_{zzz} = 0$, which can be explained by the conditional Lie–Bäcklund symmetry with character $\widetilde{U}_{zzz}$ [1,32,33]). Besides those, the nonlinear evolution equation $U_t = (UU_x)_y$ also admits the Lie group of transformation with the infinitesimal generator (28). By the similar calculations as above, we find that the operator $(UU_x)_y$ admits the invariant subspace

$\mathcal{L}\{x^2, xy, y^2\}$. In [10], the operators preserving a given invariant subspace were discussed, for instance the space $\mathcal{M} = \{x^2, xy, y^2\}$, which was regarded as a "simple" problem for the affine annihilator.

Example 6. *Consider the evolution Monge–Ampère equation:*

$$u_t = u_{xx}u_{yy} - u_{xy}^2. \tag{29}$$

It is easy to verify that this equation admits the Lie groups of transformations with infinitesimal operators:

$$X_1 = y\partial_x \pm x\partial_y, \quad X_2 = y\partial_x \pm \frac{1}{2}\partial_y, \quad X_3 = x\partial_y \pm \frac{1}{2}\partial_x.$$

We find that X_1 has invariants $\widetilde{u} = u(z,t)$, $z = x^2 \pm y^2$ and $\widetilde{t} = t$. With respect to this Lie symmetry, Equation (29) is reduced to:

$$\widetilde{u}_t = \pm(-8z\widetilde{u}_z\widetilde{u}_{zz} + 4\widetilde{u}_z^2) \equiv \widetilde{M}_\pm[\widetilde{u}].$$

The operator $\widetilde{M}_\pm[\widetilde{u}]$ admits the invariant subspace $\widetilde{W}_3 = \mathcal{L}\{1, \sqrt{z}, z^2\}$ determined by the ODE:

$$\frac{d^3w}{dz^3} + \frac{1}{2z}\frac{d^2w}{dz^2} - \frac{1}{2z^2}\frac{dw}{dz} = 0,$$

and the invariant subspace $\widetilde{W}_3 = \mathcal{L}\{1, z, z^2\}$ determined by ODE $d^3w/dz^3 = 0$. Hence, the Monge–Ampère operator $M[u] = u_{xx}u_{yy} - u_{xy}^2$ admits the invariant subspaces:

$$W_3 = \mathcal{L}\{1, \sqrt{x^2 \pm y^2}, (x^2 \pm y^2)^2\}, \text{ and } W_3 = \mathcal{L}\{1, x^2 \pm y^2, (x^2 \pm y^2)^2\}$$

Similarly, under the Lie symmetries $X_{2,3}$, we obtain the following invariant subspaces preserved by the Monge–Ampère operator:

$$W_3 = \mathcal{L}\{1, (x \pm y^2)^{\frac{3}{2}}, (x \pm y^2)^3\}, \quad W_4 = \mathcal{L}\{1, (x \pm y^2), (x \pm y^2)^2, (x \pm y^2)^3\},$$

In general, assume that nonlinear evolution Equation (6) admits the Lie group of transformation with infinitesimal generator X, which has invariants:

$$z = p(x,y), \quad \widetilde{u} = \frac{u}{r(x,y)}, \quad \widetilde{t} = t,$$

and reduces it to the one-dimensional nonlinear evolution equation:

$$\widetilde{u}_t = \widetilde{F}(z, \widetilde{u}, \widetilde{u}_z, \widetilde{u}_{zz}, \cdots) \equiv \widetilde{F}[\widetilde{u}].$$

We then obtain the following proposition.

Proposition 5. *If the nonlinear differential operator $\widetilde{F}$ admits the invariant subspaces $\widetilde{W}_n = \mathcal{L}\{f_1(z), \cdots, f_n(z)\}$, then two-dimensional nonlinear differential operator F preserves the invariant subspaces $W_n = \mathcal{L}\{r(x,y)f_1(p(x,y)), \cdots, r(x,y)f_n(p(x,y))\}$.*

The proof is similar to that of Proposition 4. Clearly, in this approach, the estimate on the dimension of invariant subspace obeys Theorem 1.

5. Concluding Remarks

In this paper, several approaches are developed to obtain invariant subspaces of the two-dimensional nonlinear operators, including two direct extensions to the invariant subspace method in $\mathbb{R}$, the method of the general change of variables and the one-dimensional invariant subspace method combined with the Lie symmetry method. In particular, we find that the subspaces W_{nm}^{xy} and W_{n+m-1}^{xy} of the two-dimensional nonlinear differential operators are extensions of the invariant subspaces for one-dimensional nonlinear differential operators, which are determined by the spaces of solutions of ODEs completely. In $\mathbb{R}^2$, the invariant subspaces admitted by the quadratic operator $A[u]$ and their applications are considered. In general, the extensions of the concept of invariant subspaces in $\mathbb{R}^N$ could be introduced. Assume that $\{f_{j1}(x_j), \cdots, f_{jm_j}(x_j)\}$ is a finite set of linearly independent functions, and $W_{m_j}^{x_j}$ denotes their linear span $W_{m_j}^{x_j} = \mathcal{L}\{f_{j1}(x_j), \cdots, f_{jm_j}(x_j)\}$, where $j = 1, \cdots, N$. The $(m_1 \cdot \cdots \cdot m_N)$-dimensional subspace:

$$\widetilde{W} = \left\{ \sum_{i_1, \cdots, i_N} C_{i_1 \cdots i_N} f_{1i_1}(x_1) \cdot \cdots \cdot f_{Ni_N}(x_N), \forall C_{i_1 \cdots i_N} \in \mathbb{R}, i_j = 1, \cdots, m_j, j = 1, \cdots, N \right\}$$

can be introduced as an extension to the subspace W_{nm}^{xy} in $\mathbb{R}^N$. Consider the N-dimensional nonlinear operator:

$$F[u] \equiv F(u, Du, D^2u, \cdots, D^k u),$$

where $Du = (u_{x_1}, \cdots, u_{x_N})$, $D^2u = (u_{x_1x_1}, \cdots, u_{x_1x_N}, u_{x_2x_2}, \cdots, u_{x_2x_N}, \cdots, u_{x_Nx_N})$, etc. Assume that the subspace $W_{m_j}^{x_j}$ is the space of solutions of the ODE:

$$L_{x_j}[v_j] \equiv \frac{d^{m_j} v_j}{dx_j^{m_j}} + a_{jm_j-1}(x_j) \frac{d^{m_j-1} v_j}{dx_j^{m_j-1}} + \cdots + a_{j0}(x_j) v_j = 0, \; j = 1, \cdots, N.$$

Then, the invariance condition of the subspace $\widetilde{W}$ preserved by the operator $F[u]$ (i.e., $F[\widetilde{W}] \subseteq \widetilde{W}$) is:

$$L_{x_j}[F[u]]|_{[\widetilde{H}]} \equiv 0, \; j = 1, \cdots, N,$$

where $[\widetilde{H}]$ denotes $L_{x_j}[u] = 0$, and their differential consequences with respect to x_i, $i, j = 1, \cdots, N$. Similarly, we assume that $\{1, f_{j1}(x_j), \cdots, f_{jm_j}(x_j)\}$ is a set of basis of solutions of the ODE system:

$$\overline{L}_{x_j}[v_j] \equiv \frac{d^{m_j} v_j}{dx_j^{m_j}} + a_{jm_j-1}(x_j) \frac{d^{m_j-1} v_j}{dx_j^{m_j-1}} + \cdots + a_{j1}(x_j) \frac{dv_j}{dx_j} = 0, \; j = 1, \cdots, N.$$

Let $\overline{W}_{m_j}^{x_j} = \mathcal{L}\{1, f_{j1}(x_j), \cdots, f_{jm_j}(x_j)\}$ denote the space of solutions of this ODE, where $j = 1, \cdots, N$. We can introduce the $(m_1 + \cdots + m_N - N + 1)$-dimensional subspace:

$$\overline{W} = \mathcal{L}\{1, f_{12}(x_1), \cdots, f_{1m_1}(x_1), \cdots, f_{N2}(x_N), \cdots, f_{Nm_N}(x_N)\}$$

as an extension of W_{n+m-1}^{xy} in $\mathbb{R}^N$. Then, the invariance condition of the subspace $\overline{W}$ preserved by the operator $F[u]$ (i.e., $F[\overline{W}] \subseteq \overline{W}$) is:

$$\overline{L}_{x_j}[F[u]]|_{[\overline{H}]} \equiv 0, \; (F[u])_{x_ix_j}|_{[\overline{H}]} \equiv 0,$$

where $[\overline{H}]$ denotes $\overline{L}_{x_j}[u] = 0$, $u_{x_ix_j} = 0$, and their differential consequences with respect to x_i, $i, j = 1, \cdots, N, i \neq j$. The invariant subspaces obtained by this method can be regarded as original subspaces and used to obtain new ones by the general changes of variables in Section 3.

To obtain more invariant subspaces of nonlinear differential operators, we adopt the direct sum of invariant subspaces, which was used by Galaktionov and Svirshchevskii [1] to obtain the new invariant subspaces preserved by a given operator. For example, in Proposition 6.1 of [1], it was shown that the direct sum of the subspaces $W_n^q = \mathcal{L}\{1, x_ix_j, 1 \leqslant i \leqslant j \leqslant N\}$ and $W_N^{\text{lin}} = \mathcal{L}\{x_1, \cdots, x_N\}$ is preserved by the operator $K[u]$. It is expected that the formulation of the direct sum can be used to obtain the invariant subspaces W_{91} and W_{92} of $J[U]$ by them. Indeed, the following result is always true.

Proposition 6. *Given a nonlinear differential operator F. If the linear subspaces W_n and W_m are preserved by the operator F and $W_n \cap W_m = \{0\}$, then the direct sum of W_n and W_m, i.e., $W_n \oplus W_m$ is invariant or partially invariant under the operator F.*

Clearly, for the "nonlinear property" of the nonlinear operator, $F[W_n \oplus W_m] \subseteq W_n \oplus W_m$ is not always true. However, in the case of $F[W_n \oplus W_m] \not\subseteq W_n \oplus W_m$, it is said to be partially invariant under the operator F (see [1]). The linear space W_n is partially invariant under the operator F, i.e., $F[W_n] \not\subseteq W_n$, but for some part M of W_n, $F[M] \subseteq W_n$. If the subspace W_n is partially invariant under a given operator, then the corresponding evolution equation can be reduced to an over-determined system of ODEs. One can verify whether the direct sum of two invariant subspaces is invariant under the given operator by a direct computation.

The following result is a further extension to Proposition 6.

Proposition 7. *Let F be a given nonlinear differential operator. If the linear subspaces $W_{n_1}^1, \cdots, W_{n_m}^m$ are preserved by the operator F, then the subspace $W_{n_1}^1 \cup \cdots \cup W_{n_m}^m$ is invariant or partially invariant under the operator F.*

Let us return to the invariant subspaces W_{91} and W_{92}. We can express:

$$W_{91} = W_3^1 \cup W_3^2 \cup W_3^3 \cup W_3^4,$$

where:

$$W_3^1 = \mathcal{L}\{1, x, y\}, \quad W_3^2 = \mathcal{L}\{x^2, xy, y^2\},$$
$$W_3^3 = \mathcal{L}\{x^2, x(x^2+y^2), (x^2+y^2)^2\}, \quad W_3^4 = \mathcal{L}\{y^2, y(x^2+y^2), (x^2+y^2)^2\}.$$

and express:

$$W_{92} = W_5^1 \cup W_3^2 \cup W_3^3 \cup W_3^4 \cup W_3^5,$$

where:

$$W_5^1 = \mathcal{L}\{1, \cos 2y, \sin 2y, \cosh 2x, \sinh 2x\},$$
$$W_3^2 = \mathcal{L}\{\cos^2 y, \cosh x \cos y, \cosh^2 x\}, \quad W_3^3 = \mathcal{L}\{\cos^2 y, \sinh x \cos y, \sinh^2 x\},$$
$$W_3^4 = \mathcal{L}\{\sin^2 y, \cosh x \sin y, \cosh^2 x\}, \quad W_3^5 = \mathcal{L}\{\sin^2 y, \sinh x \sin y, \sinh^2 x\}.$$

Note that $2\cos^2 y - 1 = -2\sin^2 y + 1 = \cos 2y$, $\sinh^2 x = (\sinh 2x - 1)/2$, $\cosh^2 x = (\cosh 2x + 1)/2$, and every component of W_{91} and W_{92} can be obtained by the knowledge of algebra and ODEs (see Sections 2 and 4). The following invariant subspace $\widehat{\widehat{W}}{}_{3+3-1}^{xy}$ of $J[U]$ can be obtained from $\widehat{W}_{3+3-1}^{xy}$ by the discrete symmetry $x \to y$, $y \to x$. Indeed, we have:

(1) $\widehat{W}_{3+3-1}^{xy} = W_3^1 \cup W_3^2$, with:

$$W_3^1 = \mathcal{L}\{(x^2+y^2)^2, (x^2+y^2)^2 \cos(b_1 \frac{x}{x^2+y^2}), (x^2+y^2)^2 \sin(b_1 \frac{x}{x^2+y^2})\},$$
$$W_3^2 = \mathcal{L}\{(x^2+y^2)^2, (x^2+y^2)^2 \exp(b_1 \frac{y}{x^2+y^2}), (x^2+y^2)^2 \exp(-b_1 \frac{y}{x^2+y^2})\};$$

(2) $\widehat{\widehat{W}}{}_{3+3-1}^{xy} = W_3^3 \cup W_3^4$, with:

$$W_3^3 = \mathcal{L}\{(x^2+y^2)^2, (x^2+y^2)^2 \cos(b_1 \frac{y}{x^2+y^2}), (x^2+y^2)^2 \sin(b_1 \frac{y}{x^2+y^2})\},$$
$$W_3^4 = \mathcal{L}\{(x^2+y^2)^2, (x^2+y^2)^2 \exp(b_1 \frac{x}{x^2+y^2}), (x^2+y^2)^2 \exp(-b_1 \frac{x}{x^2+y^2})\}.$$

Here, $W_3^i(i = 1, \cdots, 4)$ can be obtained by the method in Section 4. Similarly, we can check that both the operator $(UU_x)_y$ and $J_p[U]$ admit the invariant subspace $\mathcal{L}\{1, x, y, x^2, xy, y^2\} = \mathcal{L}\{1, x, y\} \cup \mathcal{L}\{x^2, xy, y^2\}$. Hence, the porous medium Equation (26) has the exact solution of the more general form:

$$u = (c_1(t) + c_2(t)x + c_3(t)y + c_4(t)x^2 + c_5(t)xy + c_6(t)y^2)^{\frac{1}{p}}.$$

On the other hand, it was shown that the operator $J_{-\frac{4}{3}}[U]$ admits the following invariant subspaces (see Example 5):

$$W_3^1 = \mathcal{L}\{x^2, xy, y^2\}, \quad W_3^2 = \mathcal{L}\{x^2, y^2, x\sqrt{x^2 + y^2}\}, \quad W_3^3 = \mathcal{L}\{x^2, y^2, y\sqrt{x^2 + y^2}\}$$

By direct calculation, one can check that the operator $J_{-\frac{4}{3}}[U]$ admits another invariant subspace:

$$W_5 = \mathcal{L}\{x^2, xy, y^2, x\sqrt{x^2 + y^2}, y\sqrt{x^2 + y^2}\} = W_3^1 \cup W_3^2 \cup W_3^3.$$

Hence, for $p = -4/3$, the porous medium Equation (26) has another solution of the form:

$$u = (c_1(t)x^2 + c_2(t)xy + c_3(t)y^2 + c_4(t)x\sqrt{x^2 + y^2} + c_5(t)y\sqrt{x^2 + y^2})^{-\frac{3}{4}}.$$

Finally, we would like to address some open questions. Firstly, although we have several operable approaches to obtain the invariant subspaces of two-dimensional nonlinear operators, we do not have a systematic approach to obtain the invariant subspaces of $J[U]$ as W_{91} and W_{92} and those of the Monge–Ampère operator as:

$$W_{13} = \mathcal{L}\{1, x, y, x^2, xy, y^2, x^3, x^2y, xy^2, y^3, x^4, x^2y^2, y^4\}.$$

Secondly, as mentioned in the Introduction, what is the maximal dimension of the certain types of invariant subspaces of multi-dimensional k-th order nonlinear differential operators? All of these questions will be the content of our future research.

Acknowledgments: This work was supported by the National Natural Science Foundation of China (Grant Nos. 11631007, 11301007 and 11471174) and the Natural Science Foundation of Anhui province (Grant No. 1408085QA05).

Author Contributions: Both authors C.R.Z. and C.Z.Q. make contributions to Sections 1, 2, 4, and 5. C.R.Z. make contribution to Section 3.

Conflicts of Interest: The authors declare no conflict of interest.

References

1. Galaktionov, V.A.; Svirshchevskii, S.R. *Exact Solutions and Invariant Subspaces of Nonlinear Partial Differential Equations in Mechanics and Physics;* Chapman and Hall/CRC: London, UK, 2007.
2. King, J.R. Exact polynomial solutions to some nonlinear diffusion equations. *Physica D* **1995**, *64* , 35–65.
3. Galaktionov, V.A. Invariant subspaces and new explicit solutions to evolution equations with quadratic nonlinearities. *Proc. Roy. Soc. Edinb.* **1995**, *125A*, 225–246.
4. Galaktionov, V.A.; Posashkov, S.A.; Svirshchevskii, S.R. On invariant sets and explicit solutions of nonlinear evolution equations with quadratic nonlinearities. *Differ. Integr. Equat.* **1995**, *31*, 1997–2024.
5. Svirshchevskii, S.R. Invariant linear spaces and exact solutions of nonlinear evolution equations. *J. Nonlinear Math. Phys.* **1996**, *3*, 164–169.
6. Svirshchevskii, S.R. Lie-Bäcklund symmetries of submaximal order of ordinary differential equations. *Nonlin. Dyn.* **2002**, *28*, 155–166.
7. Svirshchevskii, S.R. Nonlinear differential operator of first and second order possessing invariant linear spaces of maximal dimension. *Theor. Math. Phys.* **1995**, *105*, 1346–1353.

8. Galaktionov, V.A. On invariant subspaces for nonlinear finite-difference operators. *Proc. Roy. Soc. Edinb. Sect. A* **1998**, *125*, 1293–1308.
9. Galaktionov, V.A. Invariant solutions of two models of evolution of turbulent bursts. *Eur. J. Appl. Math.* **1999**, *10*, 237–249.
10. Karmaran, N.; Milson, R.; Olver, P.J. Invariant modules and the reduction of nonlinear partial differential equations to dynamical systems. *Adv. Math.* **2000**, *156*, 286–319.
11. Galaktionov, V.A. Groups of scalings and invariant sets for higher-order nonlinear evolution equations. *Differ. Integr. Equat.* **2001**, *14*, 913–924.
12. King, J.R. Exact multidimensional solutions to some nonlinear diffusion equations. *Quart. J. Mech. Appl. Math.* **1993**, *46*, 419–436.
13. King, J.R. Mathematical analysis of a models for substitutional diffusion. *Proc. Roy. Soc. Lond. A* **1990**, *430*, 377–404.
14. Gomez-Ullate, D.; Kamran, N.; Milson, R. Structure theorems for linear and non-linear differential operators admitting invariant polynomial subspaces. *Discr. Cont. Dyn. Syst.* **2007**, *18*, 85–106.
15. Qu, C.Z.; Zhu C.Z. Classification of coupled systems with two-component nonlinear diffusion equations by the invariant subspace method. *J. Phys. A Math. Theor.* **2009**, *42*, 475201.
16. Zhu, C.R.; Qu, C.Z. Maximal dimension of invariant subspaces admitted by nonlinear vector differential equations. *J. Math. Phys.* **2011**, *52*, 043507, doi:10.1063/1.3574534.
17. Galaktionov, V.A.; Svirshchevskii, S.R. Invariant solutions of nonlinear diffusion equations with maximal symmetry algebra. *J. Nonlinear Math. Phys.* **2011**, *18*, 107–121.
18. Kaptsov, O.V. Invariant sets of evolution equations. Nonlin. *Anal. Theor. Meth. Appl.* **1992**, *19*, 753–761.
19. Kaptsov, O.V.; Verevkin, I.V. Differential constraints and exact solutions of nonlinear diffusion equations. *J. Phys. A Theor. Meth.* **2003**, *36*, 1401–1414.
20. Shen, S.F.; Qu, C.Z.; Jin, Y.Y. Maximal dimension of invariant subspaces to systems of nonlinear evolution equations. *Chin. Ann. Math.* **2012**, *33B*, 161–178.
21. Oron, A.; Rosenau, P. Some symmetries of the nonlinear heat and wave equations. *Phys. Lett. A* **1986**, *118*, 172–176.
22. Cherniha, R.M. A constructive method for obtaining new exact solutions of nonlinear evolution equations. *Rept. Math. Phys.* **1996**, *38*, 301–312.
23. Cherniha, R.M. New non-Lie ansätze and exact solutions of nonlinear reaction-diffusion-convection equations. *J. Phys. A Math. Gen.* **1998**, *31*, 8179–8198.
24. Qu, C.Z.; Ji, L.N. Invariant subspaces and conditional Lie-Bäcklund symmetries of inhomogeneous nonlinear diffusion equations. *Sci. China Math.* **2013**, *56*, 2187–2202.
25. Ji, L.N.; Qu, C.Z. Conditional Lie-Bäcklund symmetries and invariant subspaces to nonlinear diffusion equations with convection and source. *Stud. Appl. Math.* **2010**, *131*, 266–301.
26. Ji, L.N.; Qu, C.Z. Conditional Lie-Bäcklund symmetries and invariant subspaces to nonlinear diffusion equations. *IMA J. Appl. Math.* **2011**, *76*, 610–632.
27. Ma, W.X. A refined invariant subspace method and applications to evolution equations. *Sci. China Math.* **2012**, *55*, 1769–1778.
28. King, J.R. Two generalisations of the thin film equation. *Math. Comput. Mod.* **2001**, *34*, 737–756.
29. Betelú, S.I.; King, J.R. Explicit solutions of a two-Dimensional fourth-order nonlinear diffusion equation. *Math. Comput. Mod.* **2003**, *37*, 395–403.
30. Cherniha, R.; Serov, M. Symmetries, ansätze and exact solutions of nonlinear second-order evolution equations with convection terms, II. *Eur. J. Appl. Math.* **2006**, *17*, 597–605.
31. Gazizov, R.K.; Kasatkin, A.A. Construction of exact solutions for fractional order differential equations by the invariant subspace method. Comput. *Math. Appl.* **2013**, *66*, 576–584.
32. Zhdanov, R.Z. Conditional Lie-Bäcklund symmetry and reduction of evolution equation. *J. Phys. A Math. Gen.* **1995**, *28*, 3841–3850.
33. Fokas, A.S.; Liu, Q.M. Nonlinear interaction of traveling waves of nonintegrable equations. *Phys. Rev. Lett.* **1994**, *72*, 3293–3296.
34. Pedlosky, J. *Geophysical Fluid Dynamics*; Springer: New York, NY, USA, 1979.
35. Olver, P.J. *Applications of Lie Groups to Differential Equations*, 2nd ed.; Graduate Texts in Mathematics; Springer: New York, NY, USA, 1993; Volume 107.

36. Bluman, G.W.; Kumei, S. *Symmetries and Differential Equations*; Springer: New York, NY, USA, 1989; Volume 154.
37. Bluman, G.W.; Cole, J.D. The general similarity solution of the heat equation. *J. Math. Mech.* **1969**, *18*, 1025–1042.
38. Qu, C.Z. Group classification and generalized conditional symmetry reduction of the nonlinear diffusion-convection equation with a nonlinear source. *Stud. Appl. Math.* **1997**, *99*, 107–136.
39. Hamilton, R.S. Three-manifolds with positive Ricci curvature. *J. Diff. Geom.* **1982**, *17*, 255–306.
40. Nariboli, G.A. Self-similar solutions of some nonlinear equations. *Appl. Sci. Res.* **1970**, *22*, 449–461.
41. Cherniha, R.; King, J.R. Lie and conditional symmetries of a class of nonlinear (1 + 2)-dimensional boundary value problems. *Symmetry* **2015**, *7*, 1410–1435.
42. Cimpoiasu, R.; Constantinescu, R. Symmetries and invariants for the 2D-Ricci flow model. *J. Nonlinear Math. Phys.* **2006**, *13*, 285–292.
43. Cimpoiasu, R. Conservation laws and associated Lie symmetries for the 2D Ricci flow model. *Rom. J. Phys.* **2013**, *58*, 519–528.
44. Nadjafikhah, M.; Jafari, M. Symmetry reduction of the two-dimensional Ricci flow equation. *Geometry* **2013**, *373701*, 1–6.
45. Wang, J.H. Symmetries and solutions to geometrical flows. *Sci. China Math.* **2003**, *56*, 1689–1704.
46. Liu, T.P. Compressible flow with damping and vacuum. *Japan J. Indust. Appl. Math.* **1996**, *13*, 25–32.
47. Li, T.H.; Wang, D.H. Blow-up phenomena of solutions to the Euler equations for compressible fluid flow. *J. Diff. Equat.* **2006**, *221*, 91–101.
48. Yuen, M. Perturbational blow-up solutions to the compressible 1-dimensional Euler equations. *Phys. Lett. A* **2011**, *375*, 3821–3825.
49. Yuen, M. Exact, rotational, infinite energy, blowup solutions to the 3-dimensional Euler equations. *Phys. Lett. A* **2011**, *375*, 3107–3113.

Chapter 3:

symmetry

MDPI

Article

Lie and Conditional Symmetries of a Class of Nonlinear (1 + 2)-Dimensional Boundary Value Problems

Roman Cherniha [1,2,*] and John R. King [1]

[1] School of Mathematical Sciences, University of Nottingham, University Park, Nottingham NG7 2RD, UK; E-Mail: John.King@nottingham.ac.uk

[2] Institute of Mathematics, National Academy of Science of Ukraine, 3 Tereshchenkivs'ka Street, Kyiv 01601, Ukraine

* Correspondence: r.m.cherniha@gmail.com; Tel.: +44-750-675-3939.

Academic Editor: Sergei Odintsov

Received: 30 June 2015 / Accepted: 11 August 2015 / Published: 17 August 2015

Abstract: A new definition of conditional invariance for boundary value problems involving a wide range of boundary conditions (including initial value problems as a special case) is proposed. It is shown that other definitions worked out in order to find Lie symmetries of boundary value problems with standard boundary conditions, followed as particular cases from our definition. Simple examples of direct applicability to the nonlinear problems arising in applications are demonstrated. Moreover, the successful application of the definition for the Lie and conditional symmetry classification of a class of (1 + 2)-dimensional nonlinear boundary value problems governed by the nonlinear diffusion equation in a semi-infinite domain is realised. In particular, it is proven that there is a special exponent, $k = -2$, for the power diffusivity u^k when the problem in question with non-vanishing flux on the boundary admits additional Lie symmetry operators compared to the case $k \neq -2$. In order to demonstrate the applicability of the symmetries derived, they are used for reducing the nonlinear problems with power diffusivity u^k and a constant non-zero flux on the boundary (such problems are common in applications and describing a wide range of phenomena) to (1 + 1)-dimensional problems. The structure and properties of the problems obtained are briefly analysed. Finally, some results demonstrating how Lie invariance of the boundary value problem in question depends on the geometry of the domain are presented.

Keywords: Lie symmetry; Q-conditional symmetry; nonlinear boundary-value problem; nonlinear diffusion; exact solution

1. Introduction

Nowadays, the Lie symmetry method is widely applied to study partial differential equations (including multi-component systems of multidimensional PDEs), notably for their reductions to ordinary differential equations (ODEs) and for constructing exact solutions. There are a huge number of papers and many excellent books (see, e.g., [1–5] and the papers cited therein) devoted to such applications. During recent decades, other symmetry methods, which are based on the classical Lie method, were derived. The Bluman–Cole method of non-classical symmetry (another widely-used terminology is Q-conditional symmetry, proposed in [3]) is perhaps the best known among them, and the recent book [6] summarizes results obtained by means of this approach for scalar PDEs (see also the recent papers [7,8] for some results and references in the case of nonlinear PDE systems).

However, a PDE cannot model any real process without additional condition(s) on the unknown function(s), a boundary value problem (BVP) based on the given PDE being needed to describe real processes arising in nature or society. One may note that symmetry-based methods have not been widely used for solving BVPs (we include initial value problems within this terminology defining the initial condition as a particular case of a boundary condition). The obvious reason follows from the following observation: the relevant boundary and initial conditions are usually not invariant under any transformations, *i.e.*, they do not admit any symmetry of the governing PDE. Nevertheless, there are some classes of BVPs that can be solved by means of the Lie symmetry-based algorithm. This algorithm uses the notion of Lie invariance of the BVP in question. Probably the first rigorous definition of Lie invariance for BVPs was formulated by G.W. Bluman in the 1970s [9] (the definition and several examples are summarized in the book [1]). This definition was used (explicitly or implicitly) in several papers to derive exact solutions of some BVPs. It should be noted that Ibragimov's definition of BVP invariance [10] (see also his recent paper [11]), which was formulated independently, is equivalent to Bluman's. On the other hand, one notes that Bluman's definition does not suit all types of boundary conditions. Notably, the definition does not work in the case of boundary conditions involving points at infinity. In recent papers [12–14], a new definition of Lie invariance of BVPs with a wide range of boundary conditions (including those involving points at infinity and moving surfaces) was formulated. Moreover, an algorithm for the group classification for the given class of BVPs was worked out and applied to a class of nonlinear two-dimensional and multidimensional BVPs of the Stefan type with the aim of showing their efficiency.

However, there are many realistic BVPs that cannot be solved using any definition of the Lie invariance of the BVP, for instance because the relevant governing equations do not admit any Lie symmetry (or possess a trivial one only). Hence, definitions involving more general types of symmetries should be worked out. Having this in mind, in this paper, we consider a class of (1 + 2)-dimensional nonlinear boundary value problems (BVPs) modelling heat transfer (for example) in the semi-infinite domain $\omega = \{(x_1, x_2) : -\infty < x_1 < +\infty,\ x_2 > 0\}$:

$$\frac{\partial u}{\partial t} = \nabla.(d(u)\nabla u) \quad (x_1, x_2) \in \omega,\ t \in \mathbb{R} \tag{1}$$

$$x_2 = 0 : d(u)\frac{\partial u}{\partial x_2} = q(t) \tag{2}$$

$$x_2 \to +\infty : \frac{\partial u}{\partial x_2} = 0 \tag{3}$$

where $u(t, x_1, x_2)$ is an unknown function describing a temperature field (say), $d(u)$ is the positive coefficient of thermal conductivity, $q(t)$ is a specified function describing the heat flux of energy absorbed at (or radiating from) the surface $x_2 = 0$, zero flux is prescribed at infinity (actually, one should use the condition $d(u)\frac{\partial u}{\partial x_2} = 0$, but we assume that two are equivalent, provided $d(u) \neq 0$ when $x_2 \to +\infty$; a discussion of the dependence of the admissibility of such boundary conditions at infinity on $d(u)$ is a delicate one that lies outside the scope of the current work; however, some results are presented at the end of Sections 3 and 4) and the standard notation $\nabla = \left(\frac{\partial}{\partial x_1}, \frac{\partial}{\partial x_2}\right)$ is used. Hereafter, we assume that $d(u) \neq$ constant (otherwise, the problem is linear and can be solved by the well-known classical methods) and all of the functions arising in Problem (1)–(3) are sufficiently smooth. It should be noted that we do not prescribe any initial condition assuming that the initial profile can be an arbitrary smooth function that can be specified with respect to a symmetry of BVPs (1)–(3) in question.

The paper is organised as follows. In Section 2, a theoretical background is developed, and relevant examples are presented. In Section 3, the Lie symmetry classification of BVPs of the form (1)–(3) is derived and the main result is presented in Theorem 2. In Section 4, all possible reductions of BVPs (1)–(3) with the power-law thermal conductivity u^k and a non-zero constant flux $q(t) = q_0$ that admit reductions to (1 + 1)-dimensional BVPs are constructed. In Section 5, the conditional symmetry classification of the BVP class (1)–(3) is derived, and the relevant reductions are presented. In Section 6, some results demonstrating how Lie invariance of the BVP in question depends on the geometry of the domain are presented. Finally, we discuss the results obtained and present some conclusions in the last section.

2. Theoretical Background: Definitions and Examples

Here, we restrict ourselves to the case when the basic equation of the BVP is a multidimensional evolution PDE of k-th order in space ($k \geq 2$), *i.e.*, our considerations here go well beyond the specific Equation (1). Thus, the relevant BVP may be formulated as follows:

$$u_t = F\left(t, x, u, u_x, \ldots, u_x^{(k)}\right), \; x \in \omega \subset \mathbb{R}^n, \; t > 0 \tag{4}$$

$$s_a(t,x) = 0: \; B_a\left(t, x, u, u_x, \ldots, u_x^{(k_a)}\right) = 0, \; a = 1, 2, \ldots, p, \; k_a < k \tag{5}$$

where F and B_a are smooth functions in the corresponding domains, ω is a domain with smooth boundaries and $s_a(t,x)$ are smooth curves. Hereafter, the subscripts t and $x = (x_1, \ldots, x_n)$ denote differentiation with respect to these variables and $u_x^{(j)}$, $j = 1, \ldots, k$ denotes a totality of partial derivatives of the order j with respect to the space variables, for example $u_x^{(k)} = (u_{x_1 \ldots x_1}, \ldots, u_{x_{j_1}, \ldots, x_{j_n}}, \ldots, u_{x_n \ldots x_n})$, where $u_{x_{j_1}, \ldots, x_{j_n}} = \frac{\partial^k u}{\partial x_{j_1} \ldots \partial x_{j_n}}$, $j = 1, 2, \ldots, k$; $j_1 + \ldots + j_n = k$. We assume that BVPs (4) and (5) have a classical solution (in a usual sense).

Consider the infinitesimal generator:

$$X = \xi^0(t,x)\frac{\partial}{\partial t} + \xi^i(t,x)\frac{\partial}{\partial x_i} + \eta(t,x,u)\frac{\partial}{\partial u} \tag{6}$$

Hereafter, ξ^0, ξ^i and η are known smooth functions, and summation is assumed from one to n over repeated index i in operators. Assuming that this operator defines a Lie symmetry acting both on the (t,x,u)-space and on its projection to (t,x)-space, consider the operator:

$$\underset{k}{X} = \underset{k-1}{X} + \sigma_1^k \frac{\partial}{\partial u_{x_1 \ldots x_1}} + \sigma_2^k \frac{\partial}{\partial u_{x_1 \ldots x_1 x_2}} + \ldots + \sigma_{k_n}^k \frac{\partial}{\partial u_{x_n \ldots x_n}}, \quad k \geq 2 \tag{7}$$

corresponding to the k-th prolongation of X, whose coefficients are calculated via the functions $\xi^0, \ldots, \xi^1, \eta$ and their derivatives by the well-known prolongation formulae [4,5] starting from the first prolongation of X:

$$\underset{1}{X} = X + \sigma_0^1 \frac{\partial}{\partial u_t} + \sigma_1^1 \frac{\partial}{\partial u_{x_1}} + \ldots + \sigma_n^1 \frac{\partial}{\partial u_{x_n}}$$

In Formula (7), k_n is the total number of different k-order derivatives of the function u w.r.t. the space variables (there is no need to take into account k-order derivatives involving the time variable, because (4) contains the first-order time derivative only).

Definition 1. *[1] The Lie symmetry X (6) is admitted by the boundary value Problems (4)–(5) if:*

(a) $\underset{k}{X}\left(F\left(x, u, u_x, \ldots, u_x^{(k)}\right) - u_t\right) = 0$ *when u satisfies (4);*
(b) $X(s_a(t,x)) = 0$ *when* $s_a(t,x) = 0$, $a = 1, \ldots, p$;
(c) $\underset{k_a}{X}\left(B_a\left(t, x, u, u_x, \ldots, u_x^{(k_a)}\right)\right) = 0$ *when* $s_a(t,x) = 0$ *and* $B_a|_{s_a(t,x)=0} = 0$, $a = 1, \ldots, p$.

Because BVPs (4)–(5) involve only the standard boundary conditions, Definition 1 cannot be applied to BVPs (1)–(3), which involve boundary conditions defined at infinity. Moreover, Definition 1 cannot be generalised in a straightforward way to the boundary Condition (3) (see an example in [13]). This issue was pointed out in [15], where it was suggested that an appropriate substitution be made to transform the unbounded domain to a bounded one. This idea was formalised in [12,13], where it was shown how this definition can be extended to classes of BVPs with more complicated boundaries and initial conditions. Here, we go essentially further, namely we extend the notion of BVP invariance to the case of operators of conditional symmetry; we describe what kind of transformations can be applied to transform boundary conditions at infinity to those containing no conditions at infinity; and we show that the domain geometry plays an important role in the multidimensional ($n > 1$) case.

Consider a BVP for the evolution Equation (4) involving Condition (5) and boundary conditions at infinity:

$$\gamma_c(t,x) = \infty : \gamma_c\left(t, x, u, u_x, \ldots, u_x^{(k_c)}\right) = 0, \ c = 1, 2, \ldots, p_\infty \tag{8}$$

Here, $k_c < k$ and p_∞ are given numbers and the $\gamma_c(t,x)$ are specified functions by which the domain (t,x) on which the BVP in question is defined extends to infinity in some directions. We assume that all of the functions arising in (4), (5) and (8) and the number of boundary and initial conditions are such that a classical solution still exists for this BVP.

Let us assume that the operator:

$$Q = \xi^0(t,x,u)\frac{\partial}{\partial t} + \xi^i(t,x,u)\frac{\partial}{\partial x_i} + \eta(t,x,u)\frac{\partial}{\partial u} \tag{9}$$

is a Q-conditional symmetry of PDE (4), *i.e.*, the following criterion is satisfied (see, e.g., [1]):

$$\underset{k}{Q}\left(u_t - F\left(t, x, u, u_x, \ldots, u_x^{(k)}\right)\right)\Big|_{M} = 0 \tag{10}$$

where $\underset{k}{Q}$ is the k-th prolongation of Q and the manifold $M = \{u_t - F\left(t, x, u, u_x, \ldots, u_x^{(k)}\right) = 0, \ Q(u) = 0\}$ with $Q(u) \equiv \xi^0(t,x,u)u_t + \xi^i(t,x,u)u_{x_i} - \eta(t,x,u)$.

Remark 1. *Rigorously speaking, one needs to reduce the manifold M by adding the differential consequences of equation $Q(u) = 0$ up to order k, which leads to huge technical problems in the application of the criterion obtained. However, in the case of evolution equations, the resulting symmetries will be still the same provided $\xi^0(t,x,u) \neq 0$ in Q, because each such differential consequence contains one or more mixed derivatives of the function u w.r.t. the variables t and x, while the evolution equation in question does not involve any such mixed derivatives.*

Let us consider for each $c = 1, 2, \ldots, p_\infty$ the manifold:

$$\mathrm{M} = \{\gamma_c(t,x) = \infty, \gamma_c\left(t, x, u, u_x, \ldots, u_x^{(k_c)}\right) = 0\} \tag{11}$$

in the extended space of variables $t, x, u, u_x, \ldots, u_x^{(k_c)}$ (obviously, the space dimensionality will depend on k_c and, e.g., one obtains $n+2$-dimensional space (t,x,u) in the case of Dirichlet boundary conditions). We assume that there exists a smooth bijective transformation of the form:

$$\tau = f(t,x), \quad y = g(t,x), \quad w = h(t,x,u) \tag{12}$$

where $y = (y_1, \ldots, y_n)$, $f(t,x)$ and $h(t,x,u)$ are smooth functions and $g(t,x)$ is a smooth vector function that maps the manifold M into:

$$\mathrm{M}^* = \{\gamma_c^*(\tau,y) = 0, \gamma_c^*\left(\tau, y, u, u_y, \ldots, u_y^{(k_c^*)}\right) = 0\} \tag{13}$$

of the same dimensionality in the extended space $\tau, y, w, w_y, \ldots, w_y^{(k_c)}$ (here $k_c^* \leq k_c$).

Definition 2. *BVPs (4)–(5) and (8) are Q-conditionally invariant under operator (9) if:*

(a) Criterion (10) is satisfied;
(b) $Q(s_a(t,x)) = 0$ when $s_a(t,x) = 0$, $B_a|_{s_a(t,x)=0} = 0$, $a = 1, \ldots, p$;

(c) $\underset{k}{Q}\left(B_a\left(t, x, u, u_x, \ldots, u_x^{(k_a)}\right)\right) = 0$ *when* $s_a(t,x) = 0$ *and* $B_a|_{s_a(t,x)=0} = 0, \ \ a = 1, \ldots, p$;
(d) *there exists a smooth bijective transform (12) mapping* M *into* M^* *of the same dimensionality;*
(e) $Q^*(\gamma_c^*(\tau, y)) = 0$ *when* $\gamma_c^*(\tau, y) = 0, \ c = 1, 2, \ldots, p_\infty$;
(f) $\underset{k_c^*}{Q^*}\left(\gamma_c^*\left(\tau, y, u, u_y, \ldots, u_y^{(k_c^*)}\right)\right) = 0$ *when* $\gamma_c^*(\tau, y) = 0$ *and* $\gamma_c^*|_{\gamma_c^*(\tau,y)=0} = 0, \ \ c = 1, \ldots, r,$

where γ_c^* *and* $\gamma_c^*(\tau, y)$ *are the functions* γ_c *and* $\frac{1}{\gamma_c(t,x)}$*, respectively, expressed via the new variables. Moreover, the operator* Q^**, i.e., (9) in the new variables, is defined almost everywhere (i.e., except at a finite number of points) on* M^**.*

Remark 2. *Because any Q-conditional symmetry operator can be multiplied by an arbitrary function, say* $s_a(t,x)$*, Definition 2 implies that the operator Q does not vanish provided* $s_a(t,x) = 0$*. Rigorously speaking, this restriction is valid also for Definition 1.*

This definition coincides with Definition 1 if Q is a Lie symmetry operator and there are no boundary conditions at infinity (*i.e.*, of the form (8)). In the case of BVPs involving boundary conditions at infinity, Definition 2 essentially generalises the definitions of Lie and conditional symmetry proposed in [13,16], respectively. In fact, those definitions are valid only for two-dimensional BVPs with essentially restricted forms of boundary conditions at infinity (for example, they work for the Dirichlet conditions, but cannot be applied for the Neumann conditions, as shown in Example 2 below) because they were created using the above-mentioned substitution from [15], which is a very particular case of (12) with $n = 1$, $\tau = t$, $y = \frac{1}{x}$, $w = u$.

Now, we demonstrate how this definition works using simple examples. Because each Q-conditional symmetry is automatically a Lie symmetry, we start from an example involving the Lie symmetry only and continue with a second example involving pure conditional invariance.

Example 1. *Consider the nonlinear BVP modelling heat transfer in a semi-infinite solid rod, assuming that thermal diffusivity depends on temperature and that the rod is insulated at the left endpoint. Hereafter, we neglect the initial distribution of the temperature in the rod. Thus, the nonlinear BVP reads as:*

$$\frac{\partial u}{\partial t} = \frac{\partial}{\partial x}\left(d(u)\frac{\partial u}{\partial x}\right), \ t > 0, \ 0 < x < +\infty \tag{14}$$

$$x = 0 : d(u)\frac{\partial u}{\partial x} = 0, \ t > 0 \tag{15}$$

$$x = +\infty : u = u_\infty, \ t > 0 \tag{16}$$

where $u(t,x)$ *is an unknown temperature field,* $d(u)$ *is a thermal diffusivity coefficient and* u_∞ *is a given temperature at infinity.*

The maximal algebra of invariance (MAI), i.e., the Lie algebra containing any Lie symmetry of the governing Equation (14), is well-known and is spanned by the basic operators [5] $\langle \partial_t, \partial_x, 2t\partial_t + x\partial_x \rangle$ *provided* $d(u)$ *is an arbitrary function. Obviously, BVPs (14)–(16) are invariant w.r.t. the operator* ∂_t *because the boundary conditions do not involve the time variable, while the first one affects the operator* ∂_x *(see Item (b) of Definition 2). Hence, we need to examine the third operator. Items (b)–(c) of Definition 2 are fulfilled in the case of the first boundary condition, while one needs to find an appropriate bijective transform of the form (12) to check Items (d)–(f).*

Let us consider the obvious change of variables:

$$\tau = t, \quad y = \frac{1}{x}, \quad w = u \tag{17}$$

which maps $M = \{x = \infty, u = u_\infty\}$ *into* $M^* = \{y = 0, w = u_\infty\}$*; both manifolds have the same dimensionality* $D = 1$*, because they are lines in the three-dimensional space of variables, i.e., Item (d) is fulfilled. Transform (17) maps the operator in question to the form* $2\tau\partial_\tau - y\partial_y$*, and now, one easily checks* *that this operator satisfies Items (e)–(f) of Definition 2 on M^*. Thus, BVPs (14)–(16) are invariant under the two-dimensional MAI $\langle \partial_t, 2t\partial_t + x\partial_x, \rangle$, provided $d(u)$ is an arbitrary function. Note that similarity reduction associated with the second operator of this algebra is well known (see, for example, [14]) and could of course have been identified without use of the definition developed here.*

Example 2. *Consider the reaction-diffusion-convectionequation:*

$$\frac{\partial u}{\partial t} = \frac{\partial}{\partial x}\left(u^m u_x\right) + \lambda_1 u^m u_x + \lambda_2 u^{-m} \tag{18}$$

where λ_k, $k = 1,2$ and $m \neq -1, 0$ are arbitrary constants, while $u_x = \frac{\partial u}{\partial x}$. Let us formulate a BVP with the governing Equation (18) in the domain $\omega = \{(t,x) : t > 0,\ x \in (0, +\infty)\}$ using the Neumann boundary conditions:

$$x = 0 :\ u_x = \varphi(t) \tag{19}$$

and

$$x \to +\infty :\ u_x = 0 \tag{20}$$

where $\varphi(t)$ is the specified smooth function. Therefore, (18)–(20) are nonlinear BVPs, which is the standard object for investigation. In [17], it was proven that (18) admits the Q-conditional symmetry:

$$Q = \frac{\partial}{\partial t} + \lambda_2 u^{-m}\frac{\partial}{\partial u} \tag{21}$$

which is not equivalent to a Lie symmetry provided $\lambda_2 \neq 0$.

Now, we apply Definition 2 to BVPs (18)–(20) in order to obtain correctly-specified constraints when this problem is conditionally invariant under Operator (21). Obviously, the first item is fulfilled by the correct choice of the operator. Item (b) is satisfied automatically because of the operator structure. A non-trivial result is obtained by the application of Item (c) to the boundary Condition (19). In fact, calculating the first prolongation (i.e., $k_a = 1$) of Operator (21):

$$\begin{aligned}
\underset{1}{Q} &= Q + \sigma_0^1\tfrac{\partial}{\partial u_t} + \sigma_1^1\tfrac{\partial}{\partial u_x} \\
&= Q + \left(\eta_t + u_t\eta_u - u_t(\xi_t^0 + u_t\xi_u^0) - u_x(\xi_t^1 + u_t\xi_u^1)\right)\tfrac{\partial}{\partial u_t} \\
&+ \left(\eta_x + u_x\eta_u - u_t(\xi_x^0 + u_x\xi_u^0) - u_x(\xi_x^1 + u_x\xi_u^1)\right)\tfrac{\partial}{\partial u_x} \\
&= \tfrac{\partial}{\partial t} + \lambda_2 u^{-m}\tfrac{\partial}{\partial u} - m\lambda_2 u^{-m-1}u_t\tfrac{\partial}{\partial u_t} - m\lambda_2 u^{-m-1}u_x\tfrac{\partial}{\partial u_x}
\end{aligned} \tag{22}$$

and acting on (19), one obtains the first-order ODE:

$$x = 0 :\ \dot{\varphi}(t) + m\lambda_2\varphi(t)u^{-m-1} = 0 \tag{23}$$

to find the function $\varphi(t)$. Because the BVP in question involves the condition at infinity (20), we also need to examine Items (d)–(f). Let us consider the following change of variables (substitution of (17) does not work in the case of zero Neumann conditions):

$$\tau = t, \quad y = \frac{1}{x}, \quad w = \frac{u}{x} \tag{24}$$

which maps $M = \{x = +\infty,\ u_x = 0\}$ into $M^* = \{y = 0,\ w = 0\}$. Since both manifolds have the same dimensionality $D = 1$, Item (d) is fulfilled. Transform (24) maps the operator in question to the form:

$$Q^* = \frac{\partial}{\partial \tau} + \lambda_2 y^{1+m}w^{-m}\frac{\partial}{\partial w} \tag{25}$$

and now, one easily checks that this operator satisfies Items (e)–(f) of Definition 2 on M^* provided $m \in (-1, 0)$. In the case $m \notin [-1, 0]$, one needs the additional constraint $y^{1+m}w^{-m} \to 0$ as $(y,w) \to (0,0)$ in order to satisfy Item (f) in Definition 2 (this case is not examined here, but it can be done in a similar way).

Thus, we have shown that BVPs (18)–(20) are Q-conditionally invariant under Operator (21) if and only if Condition (23) and constraint $m \in (-1, 0)$ hold.

One may note that Condition (23) corresponds to a Dirichlet condition, and generally speaking, will not be compatible with the Neumann Condition (19). Happily (but not coincidentally), there is no contradiction in this case. In fact, Operator (21) generates the ansatz:

$$u^{1+m} = f(x) + \lambda_2(m+1)t$$

where $f(x)$ is an unknown function. Substituting this ansatz into the governing Equation (18) and solving the ordinary differential equation obtained, one finds that $f(x) = C_0 + C_1 e^{-\lambda_1 x}$ (C_0 and C_1 are arbitrary constants); hence, the exact solution:

$$u = \left(C_0 + C_1 e^{-\lambda_1 x} + \lambda_2(m+1)t\right)^{\frac{1}{1+m}} \quad (26)$$

of the nonlinear Equation (18) is constructed. Now, we need to specify the function $\varphi(t)$ using (19); therefore, $\varphi(t) = -\frac{\lambda_1 C_1}{1+m}\left(C_0 + C_1 + \lambda_2(m+1)t\right)^{-\frac{m}{1+m}}$ is obtained by simple calculations. The last step is to check the additional Condition (23), which is fulfilled identically by the function $\varphi(t)$ obtained.

Note that there is a case when Constraint (23) does not produce any boundary condition, namely $\varphi(t) = 0$, i.e., the problem with the zero Neumann conditions (zero flux) on the boundary $x = 0$ and at infinity $x = +\infty$ is invariant under the Q-conditional symmetry (21) provided $m \in (-1, 0)$.

3. Lie Symmetry Classification of the BVPs Class (1)–(3)

Since the BVP class (1)–(3) contains two arbitrary functions, $d(u)$ and $q(t)$, the problem of Lie group classification arises, *i.e.*, to describe all possible Lie (or indeed conditional) symmetries that can be admitted by BVPs from this class depending on the pair (d, q). The problem of group classification for classes of partial differential equations (PDEs) was formulated by Ovsiannikov using notions of the equivalence group E_{eq} and the principal (kernel) group of invariance [5]. The relevant algorithm for solving this problem, the so called Lie–Ovsiannikov algorithm, is well known (see [5] for details). During the last few decades, this problem was further studied, and more efficient algorithms were worked out (see, e.g., [18–21] and the references cited therein). It is widely accepted that the problem of group classification is completely solved for the given PDE class if it has been proven that:

(i) The Lie symmetry algebras are the maximal algebras of invariance of the relevant PDEs from the list obtained;
(ii) All PDEs from the list are inequivalent with respect to a set of transformations, which are explicitly (or implicitly) presented and, generally speaking, may not form any group;
(iii) Any other PDE from the class that admits a non-trivial Lie symmetry algebra is reduced by transformations from the set to one of those from the list.

In [12,14], an algorithm for solving the group classification problem for BVP classes was proposed. The algorithm, which is based on the concept of the equivalence group [22] of a class of BVPs, has its origins in the Lie–Ovsiannikov algorithm. The main steps of the algorithm in the case of the BVP class (1)–(3) can be formulated as follows:

(I) To construct the equivalence group E_{eq} of local transformations that transform the governing Equation (1) into itself;
(II) To find the equivalence group E_{eq}^{BVP} of local transformations that transform the class of BVPs (1)–(3) into itself: to do this, one extends the space of the group E_{eq} action on the prolonged space, where the function q arising in the boundary condition is treated as a new variable;
(III) To perform the group classification of Equation (1) up to local transformations generated by the group E_{eq}^{BVP};
(IV) Using Definition 2, to find the principal algebra of invariance of the BVP class (1)–(3), *i.e.*, the algebra admitted by each BVP from this class;

(V) Using Definition 2 and the results obtained in Steps (III)–(IV), to describe all possible E_{eq}^{BVP}-inequivalent BVPs of the form (1)–(3) admitting MAIs of higher dimensionality (depending on the pair (d, q)) than the principal algebra.

The algorithm can also be applied when one is looking for Q-conditional symmetries, because such symmetries cannot generate any new group of transformations; hence, classification can be still carried out modulo the group E_{eq}.

Now, we carry out the group classification of BVPs of the form (1)–(3) using the definition and the algorithm presented above.

As the first step, we find the equivalence group $E_{\rm eq}$ of the class of PDEs (1) by direct calculations and obtain the following result.

Lemma 1. *The equivalence group $E_{\rm eq}$ of the PDE class (1) is formed by the transformations:*

$$\tilde{t} = \alpha t + \gamma_0, \quad \tilde{x} = \beta R(\theta) x + \gamma, \quad \tilde{u} = \delta u + \gamma_u, \quad \tilde{d} = \frac{\beta^2}{\alpha} d$$

where $\alpha, \beta, \gamma_u, \gamma_i \ (i = 0,1,2), \delta, \theta$ are arbitrary real constants obeying the conditions $\alpha\beta\delta \neq 0$ and $\theta \in [-\pi, \pi)$; $R(\theta) = \begin{pmatrix} \cos\theta & \sin\theta \\ -\sin\theta & \cos\theta \end{pmatrix}$ is the rotation matrix; and the vectors $\tilde{x} = \begin{pmatrix} \tilde{x}_1 \\ \tilde{x}_2 \end{pmatrix}$, $x = \begin{pmatrix} x_1 \\ x_2 \end{pmatrix}$ and $\gamma = \begin{pmatrix} \gamma_1 \\ \gamma_2 \end{pmatrix}$.

Note that this equivalence group can be easily extracted from paper [23], where Lie symmetries of the class of reaction-diffusion equations of the form:

$$\frac{\partial u}{\partial t} = \nabla . (d(u) \nabla u) + Q(u) \tag{27}$$

were completely described.

In the second step, we substitute the transformations from the group $E_{\rm eq}$ into (1)–(3) and require that those transformations preserve the structure of the class: hence, we find the set $E_{\rm eq}^{\rm BVP}$ of equivalence transformations that are essentially different, using the result of Lemma 1.

Lemma 2. *The equivalence group $E_{\rm eq}^{\rm BVP}$ of the class of BVPs (1)–(3) is formed by the transformations:*

$$\tilde{t} = \alpha t + \gamma_0, \quad \tilde{x}_1 = \beta x_1 + \gamma_1, \quad \tilde{x}_2 = \beta x_2, \quad \tilde{u} = \delta u + \gamma_u,$$
$$\tilde{d} = \frac{\beta^2}{\alpha} d, \quad \tilde{q} = \frac{\beta\delta}{\alpha} q$$

where $\alpha > 0, \gamma_u, \gamma_i \ (i = 0,1), \delta$ and $\beta > 0$ are arbitrary real constants obeying only the non-degeneracy condition $\delta \neq 0$.

In the third step, we have used the known results [23] (it is interesting to note that Lie symmetries of the nonlinear Equation (1) seem to have been described for the first time in paper [24], incidentally not cited so often as [23] published 13 years later) for solving the relevant group classification problem in the case of the equivalence group $E_{\rm eq}^{\rm BVP}$ and have proven the following statement.

Theorem 1. *All possible MAIs (up to the equivalent transformations from the group $E_{\rm eq}^{\rm BVP}$) of Equation (1) for any fixed non-negative function $d(u) \neq const$ are presented in Table 1. Any other equation of the form (1) is reduced by an equivalence transformation from the group $E_{\rm eq}^{\rm BVP}$ to one of those given in Table 1.*

Table 1. Result of group classification of the class of PDEs (1). MAI, maximal algebra of invariance.

Case	$d(u)$	Basic Operators of MAI
1.	$\forall$	$AE(1,2) = \langle T, X_1, X_2, D, J_{12}\rangle$
2.	$u^k, k \neq 0, -1$	$AE(1,2), D_k$
3.	u^{-1}	$AE(1,2)$, $A(x)\partial_{x_1} + B(x)\partial_{x_2} - 2A_{x_1}u\partial_u$
4.	e^u	$AE(1,2), D_e$

Remark 3. *In Table 1, the following designations of the Lie symmetry operators are used:*

$$\begin{aligned} &T = \partial_t,\ X_1 = \partial_{x_1},\ X_2 = \partial_{x_2},\ D = 2t\partial_t + x_a\partial_{x_a} \\ &J_{12} = x_1\partial_{x_2} - x_2\partial_{x_1},\ D_k = kt\partial_t - u\partial_u,\ D_e = t\partial_t - \partial_u \end{aligned} \tag{28}$$

while $A(x)$ and $B(x)$ (hereafter, $x = (x_1, x_2)$ and summation is assumed over the repeated index $a = 1,2$) are an arbitrary solution of the Cauchy–Riemann system $A_{x_1} = B_{x_2}$, $A_{x_2} = -B_{x_1}$.

Now, one needs to proceed to the final two steps of the group classification algorithm presented above. The result can be formulated in form of the main theorem (Theorem 2), which gives the complete list of the non-equivalent BVPs of the form (1)–(3) and the relevant MAIs.

Theorem 2. *All possible MAIs (up to equivalent transformations from the group $E^{\rm BVP}_{\rm eq}$) of the nonlinear BVPs (1)–(3) for any fixed pair $(d(u), q(t))$, where $d(u) \neq const$, are presented in Table 2. Any other BVP of the form (1)–(3) is reduced by an equivalence transformation from the group $E^{\rm BVP}_{\rm eq}$ from Lemma 2 to one of those listed in Table 2.*

Table 2. Result of group classification of the class of BVPs (1)–(3).

Case	$d(u)$	$q(t)$	Basic Operators of MAI	Relevant Constraints
1.	$\forall$	$\forall$	X_1	
2.	$\forall$	$q_0t^{-\frac{1}{2}}$	X_1, D	
3.	$\forall$	q_0	X_1, T	
4.	$\forall$	0	X_1, T, D	
5.	u^k	q_0t^p	X_1, D_{kp}	$k \neq -2,\ p \neq 0$
6.	u^k	$q_0e^{\pm t}$	$X_1, D_\pm$	$k \neq -2$
7.	u^k	q_0	X_1, T, D_{kp}	$p = 0$
8.	u^k	0	X_1, T, D, D_k	$k \neq -1$
9.	u^{-2}	$\forall$	$X_1, D_\pm$	$k = -2$
10.	u^{-2}	$q_0t^{-\frac{1}{2}}$	X_1, D, D_k	$k = -2$
11.	u^{-1}	0	X_1, T, D, X^∞	$\mathcal{M}$
12.	e^u	q_0t^p	X_1, D_p	$p \neq 0$
13.	e^u	$q_0e^{\pm t}$	$X_1, D_{\pm e}$	
14.	e^u	q_0	X_1, T, D_p	$p = 0$
15.	e^u	0	X_1, T, D, D_e	

Remark 4. *In Table 2 the arbitrary constant $q_0 \neq 0$ and the following designations of the Lie symmetry operators are used:*

$$\begin{aligned} &D_{kp} = (k+2)t\partial_t + [k(p+1)+1]x_a\partial_{x_a} + (2p+1)u\partial_u \\ &D_\pm = \pm(k+2)\partial_t + kx_a\partial_{x_a} + 2u\partial_u \\ &D_p = t\partial_t - (p-1)x_a\partial_{x_a} - (2p-1)u\partial_u \\ &D_{\pm e} = \pm\partial_t - x_a\partial_{x_a} - 2\partial_u \end{aligned} \tag{29}$$

In Case 11, the coefficient $B(x)$ of the operator:

$$X^{\infty} = A(x)\partial_{x_1} + B(x)\partial_{x_2} - 2A_{x_1}u\partial_u$$

must satisfy the set of conditions $\mathcal{M}$:

$$B(x_1, 0) = 0 \tag{30}$$

$$x_2 \to +\infty : \quad \frac{B(x_1, x_2)}{x_2} \neq \infty, \qquad \frac{\partial B(x_1, x_2)}{\partial x_2} \neq \infty \tag{31}$$

Proof. The proof is based at Definition 2, Lemma 2 and Theorem 1. According to the algorithm described above (see Steps (IV) and (V)), we need to examine the four different cases listed in Table 1. First of all, we should consider Case 1 with the aim to find the principal algebra of invariance, *i.e.*, the invariance algebra, admitting by each BVP of the form (1)–(3). Taking the most general form of the Lie symmetry in this case, one obtains:

$$X = (\lambda_0 + 2\lambda_3 t)\partial_t + (\lambda_1 + \lambda_3 x_1 + \lambda_4 x_2)\partial_{x_1} + (\lambda_2 - \lambda_4 x_1 + \lambda_3 x_2)\partial_{x_2} \tag{32}$$

where $\lambda_0, \ldots, \lambda_4$ are arbitrary real constants. Applying Item (a) of Definition 2 to the first part of the boundary condition (2), we immediately obtain:

$$X(x_2)|_{x_2=0} = 0 \Leftrightarrow \lambda_2 = \lambda_4 = 0 \tag{33}$$

To finish application of Item (a), we need the first prolongation of operator (32) with $\lambda_2 = \lambda_4 = 0$, because the boundary condition in question involves the derivative u_{x_2}. Hence, using the prolongation formulae (see, e.g., [4,5]), we arrive at the expression:

$$\underset{1}{X}\left(d(u)\frac{\partial u}{\partial x_2} - q(t)\right)\bigg|_{\mathrm{M}} = -\lambda_3 q(t) - (\lambda_0 + 2\lambda_3 t)\dot{q}(t) = 0 \tag{34}$$

where $\underset{1}{X} = X - \lambda_3(2u_t\partial_{u_t} + u_{x_a}\partial_{u_{x_a}})$ and:

$$\mathrm{M} = \{x_2 = 0,\ d(u)\frac{\partial u}{\partial x_2} = q(t)\} \tag{35}$$

Obviously, the zero flux case $q(t) = 0$ does not produce any constraints; hence λ_0, λ_1 and λ_3 are arbitrary, *i.e.*, the relevant BVP is invariant under a three-dimensional MAI (see Case 4 of Table 4). Rather simple analysis of the linear ODE $\lambda_3 q(t) + (\lambda_0 + 2\lambda_3 t)\dot{q}(t) = 0$ with non-zero $q(t)$ immediately leads to three different possibilities only:

(i) If $q(t)$ is an arbitrary function, then $\lambda_0 = \lambda_3 = 0$, *i.e.*, $X = X_1$;
(ii) If $q(t) = q_0/\sqrt{t + \lambda_0^*}$ with $\lambda_0^* = \lambda_0/(2\lambda_3)$, then $X = \lambda_0 T + \lambda_1 X_1 + \lambda_3 D$ (here, λ_0 and $\lambda_3 \neq 0$ are no longer arbitrary);
(iii) If $q(t) = q_0$, q_0 being a constant, then $\lambda_3 = 0$, *i.e.*, $X = \lambda_0\partial_t + \lambda_2\partial_{x_2}$.

The function $q(t)$ and the operator X arising in Item (ii) can be simplified using Lemma 2 (see the transformation for t) as follows $q(t) \rightarrowtail q_0/\sqrt{t},\ X \rightarrowtail \lambda_1 X_1 + D$ Now, we need to find an appropriate transform of the form (12). Let us consider the transformation:

$$\tau = t, \quad y_1 = x_1, \quad y_2 = x_2^{-\epsilon}, \epsilon > 0, \quad u = x_2 w \tag{36}$$

which transforms the manifold:

$$\mathrm{M} = \{x_2 \to +\infty,\ \frac{\partial u}{\partial x_2} = 0\} \tag{37}$$

into:

$$\mathrm{M}^* = \{y_2 = 0,\ w = 0\} \tag{38}$$

provided the function w is differentiable at $y_2 = 0$ (it should be noted that transforming only y (keeping the other variables the same) does not work in that sense).

Now, one realizes that Items (e)–(f) of Definition 2 are automatically fulfilled if $q(t)$ is an arbitrary function and $X = X_1$ (see Item (i) above); hence, we have found the principal algebra of invariance of the BVP class (1)–(3), and one is listed in Case 1 of Table. To examine the other two items, one needs to express the relevant operators via the new variables. In particular, the operator $X = \lambda_1 X_1 + D$ takes the form:

$$X^* = 2t\partial_t + (\lambda_1 + x_1)\partial_{x_1} - \epsilon y_2 \partial_{y_2} - w\partial_w \tag{39}$$

Obviously, Items (e)–(f) of Definition 2 are automatically fulfilled for Operator (39), provided the boundary condition is giving M^*; hence, Case 2 of Table 2 is derived. Obviously, the third possibility for the function $q(t) = q_0$ leads to Case 3 of Table 2. Thus, Case 1 of Table 1 produces the principal algebra of invariance of the BVP class (1)–(3) and three extensions depending on the function $q(t)$ (see Cases 1–4 in Table 2).

Case 2 of Table 1 can be examined in quite a similar way as we have done above for Case 1. Application of Definition 2 and Lemma 2 leads to the six different cases listed in Table 2 (see $5, \ldots, 10$). It should be stressed that the power $k = -2$ is a special one (but not one for Lie invariance of the governing PDE!) and leads to the two additional Cases 9 and 10 (see an analogous result for a (1 + 1)-dimensional BVP in [14,25]).

The most complicated examination is in Case 3 of Table 1, because one needs to analyse an infinite-dimensional Lie algebra. In this case, the MAI of Equation (1) is spanned by the following operators:

$$T = \partial_t,\ X_a = \partial_{x_a},\ J = x_2\partial_{x_1} - x_1\partial_{x_2},\ D = 2t\partial_t + x_a\partial_{x_a},\ D_{-1} = t\partial_t + u\partial_u$$
$$X^\infty = A(x_1, x_2)\partial_{x_1} + B(x_1, x_2)\partial_{x_2} - 2A_{x_1} u\partial_u$$

Hence, the most general form of Lie symmetry operator is:

$$\begin{aligned} X = {} & (\lambda_0 + (2\lambda_4 + \lambda_5)t)\partial_t + (\lambda_1 + \lambda_4 x_1 + \lambda_3 x_2 + \lambda_6 A(x_1, x_2))\partial_{x_1} \\ & + (\lambda_2 - \lambda_3 x_1 + \lambda_4 x_2 + \lambda_6 B(x_1, x_2))\partial_{x_2} - (2\lambda_6 A_{x_1} - \lambda_5)u\partial_u \end{aligned} \tag{40}$$

where $\lambda_0, \ldots, \lambda_6$ are arbitrary real constant.

Obviously, we should assume $\lambda_6 \neq 0$, otherwise particular cases of the results already derived for $d(u) = u^k$ will be obtained. First of all, we simplify Operator (40) as follows. Because the functions $A(x_1, x_2)$ and $B(x_1, x_2)$ are harmonic, then we can construct the functions $\bar{A} = \lambda_1 + \lambda_3 x_2 + \lambda_6 A(x_1, x_2)$ and $\bar{B} = \lambda_2 - \lambda_3 x_1 + \lambda_6 B(x_1, x_2)$. It can be easily seen that the functions $\bar{A}$ and $\bar{B}$ are also harmonic. Thus, without losing generality, the operator X reduces to the form:

$$X = (\lambda_0 + (2\lambda_4 + \lambda_5)t)\partial_t + (\bar{A}(x_1, x_2) + \lambda_4 x_1)\partial_{x_1} + (\bar{B}(x_1, x_2) + \lambda_4 x_2)\partial_{x_2} - (2\bar{A}_{x_1} - \lambda_5)u\partial_u \tag{41}$$

Applying Items (b)–(c) of Definition 2 to the boundary Condition (2), we obtain:

$$X(x_2)|_{x_2=0} = 0, \quad \underset{1}{X}\left(u^{-1}\frac{\partial u}{\partial x_2} - q(t)\right)\Big|_{\mathrm{M}} = 0 \tag{42}$$

where the first prolongation of the operator X has the form $\underset{1}{X} = X + \rho_0\partial_{u_t} + \rho_a\partial_{u_{x_a}}$, and M is defined in (35). We need to calculate only the coefficient ρ_2, because (42) does not involve any other derivatives. Since the known formulae mentioned above produce in the case of Operator (41):

$$\rho_2 = -2\frac{\partial^2 \bar{B}}{\partial x_2^2}u + \left(\lambda_4 + \lambda_5 - 3\frac{\partial \bar{B}}{\partial x_2}\right)u_{x_2} - \frac{\partial \bar{A}}{\partial x_2}u_{x_1} \tag{43}$$

the invariance conditions (42) are simplified to the form:

$$\bar{B}(x_1,0)=0,\quad \frac{\partial \bar{A}(x_1,0)}{\partial x_2}=0,\quad \frac{\partial^2 \bar{B}(x_1,0)}{\partial x_2^2}=0 \tag{44}$$

provided $q(t)=0$ (we remind the reader that the functions $\bar{A}$ and $\bar{B}$ satisfy the Cauchy–Riemann system). Obviously, Condition (44) is equivalent to this (30) if one takes into account that the second and third equations in (44) are direct consequences of the first equation.

To finish the examination of Case 3 of Table 1 when the nonlinear BVP in question involves zero flux $q(t)=0$, one needs to check the invariance of the boundary Condition (3). Hence, using again Transformation (36) and applying Items (f) and (e) of Definition 2, one arrives at the restrictions:

$$y_2=0:\ \bar{B}(x_1,y_2^{-1/\epsilon})y_2^{1+1/\epsilon}=0 \tag{45}$$

and

$$\left(\left(\bar{B}(x_1,y_2^{-1/\epsilon})y_2^{1/\epsilon}+2\frac{\partial \bar{A}(x_1,y_2^{-1/\epsilon})}{\partial x_1}\right)w\right)\Bigg|_{M^*}=0 \tag{46}$$

Because Transformation (36) is bijective and differentiable, Formulae (45)–(46) are equivalent to (31). Hence, we have proven that BVPs (1)–(3) with $d(u)=u^{-1}$ and $q(t)=0$ admit Operator (41) provided Restrictions (30)–(31) are fulfilled. This immediately leads to the result presented in Case 11 of Table 2 (the rotation operator J_{12} must be excluded because its coefficient $B=-x_1$ does not satisfy (30)).

The invariance Condition (42) leads to more complicated analysis if $q(t)\neq 0$. We omit here the relevant routine analysis and present the result only: the restrictions obtained on the functions $\bar{A}$ and $\bar{B}$ lead to the correctly-specified functions $q(t)$ and MAIs listed in Cases 5–7 with $k=-1$ of Table 2 only. Thus, examination of Case 3 of Table 1 when $d(u)=u^{-1}$ is completed.

Finally, Case 4 of Table 1 should be analysed. It turns out that the results obtained are very similar to Case 2 of Table 1 when $k\neq -2$; therefore, the MAIs of the same dimensionality and the fluxes $q(t)$ of the same forms were derived (see Cases 12–15 in Table 2).

The proof is now completed. □

While Restrictions (30)–(31) on the harmonic functions A and B are very strong, the MAI of the problem in Case 11 is still infinite-dimensional. The real and imaginary parts of the complex function $z^{-n}=(x_1+ix_2)^{-n}$ with arbitrary $n=1,2,3,\ldots$ generate the operator of the form X^{∞}, which is a symmetry of BVPs (1)–(3) with $d(u)=u^{-1}$ and $q(t)=0$. Note that here, we allow the singular behaviour of X^{∞} at the origin $(x_1,x_2)=(0,0)$.

Example 3. *The complex function $z^{-1}=(x_1+ix_2)^{-1}$ generates the operator:*

$$\frac{x_1}{x_1^2+x_2^2}\partial_{x_1}-\frac{x_2}{x_1^2+x_2^2}\partial_{x_2}+2\frac{x_1^2-x_2^2}{(x_1^2+x_2^2)^2}u\partial_u \tag{47}$$

Applying Items (b) and (c) of Definition 2, one obtains (42) and (43) with $\lambda_4=\lambda_5=0$ and $\bar{A}=\frac{x_1}{x_1^2+x_2^2}$, $\bar{B}=-\frac{x_2}{x_1^2+x_2^2}$. Now, one easily checks that the invariance conditions are satisfied (42) because the given functions $\bar{A}$ and $\bar{B}$ fulfil Condition (44).

Let us consider Transformation (36). As indicated above, one transforms Manifold (37) into (38). Simultaneously, Operator (47) takes the form:

$$\frac{y_1}{|y|_\epsilon^2}\partial_{y_1}+\frac{1}{|y|_\epsilon^2}(\epsilon y_2\partial_{y_2}+w\partial_w)+2\frac{y_1^2-y_2^{-2/\epsilon}}{|y|_\epsilon^4}w\partial_w \tag{48}$$

where $|y|_\epsilon^2=y_1^2+y_2^{-2/\epsilon}$.

After directly checking Items (d)–(f) of Definition 2 in the case of Manifold (38) and Operator (48), one concludes that (47) is a Lie symmetry operator of BVPs (1)–(3) with $d(u) = u^{-1}$ *and* $q(t) = 0$.

We conclude this section by presenting the following observation. Let us replace the last condition in BVPs (1)–(3) by:

$$x_2 \to +\infty : d(u)\frac{\partial u}{\partial x_2} = 0 \qquad (49)$$

which is more usually adopted in applications. It can be easily checked by direct calculations (each Lie symmetry operator generates the corresponding Lie group of transformations) that the results presented in Table 2 are still valid for BVPs of the form (1), (2) and (49). Moreover, the assumption $d(u) \neq 0$ for $x_2 \to +\infty$ is not important (though it is of course significant with respect to BVP theory). However, one should ideally show that there are no cases other than those presented in Table 2. Unfortunately, this is a non-trivial task; in particular, Transformation (36) does not work in all cases as above. For example, to examine the case of the power-law diffusivity $d(u) = u^k$, $k \neq -1$, one could use the transformation:

$$\tau = t, \quad y_1 = x_1, \quad y_2 = x_2^{-\epsilon}, \epsilon > 0, \quad u = x_2^{\frac{1}{k+1}} w$$

which transforms M into:

$$\mathsf{M}^* = \{y_2 = 0,\ w^{k+1} = 0\}$$

4. Lie Symmetry Reduction of Some BVPs of the Form (1)–(3)

First of all, it should be noted that each BVP of the form (1)–(3) reduces to a (1 + 1)-dimensional problem using the operator $X_1 = \partial_{x_1}$. However, the problem obtained is simply the corresponding (1 + 1)-dimensional one, with no dependence on x_1; hence, we do not consider such a reduction below.

Another special case arises for each BVPs (1)–(3) with $q(t) = q_0$ (Case 3 of Table 2) when the problem reduces to the stationary one using the operator $T = \partial_t$:

$$\nabla.\,(d(U)\nabla U) = 0 \qquad (50)$$

$$x_2 = 0 : d(U)\frac{\partial U}{\partial x_2} = q_0 \qquad (51)$$

$$x_2 \to +\infty : \frac{\partial U}{\partial x_2} = 0 \qquad (52)$$

where $U(x_1, x_2)$ is an unknown function. BVPs (50)–(52) are linearisable via the Kirchoff substitution $W = \int d(U)dU$, and the linear problem obtained can be treated by the classical methods for solving linear problems for the Laplace equation.

A brief analysis of Table 2 shows that seven cases when the relevant problems are invariant under the MAI of dimensionality three and higher are the most interesting, because a few different reductions to BVPs of lower dimensionality can be obtained. Obviously, the most complicated case occurs for the critical exponent $k = -1$ (see Case 11), and we are going to treat in detail this one elsewhere. On the other hand, Cases 7 and 8 seem to be the most interesting, because the power diffusivity u^k is very common in applications and describes a wide range of phenomena depending on the value of k.

Let us consider Case 7. Because the operator D_{kp} with $p = 0$ has the form:

$$D_{k0} = (k+2)t\partial_t + (k+1)x_a\partial_{x_a} + u\partial_u \qquad (53)$$

one needs to consider three different cases:

(i) $k \neq -1, -2$;
(ii) $k = -2$;
(iii) $k = -1$.

In the first case, the Lie algebra $\langle X_1, T, D_{k0}\rangle$ leads only to two essentially different reductions, via the operators $T + vX_1,\ v \in \mathbb{R}$ and D_{k0}. Obviously, the operator $T + vX_1,\ v \in \mathbb{R}$ generates the travelling wave ansatz:

$$u = \phi(y, x_2), \quad y = x_1 - vt \tag{54}$$

which reduces the nonlinear BVP:

$$\frac{\partial u}{\partial t} = \nabla.\left(u^k \nabla u\right), \ (x_1, x_2) \in \omega,\ t \in \mathbb{R} \tag{55}$$

$$x_2 = 0 : u^k \frac{\partial u}{\partial x_2} = q_0 \tag{56}$$

$$x_2 \to +\infty : \frac{\partial u}{\partial x_2} = 0 \tag{57}$$

to the (1 + 1)-dimensional elliptic problem:

$$-v\phi_y = (\phi^k \phi_y)_y + (\phi^k \phi_{x_2})_{x_2} \tag{58}$$

$$x_2 = 0 : \phi^k \phi_{x_2} = q_0 \tag{59}$$

$$x_2 \to +\infty : \phi_{x_2} = 0 \tag{60}$$

where ϕ is an unknown function (hereafter, subscripts on ϕ denote differentiation w.r.t. the relevant variables). Like other cases described below, the relevance of the function ϕ to a specific BVP will depend on the behaviour at infinity of the initial data, and we shall make no attempt to explore such matters here in detail.

The operator D_{k0} generates a more complicated ansatz:

$$u = t^{\frac{1}{k+2}} \phi(\omega_1, \omega_2), \quad \omega_a = x_a t^{-\gamma},\ \gamma = \frac{k+1}{k+2} \tag{61}$$

After substituting ansatz (61) into BVPs (55)–(57), direct calculations show that one obtains the (1 + 1)-dimensional elliptic problem:

$$\frac{1}{k+2}\phi - \gamma \omega_a \phi_{\omega_a} = (\phi^k \phi_{\omega_a})_{\omega_a} \tag{62}$$

$$\omega_2 = 0 : \phi^k \phi_{\omega_2} = q_0 \tag{63}$$

$$\omega_2 \to +\infty : \phi_{\omega_2} = 0 \tag{64}$$

In Case (ii), the Lie algebra $\langle X_1, T, -x_a\partial_{x_a} + u\partial_u\rangle$ leads to three essential different reductions, via the operators $T + vX_1,\ v \in \mathbb{R},\ \frac{1}{\lambda}T - x_a\partial_{x_a} + u\partial_u,\ \lambda \neq 0$ and $-x_a\partial_{x_a} + u\partial_u$. Obviously, the operator $T + vX_1,\ v \in \mathbb{R}$ leads to the same ansatz as in Case (i); hence, BVPs (58)–(60) with $k = -2$ are obtained.

The operator $\frac{1}{\lambda}T - x_a\partial_{x_a} + u\partial_u$ generates a new ansatz of the form:

$$u = e^{\lambda t}\phi(\omega_1, \omega_2), \quad \omega_a = x_a e^{\lambda t},\ \lambda \neq 0 \tag{65}$$

which reduces BVPs (55)–(57) with $k = -2$ to the (1 + 1)-dimensional problem:

$$\lambda\phi + \lambda\omega_a\phi_{\omega_a} = (\phi^{-2}\phi_{\omega_a})_{\omega_a} \tag{66}$$

$$\omega_2 = 0 : \phi^{-2}\phi_{\omega_2} = q_0 \tag{67}$$

$$\omega_2 \to +\infty : \phi_{\omega_2} = 0 \tag{68}$$

Because the diffusivity $\phi^{-2} \to \infty$ as $\phi \to 0$, this singular nature of the diffusivity in (66) prevents immediate physical interpretation of such reductions, but it is worth noting that the reduction (65) applies for a continuum of values of the similarity exponent λ. The one-dimensional case is instructive here, yielding the equation:

$$q_0 + \lambda \omega_2 \phi = \phi^{-2} \phi_{\omega_2} \quad (69)$$

This ODE is easily solved by setting $\psi = \omega_2 \phi$, and its general solution can be presented in the implicit form (for $q_0 = 0$ and $\lambda = \frac{q_0^2}{4}$, the solutions are obvious):

$$C = \frac{\phi}{\sqrt{\lambda(\omega_2\phi)^2 + q_0\omega_2\phi + 1}} \left(\frac{q_\lambda + q_0 + 2\lambda\omega_2\phi}{q_\lambda - q_0 - 2\lambda\omega_2\phi} \right)^{\frac{q_0}{2q_\lambda}} \quad (70)$$

where C is an arbitrary non-zero constant and $q_\lambda \equiv \sqrt{q_0^2 - 4\lambda}$. Note that we need $\lambda < q_0^2/4$ in order to obtain a real solution. Because Solution (70) should satisfy also the condition at infinity (68), we need to analyse it as $\omega_2 \to +\infty$. Indeed, it can be noted that:

$$\phi \sim \frac{\phi_\infty}{\omega_2}, \quad \omega_2 \to +\infty \quad (71)$$

where ϕ_∞ is a solution of the quadratic equation:

$$\lambda \phi_\infty^2 + q_0 \phi_\infty + 1 = 0 \quad (72)$$

(there are two roots, and which should be used depends on sign of q_0). Thus, we conclude that:

$$u \sim \frac{\phi_\infty}{x_2}, \quad x_2 \to +\infty, t = 0 \quad (73)$$

whereby the similarity exponent $\lambda \leq q_0^2/4$ is determined in terms of this initial data via (72).

The operator $-x_a\partial_{x_a} + u\partial_u$ generates the ansatz

$$u = x_1^{-1}\phi(t,z), \quad z = \frac{x_2}{x_1} \quad (74)$$

reducing the (1 + 2)-dimensional BVP in question to the (1 + 1)-dimensional parabolic problem:

$$\phi_t = (\phi^{-2}\phi_z)_z + z\left(\phi^{-1} + z\phi^{-2}\phi_z\right)_z \quad (75)$$

$$z = 0 : \phi^{-2}\phi_z = q_0 \quad (76)$$

$$z \to +\infty : \phi_z = 0 \quad (77)$$

Remark 5. *The reduced BVPs (75)–(77) were derived under assumption $x_1 > 0$. In the case $x_1 < 0$, the same problem is obtained, but $z \to +\infty$ should be replaced by $z \to -\infty$.*

Finally, we examine Case (iii), in which the Lie algebra $\langle X_1, T, t\partial_t + u\partial_u \rangle$ arises. There are only two essentially different reductions, via the operators $T + vX_1$ and $\lambda X_1 + t\partial_t + u\partial_u$ where $(v, \lambda) \in \mathbb{R}^2$. The first one again leads to a particular case of BVPs (58)–(60) with $k = -1$, while the second generates a new (1 + 1)-dimensional elliptic problem of the form:

$$\phi - \lambda\phi_w = (\phi^{-1}\phi_w)_w + (\phi^{-1}\phi_{x_2})_{x_2} \quad (78)$$

$$x_2 = 0 : \phi^{-1}\phi_{x_2} = q_0 \quad (79)$$

$$x_2 \to +\infty : \phi_{x_2} = 0 \quad (80)$$

where

$$u = t\phi(w, x_2), \quad w = x_1 - \lambda \log t \quad (81)$$

It should be noted that BVPs (78)–(80) with $\lambda = 0$ are equivalent to the problem (provided $\phi = e^{\psi} \geq 0$):

$$e^{\psi} = \Delta\psi \tag{82}$$

$$x_2 = 0 : \psi_{x_2} = q_0 \tag{83}$$

$$x_2 \to +\infty : \psi_{x_2} = 0 \tag{84}$$

where (82) is the known Liouville equation, which has been widely studied for many years (see, e.g., the books [26,27]), and its general solution (for $n = 2$) is well known.

Now, we make the following observation: while the BVP in question is a parabolic problem, all of the (1 + 1)-dimensional BVPs obtained (except (75)–(77)) are elliptic. Each of the (1 + 1)-dimensional BVPs derived above can be further analysed by symmetry-based, asymptotical and numerical methods, and we shall investigate such matters in a forthcoming paper. Here, we present an interesting example only.

Example 4. *Here, we present in Figure 1 a result of numerical simulations of (75)–(77). Note that we take the initial profile* $\phi_0(z)$ *to agree the with boundary Conditions (76)–(77), namely* $\phi(0,z) = \phi_0(z) \equiv -\frac{1}{q_0(z+z_0)}$, *where* $z_0 > 0$, $q_0 < 0$ *are arbitrary constants. It is appropriate to touch on the large-time behaviour of BVPs (75)–(77). This is best done in polar coordinates. From the symmetry point of view, it means that one uses the ansatz:*

$$u = r^{-1}v(t,\theta), \quad \theta = \arctan\frac{x_2}{x_1}, \; r^2 = x_1^2 + x_2^2 \tag{85}$$

which is equivalent to (74). As a result, the reduced BVP takes form:

$$v_t = (v^{-2}v_\theta)_\theta - v^{-1} \tag{86}$$

$$z = 0 : v^{-2}v_\theta = q_0 \tag{87}$$

$$\theta \to \frac{\pi^-}{2} : \frac{\cos\theta v_\theta}{v} \to 1 \tag{88}$$

Note that the conditions $\theta \to \frac{\pi^-}{2} : v = 0,\ \cos\theta v_\theta \to 0$, *which formally can be also used instead of (88), are inappropriate, because (86) has no smooth solutions satisfying* $v = 0$ *at any finite* θ.

The boundary Condition (88) implies that $v \to +\infty$ *as* $\theta \to \frac{\pi^-}{2}$ *(this allows* $d(u) \equiv u^{-2} = 0$ *when* $x_2 \to +\infty$*); this and (87) immediately lead to the conservation law:*

$$\frac{d}{dt}\int_0^{\frac{\pi}{2}} v\cos\theta d\theta = -q_0 \tag{89}$$

where $v(t,\theta)\cos\theta = \phi(t,\tan\theta)$ *(see (74)). Then, the following three-layer structure is a plausible description of the behaviour of* v *as* $t \to +\infty$.

(A) Outer region $\theta = O(1)$: *Here,* $v \sim \sqrt{t}\psi(\theta)$, *which reduces (86) to:*

$$\frac{1}{2}\psi = (\psi^{-2}\psi_\theta)_\theta - \psi^{-1}$$

hence, one obtains:

$$\psi \sim \frac{1}{\theta\sqrt{-\log\theta}} \quad as \quad \theta \to 0^+, \quad \psi \sim \frac{1}{(\frac{\pi}{2}-\theta)\sqrt{-\log(\frac{\pi}{2}-\theta)}} \quad as \quad \theta \to \frac{\pi^-}{2}$$

(B) Transition region $\theta = O(1)$, *where* $-\lambda < \theta < 0$: *Here, we set* $v = \theta^{-1}\sigma(t,\zeta)$, $\zeta = \log\theta$ *in order to obtain from (86) the equation (note the term* $v^{-1} = \theta\sigma^{-1}$ *is negligible under these scalings):*

$$\sigma_t = (\sigma^{-2}\sigma_\zeta)_\zeta + \sigma^{-2}\sigma_\zeta \tag{90}$$

Whereby the middle term in (90) is negligible (for a large time) and $\sigma \sim \sigma(\theta)$, $\theta = \zeta t^{-1}$, one obtains the equation:

$$-\theta\sigma_\theta = \sigma^{-2}\sigma_\theta$$

with the solution $\sigma = \frac{1}{\sqrt{-\theta}}$. The region dominates the integral in (89), whereby:

$$\int_{-\lambda}^{0} \frac{d\theta}{\sqrt{-\theta}} = -q_0$$

so that λ (which plays a crucial role in the inner region) is given by $\lambda = \frac{q_0^2}{4}$, a conclusion that also follows by other arguments.

(C) Inner region $y \equiv e^{\lambda t}\theta = O(1)$: Here, we introduce the variable:

$$v \sim e^{\lambda t} V(y)$$

(the large time solution behaviour may in fact also involve algebraic dependence on t; such refinements are of little importance here). Using these variables and neglecting the final term in (86), one obtains the equation:

$$(V^{-2}V_y)_y = \lambda V + \lambda y V_y$$

which is equivalent to the first order ODE:

$$V^{-2}V_y = \lambda y V + q_0 \tag{91}$$

if one takes into account (87). Now, we note that ODE (91) coincides with (69), which was analysed above. In particular, the exponent $\lambda = \frac{q_0^2}{4}$ is associated with a repeated-root condition (see (72)). Thus, we obtain:

$$V \sim \frac{2}{-q_0 y}, \quad as \quad y \to +\infty \tag{92}$$

which matches with Region (B) and provides an alternative route to the derivation of the value of the similarity exponent λ.

In comparing this analysis with the numerics above (see Figure1), one notes that $\phi(t,z)$ scales as $\sqrt{t}$ for large z according to the result of (A). On the other hand, $\phi(t,z)$ is exponentially growing in time for t large and z small. This is in agreement with (C)(see Formula (92)) and the boundary Condition (76).

5. Conditional Symmetry Classification of the BVPs Class (1)–(3)

Q-conditional (nonclassical) symmetries of the class of (1 + 2)-dimensional heat Equations (1) were described in paper [28]. In contrast to the (1 + 1)-dimensional case, the result is very simple: in the case of the Q-conditional symmetry Operator (6) with $\xi^0(t,x) \neq 0$, there is only a unique nonlinear equation from this class, namely (1) with $d(u) = u^{-1/2}$, admitting a conditional symmetry. Any other nonlinear heat equation admits conditional symmetry operators of the form (6), which are equivalent to the relevant Lie symmetry operators. In the case of Q-conditional symmetry Operator (6) with $\xi^0(t,x) = 0$, the system of determining equations is analysed in [28] (see System (3.30) therein), and their conclusion is as follows: each known solution of the system leads again to a Lie symmetry, and they were not able to construct any other solution.

Let us consider the equation:

$$\frac{\partial u}{\partial t} = \nabla . \left(u^{-1/2}\nabla u\right) \tag{93}$$

and its conditional symmetry:

$$Q = \frac{\partial}{\partial t} + 2h(x_1, x_2)\sqrt{u}\frac{\partial}{\partial u} \tag{94}$$

where the function h is an arbitrary solution of the nonlinear equation $\triangle h = h^2$ (in [28], these formulae have a slightly different form, because in the very beginning, the authors applied the Kirchoff transformation to (1)). Now, we examine BVPs (93), (2) with $d(u) = u^{-1/2}$ and (3) using Definition 2.

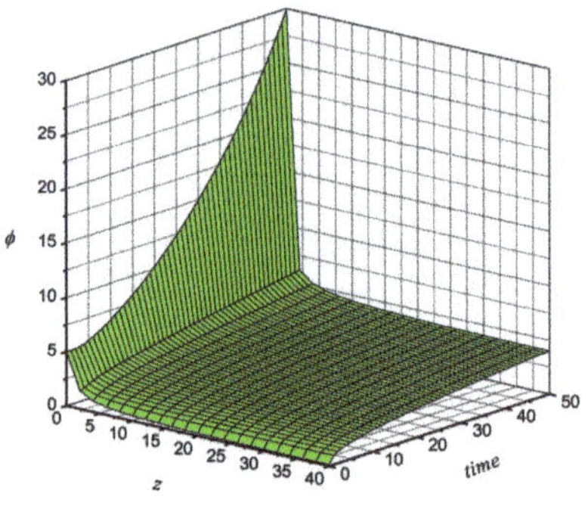

Figure 1. Numerical solution of BVPs (75)–(77) for $\phi(0,z) = \phi_0(z) \equiv -\frac{1}{q_0(z+z_0)}$ and $z_0 = 1.0, q_0 = -0.2$.

Obviously, Items (a) and (b) are automatically fulfilled. To fulfil Item (c), one needs the first prolongation of Operator (94):

$$\underset{1}{Q} = \frac{\partial}{\partial t} + 2h\sqrt{u}\frac{\partial}{\partial u} + \left(2h_{x_2}\sqrt{u} + hu^{-1/2}u_{x_2}\right)\frac{\partial}{\partial u_{x_2}} \tag{95}$$

Applying this operator to the boundary Condition (2) with $d(u) = u^{-1/2}$, we arrive at the equation:

$$x_2 = 0 : q'(t) = 2h_{x_2} \tag{96}$$

which immediately gives:

$$q(t) = q_0 + 2q_1 t, \quad h_{x_2}(x_1, 0) = q_1 \tag{97}$$

where q_0 and q_1 are arbitrary constants. Finally, we can use again Transformation (36) for the examination of Items (d)–(f), and direct checking shows that a sufficient condition is that the function h be bounded as $x_2 \to \infty$.

Thus, BVPs (93), (2) with $d(u) = u^{-1/2}$ and (3) are Q-conditionally invariant only in the case of linear flux $q(t)$ (see (97)), and the relevant conditional symmetry operator possesses the form (94), where the function h solves the initial problem:

$$\triangle h = h^2, \quad h_{x_2}(x_1, 0) = q_1 \tag{98}$$

Remark 6. *Because each conditional symmetry operator (6) multiplied by an arbitrary smooth function M is again a conditional symmetry, we have examined also the operator $M(t, x_1, x_2, u)Q$ and show that no further results are obtained.*

Now, we apply the Q-conditional symmetry (94) in reducing the nonlinear BVP with the governing Equation (93) and conditions:

$$x_2 = 0 : u^{-1/2}u_{x_2} = q_0 + 2q_1 t \tag{99}$$

$$x_2 \to +\infty : u_{x_2} = 0 \tag{100}$$

Operator (94) produces the ansatz:

$$u = (t\phi(x_1, x_2) + h(x_1, x_2))^2 \tag{101}$$

where $\phi(x_1, x_2)$ is a new unknown function. It can be noted that Ansatz (101) was proposed (and applied for finding exact solutions) in [29] without knowledge of symmetry (94). Substituting (101) into BVPs (93), (99), (100) and taking into account (98), we arrive at the two-dimensional problem for the nonlinear system of two elliptic equations:

$$\triangle\phi = \phi h, \quad \triangle h = h^2, \tag{102}$$

$$x_2 = 0 : \phi_{x_2} = q_0, \quad h_{x_2} = q_1 \tag{103}$$

$$x_2 \to +\infty : \phi_{x_2} = 0, \quad h_{x_2} = 0 \tag{104}$$

6. Some Remarks about the Domain Geometry

A natural question arises: how do Lie and conditional invariance of BVPs depend on the geometry of the domain ω? Obviously, the problem essentially depends on the space dimensionality. For example, there are only three essentially different cases for BVPs with the (1 + 1)-dimensional evolution equations, namely: ω is a finite interval, a semi-infinite interval and $\omega = \mathbb{R}$. Here, we treated (1 + 2)-dimensional BVPs with $\omega = \{(x_1, x_2) : -\infty < x_1 < +\infty,\ x_2 > 0\}$. In the general case, the domain can be any open subset $\omega \subset \mathbb{R}^2$ with a smooth boundary, *i.e.*, one is formed by differentiable (except possibly a finite number of points) curves. However, if one fixes a governing equation, then the geometric structure of ω may be predicted in advance if one is looking for Lie and conditional invariance of the relevant BVP. In the case of the governing Equation (1), all possible Lie symmetries are presented in Table 1. Let us skip the critical Case 3, because this involves an infinite-dimensional algebra. The projection of all MAIs arising in Cases 1, 2 and 4 on the (x_1, x_2)-space gives the Lie algebra with basic operators:

$$X_1 = \partial_{x_1},\ X_2 = \partial_{x_2},\ J_{12} = x_1\partial_{x_2} - x_2\partial_{x_1},\ D_{12} = x_1\partial_{x_1} + x_2\partial_{x_2} \tag{105}$$

which is nothing else but the Euclidean algebra $AE(2)$ extended by the operator of scale transformations. Now, we realize that a non-trivial result can be obtained provided ω is invariant under transformations generated by this algebra. For example, the case addressed above, namely $\omega = \{(x_1, x_2) : -\infty < x_1 < +\infty,\ x_2 > 0\}$, is invariant under x_1-translations and scale transformations generated by D_{12}; however, to note a simple such example, any triangle in the (x_1, x_2)-space does not admit any transformations generated by (105). Of course, the domain $\omega = \mathbb{R}^2$ is invariant under the extended Euclidean algebra (105); however, this domain is appropriate to initial value problems only (note that interesting symmetry-based approaches for solving such problems were proposed in [30,31]), while any boundary-value problem implies $\omega \neq \mathbb{R}^2$.

It turns out that all of the domains admitting at least one-dimensional algebra can be described using the well-known results of the classification of inequivalent (non-conjugate) subalgebras for the extended Euclidean algebra, which are presented, for example, in [32]. The corresponding list of subalgebras can be divided on subalgebras of different dimensionality. We present only those of dimensionality one and two, because subalgebras of higher dimensionality immediately lead to $\omega = \mathbb{R}^2$.

The one-dimensional subalgebras are:

$$\langle X_1 \rangle,\ \langle J_{12} \rangle,\ \langle D_{12} \rangle,\ \langle J_{12} + \beta D_{12} \rangle\ (\beta > 0)$$

and the two-dimensional ones are:

$$\langle X_1, X_2 \rangle,\ \langle X_1, D_{12} \rangle,\ \langle J_{12}, D_{12} \rangle\,.$$

Obviously, absolute invariants of each algebra can be easily calculated in explicit form (see, e.g., the relevant theory in [4]); hence, we need only to provide a geometrical interpretation for each algebra. In the case of the algebra $\langle X_1\rangle$, the absolute invariant is x_2; hence, the domain ω can be created by lines of the form $x_2 = const$. This means that there are only two generic domains, the strip $\omega_1 = \{(x_1, x_2) : -\infty < x_1 < +\infty,\ C_1 < x_2 < C_2\}$ and the half-plane $\omega_2 = \{(x_1, x_2) : -\infty < x_1 < +\infty,\ x_2 > C_2\}$ (hereafter, C_1 and C_2 are arbitrary consts). Any other domain admitting the x_1-translations can be obtained via a combination of ω_1 and ω_2.

In the case of the algebra $\langle J_{12}\rangle$, the absolute invariant is $x_1^2 + x_2^2$; hence, the domain ω can be created by circles of the form $x_1^2 + x_2^2 = const$. This means that there are only three generic domains, the interior of the circle $\omega_1 = \{(x_1, x_2) : x_1^2 + x_2^2 < C_2\}$, the exterior of the circle $\omega_2 = \{(x_1, x_2) : x_1^2 + x_2^2 > C_1 > 0\}$ and the annulus $\omega_3 = \{(x_1, x_2) : 0 < C_1 < x_1^2 + x_2^2 < C_2\}$.

In the case of the algebra $\langle D_{12}\rangle$, the absolute invariant is $\frac{x_1}{x_2}$; hence, the domain ω can be created by lines of the form $x_1 = const\, x_2$ and $x_2 = 0$. This means that there are only two generic domains, the wedge $\omega_1 = \{(x_1, x_2) : C_1 x_2 < x_1 < C_2 x_2\}$ and the half-plane $\omega_2 = \{(x_1, x_2) : -\infty < x_1 < +\infty,\ x_2 > 0\}$.

Finally, in the case of the one-dimensional algebra $\langle J_{12} + \beta D_{12}\rangle$, the absolute invariant is $\sqrt{x_1^2 + x_2^2}\exp\left(-\beta \arctan\frac{x_1}{x_2}\right)$; hence, the domain ω can be created by the curves $\sqrt{x_1^2 + x_2^2} = const \exp\left(\beta \arctan\frac{x_1}{x_2}\right)$. In the polar coordinates (r, θ), such curves are the logarithmic spirals $r = const \exp(\beta\theta)$, and one obtains only the generic domain $\omega = \{(r, \theta) : C_1 \exp(\beta\theta) < r < C_2 \exp(\beta\theta)\}$ $(0 < C_1 < C_2)$, which is the space between two spirals.

Examination of two-dimensional subalgebras listed above does not lead to any new domains. In fact, the first and the third produce $\omega = \mathbb{R}^2$, while the second leads only to the half-space $\omega = \{(x_1, x_2) : -\infty < x_1 < +\infty,\ x_2 > 0\}$, which is a particular case of the domain ω_2 obtained above for the algebra $\langle X_1\rangle$.

The above considerations provide a symmetry-based motivation for investigating half-space problems, as we have done above. The other domains just recorded should be taken into account for further application of the technique established above.

7. Conclusions

In this paper, a new definition (see Definition 2) of conditional invariance for BVPs is proposed. It is shown that Bluman's definition [1,9] for Lie invariance of BVPs, which is widely used to find Lie symmetries of BVPs with standard boundary conditions, follows as a natural particular case from Definition 2. Simple examples of the direct applicability of the definition to nonlinear (1 + 1)-dimensional BVPs, leading to both known and new results, are demonstrated.

The second main result of the paper consists of the successful application of the definition for Lie and conditional symmetry classification of BVPs of the form (1)–(3). It turns out that a wide range of possibilities arises for BVPs with the governing (1 + 2)-dimensional nonlinear heat equation if one looks for Lie symmetries. Depending on the form of the pair $(d(u); q(t))$, there are 15 different cases (see Table 2) in contrast to the four different cases only that arise for the governing Equation (1). In particular, we have proven that there is a special exponent, $k = -2$, for the power diffusivity u^k when the BVP with non-vanishing flux on the boundary admits additional Lie symmetry operators compared to the case $k \neq -2$ (see Cases 9 and 10 in Table 2). It should be stressed that the power $k = -2$ is not a special case for the governing Equation (1) with $d(u) = u^k$ in two space dimensions, though in some respects, this result reflects its exceptional status in one dimension. It is worth noting that the well-known critical power $k = -1$, leading to an infinite-dimensional invariance algebra of the (1 + 2)-dimensional nonlinear heat equation, preserves its special status only in the case of zero flux on the boundary (see Case 11 in Table 2 and Remark 4).

In the case of conditional symmetry classification of the BVP class (1)–(3), our result is modest, because the governing Equation (1) admits a Q-conditional symmetry only for the diffusivity

$d(u) = u^{-1/2}$ [28]. Hence, we have examined BVPs (1)–(3) with $d(u) = u^{-1/2}$ only and proven that this problem is conditionally invariant under Operator (94) provided restriction (97) holds.

In order to demonstrate the applicability of the symmetries derived, we used those for reducing the nonlinear BVPs (1)–(3) with power diffusivity u^k and a constant non-zero flux (such problems are common in applications and describe a wide range of phenomena depending on values of k). One motivation was to investigate the structure of the (1 + 1)-dimensional problems obtained. It turns out that all of the reduced problems (excepting the parabolic Problems (75)–(77), for which an analysis is presented in Example 4) are elliptic ones. Some of them are well known (see (82)–(84)), while others seem to be new and will be treated in a forthcoming paper.

Finally, we have described a brief analysis of a problem of independent interest, which follows in a natural way from the theoretical considerations presented in Section 2. The problem can be formulated as follows: how do Lie and conditional invariance of BVP depend on geometry of the domain, in which the given BVP is defined? We have solved this problem for BVPs with the governing Equation (1) and obtained an exhaustive list of possible domains preserving at least a one-dimensional subalgebra of MAI of Equation (1). It turns out that the geometrical interpretation of the domains obtained is rather simple. However, we foresee much more difficulties for BVPs in this regard with the governing equations in spaces of higher dimensionality.

Acknowledgments: This research was supported by a Marie Curie International Incoming Fellowship to the first author within the 7th European Community Framework Programme. The authors are grateful to V. Dutka (NAS of Ukraine, Kyiv) for numerical simulations in order to create Figure 1. The authors are also grateful to a reviewer for the useful critical comments.

Author Contributions: The authors contributed equally to this work.

Conflicts of Interest: The authors declare no conflict of interest.

References

1. Bluman, G.W.; Anco, S.C. *Symmetry and Integration Methods for Differential Equations*; Applied Mathematical Sciences; Springer-Verlag: New York, NY, USA, 2002; Volume 154.
2. Bluman, G.; Kumei, S. *Symmetries and Differential Equations*; Springer: Berlin, Germany, 1989.
3. Fushchych, W.; Shtelen, W.; Serov, M. *Symmetry Analysis and Exact Solutions of Equations of Nonlinear Mathematical Physics*; Kluwer: Dordrecht, The Nertherlands, 1993.
4. Olver, P.J. *Applications of Lie Groups to Differential Equations*, 2nd ed.; Graduate Texts in Mathematics; Springer-Verlag: New York, NY, USA, 1993; Volume 107.
5. Ovsiannikov, L. *The Group Analysis of Differential Equations*; Academic Press: New York, NY, USA, 1982.
6. Bluman, G.W.; Cheviakov, A.F.; Anco, S.C. *Applications of Symmetry Methods to Partial Differential Equations*; Applied Mathematical Sciences; Springer: New York, NY, USA, 2010; Volume 168.
7. Cherniha, R. Conditional symmetries for systems of PDEs: New definitions and their application for reaction-diffusion systems. *J. Phys. A* **2010**, *43*, doi:10.1088/1751-8113/43/40/405207.
8. Cherniha, R.; Davydovych, V. Lie and conditional symmetries of the three-component diffusive Lotka-Volterra system. *J. Phys. A* **2013**, *46*, doi:10.1088/1751-8113/46/18/185204.
9. Bluman, G. Application of the general similarity solution of the heat equation to boundary value problems. *Q. Appl. Math.* **1974**, *31*, 403–415.
10. Ibragimov, N.K. Group analysis of ordinary differential equations and the invariance principle in mathematical physics (on the occasion of the 150th anniversary of the birth of Sophus Lie). *Uspekhi Mat. Nauk* **1992**, *47*, 83–144.
11. Ibragimov, N.H. Lie group analysis of Moffatt's model in metallurgical industry. *J. Nonlinear Math. Phys.* **2011**, *18*, 143–162.
12. Cherniha, R.; Kovalenko, S. Lie symmetries and reductions of multi-dimensional boundary value problems of the Stefan type. *J. Phys. A* **2011**, *44*, 485202.
13. Cherniha, R.; Kovalenko, S. Lie symmetry of a class of nonlinear boundary value problems with free boundaries. *Banach Center Publ.* **2011**, *93*, 95–104.

14. Cherniha, R.; Kovalenko, S. Lie symmetries of nonlinear boundary value problems. *Commun. Nonlinear Sci. Numer. Simul.* **2012**, *17*, 71–84.
15. King, J.R. Exact results for the nonlinear diffusion equations $\partial u/\partial t = (\partial/\partial x)(u^{-4/3}\partial u/\partial x)$ and $\partial u/\partial t = (\partial/\partial x)(u^{-2/3}\partial u/\partial x)$. *J. Phys. A* **1991**, *24*, 5721–5745.
16. Cherniha, R. Conditional symmetries for boundary value problems: New definition and its application for nonlinear problems with Neumann conditions. *Miskolc Math. Notes* **2013**, *14*, 637–646.
17. Cherniha, R.; Pliukhin, O. New conditional symmetries and exact solutions of nonlinear reaction-diffusion-convection equations. *J. Phys. A* **2007**, *40*, 10049–10070.
18. Cherniha, R.; King, J.R. Lie symmetries of nonlinear multidimensional reaction-diffusion systems. II. *J. Phys. A* **2003**, *36*, 405–425.
19. Cherniha, R.; King, J.R. Non-linear reaction-diffusion systems with variable diffusivities: Lie symmetries, ansätze and exact solutions. *J. Math. Anal. Appl.* **2005**, *308*, 11–35.
20. Vaneeva, O.O.; Johnpillai, A.G.; Popovych, R.O.; Sophocleous, C. Enhanced group analysis and conservation laws of variable coefficient reaction-diffusion equations with power nonlinearities. *J. Math. Anal. Appl.* **2007**, *330*, 1363–1386.
21. Cherniha, R.; Serov, M.; Rassokha, I. Lie symmetries and form-preserving transformations of reaction-diffusion-convection equations. *J. Math. Anal. Appl.* **2008**, *342*, 1363–1379.
22. Akhatov, I.; Gazizov, R.; Ibragimov, N. Nonlocal symmetries. Heuristic approach. *J. Sov. Math.* **1991**, *55*, 1401–1450.
23. Dorodnitsyn, V.; Knyazeva, I.; Svirshchevskii, S. Group properties of the nonlinear heat equation with source in the two- and three-dimensional cases. *Differential'niye Uravneniya* **1983**, *19*, 1215–1223.
24. Nariboli, G. Self-similar solutions of some nonlinear equations. *Appl. Sci. Res.* **1970**, *22*, 449–461.
25. Kovalenko, S. *Group Theoretic Analysis of a Class of Boundary Value Problems for a Nonlinear Heat Equation*; The University of Cornell: Ithaca, NY, USA, 2012.
26. Dubrovin, B.A.; Fomenko, A.T.; Novikov, S.P. *Modern Geometry—Methods and Applications. Part I: The Geometry of Surfaces, Transformation Groups, and Fields*, 2nd ed.; Robert, G.B., Translator; Graduate Texts in Mathematics; Springer-Verlag: New York, NY, USA, 1992.
27. Henrici, P. *Applied and Computational Complex Analysis, Discrete Fourier analysis—Cauchy Integrals—Construction of Conformal Maps—Univalent Functions*; Pure and Applied Mathematics; John Wiley & Sons, Inc.: New York, NY, USA, 1986; Volume 3.
28. Arrigo, D.J.; Goard, J.M.; Broadbridge, P. Nonclassical solutions are non-existent for the heat equation and rare for nonlinear diffusion. *J. Math. Anal. Appl.* **1996**, *202*, 259–279.
29. King, J.R. Some non-self-similar solutions to a nonlinear diffusion equation. *J. Phys. A* **1992**, *25*, 4861–4868.
30. Hydon, P.E. Symmetry analysis of initial-value problems. *J. Math. Anal. Appl.* **2005**, *309*, 103–116.
31. Goard, J. Finding symmetries by incorporating initial conditions as side conditions. *Eur. J. Appl. Math.* **2008**, *19*, 701–715.
32. Fushchych, W.; Barannyk, L.; Barannyk, A. *Subgroup Analysis of the Galilei and Poincare Groups and Reduction of Nonlinear Equations*; Naukova Dumka: Kiev, Ukraine, 1991.

symmetry

MDPI

Article

Lie Group Method for Solving the Generalized Burgers', Burgers'–KdV and KdV Equations with Time-Dependent Variable Coefficients

Mina B. Abd-el-Malek [1,2,*] **and Amr M. Amin** [1]

1 Department of Engineering Mathematics and Physics, Faculty of Engineering, Alexandria University, Alexandria 21544, Egypt; amr.mahmoud.amin@gmail.com

2 Department of Mathematics and Actuarial Science, The American University in Cairo, New Cairo 11835, Egypt

* Author to whom correspondence should be addressed; minab@aucegypt.edu; Tel.: +20-1-000-066-496.

Academic Editor: Roman M. Cherniha
Received: 26 June 2015; Accepted: 30 September 2015; Published: 13 October 2015

Abstract: In this study, the Lie group method for constructing exact and numerical solutions of the generalized time-dependent variable coefficients Burgers', Burgers'–KdV, and KdV equations with initial and boundary conditions is presented. Lie group theory is applied to determine symmetry reductions which reduce the nonlinear partial differential equations to ordinary differential equations. The obtained ordinary differential equations were solved analytically and the solutions are obtained in closed form for some specific choices of parameters, while others are solved numerically. In the obtained results we studied effects of both the time t and the index of nonlinearity on the behavior of the velocity, and the solutions are graphically presented.

Keywords: Lie group method; Burgers' Equation; Burgers'–KdV Equation; KdV Equation

1. Introduction

Burgers' equation was formed as a model of turbulent fluid motion by Burgers in a series of several articles. These articles have been summarized in Burgers' book [1]. Burgers' equation is used in the modeling of water in unsaturated oil, dynamics of soil in water, mixing and turbulent diffusion [2], and optical-fiber communications [3]. Burgers' equation can be transformed to the standard heat equation by means of the Hopf–Cole transformation [4,5]. It is well known that Burgers' equation involves dissipation term u_{xx} and Burgers' Korteweg–de Vries Burgers' (BKdV) equation involves both dispersion term u_{xxx} and dissipation term u_{xx} [6]. Typical examples describing the behavior of a long wave in shallow water and waves in plasma, also describing the behavior of flow of liquids containing gas bubbles and the propagation of waves on an elastic tube filed with a viscous fluid, [6] are considered.

Several numerical techniques for Burgers' equation exist, such as Crank–Nicolson finite difference method applied by Kadalbajoo and Awasthi [7] to the linearized Burgers' equation. Gorguis applied the Adomian decomposition method on Burgers' equation directly [8] and Kutluay *et al.* applied the direct approach via the least square quadratic B-spline finite element method [9].

In 1985, Ames and Nucci offered the analysis of fluid equations by group methods. Their work included the solution of Burgers' equation [10]. In 1989, Hammerton and Crighton derived the generalized Burgers' equation describing the propagation of weakly nonlinear acoustic waves under the influence of geometrical spreading and thermo-viscous diffusion [11], which is, in non-dimensional variables, reducible to the form $u_t + uu_x = g(t)u_{xx}$. In 2005, Wazwaz presented an analysis to generalized forms of Burgers', Burgers'–KdV, and Burgers'–Huxley equations using the traveling wave method [12]. In 2011, Abd-el-Malek and Helal applied an analysis to the generalized forms of Burgers' and

Burgers'–KdV with variable coefficients and with initial and boundary conditions using the group theoretic approach [13].

There are several approaches for using Lie symmetries to reduce initial boundary value problems (IBVPs) of partial differential equations to those of ordinary differential equations. The classical technique is to require that the equation and both initial and boundary conditions are left invariant under the one parameter Lie group of infinitesimal transformations. The first rigorous definition of Lie's invariance for IBVPs was formulated by Bluman [14]. This definition and several examples are summarized in his book [15]. His definition was used (explicitly or implicitly) in several papers to derive exact solutions of some IBVPs. On other hand, one notes that Bluman's definition cannot be applied directly to IBVPs which contain boundary conditions involving points at infinity. A new definition of Lie invariance of IBVPs with a wide range of boundary conditions was formulated by Cherniha *et al.* [16,17]. In this paper, we have to deal with the definitions of Lie's invariance for IBVPs that were defined by Bluman and Anco [15] and Cherniha and King [17].

In the present work we apply an analysis to the generalized Burgers' equation, Burgers'–KdV, and KdV equations with time-dependent variable coefficients as well as initial and boundary conditions using the Lie group method. Hence, the obtained symmetries are used to reduce the partial differential equation to an ordinary differential equation and some exact solutions are obtained for some cases while numerical solutions are obtained for others.

2. The Generalized Burgers' Equation

We consider the generalized Burgers' equation in the form [12,13]

$$u_t + \alpha\,(u^n)_x = \beta\, g(t)\,(u^n)_{xx},\; x > 0, t > 0,\; n > 1,\; \alpha, \beta \neq 0 \tag{1}$$

Equation (1) can be written in the form

$$u_t + \alpha\, n\, u^{n-1} u_x - \beta\, n\,(n-1)\, g(t)\, u^{n-2}\, u_x^2 - \beta\, n g(t)\, u^{n-1}\, u_{xx} = 0,\; x > 0, t > 0,\; n > 1,\; \alpha, \beta \neq 0 \tag{2}$$

with initial and boundary conditions given by:

$$u(x,0) \to \infty,\; x > 0 \tag{3}$$

$$u(0,t) = \gamma\, r(t),\; t > 0,\; \gamma \neq 0 \tag{4}$$

$$\lim_{x\to\infty} u(x,t) \to \infty, t > 0 \tag{5}$$

To specify the symmetry algebra of Equation (2), we use the Lie symmetry method. For a general introduction to the subject we refer to [14,15,18,19] specifically [18].

Considering the infinitesimal generator of the symmetry group admitted by Equation (2), given by

$$X \equiv \xi^1 \frac{\partial}{\partial x} + \xi^2 \frac{\partial}{\partial t} + \eta^1 \frac{\partial}{\partial u} \tag{6}$$

Since the generalized Burgers' equation has at most second-order derivatives, we prolong the vector field X to the second order. The action of $\mathrm{Pr}^{(2)}X$ on Equation (2) must vanish, where u is the solution of Equation (2), and then we find the following determining equations:

$$\left.\begin{array}{c} \xi^1_u = \xi^1_t = \xi^1_{xx} = 0\,,\; \xi^1_x = \frac{g_t\,\xi^2}{g},\; \frac{g_t\,\xi^2}{g} = \text{constant} \\ \xi^2_u = \xi^2_x = \xi^2_{tt} = 0 \\ \eta^1_x = \eta^1_t = \eta^1_{uu} = 0,\; \eta^1 = \frac{\xi^1_x - \xi^2_t}{n-1}\, u \end{array}\right\} \tag{7}$$

We have found all possible forms of $g = g(t)$ when Equation (1) admits different Lie algebras of invariance (see Table 1 (Tables 2–4)):

Table 1. Infinitesimals for $u_t + \alpha\, (u^n)_x = \beta\, g(t)\, (u^n)_{xx}$.

No.	g	Infinitesimals
1	$\forall$	$\xi^1 = c_2, \xi^2 = 0, \eta^1 = 0$
2	t^λ	$\xi^1 = c_1\,\lambda\, x + c_2, \xi^2 = c_1\, t, \eta^1 = \frac{\lambda-1}{n-1} c_1 u$
3	e^t	$\xi^1 = c_1\, x + c_2, \xi^2 = c_1, \eta^1 = \frac{\lambda-1}{n-1} c_1 u$
4	1	$\xi^1 = c_2, \xi^2 = c_1 t + c_3, \eta^1 = \frac{-1}{n-1} c_1 u$

The Lie symmetries listed in Cases 1–3 in Table 1 are completely new; however, the Lie symmetries listed in Case 4 in Table 1 have been obtained much earlier in [20].

We consider Case 2 in Table 1. The infinitesimal generator of the symmetry group admitted by Equation (2) is given by $X \equiv\ (c_1\,\lambda\, x + c_2)\,\frac{\partial}{\partial x} + c_1\, t\,\frac{\partial}{\partial t} + \frac{\lambda-1}{n-1} c_1 u \frac{\partial}{\partial u}$.

The action of X on Equation (4) must vanish, *i.e.*,

$$X\,(u(0,t) - \gamma\, r(t) = 0) = 0 \tag{8}$$

Therefore, $c_2 = 0$, $c_1 \neq 0$ and $-t\,\frac{dr(t)}{dt} + \frac{\lambda-1}{n-1}\, r(t) = 0$, from which we get:

$$r(t)\ =\ t^{\frac{\lambda-1}{n-1}},\ n\ >\ 1 \tag{9}$$

Clearly, our initial and boundary conditions in Equations (3) and (5) include infinity. Therefore, Bluman's definition cannot be applied directly to IBVPs which contain boundary conditions with points at infinity. Hence, we need to examine the operator X according to Definition 2 in [17]. Items (b)–(c) of Definition 2 are fulfilled in the case of Equation (4). To check items (d)–(f) we need to find an appropriate bijective transform. Let us consider the transform in form the $y = \frac{1}{x}$, $t = \tau$, $u = \frac{1}{U}$ which maps $M = \{x \to \infty,\ u \to \infty\}$ to $M^* = \{y \to 0,\ U \to 0\}$, and both manifolds have the same dimensionality. Now, item (d) is fulfilled. Our transform maps the operator X to the form $X^* \equiv\ -c_1\,\lambda\, y\frac{\partial}{\partial y} + c_1\,\tau\,\frac{\partial}{\partial \tau}\ -\ \frac{\lambda-1}{n-1} c_1 U \frac{\partial}{\partial U}$, and now, one can easily check that this operator satisfies items (e)–(f) of Definition 2 on M^*.

In the previous calculation, we have shown that the symmetry operator X which is admitted by both Equation (1) and initial and boundary conditions (3)–(5) with $r(t)$ being a power function is called the dilatation operator, *i.e.*, the operator corresponding to the one-parameter Lie group of scaling of the variables x, t, and u. Also, Equation (1) admits a Lie symmetry generator which keeps the boundary conditions invariant if and only if $g(t)$ is a power function or constant.

By repeating the above procedures, the symmetry operator presented in Case 3 in Table 1 does not leave the boundary conditions invariant and the same goes for Case 4 in Table 1. Consider Case 2 with $\lambda = 0$ as a special case.

The auxiliary equation according to the symmetry operator X, which satisfies the given initial and boundary conditions, will be:

$$\frac{dx}{c_1\lambda\, x} = \frac{dt}{c_1\, t} = \frac{du}{\frac{c_1(\lambda-1)u}{n-1}} \tag{10}$$

Solving Equation (10), we get:

$$\left.\begin{aligned} \eta(x,t)\ &= x\, t^{-\lambda}\ ,\ u(x,t)\ = t^{\frac{\lambda-1}{n-1}} F(\eta) \\ r(t)\ &=\ t^{\frac{\lambda-1}{n-1}}\ ,\ g(t) = t^\lambda \end{aligned}\right\} \tag{11}$$

From Equations (9) and (11), $u(x,t)\ = t^{\frac{\lambda-1}{n-1}} F(\eta) = r(t)\ F(\eta)$, by choosing $\frac{\lambda-1}{n-1}\ <\ 0$ (*i.e.*, $\lambda < 1$, since $n > 1$). Therefore, $r(t) \to \infty$ and $u(x,0) \to \infty$ as $t \to 0$. Hence, a solution in this case exists;

otherwise, a smooth solution of the problem may not exist. Finally, when $\lambda < 1$, the obtained solution in this case will satisfy Equation (3) directly.

By substituting Equation (11) in Equation (2) we get:

$$\beta\, nF^{n-1}F'' + \beta n(n-1)F^{n-2}F'^2 - \alpha nF^{n-1}F' = \frac{\lambda-1}{n-1}F - \lambda\eta F' \tag{12}$$

By substituting Equation (11) in Equations (4) and (5) we get:

$$F(0) = \gamma \tag{13}$$

$$\lim_{\eta\to\infty} F(\eta) \to \infty \tag{14}$$

Equation (12) can be written as

$$\left(F^{n-1}F' - \frac{\alpha}{\beta n}F^n\right)' = \frac{\lambda-1}{n\,(n-1)\beta}F - \frac{\lambda}{n\,\beta}\eta\, F' \tag{15}$$

We have shown that solution of the problem exists and Equation (3) is satisfied when $\lambda < 1$. Therefore without loss of generality, we consider $\lambda = 1/n$. Hence Equation (15) reduces to:

$$\left(F^{n-1}F' - \frac{\alpha}{\beta n}F^n\right)' = -\frac{1}{n^2\,\beta}(F + \eta F') \tag{16}$$

from which its solution is:

$$F = \left[\frac{\beta}{\alpha^2(n-1)} + \frac{\eta}{\alpha\, n} + \left(\gamma^{n-1} - \frac{\beta}{\alpha^2(n-1)}\right) e^{\left(\frac{\alpha\,(n-1)\,\eta}{n\,\beta}\right)}\right]^{\frac{1}{n-1}} \tag{17}$$

under the condition $\gamma^{n-1} > \frac{\beta}{\alpha^2(n-1)}$ which should satisfy Equation (14). Therefore

$$u(x.t) = \frac{1}{t^{\frac{1}{n}}}\left[\frac{\beta}{\alpha^2(n-1)} + \frac{x}{\alpha\, n\, t^{\frac{1}{n}}} + \left(\gamma^{n-1} - \frac{\beta}{\alpha^2(n-1)}\right) e^{\left(\frac{\alpha\,(n-1)\,x}{n\,\beta\, t^{\frac{1}{n}}}\right)}\right]^{\frac{1}{n-1}} \tag{18}$$

which is the same as obtained by Abd-el-Malek and Helal [13] for the case $g(t) = t^{\lambda}$. While we succeeded to find other possible forms of $g(t)$, the only case which keeps the boundary conditions invariant is a power function or constant.

It is clear that the obtained similarity solutions which describe the behavior of the velocity field $u(x,t)$ decrease with the increase of time as shown in Figure 1a,b. Furthermore, the velocity field $u(x,t)$ decreases with the increase of the nonlinearity index "n" as it follows from Figure 1.

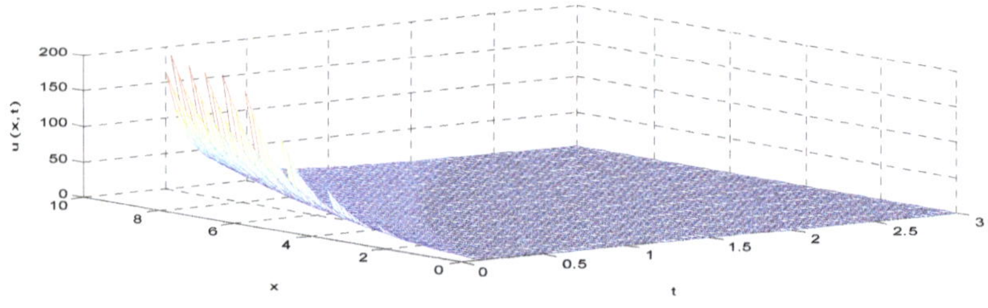

(a)

Figure 1. *Cont.*

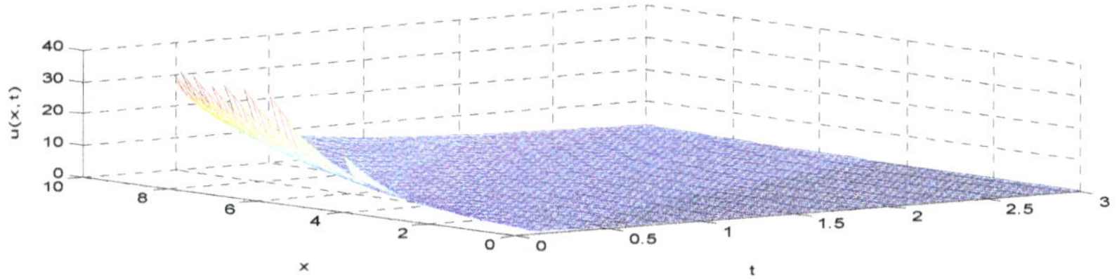

(b)

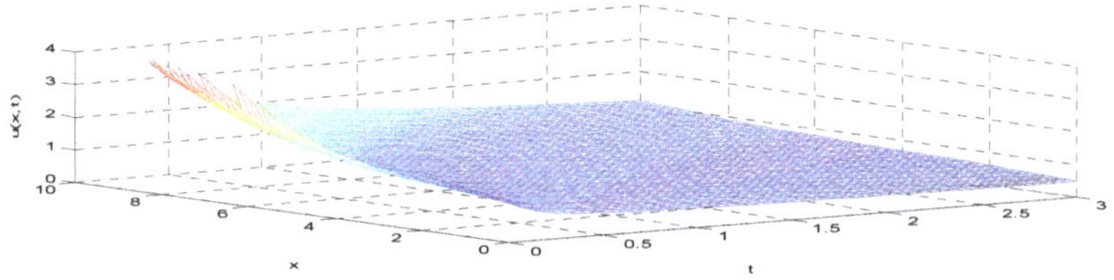

(c)

Figure 1. (**a**) Exact solution Equation (18) for $\alpha = 2$, $\beta = 2$, $\gamma = 0.6$, and $n = 2$; (**b**) Exact solution Equation (18) for $\alpha = 2$, $\beta = 2$, $\gamma = 0.6$, and $n = 3$; (**c**) Exact solution Equation (18) for $\alpha = 2$, $\beta = 2$, $\gamma = 0.6$, and $n = 5$.

3. The Generalized Burgers'–KdV Equation "GBKdV"

The generalized Burgers'–KdV equation "GBKdV" [12,13] will be studied by using the Lie group method.

Consider the generalized Burgers'–KdV equation in form:

$$u_t + \alpha\,(u^n)_x + u_{xx} = \beta\, g(t)\, u_{xxx},\ x > 0,\ t > 0,\ n > 1,\ \alpha, \beta \neq 0 \tag{19}$$

with the initial and boundary conditions

$$u(x,0) = 0,\ x > 0 \tag{20}$$

$$u(0,t) = \gamma\, r(t),\ t > 0,\ \gamma \neq 0 \tag{21}$$

$$u_x(0,t) = 0,\ t > 0 \tag{22}$$

$$u_{xx}(0,t) = 0,\ t > 0 \tag{23}$$

Apply Lie group again to Equation (19) as we did before in Equation (2) with the infinitesimal generator

$$X \equiv \xi^1 \frac{\partial}{\partial x} + \xi^2 \frac{\partial}{\partial t} + \eta^1 \frac{\partial}{\partial u} \tag{24}$$

Since the generalized Burgers'–KdV equation has at most third-order derivatives, we prolong the vector field X in Equation (24) to the third order. The action of $\mathrm{Pr}^{(3)}X$ on Equation (19) must vanish, where u is the solution of Equation (19). Clearly, the determining Equations can be split with respect to different powers of u. Special cases of splitting cases arise if $n \neq 2$ and if $n = 2$. Therefore, we investigate two cases, namely $n \neq 2$ and $n = 2$

Case 1: For $n \neq 2$, the determining equations are:

$$\left.\begin{array}{l} \xi^1_u = \xi^1_t = \xi^1_{xx} = 0\,,\ \xi^1_x = \frac{g_t\,\xi^2}{g} = \frac{1}{2}\xi^2_t,\ \frac{g_t\,\xi^2}{g} = \text{constant} \\ \xi^2_u = \xi^2_x = \xi^2_{tt} = 0,\ \eta^1_x = \eta^1_t = \eta^1_{uu} = 0,\ \eta^1 = \frac{\xi^1_x - \xi^2_t}{n-1}\,u \end{array}\right\} \tag{25}$$

We have found all possible forms of $g = g(t)$ when Equation (19) admits different Lie algebras of invariance:

Our initial and boundary conditions are satisfied only for Case 1 in Table 2, so $g = g(t) = \sqrt{t}$. Then the infinitesimal generator of the symmetry group admitted by Equation (19) is given by $X \equiv \left(\frac{1}{2}c_1\, x + c_2\right)\frac{\partial}{\partial x} + c_1\, t\,\frac{\partial}{\partial t} + \frac{-1}{2\,(n-1)}c_1 u\,\frac{\partial}{\partial u}$, and the action of X on Equation (21) must vanish, *i.e.*,

$$X\,(u(0,t) - \gamma\, r(t) = 0) = 0 \tag{26}$$

Table 2. Infinitesimals $u_t + \alpha\,(u^n)_x + u_{xx} = \beta\, g(t)\, u_{xxx}$.

No.	g	Infinitesimals
1	$\sqrt{t}$	$\xi^1 = \frac{c_1}{2}\,x + c_2\,, \xi^2 = c_1\,t,$ $\eta^1 = \frac{-1}{2\,(n-1)}c_1 u$
2	1	$\xi^1 = c_1\,, \xi^2 = c_2, \eta^1 = 0$

This gives $c_2 = 0$, $c_1 \neq 0$, and $-t\,\frac{dr(t)}{dt} - \frac{1}{2\,(n-1)}\,r(t) = 0$, from which we get:

$$r(t) = t^{\overline{-\,2\,(n-1)}^{\,1}} \tag{27}$$

The auxiliary equation will be:

$$\frac{dx}{\frac{1}{2}c_1\,x} = \frac{dt}{c_1\,t} = \frac{du}{\frac{-1}{2\,(n-1)}c_1 u} \tag{28}$$

Solving Equation (28)

$$\left.\begin{aligned} \eta(x,t) &= \frac{x}{\sqrt{t}},\ u(x,t) = t^{\frac{1}{-2(n-1)}}\,F(\eta) \\ r(t) &= t^{\frac{1}{-2(n-1)}},\ g(t) = \sqrt{t} \end{aligned}\right\} \tag{29}$$

By inserting Equation (29) in Equation (19), it will be reduced to an ODE

$$\beta\, F''' - F'' - \alpha n F^{n-1}F' + \frac{1}{2}\eta F' + \frac{1}{2n-2}F = 0 \tag{30}$$

The conditions reduce to

$$F(0) = \gamma \tag{31}$$

$$F'(0) = 0 \tag{32}$$

$$F''(0) = 0 \tag{33}$$

Equations (30)–(33) are identical with the results obtained by [13] for $g(t) = \sqrt{t}$.

We use the fourth, fifth-order Runge–Kutta method to solve Equation (30).

Solutions of Equation (30) are plotted in Figure 2 with two different values of the nonlinearity index "n" (n = 5 and 100). The figure shows the vibrations of the variable "F" increase with the increase of the nonlinearity index "n" values and the wave length is seen to decrease with increasing values of "η". The shape of these waves seems to be a bugle or the sound waves are reborn from the bugle, so we can name these waves as "Bugle-shaped waves".

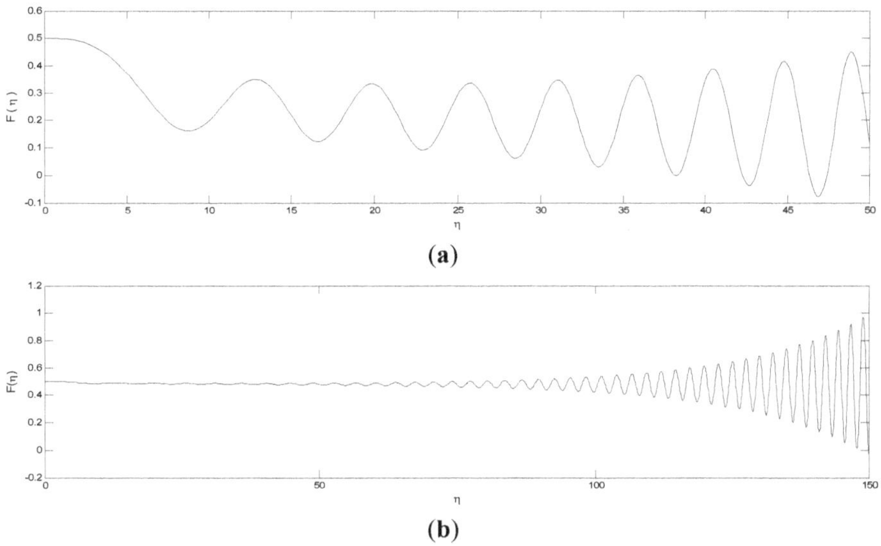

Figure 2. (**a**) Bugle-shaped wave solution of BKdV Burgers'–KdV Equation (30) for α = 1, β = 10, and n = 5; (**b**) Bugle-shaped wave solution of BKdV Burgers'–KdV Equation (30) for α = 1, β = 10, and n = 100.

Case (2): For $n = 2$, the determining equations are:

$$\left.\begin{array}{l} \xi^1_u = \xi^1_{tt} = \xi^1_{xx} = 0\,,\ \xi^1_x = \frac{g_t\,\xi^2}{g} = \frac{1}{2}\xi^2_t, \\ \xi^2_u = \xi^2_x = 0\,,\ \eta^1 = \frac{1}{2}\frac{g\,\xi^1_t - 2\,\alpha\,g_t\,\xi^2\,u}{g\,\alpha} \end{array}\right\} \tag{34}$$

We have found all possible forms of $g = g(t)$ when Equation (19) with $n = 2$ admits different Lie algebras of invariance.

One may note that $g(t) = \sqrt{t}$ or $g(t) = t$ from Table 3 also satisfy our initial and boundary conditions, and we can repeat the same procedure from Equation (26) to Equation (33).

This case is completely new and has not been considered before by Abd-el-Malek and Helal [13].

Table 3. Infinitesimals $u_t + \alpha\,(u^2)_x + u_{xx} = \beta\,g(t)\,u_{xxx}$.

No.	g	Infinitesimals
1	$\sqrt{t}$	$\xi^1 = \frac{1}{2}\,c_1\,x + c_2\,t + c_3, \xi^2 = c_1\,t, \eta^1 = \frac{c_2 - c_1\,\alpha\,u}{2\,\alpha}$
2	t	$\xi^1 = (\,c_1\,x + c_2)\,t + c_3, \xi^2 = c_1\,t^2, \eta^1 = \frac{c_1\,x + c_2 - 2\,c_1\,t\,\alpha\,u}{2\,\alpha}$
3	1	$\xi^1 = c_2\,t + c_3, \xi^2 = c_1, \eta^1 = \frac{1}{2\,\alpha}c_2$
4	$\forall$	$\xi^1 = c_2\,t + c_3, \xi^2 = 0, \eta^1 = \frac{1}{2\,\alpha}c_2$

4. The Generalized KdV Equation "GKdV"

We introduce the following generalized KdV (GKdV) equation [13,21],

$$u_t + u^n\,u_x + w(t)\,u + g(t)\,u_{xxx} = 0,\ x > 0,\ t > 0,\ n > 1 \tag{35}$$

with initial and boundary conditions

$$u(x,0) \to \infty,\ x > 0 \tag{36}$$

$$u(0,t) = \gamma\, r(t)\; t > 0,\ \gamma \neq 0 \tag{37}$$

$$u_x(0,t) = 0,\ t > 0 \tag{38}$$

$$u_{xx}(0,t) = 0,\ t > 0 \tag{39}$$

This type of GKdV equation with variable coefficients has several important physical circumstances such as coastal waves in oceans, liquid drops and bubbles, and most interestingly, the atmospheric blocking phenomenon, particularly dipole blocking [22]. The first term of Equation (35) is the evolution term, the second term represents the nonlinear term, the third term is the linear damping with time-dependent coefficient $w(t)$, and the fourth term is the dispersion term with time-dependent coefficient $g(t)$.

The term $w(t)u$ in the generalized KdV Equation (35) can be removed by the point transformation as in [23,24]:

$$\tau = \int e^{-n\int w(t)dt} dt,\ x = x,\ v(x,\tau) = e^{n\int w(t)dt}\, u,\ \tilde{g}(\tau) = e^{n\int w(t)dt}\, g(t) \tag{40}$$

By substituting Equation (40) in Equation (35) we get:

$$v_\tau + v^n\, v_x + \tilde{g}(\tau)\, v_{xxx} = 0,\ x > 0,\ \tau > 0,\ n > 1 \tag{41}$$

All results in Lie symmetries and exact solutions for Equation (35) can be derived from similar results obtained for Equation (41).

Apply Lie group again on Equation (41) as we did before in Equation (19) with the infinitesimal generator

$$X \equiv \xi^1 \frac{\partial}{\partial x} + \xi^2 \frac{\partial}{\partial \tau} + \eta^1 \frac{\partial}{\partial v} \tag{42}$$

Since the generalized KdV equation has at most third-order derivatives, we prolong the vector field X in Equation (42) to the third order. The action of $Pr^{(3)}X$ on Equation (41) vanishes where ν, the solution of Equation (41), is, and then we find the following determining equations:

$$\left.\begin{array}{c} \xi^1_u = \xi^1_\tau = \xi^1_{xx} = 0,\ \xi^1_x = \frac{1}{3}\frac{\tilde{g}_\tau\, \xi^2 + \tilde{g}\, \xi^2_\tau}{\tilde{g}} \\ \xi^2_u = \xi^2_x = \xi^2_{\tau\tau} = 0, \\ \eta^1_x = \eta^1_\tau = 0,\ \eta^1 = \frac{(\xi^1_x - \xi^2_\tau)}{n}\, v \end{array}\right\} \tag{43}$$

The general solution of Equation (43) is

$$\xi^1 = (c_1 + n\, c_4)\, x + c_2,\ \xi^2 = c_1\, \tau + c_3,\ \eta^1 = c_4\, v \tag{44}$$

with the following restriction $(c_1\, \tau + c_3)\, \tilde{g}_\tau = (2\, c_1 + 3\, n\, c_4)\, \tilde{g}$.

We have found all possible forms of $\tilde{g}(\tau)$ when Equation (41) admits different Lie algebras of invariance.

Clearly, by substituting $w(t) = 0$ in Equation (35), the latter reduces to Equation (41) with the same initial and boundary conditions in Equations (36)–(39) (*i.e.*, u = ν). The same result is also obtained by substituting $w(t) = 0$ in Equation (40).

It can be easily checked using Equation (40) that when $w(t) = \frac{\delta}{t}$, both $\tilde{g}(\tau)$ and $g(t)$ are power functions.

By substituting $w(t) = \frac{\delta}{t}$ in Equation (35), we get:

$$u_t + u^n\, u_x + \frac{\delta}{t}\, u + g(t)\, u_{xxx} = 0,\ x > 0,\ t > 0,\ n > 1 \tag{45}$$

with initial and boundary conditions in Equations (36)–(39).

By substituting $w(t) = \frac{\delta}{t}$ in Equation (40) and by using Case 2 in Table 4, Equation (45) reduces to

$$u_t + u^n\, u_x + \frac{\delta}{t}\, u + t^{\lambda(1-\delta\, n)-\delta\, n}\, u_{xxx} = 0 \tag{46}$$

which admits the Lie symmetry generators

$$X \equiv \left(\frac{\lambda+1}{3}\right) c_1\, x \frac{\partial}{\partial x} + \left(\frac{1}{1-\delta\, n}\right) c_1\, t \frac{\partial}{\partial t} + \left(\frac{\lambda-2}{3\, n} - \frac{\delta}{1-\delta\, n}\right) c_1\, u \frac{\partial}{\partial u} \tag{47}$$

The action of X on Equation (37) must vanish, *i.e.*,

$$X\,(u(0,t) - \gamma\, r(t) = 0) = 0 \tag{48}$$

This gives

$$r(t) \;=\; t^{\frac{\lambda\,(1-\delta\, n)\;-\;\delta\, n\;-\;2}{3\, n}} \tag{49}$$

Table 4. Infinitesimals of $v_\tau + v^n\, v_x + \tilde{g}(\tau)\, v_{xxx} = 0$.

No.	$\tilde{g}(\tau)$	Infinitesimals
1	$\forall$	$\xi^1 = c_2, \xi^2 = 0, \eta^1 = 0$
2	τ^λ	$\xi^1 = \frac{c_1}{3}(\lambda+1)\, x + c_2\, ,\xi^2 = c_1\, \tau, \eta^1 = \frac{\lambda-2}{3\, n} c_1 v$
3	e^τ	$\xi^1 = \frac{c_1}{3}\, x + c_2\, ,\xi^2 = c_1, \eta^1 = \frac{1}{3\, n} c_1 v$
4	1	$\xi^1 = \frac{c_1}{3}\, x + c_2\, ,\xi^2 = c_1\, \tau\, +\, c_3, \eta^1 = \frac{-2}{3\, n} c_1 v$

We have shown that Equation (45) admits the Lie symmetry generator Equation (47), which keeps the boundary condition invariant if and only if $g(t)$ is a power function or constant.

The auxiliary equation is

$$\frac{dx}{\frac{\lambda+1}{3}\, c_1\, x} = \frac{dt}{\left(\frac{1}{1-\delta\, n}\right)\, c_1\, t} = \frac{du}{\left(\frac{\lambda-2}{3\, n} - \frac{\delta}{1-\delta\, n}\right)\, c_1\, u} \tag{50}$$

Solving Equation (50), we get:

$$\left.\begin{aligned} \eta(x,t) \;&=\; \frac{x}{t^{\frac{(\lambda+1)\;(1-\delta\, n)}{3}}} \\ u(x,t) \;&= t^{\frac{\lambda\,(1-\delta\, n)-\delta\, n-2}{3\, n}}\, F(\eta) \end{aligned}\right\} \tag{51}$$

From Equations (49) and (51), $u(x,t) = t^{\frac{\lambda\,(1-\delta\, n)-\delta\, n-2}{3\, n}}\, F(\eta) = r(t)\, F(\eta)$ by choosing $\lambda(1-\delta n) - \delta n - 2 < 0$. Therefore, $r(t) \to \infty$ and $u(x,0) \to \infty$ as $t \to 0$. Hence, the solution in this case exists; otherwise, a smooth solution of the problem may not exist.

By substituting Equation (51) in Equation (35), we get:

$$F''' + \left(F^n + \frac{(\lambda+1)\;(\delta\, n-1)}{3}\eta\right)\, F' + \frac{(\lambda-2)\;(1-\delta\, n)}{3n}\, F = 0 \tag{52}$$

Equation (52) can be written in the form

$$\left(F'' + \frac{1}{n+1}F^{n+1}\right)' = -\frac{(\lambda+1)\;(\delta\, n-1)}{3}\eta F' - \frac{(\lambda-2)\;(1-\delta\, n)}{3n}F \tag{53}$$

Assume $\lambda = -\frac{n-2}{n+1}$, and by setting $\delta = 1$, then $\lambda(1 - \delta n) - \delta n - 2 < 0$. Thus, Equation (53) reduces to

$$\left(F'' + \frac{1}{n+1}F^{n+1}\right)' = -\frac{n-1}{n+1}(\eta F' + F) \tag{54}$$

Integrating Equation (54) once, we get:

$$F'' + \frac{1}{n+1}F^{n+1} = -\frac{n-1}{n+1}\eta F + k_3 \tag{55}$$

where k_3 is a constant. The condition that $F(0) = \gamma$ leads to $k_3 = \frac{\gamma^{n+1}}{n+1}$, therefore

$$F'' + \frac{1}{n+1}F^{n+1} + \frac{n-1}{n+1}\eta F = \frac{\gamma^{n+1}}{n+1} \tag{56}$$

$$F(0) = \gamma \tag{57}$$

$$F'(0) = 0 \tag{58}$$

Equation (56) as well as Equations (57) and (58) are identical with the results obtained by Abd-el-Malek and Helal [13] only for $g(t) = t^{\lambda}$ and $w(t) = t^{-1}$. We use the fourth, fifth-order Runge–Kutta method to solve Equations (56)–(58).

It is clear that the wave vibrations of F take on a blubber vase-shaped wave as in Figure 3a,d with an increase in the value of "η". Figure 3b,d display the alternations of the variable F which increases with the increase of the nonlinearity index "n" and the wavelength is seen to decrease with increasing values of "η".

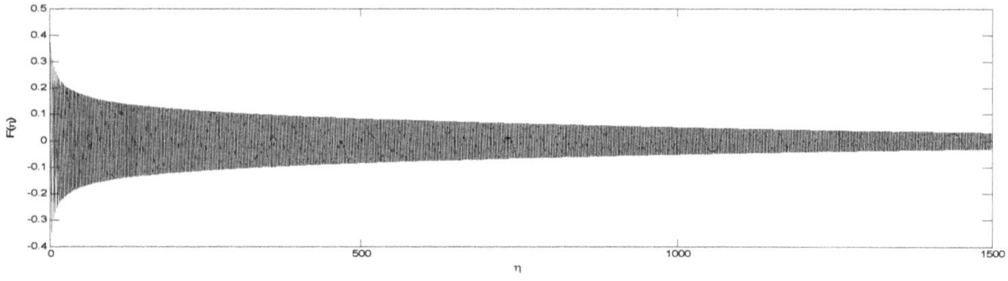

(a)

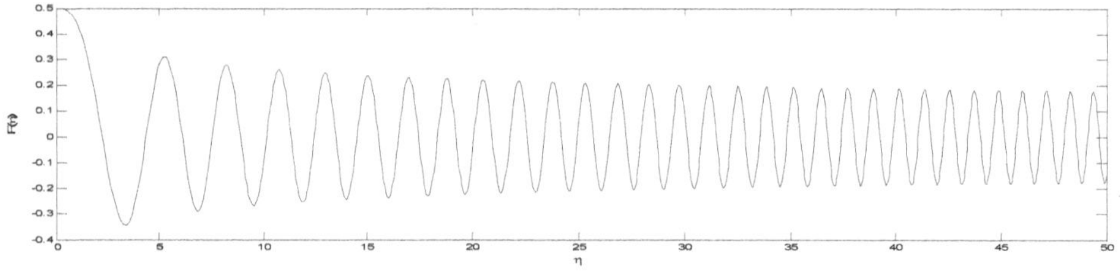

(b)

Figure 3. *Cont.*

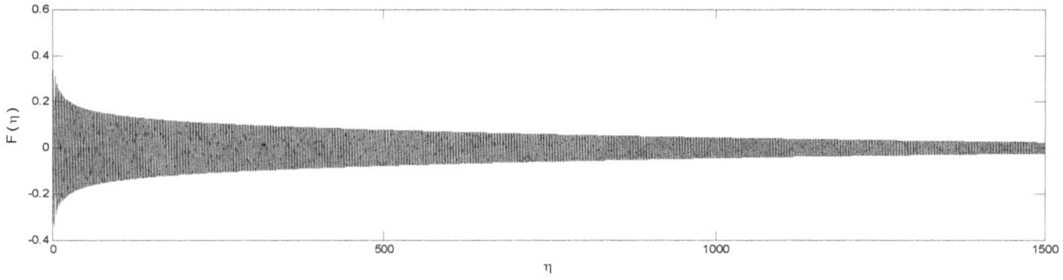

(c)

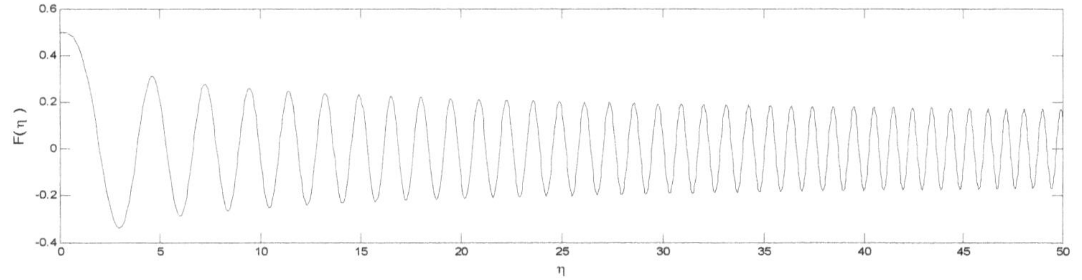

(d)

Figure 3. (**a**) Wave solution of boundary-value problem (56)–(58) for $w(t) = t^{-1}$, $n = 5$, $\gamma = 0.5$; (**b**) Wave solution of boundary-value problem (56)–(58) for $n = 5$, $\gamma = 0.5$, and $\eta \in [0, 50]$; (**c**) Wave solution of boundary-value problem (56)–(58) for $w(t) = t^{-1}$, $n = 100$, $\gamma = 0.5$; (**d**) Wave solution of boundary-value problem (56)–(58) for $n = 100$, $\gamma = 0.5$, and $\eta \in [0, 50]$.

We also succeeded to find other possible forms of $g(t)$, but the only case which keeps the boundary conditions invariant is a power function or constant.

5. Concluding Remarks

In this paper, we applied the Lie group method to generate all symmetries for the three nonlinear partial differential equations, namely generalized Burgers', Burgers'–KdV, and GKdV with initial and boundary conditions. We used the suitable symmetry which satisfies the initial and boundary conditions to reduce the partial differential equations to ordinary differential equations. According to the Lie group method, the traveling wave solutions of all the given equations can be obtained.

Also, the function $g(t)$ in the partial differential equations cannot take exponential form as our initial and boundary condition will not be satisfied. We were able to find more solutions than those that were obtained before by Abd-el-Malek and Helal [13].

The Lie group method can be applied easily to solve boundary value problems by simple procedures.

Finally, in Section 2, we succeeded in constructing the exact solution that describes the velocity field of the generalized Burgers' equation. However, in Sections 3 and 4, the reduced ordinary differential equations could not be solved analytically, only numerically. The effect of the nonlinearity index on the solutions was examined and the numerical solutions are graphically presented.

Acknowledgments: The authors would like to express their gratitude to the expert reviewers for their valuable comments that improved the paper to the present form.

Author Contributions: Mina Abd-el-Malek was responsible for suggesting the problem, reviewing the results and editing the writings; Amr Amin was responsible for the calculations and the writing of the manuscript.

Conflicts of Interest: The authors declare no conflict of interest.

References

1. Burgers, J.M. *A Mathematical Model Illustrating the Theory of Turbulence*; Academic Press: New York, NY, USA, 1948.
2. Su, N.; Watt, J.P.C.; Vincent, K.W.; Close, M.E.; Mao, R. Analysis of Turbulent Flow Patterns of Soil Water under Field Conditions Using Burgers' Equation and Porous Suction—Cup Samplers. *Aust. J. Soil Res.* **2004**, *42*, 9–16. [CrossRef]
3. Turitsyn, S.; Aceves, A.; Jones, C.K.R.T.; Zharnitsky, V. Average dynamics of the optical soliton in communication lines with dispersion management, analytical results. *Phys. Rev. E* **1998**, *58*, R48–R51. [CrossRef]
4. Cole, J.D. On a quasi linear parabolic Equation occurring in aerodynamics. *Quart. Appl. Math.* **1951**, *9*, 225–236.
5. Hopf, E. The partial differential Equation $u_t + uu_x = \mu u_{xx}$. *Comm. Pure Appl. Math.* **1950**, *33*, 201–230.
6. Jeffrey, A.; Mohamad, M.N.B. Exact solutions to the KdV–Burgers' Equation. *Motion* **1991**, *14*, 369–375. [CrossRef]
7. Kadalbajoo, M.K.; Awasthi, A. A Numerical Method Based on Crank—Nicolson scheme for Burgers' Equation. *Appl. Math. Comput.* **2006**, *182*, 1430–1442. [CrossRef]
8. Gorguis, A. A Comparison between Cole–Hopf transformation and decomposition method for solving Burgers' Equation. *Appl. Math. Comput.* **2006**, *173*, 126–136. [CrossRef]
9. Kutluay, S.; Esen, A.; Dag, I. Numerical solutions of the Burgers' Equation by the least—Squares quadratic B—Spline finite element method. *J. Comput. Appl. Math.* **2004**, *167*, 21–33. [CrossRef]
10. Ames, W.F.; Nucci, M.C. Analysis of fluid Equations by group method. *J. Eng. Math.* **1985**, *20*, 181–187. [CrossRef]
11. Hammerton, P.W.; Crighton, D.G. Approximate solution methods for nonlinear acoustic propagation over long ranges. *Proc. R. Soc. Lond. A* **1989**, *426*, 125–152. [CrossRef]
12. Wazwaz, A. Travelling wave solutions of generalized forms of Burgers, Burgers–KdV and Burgers–Huxley Equations. *Appl. Math. Comput.* **2005**, *169*, 639–656. [CrossRef]
13. Abd-el-Malek, M.B.; Helal, M.M. Group method solutions of the generalized forms of Burgers, Burgers–KdV and KdV Equations with time -dependent variable coefficients. *Acta Mech.* **2011**, *221*, 281–296. [CrossRef]
14. Bluman, G.W.; Kumei, S. *Symmetries and Differential Equations*; Springer: Berlin, Germany, 1989.
15. Bluman, G.W.; Anco, S.C. *Symmetry and Integration Methods for Differential Equations*; Applied Mathematical Sciences; Springer-Verlag: New York, NY, USA, 2002; Volume 154.
16. Cherniha, R.; Kovalenko, S. Lie symmetries of nonlinear boundary value problems. *Commun. Nonlinear Sci. Numer. Simulat.* **2012**, *17*, 71–84. [CrossRef]
17. Cherniha, R.; King, J.R. Lie and conditional symmetries of a class of nonlinear (1 + 2) dimensional boundary value problems. *Symmetry* **2015**, *7*, 1410–1435. [CrossRef]
18. Olver, P.J. *Applications of Lie Groups to Differential Equations*; Springer: Berlin, Germany, 1986.
19. Ovsiannikov, L.V. *Group Analysis of Differential Equations*; Academic Press: New York, NY, USA, 1982.
20. Cherniha, R.; Serov, M. Symmetries, Ansaetze and Exact Solutions of Nonlinear Second-order Evolution Equations with Convection Terms, II. *Eur. J. Appl. Math.* **2006**, *17*, 597–605. [CrossRef]
21. Senthilkumaran, M.; Pandiaraja, D.; Vaganan, B.M. New exact explicit solutions of the generalized KdV Equations. *Appl. Math. Comput.* **2008**, *202*, 693–699. [CrossRef]
22. Tang, X.-Y.; Huang, F.; Lou, S.-Y. Variable coefficient KdV Equation and the analytical diagnoses of a dipole blocking life cycle. *Chin. Phys. Lett.* **2006**, *23*, 887–890.
23. Vaneeva, O.O.; Papanicolaou, N.C.; Christou, M.A.; Sophocleous, C. Numerical solutions of boundary value problems for variable coefficient generalized KdV Equations using Lie symmetries. *Commun. Nonlinear Sci. Numer. Simul.* **2014**, *19*, 3074–3085. [CrossRef]
24. Popovych, R.O.; Vaneeva, O.O. More common errors in finding exact solutions of nonlinear differential Equations: Part I. *Commun. Nonlinear Sci. Numer. Simulat.* **2010**, *15*, 3887–3899. [CrossRef]

symmetry

MDPI

Article

A (1 + 2)-Dimensional Simplified Keller–Segel Model: Lie Symmetry and Exact Solutions

Maksym Didovych

Institute of Mathematics, National Academy of Science of Ukraine, 3, Tereshchenkivska Str., Kyiv 01601, Ukraine; m.didovych@gmail.com

Academic Editor: Roman M. Cherniha
Received: 29 June 2015; Accepted: 20 August 2015; Published: 24 August 2015

Abstract: This research is a natural continuation of the recent paper "Exact solutions of the simplified Keller–Segel model" (Commun Nonlinear Sci Numer Simulat **2013**, *18*, 2960–2971). It is shown that a (1 + 2)-dimensional Keller–Segel type system is invariant with respect infinite-dimensional Lie algebra. All possible maximal algebras of invariance of the Neumann boundary value problems based on the Keller–Segel system in question were found. Lie symmetry operators are used for constructing exact solutions of some boundary value problems. Moreover, it is proved that the boundary value problem for the (1 + 1)-dimensional Keller–Segel system with specific boundary conditions can be linearized and solved in an explicit form.

Keywords: Keller–Segel model; Lie symmetry; Neumann boundary-value problem; exact solution

1. Introduction

In 1970–1971, E.F. Keller and L.A. Segel published a remarkable papers [1,2], which they constructed the mathematical model for describing the chemotactic interaction of amoebae mediated by the chemical (acrasin) in. Nowadays their model is called the Keller–Segel model and used for modeling a wide range of processes in biology and medicine. The one-dimensional (with respect to the space variable) version of the Keller–Segel model reads as

$$\begin{aligned} N_t &= \left[D^1(N,P)N_x - \chi(P)NP_x\right]_x, \\ P_t &= \left[D^2(P)P_x\right]_x + \alpha(P)N - \beta(P) \end{aligned} \tag{1}$$

where unknown functions $N(t,x)$ and $P(t,x)$ describe the densities of cells (species) and chemicals, respectively, t and x denote the time and space variables, respectively, $D^1(N,P)$ and $D^2(P)$ are the diffusivities of cells (species) and chemicals, while $\alpha(P)$ and $\beta(P)$ are known non-negative smooth functions. The function $\chi(P)$ (usually a constant χ_0) is called the chemotactic sensitivity. Nowadays a wide range of simplifications of the Keller–Segel model are used for modeling processes in biology and medicine. Here we restrict ourselves on the (1 + 2)-dimensional Keller–Segel system of the form [3–6]

$$\begin{aligned} N_t(t,x,y) &= d_1 \triangle N(t,x,y) - \chi_0 \nabla(N(t,x,y)\nabla P(t,x,y)), \\ 0 &= \triangle P(t,x,y) + \alpha N(t,x,y) - \beta P(t,x,y) \end{aligned} \tag{2}$$

where the parameters d_1, χ_0, α and β are non-negative constants, moreover, $\chi_0 \alpha \neq 0$ (otherwise the model loses its biological meaning). Nowadays, System (2), including the special case $\beta = 0$, is extensively examined by means of different mathematical techniques, in particular, several talks were devoted to this model at a special session within 10th AIMS Conference [7,8].

However, to the best of our knowledge, there are no papers devoted to application of the Lie symmetry method for investigation of System (2), notably for construction of exact solutions. In this paper, we show that this nonlinear system with $\beta = 0$ is invariant with respect infinite-dimensional Lie algebra generated by the operators involving three arbitrary functions, which depend on the time variable. Moreover, the corresponding Neumann boundary-value problems also admit infinite-dimensional Lie algebras. Using these algebras we find exact solutions for (1 + 1) and (1 + 2)-dimensional BVPs. This research is a natural continuation of the recent paper [9].

The paper is organized as follows: in Section 2 maximal algebras of invariance (MAIs) of the Keller–Segel system and corresponding Neumann boundary-value problems are presented. Section 3 is devoted to the application of the Lie symmetry operators for finding exact solutions of some Neumann boundary-value problems with correctly specified parameters. It is also proved that the boundary value problem for the (1 + 1)-dimensional Keller–Segel system with specific boundary conditions can be linearized and solved in an explicit form. The results are summarized in Conclusions.

2. Lie Symmetry of the Neumann Boundary-Value Problem

First of all, one notes that all the parameters, excepting β, can be dropped in System (2) if one introduces non-dimensional variables using the standard re-scaling procedure, *i.e.*, this simplified Keller–Segel system is equivalent to

$$\begin{aligned} \rho_t(t,x,y) &= \Delta\, \rho(t,x,y) - \nabla(\rho(t,x,y)\nabla S(t,x,y)), \\ 0 &= \Delta\, S(t,x,y) + \rho(t,x,y) - \beta^* S(t,x,y) \end{aligned} \tag{3}$$

where $\beta^* = \beta d_1/\alpha$. Obviously, one may set $\beta^* = 0$ provided $\beta d_1/\alpha = \varepsilon << 1$ in (2), hence the nonlinear system

$$\begin{aligned} \rho_t(t,x,y) &= \Delta\, \rho(t,x,y) - \nabla(\rho(t,x,y)\nabla S(t,x,y)), \\ 0 &= \Delta\, S(t,x,y) + \rho(t,x,y) \end{aligned} \tag{4}$$

is obtained.

Theorem 1. *Maximal algebra of invariance (MAI) of the (1 + 2) KS System (4) is the infinite-dimensional Lie algebra generated by the operators*

$$\begin{aligned} G_1^\infty = f_1(t)\tfrac{\partial}{\partial x} + x f_1'(t)\tfrac{\partial}{\partial S},\ G_2^\infty = f_2(t)\tfrac{\partial}{\partial y} + y f_2'(t)\tfrac{\partial}{\partial S}, \\ X_S^\infty = g(t)\tfrac{\partial}{\partial S}, P_t = \tfrac{\partial}{\partial t}, J_{12} = -x\tfrac{\partial}{\partial y} + y\tfrac{\partial}{\partial x}, \\ D = 2t\tfrac{\partial}{\partial t} + x\tfrac{\partial}{\partial x} + y\tfrac{\partial}{\partial y} - 2\rho\tfrac{\partial}{\partial \rho} \end{aligned} \tag{5}$$

where $f_1(t)$, $f_2(t)$ and $g(t)$ are arbitrary function, which possess derivatives of any order.

Proof of the theorem is obtained by straightforward calculations using the well-known technique created by Sophus Lie in 80s of 19 century. Nowadays this routine can be done using computer algebra packages therefore we used Maple 16.

Remark. Maximal algebra of invariance of System (3) with $\beta^* \neq 0$ is the trivial Lie algebra with the basic Lie symmetry operators

$$P_t = \frac{\partial}{\partial t}, P_x = \frac{\partial}{\partial x}, P_y = \frac{\partial}{\partial y}, J_{12} = -x\frac{\partial}{\partial y} + y\frac{\partial}{\partial x}$$

It should be noted that the infinite-dimensional Lie algebra generated by Operators (5) contains as a subalgebra the well-known Galilei algebra $AG(1,2)$ (see, e.g., [10]) with the basic operators

$$\begin{aligned} P_t, P_x, P_y, G_x = tP_x + x\tfrac{\partial}{\partial S}, \\ G_y = tP_y + y\tfrac{\partial}{\partial S}, J_{12} \end{aligned}$$

and its extension $AG_1(1,2)$ with the additional operator D. Here the operators G_x and G_y produce the celebrated Galilei transformations.

Commutators of the MAI (5) are presented in Table 1.

Table 1. Commutators of the maximal algebras of invariance (MAI) (5).

	G_1^∞	G_2^∞	X_S^∞	P_t	J_{12}	D
G_1^∞	0	0	0	$-f_1'(t)\frac{\partial}{\partial x} - xf_1''(t)\frac{\partial}{\partial S}$	$-f_1(t)\frac{\partial}{\partial y} - yf_1'(t)\frac{\partial}{\partial S}$	G_1^*
G_2^∞		0	0	$-f_2'(t)\frac{\partial}{\partial y} - yf_2''(t)\frac{\partial}{\partial S}$	$f_2(t)\frac{\partial}{\partial x} + xf_2'(t)\frac{\partial}{\partial S}$	G_2^*
X_S^∞			0	$-g'(t)\frac{\partial}{\partial S}$	0	$-2tg'(t)\frac{\partial}{\partial S}$
P_t				0	0	$2\frac{\partial}{\partial t}$
J_{12}					0	0
D						0

$$G_1^* = (f_1(t) - 2tf_1'(t))\frac{\partial}{\partial x} + x(-f_1'(t) - 2tf_1''(t))\frac{\partial}{\partial S} = f_1^*(t)\frac{\partial}{\partial x} + xf_1^{*\prime}(t)\frac{\partial}{\partial S},$$
$$G_2^* = (f_2(t) - 2tf_2'(t))\frac{\partial}{\partial y} + y(-f_2'(t) - 2tf_2''(t))\frac{\partial}{\partial S} = f_2^*(t)\frac{\partial}{\partial y} + yf_2^{*\prime}(t)\frac{\partial}{\partial S}$$

It is well-known that a PDE (system of PDEs) cannot model any real process without additional condition(s) on unknown function(s). Thus, boundary-value problems (BVPs) based on the chemotaxis systems of the form (1) are usually studied (see [2,3,11,12] and papers cited therein). In most of these papers authors investigate Neumann problems with zero-flux boundary conditions. Here we examine the Neumann problem for System (4) in half-plane

$$\begin{aligned} \rho_t(t,x,y) &= \Delta\, \rho(t,x,y) - \nabla(\rho(t,x,y)\nabla S(t,x,y)),\\ 0 &= \Delta\, S(t,x,y) + \rho(t,x,y),\\ y = 0 &: \rho_y = q_1(t),\ S_y = q_2(t),\\ y = +\infty &: \rho_y = S_y = 0 \end{aligned} \tag{6}$$

where $q_1(t)$ and $q_2(t)$ are arbitrary functions, which possess derivatives of any order.

Obviously, Lie algebra (5) cannot be MAI of the BVP (6) for arbitrary functions $q_1(t)$ and $q_2(t)$. Moreover, BVP (6) involves conditions at infinity, so one cannot apply the definition [13,14] in order to examine Lie invariance of this problem. Here we adapt for such purpose the definition proposed in [15].

First, let us calculate the linear combination for all the operators listed in (5).

$$\begin{aligned} X = a_1G_1^\infty + a_2G_2^\infty + a_3X_S^\infty + a_4P_t + a_5J_{12} + a_6D = \\ (a_4 + 2ta_6)\frac{\partial}{\partial t} + (a_1f_1(t) + a_5y + a_6x)\frac{\partial}{\partial x} + (a_2f_2(t) - a_5x + a_6y)\frac{\partial}{\partial y} + \\ (a_3g(t) + a_1f_1'(t)x + a_2f_2'(t)y)\frac{\partial}{\partial S} - 2a_6\rho\frac{\partial}{\partial \rho} \end{aligned} \tag{7}$$

and its first prolongation

$$\underset{1}{X} = X + \sigma_0^1\frac{\partial}{\partial \rho_t} + \sigma_1^1\frac{\partial}{\partial \rho_x} + \sigma_2^1\frac{\partial}{\partial \rho_y} + \sigma_0^2\frac{\partial}{\partial S_t} + \sigma_1^2\frac{\partial}{\partial S_x} + \sigma_2^2\frac{\partial}{\partial S_y}$$

where $a_1, \ldots, a_6$ to be determined parameters.

Using Definition 2 [15] we formulate the following invariance criteria.

Definition 1. *BVP (6) is invariant w.r.t. the Lie operator (7) if:*

(a) *Operator (7) is a Lie symmetry operator of System (4);*

(b) $X(y) = 0$ *when* $y = 0$;

(c) $\overset{X}{1}\left(\rho_y - q_1(t)\right) = 0$ *when* $y = 0, \rho_y = q_1(t)$ *and* $\overset{X}{1}(S_y - q_2(t)) = 0$ *when* $y = 0, S_y = q_2(t)$;

(d) *there exists a smooth bijective transform T mapping* $M = \left\{y = +\infty, \rho_y = 0, S_y = 0\right\}$ *into* $M^* = \left\{y^* = 0, B_1\left(\rho^*, \rho^*_{y^*}\right) = 0, B_2\left(S^*, S^*_{y^*}\right) = 0\right\}$ *of the same dimensionality;*

(e) $X^*(y^*) = 0$ *when* $y^* = 0$*;*

(f) $\overset{X^*}{k}(B_1) = 0$ *when* $y^* = 0$, $B_1 = 0$ *and* $\overset{X^*}{k}(B_2) = 0$ *when* $y^* = 0$, $B_2 = 0$, $k = 0$ *or* $k = 1$. *Where* y^*, ρ^*, S^* *are new variables,* X^* *is operator X expressed via the new variables and the functions* B_1 *and* B_2 *are defined by T.*

Let us apply this definition to BVP (6).

Taking into account item (b) one immediately obtains the condition $a_2 f_2(t) - a_5 x = 0$ which means that $a_2 = a_5 = 0$.

Now we apply the operator $\underset{1}{X}$ to the manifolds $\left\{y = 0, \rho_y = q_1(t)\right\}$ and $\{y = 0, S_y = q_2(t)\}$ (item (c))

$$\underset{1}{X}\left(\rho_y - q_1(t)\right)\Big|_{y=0,\rho_y=q_1(t)} = -3a_6 q_1(t) - (a_4 + 2a_6 t)\,\dot{q}_1(t) = 0,$$
$$\underset{1}{X}(S_y - q_2(t))\Big|_{y=0,S_y=q_2(t)} = -a_6 q_2(t) - (a_4 + 2a_6 t)\,\dot{q}_2(t) = 0$$

Thus two conditions are obtained:

$$\begin{aligned} 3a_6 q_1(t) + (a_4 + 2a_6 t)\dot{q}_1(t) &= 0, \\ a_6 q_2(t) + (a_4 + 2a_6 t)\dot{q}_2(t) &= 0 \end{aligned} \tag{8}$$

Let us consider the following change of variables, which was used in [15] for the similar purposes, in order to examine items (d)–(f)

$$\tau = t, x^* = x, y^* = \frac{1}{y}, U = \frac{\rho}{y}, V = \frac{S}{y} \tag{9}$$

By direct calculations we have proved that Transform (9) maps $M = \left\{y = +\infty, \rho_y = 0, S_y = 0\right\}$ into $M^* = \{y^* = 0, U = 0, V = 0\}$. Since both manifolds have the same dimensionality, item (d) is fulfilled. Transform (9) maps Operator X (7) (here we take into account that $a_2 = a_5 = 0$) to the form

$$\begin{aligned} X^* = (a_4 + 2a_6\tau)\tfrac{\partial}{\partial\tau} + (a_1 f_1(\tau) + a_6 x^*)\tfrac{\partial}{\partial x^*} - a_6 y^* \tfrac{\partial}{\partial y^*} + \\ (a_1 f_1'(\tau) x^* y^* + a_3 g(\tau) y^* - a_6 V)\tfrac{\partial}{\partial V} - 3a_6 U \tfrac{\partial}{\partial U} \end{aligned}$$

Now it is easy to check items (e)–(f)

$$\begin{aligned} X^*(y^*)|_{y^*=0} &= -a_6 y^*|_{y^*=0} \equiv 0, \\ X^*(U)|_{y^*=0,U=0} &= -3a_6 U|_{y^*=0,U=0} \equiv 0, \\ X^*(V)|_{y^*=0,V=0} &= (a_1 f_1'(\tau) x^* y^* + a_3 g(\tau) y^* - a_6 V)|_{y^*=0,V=0} \equiv 0 \end{aligned}$$

Thus we only need to satisfy Conditions (8). It can be noted that these conditions lead to four different possibilities only:

1. if $q_1(t)$ and $q_2(t)$ are arbitrary function, which possess derivatives of any order, then $a_4 = a_6 = 0$, *i.e.*, $X = a_1 G_1^\infty + a_3 X_S^\infty$;
2. if $q_1(t) = \frac{q_1^0}{\sqrt{\left(t + \frac{a_4}{2a_6}\right)^3}}$, $q_2(t) = \frac{q_2^0}{\sqrt{t + \frac{a_4}{2a_6}}}$, where $q_1^0, q_2^0 \in \mathbb{R}$, then $X = a_1 G_1^\infty + a_3 X_S^\infty + a_4 P_t + a_6 D$ (here a_4 and $a_6 \neq 0$ are no longer arbitrary);
3. if $q_1(t) = q_1^0 = const$, $q_2(t) = q_2^0 = const$ then $a_6 = 0$, *i.e.*, $X = a_1 G_1^\infty + a_3 X_S^\infty + a_4 P_t$;
4. if $q_1(t) = q_2(t) = 0$ then $X = a_1 G_1^\infty + a_3 X_S^\infty + a_4 P_t + a_6 D$.

Let us formulate the result as follows (we set $t + \frac{a_4}{2a_6} \to t$ without losing a generality).

Theorem 2. *All possible MAIs of the (1 + 2)-dimensional Neumann boundary-value problem (6) depending on the form of the functions $q_1(t)$ and $q_2(t)$ are presented in Table 2. In Table 2 $q_1^0, q_2^0 \in \mathbb{R}$ and $\left(q_1^0\right)^2 + \left(q_2^0\right)^2 \neq 0$.*

Table 2. MAIs and restrictions for Neumann BVP (6).

	$q_1(t)$	$q_2(t)$	**MAI**
1	$\forall$	$\forall$	G_1^∞, X_S^∞
2	$\frac{q_1^0}{\sqrt{t^3}}$	$\frac{q_2^0}{\sqrt{t}}$	$G_1^\infty, X_S^\infty, D$
3	q_1^0	q_2^0	$G_1^\infty, X_S^\infty, P_t$
4	0	0	$G_1^\infty, X_S^\infty, P_t, D$

3. Exact Solutions of Neumann Problems

This section is devoted to the applying of Lie symmetry operators obtained in Theorem 2 in order to reduce the Neumann BVP (6) to BVPs of lower dimensionality and find exact solutions.

In the most general case we apply a linear combination of operators G_1^∞ and X_s^∞ (case 1, Theorem 2):

$$G_1^\infty + a_3 X_s^\infty = f_1(t)\frac{\partial}{\partial x} + \left(x f_1'(t) + a_3\, g(t)\right)\frac{\partial}{\partial S}$$

This operator generates ansatz

$$\begin{gathered} \rho(t,x,y) = \varrho(t,y), \\ S(t,x,y) = \varphi(t,y) + \tfrac{f_1'(t)}{2f_1(t)}\, x^2 + a_3 \tfrac{g(t)}{f_1(t)}\, x \end{gathered} \tag{10}$$

Ansatz (10) reduces BVP (6) to the (1 + 1)-dimensional BVP

$$\begin{gathered} \varrho_t(t,y) = \varrho_{yy}(t,y) - \left(\varrho(t,y)\varphi_y(t,y)\right)_y - \tfrac{f_1'(t)}{f_1(t)}\varrho(t,y), \\ 0 = \varphi_{yy}(t,y) + \varrho(t,y) + \tfrac{f_1'(t)}{f_1(t)}, \\ y = 0 : \varrho_y = q_1(t), \varphi_y = q_2(t), \\ y = +\infty : \varrho_y = \varphi_y = 0 \end{gathered} \tag{11}$$

Let us consider special case of BVP (11): $f_1(t) = 1$ and $q_1(t) = q_1^0$, $q_2(t) = q_2^0$. In this case the Nonlinear problem (11) can be presented as follows

$$\begin{gathered} \varrho_t(t,y) = \varrho_{yy}(t,y) - \left(\varrho(t,y)\varphi_y(t,y)\right)_y, \\ 0 = \varphi_{yy}(t,y) + \varrho(t,y) \end{gathered} \tag{12}$$

$$\begin{gathered} y = 0 : \varrho_y = q_1^0, \varphi_y = q_2^0, \\ y = +\infty : \varrho_y = \varphi_y = 0 \end{gathered} \tag{13}$$

In reality (12) and (13) is the (1 + 1)-dimensional analog of the (1 + 2)-dimensional BVP (6) with $q_k(t) = q_k^0, k = 1,2$. System (12) can be reduced to the 3-rd order PDE

$$\varphi_{ty} = \varphi_{yyy} - \varphi_{yy}\varphi_y + \vartheta(t)$$

where $\vartheta(t)$ is an arbitrary function. Setting $\vartheta(t) = 0$, using the Cole–Hopf substitution

$$\varphi_y(t,y) = -2\frac{V_y(t,y)}{V(t,y)} \tag{14}$$

and taking into account the Boundary conditions (13), we obtain BVP problem for the heat equation

$$\begin{aligned} V_t &= V_{yy}, \\ y = 0 &: V_y + \frac{q_2^0}{2} V = 0, \\ y = +\infty &: V_y = 0 \end{aligned} \tag{15}$$

In order to solve (15) by using the classical technique, we should specify an initial profile. Let us set for simplicity $V(0, y) = V_0 = const$. Now one may use Laplace transform $\mathbb{V}_L(s, y) = \int_0^{+\infty} V(t, y) e^{-st} dt$ to reduce heat equation to the 2nd order ODE

$$\mathbb{V}_L'' - s\mathbb{V}_L(s, y) + V_0 = 0 \tag{16}$$

with boundary conditions

$$\begin{aligned} y = 0 &: \mathbb{V}_L' + \frac{q_2^0}{2} \mathbb{V}_L = 0, \\ y = +\infty &: \mathbb{V}_L' = 0 \end{aligned} \tag{17}$$

The general solution of BVP (16) and (17) is

$$\mathbb{V}_L(s, y) = \frac{q_2^0 V_0}{s(2\sqrt{s} - q_2^0)} e^{-\sqrt{s}y} + \frac{V_0}{s}$$

By using the inverse Laplace transform (see for example [16]) and the relevant simplifications one obtains the general solution of the Linear BVP (15)

$$V(t, y) = V_0 \left(1 - \operatorname{erfc}\left(\frac{y}{2\sqrt{t}} \right) + e^{\frac{(q_2^0)^2}{4} t - \frac{q_2^0}{2} y} \operatorname{erfc}\left(\frac{y}{2\sqrt{t}} - \frac{q_2^0}{2} \sqrt{t} \right) \right)$$

Now, by using Cole-Hopf substitution (14), one finds the exact solution for the Nonlinear problem (12) and (13)

$$\begin{aligned} \varrho(t, y) &= q_2^0 \frac{\frac{1}{\sqrt{\pi t}} e^{-\frac{y^2}{4t}} \operatorname{erf}\left(\frac{y}{2\sqrt{t}}\right) + e^{\frac{(q_2^0)^2}{4} t - \frac{q_2^0}{2} y} \operatorname{erfc}\left(\frac{y}{2\sqrt{t}} - \frac{q_2^0}{2}\sqrt{t}\right) \left(\frac{q_2^0}{2} \operatorname{erf}\left(\frac{y}{2\sqrt{t}}\right) + \frac{1}{\sqrt{\pi t}} e^{-\frac{y^2}{4t}} \right)}{\left(\operatorname{erf}\left(\frac{y}{2\sqrt{t}}\right) + e^{\frac{(q_2^0)^2}{4} t - \frac{q_2^0}{2} y} \operatorname{erfc}\left(\frac{y}{2\sqrt{t}} - \frac{q_2^0}{2}\sqrt{t}\right) \right)^2}, \\ \varphi(t, y) &= -2 \ln \left(\operatorname{erf}\left(\frac{y}{2\sqrt{t}}\right) + e^{\frac{(q_2^0)^2}{4} t - \frac{q_2^0}{2} y} \operatorname{erfc}\left(\frac{y}{2\sqrt{t}} - \frac{q_2^0}{2}\sqrt{t}\right) \right) + h(t) \end{aligned} \tag{18}$$

where $h(t)$ is an arbitrary smooth function. Plots of Solution (18) are presented on Figure 1. It should be noted that the very similar profile of the function ρ which describes density of cells was presented in many papers (see, e.g., [2,17–19]). However, in papers [2,17,18] the traveling wave solutions were found, and in [19] the numerical ones. So the exact Solution (18) is new because it is neither traveling wave solution nor numerical. It possesses much more complicated structure. Nevertheless this profile of the function ρ represents the traveling band of cells. This phenomenon was studied by J. Adler in his experiments which were described in [20].

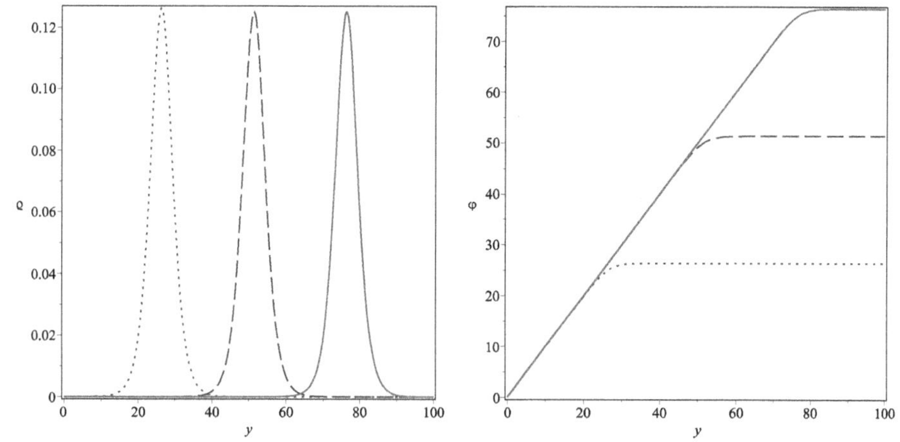

Figure 1. Plots of functions $\varrho(t,y)$ and $\varphi(t,y)$ with $q_2^0 = 1, h(t) = 0$ and $t = 50$ (dot line), $t = 100$ (dash line), $t = 150$ (solid line).

Consider Case 2 in Table 2. The linear combination of operators $a_1 G_1^\infty + a_3 X_S^\infty + D$ produces the following ansatz

$$\begin{aligned} \rho(t,x,y) &= \tfrac{1}{t}\,\psi(\omega_1,\omega_2), \\ S(t,x,y) &= \phi(\omega_1,\omega_2) + \tfrac{a_1 x}{2\sqrt{t}} \int \tfrac{f_1'(t)}{\sqrt{t}} dt + \kappa(t), \\ \omega_1 &= \tfrac{x}{\sqrt{t}} - \tfrac{a_1}{2} \int \tfrac{f_1(t)}{t^{\frac{3}{2}}} dt, \\ \omega_2 &= \tfrac{y}{\sqrt{t}} \end{aligned} \tag{19}$$

where $\kappa(t)$ is an arbitrary smooth function.

This ansatz reduces BVP (6) to the elliptic BVP

$$\begin{aligned} \triangle\, \psi - \nabla(\psi \nabla \phi) + \tfrac{\omega_1}{2} \psi_{\omega_1} + \tfrac{\omega_2}{2} \psi_{\omega_2} + \psi = 0, \\ \psi + \triangle\, \phi = 0 \end{aligned} \tag{20}$$

$$\begin{aligned} \omega_2 = 0 : \psi_{\omega_2} = q_1^0, \phi_{\omega_2} = q_2^0, \\ \omega_2 = +\infty : \psi_{\omega_2} = \phi_{\omega_2} = 0 \end{aligned} \tag{21}$$

It can be easily established that System (20) is invariant w.r.t. the 4-dimensional MAI generated by the operators

$$P_1 = \frac{\partial}{\partial \omega_1} + \frac{\omega_1}{2}\frac{\partial}{\partial \phi}, P_2 = \frac{\partial}{\partial \omega_2} + \frac{\omega_2}{2}\frac{\partial}{\partial \phi}, P_3 = \frac{\partial}{\partial \phi}, P_4 = \omega_2 \frac{\partial}{\partial \omega_1} - \omega_1 \frac{\partial}{\partial \omega_2}$$

In quite a similar way as it was done for BVP (6) we have proved that only operators P_1 and P_3 are the Lie symmetry operators of BVP (20) and (21). The linear combination of these operators $P_1 + \lambda P_3 = \frac{\partial}{\partial \omega_1} + \left(\lambda + \frac{\omega_1}{2}\right)\frac{\partial}{\partial \phi}, \lambda \in \mathbb{R}$ produces ansatz:

$$\begin{aligned} \psi(\omega_1,\omega_2) &= \psi_*(\omega_2), \\ \phi(\omega_1,\omega_2) &= \phi_*(\omega_2) + \tfrac{\omega_1^2}{4} + \lambda\, \omega_1 \end{aligned}$$

which reduces the Elliptic BVP (20) and (21) to the problem for the second-order ODEs

$$\begin{array}{c} \psi_*'' - (\psi_*\phi_*')' + \frac{\omega_2}{2}\psi_*' + \frac{1}{2}\psi_* = 0, \\ \phi_*'' + \psi_* + \frac{1}{2} = 0, \\ \omega_2 = 0 : \frac{\partial \psi_*}{\partial \omega_2} = q_1^0, \frac{\partial \phi_*}{\partial \omega_2} = q_2^0, \\ \omega_2 = +\infty : \frac{\partial \psi_*}{\partial \omega_2} = \frac{\partial \phi_*}{\partial \omega_2} = 0 \end{array} \tag{22}$$

Unfortunately we were unable to solve BVP (22) because the governing system of ODEs is non-integrable. Happily we noted that BVP (20) and (21) is invariant w.r.t. the Q-conditional symmetry operator $\frac{\partial}{\partial \omega_1}$ (in the sense of Definition 2 [15]). The ansatz generated by the operator $\frac{\partial}{\partial \omega_1}$ has the form

$$\begin{array}{c} \psi(\omega_1, \omega_2) = \hat{\psi}(\omega_2), \\ \phi(\omega_1, \omega_2) = \hat{\phi}(\omega_2) \end{array} \tag{23}$$

In contrast to the previous ansatz, this one reduces BVP (20) and (21) to the simpler system of ODEs

$$\begin{array}{c} \hat{\psi}'' - \left(\hat{\psi}\hat{\psi}'\right)' + \frac{\omega_2}{2}\hat{\psi}' + \hat{\psi} = 0, \\ \hat{\psi} + \hat{\phi}'' = 0 \end{array} \tag{24}$$

with boundary conditions

$$\begin{array}{c} \omega_2 = 0 : \hat{\psi}_{\omega_2} = q_1^0, \hat{\phi}_{\omega_2} = q_2^0, \\ \omega_2 = +\infty : \hat{\psi}_{\omega_2} = \hat{\phi}_{\omega_2} = 0 \end{array} \tag{25}$$

System (24) can be reduced to the 4-th order ODE

$$\hat{\phi}^{(4)} + \frac{\omega_2}{2}\hat{\phi}^{(3)} - \left(\hat{\phi}''\hat{\phi}'\right)' + \hat{\phi}'' = 0$$

By integrating this equation twice and then using substitution $\hat{\phi}'(\omega_2) = \mu(\omega_2)$, one can obtain the first order ODE

$$\mu' - \frac{1}{2}\mu^2 + \frac{\omega_2}{2}\mu + \mu_1^0\omega_2 + \mu_2^0 = 0 \tag{26}$$

where $\mu_0^1, \mu_0^2 \in \mathbb{R}$.

In order to construct the general solution of Equation (26), we apply the substitution (see, e.g., [21])

$$U(\omega_2) = e^{-\frac{1}{2}\int \mu(\omega_2)\, d\omega_2}$$

Now the linear ODE

$$U'' + \frac{\omega_2}{2}U' + \left(\frac{1}{2}\,\mu_1^0\,\omega_2 + \frac{1}{2}\,\mu_2^0\right)U = 0 \tag{27}$$

is obtained with the general solution:

$$\begin{array}{c} U(\omega_2) = A\,e^{-\mu_1^0\omega_2}K\left(\left(\mu_1^0\right)^2 + \frac{1}{2}\mu_2^0, \frac{1}{2}, -\frac{1}{4}\left(\omega_2 - 4\mu_1^0\right)^2\right) + \\ B\left(\omega_2 - 4\mu_1^0\right)e^{-\mu_1^0\omega_2}\,K\left(\left(\mu_1^0\right)^2 + \frac{1}{2}\mu_2^0 + \frac{1}{2}, \frac{3}{2}, -\frac{1}{4}\left(\omega_2 - 4\mu_1^0\right)^2\right) \end{array}$$

where $A, B \in \mathbb{R}$ and $K(a,b,z)$ is Kummer's function

$$K(a,b,z) = 1 + \sum_{k=1}^{\infty} \frac{(a)_k}{(b)_k}\frac{z^k}{k!}, (a)_k = a(a+1)...(a+k-1), (a)_0 = 1$$

Because Kummer's functions lead to a very cumbersome solution of BVP in question, we consider the special case $\mu_1^0 = \mu_2^0 = 0$ (let us note that more general case $\mu_2^0 = -2(\mu_1^0)^2$ leads to the same result because of the Boundary conditions (25). In this case Equation (26) has the general solution

$$\mu(\omega_2) = \frac{e^{-\frac{\omega_2^2}{4}}}{A - \frac{\sqrt{\pi}}{2}\operatorname{erf}(\frac{\omega_2}{2})}$$

From the Boundary condition (25) follows $\mu(0) = \frac{1}{A} = q_2^0$, hence

$$\mu(\omega_2) = q_2^0 \frac{e^{-\frac{\omega_2^2}{4}}}{1 - q_2^0 \frac{\sqrt{\pi}}{2}\operatorname{erf}(\frac{\omega_2}{2})}$$

Now one obtains the general solution of BVP (24) and (25)

$$\begin{aligned} \hat{\phi}(\omega_2) &= -2\ln\left(1 - \frac{q_2^0\sqrt{\pi}}{2}\operatorname{erf}(\frac{\omega_2}{2})\right), \\ \hat{\psi}(\omega_2) &= q_2^0 \frac{\frac{\omega_2}{2}e^{-\frac{\omega_2^2}{4}}\left(1-q_2^0\frac{\sqrt{\pi}}{2}\operatorname{erf}(\frac{\omega_2}{2})\right)-q_2^0 e^{-\frac{\omega_2^2}{2}}}{\left(1-q_2^0\frac{\sqrt{\pi}}{2}\operatorname{erf}(\frac{\omega_2}{2})\right)^2} \end{aligned} \tag{28}$$

Since $\psi'(0) = q_1^0$ one can calculate that $q_1^0 = \frac{q_2^0}{2} - 2(q_2^0)^3$. Thus, the exact solution of BVP (6) with $q_1(t) = \frac{\frac{q_2^0}{2} - 2(q_2^0)^3}{\sqrt{t^3}}$ and $q_2(t) = \frac{q_2^0}{\sqrt{t}}$ has the form

$$\begin{aligned} \rho(t,x,y) &= \frac{q_2^0}{t} \cdot \frac{\frac{y}{2\sqrt{t}}e^{-\frac{y^2}{4t}}\left(1-q_2^0\frac{\sqrt{\pi}}{2}\operatorname{erf}\left(\frac{y}{2\sqrt{t}}\right)\right)-q_2^0 e^{-\frac{y^2}{2t}}}{\left(1-q_2^0\frac{\sqrt{\pi}}{2}\operatorname{erf}\left(\frac{y}{2\sqrt{t}}\right)\right)^2} \frac{y}{\sqrt{t}}, \\ S(t,x,y) &= -2\ln\left(1 - \frac{q_2^0\sqrt{\pi}}{2}\operatorname{erf}\left(\frac{y}{2\sqrt{t}}\right)\right) + \frac{a_1 x}{2\sqrt{t}} \int \frac{f_1'(t)}{\sqrt{t}}dt + \kappa(t) \end{aligned} \tag{29}$$

where $\kappa(t)$ is an arbitrary smooth function. Solution (29) is continuous when $q_2^0 < \frac{2}{\sqrt{\pi}}$.

4. Conclusions

In this work we studied a simplified version of (1 + 2)-dimensional Keller–Segel model. It is well-known that Keller–Segel model is widely used for modeling a wide range of processes in biology and medicine (especially for the tumour growth modeling) therefore one is extensively examined by means of different mathematical techniques.

It was established that MAI of System (4) is the infinite-dimensional Lie algebra. Moreover we have proved that different Neumann BVPs for this system of the form (6) still admit infinite-dimensional Lie algebras depending on the form of fluxes $q_1(t)$ and $q_2(t)$. Using the definition from [15], all inequivalent problems of the form (6) were found, which admit different MAIs (see Theorem 2).

In order to construct the exact solutions of some Neumann problems, the Lie symmetry operators were applied. In particular, we have proved that the BVP for the one-dimensional (in space) Keller–Segel system in question can be linearized. As result, the exact solution of the BVP was constructed in explicit form (18). It should be stressed that this solution has a remarkable properties, which allow a biological interpretation.

Finally, the exact solution for the (1 + 2)-dimensional BVP with the correctly specified boundary conditions was found (see Formula (29)).

Acknowledgments: The author is grateful to R Cherniha who brought my attention to this problem and for his helpful comments.

Conflicts of Interest: The author declares no conflict of interest.

References

1. Keller, E.F.; Segel, L.A. Initiation of slime mold aggregation viewed as an instability. *J. Theor. Biol.* **1970**, *26*, 399–415. [CrossRef]
2. Keller, E.F.; Segel, L.A. Traveling bands of chemotactic bacteria: A theoretical analysis. *J. Theor. Biol.* **1971**, *30*, 235–248. [CrossRef]
3. Horstmann, D. From 1970 until present: the Keller–Segel model in chemotaxis and its consequences. *Jahresber. Deutsch. Math. Verein.* **2003**, *105*, 103–165.
4. Calvez, V.; Dolak-Strub, Y. Asymptotic Behavior of a Two-Dimensional Keller–Segel Model with and without Density Control. *Math. Modeling Biol. Syst.* **2007**, *2*, 323–329.
5. Nagai, T. Convergence to Self-Similar Solutions for a Parabolic–Elliptic System of Drift-Diffusion Type in R^2. *Adv. Differ. Equ.* **2011**, *9–10*, 839–866.
6. Nagai, T. Global Existence and Decay Estimates of Solutions to a Parabolic–Elliptic System of Drift-Diffusion Type in R^2. *Differ. Integral Equ.* **2011**, *1–2*, 29–68.
7. Fujie, K.; Yokota, T. Behavior of solutions to parabolic–elliptic Keller–Segel systems with signal-dependent sensitivity. In Mathematical Models of Chemotaxis, Proceeding of the 10th AIMS Conference on Dynamical Systems, Differential Equations and Applications, Madrid, Spain, 7–11 July 2014.
8. Granero-Belinchon, R.; Ascasibar, Y. On the Patlak–Keller–Segel model with a nonlocal flux. In Mathematical Models of Chemotaxis, Proceeding of the 10th AIMS Conference on Dynamical Systems, Differential Equations and Applications, Madrid, Spain, 7–11 July 2014.
9. Cherniha, R.; Didovych, M. Exact solutions of the simplified Keller–Segel model. *Commun. Nonlinear Sci. Numer. Simulat.* **2013**, *18*, 2960–2971. [CrossRef]
10. Fushchych, W.; Cherniha, R. Galilei-invariant systems of nonlinear systems of evolution equations. *J. Phys. A Math. Gen.* **1995**, *28*, 5569–5579. [CrossRef]
11. Hillen, T.; Painter, K.J. A user's guide to PDE models for chemotaxis. *J. Math. Biol.* **2009**, *58*, 183–217. [CrossRef]
12. Nagai, T. Blow-up of Radially Symmetric Solutions. *Adv. Math. Sci. Appl.* **1995**, *5*, 581–601.
13. Bluman, G.W. Application of the general similarity solution of the heat equation to boundary value problems Q. *Appl. Math.* **1974**, *31*, 403–415.
14. Bluman, W.; Kumei, S. *Symmetries and Differential Equations*; Springer: Berlin, Germany, 1989.
15. Cherniha, R.; King, J. Lie and Conditional Symmetries of a Class of Nonlinear (1 + 2)-dimensional Boundary Value Problems. *Symmetry* **2015**, *7*, 1410–1435. [CrossRef]
16. Polyanin, A.; Manzhirov, A. *Handbook of Integral Equations*; CRC Press Company: Boca Raton, FL, USA, 1998.
17. Novick-Cohen, A.; Segel, L. A Gradually Slowing Traveling Band of Chemotactic Bacteria. *J. Math. Biol.* **1984**, *19*, 125–132. [CrossRef]
18. Feltham, D.L.; Chaplain, M.A.J. Travelling Waves in a Model of Species Migration. *Appl. Math. Lett.* **2000**, *13*, 67–73. [CrossRef]
19. Wang, Q. Boundary Spikes of a Keller–Segel Chemotaxis System with Saturated Logarithmic Sensitivity. *Disc. Contin. Dyn. Syst. Ser. B* **2015**, *20*, 1231–1250. [CrossRef]
20. Adler, J. Chemotaxis in bacteria. *Ann. Rev. Biochem.* **1975**, *44*, 341–356. [CrossRef]
21. Polyanin, A.; Zaitsev, V. *Exact Solutions for Ordinary Differential Equations*; CRC Press: Boca Raton, FL, USA, 2003.

symmetry

MDPI

Article

A (1 + 2)-Dimensional Simplified Keller–Segel Model: Lie Symmetry and Exact Solutions. II

Roman Cherniha * and Maksym Didovych

Institute of Mathematics, National Academy of Sciences of Ukraine, 3 Tereshchenkivs'ka Street, Kyiv 01004, Ukraine; m.didovych@gmail.com

* Correspondence: r.m.cherniha@gmail.com; Tel.: +38-044-235-2010

Academic Editor: Hari M. Srivastava

Received: 31 October 2016; Accepted: 10 January 2017; Published: 20 January 2017

Abstract: A simplified Keller–Segel model is studied by means of Lie symmetry based approaches. It is shown that a (1 + 2)-dimensional Keller–Segel type system, together with the correctly-specified boundary and/or initial conditions, is invariant with respect to infinite-dimensional Lie algebras. A Lie symmetry classification of the Cauchy problem depending on the initial profile form is presented. The Lie symmetries obtained are used for reduction of the Cauchy problem to that of (1 + 1)-dimensional. Exact solutions of some (1 + 1)-dimensional problems are constructed. In particular, we have proved that the Cauchy problem for the (1 + 1)-dimensional simplified Keller–Segel system can be linearized and solved in an explicit form. Moreover, additional biologically motivated restrictions were established in order to obtain a unique solution. The Lie symmetry classification of the (1 + 2)-dimensional Neumann problem for the simplified Keller–Segel system is derived. Because Lie symmetry of boundary-value problems depends essentially on geometry of the domain, which the problem is formulated for, all realistic (from applicability point of view) domains were examined. Reduction of the the Neumann problem on a strip is derived using the symmetries obtained. As a result, an exact solution of a nonlinear two-dimensional Neumann problem on a finite interval was found.

Keywords: Lie symmetry; algebra of invariance; nonlinear boundary-value problem; Keller–Segel model; Cauchy problem; exact solution

MSC: 35K5; 22E70

1. Introduction

Nonlinear partial differential equations describe various processes in society and nature. The well-known principle of linear superposition cannot be applied to generate new exact solutions to nonlinear partial differential equations (PDEs). Therefore, the classical methods for solving *linear PDEs* are not applicable for solving *nonlinear PDEs*. It means that finding exact solutions of most nonlinear PDEs generally requires new methods. Finding exact solutions that have a physical, chemical or biological interpretation is of fundamental importance. The most popular method for construction of exact solutions to nonlinear PDEs is the Lie method, which was created by Sophus Lie, the famous Norwegian mathematician, in 1880s–1890s and published in his papers and books. His most important work in this direction is [1] (see also [2]). Nowadays, the Lie symmetry method is widely applied to study partial differential equations (including multi-component systems of PDEs), notably for their reductions to ordinary differential equations (ODEs) and for constructing exact solutions. There are a huge number of papers and many excellent books [3–7] devoted to such applications.

In real world applications, mathematical models are typically based on PDEs with the relevant boundary and/or initial conditions. As a result, one needs to investigate boundary value problems

(BVPs) and initial problems (Cauchy problems). In the case of nonlinear BVPs and initial problems, a fundamental difficulty arises in solving such problems using analytical methods. One may note that the Lie method has not been widely used for solving BVPs and initial problems. A natural reason follows from the following observation: the relevant boundary and initial conditions are usually not invariant under transformations, which are generated by Lie symmetry of the governing PDE. Nevertheless, there are some classes of BVPs that can be solved by means of the Lie symmetry algorithm. A brief history concerning first attempts to apply Lie symmetries for solving BVPs are discussed in the recent papers [8–11] and the relevant papers cited therein.

In this work, we continue study of a simplified (1+2)-dimensional Keller–Segel model, initiated in the first part of this work [12]. This model is a particular case of the classical Keller–Segel model [13,14] used for modeling a wide range of processes in biology, ecology, medicine, etc. The basic equations of the simplified Keller–Segel model that we are interested in have the form (see more details about these equations in [15–18]):

$$\begin{aligned} N_t(t,x,y) &= d_1 \triangle N(t,x,y) - \chi_0 \nabla.(N(t,x,y)\nabla P(t,x,y)),\\ 0 &= \triangle P(t,x,y) + \alpha N(t,x,y) - \beta P(t,x,y), \end{aligned} \tag{1}$$

where unknown functions $N(t,x)$ and $P(t,x)$ describe the densities of cells (species) and chemicals, respectively; t and x denote the time and space variables; the parameters d_1, χ_0, α and β are non-negative constants, $\chi_0\alpha \neq 0$ (otherwise, the model loses its biological meaning) and the operators $\triangle = \partial_x^2 + \partial_y^2$, $\nabla = (\partial_x, \partial_y)$. We start from the nonlinear System (1) supplied by initial profiles for unknown functions, i.e. the Cauchy problem, and continue by examination of System (1) with Neumann boundary conditions (including zero flux conditions as an important particular case).

The paper is organized as follows. In Section 2, the Lie symmetry classification of the Cauchy problem for a simplified Keller–Segel (SKS) system is derived. In Section 3, the exact solutions of the (1 + 1) and (1 + 2)-dimensional Cauchy problems were constructed including a nontrivial example of the exact solution for the correctly-specified initial profiles. In Section 4, Lie symmetry of BVPs with the Neumann boundary conditions is studied. Because Lie symmetry of BVPs essentially depends on geometry of the domain, which the problem is formulated on, all realistic (from applicability point of view) domains were examined. In Section 5, a Lie symmetry operator was used in order to reduce the (1 + 2)-dimensional Neumann problem for SKS and to construct the exact solution of the corresponding (1 + 1)-dimensional Neumann problem. The results obtained are summarized in the Conclusions section.

2. Lie Symmetry of the Cauchy Problem

It was shown in [12] that the SKS System (1) can be further simplified provided $\beta d_1/\alpha = \varepsilon \ll 1$. In this case, one may reduce SKS System (1) to the form:

$$\begin{aligned} \rho_t(t,x,y) &= \Delta\rho(t,x,y) - \nabla(\rho(t,x,y)\nabla S(t,x,y)),\\ 0 &= \Delta S(t,x,y) + \rho(t,x,y). \end{aligned} \tag{2}$$

Of course, the system derived is still nonlinear; however, one admits infinite-dimensional Lie algebra of invariance generated by the operators [12]:

$$\begin{aligned} &G_1^\infty = f_1(t)\tfrac{\partial}{\partial x} + x f_1'(t)\tfrac{\partial}{\partial S},\ G_2^\infty = f_2(t)\tfrac{\partial}{\partial y} + y f_2'(t)\tfrac{\partial}{\partial S},\\ &X_S^\infty = g(t)\tfrac{\partial}{\partial S},\ P_t = \tfrac{\partial}{\partial t},\ J_{12} = -x\tfrac{\partial}{\partial y} + y\tfrac{\partial}{\partial x},\\ &D = 2t\tfrac{\partial}{\partial t} + x\tfrac{\partial}{\partial x} + y\tfrac{\partial}{\partial y} - 2\rho\tfrac{\partial}{\partial \rho}, \end{aligned} \tag{3}$$

where $f_1(t)$, $f_2(t)$ and $g(t)$ are arbitrary smooth functions.

Now, we consider the Cauchy problem for the nonlinear System (2), which can be formulated as follows:

$$\begin{aligned} &\rho_t = \triangle\rho - \nabla(\rho\nabla S), \\ &0 = \triangle S + \rho, \\ &t = 0: \; S = \phi(x), \quad \rho = -\phi_{xx}(x), \end{aligned} \tag{4}$$

where $\phi(x)$ is an arbitrary smooth function. In the most general case, this function may depend also on the variable y; however, here we restricted ourselves in this case, in order to avoid cumbersome calculations. To guarantee existence of a classical solution of the Cauchy Problem (4), the initial profile for the component ρ should be specified as above.

Obviously, the Lie algebra (3) cannot be maximal algebra of invariance (MAI) of the Cauchy Problem (4) for the arbitrary given Function $\phi(x)$. To determine which of the operators listed in (3) are Lie symmetry operators of the Cauchy problem in question, we use the well-known criteria [4]. According to the criteria, one should examine whether initial conditions from Equation (4) are invariant under the operator in question. In order to check this for all the operators listed in Equation (3), we take their linear combination (hereafter, a_i, $i = 1, 2, 3...$ are arbitrary parameters):

$$\begin{aligned} X = {} & a_1 G_1^\infty + a_2 G_2^\infty + a_3 X_S^\infty + a_4 P_t + a_5 J_{12} + a_6 D = \\ & (a_4 + 2ta_6)\frac{\partial}{\partial t} + (a_1 f_1(t) + a_5 y + a_6 x)\frac{\partial}{\partial x} + \\ & (a_2 f_2(t) - a_5 x + a_6 y)\frac{\partial}{\partial y} + \\ & (a_3 g(t) + a_1 f_1'(t) x + a_2 f_2'(t) y)\frac{\partial}{\partial S} - 2a_6\rho\frac{\partial}{\partial \rho}. \end{aligned} \tag{5}$$

Applying operator X to the manifold $\mathrm{M} = \{t = 0,\ S = \phi(x),\ \rho = -\phi_{xx}(x)\}$ generated by the initial conditions, one arrives at the restriction and two equations:

$$a_4 = 0, \tag{6}$$

$$(a_1 f_1(0) + a_5 y + a_6 x)\phi_x(x) = a_3 g(0) + a_1 f_1'(0) x + a_2 f_2'(0) y, \tag{7}$$

$$2a_6\phi_{xx}(x) + (a_1 f_1(0) + a_5 y + a_6 x)\phi_{xxx}(x) = 0. \tag{8}$$

Because Equation (8) is a differential consequence of (7), we need to analyze (7) only. Obviously, the restriction means (6) that the Cauchy problem is not invariant w.r.t. time translation.

Equation (7) implies certain limitations on the function $\phi(x)$. This function can only be arbitrary when both sides of equation vanish. When the multiplier on the left-hand-side is non-zero, then one obtains a linear ordinary differential equation (ODE) to find $\phi(x)$. This ODE has been solved depending on the values of the parameters a_i, $i = 1, 2, 3....$ As a result, four different profiles for the function $\phi(x)$ were derived. One of them, namely, $\phi(x) = \gamma \ln|x| + \lambda_1 x + \lambda_0$ (this function springs up if $a_6 \neq 0$), was exempted from the further examination because the function $\ln|x|$ possesses singularity. The other three cases are presented in Table 1 together with the relevant MAIs.

Theorem 1. *All possible MAIs of the (1+2)-dimensional Cauchy Problem (4), depending on the form of initial profiles (up to translations w.r.t. the space variable x), are presented in Table 1.*

Remark 1. *Because G_1^∞, G_2^∞ and X_S^∞ contain arbitrary functions on time (see Formulae (3)), one notes corresponding restrictions on these functions in Table 1, which reduce their arbitrariness. For instance, MAI in case 4 differs from that in case 1 because there is only a single restriction on the function f_1, while there are two restrictions on f_1 in case 1.*

Table 1. maximal algebra of invariances (MAIs) of the Cauchy Problem (4).

	ϕ(x)	MAI
1	$\forall$	G_1^∞ with $f_1(0) = f_1'(0) = 0$, G_2^∞ with $f_2'(0) = 0$, X_S^∞ with $g(0) = 0$
2	0	G_1^∞ with $f_1'(0) = 0$, G_2^∞ with $f_2'(0) = 0$, X_S^∞ with $g(0) = 0$, J_{12}, D
3	$\lambda_1 x, \lambda_1 \neq 0$	$G_1^\infty + b_6 D$ with $f_1(0) = 0, f_1'(0) = b_6\lambda_1$, $G_2^\infty + b_5 J_{12}$ with $f_2'(0) = b_5\lambda_1$, $b_1 G_1^\infty + X_S^\infty + b_1 a_6 D$ with $g(0) = b_1\lambda_1 f_1(0)$, $f_1(0) \neq 0, f_1'(0) = a_6\lambda_1$
4	$\lambda_2 x^2 + \lambda_0, \lambda_2 \neq 0$	G_1^∞ with $f_1'(0) = 2\lambda_2 f_1(0)$, G_2^∞ with $f_2'(0) = 0$, X_S^∞ with $g(0) = 0$

3. Application of Lie Symmetry for Constructing Exact Solutions of Cauchy Problems

The Cauchy Problem (4) can be reduced to a set of (1 + 1)-dimensional ones by using the Lie symmetry operators. Nowadays, it is a standard routine in the case of application of symmetry operators to PDEs in the case of Lie algebras of low dimensionality. In the case of BVPs with a wide Lie symmetry, classification of inequivalent subalgebras of MAI and its application for reducing can be highly nontrivial (see examples in [8]). One notes that all MAIs listed in Table 1 are infinite-dimensional, hence here we restrict ourselves on the operator G_2^∞ from case 1 of Table 1.

Because the operator G_2^∞ contains an arbitrary function $f_2(t)$ with the property $f_2'(0) = 0$, we consider two cases, namely: (i) $f_2(t) = conts \neq 0$, i.e., $G_2^\infty = \partial_y$, and (ii) $f_2(t)$ is an arbitrary non-constant function.

3.1. Exact Solutions of the (1 + 1)-Dimensional Cauchy Problem

In case (i), one easily construct the ansatz:

$$S(t,x,y) = S(t,x), \quad \rho(t,x,y) = \rho(t,x). \tag{9}$$

Substituting ansatz (9) into Equation (4), the (1 + 1)-dimensional Cauchy problem,

$$\begin{aligned} &\rho_t = \rho_{xx} - (\rho S_x)_x, \\ &0 = S_{xx} + \rho, \\ &t = 0 : \rho = -\phi_{xx}(x), S = \phi(x), \end{aligned} \tag{10}$$

is obtained. Thus, the same problem is derived, however, in the case of a single spacial variable. Let us reduce the governing system to a single equation, extracting $\rho = -S_{xx}$ from the second equation of Equation (10) and substituting it into the first. Hence, the 4th order differential equation:

$$S_{txx} = S_{xxxx} - (S_{xx}S_x)_x$$

is obtained, which is equivalent to the 3rd order equation:

$$S_{tx} = S_{xxx} - S_{xx}S_x + \theta(t),$$

where $\theta(t)$ is an arbitrary function. The obvious substitution:

$$W(t,x) = S_x(t,x)$$

transforms the last equation to the nonlinear 2nd order equation:

$$W_t = W_{xx} - W_x W + \theta(t). \tag{11}$$

It can be noted that the substitution:

$$\begin{aligned} &W(t,x) = U(t,y) + \textstyle\int_0^t \theta(\tau)\,\mathrm{d}\tau, \\ &x = z + j(t), \end{aligned}$$

where $j(t) = \int_0^t (\int_0^\tau \theta(\tau_1)\,\mathrm{d}\tau_1)\mathrm{d}\tau$, reduces Equation (11) to the Burgers equation:

$$U_t = U_{zz} - UU_z.$$

The Burgers equation is linearizable via the famous Cole–Hopf substitution [19,20]:

$$U(t,z) = -2\frac{V_z(t,z)}{V(t,z)}$$

to the linear heat equation:

$$V_t = V_{zz}. \tag{12}$$

All the substitutions mentioned above could be combined as follows:

$$\begin{aligned} &S_x(t,x) = -2\tfrac{V_z(t,z)}{V(t,z)} + \textstyle\int_0^t \theta(\tau)\,\mathrm{d}\tau, \\ &x = z + j(t). \end{aligned} \tag{13}$$

It should be stressed that the Substitution (13) reduces the nonlinear Cauchy Problem (10) to the linear problem for the heat Equation (12), which can be exactly solved. In fact, having the specified initial profiles in (10), we find the initial condition for Equation (12) as follows:

$$-2\,\frac{V_z(0,z)}{V(0,z)} = (\phi(z))_z,$$

i.e.,

$$V(0,z) = a\,e^{-\frac{1}{2}\phi(z)} \equiv a\,p(z),\ a > 0. \tag{14}$$

Obviously, the exact solution of Cauchy Problems (12) and (14) is the Poisson integral:

$$V(t,z) = \frac{a}{\sqrt{4\pi t}} \int_{-\infty}^{+\infty} p(\xi)\,e^{-\frac{(z-\xi)^2}{4t}}\,\mathrm{d}\xi. \tag{15}$$

Thus, calculating the derivatives $V_z(t,z)$ and $V_{zz}(t,z)$, and using Substitution (13), one can construct the solution of the Cauchy Problem (10):

$$\begin{aligned} S(t,x) &= \int W(t,x)\,\mathrm{d}x = \int \left(U(t,z) + \int_0^t \theta(\tau)\,\mathrm{d}\tau \right) \mathrm{d}x = \\ &- 2\ln(V(t,z)) + x \cdot \int_0^t \theta(\tau)\,\mathrm{d}\tau + A(t) = \\ &- 2\ln\left(\frac{1}{\sqrt{4\pi t}} \int_{-\infty}^{+\infty} p(\xi)\,e^{-\frac{(x-j(t)-\xi)^2}{4t}}\,\mathrm{d}\xi \right) + x \cdot \int_0^t \theta(\tau)\,\mathrm{d}\tau + A(t), \end{aligned}$$

$$\rho(t,x) = -\,U_z(t,z) = 2\left(\frac{V_{zz}(t,z)}{V(t,z)} - \left(\frac{V_z(t,z)}{V(t,z)}\right)^2\right) =$$

$$-\,\frac{1}{t}\cdot\frac{\int_{-\infty}^{+\infty} p'(\xi)\,(x-j(t)-\xi)\,e^{-\frac{(x-j(t)-\xi)^2}{4t}}\,\mathrm{d}\xi}{\int_{-\infty}^{+\infty} p(\xi)\,e^{-\frac{(x-j(t)-\xi)^2}{4t}}\,\mathrm{d}\xi} -$$

$$-\,\frac{1}{2t^2}\cdot\left(\frac{\int_{-\infty}^{+\infty} p(\xi)\,(x-j(t)-\xi)\,e^{-\frac{(x-j(t)-\xi)^2}{4t}}\,\mathrm{d}\xi}{\int_{-\infty}^{+\infty} p(\xi)\,e^{-\frac{(x-j(t)-\xi)^2}{4t}}\,\mathrm{d}\xi}\right)^2.$$

Because the Cole–Hopf substitution in non-local transform (i.e., involves derivatives), we need to examine the behavior of the solution as $t \to 0$:

$$\lim_{t\to 0} S(t,x) = -2\ln\left(\lim_{t\to 0}(V(t,x))\right) + A(0) = -2\ln(e^{-\frac{1}{2}\phi(x)}) + A(0) = \phi(x) + A(0).$$

Hence, having chosen the function $A(t)$ with the property $A(0) = 0$, we obtain $\lim_{t\to 0} S(t,x) = S(0,x)$, i.e., the function $S(t,x)$ is continuous at $t = 0$.

In order to prove that $\lim_{t\to 0}\rho(t,x) = \rho(0,x)$, one needs to show that $\lim_{t\to 0} V_z(t,z) = V_z(0,z)$ and $\lim_{t\to 0} V_{zz}(t,z) = V_{zz}(0,z)$. The proof of the first equality can be found in [19]. We have proved that $\lim_{t\to 0} V_{zz}(t,z) = V_{zz}(0,z)$ under the restriction $\phi(x) = o(x^2), x \to \infty$ (here, the relevant calculations are omitted). Thus, the following statement can be formulated.

Theorem 2. *The classical solution of the Cauchy Problem (10), in the case when $\phi(x)$ is differentiable twice and $\phi(x) = o(x^2), x \to \infty$ can be presented as:*

$$S(t,x) = -\,2\ln\left(\frac{1}{\sqrt{4\pi t}}\int_{-\infty}^{+\infty} p(\xi)\,e^{-\frac{(x-j(t)-\xi)^2}{4t}}\,\mathrm{d}\xi\right) + $$
$$+\,x\cdot\int_0^t \theta(\tau)\,\mathrm{d}\tau + A(t),$$
$$\rho(t,x) = -\,\frac{1}{t}\cdot\frac{\int_{-\infty}^{+\infty} p'(\xi)\,(x-j(t)-\xi)\,e^{-\frac{(x-j(t)-\xi)^2}{4t}}\,\mathrm{d}\xi}{\int_{-\infty}^{+\infty} p(\xi)\,e^{-\frac{(x-j(t)-\xi)^2}{4t}}\,\mathrm{d}\xi} - \tag{16}$$
$$-\,\frac{1}{2t^2}\cdot\left(\frac{\int_{-\infty}^{+\infty} p(\xi)\,(x-j(t)-\xi)\,e^{-\frac{(x-j(t)-\xi)^2}{4t}}\,\mathrm{d}\xi}{\int_{-\infty}^{+\infty} p(\xi)\,e^{-\frac{(x-j(t)-\xi)^2}{4t}}\,\mathrm{d}\xi}\right)^2,$$

where $p(\xi) = e^{-\frac{1}{2}\phi(\xi)}$, $A(0) = 0$.

Obviously, the exact Solution (16) is not unique because one contains two arbitrary functions $\theta(t)$ and $A(t)$. To specify these functions, one needs additional biologically motivated restrictions. We remind the reader that function $S(t,x)$ describing the density of chemicals should be bounded in space. Therefore, functions $\theta(t)$ must vanish (in this case, function $j(t)$, which depends on $\theta(t)$, also vanishes). In order to specify both functions $\theta(t)$ and $A(t)$ one needs, for example, to assume that the quantity of the chemical $S(t,x)$ is finite in space and time, i.e., $\int_{-\infty}^{+\infty}|S(t,x)|dx < 0$ for $\forall t > 0$. This assumption immediately leads to the unique solution of Cauchy Problem (10) in the form:

$$S(t,x) = -2\ln\left(\frac{1}{\sqrt{4\pi t}}\int_{-\infty}^{+\infty} p(\xi)\,e^{-\frac{(x-\xi)^2}{4t}}\,\mathrm{d}\xi\right),$$
$$\rho(t,x) = -\,\frac{1}{t}\cdot\frac{\int_{-\infty}^{+\infty} p'(\xi)\,(x-\xi)\,e^{-\frac{(x-\xi)^2}{4t}}\,\mathrm{d}\xi}{\int_{-\infty}^{+\infty} p(\xi)\,e^{-\frac{(x-\xi)^2}{4t}}\,\mathrm{d}\xi} - \tag{17}$$
$$\frac{1}{2t^2}\cdot\left(\frac{\int_{-\infty}^{+\infty} p(\xi)\,(x-\xi)\,e^{-\frac{(x-\xi)^2}{4t}}\,\mathrm{d}\xi}{\int_{-\infty}^{+\infty} p(\xi)\,e^{-\frac{(x-\xi)^2}{4t}}\,\mathrm{d}\xi}\right)^2.$$

In order to present a non-trivial exact Solution of (10) in terms of elementary functions, we consider the initial profiles:

$$\begin{aligned} S(0,x) &= \phi(x) = -2\ln(\sin(\gamma x)+2)+2\ln(3), \\ \rho(0,x) &= -2\gamma^2\,\tfrac{1+2\sin(\gamma x)}{(\sin(\gamma x)+2)^2}, \end{aligned} \tag{18}$$

where γ is an arbitrary constant.

Using Formula (17), one obtains:

$$S(t,x) = -2\ln\left(\frac{1}{2\sqrt{\pi t}}\int_{-\infty}^{+\infty} e^{-\frac{1}{2}\phi(\xi)}\,e^{-\frac{(\xi-x)^2}{4t}}\,\mathrm{d}\xi\right) =$$
$$-2\ln\left(\frac{2}{3}+\frac{1}{6\sqrt{\pi t}}\,e^{-\frac{x^2}{4t}}\int_{-\infty}^{+\infty}\sin(\gamma\xi)\,e^{-\frac{1}{4t}\xi^2+\frac{x}{2t}\xi}\,\mathrm{d}\xi\right).$$

To calculate the integral $I(t,x) = \int_{-\infty}^{+\infty}\sin(\gamma\xi)\,e^{-\frac{1}{4t}\xi^2+\frac{x}{2t}\xi}\,\mathrm{d}\xi$, the Mellin transformation (see, e.g., [21]) has been used:

$$\begin{aligned} I(t,x) = &\int_{0}^{+\infty}\sin(\gamma\xi)\,e^{-\frac{1}{4t}\xi^2+\frac{x}{2t}\xi}\,\mathrm{d}\xi - \int_{0}^{+\infty}\sin(\gamma\xi)\,e^{-\frac{1}{4t}\xi^2-\frac{x}{2t}\xi}\,\mathrm{d}\xi = \\ &\frac{i}{2}\sqrt{2t}\,e^{\frac{x^2}{8t}-\frac{t}{2}\gamma^2}(e^{-\frac{ix\gamma}{2}}(D_{-1}(\frac{x}{\sqrt{2t}}-i\gamma\sqrt{2t})+D_{-1}(-\frac{x}{\sqrt{2t}}+i\gamma\sqrt{2t}))- \\ &e^{\frac{ix\gamma}{2}}(D_{-1}(\frac{x}{\sqrt{2t}}+i\gamma\sqrt{2t})+D_{-1}(-\frac{x}{\sqrt{2t}}-i\gamma\sqrt{2t}))). \end{aligned}$$

Here, $D_{-1}(\cdot)$ is a parabolic cylinder function. Using the known properties of such functions, the integral in question can be explicitly calculated:

$$I(t,x) = 2\sqrt{\pi t}\,\sin(\gamma x)e^{\frac{x^2}{4t}-\gamma^2 t}.$$

Now, we find the function:

$$\begin{aligned} S(t,x) = &-2\ln\left(\frac{2}{3}+\frac{1}{6\sqrt{\pi t}}\,e^{-\frac{x^2}{4t}}\,2\sqrt{\pi t}\,\sin(\gamma x)e^{\frac{x^2}{4t}-\gamma^2 t}\right) = \\ &-2\ln\left(\sin(\gamma x)\,e^{-\gamma^2 t}+2\right)+2\ln(3). \end{aligned}$$

Because $\rho(t,x) = -S_{xx}(t,x)$, one also finds the function $\rho(t,x)$. Thus, the exact solution of the Cauchy Problems (10) and (18) has the form:

$$\begin{aligned} S(t,x) &= -2\ln\left(\sin(\gamma x)\,e^{-\gamma^2 t}+2\right)+2\ln(3), \\ \rho(t,x) &= -2\gamma^2 e^{-\gamma^2 t}\,\tfrac{e^{-\gamma^2 t}+2\sin(\gamma x)}{(\sin(\gamma x)\,e^{-\gamma^2 t}+2)^2}. \end{aligned} \tag{19}$$

Plots of this solution with $\gamma = 1$ are presented in Figures 1 and 2. Both functions, $\rho(t,x)$ and $S(t,x)$, have attractive properties. For example, they are periodic, bounded and tend to some constants if $t \to +\infty$. However, one notes that the function $\rho(t,x)$, which usually describes the densities of cells or species (see Introduction), are not non-negative for $\forall x \in \mathbb{R}$. It turns out that this unrealistic behavior (for real world applications) is a natural property of each non-constant solution of the (1 + 1)-dimensional Cauchy Problem (10) with an arbitrary non-negative function $\phi(x) \neq const$.

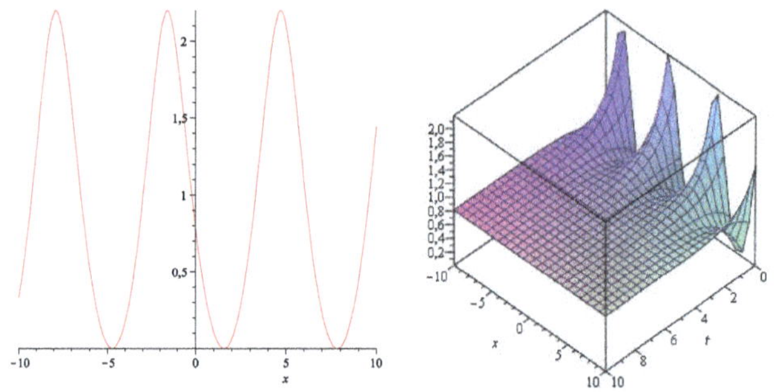

Figure 1. Plots of the functions $S(0,x)$ and $S(t,x)$ using Formulae (19) with $\gamma = 1$.

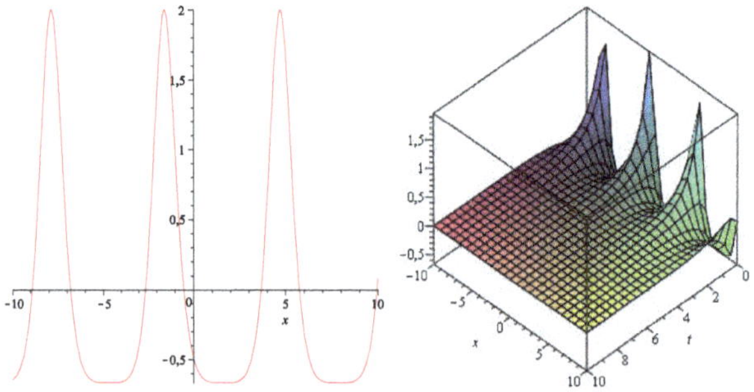

Figure 2. Plots of the functions $\rho(0,x)$ and $\rho(t,x)$ using Formulae (19) with $\gamma = 1$.

Let us show this assuming that there is a Solution of (10) with a non-negative function $\phi(x) \neq const$, such that the functions $\rho(t,x)$ and $S(t,x)$ are non-negative for $\forall (t,x) \in \mathbb{R}^+ \times \mathbb{R}$. Let us fix an arbitrary $t = t_0 > 0$. Because $0 \leq \rho(t_0,x) = -S_{xx}(t_0,x)$, the second derivative $S_{xx}(t_0,x)$ is non-positive for all x. It means the continues function $S(t_0,x)$ is convex upwards for all x (otherwise, one is a constant). Now, one realizes that any function with such property (like $-x^2 + c(t)$, $-e^{-x} + c(t)$, etc.) cannot be non-negative for all x.

3.2. *Reduction and Exact Solutions of (1 + 2)-Dimensional Cauchy Problem*

Now, we apply the operator G_2^{∞} with an arbitrary non-constant function $f_2(t)$. In order to reduce System (4), we need to construct an ansatz by solving the corresponding system of characteristic equations for this operator. After rather standard calculations, the ansatz:

$$S(t,x,y) = \frac{f_2'(t)}{f_2(t)} \cdot \frac{y^2}{2} + \psi(t,x), \quad \rho(t,x,y) = \Psi(t,x) \tag{20}$$

is obtained. Ansatz (20) reduces Cauchy Problem (4) to the (1 + 1)-dimensional Cauchy problem:

$$\begin{aligned} \Psi_t(t,x) =& \Psi_{xx}(t,x) - \Psi(t,x)\psi_{xx}(t,x) - \\ & \Psi_x(t,x)\psi_x(t,x) - \frac{f_2'(t)}{f_2(t)}\Psi(t,x), \\ 0 =& \psi(t,x) + \frac{f_2'(t)}{f_2(t)} + \Psi(t,x), \\ t = 0 :& \ \psi = \phi(x), \Psi = -\phi_{xx}(x). \end{aligned} \tag{21}$$

The function $\Psi(t,x)$ can be easily found by using the second equation:

$$\Psi(t,x) = -\frac{f_2'(t)}{f_2(t)} - \psi_{xx}(t,x).$$

Substituting this expression into the first Equation of (21), one arrives at the fourth order partial differential equation:

$$\psi_{txx} = \psi_{xxxx} - (\psi_{xx}\psi_x)_x - \frac{2f_2'(t)}{f_2(t)}\psi_{xx} - \frac{f_2''(t)}{f_2(t)},$$

which can be integrated twice w.r.t. the variable x. Thus, the second order nonlinear equation:

$$\psi_t = \psi_{xx} - \frac{1}{2}(\psi_x)^2 - \frac{2f_2'(t)}{f_2(t)}\psi - \frac{f_2''(t)}{2f_2(t)}x^2 + c_1(t)x + c_0(t), \tag{22}$$

is obtained ($c_1(t)$ and $c_0(t)$ are arbitrary smooth functions).

In contrast to the nonlinear Equation (11), we were unable to linearize Equation (22) (we remind the reader that $f_2'(t) \neq 0$). To find particular solutions, we used a non-Lie ansatz (see [22] for an example):

$$\psi(t,x) = \psi_2(t)x^2 + \psi_1(t)x + \psi_0(t). \tag{23}$$

in order to reduce the partial differential Equation (22) to a system of ODEs. In fact, Substituting (23) into Equation (22), the system of three ODEs:

$$\begin{aligned} &\psi_2' + 2\psi_2^2 + \tfrac{2f_2'}{f_2}\psi_2 + \tfrac{f_2''}{2f_2} = 0, \\ &\psi_1' + 2\psi_1\psi_2 + \tfrac{2f_2'}{f_2}\psi_1 = c_1(t), \\ &\psi_0' - 2\psi_2 + \tfrac{1}{2}\psi_1^2 + \tfrac{2f_2'}{f_2}\psi_0 = c_0(t), \end{aligned} \tag{24}$$

is obtained to find unknown functions $\psi_0(t), \psi_1(t)$ and $\psi_2(t)$. Because System (24) has the same structure as (38) [22], one is integrable. In fact, the first equation can be solved for arbitrary function $f_2(t)$. By substituting the known function $\psi_2(t)$ into the second equation, one obtains the first order linear ODE to find the function $\psi_1(t)$. Finally, having the known functions $\psi_1(t)$ and $\psi_2(t)$, the third equation of System (24) can be easily solved. As a result, the general Solution of (24) has the form:

$$\begin{aligned} \psi_2(t) &= \tfrac{1}{2(t-t_0)} - \tfrac{f_2'}{2f_2},\ t_0 \in \mathbb{R}, \\ \psi_1(t) &= \tfrac{1}{(t-t_0)f_2(t)}\left(\textstyle\int (t-t_0)f_2(t)c_1(t)\,dt + k_1\right),\ k_1 \in \mathbb{R}, \\ \psi_0(t) &= \frac{k_0}{f_2^2(t)} - \frac{1}{2} + \frac{1}{f_2^2(t)}\int \left(c_0(t)f_2^2(t) + \frac{f_2^2(t)}{t-t_0}\right) dt - \\ &\quad \frac{1}{f_2^2(t)}\int \left(\frac{\left(\int (t-t_0)f_2(t)c_1(t)\,dt + k_1\right)^2}{2(t-t_0)^2}\right) dt,\ k_0 \in \mathbb{R}. \end{aligned} \tag{25}$$

Taking into account that $c_1(t)$ and $c_0(t)$ are arbitrary smooth functions, we may simplify the general Solution (25) to the form:

$$\begin{aligned} \psi_2(t) &= \tfrac{1}{2(t-t_0)} - \tfrac{f_2'(t)}{2f_2(t)},\ t_0 \in \mathbb{R}, \\ \psi_1(t) &= c_1^*(t), \\ \psi_0(t) &= c_0^*(t), \end{aligned} \tag{26}$$

by introducing new notations $c_1^*(t)$ and $c_0^*(t)$, which are arbitrary smooth functions, while t_0 is an arbitrary parameter.

Thus, using Equations (20), (23) and (26), we construct the exact solution:

$$
\begin{aligned}
S(t,x,y) =& \frac{f_2'(t)}{2f_2(t)}y^2 - \frac{f_2'(t)}{2f_2(t)}x^2 + \frac{1}{2(t-t_0)}x^2 + \\
& c_1^*(t)x + c_0^*(t),
\end{aligned} \tag{27}
$$

$$
\rho(t,x,y) = \tfrac{1}{t_0 - t}
$$

of Cauchy Problem (4) with the correctly-specified initial profile $\phi(x) = -\frac{1}{2t_0}x^2 + c_1 x + c_0$.

Obviously, this solution blows up for the finite time $t_0 > 0$, which makes its immediate biological interpretation unlikely. On the other hand, it is in agreement with [18] (see also the references cited therein) because the global solution of this problem requires the constraint $\int_{\mathbb{R}^2} \rho(0,x,y)dxdy < 8\pi$, which does not take place for $\rho(0,x,y) = \frac{1}{t_0}$. Notably, blow-up solutions occur in some physical models, and they have been intensively studied since the 1980s (see, e.g., [23] and references therein).

4. Lie Symmetry of the Neumann Problems

In this section, we study Lie symmetry of BVPs with the Neumann boundary conditions. Because the result depends essentially on geometry of the domain Ω, where BVP in question is defined, one needs to examine different cases. In [12], the simplest case, when Ω is a half-plane, was under study. In principle, Ω can be an arbitrary (bounded or unbounded) domain with smooth boundaries. However, it was established in [10] that geometry of Ω is predicted by the Lie symmetry of the governing equations of BVP in question. In particular, all possible domains were established if the projection of MAI of the governing equations on the (x,y) space gives the Lie algebra with the basic operators:

$$
X = \partial_x,\ Y = \partial_y,\ J_{12} = -x\partial_y + y\partial_x,\ D_{12} = x\partial_x + y\partial_y. \tag{28}
$$

Now, one notes that such projection of MAI (3) is exactly the Lie algebra (28) (the projections of G_1^∞ and G_2^∞ are the operators $f_1(t)\frac{\partial}{\partial x}$ and $f_2(t)\frac{\partial}{\partial x}$, which are equivalent to X and Y because the time variable is now a parameter). It means that all generic domains leading to non-trivial Lie symmetry of any Neumann problem for the SKS System (2) are already known [10]. Here, we consider the most important (from an applicability point of view) of them, and the first one is a strip.

4.1. Neumann Problem on the Strip

The Neumann problem on the strip $\Omega = \{(x,y) : -\infty < x < +\infty,\ 0 < y < \pi\}$ can be formulated as follows:

$$
\begin{aligned}
&\rho_t(t,x,y) = \triangle\rho(t,x,y) - \nabla(\rho(t,x,y)\nabla S(t,x,y)),\\
&0 = \triangle S(t,x,y) + \rho(t,x,y),\\
&t = 0 : S = \varphi(x,y), \rho = -\triangle\varphi(x,y),\\
&y = 0 : \rho_y = q_1(t), S_y = q_2(t),\\
&y = \pi : \rho_y = 0, S_y = 0,
\end{aligned} \tag{29}
$$

where $\varphi(x,y), q_1(t)$ and $q_2(t)$ are arbitrary smooth functions. We note that BVP on an arbitrary strip $\Omega_1 = \{(x,y) : -\infty < x < +\infty,\ C_1 < y < C_2\}$ can be reduced to BVP (29) by the translation and scaling transformation w.r.t. the variable y.

In contrast to the previous problem, BVP (29) does not involve boundary conditions at infinity; hence, the standard definition of invariance can be applied in order to find Lie symmetry of Problem (29). Let us apply the operator X (5) to the manifolds $\{y = 0\}$, $\{y = \pi\}$, $\{t = 0\}$:

$$
\begin{aligned}
&X(y-0)|_{y=0} = a_2 f_2(t) - a_5 x + a_6 y = a_2 f_2(t) - a_5 x = 0,\\
&X(y-\pi)|_{y=\pi} = a_2 f_2(t) - a_5 x + a_6 y = a_2 f_2(t) - a_5 x + a_6\pi = 0,\\
&X(t-0)|_{t=0} = a_4 = 0.
\end{aligned}
$$

A simple analysis of the conditions obtained above immediately leads to $a_2 = a_4 = a_5 = a_6 = 0$. Now, one needs to apply the first prolongation $\underset{1}{X}$ to the manifolds $\{y = 0, \rho_y = q_1(t)\}$, $\{y = 0, S_y = q_2(t)\}$, $\{y = \pi, \rho_y = 0\}$ and $\{y = \pi, S_y = 0\}$. Taking into account the restrictions $a_2 = a_4 = a_5 = a_6 = 0$, one may easily check that the expressions:

$$\underset{1}{X}\left(\rho_y - q_1(t)\right)\Big|_{y=0,\ \rho_y=q_1(t)} = -a_4\, q_1'(t) = 0,$$

$$\underset{1}{X}\left(S_y - q_2(t)\right)\Big|_{y=0,\ S_y=q_2(t)} = -a_4\, q_2'(t) = 0,$$

$$\underset{1}{X}\left(\rho_y - 0\right)\Big|_{y=\pi,\ \rho_y=0} = 0,$$

$$\underset{1}{X}\left(S_y - 0\right)\Big|_{y=\pi,\ S_y=0} = 0,$$

are fulfilled automatically. Thus, there are no any restrictions on the functions $q_1(t)$ and $q_2(t)$. Finally, we apply operator X with $a_2 = a_4 = a_5 = a_6 = 0$ to the initial profiles:

$$X\left(S - \varphi(x,y)\right)\Big|_{t=0,\ S=\varphi(x,y)} = a_1 f_1'(0)x - a_1 f_1(0)\varphi_x(x,y) + a_3 g(0) = 0,$$

$$X\left(\rho + \triangle\varphi(x,y)\right)\Big|_{t=0,\ \rho=-\triangle\varphi(x,y)} = a_1 f_1(0)(\varphi_{xxx}(x,y) + \varphi_{yyx}(x,y)) = 0.$$

One notes that the second equation in these formulae is a differential consequence of the first. Thus, we need solve only the equation:

$$a_1 f_1'(0)x - a_1 f_1(0)\varphi_x(x,y) + a_3 g(0) = 0. \tag{30}$$

Obviously, it is the linear ODE for the function $\varphi(x,y)$ (variable y should be treated as a parameter). A simple analysis of Equation (30) leads two different cases, $f_1(0) = 0$ and $f_1(0) \neq 0$. As a result, the following statement can be formulated.

Theorem 3. *MAI of the Neumann Problem (29) does not depend on the form of the functions $q_1(t)$ and $q_2(t)$. If the initial profile $\varphi(x,y)$ is an arbitrary smooth function, then MAI is the infinite-dimensional Lie algebra generated by the operators:*

$$G_1^{\infty} = f_1(t)\frac{\partial}{\partial x} + x f_1'(t)\frac{\partial}{\partial S}, \quad X_S^{\infty} = g(t)\frac{\partial}{\partial S}, \tag{31}$$

with $f_1(0) = f_1'(0) = 0$ and $g(0) = 0$. In the case,

$$\varphi(x,y) = \lambda_2 x^2 + \lambda_1 x + \mu(y),$$

where $\mu(y)$ is an arbitrary smooth function, this algebra is extended by operators of the form:

$$G_1^{\infty} + bX_S^{\infty}, \tag{32}$$

with $\frac{f_1'(0)}{2f_1(0)} = \lambda_2$, $b\frac{g(0)}{f_1(0)} = \lambda_1$, $f_1(0) \neq 0$.
There are no other initial profiles $\varphi(x,y)$, leading to extensions of the Lie algebra (31).

It is worth noting that Lie symmetry of the Neumann problem on the strip (29) is essentially different than one for the same problem on the half-plane. In particular, there is no any dependence on the form of flux (i.e., the functions $q_1(t)$ and $q_2(t)$) in contrast to case of the half-plane (see Theorem 2 in [12]).

4.2. Neumann Problem on Interior/Exterior of a Circle

Now, we turn to the case of a bounded domain. As it was pointed out in [10], the interior of the circle $\Omega = \{(x,y) : x^2 + y^2 < R^2\}$ is the simplest case of such domain, which may occur for the system in question. It should be noted that the circle interior is a two-dimensional analog of the ball in the 3D space. On the other hand, the ball is a typical approximation of the domains arising in biomedical applications (for instance, when the Keller–Segel model is applied for tumour growth).

The Neumann problem in Ω can be formulated as follows:

$$\begin{aligned}
&\rho_t(t,x,y) = \triangle\rho(t,x,y) - \nabla(\rho(t,x,y)\nabla S(t,x,y)),\\
&0 = \triangle S(t,x,y) + \rho(t,x,y),\\
&t = 0 : S = \varphi(x,y), \rho = -\triangle\varphi(x,y),\\
&x^2 + y^2 = R^2 : \tfrac{\partial\rho}{\partial\mathbf{n}} = q_1(t),\ \tfrac{\partial S}{\partial\mathbf{n}} = q_2(t),
\end{aligned} \tag{33}$$

where $\varphi(x,y), q_1(t)$ and $q_2(t)$ are again arbitrary smooth functions, while $\mathbf{n}$ is the outer normal to the boundary (the circus of the radius R) of the domain. Now, we formulate the theorem presenting the Lie symmetry classification of this BVP.

Theorem 4. *MAI of the Neumann Problem (33) does not depend on the form of the functions $q_1(t)$ and $q_2(t)$. If the initial profile $\varphi(x,y)$ is an arbitrary smooth function, then MAI is the infinite-dimensional Lie algebra generated by the operator X_S^∞ with $g(0) = 0$.*

In the case:

$$\varphi(x,y) = \psi(x^2 + y^2) + h_0 \arctan\frac{y}{x}, \tag{34}$$

this algebra is extended by operator of the form:

$$J_{12} - h(t)\frac{\partial}{\partial S}, \tag{35}$$

where ψ and h are arbitrary smooth functions of their arguments, the function h additionally satisfies the condition $h(0) = h_0$ with a fixed constant h_0.

There are no other initial profiles $\varphi(x,y)$ leading to extensions of the above Lie algebras.

The proof of the theorem is similar to that of Theorem 3 and is based on applying the invariance criteria to the operator X (5). It is worth noting that the relevant calculations can be essentially simplified if one applies the polar coordinates (r,θ) instead of the Cartesian of those (x,y). In particular, one may see that Formula (34) is much simpler in the polar coordinates, and the operator J_{12} takes the form $J_{12} = -\frac{\partial}{\partial\theta}$.

Another possible bounded domain is the annulus $\Omega = \{(x,y) : R_1^2 < x^2 + y^2 < R_2^2\}$ only [10]. In the case of the annulus, the corresponding Neumann problem has the form (33) with the additional boundary conditions:

$$x^2 + y^2 = R_2^2 : \frac{\partial\rho}{\partial\mathbf{n}} = q_3(t),\ \frac{\partial S}{\partial\mathbf{n}} = q_4(t), \tag{36}$$

$q_3(t)$ and $q_4(t)$ being arbitrary smooth functions. It turns out that the result of the Lie symmetry classification for such BVP will be the same as presented in Theorem 4.

Finally, the case of the domain, which is the interior of the circle $\Omega = \{(x,y) : x^2 + y^2 > R^2\}$, is also interesting. The Neumann Problem (33) in such domain should be supplied by the boundary conditions at infinity. Typically, they have the form:

$$x^2 + y^2 \to \infty : \frac{\partial\rho}{\partial\mathbf{n}} = 0,\ \frac{\partial S}{\partial\mathbf{n}} = 0, \tag{37}$$

i.e., zero-flux at infinity. In order to establish Lie symmetry of the nonlinear BVPs (33) and (37), one needs to examine separately the boundary Conditions (37). Because Formula (37) presents conditions at infinity, we cannot apply the standard invariance criteria [3]; however, we may use Definition 2 [10]. In fact, applying the transformation:

$$r^* = \frac{1}{r}, U = \frac{\rho}{r}, V = \frac{S}{r}, \tag{38}$$

to Equation (37) in the polar coordinates, one may easily calculate that the manifold $M = \{r = +\infty, \frac{\partial\rho}{\partial r} = 0, \frac{\partial S}{\partial r} = 0\}$ corresponding to Equation (37) is mapped into $M^* = \{r^* = 0, U = 0, V = 0\}$. At the final step, one checks invariance of the manifold M^* w.r.t. the operators X_S^∞ and Equation (35), taking into account transformation (38).

Thus, we have shown that Theorem 4 is also valid for the Neumann problems on Ω, which is the annulus or the interior of the circle, and the additional boundary Conditions (36) or (37) take place, respectively.

5. Exact Solutions of the Neumann Problem

This section is devoted to the application of the Lie symmetry operators obtained in the previous section in order to reduce BVP (29) to BVPs of lower dimensionality and to find exact solutions.

Let us consider the case of the specified initial profile from Theorem 3:

$$\varphi(x,y) = \lambda_2 x^2 + \lambda_1 x + \mu(y),$$

and apply operator (32). This operator generates ansatz:

$$\begin{aligned} &\rho(t,x,y) = \varrho(t,y), \\ &S(t,x,y) = s(t,y) + \tfrac{f_1'(t)}{2f_1(t)}\, x^2 + b\tfrac{g(t)}{f_1(t)} x, \end{aligned} \tag{39}$$

which reduces BVP (29) to the (1 + 1)-dimensional BVP:

$$\begin{aligned} &\varrho_t(t,y) = \varrho_{yy}(t,y) - (\varrho(t,y)s_y(t,y))_y - \tfrac{f_1'(t)}{f_1(t)}\varrho(t,y), \\ &0 = s_{yy}(t,y) + \varrho(t,y) + \tfrac{f_1'(t)}{f_1(t)}, \\ &t = 0 : s = \mu(y), \varrho = -\mu_{yy}(y), \\ &y = 0 : \varrho_y = q_1(t), s_y = q_2(t), \\ &y = \pi : \varrho_y = 0, s_y = 0. \end{aligned} \tag{40}$$

This (1 + 1)-dimensional BVP is still nonlinear, and it is a difficult task to construct its exact solution in an explicit form. We are interested in the special case when $f_1(t) = 1$ (then, automatically, $\lambda_2 = 0$), when Problem (40) is nothing else but the (1 + 1)-dimensional case of BVP (29), i.e., the Neumann problem on the interval $[0;\pi]$ for the simplified Keller–Segel system.

Nevertheless, the corresponding BVP can be solved for arbitrary smooth functions $q_1(t), q_2(t)$, and we take $q_1(t) = q_1 = const$, $q_2(t) = q_2 = const$ in order to avoid cumbersome formulae. In this case, the nonlinear Problem (40) takes the form:

$$\begin{aligned} &\varrho_t(t,y) = \varrho_{yy}(t,y) - (\varrho(t,y)s_y(t,y))_y, \\ &0 = s_{yy}(t,y) + \varrho(t,y), \end{aligned} \tag{41}$$

$$\begin{aligned} &t = 0 : s = \mu(y), \varrho = -\mu_{yy}(y), \\ &y = 0 : \varrho_y = q_1, s_y = q_2, \\ &y = \pi : \varrho_y = 0, s_y = 0. \end{aligned} \tag{42}$$

First of all, we reduce System (41) to the 3rd order PDE in a similar way as it was done in Section 3; hence, we arrive at:

$$s_{ty} = s_{yyy} - s_{yy}s_y + \vartheta(t).$$

Setting $\vartheta(t) \equiv 0$, one notes that it is the Burgers equation w.r.t. the function $s_y(t,y)$. Thus, using the celebrated Cole–Hopf Substitution [20]:

$$s_y(t,y) = -2\frac{v_y(t,y)}{v(t,y)}, \tag{43}$$

and taking into account the Conditions (42), one obtains the Neumann problem for heat equation:

$$v_t = v_{yy}, \tag{44}$$

$$\begin{aligned} &t = 0 : v = \omega(y), \\ &y = 0 : v_y + \tfrac{q_2}{2}v = 0, \\ &y = \pi : v_y = 0, \end{aligned} \tag{45}$$

where $\omega(y) = Ce^{-\frac{1}{2}\mu(y)}$, $C \neq 0$. We will use Fourier method to solve Problems (44) and (45).

Let $v(t,y) = T(t)W(y)$, by substituting it into (44), we obtain:

$$\frac{W''}{W} = \frac{T''}{T} = -\lambda, \ \lambda > 0.$$

Thus, $T(t) = ae^{-\lambda t}$ and $W(y) = c_1\cos(\sqrt{\lambda}y) + c_2\sin(\sqrt{\lambda}y)$. Using boundary Conditions (45), we obtain the equation for λ:

$$\tan(\sqrt{\lambda}\pi) = -\frac{q_2}{2\sqrt{\lambda}}. \tag{46}$$

Now, we can calculate the general solution for Equations (44) and (45):

$$v(t,y) = \sum_{n=1}^{\infty} a_n e^{-\lambda_n t}(\cos(\sqrt{\lambda_n}y) - \frac{q_2}{2\sqrt{\lambda_n}}\sin(\sqrt{\lambda_n}y)), \tag{47}$$

where λ_n are the roots of Equation (46) and:

$$a_n = \frac{1}{||W_n||^2}\int_0^{\pi} \omega(y)(\cos(\sqrt{\lambda_n}y) - \frac{q_2}{2\sqrt{\lambda_n}}\sin(\sqrt{\lambda_n}y))\,dy,$$

$$||W_n||^2 = \int_0^{\pi}(\cos(\sqrt{\lambda_n}y) - \frac{q_2}{2\sqrt{\lambda_n}}\sin(\sqrt{\lambda_n}y))^2\,dy = \frac{\pi}{2} + \frac{q_2^2\pi - 2q_2}{8\lambda_n},$$

for $n > 0$. Thus, using Substitution (43) and the second equation from (41), one constructs the exact solution of the nonlinear BVPs (41) and (42):

$$\begin{aligned} s(t,y) = &-2\ln(v(t,y)) + B(t) = \\ &-2\ln\left(\sum_{n=1}^{\infty} a_n e^{-\lambda_n t}(\cos(\sqrt{\lambda_n}y) - \frac{q_2}{2\sqrt{\lambda_n}}\sin(\sqrt{\lambda_n}y))\right) + B(t), \end{aligned}$$

$$\varrho(t,y) = 2\left(\frac{v_{yy}}{v} - \left(\frac{v_y}{v}\right)^2\right) =$$
$$-2\left(\frac{\sum\limits_{n=1}^{\infty} a_n\lambda_n e^{-\lambda_n t}(\cos(\sqrt{\lambda_n}y) - \frac{q_2}{2\sqrt{\lambda_n}}\sin(\sqrt{\lambda_n}y))}{\sum\limits_{n=1}^{\infty} a_n e^{-\lambda_n t}(\cos(\sqrt{\lambda_n}y) - \frac{q_2}{2\sqrt{\lambda_n}}\sin(\sqrt{\lambda_n}y))} + \left(\frac{a_n\sqrt{\lambda_n}e^{-\lambda_n t}(\sin(\sqrt{\lambda_n}y) + \frac{q_2}{2\sqrt{\lambda_n}}\cos(\sqrt{\lambda_n}y))}{a_n e^{-\lambda_n t}(\cos(\sqrt{\lambda_n}y) - \frac{q_2}{2\sqrt{\lambda_n}}\sin(\sqrt{\lambda_n}y))}\right)^2\right),$$

where $B(t)$ is an arbitrary function.

Because the Cole–Hopf substitution in non-local transform (i.e., involves derivatives) we need to check whether this solution satisfies the initial and boundary conditions. Setting $t = 0$, we note that:

$$s(0,y) = -2\ln\left(\sum_{n=1}^{\infty} a_n(\cos(\sqrt{\lambda_n}y) - \frac{q_2}{2\sqrt{\lambda_n}}\sin(\sqrt{\lambda_n}y))\right) + B(0) =$$
$$\mu(y) - 2\ln|C| + B(0).$$

Therefore, $s(0,y) = \mu(y)$ provided $B(0) = 2\ln|C|$. Simple calculations show that the second initial condition $\varrho(0,y) = -\mu_{yy}(y)$ is also fulfilled.

It can be easily checked that the functions $s(t,y)$ and $\rho(t,y)$ constructed above also satisfy zero Neumann conditions from Equation (42).

Finally, our result can be formulated as the following theorem.

Theorem 5. *The exact solution of the Neumann problem for the simplified Keller–Segel Systems (41) and (42) can be presented as:*

$$s(t,y) = -2\ln\left(\sum_{n=1}^{\infty} a_n e^{-\lambda_n t}(\cos(\sqrt{\lambda_n}y) - \frac{q_2}{2\sqrt{\lambda_n}}\sin(\sqrt{\lambda_n}y))\right) + B(t),$$
$$\varrho(t,y) = -2\left(\frac{\sum\limits_{n=1}^{\infty} a_n\lambda_n e^{-\lambda_n t}(\cos(\sqrt{\lambda_n}y) - \frac{q_2}{2\sqrt{\lambda_n}}\sin(\sqrt{\lambda_n}y))}{\sum\limits_{n=1}^{\infty} a_n e^{-\lambda_n t}(\cos(\sqrt{\lambda_n}y) - \frac{q_2}{2\sqrt{\lambda_n}}\sin(\sqrt{\lambda_n}y))} + \left(\frac{\sum\limits_{n=1}^{\infty} a_n\sqrt{\lambda_n}e^{-\lambda_n t}(\sin(\sqrt{\lambda_n}y) + \frac{q_2}{2\sqrt{\lambda_n}}\cos(\sqrt{\lambda_n}y))}{\sum\limits_{n=1}^{\infty} a_n e^{-\lambda_n t}(\cos(\sqrt{\lambda_n}y) - \frac{q_2}{2\sqrt{\lambda_n}}\sin(\sqrt{\lambda_n}y))}\right)^2\right), \quad (48)$$

where λ_n are the roots of the transcendent equation $\tan(\sqrt{\lambda}\pi) = -\frac{q_2}{2\sqrt{\lambda}}$, while:

$$a_n = \frac{1}{||W_n||^2}\int_0^{\pi} \omega(y)(\cos(\sqrt{\lambda_n}y) - \frac{q_2}{2\sqrt{\lambda_n}}\sin(\sqrt{\lambda_n}y))\,dy,$$

$$||W_n||^2 = \frac{\pi}{2} + \frac{q_2^2\pi - 2q_2}{8\lambda_n},$$

and for $n > 0$, $\omega(y) = Ce^{-\frac{1}{2}\mu(y)}$, $C \neq 0$, $B(t)$ is an arbitrary function, such that $B(0) = 2\ln|C|$.

Obviously, the exact Solution (48) is not unique because one contains the arbitrary function $B(t)$ and a parameter C. To specify this function, one needs additional biologically motivated restrictions; however, it lies outside of the scope of this paper.

6. Conclusions

In this paper, the simplified Keller–Segel model has been studied by means of Lie symmetry based approaches. It is shown that (1 + 2)-dimensional Keller–Segel type System (2), together with the relevant boundary and initial conditions, is invariant with respect to infinite-dimensional Lie algebras. A classification of Lie symmetries for the Cauchy problem and the Neumann problem for this system is derived and presented in Theorems 1, 3 and 4, which say that the Cauchy (initial) problem and some Neumann problems for this system are still invariant w.r.t. infinite-dimensional Lie algebras (with the relevant restrictions on the structure of arbitrary functions arising in Equation (3)). It should be stressed that Lie symmetry of a boundary-value problem depends essentially on geometry of the domain, which the problem is formulated on. All possible domains, which may lead to nontrivial Lie symmetry of BVPs with the governing System (2), have been identified using the result of the recent paper [10]. All realistic from applicability point of view domains (a strip, an interior and exterior of the circle, an annulus) were examined (the case of a half-plane was studied earlier in [12]).

The results obtained, in particular *infinite-dimensional* Lie algebras of invariance, seem to be very interesting because initial and boundary-value problems usually demonstrate a full scale breaking Lie symmetry of the governing equation(s). For example, the classical example of the Cauchy problem for the linear heat equation says that this problem can be invariant only w.r.t. *finite-dimensional* Lie algebra [3]. *Finite-dimensional* Lie algebras of invariance occur also for boundary-value problems involving the linear heat equation [3,24] and nonlinear heat equations [8,9]. However, one cannot claim that the result obtained here is unique because infinite-dimensional Lie algebras of invariance may occur for BVPs with the governing equation(s) possessing infinite-dimensional MAI. A non-trivial example can be found in [10] (see case 11 in Table 2).

The Lie symmetries obtained are used for reduction of the problems in question to two-dimensional those. Exact solutions of some two-dimensional problems are constructed. In particular, we have proven that the Cauchy problem for the (1 + 1)-dimensional Keller–Segel type system can be linearized and solved in an explicit form (see Theorem 2). Because the exact solution involves two arbitrary functions, the relevant biologically motivated restrictions were proposed in order to obtain a unique solution. A non-trivial example of the solution in terms of elementary functions was also derived (see Formulae (19)). It should be stressed that exact solutions of Cauchy problems with nonlinear governing PDEs can be derived only in exceptional cases because there are no constructive methods for solving such nonlinear problems (in contrast to linear Cauchy problems).

Symmetry operators were applied also for reduction of the Neumann problems on the strip. As a result, the exact solution of the Neumann problem for the (1 + 1)-dimensional simplified Keller–Segel system has been constructed (see Theorem 5). The work is in progress for finding exact solutions of the Neumann problem on bounded domains.

Acknowledgments: This research was supported by a Marie Curie International Incoming Fellowship to the first author within the 7th European Community Framework Programme (Project BVP symmetry 912563).

Author Contributions: The authors contributed equally to this work.

Conflicts of Interest: The authors declare no conflict of interest.

References

1. Lie, S. Uber integration durch bestimmte integrale von einer Klasse lineare partiellen differentialgleichungen. *Arch. Math.* **1881**, *8*, 328–368.
2. Engel, F.; Heegaard, P. *Gesammelte Abhandlungen, Band 3*; Benedictus Gotthelf Teubner: Leipzig, Germany, 1922.
3. Bluman, G.W.; Anco, S.C. *Symmetry and Integration Methods for Differential Equations*; Springer: New York, NY, USA, 2002.
4. Bluman, G.W.; Kumei, S. *Symmetries and Differential Equations*; Springer: Berlin, Germany, 1989.

5. Fushchych, W.I.; Shtelen, W.M.; Serov, M.I. *Symmetry Analysis and Exact Solutions of Equations of Nonlinear Mathematical Physics*; Kluwer: Dordrecht, The Netherlands, 1993.
6. Olver, P.J. *Applications of Lie Groups to Differential Equations*; Springer: New York, NY, USA, 1993.
7. Ovsiannikov, L.V. *The Group Analysis of Differential Equations*; Academic Press: New York, NY, USA, 1982.
8. Cherniha, R.; Kovalenko, S. Lie symmetries and reductions of multi-dimensional boundary value problems of the Stefan type. *J. Phys. A Math. Theor.* **2011**, *44*, doi: 10.1088/1751-8113/44/48/485202.
9. Cherniha, R.; Kovalenko, S. Lie symmetries of nonlinear boundary value problems. *Commun. Nonlinear Sci. Numer. Simulat.* **2012**, *17*, 71–84.
10. Cherniha, R.; King, J.R. Lie and conditional symmetries of a class of nonlinear (1 + 2)-dimensional boundary value problems. *Symmetry* **2015**, *7*, 1410–1435.
11. Abd-el Malek, M.B.; Amin, A.M. Lie group method for solving the generalized Burgers', Burgers'-KdV and KdV equations with time-dependent variable coefficients. *Symmetry* **2015**, *7*, 1816–1830.
12. Didovych, M. A (1 + 2)-dimensional simplified Keller–Segel model: Lie symmetry and exact solutions. *Symmetry* **2015**, *7*, 1463–1474.
13. Keller, E.K.; Segel, L.A. Initiation of slime mold aggregation viewed as an instability. *J. Theor. Biol.* **1970**, *26*, 399–415.
14. Keller, E.K.; Segel, L.A. Traveling bands of chemotactic bacteria: A theoretical analysis. *J. Theor. Biol.* **1971**, *30*, 235–248.
15. Horstmann, D. From 1970 until present: The Keller–Segel model in chemotaxis and its consequences. *Jahresber. Deutsch. Math.-Verein* **2013**, *105*, 103–165.
16. Calvez, V.; Dolak-Strub, Y. Asymptotic behavior of a two-dimensional Keller–Segel model with and without density control. *Math. Model. Biol. Syst.* **2007**, *2*, 323–329.
17. Nagai, T. Convergence to self-similar solutions for a parabolic-elliptic system of drift-diffusion type in R^2. *Adv. Differ. Equ.* **2011**, *9*, 839–866.
18. Nagai, T. Global existence and decay estimates of solutions to a parabolic-elliptic system of drift-diffusion type in R^2. *Differ. Integral Equ.* **2011**, *1*, 29–68.
19. Hopf, E. The partial differential equation $U_t + UU_x = \mu U_{xx}$. *Comm. Pure Appl. Math.* **1950**, *3*, 201–230.
20. Burgers, J.M. *The Nonlinear Diffusion Equation*; D. Reidel Publishing Company: Dordrecht, The Netherlands, 1973.
21. Bateman, H.; Erdelyi, A. *Tables of Integral Transforms*; McGraw-Hill Book Company: New York, NY, USA, 1954.
22. Cherniha, R. New non-Lie ansatze and exact solutions of nonlinear reaction-diffusion-convection equations. *J. Phys. A Math.* **1998**, *31*, 8179–8198.
23. Samarskii, A.A.; Galaktionov, V.A.; Kurdyumov, S.P.; Mikhailov, A.P. Blow-up in quasilinear parabolic equations. In *de Gruyter Expositions in Mathematics*; Walter de Gruyter & Co.: Berlin, Germany, 1995; Volume 19.
24. Bluman, G.W. Application of the general similarity solution of the heat equation to boundary value problems. *Q. Appl. Math.* **1974**, *31*, 403–415.

Chapter 4:

symmetry

MDPI

Review

Dynamical Symmetries and Causality in Non-Equilibrium Phase Transitions

Malte Henkel

Groupe de Physique Statistique, Institut Jean Lamour (CNRS UMR 7198), Université de Lorraine Nancy, B.P. 70239, F-54506 Vandœuvre-lès-Nancy Cedex, France; E-Mail: malte.henkel@univ-lorraine.fr

Academic Editor: Roman M. Cherniha
Received: 10 July 2015 / Accepted: 29 October 2015 / Published: 13 November 2015

Abstract: Dynamical symmetries are of considerable importance in elucidating the complex behaviour of strongly interacting systems with many degrees of freedom. Paradigmatic examples are cooperative phenomena as they arise in phase transitions, where conformal invariance has led to enormous progress in equilibrium phase transitions, especially in two dimensions. Non-equilibrium phase transitions can arise in much larger portions of the parameter space than equilibrium phase transitions. The state of the art of recent attempts to generalise conformal invariance to a new generic symmetry, taking into account the different scaling behaviour of space and time, will be reviewed. Particular attention will be given to the causality properties as they follow for co-variant n-point functions. These are important for the physical identification of n-point functions as responses or correlators.

Keywords: Schrödinger algebra; conformal Galilei algebra; ageing algebra; representations; causality; parabolic sub-algebra; holography; physical ageing

1. Introduction

Improving our understanding of the collective behaviour of strongly interacting systems consisting of a large number of strongly interacting degrees of freedom is an ongoing challenge. From the point of view of the statistical physicist, paradigmatic examples are provided by systems undergoing a continuous phase transition, where fluctuation effects render traditional methods such as mean-field approximations inapplicable [1,2]. At the same time, it turns out that these systems can be effectively characterised in terms of a small number of "relevant" scaling operators, such that the net effect of all other physical quantities, the "irrelevant" ones, merely amounts to the generation of corrections to the leading scaling behaviour. From a symmetry perspective, phase transitions naturally acquire some kind of scale-invariance, and it then becomes a natural question whether further dynamical symmetries can be present.

1.1. *Conformal Algebra*

In equilibrium critical phenomena (roughly, for systems with sufficiently short-ranged, local interactions), scale-invariance can be extended to conformal invariance. In two space dimensions, the generators $\ell_n, \bar{\ell}_n$ should obey the infinite-dimensional algebra

$$[\ell_n, \ell_m] = (n-m)\ell_{n+m}, \quad [\bar{\ell}_n, \bar{\ell}_m] = (n-m)\bar{\ell}_{n+m}, \quad [\ell_n, \bar{\ell}_m] = 0 \tag{1}$$

for $n, m \in \mathbb{Z}$. The action of these generators on physical scaling operators $\phi(z, \bar{z})$, where complex coordinates $z, \bar{z}$ are used, is conventionally given by the representation [3]

$$\ell_n \phi(z, \bar{z}) \to [\ell_n, \phi(z, \bar{z})] = -\left(z^{n+1}\partial_z + \Delta(n+1)z^n\right)\phi(z, \bar{z}) \tag{2}$$

and similarly for $\bar{\ell}_n$, where the rôles of z and $\bar{z}$ are exchanged. Herein, the conformal weights Δ, $\overline{\Delta}$ are real constants, and related to the scaling dimension $x_\phi = \Delta + \overline{\Delta}$ and the spin $s_\phi = \Delta - \overline{\Delta}$ of the scaling operator ϕ. The representation (2) is an infinitesimal form of the (anti)holomorphic transformations $z \mapsto w(z)$ and $\bar{z} \mapsto \bar{w}(\bar{z})$. The maximal finite-dimensional sub-algebra of Equation (1) is isomorphic to $\mathfrak{sl}(2,\mathbb{R}) \oplus \mathfrak{sl}(2,\mathbb{R}) \cong \langle \ell_{\pm1,0}, \bar{\ell}_{\pm1,0} \rangle$. It is this conformal sub-algebra only which has an analogue in higher space dimensions $d > 2$. Denoting the Laplace operator by $\mathcal{S} := 4\partial_z\partial_{\bar{z}} = 4\ell_{-1}\bar{\ell}_{-1}$, the conformal invariance of the Laplace equation $\mathcal{S}\phi(z,\bar{z}) = 0$ is expressed through the commutator

$$[\mathcal{S}, \ell_n]\,\phi(z,\bar{z}) = -(n+1)z^n \mathcal{S}\phi(z,\bar{z}) - 4\Delta(n+1)nz^{n-1}\partial_{\bar{z}}\phi(z,\bar{z}) \tag{3}$$

and analogously for $\bar{\ell}_n$. Hence, for vanishing conformal weights $\Delta = \Delta_\phi = 0$ and $\overline{\Delta} = \overline{\Delta}_\phi = 0$, any solution of $\mathcal{S}\phi = 0$ is mapped onto another solution of the same equation. Thermal fluctuations in $2D$ classical critical points or quantum fluctuations in $1D$ quantum critical points (at temperature $T = 0$) modify the conformal algebra (1) to a pair of commuting Virasoro algebras, parametrised by the central charge c. Then Equation (2) retains its validity when the set of admissible operators ϕ is restricted to the set of primary scaling operators (a scaling operator is called quasi-primary if the transformation (2) only holds for the finite-dimensional sub-algebra $\mathfrak{sl}(2,\mathbb{R}) \cong \langle \ell_{\pm1,0} \rangle$) [4]. In turn, this furnishes the basis for the derivation of conformal Ward identities obeyed by n-point correlation functions $F_n := \langle \phi_1(z_1,\bar{z}_1)\ldots\phi_n(z_n,\bar{z}_n)\rangle$ of primary operators $\phi_1\ldots\phi_n$. Celebrated theorems provide a classification of the Virasoro primary operators from the unitary representations of the Virasoro algebra, for example through the Kac formula for central charges $c < 1$ [5,6]. Novel physical applications are continuously being discovered.

1.2. Schrödinger Algebra

When turning to time-dependent critical phenomena, the theory is far less advanced. One of the best-studied examples is the Schrödinger–Virasoro algebra $\mathfrak{sv}(d)$ in d space dimensions [7,8]

$$\begin{array}{rclcrcl}
[X_n, X_{n'}] &=& (n-n')X_{n+n'} &,& [X_n, Y_m^{(j)}] &=& \left(\frac{n}{2} - m\right) Y_{n+m}^{(j)} \\
{[X_n, M_{n'}]} &=& -n' M_{n+n'} &,& [X_n, R_{n'}^{(jk)}] &=& -n' R_{n+n'}^{(jk)} \\
{[Y_m^{(j)}, Y_{m'}^{(k)}]} &=& \delta^{j,k}\left(m-m'\right) M_{m+m'} &,& [R_n^{(jk)}, Y_m^{(\ell)}] &=& \delta^{j,\ell}\, Y_{n+m}^{(k)} - \delta^{k,\ell}\, Y_{n+m}^{(j)} \\
{[R_n^{(jk)}, R_{n'}^{(\ell i)}]} &=& \multicolumn{5}{l}{\delta^{j,i} R_{n+n'}^{(\ell k)} - \delta^{k,\ell} R_{n+n'}^{(ji)} + \delta^{k,i} R_{n+n'}^{(j\ell)} - \delta^{j,\ell} R_{n+n'}^{(ik)}}
\end{array} \tag{4}$$

(all other commutators vanish) with integer indices $n, n' \in \mathbb{Z}$, half-integer indices $m, m' \in \mathbb{Z} + \frac{1}{2}$ and $i, j, k, \ell \in \{1, \ldots, d\}$. Casting the generators of $\mathfrak{sv}(d)$ into the four families $X, Y^{(j)}, M, R^{(jk)} = -R^{(kj)}$ makes explicit (i) that the generators X_n form a conformal sub-algebra and (ii) that the families $Y^{(j)}$ and $M, R^{(jk)}$ make up Virasoro primary operators of weight $\frac{3}{2}$ and 1, respectively [7]. Non-trivial central extensions are only possible (i) either in the conformal sub-algebra $\langle X_n \rangle_{n\in\mathbb{Z}}$, where it must be of the form of the Virasoro central charge, or else (ii) in the $\mathfrak{so}(d)$-current algebra $\left\langle R_n^{(jk)} \right\rangle_{n\in\mathbb{Z}}$, where it must be a Kac–Moody central charge [5–7,9]. The maximal finite-dimensional sub-algebra of $\mathfrak{sv}(d)$ is the Schrödinger algebra $\mathfrak{sch}(d) = \left\langle X_{0,\pm1}, Y_{\pm1/2}^{(j)}, M_0, R_0^{(jk)} \right\rangle_{j,k=1,\ldots d}$, where M_0 is central. An explicit representation in terms of time-space coordinates $(t, \boldsymbol{r}) \in \mathbb{R} \times \mathbb{R}^d$, acting on a (scalar) scaling operator $\phi(t, \boldsymbol{r})$ of scaling dimension x and of mass $\mathcal{M}$, is given by [7]

$$
\begin{aligned}
X_n &= -t^{n+1}\partial_t - \frac{n+1}{2}t^n \boldsymbol{r}\cdot\boldsymbol{\nabla}_r - \frac{\mathcal{M}}{4}(n+1)nt^{n-1}\boldsymbol{r}^2 - \frac{n+1}{2}xt^n \\
Y_m^{(j)} &= -t^{m+1/2}\partial_j - \left(m+\frac{1}{2}\right)t^{m-1/2}\mathcal{M}r_j \\
M_n &= -t^n\mathcal{M} \\
R_n^{(jk)} &= -t^n\left(r_j\partial_k - r_k\partial_j\right) = -R_n^{(kj)}
\end{aligned} \tag{5}
$$

with the abbreviations $\partial_j := \partial/\partial r_j$ and $\boldsymbol{\nabla}_r = (\partial_1, \ldots, \partial_d)^{\mathrm{T}}$. These are the infinitesimal forms of the transformations $(t, \boldsymbol{r}) \mapsto (t', \boldsymbol{r}')$, where

$$
\begin{aligned}
X_n: &\quad t = \beta(t'), \quad \boldsymbol{r} = \boldsymbol{r}'\sqrt{\frac{\mathrm{d}\beta(t')}{\mathrm{d}t'}} \\
Y_m: &\quad t = t', \quad \boldsymbol{r} = \boldsymbol{r}' - \alpha(t') \\
R_n: &\quad t = t', \quad \boldsymbol{r} = \mathcal{R}(t')\boldsymbol{r}'
\end{aligned} \tag{6}
$$

where $\alpha(t)$ is an arbitrary time-dependent function, $\beta(t)$ is a non-decreasing function and $\mathcal{R}(t) \in SO(d)$ denotes a rotation matrix with time-dependent rotation angles. The generators M_n do not generate a time-space transformation, but rather produce a time-dependent "phase shift" of the scaling operator ϕ [10].

The dilatations X_0 are the infinitesimal form of the transformations $t \mapsto \lambda^z t$ and $\boldsymbol{r} \mapsto \lambda\boldsymbol{r}$, where $\lambda \in \mathbb{R}_+$ is a constant and z is called the *dynamical exponent*. In the representation (5), one has $z = 2$.

Since the work of Lie [12], and before of Jacobi [13], the Schrödinger algebra is known to be a dynamic symmetry of the the free diffusion equation (and, much later, also of the free Schrödinger equation). Define the Schrödinger operator

$$
\mathcal{S} = 2\mathcal{M}\partial_t - \boldsymbol{\nabla}_r\cdot\boldsymbol{\nabla}_r = 2M_0X_{-1} - \boldsymbol{Y}_{-1/2}\cdot\boldsymbol{Y}_{-1/2} \tag{7}
$$

Following Niederer [14], dynamical symmetries of such linear equations are analysed through the commutators of $\mathcal{S}$ with the symmetry Lie algebra. For the case of $\mathfrak{sch}(d)$, the only non-vanishing commutators with $\mathcal{S}$ are

$$
[\mathcal{S}, X_0] = -\mathcal{S}, \quad [\mathcal{S}, X_1] = -2t\mathcal{S} - (2x-d)M_0 \tag{8}
$$

Hence, any solution ϕ of the free Schrödinger/diffusion equation $\mathcal{S}\phi = 0$ with scaling dimension $x_\phi = \frac{d}{2}$ is mapped onto another solution of the free Schrödinger equation [15]. Finally, from representations such as Equation (5), one can derive Schrödinger–Ward identities in order to compute the form of covariant n-point functions $\langle\phi_1(t_1,\boldsymbol{r}_1)\ldots\phi_n(t_n,\boldsymbol{r}_n)\rangle$. With respect to conformal invariance, one has the important difference that the generator $M_0 = -\mathcal{M}$ is central in the finite-dimensional non-semi-simple Lie algebra $\mathfrak{sch}(d)$. This implies the Bargman super-selection rule [17]

$$
(\mathcal{M}_1 + \cdots + \mathcal{M}_n)\,\langle\phi_1(t_1,\boldsymbol{r}_1)\ldots\phi_n(t_n,\boldsymbol{r}_n)\rangle = 0 \tag{9}
$$

Physicists' conventions require that "physical masses" $\mathcal{M}_i \geq 0$. It it therefore necessary to define a formal "complex conjugate" ϕ^* of the scaling operator ϕ, such that its mass $\mathcal{M}^* := -\mathcal{M} \leq 0$ becomes negative. Then one may write, e.g., a non-vanishing co-variant two-point function of two quasi-primary scaling operators (up to an undetermined constant of normalisation) [7]

$$
\langle\phi_1(t_1,\boldsymbol{r}_1)\phi_2^*(t_2,\boldsymbol{r}_2)\rangle = \delta_{x_1,x_2}\delta(\mathcal{M}_1 - \mathcal{M}_2^*)\,(t_1-t_2)^{-x_1}\exp\left[-\frac{\mathcal{M}_1}{2}\frac{(\boldsymbol{r}_1-\boldsymbol{r}_2)^2}{t_1-t_2}\right] \tag{10}
$$

Here and throughout this paper, $\delta_{a,b} = 1$ if $a = b$ and $\delta_{a,b} = 0$ if $a \neq b$. While Equation (10) looks at first sight like a reasonable heat kernel, a closer inspection raises several questions:

1. Why should it be obvious that the time difference $t_1 - t_2 > 0$, to make the power-law prefactor real-valued ?
2. Given the convention that $\mathcal{M}_1 \geq 0$, the condition $t_1 - t_2 > 0$ is also required in order to have a decay of the two-point function with increasing distance $|\boldsymbol{r}| = |\boldsymbol{r}_1 - \boldsymbol{r}_2| \to \infty$.
3. In applications to non-equilibrium statistical physics, one studies indeed two-point functions of the above type, which are then interpreted as the linear response function of the scaling operator ϕ with respect to an external conjugate field $h(t, \boldsymbol{r})$

$$R(t_1, t_2; \boldsymbol{r}_1, \boldsymbol{r}_2) = \left.\frac{\delta\langle\phi(t_1, \boldsymbol{r}_1)\rangle}{\delta h(t_2, \boldsymbol{r}_2)}\right|_{h=0} = \langle\phi(t_1, \boldsymbol{r}_1)\widetilde{\phi}(t_2, \boldsymbol{r}_2)\rangle \tag{11}$$

   which in the context of the non-equilibrium Janssen–de Dominicis theory [2] can be re-expressed as a two-point function involving the scaling operator ϕ and its associate response operator $\widetilde{\phi}$. In this physical context, one has a natural interpretation of the "complex conjugate" in terms of the relationship of ϕ and $\widetilde{\phi}$.
   Then, the formal condition $t_1 - t_2 > 0$ simply becomes the causality condition, namely that a response will only arise at a later time $t_1 > t_2$ after the stimulation at time $t_2 \geq 0$.

Hence, it is necessary to inquire under what conditions the causality of Schrödinger-covariant n-point functions can be guaranteed.

1.3. Conformal Galilean Algebra

Textbooks in quantum mechanics show that the Schrödinger equation is the non-relativistic variant of relativistic wave equations, be it the Klein–Gordon equation for scalars or the Dirac equations for spinors. One might therefore expect that the Schrödinger algebra could be obtained by a contraction from the conformal algebra, but this is untrue (although there is a well-known contraction from the Poincaré algebra to the Galilei sub-algebra). Rather, applying a contraction to the conformal algebra, one arrives at a different Lie algebra, which we call here the altern-Virasoro algebra [11,18–20]. $\mathfrak{av}(d) = \left\langle X_n, Y_n^{(j)}, R_n^{(jk)} \right\rangle_{n\in\mathbb{Z}}$ with $j, k = 1, \ldots, d$, but which nowadays is often referred to as infinite conformal Galilean algebra. Its non-vanishing commutators can be given as follows

$$\begin{array}{rcl} [X_n, X_{n'}] &=& (n-n')X_{n+n'} \;,\;\; [X_n, Y_m^{(j)}] \;=\; (n-m)\,Y_{n+m}^{(j)} \\ [X_n, R_{n'}^{(jk)}] &=& -n' R_{n+n'}^{(jk)} \qquad ,\;\; [R_n^{(jk)}, Y_m^{(\ell)}] \;=\; \delta^{j,\ell} Y_{n+m}^{(k)} - \delta^{k,\ell} Y_{n+m}^{(j)} \\ [R_n^{(jk)}, R_{n'}^{(\ell i)}] &=& \delta^{j,i} R_{n+n'}^{(\ell k)} - \delta^{k,\ell} R_{n+n'}^{(ji)} + \delta^{k,i} R_{n+n'}^{(j\ell)} - \delta^{j,\ell} R_{n+n'}^{(ik)} \end{array} \tag{12}$$

An explicit representation as time-space transformation is [21]

$$\begin{array}{rcl} X_n &=& -t^{n+1}\partial_t - (n+1)t^n \boldsymbol{r}\cdot\boldsymbol{\nabla}_{\boldsymbol{r}} - n(n+1)t^{n-1}\boldsymbol{\gamma}\cdot\boldsymbol{r} - x(n+1)t^n \\ Y_n^{(j)} &=& -t^{n+1}\partial_j - (n+1)t^n\gamma_j \\ R_n^{(jk)} &=& -t^n\left(r_j\partial_k - r_k\partial_j\right) - t^n\left(\gamma_j\partial_{\gamma_k} - \gamma_k\partial_{\gamma_j}\right) \;=\; -R_n^{(kj)} \end{array} \tag{13}$$

where $\boldsymbol{\gamma} = (\gamma_1, \ldots, \gamma_d)$ is a vector of dimensionful constants, called rapidities, and x is again a scaling dimension. The dynamical exponent $z = 1$. The maximal finite-dimensional sub-algebra of $\mathfrak{av}(d)$ is the conformal Galilean algebra $\mathrm{CGA}(d) = \langle X_{\pm1,0}, Y_{\pm1,0}^{(j)}, R_0^{(jk)}\rangle_{j,k=1,\ldots,d}$ [11,18,22–27].

A more abstract characterisation of $\mathfrak{av}(1)$ can be given in terms of α-densities $\mathcal{F}_\alpha = \{u(z)(\mathrm{d}z)^\alpha\}$, with the action

$$f(z)\frac{\mathrm{d}}{\mathrm{d}z}\left(u(z)(\mathrm{d}z)^\alpha\right) = (fu' + \alpha f'u)(z)(\mathrm{d}z)^\alpha \tag{14}$$

Lemma 1. [33] *One has the isomorphism, where* $\ltimes$ *denotes the semi-direct sum*

$$\mathfrak{av}(1) \cong \mathrm{Vect}(S^1) \ltimes \mathcal{F}_{-1} \tag{15}$$

Clearly, it follows that $\mathrm{CGA}(1) \cong \mathfrak{sl}(2,\mathbb{R}) \ltimes \mathcal{F}_{-1}$.

As before, the time-space representation (13) can be used to derive conformal-Galilean Ward identities. For example, the $\mathrm{CGA}(d)$-covariant two-point function takes the form

$$\langle \phi_1(t_1, \boldsymbol{r}_1)\phi_2(t_2, \boldsymbol{r}_2)\rangle = \delta_{x_1,x_2}\delta_{\gamma_1,\gamma_2}(t_1 - t_2)^{-2x_1} \exp\left[-\frac{\gamma_1 \cdot (\boldsymbol{r}_1 - \boldsymbol{r}_2)}{t_1 - t_2}\right] \tag{16}$$

Again, at first sight this looks physically reasonable, but several questions must be raised:

1. Why should one have $t_1 - t_2 > 0$ for the time difference, as required to make the power-law prefactor real-valued ?
2. Even for a fixed vector γ_1 of rapidities, and even if $t_1 - t_2 > 0$ could be taken for granted, how does one guarantee that the scalar product $\gamma_1 \cdot (\boldsymbol{r}_1 - \boldsymbol{r}_2) > 0$, such that the two-point function decreases as $|\boldsymbol{r}| = |\boldsymbol{r}_1 - \boldsymbol{r}_2| \to \infty$?

The finite-dimensional $\mathrm{CGA}(2)$ admits a so-called "exotic" central extension [34,35]. Abstractly, this is achieved by completing the commutators (12) by the following

$$\left[Y_n^{(1)}, Y_m^{(2)}\right] = \delta_{n+m,0}\left(3\delta_{n,0} - 2\right)\Theta, \quad n,m \in \{\pm 1, 0\} \tag{17}$$

with a central generator Θ. This is called the exotic Galilean conformal algebra $\mathrm{ECGA} = \mathrm{CGA}(2) + \mathbb{C}\Theta$ in the physics literature. A representation as time-space transformation of ECGA is, with $n \in \{\pm 1, 0\}$ and $j,k \in \{1,2\}$ [21,24,36]

$$\begin{aligned}
X_n &= -t^{n+1}\partial_t - (n+1)t^n \boldsymbol{r}\cdot\boldsymbol{\nabla}_{\boldsymbol{r}} - x(n+1)t^n - (n+1)nt^{n-1}\gamma\cdot\boldsymbol{r} - (n+1)n\boldsymbol{h}\cdot\boldsymbol{r} \\
Y_n^{(j)} &= -t^{n+1}\partial_j - (n+1)t^n\gamma_j - (n+1)t^n h_j - n(n+1)\theta\varepsilon_{jk}r_k \\
R_0^{(12)} &= -(r_1\partial_2 - r_2\partial_1) - (\gamma_1\partial_{\gamma_2} - \gamma_2\partial_{\gamma_1}) - \frac{1}{2\theta}\boldsymbol{h}\cdot\boldsymbol{h}
\end{aligned} \tag{18}$$

The components of the vector $\boldsymbol{h} = (h_1, h_2)$ satisfy $[h_i, hj] = \varepsilon_{ij}\theta$, where θ is a constant, ε is the totally antisymmetric 2×2 tensor and $\varepsilon_{12} = 1$ [37]. The dynamical exponent $z = 1$. Because of Schur's lemma, the central generator Θ can be replaced by its eigenvalue $\theta \neq 0$. The ECGA-invariant Schrödinger operator is

$$\mathcal{S} = -\theta X_{-1} + \varepsilon_{ij}Y_0^{(i)}Y_{-1}^{(j)} = \theta\partial_t + \varepsilon_{ij}\left(\gamma_i + h_i\right)\partial_j \tag{19}$$

with $x = x_\phi = 1$. The requirement that these representations should be unitary gives the bound $x \geq 1$ [24]. Co-variant n-point functions and their applications have been studied in great detail.

1.4. Ageing Algebra

The common sub-algebra of $\mathfrak{sch}(d)$ and $\mathrm{CGA}(d)$ is called the ageing algebra $\mathfrak{age}(d) := \langle X_{0,1}, Y_{\pm\frac{1}{2}}^{(j)}, M_0, R_0^{(jk)}\rangle$ with $j,k = 1,\ldots,d$ and does not include time-translations. Starting from the representation (5), only the generators X_n assume a more general form [38]

$$X_n = -t^{n+1}\partial_t - \frac{n+1}{2}t^n\boldsymbol{r}\cdot\boldsymbol{\nabla}_{\boldsymbol{r}} - \frac{n+1}{2}xt^n - n(n+1)\xi t^n - \frac{n(n+1)}{4}\mathcal{M}t^{n-1}\boldsymbol{r}^2 \tag{20}$$

such that $z = 2$ is kept from Equation (5). When the generator X_n is applied to a scaling operator, the constant ξ describes a second scaling dimension, besides the habitual one denoted here by x, of that scaling operator ϕ. It is an important new aspect of extended dynamical symmetries, far from a stationary state, that at least two distinct scaling dimensions of a given scaling operator ϕ must be introduced. This will be made explicit later through concrete examples.

The invariant Schrödinger operates now becomes $\mathcal{S} = 2\mathcal{M}\partial_t - \partial_r^2 + 2\mathcal{M}t^{-1}\left(x + \xi - \frac{1}{2}\right)$, but without any constraint, neither on x nor on ξ [39]. Co-variant n-point functions can be derived as before [1,38,40], but we shall include these results with those to be derived from more general representations in the next sections. The absence of time-translations is particular appealing for application to dynamical critical phenomena, such as physical ageing, in non-stationary states far from equilibrium, see [1].

In Figure 1 (on page 19 below), the root diagrammes [41] of the Lie algebra (a) $\mathfrak{age}(1)$, (b) $\mathfrak{sch}(1)$ and (c) $\mathrm{CGA}(1)$ are shown, where the generators (roots) are represented by the black dots. This visually illustrates that the Schrödinger and conformal Galilean algebras are not isomorphic, $\mathfrak{sch}(d) \not\cong \mathrm{CGA}(d)$.

Comparing Figure 1a with Figure 1c, a different representation of $\mathrm{CGA}(1)$ can be identified. This representation is spanned by the generators $X_{0,1}, Y_{\pm 1/2}, M_0$ from Equation (5), along with a new generator V_+, and leads to a dynamic exponent $z = 2$ [11]. It is not possible to extend this to a representation of $\mathfrak{av}(1)$ [33]. Explicit expressions of V_+ will be given in Section 4.

This algebra also appears in more systematic approaches, either from a classification of non-relativistic limits of conformal symmetries [42] or else from an attempt to construct all possible infinitesimal local scale transformations [8,18].

1.5. Langevin Equation and Reduction formulæ

In non-equilibrium statistical mechanics [2], one considers often equations under the form of a stochastic Langevin equation, viz. (we use the so-called "model-A" dynamics with a non-conserved order-parameter)

$$2\mathcal{M}\partial_t\phi = \nabla_r \cdot \nabla_r\phi - \frac{\delta\mathcal{V}[\phi]}{\delta\phi} + \eta \tag{21}$$

for a physical field ϕ (called the order parameter), and where $\delta/\delta\phi$ stands for a functional derivative. Herein, $\mathcal{V}[\phi]$ is the Ginzburg–Landau potential and η is a white noise, *i.e.*, its formal time-integral is a Brownian motion. In the context of Janssen–de Dominicis theory, see [2], this can be recast as the variational equation of motion of the functional

$$\begin{aligned}
\mathcal{J}[\phi,\widetilde{\phi}] &= \mathcal{J}_0[\phi,\widetilde{\phi}] + \mathcal{J}_b[\widetilde{\phi}] \\
\mathcal{J}_0[\phi,\widetilde{\phi}] &= \int_{\mathbb{R}_+\times\mathbb{R}^d} \mathrm{d}t\mathrm{d}\boldsymbol{r}\, \widetilde{\phi}\left(2\mathcal{M}\partial_t\phi - \nabla_r\cdot\nabla_r\phi + \frac{\delta\mathcal{V}[\phi]}{\delta\phi}\right) \\
\mathcal{J}_b[\widetilde{\phi}] &= -T\int_{\mathbb{R}_+\times\mathbb{R}^d} \mathrm{d}t\mathrm{d}\boldsymbol{r}\, \widetilde{\phi}^2(t,\boldsymbol{r}) - \frac{1}{2}\int_{\mathbb{R}^{2d}} \mathrm{d}\boldsymbol{r}\mathrm{d}\boldsymbol{r}'\, \widetilde{\phi}(0,\boldsymbol{r})c(\boldsymbol{r}-\boldsymbol{r}')\widetilde{\phi}(0,\boldsymbol{r}')
\end{aligned} \tag{22}$$

where the term $\mathcal{J}_0[\phi,\widetilde{\phi}]$ contains the deterministic terms coming from the Langevin equation and $\mathcal{J}_b[\widetilde{\phi}]$ contains the stochastic terms generated by averaging over the thermal noise and the initial condition, characterised by an initial correlator $c(\boldsymbol{r})$ [43]. In particular, by adding an external source term $h(t,\boldsymbol{r})\phi(t,\boldsymbol{r})$ to the potential $\mathcal{V}[\phi]$, one can write the two-time linear response function as follows (spatial arguments are suppressed for brevity)

$$R(t,s) = \left.\frac{\delta\langle\phi(t)\rangle}{\delta h(s)}\right|_{h=0} = \int\mathcal{D}\phi\mathcal{D}\widetilde{\phi}\,\phi(t)\widetilde{\phi}(s)e^{-\mathcal{J}[\phi,\widetilde{\phi}]} = \langle\phi(t)\widetilde{\phi}(s)\rangle \tag{23}$$

with an explicit expression of the average $\langle.\rangle$ as a functional integral.

Theorem 1. [44] *If in the functional* $\mathcal{J}[\phi,\widetilde{\phi}] = \mathcal{J}_0[\phi,\widetilde{\phi}] + \mathcal{J}_b[\widetilde{\phi}]$, *the part* $\mathcal{J}_0$ *is Galilei-invariant with non-vanishing masses and* $\mathcal{J}_b[\widetilde{\phi}]$ *does not contain the field* ϕ, *then the computation of all responses and correlators can be reduced to averages which only involve the Galilei-invariant part* $\mathcal{J}_0$.

Proof. We illustrate the main idea for the calculation of the two-time response. Define the average $\langle X\rangle_0 = \int \mathcal{D}\phi\mathcal{D}\widetilde{\phi}\; X[\phi]e^{-\mathcal{J}_0[\phi,\widetilde{\phi}]}$ with respect to the functional $\mathcal{J}_0[\phi,\widetilde{\phi}]$. Then, from Equation (23)

$$R(t,s) = \left\langle \phi(t)\widetilde{\phi}(s)e^{-\mathcal{J}_b[\widetilde{\phi}]}\right\rangle_0 = \sum_{k=0}^{\infty}\frac{(-1)^k}{k!}\left\langle \phi(t)\widetilde{\phi}(s)\mathcal{J}_b[\widetilde{\phi}]^k\right\rangle_0 = \langle \phi(t)\widetilde{\phi}(s)\rangle_0$$

since the Bargman super-selection rule (9) implies that only the term with $k=0$ remains. Hence the response function $R(t,s) = R_0(t,s)$ is reduced to the expression obtained from the deterministic part $\mathcal{J}_0$ of the action.

Analogous reduction formulæ can be derived for all Galilei-covariant n-point responses and correlators [1,44]. □

This means that one may study the deterministic, noiseless truncation of the Langevin equation and its symmetries, provided that spatial translation- and Galilei-invariance are included therein, in order to obtain the form of the stochastic two-time response functions, as it will be obtained from models, simulations or experiments.

This work is organised as follows. In Section 2, we review several distinct representations of the Schrödinger and conformal Galilean algebras, discuss the associated invariant Schrödinger operators an co-covariant two-point functions. Applications to non-equilibrium statistical mechanics and the non-relativistic AdS/CFT correspondence will be indicated. In Section 3, the dual representations and the extensions to parabolic sub-algebras will be reviewed. In Section 4, it will be shown how to use these, to algebraically derive causality and long-distance properties of co-variant two-point functions. Conclusions are given in Section 5.

2. Representations

We now list several results relevant for the extension of the representations discussed in the introduction. The basic new fact, first observed in [40], is compactly stated as follows.

Proposition 1. *Let γ be a constant and $g(z)$ a non-constant function. Then the generators*

$$\ell_n = -z^{n+1}\partial_z - n\gamma z^n - g(z)z^n \tag{24}$$

obey the conformal algebra $[\ell_n, \ell_m] = (n-m)\ell_{n+m}$ for all $n, m \in \mathbb{Z}$.

The commutator is readily checked. We point out that the rapidity γ serves as a second scaling dimension and the choice of the function $g(z)$ can be helpful to include effects of corrections to scaling into the generators of time-space transformations. Next, we give an example on how these terms in the generators ℓ_n appear in the two-point function, co-variant under the maximal finite-dimensional sub-algebra $\langle \ell_{\pm 1,0}\rangle$.

Proposition 2. *If $\phi(z)$ is a quasi-primary scaling operator under the representation (24) of the conformal algebra $\langle \ell_{\pm 1,0}\rangle$, its co-variant two-point function is, where φ_0 is a normalisation constant*

$$\langle \phi_1(z_1)\phi_2(z_2)\rangle = \varphi_0\,\delta_{\gamma_1,\gamma_2}\,(z_1-z_2)^{-\gamma_1-\gamma_2}\Gamma_1(z_1)\Gamma_2(z_2),\quad \Gamma_i(z) := z^{\gamma_i}\exp\left(-\int_1^z \mathrm{d}\zeta\,\frac{g(\zeta)}{\zeta}\right) \tag{25}$$

Proof. For brevity, denote $F(z_1,z_2) = \langle \phi_1(z_1)\phi_2(z_2)\rangle$. Then the co-variance of F is expressed by the three Ward identities, with $\partial_i := \partial/\partial z_i$

$$\begin{aligned}
\ell_{-1}F &= \left(-\partial_1 - \partial_2 + \gamma_1 z_1^{-1} + \gamma_2 z_2^{-1} - g(z_1)z_1^{-1} - g(z_2)z_2^{-1}\right) F = 0 \\
\ell_0 F &= \left(-z_1\partial_1 - z_2\partial_2 - g(z_1) - g(z_2)\right) F = 0 \\
\ell_1 F &= \left(-z_1^2\partial_1 - z_2^2\partial_2 - \gamma_1 z_1 - \gamma_2 z_2 - g(z_1)z_1 - g(z_2)z_2\right) F = 0
\end{aligned}$$

Rewrite the correlator as $F(z_1, z_2) = \Gamma_1(z_1)\Gamma_2(z_2)\Psi(z_1, z_2)$. Then the function $\Psi(z_1, z_2)$ satisfies

$$\begin{aligned}
\left(-\partial_1 - \partial_2\right)\Psi &= 0 \\
\left(-z_1\partial_1 - z_2\partial_2 - \gamma_1 - \gamma_2\right)\Psi &= 0 \\
\left(-z_1^2\partial_1 - z_2^2\partial_2 - 2\gamma_1 z_1 - 2\gamma_2 z_2\right)\Psi &= 0
\end{aligned}$$

which are the standard Ward identities of the representation (2) of conformal invariance, where the γ_i take the rôle of the conformal weights. The resulting function Ψ is well-known [45]. □

One can now generalise the representation (5) of the Schrödinger–Virasoro algebra $\mathfrak{sv}(d)$.

Proposition 3. *If one replaces in the representation (5) the generator X_n as follows*

$$X_n = -t^{n+1}\partial_t - \frac{n+1}{2}t^n \boldsymbol{r}\cdot\nabla_{\boldsymbol{r}} - \frac{n+1}{2}xt^n - n(n+1)\xi t^n - \Xi(t)t^n - \frac{n(n+1)}{4}\mathcal{M}t^{n-1} \tag{26}$$

where x, ξ are constants and $\Xi(t)$ is an arbitrary (non-constant) function, then the commutators (4) of the Lie algebra $\mathfrak{sv}(d)$ are still satisfied.

This result was first obtained, for the maximal finite-dimensional sub-algebra $\mathfrak{sch}(d)$, by Minic, Vaman and Wu [40], who also further take the dependence on the mass $\mathcal{M}$ into account and write down terms of order $O(1/\mathcal{M})$ and $O(1)$ in $v(t)$ explicitly. We extend this observation to $\mathfrak{sv}(d)$, but do not trace the dependence in $\mathcal{M}$ explicitly, although one could re-introduce it, if required. The proof is immediate, since all modifications of the generator X_n merely depend on the time t and none of the other generators of $\mathfrak{sv}(d)$ changes t. For the sub-algebra $\mathfrak{age}(d) \subset \mathfrak{sch}(d)$, the representation (20) is a special case, with arbitrary ξ, but with $\Xi(t) = 0$.

It is obvious that similar extensions of the representations of time-space transformation of the other algebras, especially $\mathfrak{av}(d)$, its finite-dimensional sub-algebra CGA(d) or the exotic algebra ECGA apply.

Proposition 4. *Consider the representation (5), but with the generators X_n replaced by Equation (26), of the ageing algebra $\mathfrak{age}(d)$ and the Schrödinger algebra $\mathfrak{sch}(d)$. The invariant Schrödinger operator has the form*

$$\mathcal{S} = 2\mathcal{M}\partial_t - \nabla_r^2 + 2\mathcal{M}v(t), \quad v(t) = \frac{x + \xi - d/2}{t} + \frac{\Xi(t)}{t} \tag{27}$$

such that a solution of $\mathcal{S}\phi = 0$ is mapped onto another solution of the same equation. For the algebra $\mathfrak{age}(d)$, there is no restriction, neither on x, nor on ξ, nor on $\Xi(t)$. For the algebra $\mathfrak{sch}(d)$, one has the additional condition $x = \frac{d}{2} - 2\xi$.

Proof. To shorten the calculations, we restrict here to $d = 1$. It is enough to restrict attention to the generators $X_{\pm 1,0}$, and we must reproduce Equation (8) in this more general setting. We first look at $\mathfrak{age}(1)$. Consideration of X_0 gives $t\dot{v}(t) + v - \dot{\Xi}(t) = 0$ and considering X_1 gives $x + \xi - \frac{1}{2} + \Xi(t) + t\dot{\Xi}(t) - 2tv(t) - t^2\dot{v}(t) = 0$, where the dot denotes the derivative with respect to t. The second relation can be simplified to $x + \xi - \frac{1}{2} + \Xi(t) - tv(t) = 0$ which gives the assertion. Going over to $\mathfrak{sch}(1)$, the condition $[\mathcal{S}, X_{-1}] = 0$ leads to $\zeta/t^2 + \dot{\Xi}(t)/t - \Xi(t)/t^2 - \dot{v}(t) = 0$. This is only compatible with the result found before for $\mathfrak{age}(1)$, if $\zeta = -x - \zeta + \frac{1}{2}$, hence $x = \frac{1}{2} - 2\zeta$, as claimed. □

Example 1. *For a physical illustration of the meaning of the explicitly time-dependent terms in the Schrödinger operator (27), we consider the growth of an interface [46]. One may imagine that an interface can be created by randomly depositing particle onto a substrate. The height of this interface will be described by a function* $h(t, \boldsymbol{r})$. *One usually works in a co-moving coordinate system such that the average height* $\langle h(t, \boldsymbol{r})\rangle = 0$ *which we shall assume from now on. Then physically interesting quantities are either the interface width* $w(t) = \langle h(t, \boldsymbol{r})^2\rangle \sim t^{\beta}$, *which for sufficiently long times t defines the* growth exponent β, *or else two-time height-height correlators* $C(t, s; \boldsymbol{r}) = \langle h(t, \boldsymbol{r})h(s, \boldsymbol{0})\rangle$ *or two-time response functions* $R(t, s; \boldsymbol{r}) = \left.\frac{\delta\langle h(t,\boldsymbol{r})\rangle}{\delta j(s,\boldsymbol{0})}\right|_{j=0}$, *with respect to an external deposition rate* $j(t, \boldsymbol{r})$. *Their scaling behaviour is described by several non-equilibrium exponents [1,2]. Herein, spatial translation-invariance was assumed for the sake of simplicity of the notation.*

Physicists have identified several universality classes *of interface growth, see e.g., [2,46]. For the* Edwards–Wilkinson *universality class, h is simply assumed to be a continuous function in space. Its equation of motion for the height is just a free Schrödinger equation with an additional white noise. A distinct universality class is given by the celebrated* Kardar–Parisi–Zhang equation *which contains an additional term, quadratic in* $\nabla_{\boldsymbol{r}} h$. *A lattice realisation may be obtained by requiring that the heights only take integer values such that the height difference on two neighbouring sites, such that* $|\boldsymbol{r}_1 - \boldsymbol{r}_2| = a$ *where a is the lattice constant, is restricted to* $h(t, \boldsymbol{r}_1) - h(t, \boldsymbol{r}_2) = \pm 1$. *An intermediate universality class is the one of the* Arcetri model, *where the strong restriction of the Kardar-Parisi-Zhang model is relaxed in that h is taken to be a real-valued function, but subject to the constraint that the sum of its* slopes $\sum_{\boldsymbol{r}}\langle \nabla_{\boldsymbol{r}} h(t, \boldsymbol{r})^2\rangle \stackrel{!}{=} \mathcal{N}$, *where* $\mathcal{N}$ *is the number of sites of the lattice [47] (this is just one of the many conditions automatically satisfied in lattice realisations of the Kardar–Parisi–Zhang universality class). Schematically, in the continuum limit, the slopes* $u_a(t, \boldsymbol{r}) = \partial h(t, \boldsymbol{r})/\partial r_a$ *in the Arcetri model satisfy a Langevin equation*

$$\partial_t u_a(t, \boldsymbol{r}) = \Delta_{\boldsymbol{r}} u_a(t, \boldsymbol{r}) + \mathfrak{z}(t) u_a(t, \boldsymbol{r}) + \frac{\partial}{\partial r_a}\eta(t, \boldsymbol{r}) \tag{28}$$

$\Delta_{\boldsymbol{r}}$ *is the spatial Laplacian and* η *is a white noise. The constraint on the slopes can be cast into a simple form by defining*

$$g(t) = \exp\left(-2\int_0^t \mathrm{d}\tau\, \mathfrak{z}(\tau)\right) \tag{29}$$

which can be shown to obey a Volterra integral equation

$$dg(t) = 2f(t) + 2T\int_0^t \mathrm{d}\tau\, f(t-\tau) g(\tau), \quad f(t) = d\frac{e^{-4t} I_1(4t)}{4t}\left(e^{-4t} I_0(4t)\right)^{d-1} \tag{30}$$

where T is the "temperature" defined by the second moment of the white noise and the I_n *are modified Bessel functions. This model is exactly soluble [47] and the exponents of the (non-stationary) interface growth are distinct from the Edwards–Wilkinson (if* $d < 2$*) and the Kardar–Parisi–Zhang universality classes.*

It turns out that for all dimensions $d > 0$, *there is a "critical temperature"* $T_c(d) > 0$ *such that for* $T \leq T_c(d)$, *long-range correlations build up. For example,* $T_c(1) = 2$ *and* $T_c(2) = 2\pi/(\pi - 2)$. *For* $T \leq T_c(d)$, *the long-time solution of Equation (30) becomes* $g(t) \sim t^{-F}$ *as* $t \to \infty$. *This is compatible with the large-time behaviour* $\mathfrak{z}(t) \sim t^{-1}$ *of the Lagrange multiplier in Equation (28).*

Hence, recalling Theorem 1, it is enough to concentrate on the deterministic part. This is given by the Schrödinger operator (27). Therein, the first term in the potential $v(t)$, *of order* $1/t$, *represents the asymptotic behaviour of the Arcetri model; whereas the term described by* $\Xi(t)/t$ *takes into account the finite-time corrections to this leading scaling behaviour [48].*

Example 2. *We give a different illustration of the new representations of* $\mathfrak{age}(d)$ *with* $\zeta \neq 0$ *(and* $\Xi(t) = 0$*). Although we shall not be able to write down explicitly the invariant Schrödinger operator of the form specified in*

Equation (27), this example makes it clear that the domain of application of these representations extends beyond the context of that single differential equation.

The physical context involved will be the kinetic Ising model with Glauber dynamics. *The statistical mechanics of the Ising model can be described in terms of discrete "spins" $\sigma_i = \pm 1$, attached to each site i of a lattice. In one spatial dimension, one associates to each configuration $\sigma = \{\sigma_1, \ldots, \sigma_{\mathcal{N}}\}$ of spins an energy (hamiltonian) $\mathcal{H} = -\sum_{i=1}^{\mathcal{N}} \sigma_i \sigma_{i+1}$, with periodic boundary conditions $\sigma_{\mathcal{N}+1} = \sigma_1$. The dynamics of these spins is described in terms of a Markov process, such that the "time" $t \in \mathbb{N}$ is discrete. At each time-step, a single spin σ_i is randomly selected and is updated according to the Glauber rates (also referred to as "heat-bath rule") [52]. These are specified in terms of the probabilites*

$$P\left(\sigma_i(t+1) = \pm 1\right) = \frac{1}{2}\left[1 \pm \tanh\left(\frac{1}{T}\left(\sigma_{i-1}(t) + \sigma_{i+1}(t) + h_i(t)\right)\right)\right] \tag{31}$$

where the constant T is the temperature and $h_i(t)$ is a time-dependent external field. From these probabilites alone, the time-evolution of the average of any local observable, such as the time-dependent magnetisation or magnetic correlators, can be evaluated analytically [52]. In one spatial dimension, and at temperature $T = 0$, the model displays dynamical scaling and the exactly-known magnetic two-time correlator and response take a simple form. In the scaling limit $t, s \to \infty$ with t/s being kept fixed, one has [53–55]

$$C(t,s) = \langle \sigma_i(t)\sigma_i(s)\rangle = \frac{2}{\pi}\arctan\sqrt{\frac{2}{t/s-1}} \tag{32}$$

$$R(t,s) = \left.\frac{\delta\langle\sigma_i(t)\rangle}{\delta h_i(s)}\right|_{h=0} = \frac{1}{\sqrt{2}\,\pi}\frac{1}{\sqrt{s(t-s)}} \tag{33}$$

This is independent of the initical conditions (which merely enter into corrections to scaling), hence these results should be interpreted as being relevant to a critical point at $T_c = 0$ [1].

As a first observation, we remark that the form Equation (33) of the auto-response function $R(t,s) = R(t,s;0)$ is not compatible with the prediction Equation (10) of Schrödinger-invariance. This means that the representation (5) of the Schrödinger-algebra $\mathfrak{sch}(1)$, with time-translation-invariance included, is too restrictive to account for the phenomenology of the relaxational behaviour, far from a stationary state, of the one-dimensional Glauber–Ising model [56].

In order to explain the exact results Equations (32) and (33) in terms of the representation (20) of $\mathfrak{age}(d)$, one first generalises the prediction Equation (10) of the Schrödinger algebra as follows, up to normalisation [44]

$$\begin{aligned} R(t,s;\boldsymbol{r}) &= \langle\phi(t,\boldsymbol{r})\widetilde{\phi}(s,\boldsymbol{0})\rangle = R(t,s)\exp\left[-\frac{\mathcal{M}}{2}\frac{\boldsymbol{r}^2}{t-s}\right] \\ &= \delta_{x+2\xi,\widetilde{x}+2\widetilde{\xi}}\,\delta(\mathcal{M}+\widetilde{\mathcal{M}})\,s^{-(x+\widetilde{x})/2}\left(\frac{t}{s}\right)^{\xi+F}\left(\frac{t}{s}-1\right)^{-x-2\xi}\exp\left[-\frac{\mathcal{M}}{2}\frac{\boldsymbol{r}^2}{t-s}\right] \end{aligned} \tag{34}$$

with $F := \frac{1}{2}(\widetilde{x} - x) + \widetilde{\xi} - \xi$. Herein, ϕ denotes the order parameter, with scaling dimensions x, ξ and of mass $\mathcal{M} > 0$. The response field $\widetilde{\phi}$, with scaling dimensions $\widetilde{x}, \widetilde{\xi}$ and mass $\widetilde{\mathcal{M}} = -\mathcal{M} < 0$ takes over the rôle of the "complex conjugate" in Equation (10), but now time-translation-invariance is no longer required. Spatial translation-invariance is implicitly admitted. Comparison of the auto-response $R(t,s)$ with the exact result Equation (33) leads to the identifications $x = \frac{1}{2}$, $\widetilde{x} = 0$, $\xi = 0$ and $\widetilde{\xi} = \frac{1}{4}$. Remarkably, only the second scaling dimension $\widetilde{\xi}$ of the response scaling operator $\widetilde{\phi}$ does not vanish—a feature also observed numerically in models such as directed percolation or the Kardar–Parisi–Zhang Equation, see [57–59] for details.

On the other hand, along the lines of Theorem 1, the autocorrelator at the critical point $T = T_c$ can be expressed as an integral of a "noiseless" three-point response, up to normalisation [7]

$$C(t,s) = \int_{\mathbb{R}_+\times\mathbb{R}^d} \mathrm{d}u\mathrm{d}\boldsymbol{R}\,\left\langle \phi(t,\boldsymbol{y})\phi(s,\boldsymbol{y})\widetilde{\phi}^2(u,\boldsymbol{R}+\boldsymbol{y})\right\rangle_0 \tag{35}$$

Ageing-invariance fixes this three-point function up to a certain undetermined scaling function. Herein, one considers $\widetilde{\phi}^2$ as a new composite operator with scaling dimensions $2\widetilde{x}_2, 2\widetilde{\xi}_2$. Up to normalisation, the autocorrelator becomes (assuming $t > s$ for definiteness) [38]

$$\begin{aligned} C(t,s) &= s^{1+d/2-x-\widetilde{x}_2} \left(\frac{t}{s}\right)^{x+F} \left(\frac{t}{s}-1\right)^{\widetilde{x}_2+2\widetilde{\xi}_2-x-2\xi-d/2} \\ &\times \int_0^1 \mathrm{d}v\, v^{2\widetilde{\xi}_2-F} \left[\left(\frac{t}{s}-v\right)(1-v)\right]^{d/2-\widetilde{x}_2-2\widetilde{\xi}_2} \boldsymbol{\Upsilon}\left(\frac{t/s+1-2v}{t/s-1}\right) \\ &= C_0 \int_0^1 \mathrm{d}v\, v^{2\mu} \left[\left(\frac{t}{s}-v\right)(1-v)\right]^{2\mu} \left(\frac{t}{s}+1-2v\right)^{2\mu} \end{aligned} \tag{36}$$

where in the second line we recognised that the scaling function can be described in terms of the single parameter $\mu = \xi + \widetilde{\xi}_2$ and there remains an undetermined scaling function $\boldsymbol{\Upsilon}$. Furthermore, the autocorrelator scaling function should be non-singular as $t \to s$. This implies $\boldsymbol{\Upsilon}(w) \sim w^{\widetilde{x}_2-x-4\xi-d/2+\mu}$ for $w \gg 1$. The most simple case arises when this form remains valid for all w. Using the values of the scaling exponents identified from the autoresponse $R(t,s)$ before, the exact 1D Glauber–Ising autocorrelator Equation (32) is recovered from Equation (36), with the choice $\mu = -\frac{1}{4}$ and $C_0 = 2/\sqrt{\pi}$ [38].

Although the discrete nature of the Ising spins does not permit to recognise explicitly the continuum equation of motion in the form Equation (27) (the underlying field theory of the model is a free-fermion theory, and not a free-boson theory as in the first example [60]), this illustrates the necessity of the second scaling dimension ξ, of the representation (20) of $\mathfrak{age}(1)$. For $d \geq 2$ dimensions, there is no known analytical solution and one must turn to numerical simulations. The available evidence suggests that the second scaling dimensions $\xi + \widetilde{\xi} \neq 0$ at criticality, at least for dimensions $d < d^* = 4$, the upper critical dimension. For details and a review of further examples, see [1].

How the choice of the representation can affect the physical interpretation, is further illustrated by considering a "lattice" representation rather than the usually employed "continuum" representation of the Schrödinger algebra $\mathfrak{sch}(1)$. In Table 1, we list the generators of the "continuum" representation (5) along with the one of the "lattice" representation. Herein, the non-linear functions of the derivative ∂_r are understood to stand for their Taylor expansions. The origin of the name of a "lattice" representation can be understood when considering the generator $Y_{-1/2}$ of "spatial translations", which reads explicitly

$$Y_{-1/2} f(t,r) = -\frac{1}{a}\Big(f(t,r+a/2) - f(t,r-a/2)\Big) \tag{37}$$

It is suggestive to interpret this as a discretised symmetric lattice derivative operator, with a as a lattice constant, although the X_n, Y_m are still generators of infinitesimal transformations.

Table 1. The "lattice" representations of the Schrödinger algebra $\mathfrak{sch}(1)$, and its "continuum" representation, to which it reduces in the limit $a \to 0$ [61].

Generator	Continuum	Lattice
X_{-1}	$-\partial_t$	$-\partial_t$
X_0	$-t\partial_t - \frac{1}{2}r\partial_r$	$-t\partial_t - \frac{1}{a\cosh(\frac{a}{2}\partial_r)} r\sinh(\frac{a}{2}\partial_r)$
X_1	$-t^2\partial_t - tr\partial_r - \frac{1}{2}\mathcal{M}r^2$	$-t^2\partial_t - \frac{2t}{a\cosh(\frac{a}{2}\partial_r)} r\sinh(\frac{a}{2}\partial_r) - \frac{\mathcal{M}}{2}\left(\frac{1}{\cosh(\frac{a}{2}\partial_r)}r\right)^2$
$Y_{-1/2}$	$-\partial_r$	$-\frac{2}{a}\sinh(\frac{a}{2}\partial_r)$
$Y_{1/2}$	$-t\partial_r - \mathcal{M}r$	$-\frac{2t}{a}\sinh(\frac{a}{2}\partial_r) - \frac{\mathcal{M}}{\cosh(\frac{a}{2}\partial_r)}r$
M_0	$-\mathcal{M}$	$-\mathcal{M}$

The Schrödinger operator has, in the "lattice" representation, the following form

$$\mathcal{S} = 2\mathcal{M}\partial_t - \frac{1}{a^2}\left(e^{a\partial_r} + e^{-a\partial_r} - 2\right) \tag{38}$$

and the equation $\mathcal{S}\phi = 0$ could be viewed as a "lattice analogue" of a free Schrödinger equation.

It is also of interest to write down the co-variant two-point functions. The extension of Equation (10) reads, up to a normalisation constant [61]

$$\Phi(t,n) := \langle \phi_1(t_1,r_1)\phi_2^*(t_2,r_2)\rangle = \delta_{\mathcal{M}_1,\mathcal{M}_2}\delta_{x_1,x_2}\, t^{1/2-x_1}\, e^{-t} I_n(t) \tag{39}$$

where I_n is again a modified Bessel function, and with the abbreviations

$$t = \frac{t_1 - t_2}{\mathcal{M}_1 a^2}, \quad n = \frac{r_1 - r_2}{a} \tag{40}$$

Herein, both r_1 and r_2 must be integer multiples of the "lattice constant" a. In the limit $a \to 0$, all these results reduce to those of the "continuum" representation, discussed in Section 1. Again, although at first sight this looks as a physically reasonable Green's function on an infinite chain [62], the same questions as raised in relation with Equation (10) should be addressed. The extensions discussed in the above propositions 2–4 can be readily added, since those only concern the time-dependence of the generators.

All representations of the Schrödinger algebra discussed so far have the dynamical exponent $z = 2$, which fixes the dilatations $t \mapsto \lambda^z t$ and $r \mapsto \lambda r$. This can be changed, however, by admitting "non-local" representations. We shall write them here, for the case $z = \nu \in \mathbb{N}$, in the form given for the sub-algebra $\mathfrak{age}(1)$, when the generators read [63]

$$\begin{aligned} X_0 &= -\frac{\nu}{2}t\partial_t - \frac{1}{2}r\partial_r - \frac{x}{2} \\ X_1 &= \left(-\frac{\nu}{2}t^2\partial_t - tr\partial_r - (x+\xi)t\right)\partial_r^{\nu-2} - \frac{\mathcal{M}}{2}r^2 \\ Y_{-1/2} &= -\partial_r, \quad Y_{1/2} = -t\partial_r^{\nu-1} - \mathcal{M}r, \quad M_0 = -\mathcal{M} \end{aligned} \tag{41}$$

and reduce to Equation (5) for $z = \nu = 2$. Clearly, these generators (especially $X_1, Y_{1/2}$) cannot be interpreted as infinitesimal transformations on time-space coordinates (t,r) and cannot be seen as mimicking a finite transformation, as was still possible with the "lattice" representation given in Table 1. In [63], a possible interpretation as transformation of distribution functions of (t,r) was explored, but the issue is not definitely settled.

Proposition 5. [63] *For any $\nu \in \mathbb{N}$, the generators (41) of the algebra $\mathfrak{age}(d)$ satisfy the commutators (4) in $d = 1$ spatial dimensions, but with the only exception*

$$[X_1, Y_{1/2}] = \frac{\nu-2}{2}t^2\partial_r^{\nu-3}\mathcal{S} \tag{42}$$

where the Schrödinger operator $\mathcal{S}$ is given by

$$\mathcal{S} = \nu\mathcal{M}\partial_t - \partial_r^\nu + 2\mathcal{M}\left(x + \xi + \frac{\nu-1}{2}\right)t^{-1} \tag{43}$$

These indeed generate a dynamical symmetry on the space of solutions of the equation $\mathcal{S}\phi = 0$, since the only non-vanishing commutators of $\mathcal{S}$ with the generators (41) are

$$[\mathcal{S}, X_0] = -\frac{\nu}{2}\mathcal{S}, \quad [\mathcal{S}, X_1] = -\nu t\partial_r^{n-2}\mathcal{S} \tag{44}$$

Verifying the required commutators is straightforward (but there is no known extension to a representation of $\mathfrak{av}(1)$). It is possible to generalise this construction to dimensions $d > 1$ and to generic dynamical exponents $z \in \mathbb{R}_+$, but this would require the introduction of fractional derivatives into the generators [39,64]. Formally, one can also derive the form of co-variant two-point functions $F(t_1, t_2; r_1, r_2) = \langle \phi_1(t_1, r_1)\phi_2^*(t_2, r_2)\rangle$.

Proposition 6. [63] *For $\nu \in \mathbb{N}$, a two-point function F, covariant under the non-local representation (41) of the Lie algebra $\mathfrak{age}(1)$, defined on the solution space of $\mathcal{S}\phi = 0$, where $\mathcal{S}$ is the Schrödinger operator (43), has the form $F = \delta(\mathcal{M}_1 - \mathcal{M}_2^*)\, t_2^{-(x_1+x_2)/\nu} F(u, v, r)$, where*

$$\begin{aligned} F(u,v,r) &= (v-1)^{-\frac{2}{\nu}[(x_1+x_2)/2+\xi_1+\xi_2-\nu+2]} v^{-\frac{1}{\nu}[x_2-x_1+2\xi_2-\nu+2]} f\left(ru^{-1/\nu}\right), \quad \nu \text{ even} \\ F(u,v,r) &= (v+1)^{-\frac{2}{\nu}[(x_1+x_2)/2+\xi_1+\xi_2-\nu+2]} v^{-\frac{1}{\nu}[x_2-x_1+2\xi_2-\nu+2]} f\left(ru^{-1/\nu}\right), \quad \nu \text{ odd} \end{aligned} \tag{45}$$

the function $f(y)$ satisfies the equation $\mathrm{d}^{\nu-1} f(y)/\mathrm{d}y^{\nu-1} + \mathcal{M}_1 y f(y) = 0$, and with the variables $r = r_1 - r_2$, $v = t_1/t_2$ and

$$u = t_1 - t_2 \quad \text{if } \nu \text{ is even} \ , \quad u = t_1 + t_2 \quad \text{if } \nu \text{ is odd} \tag{46}$$

The set of admissible functions $f(y)$ will have to be restricted by imposing physically reasonable boundary conditions, especially $\lim_{y\to\infty} f(y) = 0$. The value $z = \nu$ of the dynamical exponent is obvious.

Again, one should inquire into the behaviour when $r \to \infty$. Furthermore, one observes that the interpretation of u depends on whether ν is even or odd. In the first case, the co-variant two-point functions could be a physical two-time response function, while in the second case, it looks more like a two-time correlator, since it is symmetric symmetry under the exchange of the two scaling operators.

All representations considered here are scalar. It is possible to consider multiplets of scaling operators. In the case of conformal invariance, one should formally replace the conformal weight Δ by a matrix [65–69]. New structures are only found if that matrix takes a Jordan form. Analogous representations can also be considered for the Schrödinger and conformal Galilean algebras and their sub-algebras. Then, it becomes necessary to consider simultaneously the scaling dimensions x, ξ and the rapidities γ as matrices [36,70–73]. From the Lie algebra commutators it can then be shown that these characteristic elements of the scaling operators are simultaneously Jordan [36]. Several applications to non-equilibrium relaxation phenomena have been explored in the literature [57,58,74,75], see [59] for a review.

3. Dual Representations

In order to understand how the causality and the large-distance behaviour of the co-variant two-point functions can be derived algebraically, it is helpful to go over to a dual description. The new dual coordinate ζ is related to either the scalar mass $\mathcal{M}$ for the Schrödinger algebra (this was first noted by Giulini [76] for the case of its Galilei-subalgebra) or else to the vector of rapidities γ for the conformal Galilei algebra. It will therefore be scalar or vector, respectively. The dual fields are [11,77]

$$\begin{aligned} \hat{\phi}(\zeta, t, \boldsymbol{r}) &:= \frac{1}{\sqrt{2\pi}} \int_{\mathbb{R}} \mathrm{d}\mathcal{M}\, e^{\mathrm{i}\mathcal{M}\zeta} \phi_{\mathcal{M}}(t, \boldsymbol{r}), \quad \text{for } \mathfrak{sch}(d), \mathfrak{age}(d) \\ \hat{\phi}(\zeta, t, \boldsymbol{r}) &:= \frac{1}{(2\pi)^{d/2}} \int_{\mathbb{R}^d} \mathrm{d}\gamma\, e^{\mathrm{i}\gamma\cdot\zeta} \phi_{\gamma}(t, \boldsymbol{r}), \quad \text{for CGA}(d) \end{aligned} \tag{47}$$

For the sake of notational simplicity, we shall almost always restrict to the one-dimensional case, although we shall quote some final results for a generic dimension d.

3.1. *Schrödinger Algebra*

From Proposition 3, the dual generators of the Schrödinger–Virasoro algebra take the form (with $j,k = 1,\ldots,d$)

$$\begin{aligned} X_n &= \frac{\mathrm{i}}{4}(n+1)n\,t^{n-1}r^2\partial_\zeta - t^{n+1}\partial_t - \frac{n+1}{2}t^n r\partial_r - \frac{n+1}{2}xt^n - n(n+1)\xi t^n - \Xi(t)t^n \\ Y_m &= \mathrm{i}\left(m+\frac{1}{2}\right)t^{m-1/2}r\partial_\zeta - t^{m+1/2}\partial_r \\ M_n &= \mathrm{i}t^n\partial_\zeta \end{aligned} \tag{48}$$

with $n \in \mathbb{Z}$ and $m \in \mathbb{Z}+\frac{1}{2}$. This acts on a $(d+2)$-dimensional space, with coordinates $\zeta, t, \boldsymbol{r}$. According to Proposition 3, not only the finite-dimensional sub-algebra $\mathfrak{age}(1)$, but also the finite-dimensional sub-algebra $\mathfrak{sch}(1)$ [40] generates dual dynamical symmetries of the Schrödinger operator

$$\mathcal{S} = -2\mathrm{i}\partial_\zeta\partial_t - \partial_r^2 - 2\mathrm{i}\left(x+\xi-\frac{1}{2}\right)t^{-1}\partial_\zeta \tag{49}$$

Co-variant dual three-point functions have been derived explicitly [40].

In the context of the non-relativistic AdS/CFT correspondence, also referred to as non-relativistic holography by string theorists, see [74,78] and references therein, one rather considers a $(d+3)$-dimensional space, with coordinates $Z, \zeta, t, \boldsymbol{r}$. The time-space transforming parts of the Schrödinger–Virasoro generators read (generalising Son [79], who restricted himself to the finite-dimensional sub-algebra $\mathfrak{sch}(d)$)

$$\begin{aligned} X_n &= \frac{\mathrm{i}}{4}(n+1)n\,t^{n-1}\left(\boldsymbol{r}^2+Z^2\right) - t^{n+1}\partial_t - \frac{n+1}{2}t^n\left(\boldsymbol{r}\cdot\boldsymbol{\nabla}_{\boldsymbol{r}} + Z\partial_Z\right) \\ Y_m^{(j)} &= \mathrm{i}\left(m+\frac{1}{2}\right)t^{m-1/2}r_j\partial_\zeta - t^{m+1/2}\partial_{r_j} \\ M_n &= \mathrm{i}t^n\partial_\zeta \\ R_n^{(jk)} &= -t^n\left(r_j\partial_{r_k} - r_k\partial_{r_j}\right) \end{aligned} \tag{50}$$

Clearly, the variable Z distinguishes the bulk from the boundary at $Z=0$. Heuristically, if one replaces $Z\partial_Z \mapsto x$ and then sets $Z=0$, one goes back from Equation (50) to Equation (48), with $\xi = 0$ and $\Xi(t) = 0$.

Following Aizawa and Dobrev [78,80], the passage between the boundary and the bulk is described in terms of the eigenvalues of the quartic Casimir operator of the Schrödinger algebra $\mathfrak{sch}(1)$ [81]

$$C_4 = (4M_0X_0 - Y_{-1/2}Y_{1/2} - Y_{1/2}Y_{-1/2})^2 - 2\left\{2M_0X_{-1} - Y_{-1/2}^2, 2M_0X_1 - Y_{1/2}^2\right\} \tag{51}$$

such that in the representation Equation (5), which lives on the boundary $Z=0$, one has the eigenvalue $c_4 = c_4(x) := \mathcal{M}^2(2x-1)(2x-5)$. Since $c_4(x) = c_4(3-x)$, two scaling operators with scaling dimensions x and $3-x$ will be related. In order to formulate the holographic principle, which prescribes the mapping of a boundary scaling operators φ to a bulk scaling operator ϕ, a necessary condition is the eigenvalue equation (in the bulk)

$$C_4\phi(Z,\zeta,t,r) = c_4(x)\phi(Z,\zeta,t,r) \tag{52}$$

The other condition is the expected limiting behaviour when the boundary is approached

$$\phi(Z,\zeta,t,r) \stackrel{Z\to 0}{\longrightarrow} Z^\alpha\varphi(\zeta,t,r), \quad \alpha = x, 3-x \tag{53}$$

Lemma 2. [80] *For the Schrödinger algebra in $d = 1$ space dimension, the holographic principle takes the form*

$$\phi(Z,\chi) = \int \mathrm{d}^3\chi'\, S_\alpha(Z,\chi-\chi')\varphi(\chi') \tag{54}$$

where $\chi = (\zeta, t, r)$ is a label for a three-dimensional coordinate, $\mathrm{d}^3\chi = \mathrm{d}\zeta \mathrm{d}t \mathrm{d}r$ and

$$S_\alpha(Z,\chi) = \left[\frac{4Z}{-2\zeta t + r^2}\right]^\alpha \tag{55}$$

and where $\alpha = x$ or $\alpha = 3 - x$.

Proof. We merely outline the main ideas. First, construct the Green's function in the bulk, by solving

$$\left(C_4 - c_4(x)\right) G(Z,\chi;Z',\chi') = Z'^4\, \delta(Z-Z')\delta^3(\chi-\chi')$$

In terms of the invariant variable

$$u := \frac{4ZZ'}{(Z+Z')^2 - 2(\zeta-\zeta')(t-t') + (r-r')^2}$$

the Casimir operator becomes $C_4 = 4u^2(1-u)\partial_u^2 - 8u\partial_u + 5$, hence $G = G(u)$. Next, the ansatz $G(u) = u^\alpha \bar{G}(u)$ reduces the eigenvalue equation to a standard hyper-geometric equation, with solutions expressed in terms of the hyper-geometric function ${}_2F_1$. Finally, $S_\alpha(Z,\chi-\chi') = \lim_{Z'\to 0} Z'^{-\alpha} G(u)$ leads to the assertion. □

We refer to the literature for the non-relativistic reduction and the derivation of invariant differential equations [78,80]. The consequences of passing to the more general representations with $\xi \neq 0$ and $\Xi(t) \neq 0$ [40] remain to be studied.

3.2. Conformal Galilean Algebra I

Starting from Equation (48), a dual representation of the conformal Galilean algebra $\mathrm{CGA}(1)$ with $z = 2$ is found if (i) the generator X_{-1} is dropped, (ii) the generator X_1 is taken as follows and (iii) and adds a new generator V_+ [11]

$$\begin{aligned} X_1 &= \mathrm{i}r^2\partial_\zeta - t^2\partial_t - tr\partial_r - (x+\xi)\,t \\ V_+ &= -\zeta r\partial_\zeta - tr\partial_t - \left(\mathrm{i}\zeta t + \frac{r^2}{2}\right)\partial_r - (x+\xi)\,r \end{aligned} \tag{56}$$

They are dynamical symmetries of the dual Schrödinger operator Equation (49).

3.3. Conformal Galilean Algebra II

Another dual representation of the algebra $\mathrm{CGA}(d)$ is given by (with $j,k = 1,\ldots,d$)

$$\begin{aligned} X_n &= +\mathrm{i}(n+1)nt^{n-1}\boldsymbol{r}\cdot\partial_{\boldsymbol{\zeta}} - t^{n+1}\partial_t - (n+1)t^n\boldsymbol{r}\cdot\partial_{\boldsymbol{r}} - (n+1)xt^n \\ Y_n^{(j)} &= -t^{n+1}\partial_{r_j} + \mathrm{i}(n+1)t^n\partial_{\zeta_j} \\ R_n^{(jk)} &= -t^n\left(r_j\partial_{r_k} - r_k\partial_{r_j}\right) - t^n\left(\zeta_j\partial_{\zeta_k} - \zeta_k\partial_{\zeta_j}\right) \end{aligned} \tag{57}$$

In contrast with the representations studied so far, there are no central generators $\sim \partial_{\zeta_j}$.

The dualisation of the "lattice" representation and the non-local representations discussed in Section 2 proceeds analogously and will not be spelt out in detail here.

3.4. Parabolic Sub-Algebras

The other important ingredient is understood by considering the root diagrams of these non-semi-simple Lie algebras, see Figure 1. Therein, it is in particular illustrated that the complexified versions of these algebras are all sub-algebras of the complex Lie algebra B_2, in Cartan's notation [11]. In particular, it is possible to add further generators in the Cartan sub-algebra in order to obtain an extension to a *maximal parabolic sub-algebra*. A parabolic sub-algebra is the sub-algebra of "positive" generators, which from a root diagramme can be identified by simply placing a straight line through the center (a.k.a. the Cartan sub-algebra). By definition, all generators which are not on the left of that line are called "positive" [41]. In Figure 1, we illustrate this for the three maximal parabolic sub-algebras. The notion of "maxima" does depend here on the precise definition of "positivity". For a generic slope, see Figure 1a, both the generators X_{-1} and V_+ are non-positive, and one has the maximal parabolic sub-algebra $\widetilde{\mathfrak{age}}(1) = \mathfrak{age}(1) + \mathbb{C}N$. This sub-algebra is indeed maximal as a *parabolic* sub-algebra: for example an extension to a Schrödinger algebra by including the time-translations X_{-1} would no longer be parabolic, according to the specific definition of "positivity" used in this specific context. If a different definition of "positivity" is used, and the slope is now taken to be exactly unity, X_{-1} is included into the positive generators, see Figure 1b, and we have the maximal parabolic sub-algebra $\widetilde{\mathfrak{sch}}(1) = \mathfrak{sch}(1) + \mathbb{C}N$. Finally, and with yet a different definition of "positivity", where the slope is now infinite, see Figure 1c, one has the maximal parabolic sub-algebra $\widetilde{\mathrm{CGA}}(1) = \mathrm{CGA}(1) + \mathbb{C}N$. The Weyl symmetries of the root diagramme of B_2 [41] imply that any other maximal and non-trivial sub-algebra of B_2 is isomorphic to one of the three already given. For a formal proof, see [11].

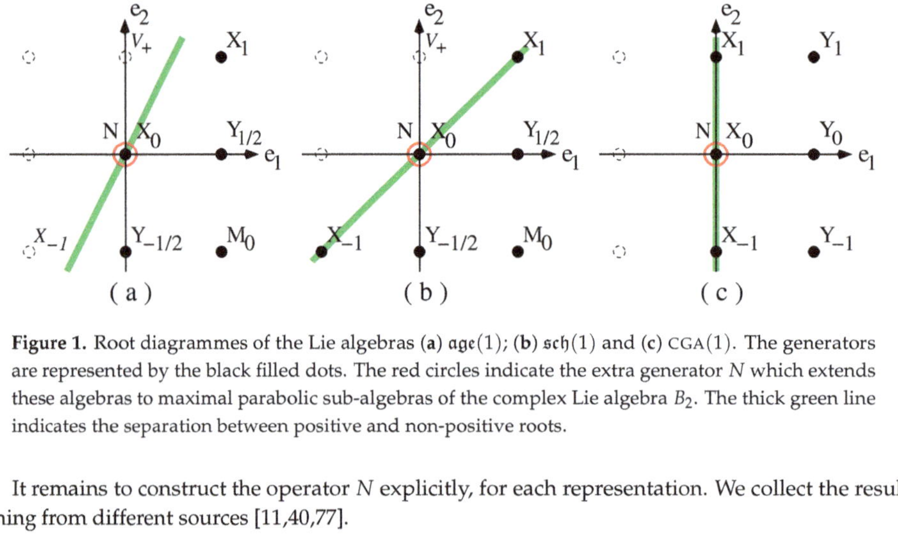

Figure 1. Root diagrammes of the Lie algebras (**a**) $\mathfrak{age}(1)$; (**b**) $\mathfrak{sch}(1)$ and (**c**) $\mathrm{CGA}(1)$. The generators are represented by the black filled dots. The red circles indicate the extra generator N which extends these algebras to maximal parabolic sub-algebras of the complex Lie algebra B_2. The thick green line indicates the separation between positive and non-positive roots.

It remains to construct the operator N explicitly, for each representation. We collect the results, coming from different sources [11,40,77].

Proposition 7. *Consider the dual representations (48) of the Schrödinger–Virasoro algebra, the $z = 2$ dual representation (56) of the conformal Galilean algebra* $\mathrm{CGA}(1)$, *the $z = 1$ dual representation (57) of* $\mathrm{CGA}(d)$ *and the dualisation of the non-local representation (41) of* $\mathfrak{age}(1)$, *dualised with respect to the mass $\mathcal{M}$. There is a generator N which extends these representations to representations of the associated maximal parabolic sub-algebra. The explicit form of the generator N is as follows:*

$$N = \begin{cases} \zeta\partial_\zeta - t\partial_t + \xi' & \text{representation (48) of } \mathfrak{sch}(d) \\ \zeta\partial_\zeta - t\partial_t + \xi & \text{representation (56) of } \mathrm{CGA}(1) \\ -\zeta\partial_\zeta - r\partial_r - \xi & \text{representation of } \mathrm{CGA}(d) \text{ as constructed in (57)} \\ \zeta\partial_\zeta - t\partial_t + \xi' & \text{dualised non-local representation (41) of } \mathfrak{age}(d) \end{cases} \tag{58}$$

Herein, ξ is the second scaling dimension and ξ' is a constant. These generators give dynamical symmetries of the Schrödinger operators $\mathcal{S}$ associated with each representation.

4. Causality

It turns out that the maximal parabolic sub-algebras are the smallest Lie algebras which permit unambiguous statements on the causality of co-variant two-point functions. For illustration, we shall concentrate on the dual representations (48) of $\mathfrak{sch}(d)$ and (57) of $\mathrm{CGA}(d)$.

Proposition 8. [11,77] *Consider the co-variant dual two-point functions. For the dual representation (48) of $\widetilde{\mathfrak{sch}}(d)$, it has the form, up to a normalisation constant*

$$\widehat{F}(\zeta,t,\boldsymbol{r}) = \langle \widehat{\phi}(\zeta,t,\boldsymbol{r})\widehat{\phi}^*(0,0,\boldsymbol{0})\rangle = \delta_{x_1,x_2}\,|t|^{-x_1}\left(\frac{2\zeta t + \mathrm{i}\boldsymbol{r}^2}{|t|}\right)^{-x_1-\xi_1'-\xi_2'} \tag{59}$$

and where translation-invariance in $\zeta, t, \boldsymbol{r}$ was used. For the dual representation (57) of $\widetilde{\mathrm{CGA}}(1)$, one has, up to a normalisation constant

$$\widehat{F}(\zeta_+,t,r) = \langle \widehat{\phi}_1(\zeta_1,t,r)\widehat{\phi}_2(\zeta_2,0,0)\rangle = \delta_{x_1,x_2}|t|^{-2x_1}\left(\zeta_+ + \frac{\mathrm{i}r}{t}\right)^{-\xi_1-\xi_2} \tag{60}$$

and where $\zeta_+ = \frac{1}{2}(\zeta_1+\zeta_2)$.

This is easily verified by insertion into the respective Ward identities which express the co-variance. Finally, we formulate precisely the spatial long-distance and co-variance properties of these two-point functions.

Theorem 2. [11] *With the convention that masses $\mathcal{M} \geq 0$ of scaling operators ϕ should be non-negative, and if $\frac{1}{2}(x_1+x_2)+\xi_1'+\xi_2' > 0$, the full two-point function, co-variant under the representation (5) of the parabolically extended Schrödinger algebra $\widetilde{\mathfrak{sch}}(d)$, has the form*

$$\langle \phi(t,\boldsymbol{r})\phi^*(0,\boldsymbol{0})\rangle = \delta(\mathcal{M}-\mathcal{M}^*)\,\delta_{x_1,x_2}\,\Theta(t)\,t^{-x_1}\,\exp\left[-\frac{\mathcal{M}}{2}\frac{\boldsymbol{r}^2}{t}\right] \tag{61}$$

where the Θ-function expresses the causality condition $t>0$, and up to a normalisation constant which depends only the mass $\mathcal{M} \geq 0$.

Proof. This follows directly from Equation (59). Carrying out the inverse Fourier transform and using the translation-invariance in the dual coordinate ζ, one recovers the habitual two-point function multiplied by an integral representation of the Θ-function. □

The treatment of the conformal Galilean algebra requires some further preparations, following Akhiezer [82] (Chapter 11).

Definition 1. *Let $\mathbb{H}_+$ be the upper complex half-plane $w = u + \mathrm{i}v$ with $v > 0$. A function $g : \mathbb{H}_+ \to \mathbb{C}$ is said to be in the* Hardy class H_2^+, *written as $g \in H_2^+$, if (i) $g(w)$ is holomorphic in $\mathbb{H}_+$ and (ii) if it satisfies the bound*

$$M^2 := \sup_{v>0}\int_{\mathbb{R}}\mathrm{d}u\;|g(u+\mathrm{i}v)|^2 < \infty \tag{62}$$

Analogously, for functions $g : \mathbb{H}_- \to \mathbb{C}$ one defines the Hardy class H_2^-, where $\mathbb{H}_-$ is the lower complex half-plane and the supremum in Equation (62) is taken over $v < 0$.

Lemma 3. [82] *If $g \in H_2^{\pm}$, then there are square-integrable functions $\mathcal{G}_{\pm} \in L^2(0,\infty)$ such that for $v > 0$ one has the integral representation*

$$g(w) = g(u \pm \mathrm{i}v) = \frac{1}{\sqrt{2\pi}} \int_0^{\infty} \mathrm{d}\gamma\, e^{\pm \mathrm{i}\gamma w}\, \mathcal{G}_{\pm}(\gamma) \tag{63}$$

We shall use Equation (63) as follows. First, consider the case $d = 1$. Fix $\lambda := r/t$. Now, recall Equation (60) and write $\widehat{F} = |t|^{-2x_1} \widehat{f}(u)$, with $u = \zeta_+ + \mathrm{i}r/t$. We shall re-write this as follows:

$$\widehat{f}(\zeta_+ + \mathrm{i}\lambda) =: f_\lambda(\zeta_+) \tag{64}$$

and concentrate on the dependence on ζ_+.

Proposition 9. [77] *Let $\xi := \frac{1}{2}(\xi_1 + \xi_2) > \frac{1}{4}$. If $\lambda > 0$, then $f_\lambda \in H_2^+$ and if $\lambda < 0$, then $f_\lambda \in H_2^-$.*

Proof. The holomorphy of f_λ being obvious, we merely must verify the bound (62). Let $\lambda > 0$. Clearly, $|f_\lambda(u + \mathrm{i}v)| = \left|(u + \mathrm{i}(v + \lambda))^{-2\xi}\right| = \left(u^2 + (v + \lambda)^2\right)^{-\xi}$. Hence, computing explicitly the integral,

$$M^2 = \sup_{v>0} \int_{\mathbb{R}} du\, |f_\lambda(u + \mathrm{i}v)|^2 = \frac{\sqrt{\pi}\,\Gamma(2\xi - \frac{1}{2})}{\Gamma(2\xi)} \sup_{v>0} (v + \lambda)^{1-4\xi} < \infty$$

since the integral converges for $\xi > \frac{1}{4}$. For $\lambda < 0$, the argument is similar. □

We can now formulate the second main result.

Theorem 3. [77] *The full two-point function, co-variant under the representation (13) of the parabolically extended conformal Galilean algebra $\widetilde{\mathrm{CGA}}(d)$, has the form*

$$\langle \phi_1(t, \boldsymbol{r}) \phi_2(0, \boldsymbol{0}) \rangle = \delta_{x_1, x_2} \delta(\gamma_1 - \gamma_2)\, |t|^{-2x_1} \exp\left[-2\left|\frac{\gamma_1 \cdot \boldsymbol{r}}{t}\right|\right] \tag{65}$$

where the normalisation constant only depends on the absolute value of the rapidity vector γ_1.

Proof. Since the final result is rotation-invariant, because of the representation (13), it is enough to consider the case $d = 1$. Let $\lambda > 0$. From Equation (63) of Lemma 3 we have

$$\sqrt{2\pi}\widehat{f}(\zeta_+ + \mathrm{i}\lambda) = \int_0^{\infty} \mathrm{d}\gamma_+\, e^{\mathrm{i}(\zeta_+ + \mathrm{i}\lambda)\gamma_+} \widehat{\mathcal{F}_+}(\gamma_+) = \int_{\mathbb{R}} \mathrm{d}\gamma_+\, \Theta(\gamma_+)\, e^{\mathrm{i}(\zeta_+ + \mathrm{i}\lambda)\gamma_+} \widehat{\mathcal{F}_+}(\gamma_+)$$

Now return from the dual two-point function $\widehat{F}$ to the original one. Let $\zeta_\pm := \frac{1}{2}(\zeta_1 \pm \zeta_2)$. We find, using also that $x_1 = x_2$

$$\begin{aligned}
F &= \frac{|t|^{-2x_1}}{\pi\sqrt{2\pi}} \int_{\mathbb{R}^2} \mathrm{d}\zeta_+ \mathrm{d}\zeta_-\, e^{-\mathrm{i}(\gamma_1 + \gamma_2)\zeta_+} e^{-\mathrm{i}(\gamma_1 - \gamma_2)\zeta_-} \int_{\mathbb{R}} \mathrm{d}\gamma_+\, \Theta(\gamma_+) \widehat{\mathcal{F}_+}(\gamma_+) e^{-\gamma_+\lambda} e^{\mathrm{i}\gamma_+\zeta_+} \\
&= \frac{|t|^{-2x_1}}{\pi\sqrt{2\pi}} \int_{\mathbb{R}} \mathrm{d}\gamma_+\, \Theta(\gamma_+) \widehat{\mathcal{F}_+}(\gamma_+) e^{-\gamma_+\lambda} \int_{\mathbb{R}} \mathrm{d}\zeta_-\, e^{-\mathrm{i}(\gamma_1 - \gamma_2)\zeta_-} \int_{\mathbb{R}} \mathrm{d}\zeta_+\, e^{\mathrm{i}(\gamma_+ - \gamma_1 - \gamma_2)\zeta_+} \\
&= \delta(\gamma_1 - \gamma_2) \Theta(\gamma_1) F_{0,+}(\gamma_1) e^{-2\gamma_1\lambda} |t|^{-2x_1}
\end{aligned}$$

where in the last line, two δ-functions were used and $F_{0,+}$ contains the unspecified dependence on the positive constant γ_1. An analogous argument applies for $\lambda < 0$. □

5. Conclusions

Results on relaxation phenomena in non-equilibrium statistical physics and the associated dynamical symmetries, scattered over many sources in the literature, have been reviewed. By analogy with conformal invariance which applies to equilibrium critical phenomena, it is tempting to try to extend the generically satisfied dynamical scaling to a larger set of dynamical symmetries. If this is possible, one should obtain a set of co-variance conditions, to be satisfied by physically relevant n-point functions. In contrast to equilibrium critical phenomena, it turned out that in non-equilibrium systems, each scaling operator must be characterised at least in terms of two independent scaling dimensions.

A straightforward realisation of this programme was seen to lead to difficulties for a consistent physical interpretation, related to the requirement of a physically sensible large-distance behaviour. This is related to the fact that writing down simple Ward identities for the n-point functions, one implicitly assumes that these n-point functions depend holomorphically on their time-space arguments, see e.g., [83]. However, the constraint of causality, required for a reasonable two-time response function $R(t_1, t_2)$, renders $R(t_1, t_2)$ non-holomorphic in the time difference $t_1 - t_2$. As a possible solution of this difficulty, we propose to go over to dual representations with respect to either the "masses" or the "rapidities", which are physically dimensionful parameters of the representations of the dynamical symmetry algebras considered. If, furthermore, the dynamical symmetry algebras can be extended to a maximal parabolic sub-algebra of a semi-simple complex Lie algebra, then causality conditions, which also guarantee the requested fall-off at large distances, can be derived.

This suggests that the dual scaling operators, rather than the original ones, might possess interesting holomorphic properties which should be further explored. This observation might also become of interest in further studies of the holographic principle.

Specifically, we considered representations of (i) the Schrödinger algebra $\mathfrak{sch}(d)$, where the co-variant two-point functions Equation (61) have the causality properties of two-time linear response functions and also representations of (ii) the conformal Galilean algebra $\text{CGA}(d)$, where the two-point functions Equation (65) have the symmetry properties of a two-time correlator [84].

Although this has not yet been done explicitly, we expect that the techniques reviewed here can be readily extended to several physically distinct representations of these algebras, see e.g., [85–88] for examples [89].

Acknowledgments: The author warmly thanks R. Cherniha, X. Durang, S. Stoimenov and J. Unterberger for discussions which helped him to slowly arrive at the synthesis presented here. This work was partly supported by the Collège Doctoral Nancy–Leipzig–Coventry (Systèmes complexes à l'équilibre et hors équilibre) of UFA-DFH.

Conflicts of Interest: The author declares no conflict of interest.

References

1. Henkel, M.; Pleimling, M. *Non-Equilibrium Phase Transitions Volume 2: Ageing and Dynamical Scaling Far from Equilibrium*; Springer: Heidelberg, Germany, 2010.
2. Täuber, U.C. *Critical Dynamics: A Field Theory Apporach to Equilibrium and Non-Equilibrium Scaling Behaviour*; Cambridge University Press: Cambridge, UK, 2014.
3. Cartan, É. Les groupes de transformation continus, infinis, simples. *Ann. Sci. Ecole Norm. S.* **1909**, *26*, 93–161.
4. Belavin, A.A.; Polyakov, A.M.; Zamolodchikov, A.B. Infinite conformal symmetry in two-dimensional quantum field-theory. *Nucl. Phys. B* **1984**, *241*, 333–380.
5. Di Francesco, P.; Mathieu, P.; Sénéchal, D. *Conformal Field-Theory*; Springer: Heidelberg, Germany, 1997.
6. Unterberger, J.; Roger, C. *The Schrödinger–Virasoro Algebra*; Springer: Heidelberg, Germany, 2011.
7. Henkel, M. Schrödinger-invariance and strongly anisotropic critical systems. *J. Stat. Phys.* **1994**, *75*, 1023–1061.
8. Henkel, M. Phenomenology of local scale invariance: From conformal invariance to dynamical scaling. *Nucl. Phys. B* **2002**, *641*, 405–486.
9. Roger, C.; Unterberger, J. The Schrödinger–Virasoro Lie group and algebra: From geometry to representation theory. *Ann. Henri Poincare* **2006**, *7*, 1477–1529.
10. To see this explicitly, one should exponentiate these generators to create their corresponding finite transformations, see [11].

11. Henkel, M.; Unterberger, J. Schrödinger invariance and space-time symmetries. *Nucl. Phys. B* **2003**, *660*, 407–435.
12. Lie, S. Über die Integration durch bestimmte Integrale von einer Klasse linearer partieller Differentialgleichungen. *Arch. Math. Nat.* **1881**, *6*, 328–368.
13. Jacobi, C.G. Vorlesungen über Dynamik (1842/43), 4. Vorlesung. In *Gesammelte Werke*; Clebsch, A., Lottner, E., Eds; Akademie der Wissenschaften: Berlin, Germany, 1866.
14. Niederer, U. The maximal kinematical invariance group of the free Schrödinger equation. *Helv. Phys. Acta* **1972**, *45*, 802–810.
15. Unitarity of the representation implies the bound $x \geq \frac{d}{2}$ [16].
16. Lee, K.M.; Lee, Sa.; Lee, Su. Nonrelativistic Superconformal M2-Brane Theory. *J. High Energy Phys.* **2009**, doi:10.1088/1126-6708/2009/09/030.
17. Bargman, V. On unitary ray representations of continuous groups. *Ann. Math.* **1954**, *56*, 1–46.
18. Henkel, M. Extended scale-invariance in strongly anisotropic equilibrium critical systems. *Phys. Rev. Lett.* **1997**, *78*, 1940–1943.
19. Ovsienko, V.; Roger, C. Generalisations of Virasoro group and Virasoro algebras through extensions by modules of tensor-densities on S^1. *Indag. Math.* **1998**, *9*, 277–288.
20. The name was originally given since at that time, relationships with physical ageing (*altern* in German) were still expected.
21. Cherniha, R.; Henkel, M. The exotic conformal Galilei algebra and non-linear partial differential equations. *J. Math. Anal. Appl.* **2010**, *369*, 120–132.
22. Bagchi, A.; Mandal, I. On representations and correlation functions of Galilean conformal algebra. *Phys. Lett. B* **2009**, *675*, 393–397.
23. Havas, P.; Plebanski, J. Conformal extensions of the Galilei group and their relation to the Schrödinger group. *J. Math. Phys.* **1978**, *19*, 482–488.
24. Martelli, D.; Tachikawa, Y. Comments on Galiean conformal field-theories and their geometric realisation. *J. High Energy Phys.* **2010**, 1005:091, doi:10.1007/JHEP05(2010)091.
25. Negro, J.; del Olmo, M.A.; Rodríguez-Marco, A. Nonrelativistic conformal groups. *J. Math. Phys.* **1997**, *38*, 3786–3809.
26. Negro, J.; del Olmo, M.A.; Rodríguez-Marco, A. Nonrelativistic conformal groups II. *J. Math. Phys.* **1997**, *38*, 3810–3831.
27. In the context of asymptotically flat 3D gravity, an isomorphic Lie algebra is known as BMS algebra, $\mathfrak{bms}_3 \cong \mathrm{CGA}(1)$ [28–32].
28. Bagchi, A.; Detournay, S.; Grumiller, D. Flat-space chiral gravity. *Phys. Rev. Lett.* **2012**, *109*, doi:10.1103/PhysRevLett.109.151301.
29. Bagchi, A.; Detournay, S.; Fareghbal, R.; Simón, J. Holographies of 3D flat cosmological horizons. *Phys. Rev. Lett.* **2013**, *110*, 141302:1–141302:5.
30. Barnich, G.; Compère, G. Classical central extension for asymptotic symmetries at null infinity in three spacetime dimensions. *Class. Quantum Grav.* **2007**, *24*, doi:10.1088/0264-9381/24/5/F01.
31. Barnich, G.; Compère, G. Classical central extension for asymptotic symmetries at null infinity in three spacetime dimensions (corrigendum). *Class. Quant. Grav.* **2007**, *24*, 3139.
32. Barnich, G.; Gomberoff, A.; González, H.A. Three-dimensional Bondi-Metzner-Sachs invariant two-dimensional field-theories as the flat limit of Liouville theory. *Phys. Rev. D* **2007**, *87*, doi:10.1103/PhysRevD.87.124032.
33. Henkel, M.; Schott, R.; Stoimenov, S.; Unterberger, J. The Poincaré algebra in the context of ageing systems: Lie structure, representations, Appell systems and coherent states. *Conflu. Math.* **2012**, *4*, 1250006:1–1250006:19.
34. Lukierski, J.; Stichel, P.C.; Zakrewski, W.J. Exotic galilean conformal symmetry and its dynamical realisations. *Phys. Lett. A* **2006**, *357*, 1–5.
35. Lukierski, J.; Stichel, P.C.; Zakrewski, W.J. Acceleration-extended galilean symmetries with central charges and their dynamical realizations. *Phys. Lett. B* **2007**, *650*, 203–207.
36. Henkel, M.; Hosseiny, A.; Rouhani, S. Logarithmic exotic conformal galilean algebras. *Nucl. Phys. B* **2014**, *879*, 292–317.
37. An infinite-dimensional extension of ECGA does not appear to be possible.

38. Henkel, M.; Enss, T.; Pleimling, M. On the identification of quasiprimary operators in local scale-invariance. *J. Phys. A Math. Gen.* **2006**, *39*, L589–L598.
39. Stoimenov, S.; Henkel, M. Non-local representations of the ageing algebra in higher dimensions. *J. Phys. A Math. Theor.* **2013**, *46*, doi:10.1088/1751-8113/46/24/245004.
40. Minic, D.; Vaman, D.; Wu, C. Three-point function of aging dynamics and the AdS-CFT correspondence. *Phys. Rev. Lett.* **2012**, *109*, doi:10.1103/PhysRevLett.109.131601.
41. Knapp, A.W. *Representation Theory of Semisimple Groups: An Overview Based on Examples*; Princeton University Press: Princeton, NJ, USA, 1986.
42. Duval, C.; Horváthy, P.A. Non-relativistic conformal symmetries and Newton–Cartan structures. *J. Phys. A Math. Theor.* **2009**, *42*, doi:10.1088/1751-8113/42/46/465206.
43. Although it might appear that $z = 2$, the renormalisation of the interactions, required in interacting field-theories, can change this and produce non-trivial values of z, see e.g., [2].
44. Picone, A.; Henkel, M. Local scale-invariance and ageing in noisy systems. *Nucl. Phys. B* **2004**, *688*, 217–265.
45. Polyakov, A.M. Conformal symmetry of critical fluctuations. *Sov. Phys. JETP Lett.* **1970**, *12*, 381–383.
46. Barabási, A.-L.; Stanley, H.E. *Fractal Concepts in Surface Growth*; Cambridge University Press: Cambridge, UK, 1995.
47. Henkel, M.; Durang, X. Spherical model of interface growth. *J. Stat. Mech.* **2015**, doi:10.1088/1742-5468/2015/05/P05022.
48. For $d = 1$, the dynamics of the Arcetri model is identical [47] to the one of the spherical Sherrington–Kirkpatrick model. The model is defined by the classical hamiltonian $\mathcal{H} = -\frac{1}{2}\sum_{i,j=1}^{\mathcal{N}} J_{i,j}s_i s_j$, where the $J_{i,j}$ are independent centred gaussian variables, of variance $\sim \mathrm{O}(1/\mathcal{N})$, and the s_i satisfy the spherical constraint $\sum_{i=1}^{\mathcal{N}} s_i^2 = \mathcal{N}$. As usual, the dynamics if given by a Langevin equation [49]. This problem is also equivalent to the statistics of the gap to the largest eigenvalue of a $\mathcal{N} \times \mathcal{N}$ gaussian unitary matrix [50,51], for $\mathcal{N} \to \infty$. Work is in progress on identifying interface growth models with $\Xi(t) \neq 0$.
49. Cugliandolo, L.F.; Dean, D. Full dynamical solution for a spherical spin-glass model. *J. Phys. A Math. Gen.* **1995**, *28*, doi:10.1088/0305-4470/28/15/003.
50. Fyodorov, Y.V.; Perret, A.; Schehr, G. Large-time zero-temperature dynamics of the spherical $p = 2$ spin model of finite size. Available online: http://arxiv.org/pdf/1507.08520.pdf (accessed on 4 November 2015).
51. Perret, A. Statistique D'extrêmes de Variables Aléatoires Fortement Corréées. Ph.D. Thesis, Université Paris Sud, Orsay, France, 2015.
52. Glauber, R.J. Time-dependent statistics of the Ising model. *J. Math. Phys.* **1963**, *4*, 294, doi:10.1063/1.1703954.
53. Godrèche, C.; Luck, J.-M. Response of non-equilibrium systems at criticality: Exact results for the Glauber-Ising chain. *J. Phys. A Math. Gen.* **2000**, *33*, doi:10.1088/0305-4470/33/6/305.
54. Henkel, M.; Schütz, G.M. On the universality of the fluctuation-dissipation ratio in non-equilibrium critical dynamics. *J. Phys. A Math. Gen.* **2004**, *37*, 591–604.
55. Lippiello, E.; Zannetti, M. Fluctuation-dissipation ratio in the one-dimensional kinetic Ising model. *Phys. Rev. E* **2000**, *61*, 3369–3374.
56. A historical comment: We have been aware of this since the very beginning of our investigations, in the early 1990s. The exact result Equation (33) looked strange, since the time-space response of the Glauber–Ising model does have the nice form $R(t,s;r) = R(t,s)\exp\left[-\frac{1}{2}\mathcal{M}r^2/(t-s)\right]$, as expected from Galilei-invariance. Only several years later, we saw how the representations of the Schrödinger algebra had to be generalised, which was only possible by giving up explicitly time-translation-invariance [38,44].
57. Henkel, M. On logarithmic extensions of local scale-invariance. *Nucl. Phys. B* **2013**, *869*, 282–302.
58. Henkel, M.; Noh, J.D.; Pleimling, M. Phenomenology of ageing in the Kardar–Parisi–Zhang equation. *Phys. Rev. E* **2012**, *85*, doi:10.1103/PhysRevE.85.030102.
59. Henkel, M.; Rouhani, S. Logarithmic correlators or responses in non-relativistic analogues of conformal invariance. *J. Phys. A Math. Theor.* **2013**, *46*, doi:10.1088/1751-8113/46/49/494004.
60. The specific structure of the dynamical functional $\mathcal{J}[\phi,\widetilde{\phi}]$, see Equation (22), of the Arcetri model (and, more generally, of the kinetic spherical model [44]) leads to $\xi + \widetilde{\xi} = 0$, such that time-translation-invariance appears to be formally satisfied, in contrast to the $1D$ Glauber–Ising model, where $\xi + \widetilde{\xi} = \frac{1}{4}$.
61. Henkel, M.; Schütz, G. Schrödinger invariance in discrete stochastic systems. *Int. J. Mod. Phys. B* **1994**, *8*, 3487–3499.
62. The scaling from Equation (39) is indeed recovered in several simple lattice models, see [61] for more details.

63. Stoimenov, S.; Henkel, M. On non-local representations of the ageing algebra. *Nucl. Phys. B* **2011**, *847*, 612–627.
64. See [39] for an application to the kinetics of the phase-separating (model-B dynamics) spherical model.
65. Gurarie, V. Logarithmic operators in conformal field theory. *Nucl. Phys. B* **1993**, *410*, 535–549.
66. Mathieu, P.; Ridout, D. From Percolation to Logarithmic Conformal Field Theory. *Phys. Lett. B* **2007**, *657*, 120–129.
67. Mathieu, P.; Ridout, D. Logarithmic $\mathcal{M}(2,p)$ minimal models, their logarithmic coupling and duality. *Nucl. Phys. B* **2008**, *801*, 268–295.
68. Rahimi Tabar, M.R.; Aghamohammadi, A.; Khorrami, M. The logarithmic conformal field theories. *Nucl. Phys. B* **1997**, *497*, 555–566.
69. Saleur, H. Polymers and percolation in two dimensions and twisted $N = 2$ supersymmetry. *Nucl. Phys. B* **1992**, *382*, 486–531.
70. Hosseiny, A.; Rouhani, S. Logarithmic correlators in non-relativistic conformal field-theory. *J. Math. Phys.* **2010**, *51*, doi:10.1063/1.3482008.
71. Hosseiny, A.; Rouhani, S. Affine extension of galilean conformal algebra in $2+1$ dimensions. *J. Math. Phys.* **2010**, *51*, doi:10.1063/1.3371191.
72. Hosseiny, A.; Naseh, A. On holographic realization of logarithmic Galilean conformal algebra. *J. Math. Phys.* **2011**, *52*, doi:10.1063/1.3637632.
73. Moghimi-Araghi, S.; Rouhani, S.; Saadat, M. Correlation functions and AdS/LCFT correspondence. *Nucl. Phys. B* **2000**, *599*, 531–546.
74. Gray, N.; Minic, D.; Pleimling, M. On non-equilibrium physics and string theory. *Int. J. Mod. Phys. A* **2013**, *28*, doi:10.1142/S0217751X13300093.
75. Hyun, S.; Jeong, J.; Kim, B.S. Aging logarithmic conformal field theory: A holographic view. *Nucl. Phys. B* **2013**, *874*, doi:10.1007/JHEP01(2013)141.
76. Giulini, D. On Galilei-invariance in quantum mechanics and the Bargmann superselection rule. *Ann. Phys.* **1996**, *249*, 222–235.
77. Henkel, M.; Stoimenov, S. Physical ageing and Lie algebras of local scale-invariance. In *Lie Theory and Its Applications in Physics*; Dobrev, V., Ed.; Springer: Heidelberg, Germany, 2015; Volume 111, pp. 33–50.
78. Dobrev, V.K. Non-relativistic holography: A group-theoretical perspective. *Int. J. Mod. Phys. A* **2013**, *29*, doi:10.1142/S0217751X14300014.
79. Son, D.T. Towards an AdS/cold atom correspondence: A geometric realisation of the Schrödinger symmetry. *Phys. Rev. D* **2008**, *78*, doi:10.1103/PhysRevD.78.046003.
80. Aizawa, N.; Dobrev, V.K. Intertwining Operator Realization of Non-Relativistic Holography. *Nucl. Phys. B* **2010**, *828*, 581–593.
81. Perroud, M. Projective representations of the Schrödinger group. *Helv. Phys. Acta* **1977**, *50*, 233–252.
82. Akhiezer, N.I. *Lectures on Integral Transforms (Translations of Mathematical Monographs)*; American Mathematical Society: Providence, RI, USA, 1988.
83. Hille, E. *Ordinary Differential Equations in the Complex Domain*; Wiley: New York, NY, USA, 1976.
84. In the numerous numerical tests of Schrödinger-invariance, the causality of the response function is simply taken for granted in the physics literature; for a review see e.g., [1]. For more recent applications and extensions, see [59].
85. Ivashkevich, E.V. Symmetries of the stochastic Burgers equation. *J. Phys. A Math. Gen.* **1997**, *30*, L525–L533.
86. Hartong, J.; Kiritsis, E.; Obers, N. Schrödinger-invariance from Lifshitz isometries in holography and field-theory. *Phys. Rev. D* **2015**, *92*, doi:10.1103/PhysRevD.92.066003.
87. Setare, M.R.; Kamali, V. Anti-de Sitter/boundary conformal field theory correspondence in the non-relativistic limit. *Eur. Phys. J. C* **2012**, *72*, doi:10.1140/epjc/s10052-012-2115-x .
88. Stoimenov, S.; Henkel, M. From conformal invariance towards dynamical symmetries of the collisionless Boltzmann equation. *Symmetry* **2015**, *7*, 1595–1612.
89. How should one dualise in the ECGA? With respect to θ or to the rapidity vector γ?

symmetry

MDPI

Article

Centrally Extended Conformal Galilei Algebras and Invariant Nonlinear PDEs

Naruhiko Aizawa * and Tadanori Kato

Department of Mathematics and Information Sciences, Graduate School of Science, Osaka Prefecture University, Nakamozu Campus, Sakai, Osaka 599-8531, Japan; E-Mail: wegci290@yahoo.co.jp

* Author to whom correspondence should be addressed; E-Mail: aizawa@mi.s.osakafu-u.ac.jp.

Academic Editor: Roman M. Cherniha

Received: 13 June 2015 / Accepted: 28 October 2015 / Published: 3 November 2015

Abstract: We construct, for any given $\ell = \frac{1}{2} + \mathbb{N}_0$, second-order *nonlinear* partial differential equations (PDEs) which are invariant under the transformations generated by the centrally extended conformal Galilei algebras. This is done for a particular realization of the algebras obtained by coset construction and we employ the standard Lie point symmetry technique for the construction of PDEs. It is observed that the invariant PDEs have significant difference for $\ell > \frac{3}{2}$.

Keywords: nonlinear PDEs; lie symmetry; conformal Galilei algebras

1. Introduction

The purpose of the present work is to construct partial differential equations (PDEs) which are invariant under the transformations generated by the conformal Galilei algebra (CGA). We consider a particular realization, which is given in [1], of CGAs with the central extension for the parameters $(d, \ell) = (1, \frac{1}{2} + \mathbb{N}_0)$, where $\mathbb{N}_0$ denotes the set of non-negative integers. We also restrict ourselves to the second-order PDEs for computational simplicity. Our main focus is on nonlinear PDEs since linear ones have already been discussed in the literatures [1–4]. CGA is a Lie algebra which generates conformal transformations in $d + 1$ dimensional *nonrelativistic* spacetime [5–8]. Even in the fixed dimension of spacetime one has infinite number of inequivalent conformal algebras. For a fixed value of d each inequivalent CGA is labelled by a parameter ℓ taking the spin value, *i.e.*, $\ell = \frac{1}{2}, 1, \frac{3}{2}, 2, \ldots$. Each CGA has an Abelian ideal (namely, CGA is a non-semisimple Lie algebra) so that it would be deformed. Indeed, it has a central extension depending the value of the parameters. More precisely, there exist two different types of central extensions. One of them exists for any values of d and half-integer ℓ, another type of extension exists for $d = 2$ and integer ℓ. Simple explanation of this fact is found in [9].

It has been observed that CGAs for $\ell = \frac{1}{2}$ and $\ell = 1$ play important roles in various kind of problems in physics and mathematics. The simplest $\ell = \frac{1}{2}$ member of CGAs is called the Schrödinger algebra which was originally discussed by Sophus Lie and Jacobi in 19th century [10,11] and reintroduced later by many physicists [12–17]. Recent renewed interest in CGAs is mainly due to the AdS/CFT correspondence. The Schrödinger algebra and $\ell = 1$ member of CGA were used to formulate nonrelativistic analogues of AdS/CFT correspondence [9,18,19]. One may find a nice review of various applications of the Schrödinger algebras in [20] and see [21] for more references on the Schrödinger algebras and $\ell = 1$ CGAs. Physical applications of $\ell = 2$ CGA is found in [22].

Now one may ask a question whether the CGAs with $\ell > 1$ are relevant structures to physical or mathematical problems. To answer this question one should find classical or quantum dynamical systems relating to CGAs and develop representation theory of CGAs (see [21,23,24] for classification of irreducible modules over $d = 1, 2$ CGAs). This is the motivation of the present work. We choose a particular differential realization of CGAs then look for PDEs invariant under the transformation generated by the realization. Investigation along this line for the Schrödinger algebras is found in [2,3,25–27] and for $\ell = 1$ CGAs in [28] and for related algebraic structure in [29,30]. For higher values of ℓ use of the representation theory such as Verma modules, singular vectors allows us to derive *linear* PDEs invariant under CGAs [1,4]. More physical applications of CGAs with higher value ℓ are found in the literatures [31–45].

The paper is organized as follows. In the next section the definition of CGA for $(d,\ell) = (1, \frac{1}{2} + \mathbb{N}_0)$ and its differential realization are given. Then symmetry of PDEs under a subset of the generators is considered. It is shown that there is a significant distinction of the form of invariant PDEs for $\ell > \frac{3}{2}$. Invariant PDEs for $\ell = \frac{3}{2}$ CGA are obtained in Section 3 For $\ell \geq \frac{5}{2}$ we first derive PDEs invariant under a subalgebra of the CGA in Section 4, then derive invariant PDEs under full CGA in Section 5. Section 6 is devoted to concluding remarks.

2. CGAs and Preliminary Consideration

The CGA for $d = 1$ and any half-integer ℓ consists of $sl(2,\mathbb{R}) \simeq so(2,1) = \langle H, D, C \rangle$ and $\ell + 1/2$ copies of the Heisenberg algebra $\langle P^{(n)}, M \rangle_{n=1,2,\dots,2\ell+1}$. Their nonvanishing commutators are given by

$$
\begin{aligned}
&[D,H] = -2H, \quad [D,C] = 2C, \quad [H,C] = D, \\
&[H,P^{(n)}] = (n-1)P^{(n-1)}, \quad [D,P^{(n)}] = 2(n-1-\ell)P^{(n)}, \\
&[C,P^{(n)}] = (n-1-2\ell)P^{(n+1)}, \quad [P^{(m)},P^{(n)}] = -\delta_{m+n,2\ell+2} I_{m-1} M
\end{aligned}
\tag{1}
$$

where the structure constant I_m is taken to be $I_m = (-1)^{m+\ell+\frac{1}{2}}(2\ell - m)!m!$ and M is the centre of the algebra. We denote this algebra by $\mathfrak{g}_\ell$. The subset $\langle P^{(n)}, M, H \rangle_{n=1,2,\dots,2\ell+1}$ forms a subalgebra of $\mathfrak{g}_\ell$ and we denote it by $\mathfrak{h}_\ell$.

We employ the following realization of $\mathfrak{g}_\ell$ on the space of functions of the variables $t = x_0, x_1, \dots, x_{\ell+\frac{1}{2}}$ and U [1]:

$$
\begin{aligned}
M &= U\partial_U, \qquad D = 2t\partial_t + \sum_{k=1}^{\ell+\frac{1}{2}} 2(\ell+1-k)x_k\partial_{x_k}, \qquad H = \partial_t, \\
C &= t^2\partial_t + \sum_{k=1}^{\ell+\frac{1}{2}} 2(\ell+1-k)tx_k\partial_{x_k} + \sum_{k=1}^{\ell-\frac{1}{2}} (2\ell+1-k)x_k\partial_{x_{k+1}} - \frac{1}{2}\Big((\ell+\frac{1}{2})!\Big)^2 x_{\ell+\frac{1}{2}}^2 U\partial_U, \\
P^{(n)} &= \sum_{k=1}^{n} \binom{n-1}{k-1} t^{n-k}\partial_{x_k}, \quad 1 \leq n \leq \ell + \frac{1}{2}, \\
P^{(n)} &= \sum_{k=1}^{\ell+\frac{1}{2}} \binom{n-1}{k-1} t^{n-k}\partial_{x_k} - \sum_{k=\ell+\frac{3}{2}}^{n} \binom{n-1}{k-1} I_{k-1} t^{n-k} x_{2\ell+2-k} U\partial_U, \quad \ell+\frac{3}{2} \leq n \leq 2\ell+1
\end{aligned}
\tag{2}
$$

where $\binom{n}{k}$ is the binomial coefficient and I_{k-1} is the structure constant appearing in Equation (1). This is in fact a realization of $\mathfrak{g}_\ell$ on the Borel subgroup of the conformal Galilei group generated by $\mathfrak{g}_\ell$ (we made some slight changes from [1]). Let us introduce the sets of indices for later convenience:

$$
\mathcal{I}_\mu = \{ \mu, \mu+1, \dots, \ell + \frac{1}{2} \}, \qquad \mu = 0, 1, 2, \dots \tag{3}
$$

Now we take x_μ, $\mu \in \mathcal{I}_0$ as independent variables and U as dependent variable: $U = U(x_\mu)$. Our aim is to find second order PDEs which are invariant under the point transformations generated by Equation (2) for $\ell > 1/2$ ($\ell = 1/2$ corresponds to Schrödinger algebra). Such a PDE is denoted by

$$
F(x_\mu, U, U_\mu, U_{\mu\nu}) = 0, \quad U_\mu = \frac{\partial U}{\partial x_\mu}, \quad U_{\mu\nu} = \frac{\partial^2 U}{\partial x_\mu x_\nu} \tag{4}
$$

We use the shorthand notation throughout this article. The left hand side of Equation (4) means that F is a function of all independent variables x_μ, $\mu \in \mathcal{I}_0$, dependent variable U and all first and second order derivatives of U. As found in the standard textbooks (e.g., [46–48]) the symmetry condition is expressed in terms of the prolonged generators:

$$\hat{X}F = 0 \; (\operatorname{mod} F = 0) \tag{5}$$

where $\hat{X}$ is the prolongation of the symmetry generator X up to second order:

$$\hat{X} = X + \sum_{\mu=0}^{\ell+\frac{1}{2}} \rho^\mu \frac{\partial}{\partial U_\mu} + \sum_{\mu \le \nu} \sigma^{\mu\nu} \frac{\partial}{\partial U_{\mu\nu}}, \qquad X = \sum_{\mu=0}^{\ell+\frac{1}{2}} \xi^\mu \frac{\partial}{\partial x_\mu} + \eta \frac{\partial}{\partial U} \tag{6}$$

The quantities $\rho^\mu, \sigma^{\mu\nu}$ are defined by

$$\rho^\mu = \eta_\mu + \eta_U U_\mu - \sum_{\nu=0}^{\ell+\frac{1}{2}} U_\nu (\xi^\nu_\mu + \xi^\nu_U U_\mu), \tag{7}$$

$$\begin{aligned} \sigma^{\mu\nu} &= \eta_{\mu\nu} + \eta_{\mu U} U_\nu + \eta_{\nu U} U_\mu + \eta_U U_{\mu\nu} + \eta_{UU} U_\mu U_\nu \\ &- \sum_{\tau=0}^{\ell+\frac{1}{2}} \xi^\tau_{\mu\nu} U_\tau - \sum_{\tau=0}^{\ell+\frac{1}{2}} (\xi^\tau_\mu U_{\nu\tau} + \xi^\tau_\nu U_{\mu\tau}) - \sum_{\tau=0}^{\ell+\frac{1}{2}} \xi^\tau_U (U_\tau U_{\mu\nu} + U_\mu U_{\nu\tau} + U_\nu U_{\mu\tau}) \\ &- \sum_{\tau=0}^{\ell+\frac{1}{2}} (\xi^\tau_{\mu U} U_\nu + \xi^\tau_{\nu U} U_\mu + \xi^\tau_{UU} U_\mu U_\nu) U_\tau \end{aligned} \tag{8}$$

In this section we consider the symmetry condition Equation (5) for M, H and $P^{(n)}$ with $n = 1, 2, \dots, 2\ell + 1$.

Lemma 1. *(i) Equation (4) is invariant under $H, P^{(1)}$ and M if it has the form*

$$F\left(x_a, \frac{U_\mu}{U}, \frac{U_{\mu\nu}}{U}\right) = 0, \quad a \in \mathcal{I}_2, \quad \mu, \nu \in \mathcal{I}_0 \tag{9}$$

(ii) For $\ell > \frac{3}{2}$, a necessary condition for the symmetry of the Equation (9) under $P^{(n)}$ with $n \in \mathcal{I}_2$ is that the function F is independent of U_{0m}, $m \in \mathcal{I}_3$.

Proof of Lemma 1. (i) It is obvious from that the generators H and $P^{(1)}$ have no prolongation, while the prolongation of M is given by

$$\hat{M} = U\partial_U + \sum_{\mu=0}^{\ell+\frac{1}{2}} U_\mu \partial_{U_\mu} + \sum_{\mu \le \nu} U_{\mu\nu} \partial_{U_{\mu\nu}} \tag{10}$$

(ii) The lemma is proved by the formula of the prolongation of $P^{(n)}$. For $n \in \mathcal{I}_2$ the generator $P^{(n)}$ is given by

$$P^{(n)} = \sum_{k=1}^{n} \binom{n-1}{k-1} t^{n-k} \partial_{x_k}$$

and its prolongation yields

$$\hat{P}^{(n)} = \sum_{a=2}^{n} \binom{n-1}{n-a} t^{n-a} \left[\tilde{P}^{(a)} - \sum_{k=a+1}^{n} (a-1) U_{a-1\,k} \partial_{U_{0k}} \right] \tag{11}$$

where

$$\tilde{P}^{(n)} = \hat{P}^{(n)}\Big|_{t=0} = \partial_{x_n} - (n-1)\left\{ U_{n-1}\partial_{U_0} + ((n-2)U_{n-2} + 2U_{0\,n-1})\partial_{U_{00}} + \sum_{k=1}^{n} U_{n-1\,k}\partial_{U_{0k}} \right\} \quad (12)$$

and the terms containing ∂_{x_1} are omitted. We give the explicit expressions for small values of n which will be helpful to see the structure of Equation (11):

$$\begin{aligned}
\hat{P}^{(2)} &= \tilde{P}^{(2)}, \\
\hat{P}^{(3)} &= \tilde{P}^{(3)} + 2t(\tilde{P}^{(2)} - U_{13}\partial_{U_{03}}), \\
\hat{P}^{(4)} &= \tilde{P}^{(4)} + 3t(\tilde{P}^{(3)} - 2U_{24}\partial_{U_{04}}) + 3t^2(\tilde{P}^{(2)} - U_{13}\partial_{U_{03}} - U_{14}\partial_{U_{04}})
\end{aligned}$$

Since $\tilde{P}^{(n)}$ is independent of t, each symmetry condition $\hat{P}^{(n)}F = 0$ decouples into some independent equations. For example, $\hat{P}^{(4)}F = 0$ decouples into the following equations:

$$\tilde{P}^{(4)}F = 0, \qquad (\tilde{P}^{(3)} - 2U_{24}\partial_{U_{04}})F = 0, \qquad (\tilde{P}^{(2)} - U_{13}\partial_{U_{03}} - U_{14}\partial_{U_{04}})F = 0$$

The condition $\hat{P}^{(2)}F = 0$ is equivalent to the condition $\tilde{P}^{(2)}F = 0$. It follows that the condition $\hat{P}^{(3)}F = 0$ yields two independent conditions $\tilde{P}^{(3)}F = 0$ and $U_{13}\partial_{U_{03}}F = 0$. The second condition means that F is independent of U_{03}. Repeating this for $\hat{P}^{(n)}F = 0$ for $n = 4, 5, \ldots, \ell + \frac{1}{2}$ one may prove the lemma.

Now we show the Equation (11). Set $\xi^k(n) = \begin{pmatrix} n-1 \\ k-1 \end{pmatrix} t^{n-k}$ then $P^{(n)} = \sum_{k=2}^{n} \xi^k(n)\partial_{x_k}$ (recall that we omit ∂_{x_1}). By the Equations (6)–(8) we have

$$\hat{P}^{(n)} = P^{(n)} - \sum_{k=1}^{n}\{\xi_0^k(n)U_k\partial_{U_0} + (\xi_{00}^k(n)U_k + 2\xi_0^k(n)U_{0k})\partial_{U_{00}} + \sum_{m=1}^{n}\xi_0^m(n)U_{mk}\partial_{U_{mk}}\} \quad (13)$$

Thus the maximal degree of t in $\hat{P}^{(n)}$ is $n-2$. The following relation is easily verified:

$$\frac{\partial^a}{\partial t^a}\xi^k(n) = \begin{cases} \frac{(n-1)!}{(n-a-1)!}\xi^k(n-a), & (1 \le k \le n-a) \\ 0, & (n-a < k) \end{cases} \quad (14)$$

Using this one may calculate the higher order derivatives of Equation (13):

$$\frac{\partial^a}{\partial t^a}\hat{P}^{(n)} = \frac{(n-1)!}{(n-a-1)!}\{P^{(n-a)} - \sum_{k=n-a+1}^{n}\sum_{m=1}^{n-a}\xi_0^m(n-a)U_{mk}\partial_{U_{0k}}\}$$

It follows that

$$\begin{aligned}
\hat{P}^{(n)} &= \sum_{a=0}^{n-2}\frac{1}{a!}\left(\frac{\partial^a}{\partial t^a}\hat{P}^{(n)}\right)_{t=0} t^a \\
&= \sum_{a=0}^{n-2}\begin{pmatrix} n-1 \\ a \end{pmatrix} t^a\left[\tilde{P}^{(n-a)} - \sum_{k=n-a+1}^{n}(n-a-1)U_{n-a-1\,k}\partial_{U_{0k}}\right]
\end{aligned}$$

By replacing $n-a$ with a we obtain the Equation (11). The Equation (12) is readily obtained by setting $t = 0$ in the Equation (13). □

Remark 1. *By Lemma 1 the symmetry condition for* $M, H, P^{(n)}$ *with* $n \in \mathcal{I}_1$ *is summarized as*

$$\tilde{P}^{(n)} F\left(x_a, \frac{U_\mu}{U}, \frac{U_{00}}{U}, \frac{U_{01}}{U}, \frac{U_{02}}{U}, \frac{U_{km}}{U},\right) = 0, \quad a \in \mathcal{I}_2,\ \mu \in \mathcal{I}_0,\ k, m \in \mathcal{I}_1 \tag{15}$$

where $\tilde{P}^{(n)}$ is given by Equation (12).

The condition Equation (15) implies that F is independent of U_{00} if $\ell \geq 7/2$, since $\tilde{P}^{(n)}$ has the term $U_{0k}\partial_{U_{00}}$ with $k \geq 3$. In fact one can make a stronger statement by looking at the symmetry conditions for $P^{(n)}$ with $\ell + \frac{3}{2} \leq n \leq 2\ell + 1$.

Lemma 2. *F given in Equation (15) is independent of U_{00} if $\ell \geq 5/2$.*

Proof of Lemma 2. We calculate the prolongation of $P^{(n)}$ for $\ell + \frac{3}{2} \leq n \leq 2\ell + 1$. The derivatives $\partial_t, \partial_{x_1}, \partial_{U_{0k}} (k \in \mathcal{I}_3)$ are ignored in the computation. Then

$$P^{(n)} = \sum_{m=2}^{\ell+\frac{1}{2}} \xi^m(n)\partial_{x_m} + \eta(n)\partial_U, \tag{16}$$

$$\xi^m(n) = \binom{n-1}{m-1} t^{n-m}, \qquad \eta(n) = -\sum_{m=\ell+\frac{3}{2}}^{n} \xi^m(n) I_{m-1} x_{2\ell+2-m} U \tag{17}$$

One may calculate derivatives of $\eta(n)$ easily

$$\eta_k(n) = \frac{\partial \eta(n)}{\partial x_k} = \begin{cases} -\xi^{2\ell+2-k}(n) I_{2\ell+1-k} U & 2\ell+2-n \leq k \leq \ell + \frac{1}{2} \\ 0 & k < 2\ell+2-n \end{cases}$$

First and second order derivatives need some care:

$$\eta_1(n) = \begin{cases} -I_{2\ell} U & n = 2\ell+1 \\ 0 & \text{otherwise} \end{cases}$$

$$\eta_2(n) = \begin{cases} -2\ell t I_{2\ell-1} U & n = 2\ell+1 \\ -I_{2\ell-1} U & n = 2\ell \\ 0 & \text{otherwise} \end{cases}$$

Then a lengthy but straightforward computation gives the following expression for the prolongation of $P^{(n)}$ up to second order:

$$\begin{aligned} \hat{P}^{(n)} &= \eta_U \hat{M} + \sum_{k=2}^{\ell+\frac{1}{2}} \xi^k(n)\partial_{x_k} + \Big(\eta_0(n) - \sum_{k=1}^{\ell+\frac{1}{2}} \xi_0^k(n) U_k\Big)\partial_{U_0} + \sum_{k=2\ell+2-n}^{\ell+\frac{1}{2}} \eta_k(n)\partial_{U_k} \\ &+ \Big(\eta_{00}(n) + 2\eta_{0U}(n)U_0 - \sum_{k=1}^{\ell+\frac{1}{2}} (\xi_{00}^k(n)U_k + 2\xi_0^k(n)U_{0k})\Big)\partial_{U_{00}} \\ &+ \sum_{k=1,2} \Big(\eta_{0U}(n)U_k - \sum_{m=1}^{\ell+\frac{1}{2}} \xi_0^m(n)U_{km}\Big)\partial_{U_{0k}} - \delta_{n,2\ell} I_{2\ell-1} U_0 \partial_{U_{02}} \\ &- \delta_{n,2\ell+1}\big(I_{2\ell}U_0\partial_{U_{01}} + 2\ell I_{2\ell-1}(U + tU_0)\partial_{U_{02}}\big) \\ &+ \sum_{k=2\ell+2-n}^{\ell+\frac{1}{2}} \sum_{m=1}^{\ell+\frac{1}{2}} \eta_{kU} U_m \partial_{U_{km}} + \sum_{k=2\ell+2-n}^{\ell+\frac{1}{2}} \eta_{kU} U_k \partial_{U_{kk}} \end{aligned} \tag{18}$$

We have already taken into account the invariance under $\hat{M}$ so that the first term of Equation (18) is omitted in the following computations. It is an easy exercise to verify that

$$\frac{\partial^a}{\partial t^a}\eta(n) = \frac{(n-1)!}{(n-a-1)!} \times \begin{cases} 0 & n-\ell-\frac{1}{2} \le a \\ \eta(n-a) & 0 \le a \le n-\ell-\frac{3}{2} \end{cases}$$

and

$$\frac{\partial^a}{\partial t^a}\sum_{k=1}^{\ell+\frac{1}{2}} \xi^k(n) = \frac{(n-1)!}{(n-a-1)!} \times \begin{cases} 0 & n \le a \\ \sum_{k=1}^{n-a} \xi^k(n-a) & n-\ell-\frac{1}{2} \le a \le n-1 \\ \sum_{k=1}^{\ell+\frac{1}{2}} \xi^k(n-a) & 0 \le a \le n-\ell-\frac{3}{2} \end{cases}$$

It follows that for $0 \le a \le n-\ell-\frac{3}{2}$

$$\frac{\partial^a}{\partial t^a}\hat{P}^{(n)} = \frac{(n-1)!}{(n-a-1)!}\hat{P}^{(n-a)} \tag{19}$$

For $n-\ell-\frac{1}{2} \le a \le n-2$ (*i.e.*, $2 \le n-a \le \ell+\frac{1}{2}$) all the derivatives of $\eta(n)$ vanishes and Equation (13) is recovered. Therefore for all values of a from 0 to $n-2$ the relation Equation (19) holds true. Thus we have

$$\hat{P}^{(n)} = \sum_{a=0}^{n-2} \frac{1}{a!}\left(\frac{\partial^a}{\partial t^a}\hat{P}^{(n)}\right)_{t=0} t^a = \sum_{a=0}^{n-2} \binom{n-1}{a} \tilde{P}^{(n-a)} t^a$$

where $\tilde{P}^{(n)} = \hat{P}^{(n)}\Big|_{t=0}$. This means that the symmetry condition under $\hat{P}^{(n)}$ is reduced to

$$\tilde{P}^{(n)}F = 0, \qquad \ell+\frac{3}{2} \le n \le 2\ell+1 \tag{20}$$

Now we look at the part containing $\partial_{U_{00}}$ in Equation (18), namely, the second line of the equation. The contribution to $\tilde{P}^{(n)}$ from the term $\sum_k \xi_0^k(n) U_{0k}\partial_{U_{00}}$ is $U_{0\,\ell+\frac{1}{2}}\partial_{U_{00}}$. Since $\ell+\frac{1}{2} \ge 3$ for $\ell \ge \frac{5}{2}$ the condition Equation (20) gives $\partial_{U_{00}}F = 0$ for this range of ℓ. Thus F has U_{00} dependence only for $\ell = \frac{3}{2}$. □

Lemma 2 requires a separate treatment of the case $\ell = \frac{3}{2}$. In the following sections we solve the symmetry conditions Equations (15) and (20) explicitly for $\ell = \frac{3}{2}$ and for $\ell > \frac{3}{2}$ separately. Before proceeding further we here present the formulae of prolongation of D which is not difficult to verify:

$$\begin{aligned} \hat{D} &= \sum_{k=2}^{\ell+\frac{1}{2}} 2(\ell+1-k)x_k\partial_{x_k} - 2U_0\partial_{U_0} - \delta_{\ell,\frac{3}{2}} 4U_{00}\partial_{U_{00}} - 2\sum_{k=1,2}(\ell+2-k)U_{0k}\partial_{U_{0k}} \\ &- 2\sum_{k=1}^{\ell+\frac{1}{2}}\Big[(\ell+1-k)U_k\partial_{U_k} + \sum_{m=k}^{\ell+\frac{1}{2}}(2\ell+2-k-m)U_{km}\partial_{U_{km}}\Big] \end{aligned} \tag{21}$$

The prolongation of C is more involved so we present it in the subsequent sections separately for $\ell = \frac{3}{2}$ and for other values of ℓ.

3. The Case of $\ell = \frac{3}{2}$

The goal of this section is to derive the PDEs invariant under the group generated by $\mathfrak{g}_{\frac{3}{2}}$. First we solve the conditions Equations (15) and (20). We have from Equation (12)

$$\tilde{P}^{(2)} = \partial_{x_2} - (U_1\partial_{U_0} + 2U_{01}\partial_{U_{00}} + \sum_{k=1,2} U_{1k}\partial_{U_{0k}}) \tag{22}$$

and collecting the $t = 0$ terms of Equation (18)

$$\begin{aligned}
\tilde{P}^{(3)} &= -2\left[U_2\partial_{U_0} + U\partial_{U_2} + (U_1 + 2U_{02})\partial_{U_{00}} + U_{12}\partial_{U_{01}} + (U_0 + U_{22})\partial_{U_{02}} + U_1\partial_{U_{12}} + 2U_2\partial_{U_{22}}\right], \\
\tilde{P}^{(4)} &= -6\left[x_2U\partial_{U_0} - U\partial_{U_1} + (U_2 + 2x_2U_0)\partial_{U_{00}} + (x_2U_1 - U_0)\partial_{U_{01}} + (x_2U_2 + U)\partial_{U_{02}}\right. \\
&\quad \left. -(2U_1\partial_{U_{11}} + U_2\partial_{U_{12}})\right]
\end{aligned} \tag{23}$$

The symmetry conditions Equations (15) and (20) are the system of first order PDEs so that it can be solved by the standard method of characteristic equation (e.g., [49]). It is not difficult to verify that the following functions are the solutions to Equations (15) and (20).

$$\begin{aligned}
\phi_1 &= \frac{U_{11}}{U} - \left(\frac{U_1}{U}\right)^2, \qquad \phi_2 = \frac{U_{22}}{U} - \left(\frac{U_2}{U}\right)^2, \\
\phi_3 &= \frac{U_{12}}{U} - \frac{U_1U_2}{U^2}, \qquad \phi_4 = \frac{U_0}{U} + \frac{x_2U_1}{U} - \frac{U_{22}}{2U}, \\
\phi_5 &= \frac{U_{01}}{U} - \frac{U_0U_1}{U^2} + x_2\phi_1 - \frac{U_2}{U}\phi_3, \\
\phi_6 &= \frac{U_{02}}{U} + \frac{U_1}{U} - \frac{U_0U_2}{U^2} - \frac{U_2}{U}\phi_2 + x_2\phi_3, \\
\phi_7 &= \frac{U_{00}}{U} - \left(\frac{U_0}{U}\right)^2 - \left(\frac{U_1}{U} + \frac{2U_{02}}{U}\right)\frac{U_2}{U} + \left(\frac{2U_0}{U} + \frac{U_{22}}{U}\right)\left(\frac{U_2}{U}\right)^2 \\
&- \left(\frac{U_2}{U}\right)^4 - x_2^2\phi_1 + 2x_2\phi_5
\end{aligned} \tag{24}$$

Thus we have proved the following lemma:

Lemma 3. *Equation (4) is invariant under* $\mathfrak{h}_{\frac{3}{2}} = \langle\, M, H, P^{(n)}\, \rangle_{n=1,2,3,4}$ *if it has the form*

$$F(\phi_1, \phi_2, \ldots, \phi_7) = 0 \tag{25}$$

Next we consider the further invariance under D and C. The computation of the second order prolongation of C for $\ell = \frac{3}{2}$ is straightforward based on Equations (6)–(8). It has the form

$$\begin{aligned}
\hat{C} &= -2x_2^2\hat{M} + t\hat{D} + 3x_1\tilde{P}^{(2)} - \tilde{C}, \\
\tilde{C} &= x_2U_2\partial_{U_0} + 3U_2\partial_{U_1} + 4x_2U\partial_{U_2} + 2(U_0 + x_2U_{02})\partial_{U_{00}} + (3U_1 + 3U_{02} + x_2U_{12})\partial_{U_{01}} \\
&+ (U_2 + 4x_2U_0 + x_2U_{22})\partial_{U_{02}} + 6U_{12}\partial_{U_{11}} + (4x_2U_1 + 3U_{22})\partial_{U_{12}} + 4(U + 2x_2U_2)\partial_{U_{22}}
\end{aligned} \tag{26}$$

It is an easy exercise to see the action of $\tilde{C}$ on ϕ_k :

$$\begin{aligned}
&\tilde{C}\phi_1 = 2\phi_3, \quad \tilde{C}\phi_2 = \frac{4}{3}, \quad \tilde{C}\phi_3 = \phi_2, \quad \tilde{C}\phi_4 = -\frac{2}{3}, \quad \tilde{C}\phi_5 = \phi_6, \\
&\tilde{C}\phi_6 = 0, \quad \tilde{C}\phi_7 = \frac{1}{3}\phi_2 + \frac{2}{3}\phi_4
\end{aligned}$$

It follows that the following combinations of ϕ_k are invariant of $\tilde{C}$:

$$w_1 = \frac{1}{2}\phi_2 + \phi_4, \qquad w_2 = 2\phi_3 - \frac{3}{4}\phi_2^2, \qquad w_3 = \frac{1}{2\sqrt{2}}\phi_6, \qquad w_4 = \phi_1 - \frac{3}{2}\left(w_2\phi_2 + \frac{1}{8}\phi_2^3\right),$$
$$w_5 = \frac{1}{\sqrt{2}}\phi_5 - \frac{3}{4\sqrt{2}}w_3\phi_2, \qquad w_6 = \phi_7 - \frac{1}{2}w_1\phi_2 \tag{27}$$

On the other hand $\hat{D}$ generates the scaling of w_k :

$$\begin{array}{lll} w_1 \to e^{2\epsilon}w_1, & w_2 \to e^{4\epsilon}w_2, & w_3 \to e^{3\epsilon}w_3, \\ w_4 \to e^{6\epsilon}w_4, & w_5 \to e^{5\epsilon}w_5, & w_6 \to e^{4\epsilon}w_6 \end{array} \tag{28}$$

With these observations one may construct all invariants of the group which is generated by $\mathfrak{g}_{\frac{3}{2}}$:

$$\psi_1 = \frac{w_2}{w_1^2} = \frac{\Psi_1}{\Phi^2}, \qquad \psi_2 = \frac{w_3^2}{w_1^3} = \frac{\Psi_2^2}{\Phi^3}, \qquad \psi_3 = \frac{w_4}{w_3^2} = \frac{\Psi_3}{\Psi_2^2},$$
$$\psi_4 = \frac{w_5^2}{w_1^5} = \frac{\Psi_4^2}{\Phi^5}, \qquad \psi_5 = \frac{w_6}{w_2} = \frac{\Psi_5}{\Psi_1} \tag{29}$$

where

$$\begin{aligned} \Phi &= 2(U_0 + x_2 U_1)U - U_2^2, \\ \Psi_1 &= 8(U_{12}U - U_1U_2)U^2 - 3(U_{22}U - U_2^2)^2, \\ \Psi_2 &= (U_1 + U_{02})U^2 - U_2\big((U_0 + U_{22})U - U_2^2\big) + x_2(U_{12}U - U_1U_2)U, \\ \Psi_3 &= 8U_{11}U^5 - 8U_1^2U^4 - 12(U_{22}U - U_2^2)(U_{12}U - U_1U_2)U^2 + 3(U_{22}U - U_2^2)^3, \\ \Psi_4 &= 4U_{01}U^4 - 4U_0U_1U^3 - 3\big((U_1 + U_{02})U - U_0U_2\big)(U_{22}U - U_2^2)U + 3U_2(U_{22}U - U_2^2)^2 \\ &+ x_2\big(4U_{11}U^3 - 4U_1^2U^2 - 3(U_{22}U - U_2^2)(U_{12}U - U_1U_2)\big)U, \\ \Psi_5 &= 4U_{00}U^3 - 2\big(2U_0^2 + U_0U_{22} + 2(U_1 + 2U_{02})U_2\big)U^2 + 5(2U_0 + U_{22})U_2^2U - 5U_2^4 \\ &+ 2x_2\big(4U_{01}U^2 - (4U_0U_1 + 4U_2U_{12} + U_1U_{22})U + 5U_1U_2^2\big)U + 4x_2^2(U_{11}U - U_1^2)U^2 \end{aligned} \tag{30}$$

Thus we obtain the PDEs with the desired symmetry.

Theorem 1. *The PDE invariant under the Lie group generated by the realization Equation (2) of* $\mathfrak{g}_{\frac{3}{2}}$ *is given by*

$$F(\psi_1, \psi_2, \psi_3, \psi_4, \psi_5) = 0 \tag{31}$$

where F is an arbitrary differentiable function and ψ_i *is given in Equation (29). Explicit form of the symmetry generators are as follows:*

$$\begin{aligned} &M = U\partial_U, \qquad D = 2t\partial_t + 3x_1\partial_{x_1} + x_2\partial_{x_2}, \qquad H = \partial_t, \\ &C = t(t\partial_t + 3x_1\partial_{x_1} + x_2\partial_{x_2}) + 3x_1\partial_{x_2} - 2x_2^2U\partial_U, \\ &P^{(1)} = \partial_{x_1}, \qquad P^{(2)} = t\partial_{x_1} + \partial_{x_2}, \\ &P^{(3)} = t^2\partial_{x_1} + 2t\partial_{x_2} - 2x_2U\partial_U, \\ &P^{(4)} = t^3\partial_{x_1} + 3t^2\partial_{x_2} - 6(tx_2 - x_1)U\partial_U \end{aligned} \tag{32}$$

4. The Case of $\ell \geq \frac{5}{2}$: $\mathfrak{h}_\ell$-Symmetry

As shown in Lemma 2 the function F is independent of U_{00} so that the PDE which we have at this stage is of the form

$$F\left(x_a, \frac{U_\mu}{U}, \frac{U_{01}}{U}, \frac{U_{02}}{U}, \frac{U_{km}}{U},\right) = 0, \quad a \in \mathcal{I}_2,\ \mu \in \mathcal{I}_0,\ k,m \in \mathcal{I}_1 \tag{33}$$

We wants to make the PDE Equation (33) invariant under all the generators of $\mathfrak{h}_\ell$. Invariance under M and $P^{(1)}$ has been completed. We need to consider the invariance under $P^{(n)}$ for $n \in \mathcal{I}_2$. The symmetry conditions are Equations (15) and (20). We give $\tilde{P}^{(n)}$ more explicitly. From Equation (12) we have

$$\tilde{P}^{(n)} = \partial_{x_n} - (n-1)(U_{n-1}\partial_{U_0} + U_{1\,n-1}\partial_{U_{01}} + U_{2\,n-1}\partial_{U_{02}}), \quad 2 \leq n \leq \ell + \frac{1}{2} \tag{34}$$

For $n \geq \ell + \frac{3}{2}$ the generator $\tilde{P}^{(n)}$ is obtained by collecting $t = 0$ terms of Equation (18). It has a slightly different form depending on the value of n. For $n = \ell + \frac{3}{2}$ it is given by

$$\tilde{P}^{(\ell+\frac{3}{2})} = -(\ell+\frac{1}{2})\Big[U_{\ell+\frac{1}{2}}\partial_{U_0} + \sum_{k=1,2} U_{k\,\ell+\frac{1}{2}}\partial_{U_{0k}} + a_\ell\Big(U\partial_{U_{\ell+\frac{1}{2}}} + \sum_{m=1}^{\ell+\frac{1}{2}} U_m\partial_{U_{m\,\ell+\frac{1}{2}}} + U_{\ell+\frac{1}{2}}\partial_{U_{\ell+\frac{1}{2}\,\ell+\frac{1}{2}}}\Big)\Big] \tag{35}$$

where

$$a_\ell = \Big((\ell - \frac{1}{2})!\Big)^2$$

For other values of n they are given by

$$\begin{aligned}
\tilde{P}^{(n)} &= -I_{n-1}\Big[-(2\ell+2-n)x_{2\ell+3-n}\Big(U\partial_{U_0} + \sum_{k=1,2} U_k\partial_{U_{0k}}\Big) + U\partial_{U_{2\ell+2-n}} \\
&+ \sum_{k=1}^{\ell+\frac{1}{2}} U_k\partial_{U_{k\,2\ell+2-n}} + U_{2\ell+2-n}\partial_{U_{2\ell+2-n\,2\ell+2-n}}\Big], \qquad \ell + \frac{5}{2} \leq n \leq 2\ell - 1 \\
\tilde{P}^{(2\ell)} &= -I_{2\ell-1}\big[-2x_3(U\partial_{U_0} + U_1\partial_{U_{01}}) + U\partial_{U_2} + (U_0 - 2x_3U_2)\partial_{U_{02}} + \sum_{k=1}^{\ell+\frac{1}{2}} U_k\partial_{U_{2k}} + U_2\partial_{U_{22}}\big]
\end{aligned}$$

and

$$\tilde{P}^{(2\ell+1)} = -I_{2\ell}\big[-x_2U\partial_{U_0} + U\partial_{U_1} + (U_0 - x_2U_1)\partial_{U_{01}} - (U + x_2U_2)\partial_{U_{02}} + \sum_{k=1}^{\ell+\frac{1}{2}} U_k\partial_{U_{1k}} + U_1\partial_{U_{11}}\big]$$

The best way to solve the symmetry condition is to start from the larger values of n. We first investigate the symmetry conditions for $P^{(2\ell+1)}$ to $P^{(\ell+\frac{3}{2})}$ in this order. They are separated in three cases (two cases for $\ell = \frac{5}{2}$).

Lemma 4. *(i) Equation (33) is invariant under $P^{(2\ell+1)}$ and $P^{(2\ell)}$ if it has the form*

$$F\left(x_a, \frac{U_0}{U}, \tilde{\phi}, \frac{U_k}{U}, \phi_{01}, \phi_{02}, \phi_{1b}, \phi_{2b}, U_{km}\right) = 0, \quad a \in \mathcal{I}_2,\ k,m \in \mathcal{I}_3,\ b \in \mathcal{I}_1 \tag{36}$$

where

$$\begin{aligned}
\tilde{\phi} &= \frac{U_0}{U} + x_2\frac{U_1}{U} + 2x_3\frac{U_2}{U}, \\
\phi_{01} &= \frac{U_{01}}{U} - \frac{U_0U_1}{U^2}, \qquad \phi_{02} = \frac{U_{02}}{U} + \frac{U_1}{U} - \frac{U_0U_2}{U^2}, \\
\phi_{\alpha k} &= \frac{U_{\alpha k}}{U} - \frac{U_\alpha U_k}{U^2}, \quad \alpha = 1,2
\end{aligned} \tag{37}$$

(ii) *Equation (36) is invariant under* $P^{(n)}$, $\ell + \frac{5}{2} \leq n \leq 2\ell - 1$ *if it has the form*

$$F\left(x_a, \phi, \frac{U_{\ell+\frac{1}{2}}}{U}, \phi_{01}, \phi_{02}, \phi_{km}\right) = 0, \quad a \in \mathcal{I}_2,\ k,m \in \mathcal{I}_1 \tag{38}$$

where ϕ_{01}, ϕ_{02} *are given in Equation (37) and*

$$\phi = \frac{U_0}{U} + \sum_{j=1}^{\ell-\frac{1}{2}} jx_{j+1}\frac{U_j}{U}, \qquad \phi_{km} = \frac{U_{km}}{U} - \frac{U_kU_m}{U^2} \tag{39}$$

(iii) *Equation (38) is invariant under* $P^{(\ell+\frac{3}{2})}$ *if it has the form*

$$F(x_a, w, w_{01}, w_{02}, \phi_{km}) = 0, \quad a \in \mathcal{I}_2,\ k,m \in \mathcal{I}_1 \tag{40}$$

where ϕ_{km} *is given in Equation (39) and*

$$w = \phi - \frac{U^2_{\ell+\frac{1}{2}}}{2a_\ell U^2}, \qquad w_{0\alpha} = \phi_{0\alpha} - \frac{\phi_{\alpha\,\ell+\frac{1}{2}}}{a_\ell}\frac{U_{\ell+\frac{1}{2}}}{U}, \quad \alpha = 1,2 \tag{41}$$

The constant a_ℓ *is defined below the Equation (35).*
For $\ell = \frac{5}{2}$ *we have the cases (i) and (iii).*

Proof of Lemma 4. (i) The symmetry conditions $\tilde{P}^{(2\ell+1)}F = \tilde{P}^{(2\ell)}F = 0$ is a system of first order PDEs. They are solved by the standard technique and it is not difficult to see that the $\tilde{\phi}$ and ϕ's given in Equation (37) are solutions to the system of PDEs; (ii) It is immediate to verify that ϕ_{01}, ϕ_{02} solve the symmetry conditions $\tilde{P}^{(n)}F = 0$ for $\ell + \frac{5}{2} \leq n \leq 2\ell - 1$. Rewriting the symmetry conditions in terms of the variables given in Equation (37) it is not difficult to solve them and find ϕ and ϕ_{km} in Equation (39) are the solutions; (iii) It is immediate to see that all ϕ_{km}, $k,m \in \mathcal{I}_1$ solves the symmetry condition $\tilde{P}^{(\ell+\frac{3}{2})}F = 0$, however, ϕ, ϕ_{01} and ϕ_{02} do not. Rewriting the symmetry condition in terms of ϕ's then solving the condition is an easy task. One may see that the variables in Equation (41) are solution of it. □

Theorem 2. *The PDE invariant under the group generated by* $\mathfrak{h}_\ell$ *with* $\ell \geq \frac{5}{2}$ *is given by*

$$F(w, \alpha_n, \beta_n, \phi_{km}) = 0, \quad n \in \mathcal{I}_2,\ k,m \in \mathcal{I}_1 \tag{42}$$

where F is an arbitrary differentiable function and

$$
\begin{aligned}
\alpha_n &= w_{01} + (n-1)x_n\phi_{1\,n-1} \\
&= \frac{U_{01}}{U} - \frac{U_0U_1}{U^2} - \frac{U_{\ell+\frac{1}{2}}}{a_\ell U}\left(\frac{U_{1\,\ell+\frac{1}{2}}}{U} - \frac{U_1U_{\ell+\frac{1}{2}}}{U^2}\right) + (n-1)x_n\left(\frac{U_{1\,n-1}}{U} - \frac{U_1U_{n-1}}{U^2}\right), \\
\beta_n &= w_{02} + (n-1)x_n\phi_{2\,n-1} \\
&= \frac{U_{02}}{U} + \frac{U_1}{U} - \frac{U_0U_2}{U^2} - \frac{U_{\ell+\frac{1}{2}}}{a_\ell U}\left(\frac{U_{2\,\ell+\frac{1}{2}}}{U} - \frac{U_2U_{\ell+\frac{1}{2}}}{U^2}\right) + (n-1)x_n\left(\frac{U_{2\,n-1}}{U} - \frac{U_2U_{n-1}}{U^2}\right)
\end{aligned}
\tag{43}
$$

Proof of Theorem 2. Theorem is proved by making the Equation (40) invariant under $P^{(n)}$ with $2 \le n \le \ell + \frac{1}{2}$. It is easy to see that w and all ϕ_{km} solve the symmetry conditions $\tilde{P}^{(n)}F = 0$ with $\tilde{P}^{(n)}$ given by Equation (34). Thus the symmetry conditions are written in terms of only x_n and $w_{0\alpha}$:

$$
\left(\partial_{x_n} - (n-1)(\phi_{1\,n-1}\partial_{w_{01}} + \phi_{2\,n-1}\partial_{w_{02}})\right)F = 0, \quad n \in \mathcal{I}_2
$$

It is easily verified that the solutions of this system of equations are given by α_n and β_n. Thus we have proved the theorem. □

5. The Case of $\ell \ge \frac{5}{2}$: $\mathfrak{g}_\ell$-Symmetry

Our next task is to make the Equation (42) invariant under D and C. From the Equation (21) one may see that D generates the following scaling:

$$
w \to e^{-2\epsilon}w, \quad \alpha_n \to e^{-2(\ell+1)\epsilon}\alpha_n, \quad \beta_n \to e^{-2\ell\epsilon}\beta_n, \quad \phi_{km} \to e^{-2(2\ell+2-k-m)\epsilon}\phi_{km} \tag{44}
$$

Now we need the prolongation of C up to second order. After lengthy but straightforward computation one may obtain the formula:

$$
\begin{aligned}
\hat{C} &= -\frac{b_\ell}{2}x^2_{\ell+\frac{1}{2}}\hat{M} + t\hat{D} + 2\ell x_1\tilde{P}^{(2)} - \tilde{C}, \\
\tilde{C} &= -\sum_{k=2}^{\ell-\frac{1}{2}}\lambda_k x_k\partial_{x_{k+1}} + \sum_{k=2}^{\ell+\frac{1}{2}}2(\ell+1-k)x_kU_k\,\partial_{U_0} \\
&+ \sum_{k=1}^{\ell-\frac{1}{2}}\left[\lambda_kU_{k+1}\partial_{U_k} + \sum_{m=k}^{\ell-\frac{1}{2}}(\lambda_kU_{k+1\,m} + \lambda_mU_{k\,m+1})\partial_{U_{km}} + (\lambda_kU_{k+1\,\ell+\frac{1}{2}} + b_\ell x_{\ell+\frac{1}{2}}U_k)\partial_{U_{k\,\ell+\frac{1}{2}}}\right] \\
&+ \sum_{k=1,2}\left[2(\ell+1-k)U_k + \sum_{m=2}^{\ell+\frac{1}{2}}2(\ell+1-m)x_mU_{km} + \lambda_kU_{0\,k+1}\right]\partial_{U_{0k}} \\
&+ b_\ell\left[x_{\ell+\frac{1}{2}}U\,\partial_{U_{\ell+\frac{1}{2}}} + (U + 2x_{\ell+\frac{1}{2}}U_{\ell+\frac{1}{2}})\partial_{U_{\ell+\frac{1}{2}\,\ell+\frac{1}{2}}}\right],
\end{aligned}
\tag{45}
$$

where

$$
b_\ell = \left((\ell+\frac{1}{2})!\right)^2, \qquad \lambda_k = 2\ell+1-k
$$

One may ignore $\hat{M}$ and $\tilde{P}^{(2)}$ since we have already taken them into account. $\tilde{C}$ is independent of t so that the invariance under D and C is reduced to the one under D and $\tilde{C}$. An immediate consequence of the Equation (45) of $\tilde{C}$ is that F may not depend on α_n and β_n :

Lemma 5. *A necessary condition for the invariance of the Equation (42) under C is that the function F is independent of α_n and β_n.*

Proof of Lemma 5. $\tilde{C}$ has the term $U_{03}\partial_{U_{02}}$ and this is the only term having U_{03}. On the other hand F is independent of U_{03} so that we have the condition $\partial_{U_{02}} F = 0$. This means that F is independent of U_{02}, *i.e.*, independent of β_n. $\tilde{C}$ also has the term $U_{02}\partial_{U_{01}}$ and this is the only term having U_{02}. Thus by the same argument F is not able to depend on U_{01}, *i.e.*, α_n. □

Now we turn to the variables w and ϕ_{km}. It is immediate to see that w is an invariant of $\tilde{C}$, however, ϕ_{km}'s are not:

$$\begin{aligned} \tilde{C}w &= 0, \\ \tilde{C}\phi_{km} &= \lambda_k \phi_{k+1\,m} + \lambda_m \phi_{k\,m+1}, \qquad \tilde{C}\phi_{k\,\ell+\frac{1}{2}} = \lambda_k \phi_{k+1\,\ell+\frac{1}{2}}, \quad 1 \le k, m \le \ell - \frac{1}{2} \\ \tilde{C}\phi_{\ell+\frac{1}{2}\,\ell+\frac{1}{2}} &= b_\ell \end{aligned} \tag{46}$$

Thus the generator $\tilde{C}$ has the simpler form in terms of ϕ_{km} (we omit $\partial_{U_{0k}}$):

$$\tilde{C} = \sum_{k=1}^{\ell-\frac{1}{2}} \sum_{m=k}^{\ell-\frac{1}{2}} (\lambda_k \phi_{k+1\,m} + \lambda_m \phi_{k\,m+1}) \partial_{\phi_{km}} + \sum_{k=1}^{\ell-\frac{1}{2}} \lambda_k \phi_{k+1\,\ell+\frac{1}{2}} \partial_{\phi_{k\ell+\frac{1}{2}}} + b_\ell \partial_{\phi_{\ell+\frac{1}{2}\,\ell+\frac{1}{2}}} \tag{47}$$

The characteristic equation of the symmetry condition $\tilde{C}F = 0$ is a system of the first order PDEs given by

$$\frac{d\phi_{km}}{\lambda_k \phi_{k+1\,m} + \lambda_m \phi_{k\,m+1}} = \frac{d\phi_{\ell+\frac{1}{2}\,\ell+\frac{1}{2}}}{b_\ell}, \qquad 1 \le k \le m \le \ell - \frac{1}{2} \tag{48}$$

$$\frac{d\phi_{k\,\ell+\frac{1}{2}}}{\lambda_k \phi_{k+1\,\ell+\frac{1}{2}}} = \frac{d\phi_{\ell+\frac{1}{2}\,\ell+\frac{1}{2}}}{b_\ell}, \qquad 1 \le k \le \ell - \frac{1}{2} \tag{49}$$

One may solve it recursively by starting with Equation (49) for $k = \ell - \frac{1}{2}$:

$$\frac{d\phi_{\ell-\frac{1}{2}\,\ell+\frac{1}{2}}}{d\phi} = \frac{\lambda_{\ell-\frac{1}{2}}}{b_\ell}\phi, \qquad \phi = \phi_{\ell+\frac{1}{2}\,\ell+\frac{1}{2}}$$

This gives the invariant of $\tilde{C}$:

$$w_{\ell-\frac{1}{2}\,\ell+\frac{1}{2}} = \phi_{\ell-\frac{1}{2}\,\ell+\frac{1}{2}} - \frac{\lambda_{\ell-\frac{1}{2}}}{b_\ell}\frac{\phi^2}{2} \tag{50}$$

Next we rewrite the Equation (49) for $k = \ell - \frac{3}{2}$ in the following way:

$$\frac{d\phi_{\ell-\frac{3}{2}\,\ell+\frac{1}{2}}}{d\phi} = \frac{\lambda_{\ell-\frac{3}{2}}}{b_\ell}\phi_{\ell-\frac{1}{2}\,\ell+\frac{1}{2}} = \frac{\lambda_{\ell-\frac{3}{2}}}{b_\ell}\left(w_{\ell-\frac{1}{2}\,\ell+\frac{1}{2}} + \frac{\lambda_{\ell-\frac{1}{2}}}{b_\ell}\frac{\phi^2}{2} \right)$$

Then we find an another invariant:

$$w_{\ell-\frac{3}{2}\,\ell+\frac{1}{2}} = \phi_{\ell-\frac{3}{2}\,\ell+\frac{1}{2}} - \frac{\lambda_{\ell-\frac{3}{2}}}{b_\ell} w_{\ell-\frac{1}{2}\,\ell+\frac{1}{2}}\,\phi - \frac{\lambda_{\ell-\frac{3}{2}}\lambda_{\ell-\frac{1}{2}}}{b_\ell^2}\frac{\phi^3}{3!} \tag{51}$$

The complete list of invariants of $\tilde{C}$ is given as follows:

Lemma 6. *Solutions of the Equations (48) and (49) are given by*

$$w_{km} = \phi_{km} - \sum_{a+b\geq 1} c_{ab}(k,m) w_{k+a\,m+b} \frac{\phi^{a+b}_{\ell+\frac{1}{2}\,\ell+\frac{1}{2}}}{(a+b)!} - \gamma(k,m) \frac{\phi^{2\ell+2-k-m}_{\ell+\frac{1}{2}\,\ell+\frac{1}{2}}}{(2\ell+2-k-m)!}, \qquad 1 \leq k \leq \ell - \frac{1}{2},\ k \leq m \leq \ell + \frac{1}{2} \tag{52}$$

where a, b *run over nonnegative integers such that* $a \leq \ell - \frac{1}{2} - k$, $b \leq \ell + \frac{1}{2} - m$ *and* $k + a \leq m + b$. *The coefficient* $\gamma(k,m)$ *depends on* $c_{ab}(k,m)$ *with the maximal value of* a *and* b :

$$\gamma(k,m) = \begin{cases} \dfrac{\lambda_{\ell-\frac{1}{2}}}{b_\ell}, & (k,m) = (\ell - \frac{1}{2}, \ell + \frac{1}{2}) \\ \dfrac{\lambda_{\ell-\frac{1}{2}}}{b_\ell} c_{\max(a)\,\max(b)}(k,m), & \text{otherwise} \end{cases} \tag{53}$$

The coefficients $c_{ab}(k,m)$ *are calculated by the algorithm given below.*

Algorithm. We borrow the terminology of graph theory.

(1) For a given w_{km}, draw a rooted tree according to the branching rules given in Figure 1. Each vetex and each edge of this tree are labelled. The root is labelled by w_{km}. Other vertices and edges are labelled as indicaed in Figure 1. Each vertex has at most two children according to its label. The vertex has no children if its label is $w_{\ell-\frac{1}{2}\,\ell+\frac{1}{2}}$. Thus the hight of the tree is $2\ell - k - m$. An example for $\ell = \frac{7}{2}$ is indicated in Figure 2.

(2) Take a directed path from the root to one of the verticies with label $w_{k+a,m+b}$ and multiply all the edge labels on this path. For instance, take the path (w_{13}, w_{14}, w_{24}) in Figure 2. Then the multiplication of the labels is $\lambda_1 \lambda_3 b_{7/2}^{-2}$.

(3) If there exit other vertices whose label is also $w_{k+a,m+b}$ (same label as (2)), then repeat the same computation as (2) for the direct paths to such vertices. In Figure 2 there is one more vertex whose label is w_{24} and the path is (w_{13}, w_{23}, w_{24}). We have $\lambda_1 \lambda_3 b_{7/2}^{-2}$ for this path, too.

(4) Take summation of all such multiplication for the paths to the vertices whose label is $w_{k+a\,m+b}$, then this summation gives the coefficient $c_{ab}(k,m)$. For the tree in Figure 2 the coefficient of w_{24} is obtained by adding the quantities calculated in (2) and (3): $c_{11}(1,3) = 2\lambda_1 \lambda_3 b_{7/2}^{-2}$.

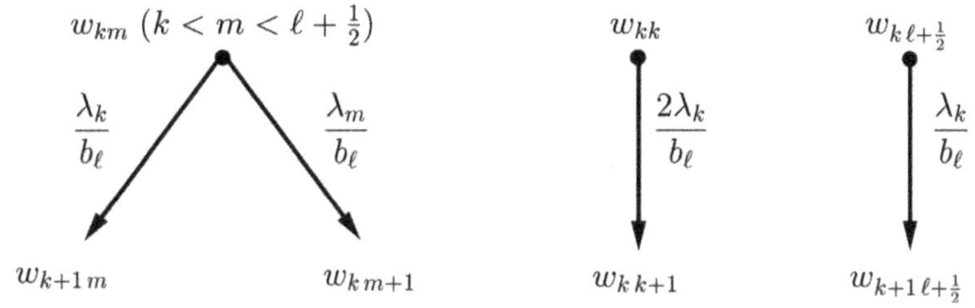

Figure 1. Vertices and edges.

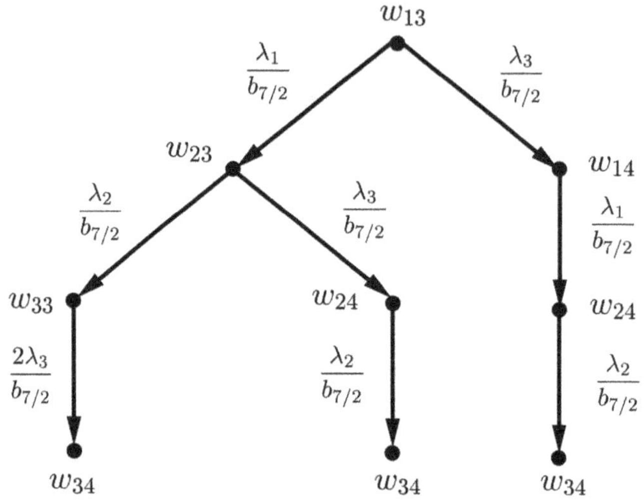

Figure 2. Example of rooted tree: $\ell = \frac{7}{2}$.

Proof of Lemma 6. The lemma is proved by induction on height of the trees. We have a tree of height zero only when label of the root is $w_{\ell-\frac{1}{2}\,\ell+\frac{1}{2}}$. In this case no c_{ab} appears so that Equation (52) yields

$$w_{\ell-\frac{1}{2}\,\ell+\frac{1}{2}} = \phi_{\ell-\frac{1}{2}\,\ell+\frac{1}{2}} - \gamma\Big(\ell-\frac{1}{2},\ell+\frac{1}{2}\Big)\frac{\phi^2_{\ell+\frac{1}{2}\,\ell+\frac{1}{2}}}{2}$$

This coincide with Equation (50). To verify the legitimacy of the algorithm calculating $c_{ab}(k,m)$ we need to start with a tree of height one. There are two possible labels of the root to obtain a tree of height one. They are $w_{\ell-\frac{3}{2}\,\ell+\frac{1}{2}}$ and $w_{\ell-\frac{1}{2}\,\ell-\frac{1}{2}}$. Let us start with the label $w_{\ell-\frac{3}{2}\,\ell+\frac{1}{2}}$

It is not difficult to verify, by employing the algorithm, that we obtain the Equation (51) for this case. For the label $w_{\ell-\frac{1}{2}\,\ell-\frac{1}{2}}$, the algorithm gives the following result:

$$w_{\ell-\frac{1}{2}\,\ell-\frac{1}{2}} = \phi_{\ell-\frac{1}{2}\,\ell-\frac{1}{2}} - \frac{2\lambda_{\ell-\frac{1}{2}}}{b_\ell} w_{\ell-\frac{1}{2}\,\ell+\frac{1}{2}}\phi_{\ell+\frac{1}{2}\,\ell+\frac{1}{2}} - 2\left(\frac{2\lambda_{\ell-\frac{1}{2}}}{b_\ell}\right)^2 \frac{\phi^3_{\ell+\frac{1}{2}\,\ell+\frac{1}{2}}}{3!} \tag{54}$$

It is easy to see that $\tilde{C}$ annihilates Equation (54). Thus the lemma is true for trees of height one.

Now we consider trees of height $h > 1$. If label of the root is w_{km} $(k < m < \ell + \frac{1}{2})$, then the tree has two rooted subtrees (height $h-1$) such that one of then has the root whose label is $w_{k+1\,m}$ and another has the root whose label is $w_{k\,m+1}$. On the other hand, if label of the root is w_{kk} or $w_{k\,\ell+\frac{1}{2}}$, then the tree has only one rooted subtree (height $h-1$) such that the subtree has the root whose label is $w_{k\,k+1}$ or $w_{k+1\,\ell+\frac{1}{2}}$. By the algorithm one may find relations between the coefficients c_{ab}, γ for the tree of height h and the subtrees of height $h-1$:

$$\begin{aligned}
c_{ab}(k,m) &= \frac{\lambda_k}{b_\ell} c_{a-1\,b}(k+1,m) + \frac{\lambda_m}{b_\ell} c_{a\,b-1}(k,m+1), \\
\gamma(k,m) &= \frac{\lambda_k}{b_\ell}\gamma(k+1,m) + \frac{\lambda_m}{b_\ell}\gamma(k,m+1), \\
c_{ab}(k,k) &= \frac{2\lambda_k}{b_\ell} c_{a\,b-1}(k,k+1), \qquad \gamma(k,k) = \frac{2\lambda_k}{b_\ell}\gamma(k,k+1)
\end{aligned} \tag{55}$$

We understand that c_{ab} and γ are zero if their indices or arguments have a impossible value.

Assumption of the induction is that the lemma is true for any rooted subtrees whose height is smaller than h. Namely, we assume that $\tilde{C}w_{k+a\,m+b} = 0$ for $a+b \geq 1$ and what we need to show is that $\tilde{C}w_{km} = 0$. We separate out $a+b=1$ terms from the summation in Equation (52) and use Equation (46) to calculate the action of $\tilde{C}$ on w_{km}. For $k < m < \ell + \frac{1}{2}$ we have

$$\begin{aligned}
\tilde{C}w_{km} &= \tilde{C}\phi_{km} - \lambda_k w_{k+1\,m} - \lambda_m w_{k\,m+1} \\
&- b_\ell \sum_{a+b\geq 2} c_{ab}(k,m) w_{k+a\,m+b} \frac{\phi^{a+b-1}_{\ell+\frac{1}{2}\,\ell+\frac{1}{2}}}{(a+b-1)!} - b_\ell \gamma(k,m) \frac{\phi^{2\ell+1-k-m}_{\ell+\frac{1}{2}\,\ell+\frac{1}{2}}}{(2\ell+1-k-m)!} \\
&= \tilde{C}\phi_{km} - \lambda_k w_{k+1\,m} - \lambda_m w_{k\,m+1} \\
&- \sum_{a+b\geq 1} \left(\lambda_k c_{ab}(k+1,m) w_{k+1+a\,m+b} + \lambda_m c_{ab}(k,m+1) w_{k\,m+1}\right) \frac{\phi^{a+b}_{\ell+\frac{1}{2}\,\ell+\frac{1}{2}}}{(a+b)!} \\
&- \left(\lambda_k \gamma(k+1,m) + \lambda_m \gamma(k,m+1)\right) \frac{\phi^{2\ell+1-k-m}_{\ell+\frac{1}{2}\,\ell+\frac{1}{2}}}{(2\ell+1-k-m)!}
\end{aligned}$$

The second equality is due to the relations Equation (55) and the replacement $a-1$ (resp. $b-1$) with
a (resp. b). By the assumption of the induction one may use Equation (52) to obtain:

$$\tilde{C}w_{km} = \tilde{C}\phi_{km} - (\lambda_k \phi_{k+1\,m} + \lambda_m \phi_{k\,m+1}) = 0$$

The second equality is due to Equation (46).

The proof of $\tilde{C}w_{kk} = \tilde{C}w_{k\,\ell+\frac{1}{2}} = 0$ is done in a similar way. This completes the proof of Lemma 6. □

Corollary 1. *The variables* $w_{k\,\ell+\frac{1}{2}}$ $(1 \leq k \leq \ell - \frac{1}{2})$ *are easily calculated by this method.*

$$w_{k\,\ell+\frac{1}{2}} = \phi_{k\,\ell+\frac{1}{2}} - \sum_{n=1}^{\ell-\frac{1}{2}-k} \binom{2\ell+1-k}{n} w_{k+n\,\ell+\frac{1}{2}} \left(\frac{\phi_{\ell+\frac{1}{2}\,\ell+\frac{1}{2}}}{b_\ell}\right)^n - \binom{2\ell+1-k}{\ell+\frac{1}{2}} \frac{1}{b_\ell^{\ell+\frac{1}{2}-k}} \frac{\phi^{\ell+\frac{3}{2}-k}_{\ell+\frac{1}{2}\,\ell+\frac{1}{2}}}{\ell+\frac{3}{2}-k}$$

Proof of Corollary 1. The rooted tree used for this computation is indicated in Figure 3. It follows that the coefficient of $w_{k+a\,\ell+\frac{1}{2}}$ is given by

$$c_{a0}(k,\ell+\tfrac{1}{2}) = \frac{\lambda_k \lambda_{k+1} \cdots \lambda_{k+a-1}}{b_\ell^a} = \binom{2\ell+1-k}{a} \frac{a!}{b_\ell^a}$$

By Equation (53) the coefficient $\gamma(k,\ell+\frac{1}{2})$ is calculated as $\gamma(k,\ell+\frac{1}{2}) = \lambda_{\ell-\frac{1}{2}} b_\ell^{-1} c_{\ell-\frac{1}{2}-k0}(k,\ell+\frac{1}{2})$. Thus we obtain the expression of $w_{k\,\ell+\frac{1}{2}}$ given in the corollary. □

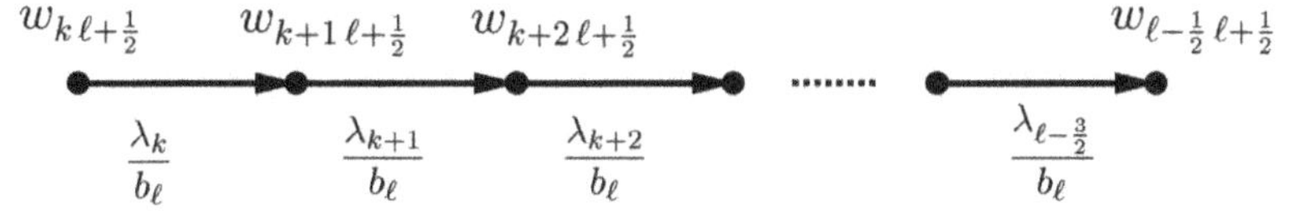

Figure 3. Rooted tree for the computation of $w_{k\,\ell+\frac{1}{2}}$.

Our final task is to consider the invariance under D. It is immediate to see that $\hat{D}$ scales w_{km} as

$$w_{km} \to e^{-2(2\ell+2-k-m)\epsilon}\, w_{km}$$

Together with the scaling law Equation (44) we arrive at the final theorem.

Theorem 3. *The PDE invariant under the group generated by $\mathfrak{g}_\ell$ with $\ell \geq \frac{5}{2}$ is given by*

$$F\left(\frac{w_{km}}{w^{2\ell+2-k-m}}\right) = 0, \quad 1 \leq k \leq \ell - \frac{1}{2},\ k \leq m \leq \ell + \frac{1}{2} \tag{56}$$

where F is an arbitrary differentiable function. The variables w and w_{km} are given in Equations (41) and (52), respectively. This is the PDE with $\ell + \frac{3}{2}$ independent and one dependent variables. The function F has $\frac{1}{2}\left(\ell - \frac{1}{2}\right)\left(\ell + \frac{5}{2}\right)$ arguments.

Example 1. *Invariant PDE for $\ell = \frac{5}{2}$.*

$$F\left(\frac{w_{11}}{w^5}, \frac{w_{12}}{w^4}, \frac{w_{13}}{w^3}, \frac{w_{22}}{w^3}, \frac{w_{23}}{w^2}\right) = 0$$

where

$$\begin{aligned}
w &= \frac{U_0}{U} + x_2\frac{U_1}{U} + 2x_3\frac{U_2}{U} - \frac{U_3^2}{8U^2}, \\
w_{23} &= \phi_{23} - \frac{1}{18}\phi_{33}^2, \\
w_{22} &= \phi_{22} - \frac{2}{9}\phi_{23}\phi_{33} + \frac{2}{3^5}\phi_{33}^3, \\
w_{13} &= \phi_{13} - \frac{5}{36}\phi_{23}\phi_{33} + \frac{5}{2^2 3^5}\phi_{33}^3, \\
w_{12} &= \phi_{12} - \frac{1}{9}\phi_{13}\phi_{33} - \frac{5}{36}\phi_{22}\phi_{33} + \frac{5}{2^3 3^3}\phi_{23}\phi_{33}^2 - \frac{5}{2^5 3^5}\phi_{33}^4, \\
w_{11} &= \phi_{11} - \frac{5}{18}\phi_{12}\phi_{33} + \frac{5}{2^2 3^4}\phi_{13}\phi_{33}^2 - \frac{25}{2^4 3^4}\phi_{22}\phi_{33}^2 - \frac{25}{2^4 3^6}\phi_{23}\phi_{33}^3 - \frac{5}{2^4 3^8}\phi_{33}^5
\end{aligned}$$

The symmetry generators are given by

$$\begin{aligned}
&M = U\partial_U, \qquad D = 2t\partial_t + 5x_1\partial_{x_1} + 3x_2\partial_{x_2} + x_3\partial_{x_3}, \qquad H = \partial_t, \\
&C = t(t\partial_t + 5x_1\partial_{x_1} + 3x_2\partial_{x_2} + x_3\partial_{x_3}) + 5x_1\partial_{x_2} + 4x_2\partial_{x_3} - 18x_3^2 U\partial_U, \\
&P^{(1)} = \partial_{x_1}, \qquad P^{(2)} = t\partial_{x_1} + \partial_{x_2}, \qquad P^{(3)} = t^2\partial_{x_1} + 2t\partial_{x_2} + \partial_{x_3}, \\
&P^{(4)} = t^3\partial_{x_1} + 3t^2\partial_{x_2} + 3t\partial_{x_3} - 12x_3 U\partial_U, \\
&P^{(5)} = t^4\partial_{x_1} + 4t^3\partial_{x_2} + 4t^2\partial_{x_3} - 24(2tx_3 + x_2)U\partial_U, \\
&P^{(6)} = t^5\partial_{x_1} + 5t^4\partial_{x_2} + 10t^3\partial_{x_3} - 120(t^2x_3 - tx_2 + x_1)U\partial_U
\end{aligned}$$

6. Concluding Remarks

We have constructed nonlinear PDEs invariant under the transformations generated by the realization of CGA given in Equation (2). This was done by obtaining the general solution of the symmetry conditions so that the PDEs constructed in this work are the most general ones invariant under Equation (2). A remarkable property of the PDEs is that they do not contain the second order derivative in t if $\ell > \frac{3}{2}$. It means that there exist no invariant PDEs of wave or Klein–Gordon type for $\ell > \frac{3}{2}$. This type of ℓ-dependence does not appear in the linear PDEs constructed in [1,4] based on the representation theory of $\mathfrak{g}_\ell$. This will be changed if one start with a realization of CGA which is different from Equation (2).

The CGAs considered in this work are only $d = 1$ members. Extending the present computation to higher values of d would be an interesting future work. Because the $d = 2$ CGA has a distinct central extension so that we will have different types of invariant PDEs. For $d \geq 3$ CGAs have $so(d)$ as a subalgebra. This will also cause a significant change in invariant PDEs.

Acknowledgments: The authors are grateful to Y. Uno for helpful discussion. N.A. is supported by the grants-in-aid from JSPS (Contract No. 26400209).

Author Contributions: Tadanori Kato performed the whole computation of Section 3. Naruhiko Aizawa designed the idea of this research and was responsible for the computations in other sections and manuscript writing.

Conflicts of Interest: The authors declare no conflict of interest.

References

1. Aizawa, N.; Kimura, Y.; Segar, J. Intertwining operators for ℓ-conformal Galilei algebras and hierarchy of invariant equations. *J. Phys. A Math. Theor.* **2013**, *46*, doi:10.1088/1751-8113/46/40/405204.
2. Aizawa, N.; Dobrev, V.K.; Doebner, H.D. Intertwining operators for Schrödinger algebras and hierarchy of invariant equations. In *Quantum Theory and Symmetries*; Kapuścik, E., Horzela, A., Eds.; World Scientific: Singapore, 2002; pp. 222–227.
3. Aizawa, N.; Dobrev, V.K.; Doebner, H.D.; Stoimenov, S. Intertwining operators for the Schrödinger algebra in $n \geq 3$ space dimension. In *Proceedings of the VII International Workshop on Lie Theory and Its Applications in Physics*; Doebner, H.D., Dobrev, V.K., Eds.; Heron Press: Sofia, Bulgaria, 2008; pp. 372–399.
4. Aizawa, N.; Chandrashekar, R.; Segar, J. Lowest weight representations, singular vectors and invariant equations for a class of conformal Galilei algebras. *SIGMA* **2015**, *11*, doi:10.3842/SIGMA.2015.002.
5. Negro, J.; del Olmo, M.; Rodrıguez-Marco, A. Nonrelativistic conformal groups. *J. Math. Phys.* **1997**, *38*, 3786–3809.
6. Negro, J.; del Olmo, M.; Rodrıguez-Marco, A. Nonrelativistic conformal groups. II. Further developments and physical applications. *J. Math. Phys.* **1997**, *38*, 3810–3831.
7. Havas, P.; Plebański, J. Conformal extensions of the Galilei group and their relation to the Schrödinger group. *J. Math. Phys.* **1978**, *19*, 482–488.
8. Henkel, M. Local scale invariance and strongly anisotropic equilibrium critical systems. *Phys. Rev. Lett.* **1997**, *78*, doi:10.1103/PhysRevLett.78.1940.
9. Martelli, D.; Tachikawa, Y. Comments on Galilean conformal field theories and their geometric realization. *JHEP* **2010**, *5*, 1–31.
10. Lie, S. *Theorie der Transformationsgruppen*; Chelsea: New York, NY, USA, 1970.
11. Jacobi, C.G.J.; Borchardt, C.W. *Vorlesungen über Dynamik*; Reimer, G., Ed.; University of Michigan Library: Ann Arbor, MI, USA, 1866.
12. Niederer, U. The maximal kinematical invariance group of the free Schrodinger equation. *Helv. Phys. Acta* **1972**, *45*, 802–810.
13. Niederer, U. The maximal kinematical invariance group of the harmonic oscillator. *Helv. Phys. Acta* **1973**, *46*, 191–200.
14. Niederer, U. The maximal kinematical invariance groups of schroedinger equations with arbitrary potentials. *Helv. Phys. Acta* **1974**, *47*, 167–172.
15. Hagen, C.R. Scale and conformal transformations in Galilean-covariant field theory. *Phys. Rev. D* **1972**, *5*, 377–388.
16. Jackiw, R. Introducing scale symmetry. *Phys. Today* **1972**, *25*, 23–27.
17. Burdet, G.; Perrin, M. Many-body realization of the Schrödinger algebra. *Lett. Nuovo Cim.* **1972**, *4*, 651–655.
18. Son, D.T. Toward an AdS/cold atoms correspondence: A geometric realization of the Schroedinger symmetry. *Phys. Rev. D* **2008**, *78*, doi:10.1103/PhysRevD.78.046003.
19. Balasubramanian, K.; McGreevy, J. Gravity duals for nonrelativistic conformal field theories. *Phys. Rev. Lett.* **2008**, *101*, doi:10.1103/PhysRevLett.101.061601.
20. Unterberger, J.; Roger, C. *The Schrödinger-Virasoro Algebra: Mathematical Structure and Dynamical Schrödinger Symmetries*; Springer: Berlin/Heidelberg, Germany, 2012.

21. Aizawa, N.; Isaac, P.S.; Kimura, Y. Highest weight representations and Kac determinants for a class of conformal Galilei algebras with central extension. *Int. J. Math.* **2012**, *23*, 1250118:1-1250118:25.
22. Henkel, M. Phenomenology of local scale invariance: From conformal invariance to dynamical scaling. *Nucl. Phys. B* **2002**, *641*, 405–486.
23. Aizawa, N.; Isaac, P.S. On irreducible representations of the exotic conformal Galilei algebra. *J. Phys. A Math. Theor.* **2011**, *44*, doi:10.1088/1751-8113/44/3/035401.
24. Lü, R.; Mazorchuk, V.; Zhao, K. On simple modules over conformal Galilei algebras. *J. Pure Appl. Algebra* **2014**, *218*, 1885–1899.
25. Fushchich, W.I.; Cherniha, R.M. The Galilean relativistic principle and nonlinear partial differential equations. *J. Phys. A Math. Gen.* **1985**, *18*, 3491–3503.
26. Fushchich, W.I.; Cherniha, R.M. Galilei invariant non-linear equations of Schrödinger type and their exact solutions. I. *Ukr. Math. J.* **1989**, *41*, 1161–1167.
27. Rideau, G.; Winternitz, P. Evolution equations invariant under two-dimensional space-time Schrödinger group. *J. Math. Phys.* **1993**, *34*, 558–570.
28. Cherniha, R.M.; Henkel, M. The exotic conformal Galilei algebra and nonlinear partial differential equations. *J. Math. Anal. Appl.* **2010**, *369*, 120–132.
29. Fushchych, W.I.; Cherniha, R.M. Galilei-invariant nonlinear systems of evolution equations. *J. Phys. A Math. Gen.* **1995**, *28*, 5569–5579.
30. Cherniha, R.M.; Henkel, M. On non-linear partial differential equations with an infinite-dimensional conditional symmetry. *J. Math. Anal. Appl.* **2004**, *298*, 487–500.
31. Duval, C.; Horvathy, P.A. Non-relativistic conformal symmetries and Newton–Cartan structures. *J. Phys. A Math. Theor.* **2009**, *42*, doi:10.1088/1751-8113/42/46/465206.
32. Duval, C.; Horvathy, P.A. Conformal Galilei groups, Veronese curves and Newton–Hooke spacetimes. *J. Phys. A Math. Theor.* **2011**, *44*, 335203:1–335203:21.
33. Gomis, J.; Kamimura, K. Schrödinger equations for higher order nonrelativistic particles and N-Galilean conformal symmetry. *Phys. Rev. D* **2012**, *85*, doi:10.1103/PhysRevD.85.045023.
34. Galajinsky, A.; Masterov, I. Remarks on l-conformal extension of the Newton–Hooke algebra. *Phys. Lett. B* **2011**, *702*, 265–267.
35. Galajinsky, A.; Masterov, I. Dynamical realization of l-conformal Galilei algebra and oscillators. *Nucl. Phys. B* **2013**, *866*, 212–227.
36. Galajinsky, A.; Masterov, I. Dynamical realizations of l-conformal Newton–Hooke group. *Phys. Lett. B* **2013**, *723*, 190–195.
37. Andrzejewski, K.; Galajinsky, A.; Gonera, J.; Masterov, I. Conformal Newton–Hooke symmetry of Pais–Uhlenbeck oscillator. *Nucl. Phys. B* **2014**, *885*, 150–162.
38. Galajinsky, A.; Masterov, I. On dynamical realizations of l-conformal Galilei and Newton–Hooke algebras. *Nucl. Phys. B* **2015**, *896*, 244–254.
39. Andrzejewski, K.; Gonera, J.; Kosiński, P.; Maślanka, P. On dynamical realizations of l-conformal Galilei groups. *Nucl. Phys. B* **2013**, *876*, 309–321.
40. Andrzejewski, K.; Gonera, J.; Kijanka-Dec, A. Nonrelativistic conformal transformations in Lagrangian formalism. *Phys. Rev. D* **2013**, *87*, doi:10.1103/PhysRevD.87.065012.
41. Andrzejewski, K.; Gonera, J. Dynamical interpretation of nonrelativistic conformal groups. *Phys. Lett. B* **2013**, *721*, 319–322.
42. Andrzejewski, K.; Gonera, J. Unitary representations of N-conformal Galilei group. *Phys. Rev. D* **2013**, *88*, doi:10.1103/PhysRevD.88.065011.
43. Andrzejewski, K. Conformal Newton–Hooke algebras, Niederer's transformation and Pais–Uhlenbeck oscillator. *Phys. Lett. B* **2014**, *738*, 405–411.
44. Andrzejewski, K.; Gonera, J.; Maślanka, P. Nonrelativistic conformal groups and their dynamical realizations. *Phys. Rev. D* **2012**, *86*, doi:10.1103/PhysRevD.86.065009.
45. Aizawa, N.; Kuznetsova, Z.; Toppan, F. ℓ-oscillators from second-order invariant PDEs of the centrally extended Conformal Galilei Algebras. *J. Math. Phys.* **2015**, *56*, 031701:1–031701:14.
46. Olver, P. *Applications of Lie Groups to Differential Equations*; Springer: New York, NY, USA, 2000.

47. Bluman, G.W.; Kumei, S. *Symmetries and Differential Equations*; Springer: New York, NY, USA, 1989.
48. Stephani, H. *Differential Equations: Their Solution Using Symmetries*; Cambridge University Press: New York, NY, USA, 1989.
49. Courant, R.; Hilbert, D. *Methods of Mathematical Physics*; CUP Archive: New York, NY, USA, 1966; Volume 1.

symmetry

MDPI

Article

From Conformal Invariance towards Dynamical Symmetries of the Collisionless Boltzmann Equation

Stoimen Stoimenov [1,*] and Malte Henkel [2]

1 Institute of Nuclear Research and Nuclear Energy, Bulgarian Academy of Sciences, 72 Tsarigradsko Chaussee, Blvd., BG–1784 Sofia, Bulgaria
2 Groupe de Physique Statistique, Institut Jean Lamour (CNRS UMR 7198), Université de Lorraine Nancy, B.P. 70239, F – 54506 Vandœuvre-lès-Nancy Cedex, France; malte.henkel@univ-lorraine.fr
* Author to whom correspondence should be addressed; spetrov@inrne.bas.bg; Tel.: +359-882-114551.

Academic Editor: Roman M. Cherniha
Received: 30 June 2015; Accepted: 18 August 2015; Published: 7 September 2015

Abstract: Dynamical symmetries of the collisionless Boltzmann transport equation, or Vlasov equation, but under the influence of an external driving force, are derived from non-standard representations of the 2D conformal algebra. In the case without external forces, the symmetry of the conformally-invariant transport equation is first generalized by considering the particle momentum as an independent variable. This new conformal representation can be further extended to include an external force. The construction and possible physical applications are outlined.

Keywords: conformal invariance; conformal Galilean algebra; Boltzmann equation

1. Introduction

The Boltzmann transport equation (BTE) [1–4] furnishes a semi-classical description of the effects of particle transport, including the influence of external forces on the effective single-particle distribution function $f = f(t, r, p)$ of a small cell in phase phase, centered at position r and momentum p. For a system with identical particles of mass m, the Boltzmann equation reads:

$$\frac{\partial f}{\partial t} + \frac{p}{m} \cdot \frac{\partial f}{\partial r} + F \cdot \frac{\partial f}{\partial p} = \left(\frac{\partial f}{\partial t} \right)_{\text{coll}} . \tag{1}$$

Here, $\mathrm{d}N = f(t, r, p,)\mathrm{d}r\, \mathrm{d}p$ is the number of particles in a cell of phase volume $\mathrm{d}r\, \mathrm{d}p$, centered at position r and momentum p [3]. In addition, $F = F(t, r)$ is the force field acting on the particles in the fluid. The term on the right-hand side is added to describe the effect of collisions between particles. It is a statistical term and requires knowledge of the statistics that the particles obey, like the Maxwell–Boltzmann, Fermi–Dirac or Bose–Einstein distributions. In his famous "Stoßzahlansatz" (or hypothesis of molecular chaos), Boltzmann obtained an explicit form for it. In modern notation, for example for an interacting Fermi gas, where a particle from a state with momentum p is scattered to a state with momentum p', whereas a second particle is scattered from a momentum q to a momentum q', the collision term reads:

$$\begin{aligned} \left(\tfrac{\partial f}{\partial t} \right)_{\text{coll}} &= -\int \mathrm{d}p' \mathrm{d}q \mathrm{d}q'\, w(\{p, q\} \to \{p', q'\}) \\ &\times [f(p)f(q)(1 - f(p'))(1 - f(q')) - f(p')f(q')(1 - f(p))(1 - f(q))] \end{aligned}$$

where $w(\{pq\} \to \{p'q'\})$ is the normalized transition probability from the two-particle state with momenta $\{p, q\}$ to the state labeled by $\{p', q'\}$. Clearly, solving this widely-studied equation is a very difficult task. It might be hoped that symmetries could be helpful. The equation without the collision term is known as the Vlasov equation [5]. The relationship with Landau damping and a physicists' derivation can be found in [6,7]. In this work, we shall explore a class of symmetries of the (collisionless) BTE.

Throughout, we shall restrict to the $d = 1$ space dimension. (By analogy with other constructions of local scale symmetries (see [8–11] and especially [12] and references therein), we expect a straightforward extension of the results reported here to $d > 1$. Since we shall construct here a finite-dimensional Lie algebra of dynamical conformal symmetries of the $1D$ collisionless BTE, one should indeed expect that an extension to $d > 1$ exists. That symmetry algebra should contain three generators $X_{\pm 1,0}$, along with a vector of generators Y_n and also spatial rotations.) We start from a non-standard representation, isomorphic to the infinite-dimensional Lie algebra of conformal transformations in $d = 2$ dimensions. (For the sake of clarity, we shall adopt the following convention of terminology: the infinite-dimensional Lie algebra $\langle X_n, Y_n \rangle_{n\in\mathbb{Z}}$ will be called a (centerless) "conformal Virasoro algebra". Its maximal finite-dimensional sub-algebra $\langle X_n, Y_n \rangle_{n\in\{-1,0,1\}}$ will be called a "conformal algebra") This Lie algebra is spanned by the generators $\langle X_n, Y_n \rangle_{n\in\mathbb{Z}}$ and can be defined from the commutators [9,12]:

$$[X_n, X_m] = (n-m)X_{n+m}, [X_n, Y_m] = (n-m)Y_{n+m}, [Y_n, Y_m] = \mu(n-m)Y_{n+m} \tag{2}$$

where μ is a parameter. An explicit realization in terms of time-space transformation is [9,12]:

$$X_n = -t^{n+1}\partial_t - \mu^{-1}[(t+\mu r)^{n+1} - t^{n+1}]\partial_r - (n+1)xt^n - (n+1)\tfrac{\gamma}{\mu}[(t+\mu r)^n - t^n]$$

$$Y_n = -(t+\mu r)^{n+1}\partial_r - (n+1)\gamma(t+\mu r)^n \tag{3}$$

such that μ^{-1} can be interpreted as a velocity ("speed of light/sound") and where x, γ are constants. (The contraction $\mu \to 0$ of Equation (3) produces the non-semi-simple "altern-Virasoro algebra" $\mathfrak{altv}(1)$ (but without central charges). Its maximal finite-dimensional sub-algebra is the conformal Galilean algebra $\mathfrak{alt}(1) \equiv$ CGA(1) [9,13]; see also [8,11]. The CGA(d) is non-isomorphic to either the standard Galilei algebra or else the Schrödinger algebra.) Writing $X_n = \ell_n + \bar{\ell}_n$ and $Y_n = \mu^{-1}\bar{\ell}_n$, where the generators $\left\langle \ell_n, \bar{\ell}_n \right\rangle_{n\in\mathbb{Z}}$ satisfy $[\ell_n, \ell_m] = (n-m)\ell_{n+m}$, $\left[\bar{\ell}_n, \bar{\ell}_m\right] = (n-m)\bar{\ell}_{n+m}$, $\left[\ell_n, \bar{\ell}_m\right] = 0$, it can be seen that, provided $\mu \neq 0$, the above Lie algebra Equation (2) is isomorphic to a pair of Virasoro algebras $\mathfrak{vect}(S^1) \oplus \mathfrak{vect}(S^1)$ with a vanishing central charge. However, this isomorphism does not imply that physical systems described by two different representations of the conformal Virasoro algebra, or the conformal algebra, with commutators Equation (2), were trivially related. For example, it is well known that if one uses the generators of the standard representation of conformal invariance or else the non-standard representation Equation (4) in order to find co-variant two-point functions, the resulting scaling forms are different [9].

Now, consider the maximal finite-dimensional sub-algebra $\langle X_{\pm 1,0}, Y_{\pm 1,0} \rangle$, which for $\mu \neq 0$, in turn, is isomorphic to the direct sum $\mathfrak{sl}(2,\mathbb{R}) \oplus \mathfrak{sl}(2,\mathbb{R})$. The explicit realization follows from from Equation (3):

$$X_{-1} = -\partial_t, X_0 = -t\partial_t - r\partial_r - x, \qquad X_1$$

$$Y_{-1} = -\partial_r, Y_0 = -t\partial_r - \mu r\partial_r - \gamma, Y_1 = -t^2\partial_r - 2\mu tr\partial_r - \mu^2 r^2\partial_r - 2\gamma t - 2\gamma\mu r \tag{4}$$

Here, the generators X_{-1}, Y_{-1} describe time- and space-translations, Y_0 is a (conformal) Galilei transformation (since the commutator $[Y_0, Y_{-1}]$ does not vanish and does not give a central element of the Lie algebra Equation (2), its structure is fundamentally different from algebras containing the usual Galilei algebra as a sub-algebra), X_0 gives the dynamical scaling $t \mapsto \lambda t$ of $r \mapsto \lambda r$ (with $\lambda \in \mathbb{R}$), such that the so-called "dynamical exponent" $z = 1$, since both time and space are re-scaled in the same way, and, finally, X_{+1}, Y_{+1} give "special" conformal transformations. In the context of statistical mechanics of conformally-invariant phase transitions, one characterizes co-variant quasi-primary scaling operators through the invariant parameters (x, μ, γ), where x is the scaling dimension.

Finally, the finite-dimensional representation Equation (4) acts as a dynamical symmetry on the equation of motion:

$$\hat{S}\phi(t,r) = (-\mu\partial_t + \partial_r)\phi(t,r) = 0. \tag{5}$$

in the sense that a solution ϕ of $\hat{S}\phi = 0$ is mapped onto another solution of the same equation. Indeed, it is easily checked that: $[\hat{S}, Y_{\pm 1,0}] = [\hat{S}, X_{-1}] = 0$ and

$$[\hat{S}, X_0] = -\hat{S}, [\hat{S}, X_1] = -2t\hat{S} + 2(\mu x - \gamma) \tag{6}$$

It follows that for fields ϕ with scaling dimensions $x_\phi = x = \gamma/\mu$, the algebra Equation (4) really leaves the solution space of Equation (5) invariant.

In order to return to the Boltzmann equation, we consider Equation (5) in the form:

$$\hat{L}f = (\mu\partial_t + v\partial_r)f(t,r,v) = 0 \tag{7}$$

where $f = f(t,r,v)$ is interpreted as a single-particle distribution function and where we consider v as an additional variable. Equation (7) is a simple Boltzmann equation, without an external force, without a collision term and in one space dimension. From Equation (6), with v fixed (and normalized to $v = 1$), its solution space is conformally invariant. (With respect to Equation (5), $\mu \mapsto -\mu$ was replaced. This change must also be made in the generators Equation (4) and commutators Equation (2)). In Section 2, we shall generalize the above representation of the conformal algebra to the situation with v as a further variable. In Section 3, we shall further extend this to the case when an external force $F = F(t,r,v)$, possibly depending on time, spatial position and velocity, is included. The aim of these calculations is to determine which situations of potential physical interest with a non-trivial conformal symmetry might be identified. This explorative study aims at identifying lines for further study, which might lead later to a more comprehensive understanding of the possible symmetries of Boltzmann equations. Taking into account the collision term is left for future work. We shall concentrate on the $d = 1$ space dimension throughout. Conclusions and final comments are given in Section 4.

2. Collisionless Boltzmann Equation without External Forces

In our construction of conformal dynamical symmetries of the 1D collisionless BTE, we shall often meet Lie algebras of a certain structure. These will be isomorphic to the two-dimensional conformal algebra.

Proposition 1. *The Lie algebra $\langle X_n, Y_n \rangle_{n\in\mathbb{Z}}$ defined by the commutators:*

$$[X_n, X_m] = (n-m)X_{n+m}, [X_n, Y_m] = (n-m)Y_{n+m}, [Y_n, Y_m] = (n-m)(kX_{n+m} + qY_{n+m}) \tag{8}$$

where k, q are constants, is isomorphic to the pair of centerless Viraso algebras $\mathfrak{vect}(S^1) \oplus \mathfrak{vect}(S^1)$.

Proof. For either $k = 0$ or $q = 0$, this is either evident or else has already been seen in Section 1. In the other case, consider the change of basis $X_n = \ell_n + \bar{\ell}_n$ and $Y_n = \alpha\ell_n - \beta\bar{\ell}_n$, where $\ell_n, \bar{\ell}_n$ are two families of commuting generators of $\mathfrak{vect}(S^1)$ and α and β are constants, such that $\alpha + \beta \neq 0$. It then follows $k = \alpha\beta$ and $q = \alpha - \beta$. □

This implies in particular the isomorphism of the maximal finite-dimensional sub-algebras, or "conformal algebras" in the terminology chosen here. By definition, this "conformal algebra" obeys the commutators Equation (8), but with $n, m \in \{-1, 0, 1\}$.

Our construction of dynamical symmetries of the Equation (7) follows the lines of the construction of local scale-invariance in time-dependent critical phenomena [9]. The physically-motivated requirements are: First of all it, is clear that the equation is invariant under time-translations:

$$X_{-1} = -\partial_t, [\hat{L}, X_{-1}] = 0 \tag{9}$$

Some kind of dynamical scaling must be present, as well. Its most general form is:

$$X_0 = -t\partial_t - \frac{r}{z}\partial_r - \frac{1-z}{z}v\partial_v - x, [\hat{L}, X_0] = -\hat{L}. \tag{10}$$

Whenever the dynamical exponent $z \neq 1$, we shall find an explicit dependence on v. In general, we look for a family of generators X_n, for which we make the ansatz:

$$X_n = -a_n(t,r,v)\partial_t - b_n(t,r,v)\partial_r - c_n(t,r,v)\partial_v - d_n(t,r,v). \tag{11}$$

We shall find X_n from the following three conditions (throughout, we use the notations $\partial_t f = \dot{f}, \partial_r f = f'$):

1. X_n must be a symmetry for the Equation (7); hence, $[\hat{L}, X_n] = \lambda_n \hat{L}$. This gives:

$$\mu\dot{a}_n + va'_n + \mu\lambda_n = 0, \mu\dot{b}_n + vb'_n - c_n + \lambda_n v = 0 \tag{12}$$

$$\mu\dot{c}_n + vc'_n = 0, \mu\dot{d}_n + vd'_n = 0.$$

2. The generator X_0 is assumed to be in the Cartan sub-algebra; hence, $[X_n, X_0] = \alpha_{n,0}X_n$. It follows:

$$(1+\alpha_{n,0})a_n - t\dot{a}_1 - \tfrac{r}{z}a'_n - \tfrac{1-z}{z}v\partial_v a_n = 0 \tag{13}$$

$$(1/z+\alpha_{n,0})b_n - t\dot{b}_n - \tfrac{r}{z}b'_n - \tfrac{1-z}{z}v\partial_v b_n = 0 \tag{14}$$

$$((1-z)/z+\alpha_{n,0})c_n - t\dot{c}_1 - \tfrac{r}{z}c'_n - \tfrac{1-z}{z}v\partial_v c_n = 0 \tag{15}$$

$$\alpha_{n,0}d_n - t\dot{d}_n - \tfrac{r}{z}d'_n - \tfrac{1-z}{z}v\partial_v d_n = 0. \tag{16}$$

3. The action of X_{-1} is as a lowering operator; hence, $[X_n, X_{-1}] = \alpha_{n,-1}X_{n-1}$. It follows:

$$\dot{a}_n = \alpha_{n,-1}t, \dot{b}_n = \alpha_{n,-1}r/z \tag{17}$$

$$\dot{c}_n = \alpha_{n,-1}v(1-z)/z, \dot{d}_n = \alpha_{n,-1}x/z.$$

These conditions, combined with the following initial conditions:

$$a_0 = t, b_0 = \tfrac{r}{z}, c_0 = \tfrac{1-z}{z}v, d_0 = x$$

$$a_{-1} = 1, b_{-1} = 0, c_{-1} = 0, d_{-1} = 0. \tag{18}$$

must be sufficient for the determination of all admissible forms of X_n.

In the special case $n = 1$, we have $\alpha_{1,0} = 1$ and find the most general form of X_1 as a symmetry of Equation (7) as follows (the requirement that $\langle X_{\pm 1,0}\rangle$ close into the Lie algebra $\mathfrak{sl}(2,\mathbb{R})$ fixes $\alpha_{1,-1} = 2$):

$$X_1 = -a_1(t,r,v)\partial_t - b_1(t,r,v)\partial_r - c_1(t,r,v)\partial_v - d_1(t,r,v) \tag{19}$$

and

$$a_1(t,r,v) = t^2 + A_{12}r^2v^{-2} + A_{110}rv^{\frac{2z-1}{1-z}} + A_{100}v^{\frac{2z}{1-z}} \tag{20}$$

$$b_1(t,r,v) = \tfrac{2}{z}tr + \left(\tfrac{A_{12}}{\mu} + \tfrac{z-2}{z}\mu\right)r^2v^{-1} + B_{110}rv^{\frac{z}{1-z}} + B_{100}v^{\frac{z+1}{1-z}} \quad (21)$$

$$c_1(t,r,v) = \tfrac{2}{z}(1-z)(vt-\mu r) + (B_{110} - \tfrac{A_{110}}{\mu})v^{\frac{z}{1-z}} \quad (22)$$

$$d_1(t,r,v) = \tfrac{2}{z}xt - \tfrac{2}{z}\mu xrv^{-1} + D_0v^{\frac{z}{1-z}} \quad (23)$$

with a certain set of undetermined constants.

For conformal invariance, a family of generators Y_n must also be found. Its construction is straightforward if the explicit form of Y_{-1} is known. Really, X_1 must act as a raising operator, in both hierarchies, such that [9]:

$$[X_1, Y_{-1}] \sim Y_0, [X_1, Y_0] \sim Y_1. \quad (24)$$

which implies that $[Y_{-1}, [Y_{-1}, X_1]] \sim Y_{-1}$. However, the usual realization of $Y_{-1} = -\partial_r$ as space translations does not work, since if we set all undetermined constants in Equation (19) to zero, one would have $[Y_{-1}, [Y_{-1}, X_1]] \sim v^{-1}Y_{-1}$. It is better to work with the form:

$$Y_{-1} = -v\partial_r. \quad (25)$$

as we shall do from now on.

We first consider the special case, when all of the constants in the expression Equation (19) for X_1 vanish:

Case A: $A_{12} = A_{110} = A_{100} = B_{110} = B_{100} = D_0 = 0$.

Proposition 2. *The six generators:*

$$\begin{aligned}
X_{-1} &= -\partial_t, X_0 = -t\partial_t - \tfrac{r}{z}\partial_r - \tfrac{1-z}{z}v\partial_v - \tfrac{x}{z} \\
X_1 &= -t^2\partial_t - (\tfrac{2}{z}tr + \tfrac{z-2}{z}\mu r^2v^{-1})\partial_r - \tfrac{2(1-z)}{z}(vt-\mu r)\partial_v - \tfrac{2}{z}xt + \tfrac{2}{z}\mu xrv^{-1} \\
Y_{-1} &= -v\partial_r, Y_0 = -(tv - \tfrac{\mu}{z}r)\partial_r - \tfrac{z-1}{z}\mu v\partial_v + \mu\tfrac{x}{z} \\
Y_1 &= -(t^2v - \tfrac{2}{z}\mu tr - \tfrac{z-2}{z}\mu^2r^2v^{-1})\partial_r - \tfrac{2}{z}(z-1)\mu(vt-\mu r)\partial_v \\
&\quad + \tfrac{2}{z}\mu xt - \tfrac{2}{z}\mu^2xrv^{-1}
\end{aligned} \quad (26)$$

span a representation of the conformal algebra Equation (2), which acts as dynamical symmetry algebra of the Equation (7), for arbitrary dynamical exponent z.

Proof. It is readily checked that the generator Equation (26) satisfies the commutation relations (2), with $\mu \mapsto -\mu$. On the other hand, for any $f = f(t,r,v)$, one has:

$$[\hat{L}, X_{-1}] = [\hat{L}, Y_{-1}] = [\hat{L}, Y_0] = [\hat{L}, Y_1] = 0$$

$$[\hat{L}, X_0] = -\hat{L}, [\hat{L}, X_1] = -2t\hat{L},$$

which establishes the asserted dynamical symmetry. □

Next, we treat the general case, when all of the constants are non-zero:

Case B: $A_{12} \neq 0, A_{110} \neq 0, A_{100} \neq 0, B_{110} \neq 0, B_{100} \neq 0, D_0 \neq 0$.

Then, the generators are modified as follows:

$$\overline{X}_1 = X_1 + \widetilde{X}_1$$

$$\begin{aligned}
\widetilde{X}_1 &= -\left(A_{12}r^2v^{-2} + A_{110}rv^{\frac{2z-1}{1-z}} + A_{100}v^{\frac{2z}{1-z}}\right)\partial_t \\
&\quad -\left(\tfrac{A_{12}}{\mu}r^2v^{-1} + B_{110}rv^{\frac{z}{1-z}} + B_{100}v^{\frac{z+1}{1-z}}\right)\partial_r
\end{aligned}$$

$$-(B_{110} - \tfrac{A_{110}}{\mu})v^{\frac{z}{1-z}}\partial_v - D_0 v^{\frac{z}{1-z}}, \tag{27}$$

$$\overline{Y}_0 = Y_0 + \widetilde{Y}_0$$

$$\widetilde{Y}_0 = \tfrac{1}{2}[\widetilde{X}_1, Y_{-1}]$$

$$= -(A_{12} r v^{-1} + \tfrac{1}{2} A_{110} v^{-1+1/(1-z)})\partial_t - \tfrac{1}{2\mu}(2A_{12} r + A_{110} v^{1/(1-z)})\partial_r. \tag{28}$$

Now, computing:

$$\left[\overline{Y}_0, Y_{-1}\right] = -\mu Y_{-1} + A_{12} X_{-1} + \frac{A_{12}}{\mu} Y_{-1} \tag{29}$$

we conclude that the cases $A_{12} = 0$ and $A_{12} \neq 0$ must be treated separately.

Case B1: $A_{12} = 0$. It follows that the constants in Equation (19) are given by:

$$B_{110} = A_{110}/\mu, A_{100} = \frac{A_{110}^2}{4\mu^2}, B_{100} = \frac{A_{100}}{\mu} = \frac{A_{110}^2}{4\mu^3}, D_0 = 0. \tag{30}$$

Proposition 3. *Let $z \neq 1$ and A_{110} be arbitrary constants. Then, the six generators:*

$$\overline{X}_{-1} = -\partial_t, \overline{X}_0 = -t\partial_t - \tfrac{r}{z}\partial_r - \tfrac{1-z}{z} v\partial_v - \tfrac{x}{z}$$

$$\overline{X}_1 = -(t^2 + A_{110} r v^{(2z-1)/(1-z)} + \tfrac{A_{110}^2}{4\mu^2} v^{2z/(1-z)})\partial_t$$

$$-\left(\tfrac{2}{z} tr + \tfrac{z-2}{z}\mu r^2 v^{-1} + \tfrac{A_{110}}{\mu} r v^{z/(1-z)} + \tfrac{A_{110}^2}{4\mu^3} v^{(z+1)/(1-z)}\right)\partial_r$$

$$-\tfrac{2(1-z)}{z}(vt - \mu r)\partial_v - \tfrac{2}{z} xt + \tfrac{2}{z}\mu x r v^{-1}$$

$$\overline{Y}_{-1} = -v\partial_r$$

$$\overline{Y}_0 = -\tfrac{A_{110}}{2} v^{z/(1-z)}\partial_t - (tv - \tfrac{\mu}{z} r + \tfrac{A_{110}}{2\mu} v^{1/(1-z)})\partial_r - \tfrac{z-1}{z}\mu v\partial_v + \mu\tfrac{x}{z}$$

$$\overline{Y}_1 = -A_{110}(t v^{z/(1-z)} - \mu r v^{(2z-1)/(1-z)})\partial_t$$

$$-\left(t^2 v - \tfrac{2}{z}\mu t r - \tfrac{z-2}{z}\mu^2 r^2 v^{-1} + \tfrac{A_{110}}{\mu}\left(t v^{1/(1-z)} - \mu r v^{z/(1-z)}\right)\right)\partial_r$$

$$-\tfrac{2}{z}(z-1)\mu(vt - \mu r)\partial_v + \tfrac{2}{z}\mu x t - \tfrac{2}{z}\mu^2 x r v^{-1} \tag{30}$$

span a representation of the conformal algebra (the above result of Case A is recovered upon setting $A_{110} = 0$). These generators give more symmetries of Equation (7).

Proof. From the above, the commutator Equation (2) is readily verified, with $\mu \mapsto -\mu$. For the dynamical symmetries, one checks the commutators:

$$[\hat{L}, X_{-1}] = [\hat{L}, Y_{\pm 1,0}] = 0$$

$$[\hat{L}, X_0] = -\hat{L}, [\hat{L}, X_1] = -(2t + \tfrac{A_{110}}{\mu} v^{z/(1-z)})\hat{L}.$$

which proves the assertion. □

In contrast to the previous Case A, the representation acting only on (t, r), but keeping v as a constant parameter, can no longer be obtained by simply setting $z = 1$. Rather, one must set $A_{110} = 0$ first, and only then, the limit $z \to 1$ is well-defined.

Case B2: $A_{12} \neq 0, A_{110} \neq 0, B_{110} \neq 0, A_{100} \neq 0, B_{100} \neq 0, D_0 \neq 0$.

It turns out that for $A_{12} \neq 0$, the algebra also can be closed, but only if $A_{12} = \mu$ and $A_{110} = 0$ (then, all other constants also vanish).

Proposition 4. *Let z be an arbitrary constant. Then, the generators* $\langle \mathcal{X}_{\pm1,0}, \mathcal{Y}_{\pm1,0}\rangle$*, where:*

$$
\begin{aligned}
\mathcal{X}_{-1} &= -\partial_t, \mathcal{X}_0 = -t\partial_t - \tfrac{r}{z}\partial_r - \tfrac{1-z}{z}v\partial_v - \tfrac{x}{z} \\
\mathcal{X}_{-1} &= X_{-1}, \mathcal{X}_0 = X_0 \\
\mathcal{X}_1 &= -(t^2+\mu r^2 v^{-2})\partial_t - \left(\tfrac{2}{z}tr + \tfrac{z+\mu(z-2)}{z}r^2v^{-1}\right)\partial_r \\
&\quad -\tfrac{2(1-z)}{z}(vt-\mu r)\partial_v - \tfrac{2}{z}xt + \tfrac{2}{z}\mu x r v^{-1} \\
\mathcal{Y}_{-1} &= -v\partial_r \\
\mathcal{Y}_0 &= -\mu r v^{-1}\partial_t - (tv - (\tfrac{\mu}{z}-1)r)\partial_r - \tfrac{z-1}{z}\mu v\partial_v + \mu\tfrac{x}{z} \\
\mathcal{Y}_1 &= -\mu(2trv^{-1} + (1-\mu)r^2v^{-2})\partial_t - \left(t^2v - \tfrac{2}{z}(z-\mu)tr + \tfrac{z(1-\mu)-(z-2)\mu^2}{z}r^2v^{-1}\right)\partial_r \\
&\quad -\tfrac{2}{z}(z-1)\mu(vt-\mu r)\partial_v + \tfrac{2}{z}\mu x t - \tfrac{2}{z}\mu^2 x r v^{-1}
\end{aligned}
\tag{31}
$$

close into a Lie algebra, with the following non-zero commutation relations:

$$
\begin{aligned}
&[\mathcal{X}_n, \mathcal{X}_{n'}] = (n-n')\mathcal{X}_{n+n'}, [\mathcal{X}_n, \mathcal{Y}_m] = (n-m)\mathcal{Y}_{n+m} \\
&[\mathcal{Y}_m, \mathcal{Y}_{m'}] = (m-m')(\mu\mathcal{X}_{m+m'} + (1-\mu)\mathcal{Y}_{m+m'}),
\end{aligned}
\tag{32}
$$

with $n, n', m, m' \in \{-1, 0, 1\}$*. The algebra is isomorphic to the usual conformal algebra Equation (2) and further extends the dynamical symmetries of Equation (7).*

Proof. The commutation relation is directly verified. The isomorphism with the conformal algebra follows from Proposition 1. The requirement to have a symmetry algebra of Equation (7) implies a relation between the constants k, q (called α, β in Proposition 1) and μ, namely $q = (k-\mu^2)/\mu$. In this case at hand, we have $k = \mu$, $q = 1-\mu$. It is then verified that $[\hat{L}, \mathcal{X}_{-1}] = [\hat{L}, \mathcal{Y}_{-1}] = 0$ and:

$$
\begin{aligned}
&[\hat{L}, \mathcal{X}_0] = -\hat{L} \\
&[\hat{L}, \mathcal{X}_1] = -2(t + \tfrac{r}{z}v^{-1})\hat{L} \\
&[\hat{L}, \mathcal{Y}_0] = -(k/\mu)\hat{L} = -\hat{L} \\
&[\hat{L}, \mathcal{Y}_1] = -2\left(\tfrac{k}{\mu}t + \tfrac{k}{z\mu^2}rv^{-1}\right)\hat{L} = -2\left(t + \tfrac{1}{z\mu}rv^{-1}\right)\hat{L}.
\end{aligned}
$$

which proves that these are dynamical symmetries of (7). □

We now ask whether the finite-dimensional representation Equations (26), (30) and (31), with $\mu \neq 0$, acting on functions $f = f(t, r, v)$, and having a dynamical exponent $z \neq 1$, can be extended to representations of an infinite-dimensional conformal Virasoro algebra. The answer turns out to be negative:

Proposition 5. *The representation Equations (26), (30) and (31) of the finite-dimensional conformal algebra* $\langle X_n, Y_n\rangle_{n\in\{\pm1,0\}}$ *with commutator Equation (8) cannot be extended to representations of an infinite-dimensional conformal Virasoro algebra with commutator Equation (8) when* $z \neq 1$*.*

Similar no-go results have been found before for variants of representations of the Schrödinger and conformal Galilean algebras [14]. On the other hand, for $\mu = 0$, extensions to a representation of a conformal Virasoro algebra with $z \neq 1$ exist [15].

Proof. Since for the finite-dimensional representations Equations (26), (30) and (31), we have:

$$[X_n, X_{n'}] = (n-n')X_{n+n'}, [X_n, Y_m] = (n-m)Y_{n+n'}, n, n', m = 0, \pm 1$$

we suppose that this must be valid for all admissible $n, m \in \mathbb{Z}$. Now, using the condition Equation (17) for $n = 2$, a conformal Virasoro algebra should contain a new generator X_2. Starting from the most general form, $X_2 = -a_2(t,r,v)\partial_t - b_2(t,r,v)\partial_r - c_2(t,r,v)\partial_v - d_2(t,r,v)$ we find that the coefficients are obtained from:

$$\begin{aligned} a_2 &= t^3 + a_{21}(r,v), b_2 = \tfrac{3}{z}t^2 r + 3\tfrac{z-2}{z}\mu t r^2 v^{-1} + b_{21}(r,v) \\ c_2 &= 3\tfrac{1-z}{z}(vt^2/2 - \mu r t) + c_{21}(r,v), d_2 = \tfrac{3}{z}xt^2 - \tfrac{6}{z}\mu x + d_{21}(r,v), \end{aligned}$$

where $a_{21}(r,v), b_{21}(r,v), c_{21}(r,v), d_{21}(r,v)$ are unknown functions of their arguments, but do no longer depend on the time t. We want to satisfy $[X_2, Y_{-1}] = 3Y_1$. However, when calculating:

$$\begin{aligned} [X_2, Y_{-1}] &= [-a_2\partial_t - b_2\partial_r - c_2\partial_v - d_2, -v\partial_r] = \\ &= 3Y_1 - va'_{21}\partial_t - (3\tfrac{1-z}{2z}t^2 v - \tfrac{3}{z}(1-z)\mu t r + vb'_{21} - c_{21} + 3\tfrac{z-2}{z}\mu^2 r^2 v^{-1})\partial_r \\ &\quad -(vc'_{21} + 3\tfrac{1-z}{z}\mu(tv - 2\mu r))\partial_v - vd'_{21} - \tfrac{6}{z}\mu\gamma r v^{-1} \end{aligned}$$

we see that closure is not possible for $z \neq 1$. Indeed, although the dependence on r, v of the functions $a_{21}, b_{21}, c_{21}, d_{21}$ can be chosen to satisfy the above closure condition, the t-dependence cannot be absorbed into these functions. Hence, our new representation Equations (26), (30) and (31) of the conformal algebra Equation (8) are necessarily finite-dimensional. □

3. Symmetry Algebra of Collisionless Boltzmann Equation with an Extra Force Term

We write the collisionless Boltzmann equation in the form:

$$\hat{B}f(t,r,v) = (\mu\partial_t + v\partial_r + F(t,r,v)\partial_v)f(t,r,v) = 0. \tag{33}$$

We want to determine the admissible forms of an external force $F(t,r,v)$, such that Equation (33) is invariant under a representation of the conformal algebra Equation (8). The unknown representation must include the "force" term and, in particular, for $F(t,r,v) = 0$, it should coincide with the representations of conformal algebra obtained in the previous section.

The idea of the construction is similar to the one used in Section 2. First, we impose invariance under basic symmetries:

- From invariance under time translation $X_{-1} = -\partial_t$, it follows:

$$[X_{-1}, \hat{B}] = -\dot{F} = 0 \to F = F(r,v) \tag{34}$$

- From invariance under dynamical scaling $X_0 = -t\partial_t - \frac{r}{z}\partial_r - \frac{1-z}{z}v\partial_v - \frac{x}{z}$, we obtain that:

$$[\hat{B}, X_0] = -\hat{B}, \tag{35}$$

if $F(r,v)$ satisfies the equation $(r\partial_r + (1-z)v\partial_v - (1-2z))F(r,v) = 0$, with solution:

$$F(r,v) = r^{1-2z}\varphi\left(r^{z-1}v\right), \tag{36}$$

where $\varphi(u)$ is an arbitrary function of the scaling variable $u := r^{z-1}v$.

It turns out that for the following calculations, it is more convenient to make a change of independent variables $(t,r,v) \mapsto (t,r,u)$. In the new variables, the generator of dynamical scaling just reads:

$$X_0 = -t\partial_t - \frac{r}{z}\partial_r - \frac{x}{z}. \tag{37}$$

Next, in order to be specific, we make the following ansatz for the analogue of space translations. (Indeed, we might also require to find Y_{-1} from the conditions to be (i) a symmetry of Boltzmann equation and (ii) to form a closed Lie algebra with the other basic symmetries $X_{-1,0}$. Such requirements lead to a system of differential equations, and the ansatz Equation (38) is a particular solution of this system, which has the special property that the Boltzmann operator can be linearly expressed $\hat{B} = -\mu X_{-1} - Y_{-1}$ by the generators. We believe this to be a natural auxiliary hypothesis):

$$Y_{-1} = -r^{1-z}u\partial_r - r^{-z}\Phi(u)\partial_u, \Phi(u) = (z-1)u^2 + \varphi(u). \tag{38}$$

In the same coordinate system, the collisionless Boltzmann equation becomes:

$$\hat{B}f(t,r,u) = \left(\mu\partial_t + r^{1-z}u\partial_r + r^{-z}\Phi(u)\partial_u\right)f(t,r,u) = 0. \tag{39}$$

Here, some comments are in order. In the structure of Boltzmann Equation (39), as well as in the form Equation (38) of the modified space translations, Y_{-1} enters an unknown function $\Phi(t,r,u)$. Therefore, the form of X_1 cannot be found only from its commutator with the other generators X_n, but the constraints form the entire conformal algebra must be used, as well as the requirement that X_1 and $Y_{0,1}$ are dynamical symmetries of Equation (39):

$$[\hat{B}, X_1] = \lambda_{X_1}(t,r,v)\hat{B}, [\hat{B}, Y_0] = \lambda_{Y_0}(t,r,v)\hat{B}, [\hat{B}, Y_1] = \lambda_{Y_1}(t,r,v)\hat{B}. \tag{40}$$

In fact, commuting the unknown generators X_1, Y_0, Y_1 with X_{-1} and X_0, we can fix the t- and r-dependence of the yet undetermined functions that occur in them:

$$\begin{aligned}
Y_0 &= -r^z a_0(u)\partial_t - (r^{1-z}u + rb_0(u))\partial_r - (r^{-z}\Phi(u)t + c_0(u))\partial_u - d_0(u) \\
X_1 &= -(t^2 + r^{2z}a_{12}(u))\partial_t - ((2/z)tr + r^{z+1}b_{12}(u))\partial_r \\
&\quad -r^z c_{12}(u)\partial_u - (2/z)xt - r^z d_{12}(u) \\
Y_1 &= -(2tr^z a_0(u) + r^{2z}A(u))\partial_t - (t^2 r^{1-z}u + 2trb_0(u) + r^{z+1}B(u))\partial_r \\
&\quad -(t^2 r^{-z}\Phi(u) + 2tc_0(u) + r^z C(u))\partial_u + (2/z)\mu xt - r^z D(u),
\end{aligned} \tag{41}$$

with the four functions:

$$\begin{aligned}
A(u) &= 2zb_0a_{12} + c_0a_{12}' - za_0b_{12} - a_0'c_{12}, C(u) = zb_0c_{12} + c_0c_{12}' - c_0'c_{12} - a_{12}\Phi \\
B(u) &= \tfrac{2}{z}a_0 + zb_0b_{12} + c_0b_{12}' - ua_{12}' - b_0'c_{12}, D(u) = \tfrac{2}{z}xa_0 + zb_0d_{12} + c_0d_{12}'.
\end{aligned} \tag{42}$$

In particular, looking for a representation of the analog of the extended Galilei algebra $\langle X_{-1}, X_0, Y_{-1}, Y_0\rangle$, we find that the unknown functions $a_0(u), b_0(u), c_0(u), d_0(u)$ must satisfy the system:

$$zua_0(u) + \Phi(u)a_0'(u) - k = 0 \tag{43}$$

$$zub_0(u) + \Phi(u)b_0'(u) - c_0(u) - qu = 0 \tag{44}$$

$$\Phi'(u)c_0 - \Phi(u)c_0'(u) + (q - zb_0)\Phi = 0 \tag{45}$$

$$\Phi(u)d_0'(u) = 0 \tag{46}$$

Because of Equation (46), one must distinguish two cases:

1. $\Phi(u) = 0$, when $d_0(u)$ can be arbitrary
2. $\Phi(u) \neq 0$, when $d_0(u) = d_0 = $ cste. is a constant.

In the second case, taking Equations (44) and (45) together, we obtain an equation for $b_0(u)$. It is:

$$\Phi^2(u)b_0''(u) + zu\Phi(u)b_0'(u) + \big(2z\Phi(u) - zu\Phi'(u)\big)b_0(u) - 2s\Phi(u) = 0, \tag{47}$$

and has in general two independent solution: $b_{01}(u), b_{02}(u)$. It follows that, for a given arbitrary value of $\Phi(u) \neq 0$, we have in general two distinct realizations of the analogue of Galilei transformation; and consequently, also two realizations of the analogue of the Galilei algebra. By construction, these are Lie algebras of symmetries of the collisionless Boltzmann Equation (39) (with $\lambda_{Y_0} = -k/\mu = -(\mu + q)$):

$$[Y_0, X_{-1}] = Y_{-1}, [X_0, X_{-1}] = X_{-1},$$

$$[Y_0, Y_{-1}] = \tfrac{k-\mu^2}{\mu} Y_{-1} + kX_{-1}. \tag{48}$$

Next, we include the generators of special conformal transformation X_1 and Y_1 to the extended Galilei algebras Equation (48) just constructed. We must also satisfy the other commutators of the conformal algebra Equation (8). Furthermore, the generators of the representation we are going to construct are dynamical symmetries of the collisionless Boltzmann equation (we use the commutators $[Y_1, Y_0] = KX_1 + QY_1$ and $[Y_1, Y_{-1}] = k_0X_0 + q_0Y_0$ in order to establish a relation between the constants k, q and K, Q, k_0, q_0). We find:

$$\lambda_{X_1}(t, r, u) = -2t - (r^z/\mu)\big(2zua_{12} + \Phi(u)a_{12}'(u)\big) = -2t - 2r^z a_0(u)/\mu \tag{49}$$

for the eigenvalue and:

$$c_{12}(u) = (2/z)\mu - (u/\mu)\big(2za_{12}(u) + \Phi(u)a_{12}'(u)\big) + \big(2zub_{12} + \Phi(u)b_{12}'(u)\big) \tag{50}$$

$$zuc_{12}(u) + \Phi c_{12}'(u) - c_{12}(u)\Phi'(u) + zb_{12}(u)\Phi(u) - 2c_0(u) = 0 \tag{51}$$

$$zud_{12}(u) + \Phi(u)d_{12}'(u) + (2/z)\mu x = 0 \tag{52}$$

$$\Phi^2(u)b_{12}''(u) + 3zu\Phi(u)b_{12}'(u) + z\big[2zu^2 + 3\Phi(u) - 2u\Phi'(u)\big]b_{12}(u) \tag{53}$$

$$2zua_{12}(u) + \Phi(u)a_{12}'(u) - 2a_0(u) = 0 \tag{54}$$

$$2zub_{12}(u) + \Phi(u)b_{12}'(u) - c_{12}(u) - 2b_0(u) = 0 \tag{55}$$

$$b_0(u) = (u/\mu)a_0(u) - \mu/z \tag{56}$$

$$c_0(u) = (\Phi/\mu)a_0(u) \tag{57}$$

$$d_0(u) = \text{cste.} = -\mu x/z. \tag{58}$$

$$k_0 = \alpha_0 k = 2k, q_0 = \alpha_0 q = 2q$$

$$2zuA(u) + \Phi(u)A'(u) - 2qa_0(u) = 0 \tag{59}$$

$$2zuB(u) + \Phi(u)B'(u) - C(u) - 2(k/z + qb_0(u)) = 0 \tag{60}$$

$$zuC(u) + \Phi(u)C'(u) - \Phi'(u)C(u) + z\Phi(u)B(u) - 2qc_0(u) = 0 \tag{61}$$

$$zuD(u) + \Phi(u)D'(u) - (2x/z)(k - \mu q) = 0. \tag{62}$$

$$K = k, Q = q \tag{63}$$

$$(q - 2zb_0)A(u) - c_0 A'(u) + za_0(u)B(u) + a_0'(u)C(u) + ka_{12}(u) - 2a_0^2 = 0 \tag{64}$$

$$(q - zb_0)B(u) - c_0 B'(u) + uA(u) + b_0'C(u) + kb_{12}(u) - 2a_0(u)b_0(u) = 0 \tag{65}$$

$$(q - zb_0 + c_0'(u))C(u) - c_0 C'(u) + \Phi(u)A(u) + kc_{12}(u) - 2a_0(u)c_0(u) = 0 \tag{66}$$

$$(q - zb_0)D(u) - c_0 D'(u) + kd_{12}(u) + \tfrac{2a_0(u)\mu x}{z} = 0 \tag{67}$$

$$2z(b_{12}(u)A(u) - a_{12}(u)B(u)) + c_{12}(u)A'(u) - a_{12}'(u)C(u) + 2a_0(u)a_{12}(u) = 0 \tag{68}$$

$$(2/z)A(u) - c_{12}(u)B'(u) + b_{12}'C(u) - 2b_0(u)a_{12}(u) = 0 \tag{69}$$

$$(zb_{12}(u) - c_{12}'(u))C(u) + c_{12}C'(u) - zc_{12}(u)B(u) + 2c_0(u)a_{12}(u) = 0 \tag{70}$$

$$(2x/z)(\mu a_{12}(u) + A(u)) + zd_{12}(u)B(u) + d_{12}'(u)C(u) \tag{71}$$

The system of Equations (43)–(45) and (50)–(71) must give a solution for the unknown functions $a_0(u), b_0(u), c_0(u), d_0(u), a_{12}(u), b_{12}(u), c_{12}(u), d_{12}(u)$. Of course, it is possible that several of the above equations are equivalent. Because of this fact, although the above system might look to be over-determined, we have not yet been able to produce an explicit solution without making an auxiliary assumption. A classification of all solutions of the above system is left as an open problem. We shall now describe some examples of solutions of this large system.

Example 1: Let $\Phi(u) = 0$. This case seems to be quite simple, provided it is compatible with our system. From Equation (43), we obtain:

$$a_0(u) = \frac{k}{z}u^{-1} \tag{72}$$

Using this value of $a_0(u)$ from Equations (52)–(54) and (56)–(58), we directly obtain:

$$b_0 = \text{cste.} = \tfrac{k}{z\mu} - \tfrac{\mu}{z}, c_0(u) = 0, d_0 = \text{cste.} = -\tfrac{\mu}{z}x, \tag{73}$$

$$a_{12}(u) = \tfrac{k}{z^2}u^{-2}, b_{12}(u) = \tfrac{1}{\mu z^2}(k - \mu^2)u^{-1}, c_{12}(u) = 0, d_{12}(u) = -\tfrac{2\mu x}{z^2}u^{-1}. \tag{74}$$

When we substitute the above results in relation Equation (42), we also find:

$$A(u) = \tfrac{k}{\mu z^2}(k - \mu^2)u^{-2}, B(u) = \tfrac{1}{\mu^2 z^2}(k(k - \mu^2) + \mu^4)u^{-1}$$

$$C(u) = 0, D(u) = \tfrac{2\mu^2 x}{z^2}u^{-1}. \tag{75}$$

One can now verify that the above results for the functions $a_0(u), b_0(u), c_0(u), d_0(u)$ and $a_{12}(u), b_{12}(u), c_{12}(u), d_{12}(u), A(u), B(u), C(u), D(u)$ satisfy all equations of the above system. Now, we can finally write the algebra generators:

$$X_{-1} = -\partial_t, X_0 = -t\partial_t - \tfrac{r}{z}\partial_r - \tfrac{x}{z}$$

$$X_1 = -\left(t^2 + \tfrac{k}{z^2}r^{2z}u^{-2}\right)\partial_t - \left(\tfrac{2}{z}tr + \tfrac{k-\mu^2}{z^2\mu}r^{z+1}u^{-1}\right)\partial_r - \tfrac{2}{z}xt + \tfrac{2\mu x}{z^2}r^z u^{-1},$$

$$Y_{-1} = -r^{1-z}u\partial_r,$$

$$Y_0 = -\tfrac{k}{z}r^z u^{-1}\partial_t - \left(tr^{1-z}u + \tfrac{k-\mu^2}{z\mu}r\right)\partial_r + \tfrac{\mu x}{z},$$

$$Y_1 = -\left(\tfrac{2k}{z}tr^z u^{-1} + \tfrac{k(k-\mu^2)}{z^2\mu}r^{2z}u^{-2}\right)\partial_t$$

$$-\left(t^2 r^{1-z} u + 2\tfrac{k-\mu^2}{z\mu} tr + \tfrac{k(k-\mu^2)+\mu^4}{z^2\mu^2} r^{z+1} u^{-1}\right)\partial_r + \tfrac{2}{z}\mu x t - \tfrac{2\mu^2 x}{z^2} r^z u^{-1}. \quad (76)$$

We return to the original variables via the change $(t,r,u) \mapsto (t,r,v)$, done through the substitutions $u \rightarrow r^{z-1}v$ and $\partial_r \rightarrow \partial_r + (1-z)r^{-1}v\partial_v$. Finally, we have the following representation of a conformal symmetry algebra of the collisionless Boltzmann Equation (33):

$$\begin{aligned}
X_{-1} &= -\partial_t, X_0 = -t\partial_t - \tfrac{r}{z}\partial_r - \tfrac{1-z}{z}v\partial_v - \tfrac{x}{z} \\
X_1 &= -\left(t^2 + \tfrac{k}{z^2}r^2v^{-2}\right)\partial_t - \left(\tfrac{2}{z}tr + \tfrac{k-\mu^2}{z^2\mu}r^2v^{-1}\right)\partial_r \\
&- (1-z)\left(\tfrac{2}{z}tv + \tfrac{k-\mu^2}{z^2\mu}r\right)\partial_v - \tfrac{2}{z}xt + \tfrac{2\mu x}{z^2}rv^{-1}, \\
Y_{-1} &= -v\partial_r - (1-z)r^{-1}v^2\partial_v, \\
Y_0 &= -\tfrac{k}{z}rv^{-1}\partial_t - \left(tv + \tfrac{k-\mu^2}{z\mu}r\right)\partial_r - (1-z)\left(tr^{-1}v^2 + \tfrac{k-\mu^2}{z\mu}v\right)\partial_v + \tfrac{\mu x}{z}, \\
Y_1 &= -\left(\tfrac{2k}{z}trv^{-1} + \tfrac{k(k-\mu^2)}{z^2\mu}r^2v^{-2}\right)\partial_t - \left(t^2v + 2\tfrac{k-\mu^2}{z\mu}tr + \tfrac{k(k-\mu^2)+\mu^4}{z^2\mu^2}r^2v^{-1}\right)\partial_r \\
&-(1-z)\left(t^2r^{-1}v^2 + 2\tfrac{k-\mu^2}{z\mu}tv + \tfrac{k(k-\mu^2)+\mu^4}{z^2\mu^2}r\right)\partial_v + \tfrac{2}{z}\mu x t - \tfrac{2\mu^2 x}{z^2}rv^{-1}.
\end{aligned} \quad (77)$$

Proposition 6. *The generator Equation (77) close into the following Lie algebra:*

$$[X_n, X_{n'}] = (n-n')X_{n+n'}, [X_n, Y_m] = (n-m)Y_{n+m}$$

$$[Y_m, Y_{m'}] = (m-m')\left(kX_{m+m'} + \tfrac{k-\mu^2}{\mu}Y_{m+m'}\right), \quad (78)$$

for $n, n', m, m' \in \{-1, 0, 1\}$ and for an arbitrary dynamical exponent z. They give a representation of the finite-dimensional conformal algebra, which acts as a dynamical symmetry algebra of the Boltzmann equation in the form:

$$\hat{B}f(t,r,v) = \left(\mu\partial_t + v\partial_r + (1-z)r^{-1}v^2\partial_v\right)f(t,r,v) = 0. \quad (79)$$

Proof. The commutation relation Equation (78) is directly checked. From the commutators $[\hat{B}, X_{-1}] = [\hat{B}, Y_{-1}] = 0$ and:

$$[\hat{B}, X_0] = -\hat{B}, [\hat{B}, X_1] = -2\left(t + \tfrac{k}{z\mu}rv^{-1}\right)\hat{B}$$

$$[\hat{B}, Y_0] = -(k/\mu)\hat{B}, [\hat{B}, Y_1] = -2\left(\tfrac{k}{\mu}t + \tfrac{k}{z\mu^2}rv^{-1}\right)\hat{B}.$$

it is seen that they generate dynamical symmetries. □

Example 2: Let $k = 0$. In this case, $\Phi(u)$ left arbitrary, which leads to $a_0 = 0$ from Equation (43) and:

$$b_0 = \text{cste.} = -\mu/z, c_0 = 0, d_0 = -\mu x/z \quad (80)$$

Then, from Equation (42), we obtain:

$$\begin{aligned}
A(u) &= -2\mu a_{12}(u), B(u) = -\mu b_{12}(u) - u a'_{12}(u) \\
C(u) &= -\mu c_{12}(u) - a_{12}(u)\Phi(u), D(u) = -\mu d_{12}.
\end{aligned} \quad (81)$$

However, when substituting in Equations (64)–(67), taking also into account that $q = -\mu$, we find that $A(u) = a_{12}(u) = 0$. Then, it is easy to check that the condition Equations (68)–(71) are fulfilled. This allows us to formulate:

Proposition 7. *Let $\Phi(u) = (z-1)u^2 + \varphi(u)$. Consider the generators:*

$$
\begin{aligned}
X_{-1} &= -\partial_t, X_0 = -t\partial_t - \tfrac{r}{z}\partial_r - \tfrac{1-z}{z}v\partial_v - \tfrac{x}{z} \\
X_1 &= -t^2\partial_t - \left(\tfrac{2}{z}tr + r^{z+1}b_{12}(u)\right)\partial_r \\
&- (1-z)\left(\tfrac{2}{z}tv + r^z v b_{12}(u) + \tfrac{r^{1-2z}}{1-z}c_{12}(u)\right)\partial_v - \tfrac{2}{z}xt - r^z d_{12}(u), \\
Y_{-1} &= -v\partial_r - (1-z)\left(r^{-1}v^2 + \tfrac{r^{1-2z}}{1-z}\Phi(u)\right)\partial_v = -v\partial_r - r^{1-2z}\varphi(u)\partial_v, \\
Y_0 &= -\left(tv - \tfrac{\mu}{z}r\right)\partial_r - (1-z)\left(\tfrac{r^{1-2z}}{1-z}\varphi(u)t - \tfrac{\mu}{z}v\right)\partial_v + \tfrac{\mu x}{z}, \\
Y_1 &= -\left(t^2 v - 2\tfrac{\mu}{z}tr - \mu r^{z+1}b_{12}(u)\right)\partial_r + \tfrac{2}{z}\mu xt + \mu r^z d_{12}(u) \qquad (82) \\
&-(1-z)\left(t^2\tfrac{r^{1-2z}}{1-z}\varphi(u) - \tfrac{2}{z}\mu tv - n\mu r^z v b_{12}(u) - \mu\tfrac{r^{1-2z}}{1-z}c_{12}(u)\right)\partial_v,
\end{aligned}
$$

where $c_{12}(u) = 2zub_{12}(u) + ((z-1)u^2 + \varphi(u))b'_{12}(u) + 2\mu/z$ and $\varphi(u), b_{12}(u), d_{12}(u)$ satisfy:

$$
\begin{aligned}
&\left[(z-1)u^2 + \varphi(u)\right]^2 b''_{12}(u) + 3zu\left[(z-1)u^2 + \varphi(u)\right]b'_{12}(u) \\
&+z\left[(z+1)u^2 - 2u\varphi'(u) + 3\varphi(u)\right]b_{12}(u) + \left[(2-z)u - \varphi'(u)\right]2\mu/z = 0 \qquad (83) \\
&zud_{12}(u) + \left[(z-1)u^2 + \varphi(u)\right]d'_{12}(u) + 2\mu x/z = 0. \qquad (84)
\end{aligned}
$$

For any triplet $(\varphi(u), b_{12}(u), d_{12}(u))$, which gives a solution of the system Equations (83) and (84), the generator Equation (82) close into the following Lie algebra:

$$
\begin{aligned}
&[X_n, X_{n'}] = (n-n')X_{n+n'}, [X_n, Y_m] = (n-m)Y_{n+m} \\
&[Y_m, Y_{m'}] = -\mu(m-m')Y_{m+m'}, \qquad (85)
\end{aligned}
$$

for $n, n', m, m' \in \{-1, 0, 1\}$ and for an arbitrary constant z. Equation (82) is a representation of the finite-dimensional conformal algebra and acts as the dynamical symmetry algebra of the Vlasov–Boltzmann equation, with a quite general "force" term:

$$
\hat{B}f(t,r,v) = \left(\mu\partial_t + v\partial_r + r^{1-2z}\varphi(u)\partial_v\right)f(t,r,v) = 0.
$$

Proof. The commutators are satisfied for $k = 0$ and $q = -\mu$ if condition Equations (83) and (84) are fulfilled. Under the same conditions, the symmetries are proven by the relations:

$$
\begin{aligned}
&[\hat{B}, X_{-1}] = [\hat{B}, Y_{-1}] = [\hat{B}, Y_0] = [\hat{B}, Y_1] = 0 \\
&[\hat{B}, X_0] = -\hat{B}, [\hat{B}, X_1] = -2t\hat{B}.
\end{aligned}
$$

□

In particular, if we implement the physical requirement that the "force" term should depend only on the positions r, that is $\varphi(u) = \varphi_0 =$ cste., we can compute explicitly the representation of the algebra Equation (82). To do this, one must find a solution of the system:

$$
\begin{aligned}
&\left[(z-1)u^2 + \varphi_0\right]^2 b''_{12}(u) + 3zu\left[(z-1)u^2 + \varphi_0\right]b'_{12}(u) \\
&+z\left[(z+1)u^2 + 3\varphi_0\right]b_{12}(u) + 2\mu\tfrac{2-z}{z}u = 0 \qquad (86) \\
&zud_{12}(u) + \left[(z-1)u^2 + \varphi_0\right]d'_{12}(u) + 2\mu x/z = 0. \qquad (87)
\end{aligned}
$$

The solution of the second equation is relatively simple, even for an arbitrary z:

$$d_{12}(u) = -\delta_0\Big[(z-1)u^2+\varphi_0\Big]^{\frac{z}{2(1-z)}}\int_{\mathbb{R}} \mathrm{d}u\,\Big[(z-1)u^2+\varphi_0\Big]^{\frac{z-2}{2(1-z)}}, \delta_0 = \text{cste.} \tag{88}$$

The solution of the Equation (86) for an arbitrary z can be expressed in terms of hypergeometric functions, but we shall not give its explicit form here. However, for $z = 2$, the system Equations (86) and (87) have an elementary solution:

$$\begin{aligned} b_{12}(u) &= b_{120}\tfrac{u}{(u^2+\varphi_0)^2} + b_{121}\tfrac{u^2-\varphi_0}{(u^2+\varphi_0)^2}, b_{120} = \text{cste.}, \mathrm{b}_{121} = \text{cste.} \\ d_{12}(u) &= -\mu x\tfrac{u}{u^2+\varphi_0}. \end{aligned} \tag{89}$$

Substituting this into the generator Equation (82) for $z = 2$ gives a finite-dimensional representation of the dynamical conformal symmetry of a collisionless Boltzmann equation of the form:

$$\hat{B}f(t,r,v) = \Big(\mu\partial_t + v\partial_r + \varphi_0 r^{-3}\partial_v\Big)f(t,r,v) = 0. \tag{90}$$

4. Conclusions

In this work, we have described the results of the first exploration of dynamical symmetries of collisionless Vlasov–Boltzmann transport equations. Our main finding is that these equations admit conformal dynamical symmetries, although it does not seem to be possible to extend this to infinite-dimensional conformal Virasoro symmetries, not even in the case of $d = 1$ space dimensions. These conformal symmetries are new representations of the conformal algebra and are inequivalent to the standard representation, which is habitually used in conformal field-theory descriptions of equilibrium critical phenomena. Our first class of new symmetries was found by admitting the momentum p (or equivalently, the velocity $v = p/\mu$) as an additional independent variable, leading to the representations Equations (26), (30) and (31). The second class of symmetries also allowed for external driving forces $F(t,r,v)$, and it has been one of the questions of which types of forces should be compatible with conformal invariance. As an example, we have seen that time-independent forces $F(r,v) = r^{1-2z}\varphi(r^{z-1}v)$, with an arbitrary scaling function φ, are admissible and lead to the general representation Equation (82). However, the solutions of the associated system of equations for the coefficients have not yet been classified and the complete content of these representations remains to be worked out in the future.

Some intuition can be gleaned from some examples. We have written down the explicit representations for the force $F(r,v) = (1-z)r^{-1}v^2$, with an $z > 1$ arbitrary Equation (77) and for $F(r,v) = \varphi_0\, r^{1-2z}$ Equations (82), (86), (87) with an arbitrary $z > 1$. In the later case, which could be related to physical situations, we have given the explicit representation of the conformal algebra for $z = 2$, when $F(r,v) = \varphi_0 r^{-3}$, Equations (82) and (89). Having identified these symmetries, the next step would be to use these to find either exact solutions [16] or else to use the algebra representations for fixing the form of co-variant n-point correlation functions, in analogy to time-dependent critical phenomena; see, e.g., [12].

The results derived here can be used as a starting point to derive forms of the transition rates w in the collision terms, which would be compatible with the dynamical symmetries of the collision-free equations. This kind of approach would be analogous to the one used for finding dynamical symmetries of non-linear Schrödinger equations; see, e.g., [17,18]. We hope to return to this elsewhere.

Acknowledgments: Most of the work in this paper was done during the visits of S.S. at Université de Lorraine Nancy and of M.H. at the XIth International workshop "Lie theories and its applications in physics". These visits were supported by PHC Rila. M.H. was partly supported by the Collège Doctoral Nancy-Leipzig-Coventry (Systèmes complexes à l'équilibre et hors équilibre) of UFA-DFH.

Author Contributions: This work was performed in close scientific collaboration of S.S. and M.H., during the scientific visits mentioned above.

Conflicts of Interest: The authors declare no conflict of interest.

References

1. Boltzmann, L. Weitere Studien über das Wärmegleichgewicht unter Gasmolekülen. *Wien. Ber.* **1872**, *66*, 275–370.
2. Haug, H. *Statistische Physik*; Vieweg: Braunschweig, Germany, 1997.
3. Huang, K. *Statistical Mechanics*, 2nd ed.; Wiley: New York, NY, USA, 1987; p. 53ff.
4. Kreuzer, H.J. *Nonequilibrium Thermodynamics and Its Statistical Foundations*; Oxford University Press: Oxford, UK, 1981; Chapter 7.
5. Vlasov, A.A. On vibration properties of electron gas. *JETP* **1938**, *8*, 291–318. (in Russian).
6. Elskens, Y.; Escande, D.; Doveil, F. Vlasov equation and N-body dynamics. *Eur. Phys. J.* **2014**, *D68*. [CrossRef]
7. Vilani, C. Particle systems and non-linear Landau damping. *Phys. Plasmas* **2014**, *21*, 030901:1–030901:9.
8. Duval, C.; Horváthy, P.A. Non-relativistic conformal symmetries and Newton-Cartan structures. *J. Phys. A* **2009**, *42*, 465206:1–465206:32. [CrossRef]
9. Henkel, M. Phenomenology of local scale-invariance: From conformal invariance to dynamical scaling. *Nucl. Phys.* **2002**, *B641*, 405–486. [CrossRef]
10. Henkel, M.; Hosseiny, A.; Rouhani, S. Logarithmic exotic conformal Galilean algebras. *Nucl. Phys.* **2014**, *B879*, 292–317. [CrossRef]
11. Martelli, D.; Tachikawa, Y. Comments on Galilean conformal field theories and their geometric realization. *J. High Energy Phys.* **2010**, *1005*, 091:1–091:31. [CrossRef]
12. Henkel, M.; Pleimling, M. *Non-Equilibrium Phase Transitions Volume 2: Ageing and Dynamical Scaling Far from Equilibrium*; Springer: Heidelberg, Germany, 2010.
13. Henkel, M.; Unterberger, J. Schrödinger invariance and space-time symmetries. *Nucl. Phys.* **2003**, *B660*, 407–435. [CrossRef]
14. Henkel, M.; Schott, R.; Stoimenov, S.; Unterberger, J. The Poincaré algebra in the context of ageing systems: Lie structure, representations, Appell systems and coherent states. *Confluentes Math.* **2012**, *4*, 1250006:1–1250006:23. [CrossRef]
15. Cherniha, R.; Henkel, M. On nonlinear partial differential equations with an infinite-dimensional conditional symmetry. *J. Math. Anal. Appl.* **2004**, *298*, 487–500. [CrossRef]
16. Fushchych, W.I.; Shtelen, W.M.; Serov, M.I. *Symmetry Analysis and Exact Solutions of Equations of Nonlinear Mathematical Physics*; Kluwer: Dordrecht, The Netherlands, 1993.
17. Boyer, C.D.; Sharp, R.T.; Winternitz, P. Symmetry-breaking interactions for the time-dependent Schrödinger equation. *J. Math. Phys.* **1976**, *17*, 1439–1451. [CrossRef]
18. Stoimenov, S.; Henkel, M. Dynamical symmetries of semi-linear Schrödinger and diffusion equations. *Nucl. Phys.* **2005**, *B723*, 205–233. [CrossRef]

symmetry

MDPI

Article

Non-Local Meta-Conformal Invariance, Diffusion-Limited Erosion and the XXZ Chain

Malte Henkel [1,2]

[1] Rechnergestützte Physik der Werkstoffe, Institut für Baustoffe (IfB), ETH Zürich, Stefano-Franscini-Platz 3, CH-8093 Zürich, Switzerland

[2] Groupe de Physique Statistique, Institut Jean Lamour (CNRS UMR 7198), Université de Lorraine Nancy, B.P. 70239, F-54506 Vandœuvre-lès-Nancy CEDEX, France; malte.henkel@univ-lorraine.fr

Academic Editor: Roman M. Cherniha

Received: 16 October 2016; 14 December 2016; Published: 24 December 2016

Abstract: Diffusion-limited erosion is a distinct universality class of fluctuating interfaces. Although its dynamical exponent $z = 1$, none of the known variants of conformal invariance can act as its dynamical symmetry. In $d = 1$ spatial dimensions, its infinite-dimensional dynamic symmetry is constructed and shown to be isomorphic to the direct sum of three loop-Virasoro algebras. The infinitesimal generators are spatially non-local and use the Riesz-Feller fractional derivative. Co-variant two-time response functions are derived and reproduce the exact solution of diffusion-limited erosion. The relationship with the terrace-step-kind model of vicinal surfaces and the integrable XXZ chain are discussed.

Keywords: meta-conformal invariance; representations; loop-Virasoro algebra; physical ageing; diffusion-limited erosion; terrace-step-kink model

1. Introduction

Symmetries have since a long time played an important role in the analysis of physical systems. The insight gained can be either calculational, in that a recognised symmetry becomes useful in simplifying calculations, or else conceptual, in that the identification of symmetries can lead to new level of understanding. In the statistical physics of equilibrium second-order phase transitions in two dimensions, *conformal invariance* has ever since the pioneering work of Belavin, Polyakov and Zamolodchikov [1] created considerable progress, both computationally as well as conceptually. It then appears natural to ask if one might find extensions of conformal invariance which apply to time-dependent phenomena. Here, we shall inquire about dynamical symmetries of the following stochastic Langevin equation, to be called *diffusion-limited erosion* (DLE) *Langevin equation*, which reads in momentum space [2]

$$\mathrm{d}\widehat{h}(t,\boldsymbol{q}) = -\nu|\boldsymbol{q}|\widehat{h}(t,\boldsymbol{q})\mathrm{d}t + \widehat{\jmath}(t,\boldsymbol{q})\mathrm{d}t + (2\nu T)^{1/2}\,\mathrm{d}\widehat{B}(t,\boldsymbol{q}) \tag{1}$$

and describes the Fourier-transformed height $\widehat{h}(t,\boldsymbol{q}) = (2\pi)^{-d/2}\int_{\mathbb{R}^d}\mathrm{d}\boldsymbol{r}\, e^{-\mathrm{i}\boldsymbol{q}\cdot\boldsymbol{r}}h(t,\boldsymbol{r})$. Because of the (Fourier-transformed) standard brownian motion $\widehat{B}$, with the variance $\langle\widehat{B}(t,\boldsymbol{q})\widehat{B}(t',\boldsymbol{q}')\rangle = \min(t,t')\delta(\boldsymbol{q}+\boldsymbol{q}')$, this is a stochastic process, called *diffusion-limited erosion* (DLE) *process*. Herein, ν, T are non-negative constants and $\delta(\boldsymbol{q})$ is the Dirac distribution. Since we shall be interested in deriving linear responses, an external infinitesimal source term $\widehat{\jmath}(t,\boldsymbol{q})$ is also included, to be set to zero at the end. Inverting the Fourier transform in order to return to direct space, Equation (1) implies spatially long-range interactions. The conformal invariance of equilibrium critical systems with long-range interactions has been analysed recently [3]. Equation (1) arises in several distinct physical contexts.

Example 1. *For the original definition of the* (DLE) *process [2], one considers how an initially flat interface is affected by the diffusive motion of corrosive particles. A single corrosive particle starts initially far away from the interface. After having undergone diffusive motion until the particle finally arrives at the interface, it erodes a particle from that interface. Repeating this process many times, an eroding interface forms which is described in terms of a fluctuating height $h(t, \mathbf{r})$, see Figure 1. It can be shown that this leads to the* DLE *Langevin Equation (1) [2,4].*

Several lattice formulations of the model [2,5–7] confirm the dynamical exponent $z = 1$.

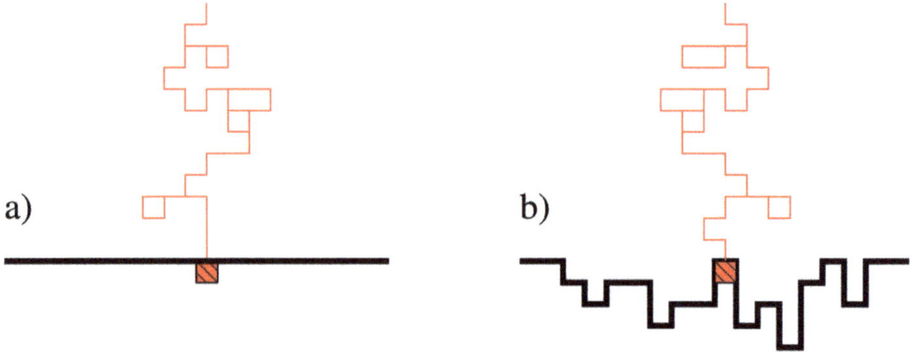

Figure 1. Schematics of the genesis of an eroding surface through the DLE process. (**a**) Initial state: a diffusing particle (red path) arrives on a flat surface (full black line) and erodes a small part of it; (**b**) Analogous process at a later time, when the surface has been partially eroded.

Example 2. *A different physical realisation of Equation (1) invokes vicinal surfaces. Remarkably, for $d = 1$ space dimension, the Langevin Equation (1) has been argued [8] to be related to a system of non-interacting fermions, conditioned to an a-typically large flux. Consider the* terrace-step-kink model *of a vicinal surface, and interpret the steps as the world lines of fermions, see Figure 2. Its transfer matrix is the matrix exponential of the quantum hamiltonian H of the asymmetric XXZ chain [8]. Use Pauli matrices $\sigma_n^{\pm,z}$, attached to each site n, such that the particle number at each site is $\varrho_n = \frac{1}{2}(1 + \sigma_n^z) = 0, 1$. On a chain of N sites, consider the quantum hamiltonian [8–10]*

$$H = -\frac{w}{2} \sum_{n=1}^{N} \left[2v\sigma_n^+ \sigma_{n+1}^- + 2v^{-1}\sigma_n^- \sigma_{n+1}^+ + \Delta \left(\sigma_n^z \sigma_{n+1}^z - 1 \right) \right] \tag{2}$$

where $w = \sqrt{pq}\, e^{\mu}$, $v = \sqrt{p/q}\, e^{\lambda}$ and $\Delta = 2\left(\sqrt{p/q} + \sqrt{q/p}\right) e^{-\mu}$. Herein, p, q describe the left/right bias of single-particle hopping and λ, μ are the grand-canonical parameters conjugate to the current and the mean particle number. In the continuum limit, the particle density $\varrho_n(t) \to \varrho(t, r) = \partial_r h(t, r)$ is related to the height h which in turn obeys (1), with a gaussian white noise *η [8]. This follows from the application of the theory of fluctuating hydrodynamics, see [11,12] for recent reviews. The low-energy behaviour of H yields the dynamical exponent $z = 1$ [8–10]. If one conditions the system to an a-typically large current, the large-time, large-distance behaviour of (2) has very recently been shown [10] (i) to be described by a conformal field-theory with central charge $c = 1$ and (ii) the time-space scaling behaviour of the stationary structure function has been worked out explicitly, for $\lambda \to \infty$. Therefore, one may conjecture that the so simple-looking Equation (1) should furnish an effective continuum description of the large-time, long-range properties of quite non-trivial systems, such as (2).*

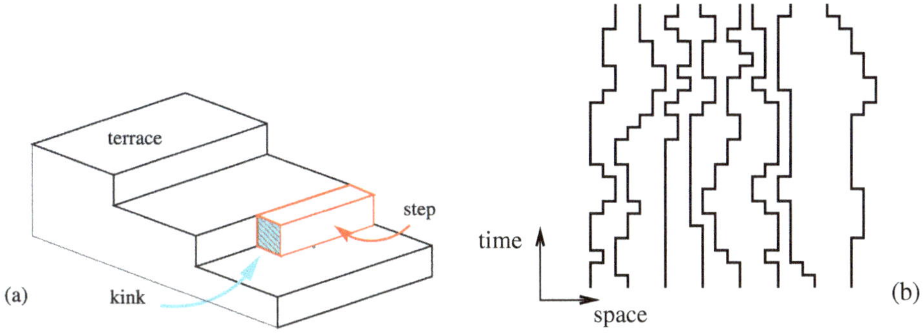

Figure 2. (**a**) Schematic illustration of a vicinal surface, formed by terraces. Fluctuations between terraces are described by steps and kinks; (**b**) Reinterpretation of the steps of a vicinal surface as non-intersecting world lines in $1+1$ dimensions of an ensemble of fermionic particles.

The physical realisation of Equation (1) in terms of the DLE process makes it convenient to discuss the results in terms of the physics of the growth of interfaces [13–15], which can be viewed as a paradigmatic example of the emergence of non-equilibrium collective phenomena [16,17]. Such an interface can be described in terms of a time-space-dependent height profile $h(t,\boldsymbol{r})$. This profile depends also on the eventual fluctuations of the set of initial states and on the noise in the Langevin equation, hence h should be considered as a random variable. The degree of fluctuations can be measured through the interface width. If the model is formulated first on a hyper-cubic lattice $\mathscr{L} \subset \mathbb{Z}^d$ of $|\mathscr{L}| = L^d$ sites, the interface width is defined by

$$w^2(t;L) := \frac{1}{L^d}\sum_{\boldsymbol{r}\in\mathscr{L}}\left\langle\left(h(t,\boldsymbol{r})-\overline{h}(t)\right)\right\rangle^2 = L^{2\beta z} f_w\left(tL^{-z}\right) \sim \left\{\begin{array}{ll} t^{2\beta} & ;\ \text{if } tL^{-z} \ll 1 \\ L^{2\beta z} & ;\ \text{if } tL^{-z} \gg 1 \end{array}\right. \tag{3}$$

where the generically expected scaling form, for large times/lattice sizes $t\to\infty$, $L\to\infty$, is also indicated. Physicists call this *Family-Vicsek scaling* [18]. Implicitly, it is assumed here that one is *not* at the 'upper critical dimension d^*', where this power-law scaling is replaced by a logarithmic scaling form, see also below. Herein, $\langle.\rangle$ denotes an average over many independent samples and $\overline{h}(t) := L^{-d}\sum_{\boldsymbol{r}\in\mathscr{L}} h(t,\boldsymbol{r})$ is the spatially averaged height. Furthermore, β is called the *growth exponent*, $z>0$ is the *dynamical exponent* and $\alpha := \beta z$ is the *roughness exponent*. When $tL^{-z} \gg 1$, one speaks of the *saturation regime* and when $tL^{-z} \ll 1$, one speaks of the *growth regime*. We shall focus on the growth regime from now on.

Definition 1. *On a spatially infinite substrate, an interface with a width $w(t)\nearrow\infty$ for large times $t\to\infty$ is called* rough. *If $\lim_{t\to\infty} w(t)$ is finite, the interface is called* smooth.

This definition permits a first appreciation of the nature of the interface: if in (3) $\beta>0$, the interface is rough.

In addition, dynamical properties of the interface can be studied through the two-time correlators and responses. In the growth regime (where effectively $L\to\infty$), one considers the double scaling limit $t,s\to\infty$ with $y := t/s > 1$ fixed and expects the scaling behaviour

$$C(t,s;\boldsymbol{r}) := \left\langle\left(h(t,\boldsymbol{r})-\left\langle\overline{h}(t)\right\rangle\right)\left(h(s,\boldsymbol{0})-\left\langle\overline{h}(s)\right\rangle\right)\right\rangle \quad = s^{-b}F_C\left(\frac{t}{s};\frac{\boldsymbol{r}}{s^{1/z}}\right) \tag{4a}$$

$$R(t,s;\boldsymbol{r}) := \left.\frac{\delta\left\langle h(t,\boldsymbol{r})-\overline{h}(t)\right\rangle}{\delta j(s,\boldsymbol{0})}\right|_{j=0} = \left\langle h(t,\boldsymbol{r})\widetilde{h}(s,\boldsymbol{0})\right\rangle = s^{-1-a}F_R\left(\frac{t}{s};\frac{\boldsymbol{r}}{s^{1/z}}\right) \tag{4b}$$

where j is an external field conjugate to the height h. Throughout, all correlators are calculated with $j = 0$. In the context of Janssen-de Dominicis theory, $\widetilde{h}$ is the conjugate response field to h, see [17]. Spatial translation-invariance was implicitly admitted in (4). This defines the *ageing exponents* a, b. The *autocorrelation exponent* λ_C and the *autoresponse exponent* λ_R are defined from the asymptotics $F_{C,R}(y, \mathbf{0}) \sim y^{-\lambda_{C,R}/z}$ as $y \to \infty$. For these non-equilibrium exponents, one has $b = -2\beta$ [15] and the bound $\lambda_C \geq (d + zb)/2$ [19,20].

For the DLE process, these exponents are readily found form the exact solution of (1) [2,4,21]. For an initially flat interface $h(0, \mathbf{r}) = 0$, the two-time correlator and response are in Fourier space

$$\widehat{C}(t, s; \mathbf{q}, \mathbf{q}') := \left\langle \widehat{h}(t, \mathbf{q}) \widehat{h}(s, \mathbf{q}') \right\rangle = \frac{T}{|\mathbf{q}|} \left[e^{-\nu |\mathbf{q}| |t-s|} - e^{-\nu |\mathbf{q}| (t+s)} \right] \delta(\mathbf{q} + \mathbf{q}'), \tag{5a}$$

$$\widehat{R}(t, s; \mathbf{q}, \mathbf{q}') := \left. \frac{\delta \langle \widehat{h}(t, \mathbf{q}) \rangle}{\delta \widehat{j}(s, \mathbf{q}')} \right|_{j=0} = \Theta(t - s)\, e^{-\nu |\mathbf{q}| (t-s)}\, \delta(\mathbf{q} + \mathbf{q}'). \tag{5b}$$

In direct space, this becomes, for $d \neq 1$ and with $C_0 := \Gamma((d+1)/2)/(\Gamma(d/2)\pi^{(d+1)/2})$

$$C(t, s; \mathbf{r}) = \frac{T C_0}{d-1} \left[\left(\nu^2 (t-s)^2 + \mathbf{r}^2 \right)^{-(d-1)/2} - \left(\nu^2 (t+s)^2 + \mathbf{r}^2 \right)^{-(d-1)/2} \right] \tag{6a}$$

$$R(t, s; \mathbf{r}) = C_0\, \Theta(t-s)\, \nu (t-s) \left(\nu^2 (t-s)^2 + \mathbf{r}^2 \right)^{-(d+1)/2} \tag{6b}$$

where the Heaviside function Θ expresses the causality condition $t > s$. In particular, in the growth regime, the interface width reads (where $C_1(\Lambda)$ is a known constant and a high-momentum cut-off Λ was used for $d > 1$)

$$w^2(t) = C(t, t; \mathbf{0}) = \frac{T C_0}{1-d} \left[(2\nu t)^{1-d} - C_1(\Lambda) \right] \stackrel{t \to \infty}{\simeq} \begin{cases} T C_0 C_1(\Lambda)/(d-1) & ; \text{ if } d > 1 \\ T C_0 \ln(2\nu t) & ; \text{ if } d = 1 \\ T C_0 (2\nu)^{1-d}/(1-d) \cdot t^{1-d} & ; \text{ if } d < 1 \end{cases} \tag{7}$$

Hence $d^* = 1$ is the upper critical dimension of the DLE process. It follows that at late times the DLE-interface is smooth for $d > 1$ and rough for $d \leq 1$. On the other hand, one may consider the *stationary* limit $t, s \to \infty$ with the time difference $\tau = t - s$ being kept fixed. Then one finds a fluctuation-dissipation relation $\partial C(s + \tau, s; \mathbf{r})/\partial \tau = -\nu T R(s + \tau, s; \mathbf{r})$. The similarity of this to what is found for equilibrium systems is unsurprising, since several discrete lattice variants of the DLE process exist and are formulated as an equilibrium system [5]. Lastly, the exponents defined above are read off by taking the scaling limit, and are listed in Table 1. In contrast to the interface width $w(t)$, which shows a logarithmic growth at $d = d^* = 1$, logarithms cancel in the two-time correlator C and response R, up to *additive* logarithmic corrections to scaling. This is well-known in the physical ageing at $d = d^*$ of simple magnets [22,23] or of the Arcetri model [20].

For comparison, we also list in Table 1 values of the non-equilibrium exponents for several other universality classes of interface growth. In particular, one sees that for the Edwards-Wilkinson (EW) [24] and Arcetri classes, the upper critical dimension $d^* = 2$, while it is still unknown if a finite value of d^* exists for the Kardar-Parisi-Zhang (KPZ) class, see [13,14,25–28]. Clearly, the stationary exponents a, b, z are the same in the EW and Arcetri classes, but the non-equilibrium relaxation exponents λ_C, λ_R are different for dimensions $d < d^*$. This illustrates the independence of λ_C, λ_R from those stationary exponents, in agreement with studies in the non-equilibrium critical dynamics of relaxing magnetic systems. On the other hand, for the KPZ class, a perturbative renormalisation-group analysis shows that $\lambda_C = d$ for $d < 2$ [29]. For $d > 2$, a new strong-coupling fixed point arises and the relaxational properties are still unknown. Even for $d = 2$, the results of different numerical studies in the KPZ class are not yet fully consistent, but recent simulations suggest that precise information on the shape of the scaling function, coming from a dynamical symmetry [30], may improve the quality of the extracted exponents [31].

Table 1. Exponents of growing interfaces in the Kardar-Parisi-Zhang (KPZ), Edwards-Wilkinson (EW), Arcetri (for both $T = T_c$ and $T < T_c$) and DLE universality classes. The numbers in bracket give the estimated error in the last digit(s).

Model	d	z	β	a	b	λ_C	λ_R	References
KPZ	1	$3/2$	$1/3$	$-1/3$	$-2/3$	1	1	[25,29,32]
	2	1.61(2)	0.2415(15)	0.30(1)	−0.483(3)	1.97(3)	2.04(3)	[33,34]
	2	1.61(2)	0.241(1)	-	−0.483	1.91(6)	-	[35]
	2	1.61(5)	0.244(2)	-	-	-	-	[26]
	2	1.627(4)	0.229(6)	-	-	-	-	[36]
	2	1.61(2)	0.2415(15)	0.24(2)	−0.483(3)	1.97(3)	2.00(6)	[31,33]
EW	<2	2	$(2-d)/4$	$d/2-1$	$d/2-1$	d	d	
	2	2	0(log) #	0	0	2	2	[24,37]
	>2	2	0	$d/2-1$	$d/2-1$	d	d	
Arcetri $T = T_c$	<2	2	$(2-d)/4$	$d/2-1$	$d/2-1$	$3d/2-1$	$3d/2-1$	
	2	2	0(log) #	0	0	2	2	[20]
	>2	2	0	$d/2-1$	$d/2-1$	d	d	
$T < T_c$	d	2	$1/2$	$d/2-1$	-1	$d/2-1$	$d/2-1$	
DLE	<1	1	$(1-d)/2$	$d-1$	$d-1$	d	d	
	1	1	0(log) #	0	0	1	1	[4,21]
	>1	1	0	$d-1$	$d-1$	d	d	

For $d = d^*$, one has logarithmic scaling $w(t;L)^2 \sim \ln t\, f_w(\ln L/\ln t)$.

Here, we are concerned with the dynamical symmetries of the DLE process. Our main results are as follows.

Theorem 1. *The* dynamical symmetry *of the* DLE *process, in $d = 1$ space dimension and with $j = 0$, is a meta-conformal algebra, in a sense to be made more precise below, and is isomorphic to the direct sum of three Virasoro algebras without central charge (or* loop-Virasoro algebra*). The Lie algebra generators will be given below in Equation (29), they are non-local in space. The general form of the co-variant two-time response function is (with $t > s$)*

$$R(t,s;r) = F_A\,(t-s)^{1-2x}\,\frac{\nu(t-s)}{\nu^2(t-s)^2+r^2}$$
$$+F_B\,(t-s)^{1+\psi-2x}\left(\nu^2(t-s)^2+r^2\right)^{-(\psi+1)/2}\cos\left((\psi+1)\arctan\left(\frac{r}{\nu(t-s)}\right)-\frac{\pi\psi}{2}\right) \tag{8}$$

where x, ψ are real parameters and $F_{A,B}$ are normalisation constants.

Remark 1. *The exact solution (6b) of the* DLE*-response in $(1+1)D$ is reproduced by (8) if one takes $x = \frac{1}{2}$, $\nu > 0$, $F_A = C_0$ and $F_B = 0$. This illustrates the importance of non-local generators in a specific physical application.*

Remark 2. *The symmetries so constructed are only dynamical symmetries of the so-called* 'deterministic part' *of Equation (1), which is obtained by setting $T = 0$. We shall see that the co-variant two-time correlator $C(t,s;r) = 0$. This agrees with the vanishing of the exact* DLE*-correlator (6a) in the $T \to 0$ limit (fix $d \neq 1$ and let first $T \to 0$ and only afterwards $d \to 1$).*

This paper presents an exploration of the dynamical symmetries of DLE process for $d = 1$ and is organised as follows. In Section 2, we introduce the distinction of ortho-conformal and meta-conformal invariance and illustrate these notions by several examples, see Table 2. In Section 3, we explain why none of these local symmetries can be considered as a valid candidate of the dynamical symmetry of the DLE process. Section 4 presents some basic properties on the Riesz-Feller fractional derivative which are used in Section 5 to explicitly construct the *non-local* dynamical symmetry of the DLE process, thereby generalising and extending earlier results [21]. Section 6 outlines the formulation of time-space Ward identities for the computation of covariant n-point functions and in Section 7 the two-point correlator and response are found for the dynamical symmetry of the DLE process. The propositions proven in Sections 5 and 7 make the Theorem 1 more precise and constitute its proof. The Lie algebra contraction, in the limit $\nu \to \infty$, and its relationship with the conformal Galilean algebra is briefly mentioned. This is summarised in Table 3.

2. Local Conformal Invariance

Can one explain the form of the two-time scaling functions of the DLE process in terms of a dynamical symmetry? To answer such a question, one must first formulate it more precisely.

Definition 2. *The* deterministic part *of the Langevin Equation (1) is obtained when formally setting* $\widehat{B} = 0$.

Our inspiration comes from Niederer's treatment [38] of the dynamical symmetries of the free diffusion equation. The resulting Lie algebra, called *Schrödinger algebra* by physicists, was found by Lie (1882) [39]. The corresponding continuous symmetries, however, were already known to Jacobi (1842/43) [40]. For growing interfaces, the Langevin equation of the EW class is the noisy diffusion equation. Hence its deterministic part, the free diffusion equation, is obviously Schrödinger-invariant. In this work, we seek dynamical symmetries of the deterministic part of the DLE process, that is, we look for dynamical symmetries of the non-local equation $(\mu\partial_t - \nabla_r)\varphi = 0$, where the non-local Riesz-Feller derivative ∇_r will be defined below, in Section 4.

Since we see from Equation (1), or the explicit correlators and responses (5), that the dynamical exponent $z = 1$, conformal invariance appears as a natural candidate, where one spatial direction is re-labelled as 'time'. However, one must sharpen the notion of conformal invariance. For notational simplicity, we now restrict to the case of $1+1$ time-space dimensions, labelled by a 'time coordinate' t and a 'space coordinate' r. Our results on the dynamical symmetries of the DLE process, see Propositions 3 and 4, require us to present here a more flexible definition than given in [21,41].

Definition 3. *(a) A set of* meta-conformal transformations $\mathscr{M}$ *is a set of maps* $(t,r) \mapsto (t',r') = \mathscr{M}(t,r)$, *which may depend analytically on several parameters and form a Lie group. The corresponding Lie algebra is isomorphic to the conformal algebra such that the maximal finite-dimensional Lie sub-algebra is semi-simple and contains at least a Lie algebra isomorphic to* $\mathfrak{sl}(2,\mathbb{R}) \oplus \mathfrak{sl}(2,\mathbb{R})$. *A physical system is* meta-conformally invariant *if its n-point functions transform covariantly under meta-conformal transformations; (b) A set of* ortho-conformal transformations $\mathscr{O}$ *is a set of meta-conformal transformations* $(t,r) \mapsto (t',r') = \mathscr{O}(t,r)$, *such that (i) the maximal finite-dimensional Lie algebra is isomorphic to* $\mathfrak{sl}(2,\mathbb{R}) \oplus \mathfrak{sl}(2,\mathbb{R})$ *and that (ii) angles in the coordinate space of the points* (t,r) *are kept invariant. A physical system is* ortho-conformally invariant *if its n-point functions transform covariantly under ortho-conformal transformations.*

The names ortho- and meta-conformal are motivated by the greek prefixes $o\varrho\theta o$: right, standard and $\mu\varepsilon\tau\alpha$: of secondary rank. Ortho-conformal transformations are usually simply called 'conformal transformations'. We now recall simple examples to illustrate these definitions. See Table 2 for a summary.

Table 2. Comparison of local ortho-conformal, conformal Galilean and meta-1 conformal invariance, in $(1+1)D$. The non-vanishing Lie algebra commutators, the defining equation of the generators, the invariant differential operator $\mathcal{S}$ and the covariant two-point function is indicated, where applicable. Physically, the co-variant quasiprimary two-point function $\mathscr{C}_{12} = \langle \varphi_1(t,r)\varphi_2(0,0)\rangle$ is a correlator, with the constraints $x_1 = x_2$ and $\gamma_1 = \gamma_2$.

	Ortho	Galilean	Meta-1
Lie algebra	$[X_n, X_m] = (n-m)X_{n+m}$ $[X_n, Y_m] = (n-m)Y_{n+m}$ $[Y_n, Y_m] = (n-m)X_{n+m}$	$[X_n, X_m] = (n-m)X_{n+m}$ $[X_n, Y_m] = (n-m)Y_{n+m}$ $[Y_n, Y_m] = 0$	$[X_n, X_m] = (n-m)X_{n+m}$ $[X_n, Y_m] = (n-m)Y_{n+m}$ $[Y_n, Y_m] = \mu(n-m)Y_{n+m}$
generators	(9)	(15)	(13)
$\mathcal{S}$	$\partial_t^2 + \partial_r^2$	-	$-\mu\partial_t + \partial_r$
$\mathscr{C}_{12}$	$t^{-2x_1}\left(1+\left(\frac{r}{t}\right)^2\right)^{-x_1}$	$t^{-2x_1}\exp\left(-2\left\lvert\frac{\gamma_1 r}{t}\right\rvert\right)$	$t^{-2x_1}\left(1+\frac{\mu}{\gamma_1}\left\lvert\frac{\gamma_1 r}{t}\right\rvert\right)^{-2\gamma_1/\mu}$

Example 3. *In $(1+1)D$, ortho-conformal transformations are analytic or anti-analytic maps, $z \mapsto f(z)$ or $\bar{z} \mapsto \bar{f}(\bar{z})$, of the complex variables $z = t + \mathrm{i}r$, $\bar{z} = t - \mathrm{i}r$. The Lie algebra generators are $\ell_n = -z^{n+1}\partial_z$ and $\bar{\ell}_n = -\bar{z}^{n+1}\partial_{\bar{z}}$ with $n \in \mathbb{Z}$. The conformal Lie algebra is a pair of commuting Virasoro algebras with vanishing central charge [42,43], viz. $[\ell_n, \ell_m] = (n-m)\ell_{n+m}$. In an ortho-conformally invariant physical system, the $\ell_n, \bar{\ell}_n$ act on physical 'quasi-primary' [1] scaling operators $\phi = \phi(z,\bar{z}) = \varphi(t,r)$ and contain terms describing how these quasi-primary operators should transform, namely*

$$\ell_n = -z^{n+1}\partial_z - \Delta(n+1)z^n \ , \ \bar{\ell}_n = -\bar{z}^{n+1}\partial_{\bar{z}} - \bar{\Delta}(n+1)\bar{z}^n \tag{9}$$

where $\Delta, \bar{\Delta} \in \mathbb{R}$ are the conformal weights of the scaling operator ϕ. The scaling dimension is $x := x_\phi = \Delta + \bar{\Delta}$. Laplace's equation $\mathcal{S}\phi = 4\partial_z\partial_{\bar{z}}\phi = \left(\partial_t^2 + \partial_r^2\right)\varphi = 0$ is a simple example of an ortho-conformally invariant system, because of the commutator

$$[\mathcal{S}, \ell_n]\,\phi(z,\bar{z}) = -(n+1)z^n\mathcal{S}\phi(z,\bar{z}) - 4\Delta n(n+1)z^{n-1}\partial_{\bar{z}}\phi(z,\bar{z}). \tag{10}$$

This shows that for a scaling operator ϕ with $\Delta = \bar{\Delta} = 0$, the space of solutions of the Laplace equation $\mathcal{S}\phi = 0$ is conformally invariant, since any solution ϕ is mapped onto another solution $\ell_n\phi$ (or $\bar{\ell}_n\phi$) in the transformed coordinates. The maximal finite-dimensional sub-group is given by the projective conformal transformations $z \mapsto \frac{\alpha z+\beta}{\gamma z+\delta}$ with $\alpha\delta - \beta\gamma = 1$; its Lie algebra is $\mathfrak{sl}(2,\mathbb{R}) \oplus \mathfrak{sl}(2,\mathbb{R})$. Two-point functions of quasi-primary scaling operators read

$$\mathscr{C}_{12}(t_1,t_2;r_1,r_2) := \langle\phi_1(z_1,\bar{z}_1)\phi_2(z_2,\bar{z}_2)\rangle = \langle\varphi_1(t_1,r_1)\varphi_2(t_2,r_2)\rangle. \tag{11}$$

Their ortho-conformal covariance implies the projective Ward identities $X_n\mathscr{C}_{12} = Y_n\mathscr{C}_{12} = 0$ for $n = \pm 1, 0$ [1]. For scalars, such that $\Delta_i = \bar{\Delta}_i = x_i$, this gives, up to the normalisation C_0 [44]

$$\mathscr{C}_{12}(t_1,t_2;r_1,r_2) = C_0\,\delta_{x_1,x_2}\left((t_1-t_2)^2 + (r_1-r_2)^2\right)^{-x_1}. \tag{12}$$

Below, we often use the basis $X_n := \ell_n + \bar{\ell}_n$ and $Y_n := \ell_n - \bar{\ell}_n$, see also Table 2.

Example 4. *An example of meta-conformal transformations in $(1+1)D$ reads [45]*

$$\begin{aligned} X_n &= -t^{n+1}\partial_t - \mu^{-1}[(t+\mu r)^{n+1} - t^{n+1}]\partial_r - (n+1)xt^n - (n+1)\frac{\gamma}{\mu}[(t+\mu r)^n - t^n] \\ Y_n &= -(t+\mu r)^{n+1}\partial_r - (n+1)\gamma(t+\mu r)^n \end{aligned} \tag{13}$$

with $n \in \mathbb{Z}$. Herein, x, γ are the scaling dimension and the 'rapidity' of the scaling operator $\varphi = \varphi(t,r)$ on which these generators act. The constant $1/\mu$ has the dimensions of a velocity. The Lie algebra $\langle X_n, Y_n\rangle_{n\in\mathbb{Z}}$ is isomorphic to the conformal Lie algebra [46], see Table 2, where it is called meta-1 conformal invariance. *If $\gamma = \mu x$, the generators (13) act as dynamical symmetries on the equation $\mathcal{S}\varphi = (-\mu\partial_t + \partial_r)\varphi = 0$. This follows from the only non-vanishing commutators of the Lie algebra with $\mathcal{S}$, namely $[\mathcal{S}, X_0]\,\varphi = -\mathcal{S}\varphi$ and $[\mathcal{S}, X_1]\,\varphi = -2t\mathcal{S}\varphi + 2(\mu x - \gamma)\varphi$. The formulation of the meta-1 conformal Ward identities does require some care, since already the two-point function turns out to be a non-analytic function of the time- and space-coordinates. It can be shown that the covariant two-point correlator is [41]*

$$\mathscr{C}_{12}(t_1,t_2;r_1,r_2) = \mathcal{C}_0\,\delta_{x_1,x_2}\delta_{\gamma_1,\gamma_2}\,|t_1-t_2|^{-2x_1}\left(1+\frac{\mu}{\gamma_1}\left|\gamma_1\frac{r_1-r_2}{t_1-t_2}\right|\right)^{-2\gamma_1/\mu}. \tag{14}$$

Although both examples have $z=1$ and isomorphic Lie algebras, the explicit two-point functions (12) and (14), as well as the invariant equations $\mathcal{S}\varphi = 0$, are different, see also Table 2. That the form of two-point functions depends mainly on the representation and not so much on the Lie algebra, is not a phenomenon restricted to the conformal algebra. Similarly, for the so-called *Schrödinger algebra* at least three distinct representations with different forms of the two-point function are known [47].

The representation (13) can be extended to produce dynamical symmetries of the $(1+1)D$ Vlasov equation [48].

Example 5. *Taking the limit $\mu \to 0$ in the meta-conformal representation (13) produces the generators*

$$\begin{aligned} X_n &= -t^{n+1}\partial_t - (n+1)t^n r\partial_r - (n+1)xt^n - (n+1)n\gamma t^{n-1}r \\ Y_n &= -t^{n+1}\partial_r - (n+1)\gamma t^n \end{aligned} \tag{15}$$

of the conformal Galilean algebra (CGA) *in $(1+1)D$ [49–60]. Its Lie algebra is obtained by standard contraction of the conformal Lie algebra, see Table 2. Hence the* CGA *is* not *a meta-conformal algebra, although $z=1$. About* CGA*-covariant equations, see [61]. The co-variant two-point correlator can either be obtained from the generators (15), using techniques similar to those applied in the above example of meta-conformal invariance [46,68], or else by letting $\mu \to 0$ in (14). Both approaches give*

$$\mathscr{C}_{12}(t_1,t_2;r_1,r_2) = \mathcal{C}_0\,\delta_{x_1,x_2}\delta_{\gamma_1,\gamma_2}\,|t_1-t_2|^{-2x_1}\exp\left(-2\left|\gamma_1\frac{r_1-r_2}{t_1-t_2}\right|\right) \tag{16}$$

Clearly, this form is different from both ortho- and meta-1-conformal invariance.

The non-analyticity of the correlators (14), and especially (16), in general overlooked in the literature, is *required* in order to achieve $\mathscr{C}_{12} \to 0$ for large time- or space-separations, viz. $t_1 - t_2 \to \pm\infty$ or $r_1 - r_2 \to \pm\infty$.

All two-point functions (12), (14) and (16) have indeed the symmetries $\mathscr{C}_{12}(t_1,t_2;r,r) = \mathscr{C}_{21}(t_2,t_1;r,r)$ and $\mathscr{C}_{12}(t,t;r_1,r_2) = \mathscr{C}_{21}(t,t;r_2,r_1)$, under permutation $\varphi_1 \leftrightarrow \varphi_2$ of the two scaling operators, as physically required for a correlator. The shape of the scaling function of these three two-point function is compared in Figure 3. In particular, the non-analyticity of the meta- and Galilean conformal invariance at $u=0$ is clearly seen, in contrast to ortho-conformal invariance, while for $u \to \infty$, the slow algebraic decay of ortho- and meta-conformal invariance is distinct from the exponential decay of conformal Galilean invariance. This illustrates the variety of possible forms already for $z=1$. Below, we shall find another form of (meta-)conformal invariance, different from all forms displayed in Figure 3.

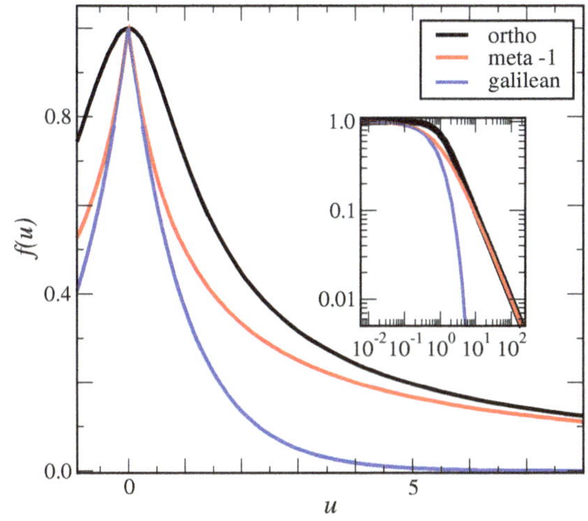

Figure 3. Scaling function $f(u)$ of the covariant two-point correlator $\mathscr{C}(t,r) = t^{-2x_1} f(r/t)$, over against the scaling variable $u = r/t$, for ortho-, meta-1- and Galilean-conformal invariance, Equations (12), (14) and (16) respectively, where $x_1 = \gamma_1 = \frac{1}{2}$ and $\mu = 1$.

3. Impossibility of a Local Meta-Conformal Invariance of the DLE Process

Can one consider these several variants of conformal invariance, which have $z = 1$ and are realised in terms of *local* first-order differential operators, as a valid dynamical symmetry of the DLE process in $1+1$ dimensions ? The answer turns out to be negative:

1. The deterministic part of the DLE Langevin Equation (1) is distinct from the simple invariant equations $\mathcal{S}\varphi = 0$ of either ortho- or meta-1-conformal invariance.
   For analogy, consider briefly *Schrödinger-invariant systems* with a Langevin equation of the form $\mathcal{S}\varphi = (2\nu T)^{1/2}\eta$, where $\eta = \frac{dB}{dt}$ is a white noise of unit variance, and such that the Schrödinger algebra is a dynamical symmetry of the noise-less equation (deterministic part) $\mathcal{S}\varphi_0 = 0$. Then, the Bargman super-selection rules [69] which follow from the combination of spatial translation-invariance and Galilei-invariance with $z = 2$, imply exact relations between averages of the full noisy theory and the averages calculated from its deterministic part [70]. In particular, the two-time response function of the full noisy equation $R(t,s;\boldsymbol{r}) = R_0(t,s;\boldsymbol{r})$, is identical to the response R_0 found when the noise is turned off and computed from the dynamical Schrödinger symmetry [16,70].
   We shall assume here that an analogous result can be derived also for the DLE Langevin equation, although this has not yet been done. It seems plausible that such a result should exist, since in the example (5b) and (6b) of the DLE process, the two-time response R is independent of T (which characterises the white noise), as it is the case for Schrödinger-invariance.
2. The explicit response function (6b) of the DLE process is distinct from the predictions (12) , (14) and (16), see also Table 2. The form of the meta-1 conformal two-point function (14), is clearly different for finite values of the scaling variable $v = (r_1 - r_2)/(t_1 - t_2)$, and similarly for the conformal Galilean case (16). The ortho-conformal two-point function (12) looks to be much closer, with the choice $x_1 = \frac{1}{2}$ and the scale factor fixed to $\nu = 1$, were it not for the extra factor $\nu(t-s)$. On the other hand, the two-time DLE-correlator (6a) does not agree with (12) either, but might be similar to a two-point function computed in a semi-infinite space $t \geq 0$, $r \in \mathbb{R}$ with a boundary at $t = 0$.

Looking for dynamical symmetries of the deterministic part of the DLE Langevin Equation (1), in $1+1$ dimensions, the first test will be the computation of the two-time response function $R(t,s;\boldsymbol{r})$.

By contrast, we shall show that the two-time correlator C cannot be found in this way. Indeed, its 'deterministic' contribution vanishes: $C_0(t,s;\boldsymbol{r}) = 0$.

4. Riesz-Feller Fractional Derivative

Formulating (1) in direct space requires the Riesz-Feller fractional derivative [71–73] of order α.

Definition 4. *For functions $f(r)$ of a single variable $r \in \mathbb{R}$, the* Riesz-Feller derivative, *of order α, is*

$$\nabla_r^\alpha f(r) := \frac{\mathrm{i}^\alpha}{\sqrt{2\pi}} \int_{\mathbb{R}} \mathrm{d}k\, |k|^\alpha\, e^{\mathrm{i}kr}\, \widehat{f}(k) = \frac{\mathrm{i}^\alpha}{2\pi} \int_{\mathbb{R}^2} \mathrm{d}k\mathrm{d}x\, |k|^\alpha\, e^{\mathrm{i}k(r-x)}\, f(x) \tag{17}$$

where $\widehat{f}(k)$ denotes the Fourier transform of $f(r)$. For brevity, we often write $\nabla_r = \nabla_r^1$ and distinguish it from the standard derivative ∂_r.

Lemma 1. *([72], Prop. 3.6) Let $f \in H^{\alpha/2}(\mathbb{R}) = \left\{ f \in L^2(\mathbb{R}) \,\middle|\, \int_{\mathbb{R}} \mathrm{d}k\, |\widehat{f}(k)|^2 (1+|k|^2)^{\alpha/2} < \infty \right\}$, a fractional Sobolev space. For $0 < \alpha < 2$, the Riesz-Feller derivative $\nabla_r^\alpha f(r)$ exists.*

Lemma 2. *([16] app. J.2, [21]) The following formal properties hold true, where α, β, q, μ are constants*

$$\begin{array}{rclcl}
\nabla_r^\alpha \nabla_r^\beta f(r) = \nabla_r^{\alpha+\beta} f(r) & , & [\nabla_r^\alpha, r]\, f(r) = \alpha \partial_r \nabla_r^{\alpha-2} f(r) & , & \nabla_r^\alpha f(\mu r) = |\mu|^\alpha \nabla_{\mu r}^\alpha f(\mu r) \\
\nabla_r^\alpha e^{\mathrm{i}qr} = (\mathrm{i}|q|)^\alpha\, e^{\mathrm{i}qr} & , & \left(\widehat{\nabla_r^\alpha f(r)}\right)(q) = (\mathrm{i}|q|)^\alpha\, \widehat{f}(q) & , & \nabla_r^2 f(r) = \partial_r^2 f(r)
\end{array} \tag{18}$$

Lemma 2 follows directly from the definition (17). The analogy with the rules of the ordinary derivative ∂_r^n, with $n \in \mathbb{N}$, applied to exponentials $e^{\mathrm{i}qr}$ and to Fourier transforms, motivated our choice of (complex) normalisation in (17). Later, we shall also need the object $\partial_r \nabla_r^{-1}$, which is formally written as

$$\partial_r \nabla_r^{-1} f(r) = \frac{1}{\sqrt{2\pi}} \int_{\mathbb{R}} \mathrm{d}k\, e^{\mathrm{i}kr}\, \mathrm{sign}\,(k)\, \widehat{f}(k) \tag{19}$$

but is best considered via its Fourier transform, viz. $\left(\widehat{\partial_r \nabla_r^{-1} f(r)}\right)(q) = \mathrm{sign}\,(q) \widehat{f}(q)$. This is well-defined, since $f \in H^{\alpha/2}(\mathbb{R}) \subset L^2(\mathbb{R})$ and because of Plancherel's theorem. The Fourier transform of $\mathrm{sign}\,(q)$ is a distribution ([74] [Equation (2.3.17)]).

Corollary 1. *One has the following formal commutator identities*

$$\begin{array}{l}
[\nabla_r, r^n] = n r^{n-1} \partial_r \nabla_r^{-1} \;,\; \left[r^2 \nabla_r, r\partial_r\right] = -r^2 \nabla_r \;,\; \left[r\partial_r \nabla_r^{-1}, r\nabla_r\right] = -r \;, \\
[\nabla_r, \partial_r] = [r\partial_r, r\nabla_r] = \left[r, \partial_r \nabla_r^{-1}\right] = \left[r^n \partial_r, \partial_r \nabla_r^{-1}\right] = \left[r^n \nabla_r, \partial_r \nabla_r^{-1}\right] = 0 \;.
\end{array} \tag{20}$$

Proof. Most of these identities are immediate consequences of (17,18). The first one is proven by induction, for all $n \in \mathbb{N}$. We only detail here the computation of the third one. Formally, one has $\left[r\partial_r \nabla_r^{-1}, r\nabla_r\right] = -r\left(\partial_r \nabla_r^{-1}\right)^2$. Using (19), and its Fourier transform, gives

$$\left(\partial_r \nabla_r^{-1}\right)^2 f(r) = \int_{\mathbb{R}} \frac{\mathrm{d}k}{\sqrt{2\pi}}\, e^{\mathrm{i}kr} \mathrm{sign}\,(k) \left(\widehat{\partial_r \nabla_r^{-1} f(r)}\right)(k) = \int_{\mathbb{R}} \frac{\mathrm{d}k}{\sqrt{2\pi}}\, e^{\mathrm{i}kr} \left(\mathrm{sign}\,(k)\right)^2 \widehat{f}(k) = f(r)$$

which establishes the assertion. □

Lemma 3. *[72,75] If $0 < \alpha < 2$, one has for either $f \in H^{\alpha/2}(\mathbb{R})$ or else $f \in \mathscr{S}(\mathbb{R})$, the Schwartz space of smooth, rapidly decreasing functions, for almost all $r \in \mathbb{R}$*

$$\nabla_r^\alpha f(r) = \frac{1 - e^{\mathrm{i}\pi\alpha}}{4\pi\mathrm{i}} \Gamma(\alpha+1) \int_{\mathbb{R}} \frac{\mathrm{d}y}{|y|^{\alpha+1}} \left[f(r+y) - 2f(r) + f(r-y)\right] \tag{21}$$

If $f \in \mathscr{S}(\mathbb{R})$, $\nabla_r^{\alpha} f(r)$ can be defined beyond the interval $0 < \alpha < 2$.

5. Non-Local Meta-Conformal Generators

The deterministic part of (1) becomes $\mathcal{S}\varphi = (-\mu\partial_r + \nabla_r)\,\varphi = 0$, where $\mu^{-1} = \mathrm{i}\nu$. Following earlier studies [45], it seems physically reasonable that the Lie algebra of dynamical symmetries should at least contain the generators of time translations $X_{-1} = -\partial_t$, dilatations $X_0 = -t\partial_t - \frac{r}{z}\partial_r - \frac{x}{z}$ and space translations $Y_{-1} = -\partial_r$. It turns out, however, if one wishes to construct a generator of generalised Galilei transformations as a dynamical symmetry, the non-local generator $-\nabla_r$ automatically arises, see below and ([16], Chapter 5.3). It is still an open problem how to close these generators into a Lie algebra, for $z \neq 2$ and beyond the examples listed above in Section 2.

This difficulty motivates us to start with the choice of a *non-local* spatial translation operator $Y_{-1} = -\mu^{-1}\nabla_r$. Here indeed, a closed Lie algebra can be found.

Proposition 1. *[21] Define the following generators*

$$\begin{aligned} X_{-1} &= -\partial_t \ , \ X_0 = -t\partial_t - r\partial_r - x \ , \ X_1 = -t^2\partial_t - 2tr\partial_r - \mu r^2\nabla_r - 2xt - 2\gamma r\partial_r\nabla_r^{-1} \\ Y_{-1} &= -\frac{1}{\mu}\nabla_r \ , \ Y_0 = -\frac{1}{\mu}t\nabla_r - r\partial_r - \frac{\gamma}{\mu} \ , \ Y_1 = -\frac{1}{\mu}t^2\nabla_r - 2tr\partial_r - \mu r^2\nabla_r - 2\frac{\gamma}{\mu}t - 2\gamma r\partial_r\nabla_r^{-1} \end{aligned} \tag{22}$$

where the constants $x = x_\varphi$ and $\gamma = \gamma_\varphi$, respectively, are the scaling dimension and rapidity of the scaling operator $\varphi = \varphi(t,r)$ on which these generators act. The six generators (22) obey the commutation relations of a meta-conformal Lie algebra, isomorphic to $\mathfrak{sl}(2,\mathbb{R}) \oplus \mathfrak{sl}(2,\mathbb{R})$

$$[X_n, X_m] = (n-m)X_{n+m} \ , \ [X_n, Y_m] = (n-m)Y_{n+m} \ , \ [Y_n, Y_m] = (n-m)Y_{n+m} \tag{23}$$

Proposition 2. *[21] The generators (22) obey the commutators*

$$[\mathcal{S}, Y_n]\,\varphi = [\mathcal{S}, X_{-1}]\,\varphi = 0 \ , \ [\mathcal{S}, X_0]\,\varphi = -\mathcal{S}\varphi \ , \ [\mathcal{S}, X_1]\,\varphi = -2t\mathcal{S}\varphi + 2(\mu x - \gamma)\varphi \tag{24}$$

with the operator $\mathcal{S} = -\mu\partial_t + \nabla_r$ and thus form a Lie algebra of meta-conformal dynamical symmetries (of the deterministic part) $\mathcal{S}\varphi = 0$ of the DLE *Langevin Equation (1), if only $\gamma = x\mu$.*

The non-local generators $X_1, Y_{0,1}$ in (22) do not generate simple local changes of the coordinates (t,r), in contrast to all examples of Section 2. Finding a clear geometrical interpretation of the generators (22) remains an open problem.

This meta-conformal symmetry algebra can be considerably enlarged.

Proposition 3. *Consider the generators (22) and furthermore define*

$$Z_{-1} = -\frac{1}{\mu}\partial_r \ , \ Z_0 = -\frac{1}{\mu}t\partial_r - r\nabla_r - \frac{\gamma}{\mu}\partial_r\nabla_r^{-1} \ , \ Z_1 = -\frac{1}{\mu}t^2\partial_r - 2tr\nabla_r - \mu r^2\partial_r - 2\frac{\gamma}{\mu}t\partial_r\nabla_r^{-1} - 2\gamma r \tag{25}$$

These generators are dynamical symmetries of the DLE *Langevin equation, since $[\mathcal{S}, Z_n] = 0$ and they extend the meta-conformal Lie algebra (23) as follows*

$$[X_n, Z_m] = (n-m)Z_{n+m} \ , \ [Y_n, Z_m] = (n-m)Z_{n+m} \ , \ [Z_n, Z_m] = (n-m)Y_{n+m} \tag{26}$$

Although Z_{-1} generates local spatial translations, the transformations obtained from $Z_{0,1}$ are non-local. In what follows, we write $\xi := \gamma/\mu$ for the second, independent scaling dimension of φ.

Corollary 2. *Define the generators $B_n^{\pm} = \frac{1}{2}(Y_n \pm Z_n)$, $n \in \{-1,0,1\}$. Then the non-vanishing commutators of the Lie algebra (23) and (26) take the form*

$$[X_n, X_m] = (n-m)X_{n+m} \ , \ [X_n, B_m^{\pm}] = (n-m)B_{n+m}^{\pm} \ , \ [B_n^{\pm}, B_m^{\pm}] = (n-m)B_{n+m}^{\pm} \tag{27}$$

The $B_n^{\pm}$ are dynamical symmetries of the DLE *process, since* $[\mathcal{S}, B_n^{\pm}] = 0$.

Corollary 3. *Define the generators* $A_n = X_n - (B_n^+ + B_n^-) = X_n - Y_n$, $n \in \{-1,0,1\}$. *Then the non-vanishing commutators of the Lie algebra (27) are*

$$[A_n, A_m] = (n-m)A_{n+m} \;,\;\; [B_n^{\pm}, B_m^{\pm}] = (n-m)B_{n+m}^{\pm} \tag{28}$$

This Lie algebra of dynamical symmetries of the deterministic part of the DLE *Langevin Equation (1) is isomorphic to the direct sum* $\mathfrak{sl}(2,\mathbb{R}) \oplus \mathfrak{sl}(2,\mathbb{R}) \oplus \mathfrak{sl}(2,\mathbb{R})$.

In this last choice of basis, all generators contain non-local terms. Their form, in Corollary 3, is suggestive for the explicit construction of an infinite-dimensional extension of the above Lie algebra.

Proposition 4. *Construct the generators, for all* $n \in \mathbb{Z}$ *and* x, ξ *constants*

$$\begin{aligned} A_n &= -t^{n+1}\left(\partial_t - \nabla_r\right) - (n+1)\left(x-\xi\right)t^n \\ B_n^{\pm} &= -\frac{1}{2}\left(t \pm r\right)^{n+1}\left(\nabla_r \pm \partial_r\right) - \frac{n+1}{2}\xi\left(t \pm r\right)^n\left(1 \pm \partial_r \nabla_r^{-1}\right) \end{aligned} \tag{29}$$

Their non-vanishing commutators are given by (28), for $n, m \in \mathbb{Z}$. *Their Lie algebra is isomorphic to the direct sum of three Virasoro algebras with vanishing central charges. They are also dynamic symmetries of the deterministic equation* $\mathcal{S}\varphi = \left(-\partial_t + \nabla_r\right)\varphi = 0$ *of* DLE *process, provided that* $x = \xi$, *because of the commutators*

$$[\mathcal{S}, A_n] = -(n+1)t^n\mathcal{S} + (n+1)n\left(x-\xi\right)t^{n-1} \;,\;\; [\mathcal{S}, B_n^{\pm}] = 0 \tag{30}$$

Proof. For $n = \pm 1, 0$, the generators (29) are those given above in (22) and (25), using $\xi = \gamma/\mu$ and rescaling $\mu \mapsto 1$. One generalises the first identity (20) in the Corollary 1 to the following form, with $n \in \mathbb{N}$

$$\left[\nabla_r, (\alpha \pm r)^n\right] = \pm n\left(\alpha \pm r\right)^{n-1}\partial_r \nabla_r^{-1}$$

where α is a constant. The assertions now follow by direct formal calculations, using (18) and (20). □

This is the DLE-analogue of the ortho- and meta-1 conformal invariances, respectively, of the Laplace equation and of simple ballistic transport, as treated in Examples 3 and 4. In Table 3, it is called "meta-2 conformal". It clearly appears that both local and non-local spatial translations are needed for realising the full dynamical symmetry of the DLE process, which we call *erosion-Virasoro algebra* and denote by $\mathfrak{ev}$. The infinite-dimensional Lie algebra $\mathfrak{ev}$ is built from *three* commuting Virasoro algebras (obviously, the maximal finite-dimensional Lie sub-algebra is $\mathfrak{sl}(2,\mathbb{R}) \oplus \mathfrak{sl}(2,\mathbb{R}) \oplus \mathfrak{sl}(2,\mathbb{R})$). The scaling operators $\varphi = \varphi(t,r)$ on which these generators act are characterised by two independent *scaling dimensions* $x = x_\varphi$ and $\xi = \xi_\varphi$. By analogy with conformal Galilean invariance [76], one expects that three independent central charges of the Virasoro type should appear if the algebra (28) will be quantised. Additional physical constraints (e.g. unitarity) may reduce the number of independent central charges.

6. Ward Identities for Co-Variant Quasi-Primary n-Point Functions

A basic application of dynamic time-space symmetries is the derivation of co-variant n-point functions. Adapting the corresponding definition from (ortho-)conformal invariance [1], a scaling operator $\varphi = \varphi(t,r)$ is called *quasi-primary*, if it transforms co-variantly under the action of the generators of the maximal finite-dimensional sub-algebra of $\mathfrak{ev}$. A *primary* scaling operator transforms co-variantly under the action of all generators of $\mathfrak{ev}$. In this work, we consider examples of n-point functions of quasi-primary scaling operators.

In the physical context of non-equilibrium dynamics, such n-point functions can either be correlators, such as $\langle\varphi(t,r)\varphi(t',r')\rangle$, or response functions $\langle\varphi(t,r)\widetilde{\varphi}(t',0)\rangle = \left.\frac{\delta\langle\varphi(t,r)\rangle}{\delta j(t',0)}\right|_{j=0}$, which can be formally rewritten as a correlator by using the formalism of Janssen-de Dominicis theory [17] which defines the response operator $\widetilde{\varphi}$, conjugate to the scaling operator φ.

Proceeding in analogy with ortho-conformal and Schrödinger-invariance [1,16,43,44,77], the quasi-primary $\mathfrak{cv}$-Ward identities are obtained from the explicit form of the Lie algebra generators (22) and (25), generalised to n-body generators. In order to do so, we assign a *signature* $\varepsilon = \pm 1$ to each scaling operator [21]. We choose the convention that $\varepsilon_i = +1$ for scaling operators φ_i and $\varepsilon_i = -1$ for response operators $\widetilde{\varphi}_i$. In order to prepare a later application to the conformal Galilean algebra, to be obtained from a Lie algebra contraction, we also multiply the generators Y_i, Z_i by the scale factor μ. The n-body generators then read

$$
\begin{aligned}
X_{-1} = X_{-1}^{[n]} &= \sum_i [-\partial_i] \quad , \qquad X_0 = X_0^{[n]} = \sum_i [-t_i\partial_i - r_i D_i - x_i] \\
X_1 = X_1^{[n]} &= \sum_i \left[-t_i^2\partial_i - 2t_i r_i D_i - \mu\varepsilon_i r_i^2 \nabla_i - 2x_i t_i - 2\mu\xi_i\varepsilon_i r_i D_i \nabla_i^{-1}\right] \\
Y_{-1} = Y_{-1}^{[n]} &= \sum_i [-\varepsilon_i\nabla_i] \quad , \qquad Y_0 = Y_0^{[n]} = \sum_i [-\varepsilon_i t_i \nabla_i - \mu r_i D_i - \mu\xi_i] \\
Y_1 = Y_1^{[n]} &= \sum_i \left[-\varepsilon_i\left(t_i^2 + \mu^2 r_i^2\right)\nabla_i - 2\mu t_i r_i D_i - 2\mu\xi_i t_i - 2\mu^2\xi_i\varepsilon_i r_i D_i \nabla_i^{-1}\right] \\
Z_{-1} = Z_{-1}^{[n]} &= \sum_i [-D_i] \quad , \qquad Z_0 = Z_0^{[n]} = \sum_i \left[-t_i D_i - \varepsilon_i r_i \nabla_i - \mu\xi_i D_i \nabla_i^{-1}\right] \\
Z_1 = Z_1^{[n]} &= \sum_i \left[-\left(t_i^2 + \mu^2 r_i^2\right) D_i - 2\varepsilon_i \mu t_i r_i \nabla_i - 2\mu\xi_i r_i - 2\mu\xi_i\varepsilon_i t_i D_i \nabla_i^{-1}\right]
\end{aligned}
\tag{31}
$$

with the short-hands $\partial_i = \frac{\partial}{\partial t_i}$, $D_i = \frac{\partial}{\partial r_i}$ and $\nabla_i = \nabla_{r_i}$. It can be checked that the generators (31) obey the meta-conformal Lie algebra of the DLE process. Define the $(n+m)$-point function

$$
\begin{aligned}
\mathscr{C}_{n,m} &= \mathscr{C}_{n,m}(t_1,\ldots,t_{n+m}; r_1,\ldots,r_{n+m}) \\
&= \langle\varphi_1(t_1,r_1)\cdots\varphi_n(t_n,r_n)\widetilde{\varphi}_{n+1}(t_{n+1},r_{n+1})\cdots\widetilde{\varphi}_{n+m}(t_{n+m},r_{n+m})\rangle
\end{aligned}
\tag{32}
$$

of quasi-primary scaling and response operators. Their co-variance is expressed through the quasi-primary Ward identities, for $k = \pm 1, 0$

$$
X_k^{[n+m]}\mathscr{C}_{n,m} = Y_k^{[n+m]}\mathscr{C}_{n,m} = Z_k^{[n+m]}\mathscr{C}_{n,m} = 0,
\tag{33}
$$

The solution of this set of (linear) differential equations gives the sought $(n+m)$-point function $\mathscr{C}_{n,m}$.

7. Co-Variant Two-Time Correlators and Responses

In order to illustrate the procedure outlined in section 6, we shall apply it to the two-point functions.

Proposition 5. *Any two-point correlator $\mathscr{C}_{2,0}(t_1,t_2;r_1,r_2) = \langle\varphi_1(t_1,r_1)\varphi_2(t_2,r_2)\rangle$, built from $\mathfrak{cv}$-quasi-primary scaling operators φ_i, vanishes.*

Proof. Time-translation-invariance, expressed by $X_{-1}\mathscr{C}_{2,0} = 0$, implies that $\mathscr{C}_{2,0} = \mathscr{C}_{2,0}(t; r_1, r_2)$, with $t = t_1 - t_2$. Invariance under both non-local and local space-translations gives $Y_{-1}\mathscr{C}_{2,0} = Z_{-1}\mathscr{C}_{2,0} = 0$. In Fourier space, this becomes

$$
(\varepsilon_1|q_1| + \varepsilon_2|q_2|)\,\widehat{\mathscr{C}}_{2,0}(t;q_1,q_2) = 0 \ , \quad (q_1+q_2)\,\widehat{\mathscr{C}}_{2,0}(t;q_1,q_2) = 0
$$

where the signatures are both positive, viz. $\varepsilon_1 = \varepsilon_2 = +1$. The only solution is $\widehat{\mathscr{C}}_{2,0}(t; q_1, q_2) = 0$. □

Recall that the dynamical symmetry of the $\mathfrak{ev}$ algebra is only a symmetry of the *deterministic part* of the DLE Langevin equation (1), which corresponds to $T = 0$. The vanishing of $\mathscr{C}_{2,0}$ is seen explicitly in the exact DLE-correlator (5a,6a), which indeed vanishes as $T \to 0$. This result of the DLE process is analogous to what is found for Schrödinger-invariant systems [16,77], where it follows from a Bargman superselection rule [69]. Still, this does not mean that symmetry methods could only predict vanishing correlators. For example, in Schrödinger-invariant systems, correlators with $T \neq 0$ can be found from certain integrals of higher n-point responses [16,70]. For a simple illustration in the noisy Edwards-Wilkinson equation, see [78]. We conjecture that an analogous procedure might work for the DLE process and hope to return to this elsewhere.

We now concentrate on the two-time *response function* $\mathscr{R} = \mathscr{R}(t_1, t_2; r_1, r_2) = \mathscr{C}_{1,1}(t_1, t_2; r_1, r_2)$. Time-translation-invariance, which imposes $X_{-1}\mathscr{R} = 0$, implies that $\mathscr{R} = \mathscr{R}(t; r_1, r_2)$, with $t = t_1 - t_2$. Invariance under non-local and local space-translations now give (in Fourier space)

$$\varepsilon_1 \left(|q_1| - |q_2|\right) \widehat{\mathscr{R}}(t; q_1, q_2) = 0 \ , \ \ (q_1 + q_2)\, \widehat{\mathscr{R}}(t; q_1, q_2) = 0$$

since the signatures are now $\varepsilon_1 = -\varepsilon_2 = +1$. Here, a non-vanishing solution is possible and we can write $\mathscr{R} = F(t, r)$, with $r = r_1 - r_2$.

Proposition 6. *The $\mathfrak{ev}$-covariant two-point response function $\mathscr{R} = \mathscr{C}_{1,1}$ from (32) satisfies the scaling form $\mathscr{R} = \langle \varphi_1(t, r)\widetilde{\varphi}_2(0, 0)\rangle = t^{-2x} f(v)$, with the scaling variable $v = r/t$. If the scaling function $f(v)$ obeys the following two conditions, with the abbreviations $x = \frac{1}{2}(x_1 + x_2)$ and $\xi = \frac{1}{2}(\xi_1 + \xi_2)$,*

$$\left(\varepsilon_1 \nabla_v + \mu v \partial_v + 2\mu\xi\right) f(v) = 0 \ , \ \ (x_1 - x_2)\left(\varepsilon_1 \nabla_v + \mu v \partial_v + \mu\right) f(v) = 0. \tag{34}$$

and the constraint $\xi_1 - \xi_2 = x_1 - x_2$ holds true, then all quasi-primary Ward identities are satisfied.

The conditions (34) come from the deterministic part of the DLE Langevin Equation (1) and do not contain T. This is consistent with the T-independence of the exact DLE-response function (5b) and (6b). A fuller justification, analogous to the derivation of the Bargman superselection rules of Schrödinger-invariance [70,77], is left as an open problem, for future work.

Proof. Denote by x_i and ξ_i (with $i = 1, 2$), the two scaling dimensions of the scaling operator φ_1 and of the response operator $\widetilde{\varphi}_2$, respectively. Time-translation-invariance and non-local and local space-translation-invariances produced the form $\mathscr{R} = F(t, r)$, with $t = t_1 - t_2$, $r = r_1 - r_2$ and the signatures $\varepsilon_1 = -\varepsilon_2 = +1$. The other six Ward identities lead to the conditions, using (18)

$$\left[-t\partial_t - r\partial_r - x_1 - x_2\right] F = 0 \tag{35a}$$

$$\left[-t\varepsilon_1\nabla_r - \mu r\partial_r - \mu\xi_1 - \mu\xi_2\right] F = 0 \tag{35b}$$

$$\left[-t\partial_r - \mu\varepsilon_1 r\nabla_r - \mu(\xi_1 + \xi_2)\varepsilon_1\partial_r\nabla_r^{-1}\right] F = 0 \tag{35c}$$

$$\left[-t^2\partial_t - 2tr\partial_r - \mu r^2\varepsilon_1\nabla_r - 2x_1 t - 2\mu\xi_1\varepsilon_1 r\partial_r\nabla_r^{-1}\right] F = 0 \tag{35d}$$

$$\left[-t^2\varepsilon_1\nabla_r - 2\mu tr\partial_r - \mu^2\varepsilon_1 r^2\nabla_r - 2\mu\xi_1 t - 2\mu^2\xi_1\varepsilon_1 r\partial_r\nabla_r^{-1}\right] F = 0 \tag{35e}$$

$$\left[-t^2\partial_r - 2\mu\varepsilon_1 tr\nabla_r - \mu^2 r^2\partial_r - 2\mu^2\xi_1 r - 2\mu\xi_1\varepsilon_1 t\partial_r\nabla_r^{-1}\right] F = 0 \tag{35f}$$

Herein, Equation (35d) is obtained by using Equations (35a) and (35c), and Equations (35e) and (35f) are obtained by using (35b) and (35c). Actually, because of the identity

$$\begin{aligned}\left[-t\partial_r - \mu\varepsilon_1 r\nabla_r - 2\mu\xi\varepsilon_1\partial_r\nabla_r^{-1}\right]F &= \varepsilon_1\left[-t\varepsilon_1\nabla_r\partial_r\nabla_r^{-1} - \mu r\nabla_r^2\nabla_r^{-1} - 2\mu\xi\partial_r\nabla_r^{-1}\right]F \\ &= \varepsilon_1\partial_r\nabla_r^{-1}\left[-\varepsilon_1 t\nabla_r - \mu r\partial_r - 2\mu\xi\right]F\end{aligned}$$

the condition $Y_0F = 0$, Equation (35b), implies $Z_0F = 0$, Equation (35c). Since $\left(\partial_r\nabla_r^{-1}\right)^2 f(r) = f(r)$, see the Corrollary 1, the converse also holds true. Next, Equation (35d) can be simplified further: multiply Equation (35a) with t and subtract it from (35d), which gives

$$\left[-tr\partial_r - \mu\varepsilon_1 r^2\nabla_r - (x_1 - x_2)t - 2\mu\xi_1\varepsilon_1 r\partial_r\nabla_r^{-1}\right]F = 0 \tag{36}$$

Then multiply (35c) with r and substract it from (36). This gives the condition

$$\left[(x_1 - x_2)\,t + \mu\,(\xi_1 - \xi_2)\,\varepsilon_1 r\partial_r\nabla_r^{-1}\right]F = 0 \tag{37}$$

Similarly, simplify Equation (35e): multiply (35b) by t and subtract from (35e), then multiply (36) by μ and subtract as well. This gives

$$\mu\left[(\xi_1 - \xi_2) - (x_1 - x_2)\right]tF = 0$$

Unless $F \sim \delta(t)$ is a distribution, this gives the constraint $\xi_1 - \xi_2 = x_1 - x_2$. Finally, Equation (35f) is simplified by multiplying first (35c) with t and subtracting and then multiplying (35b) with r and subtracting as well. This leads to $(\xi_1 - \xi_2)\left[\mu r + \varepsilon_1 t\partial_r\nabla_r^{-1}\right]F = 0$. Since in the proof of the Corollary 1, we have seen that $\left(\partial_r\nabla_r^{-1}\right)^2 f(r) = f(r)$, this can be rewritten as follows: $(\xi_1 - \xi_2)\varepsilon_1\left(\partial_r\nabla_r^{-1}\right)\left[\varepsilon_1\mu r\partial_r\nabla_r^{-1} + t\right]F = 0$. Taking the constraint into account, the last condition can be combined with (37) into the single equation

$$(x_1 - x_2)\left[t + \mu\varepsilon_1 r\partial_r\nabla_r^{-1}\right]F = 0 \tag{38}$$

The form of F is now fixed by the three equations (35a,35b,38) and the constraint has to be obeyed.

Equation (35a) implies the scaling form $F = t^{-2x}f(r/t)$. Inserting this into (35b) produces, with the help of (18), the first of the Equations (34). Finally, inserting the scaling form for F into (38) gives $(x_1 - x_2)\left(1 + \mu\varepsilon_1 v\partial_v\nabla_v^{-1}\right)f(v) = 0$. Since it is not immediately obvious if that condition is consistent with the first Equation (34), we rephrase it as follows: use the commutator $\left[v\partial_v, \nabla_v^{-1}\right] = \nabla_v^{-1}$ to write formally $v\partial_v\nabla_v^{-1} = \nabla_v^{-1} + \nabla_v^{-1}(v\partial_v)$. Then, apply ∇_v to the last condition on $f(v)$ derived from (38), in order to rewrite it as follows

$$\nabla_v\nabla_v^{-1}\left[(x_1 - x_2)\left(\varepsilon_1\nabla_v + \mu v\partial_v + \mu\right)\right]f(v) = 0$$

and this equation is obeyed if the second Equation (34) holds true. We have found a sufficient set of conditions to satisfy all nine DLE-quasi-primary Ward identities for $\mathscr{R} = \mathscr{C}_{1,1}$. □

The two conditions in Equation (34) are compatible in two distinct cases:

Case A: $2\xi = 1$. Then $(\varepsilon_1\nabla_v + \mu v\partial_v + \mu)\,f(v) = 0$ and $x_1 \neq x_2$ is still possible.
Case B: $x_1 = x_2$. Then $\xi_1 = \xi_2$ and $(\varepsilon_1\nabla_v + \mu v\partial_v + 2\mu\xi)\,f(v) = 0$.

We must also compare the differential operator $\mathcal{S} = -\mu\partial_t + \nabla_r$ with the DLE Langevin Equation (1). Taking into account the normalisation in the definition of the Riesz-Feller derivative, we find $\mu^{-1} = \mathrm{i}\nu$. Physically, one should require $\nu > 0$ in order that the correlators and responses vanish for large momenta $|q| \to \infty$.

Proposition 7. *The $\mathfrak{ev}$-co-variant two-time response function $\mathscr{R}_{12}(t,r) = F(t,r)$ has the form*

$$\begin{aligned} F(t,r) &= F_A\, \delta_{\xi_1+\xi_2,1}\, \delta_{\xi_1-\xi_2,x_1-x_2}\, t^{1-x_1-x_2} \frac{\nu t}{\nu^2 t^2 + r^2} \\ &\quad + F_B\, \delta_{x_1,x_2}\, \delta_{\xi_1,\xi_2}\, t^{1+\psi-2x_1} \left(\nu^2 t^2 + r^2\right)^{-(\psi+1)/2} \cos\left((\psi+1) \arctan\left(\frac{r}{\nu t}\right) - \frac{\pi\psi}{2}\right) \end{aligned} \tag{39}$$

where $\psi = (\xi_1 + \xi_2) - 1$ is assumed real, $F_{A,B}$ are normalisation constants and the convention $\varepsilon_1 = +1$ is admitted.

Proof. Both cases can be treated in the same way. The first Equation (34) becomes in Fourier space

$$\left[\mathrm{i}\varepsilon_1 |q| - \mu q \partial_q + \mu(2\xi - 1)\right] \widehat{f}(q) = 0$$

In case A, the constant term vanishes, while it is non-zero in case B. The solution reads

$$\widehat{f}(q) = \widehat{f}_0 q^{2\xi-1} \exp\left(\mathrm{i}\varepsilon_1 |q|/\mu\right) = \widehat{f}_0 q^{2\xi-1} \exp\left(-\varepsilon_1 \nu |q|\right)$$

where $\widehat{f}_0$ is a normalisation constant and we can now adopt $\varepsilon_1 = +1$. We also introduced the constant ν from the DLE Langevin Equation (1) to illustrate that $\widehat{f}(q) \to 0$ for $|q|$ large when ν is positive. Both cases A and B produce valid solutions of the linear Equations (34). Therefore, the general solution should be a linear superposition of both cases. Carrying out the inverse Fourier transforms is straightforward. □

Remark 3. *Propositions 4 and 7 contain the assertions in the Theorem 1, which are also listed in Table 3. Proposition 5 proves the statement in Remark 2. We had already mentioned in Section 1 (Remark 1), that if we restrict to case A and take $x = x_1 = x_2 = \frac{1}{2}$ and $\nu > 0$, the resulting two-time response $F(t,r) = F_0\, t^{1-2x}\, \varepsilon_1 \nu t/(\nu^2 t^2 + r^2)$, with $t = t_1 - t_2$ and $r = r_1 - r_2$, reproduces the exact solution (6b). We stress that no choice of x_1 will make the ortho-conformal prediction (12) compatible with (6b).*

This is the main conceptual point of this work: The non-local representation (29) of the meta-conformal algebra $\mathfrak{ev}$ is necessary to reproduce the correct scaling behaviour of the non-stationary response of the DLE process.

The non-local meta-2 conformal invariance produces the response function $\mathscr{R} = \mathscr{C}_{1,1}$, whereas all local ortho-, Galilean and meta-conformal invariances yielded a correlator $\mathscr{C}_{2,0}$.

The main result (8) and (39) on the shape of the meta-2-conformal response can be cast into the scaling form $t^{x_1+x_2}\mathscr{R}_{12}(t,r) = f(r/t)$, with the explicit scaling function

$$f(u) = \frac{\left(1+u^2\right)^{-1} + \rho\left(1+u^2\right)^{-\xi_1} \sin\left(\pi\xi_1 - 2\xi_1 \arctan u\right)}{1 + \rho \sin \pi\xi_1}. \tag{40}$$

We see that the first scaling dimensions x_1, x_2 merely arrange the data collapse, while the form of the scaling functions only depends on the second scaling dimension $\xi_1 = \xi_2$ and the amplitude ratio ρ (the exact solution (6b) of the DLE-process corresponds to $\rho = 0$). The normalisation is chosen such that $f(0) = 1$. For $\xi_1 = \frac{1}{2}$, we simply have $f(u) = (1+u^2)^{-1}$. In Figure 4, several examples of the shape of $f(u)$ are shown. Clearly, these are quite distinct from all the examples of ortho-, meta-1- and Galilean-conformal invariance, displayed above in Figure 3.

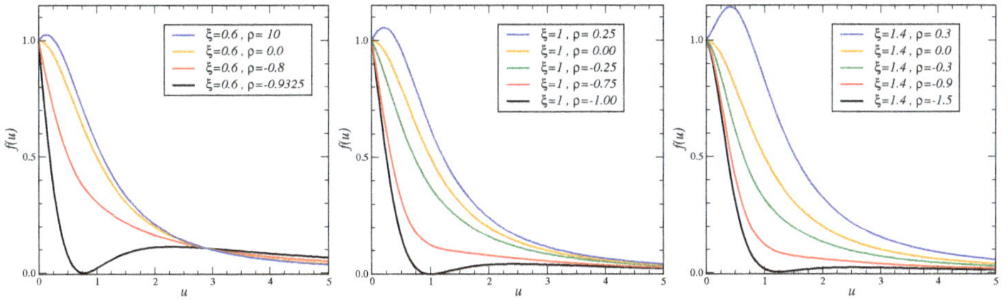

Figure 4. Scaling function $f(u)$ of the covariant meta-2-conformal two-point response $\mathscr{R}(t,r) = t^{-x_1-x_2} f(r/t)$, over against the scaling variable $u = r/t$, for $\xi = \xi_1 = [0.6, 1.0, 1.4]$ in the left, middle and right panels, respectively, and several values of the amplitude ratio ρ.

Remark 4. *By analogy with Schrödinger-invariance, we conjecture that the fluctuation-dominated correlators should be obtained from certain time-space integrals of higher n-point responses [16,70]. Working with the quantum chain representation of the terrace-step-kink model, Karevski and Schütz have calculated the stationary two-point correlator of the densities, which in our terminology correspond to the slopes $u(t,r) = \partial_r h(t,r)$. They find [10]*

$$C(t,r) = C_A t^{-2} \frac{1-\zeta^2}{(1+\zeta^2)^2} + C_B t^{-\psi} \frac{\cos[2(q^* r - \omega t)]}{(1+\zeta^2)^{\psi}} \tag{41}$$

with the scaling variable $\zeta = (r - v_c t)/(\nu t)$, where v_c is the global velocity of the interface, $\psi \geq \frac{1}{2}$ is a real parameter and $C_{A,B}$ are normalisation constants. The structure of their result is qualitatively very close to the form (8) and (39) for the two-time response of the $\mathfrak{ev}$-algebra, in the sense that it contains a dominant and monotonous term and a non-dominant and oscillatory one. Indeed, it can be checked that from the exact height-height correlator (5a) and (6a) this first term in (41) is recovered by computing the correlator $\langle u(t,r)u(0,0)\rangle$ of the densities $u = \partial_r h$. We interpret this as an encouraging signal that it should be possible to find the correlators from the $\mathfrak{ev}$ as well, by drawing on the analogies with Schrödinger-invariance. The first step in this direction would be the derivation of an analogue of a Bargman superselection rule, which is work in progress.

Remark 5. *The consequences of the choice of the fractional derivative are difficult to appreciate in advance and largely remain a matter of try and error. Our choice of the Riesz-Feller derivative was suggested that in this way the Lie algebra becomes a dynamical symmetry of the* DLE *process. In the past, we had also worked [16,45] with an extension of the Riemann-Liouville derivative by distributional terms [74]. For dynamical exponents $z \neq 1,2$, this leads to a strong oscillatory behaviour of the response functions which appears to be physically undesirable. We consider the success of the simple case study of the* DLE *process treated here as suggestive for future investigations.*

Corollary 4. *In the limit $\mu \to 0$ (or $\nu \to \infty$) the Lie algebra $\mathfrak{ev}$ can be contracted into the algebra*

$$[X_n, X_m] = (n-m)X_{n+m} \;,\; [X_n, B_m^{\pm}] = (n-m)B_{n+m}^{\pm} \tag{42}$$

with $n, m \in \mathbb{Z}$ and with the explicit generators

$$\begin{aligned} X_n &= -t^{n+1}\partial_t - (n+1)t^n r\partial_r - (n+1)xt^n - n(n+1)\varepsilon\gamma t^{n-1} r\partial_r \nabla_r^{-1} \\ B_n^{\pm} &= -\frac{1}{2}t^{n+1}\left(\varepsilon\nabla_r \pm \partial_r\right) - \frac{1}{2}(n+1)\gamma t^n \left(1 \pm \varepsilon\partial_r \nabla_r^{-1}\right) \end{aligned} \tag{43}$$

where $\varepsilon = \pm 1$ is the signature and $x = x_\varphi$ and $\gamma = \gamma_\varphi$ are the scaling dimension and the rapidity of the scaling operator on which these generators act. The co-variant quasi-primary two-point correlator $\mathscr{C}_{2,0} = 0$, whereas the co-variant quasi-primary two-point response $\mathscr{R} = \mathscr{C}_{1,1}$ is, with the normalisation constant $\mathscr{R}_0$

$$\mathscr{R}(t,r) = \delta_{x_1,x_2}\delta_{\gamma_1,\gamma_2}\mathscr{R}_0\, t^{-2x_1}\exp\left(-2\left|\frac{\gamma_1 r}{t}\right|\right). \tag{44}$$

The algebra (42), which one might call *meta-conformal Galilean algebra*, contains the conformal Galilean algebra as a sub-algebra, although the generators (43) are in general non-local, in contrast with those in Equation (15). However, the co-variant two-point function is here a response, and *not* a correlator.

Proof. In order to carry out the contraction on the generators (29), where $X_n = A_n + B_n^+ + B_n^-$, we first change coordinates $r \mapsto \mu r$, let $\xi_i = \gamma_i/\mu$ and rescale the generators $B_n^\pm \mapsto \mu B_n^\pm$. Then the last commutator in (27) becomes $[B_n^\pm, B_m^\pm] = \mu(n-m)B_{n+m}^\pm$. Taking the limit $\mu \to 0$ produces the generators (43) and the commutators (42) immediately follow. The Ward identities for the finite-dimensional sub-algebra are written down as before and $\mathscr{C}_{2,0} = 0$ follows. For the response function, going again through the proof of the proposition 7 and recalling that $\mu^{-1} = \mathrm{i}\nu$, we see that case A in (39) does not have a non-vanishing limit as $\nu \to \infty$. For case B, consider the scaling form $\mathscr{R}(t,r) = t^{-2x_1}f(r/t)$, with the scaling function $f(v)$ written as

$$f(v) = f_0\left[\left(\frac{\varepsilon_1}{\mathrm{i}\mu} - iv\right)^{-\psi-1} + e^{\mathrm{i}\pi\psi}\left(\frac{\varepsilon_1}{\mathrm{i}\mu} + iv\right)^{-\psi-1}\right]$$

and $\psi + 1 = 2\gamma_1/\mu$. If $\gamma_1 > 0$, the first term $(-\mathrm{i}/\mu - \mathrm{i}v)^{-\psi-1} = (\mathrm{i}\mu)^{2\gamma_1/\mu}(1+\mu v)^{-2\gamma_1/\mu} \stackrel{\mu\to 0}{=} \mu_0 e^{-2\gamma_1 v}$, and where μ_0 is a constant, to be absorbed into the overall normalisation. The second term vanishes, since $e^{\mathrm{i}\pi\psi} = e^{\mathrm{i}\pi(-1+2\mathrm{i}\nu\gamma_1)} = -e^{-2\pi\nu\gamma_1} \to 0$ as $\nu \to \infty$. On the other hand, if $\gamma_1 < 0$, one divides $f(v)$ by $e^{\mathrm{i}\pi\psi}$, and redefines the normalisation constant. Now, the second term produces $\sim e^{+2\gamma_1 v}$ and the first one vanishes in the $\nu \to \infty$ limit. Both cases are combined into $f(v) = \bar{f}_0 e^{-2|\gamma_1 v|}$. Alternatively, one derives from the Ward identities the two constraints $x_1 = x_2$ and $\gamma_1 = \gamma_2$. Global dilation-invariance gives the scaling form $\mathscr{R}(t,r) = t^{-2x_1}f(r/t)$ where the scaling function $f(v)$ must satisfy the equation $f'(v) + 2\gamma_1 \mathrm{sign}\,(t) f(v) = 0$ which leads to the asserted form, modulo a dualisation procedure, analogous to [41,46] to guarantee the boundedness for large separations. □

Our results on non-local meta-conformal algebras are summarised in Table 3.

Table 3. Comparison of non-local meta-2 conformal invariance, and meta-conformal galilei invariance in $(1+1)D$. The non-vanishing Lie algebra commutators, the defining equation of the generators and the invariant differential operator $\mathcal{S}$ are indicated. The usual generators are $X_n = A_n + B_n^+ + B_n^-$, $Y_n = B_n^+ + B_n^-$ and $Z_n = B_n^+ - B_n^-$, see also Table 2. Physically, the co-variant quasiprimary two-point function $\mathscr{R}_{12} = \langle\varphi_1(t,r)\widetilde{\varphi}_2(0,0)\rangle$ is a response function. In case B, one has $\psi = 2\xi_1 - 1$.

	Meta-2 Conformal	Meta-Conformal Galilean	Constraints	
Lie algebra	$[A_n, A_m] = (n-m)A_{n+m}$ $[B_n^\pm, B_m^\pm] = (n-m)B_{n+m}^\pm$	$[X_n, X_m] = (n-m)X_{n+m}$ $[X_n, B_m^\pm] = (n-m)B_{n+m}^\pm$		
generators	(29)	(43)		
$\mathcal{S}$	$-\mu\partial_t + \nabla_r$	-		
$\mathscr{R}_{12}$	$t^{1-x_1-x_2}\cdot \nu t\left(\nu^2t^2 + r^2\right)^{-1}$	-	$\xi_1 + \xi_2 = 1$ $x_1 - \xi_1 = x_2 - \xi_2$	(A)
	$t^{2\xi_1 - 2x_1}\left(\nu^2t^2 + r^2\right)^{-\xi_1}$ $\cdot\sin\left[\pi\xi_1 - 2\xi_1\arctan\left(\frac{r}{\nu t}\right)\right]$	$t^{-2x_1}\exp\left(-2\left|\gamma_1 r/t\right|\right)$	$x_1 = x_2$ $\xi_1 = \xi_2$, or $\gamma_1 = \gamma_2$	(B)

Acknowledgments: I thank Xavier Durang, Jeffrey Kelling and Geza Ódor for useful discussions or correspondence. This work was partly supported by the Collège Doctoral Nancy-Leipzig-Coventry (Systèmes complexes à l'équilibre et hors équilibre) of UFA-DFH.

Conflicts of Interest: The author declares no conflict of interest.

References

1. Belavin, A.A.; Polyakov, A.M.; Zamolodchikov, A.B. Infinite conformal symmetry in two-dimensional quantum field-theory. *Nuclear Phys. B* **1984**, *241*, 330–338.
2. Krug, J.; Meakin, P. Kinetic roughening of laplacian fronts. *Phys. Rev. Lett.* **1991**, *66*, 703–706.
3. Paulos, M.F.; Rychkov, S.; van Rees, B.C.; Zan, B. Conformal Invariance in the Long-Range Ising Model. *Nuclear Phys. B* **2016**, *902*, 249–291.
4. Krug, J. Statistical physics of growth processes. In *Scale invariance, Interfaces and Non-Equilibrium Dynamics*; McKane, A., Droz, M., Vannimenus, J., Wolf, D., Eds.; NATO ASI Series; Plenum Press: London, UK, 1994; Volume B344, p. 1.
5. Yoon, S.Y.; Kim, Y. Surface growth models with a random-walk-like nonlocality. *Phys. Rev. E* **2003**, *68*, 036121.
6. Aarão Reis, F.D.A.; Stafiej, J. Crossover of interface growth dynamics during corrosion and passivation. *J. Phys. Cond. Matt.* **2007**, *19*, 065125.
7. Zoia, A.; Rosso, A.; Kardar, M. Fractional Laplacian in Bounded Domains. *Phys. Rev. E* **2007**, *76*, 021116.
8. Spohn, H. Bosonization, vicinal surfaces, and hydrodynamic fluctuation theory. *Phys. Rev. E* **1999**, *60*, 6411–6420.
9. Popkov, V.; Schütz, G.M. Transition probabilities and dynamic structure factor in the ASEP conditioned on strong flux. *J. Stat. Phys.* **2011**, *142*, 627–639.
10. Karevski, D.; Schütz, G.M. Conformal invariance in driven diffusive systems at high currents. *arXiv* **2016**, arxiv:1606.04248.
11. Spohn, H. Nonlinear fluctuating hydrodynamics for anharmonic chains. *J. Stat. Phys.* **2014**, *154*, 1191–1227.
12. Bertini, L.; De Sole, A.; Gabrielli, D.; Jona-Lasinio, G.; Landim, C. Macroscopic fluctuation theory. *Rev. Mod. Phys.* **2015**, *87*, 593–636.
13. Barabási, A.-L.; Stanley, H.E. *Fractal Concepts in Surface Growth*; Cambridge University Press: Cambridge, UK, 1995.
14. Halpin-Healy, T.; Zhang, Y.-C. Kinetic roughening phenomena, stochastic growth, directed polymers and all that. *Phys. Rep.* **1995**, *254*, 215–414.
15. Krug, J. Origins of scale-invariance in growth processes. *Adv. Phys.* **1997**, *46*, 139–282.
16. Henkel, M.; Pleimling, M. *Non-Equilibrium Phase Transitions Volume 2: Ageing and Dynamical Scaling Far from Equilibrium*; Springer: Heidelberg, Germany, 2010.
17. Täuber, U.C. *Critical Dynamics: A Field-Theory Approach to Equilibrium and Non-Equilibrium Scaling Behaviour*; Cambridge University Press: Cambridge, UK, 2014.
18. Family, F.; Vicsek, T. Scaling of the active zone in the Eden process on percolation networks and the ballistic deposition model. *J. Phys. A Math. Gen.* **1985**, *18*, L75–L81.
19. Yeung, C.; Rao, M.; Desai, R.C. Bounds on the decay of the auto-correlation in phase ordering dynamics. *Phys. Rev. E* **1996**, *53*, 3073–3077.
20. Henkel, M.; Durang, X. Spherical model of interface growth. *J. Stat. Mech.* **2015**, *2015*, P05022.
21. Henkel, M. Non-local meta-conformal invariance in diffusion-limited erosion. *J. Phys. A Math. Theor.* **2016**, *49*, 49LT02.
22. Hase, M.O.; Salinas, S.R. Dynamics of a mean spherical model with competing interactions. *J. Phys. A Math. Gen.* **2006**, *39*, 4875–4899.
23. Ebbinghaus, M.; Grandclaude, H.; Henkel, M. Absence of logarithmic scaling in the ageing behaviour of the 4*D* spherical model. *Eur. Phys. J. B* **2008**, *63*, 85–91.
24. Edwards, S.F.; Wilkinson, D.R. The surface statistics of a granular aggregate. *Proc. R. Soc. Lond. A* **1982**, *381*, 17–31.
25. Kardar, M.; Parisi, G.; Zhang, Y.-C. Dynamic scaling of growing interfaces. *Phys. Rev. Lett.* **1986**, *56*, 889–892.
26. Rodrigues, E.A.; Mello, B.A.; Oliveira, F.A. Growth exponents of the etching model in high dimensions. *J. Phys. A Math. Theor.* **2015**, *48*, 035001.
27. Rodrigues, E.A.; Oliveira, F.A.; Mello, B.A. On the existence of an upper critical dimension for systems within the KPZ universality class. *Acta. Phys. Pol. B* **2015**, *46*, 1231–1237, doi:10.5506/APhysPolB.46.1231.

28. Alves, W.S.; Rodrigues, E.A.; Fernades, H.A.; Mello, B.A.; Oliveira, F.A.; Costa, I.V.L. Analysis of etching at a solid-solid interface. *Phys. Rev. E* **2016**, *94*, 042119.
29. Krech, M. Short-time scaling behaviour of growing interfaces. *Phys. Rev. E* **1997**, *55*, 668–679; Erratum in **1997**, *56*, 1285.
30. Henkel, M. On logarithmic extensions of local scale-invariance. *Nuclear Phys. B* **2013**, *869*, 282–302.
31. Kelling, J.; Ódor, G.; Gemming, S. Local scale-invariance of the $(2+1)$-dimensional Kardar-Parisi-Zhang model. *arXiv* **2016**, arxiv:1609.05795.
32. Henkel, M.; Noh, J.D.; Pleimling, M. Phenomenology of ageing in the Kardar-Parisi-Zhang equation. *Phys. Rev. E* **2012**, *85*, 030102(R).
33. Ódor, G.; Kelling, J.; Gemming, S. Ageing of the $(2+1)$-dimensional Kardar-Parisi-Zhang model. *Phys. Rev. E* **2014**, *89*, 032146.
34. Kelling, J.; Ódor, G.; Gemming, S. Universality of $(2+1)$-dimensional restricted solid-on-solid models. *Phys. Rev. E* **2016**, *94*, 022107.
35. Halpin-Healy, T.; Palansantzas, G. Universal correlators and distributions as experimental signatures of (2 + 1)-dimensional Kardar-Parisi-Zhang growth, *Europhys. Lett.* **2014**, *105*, 50001.
36. Kloss, T.; Canet, L.; Wschebor, N. Nonperturbative renormalization group for the stationary Kardar-Parisi-Zhang equation: Scaling functions and amplitude ratios in $1+1$, $2+1$ and $3+1$ dimensions. *Phys. Rev. E* **2012**, *86*, 051124.
37. Röthlein, A.; Baumann, F.; Pleimling, M. Symmetry-based determination of space-time functions in nonequilibrium growth processes. *Phys. Rev. E* **2006**, *74*, 061604; Erratum in **2007**, *76*, 019901(E).
38. Niederer, U. The maximal kinematical invariance group of the free Schrödinger equation. *Helv. Phys. Acta* **1972**, *45*, 802–810.
39. Lie, S. Über die Integration durch bestimmte Integrale von einer Klasse linearer partieller Differentialgleichungen. *Arch. Mathematik og Naturvidenskab* **1882**, *6*, 328-368.
40. Jacobi, C.G. Vorlesungen über Dynamik (1842/43), 4. Vorlesung. In *"Gesammelte Werke"*; Clebsch, A., Lottner, E., Eds.; Akademie der Wissenschaften: Berlin, Germany, 1866, ^{2}1884.
41. Henkel, M.; Stoimenov, S. Meta-conformal invariance and the boundedness of two-point correlation functions. *J. Phys. A Math. Theor.* **2016**, *49*, 47LT01.
42. Cartan, É. Les groupes de transformation continus, infinis, simples. *Annales Scientifiques de l'École Normale Supérieure (3e série)* **1909**; *26*, 93–161.
43. Di Francesco, P.; Mathieu, P.; Sénéchal, D. *Conformal Field-Theory*; Springer: Heidelberg, Germany, 1997.
44. Polyakov, A.M. Conformal symmetry of critical fluctuations. *Sov. Phys. JETP Lett.* **1970**, *12*, 381–383.
45. Henkel, M. Phenomenology of local scale invariance: From conformal invariance to dynamical scaling. *Nuclear Phys. B* **2002**, *641*, 405–486.
46. Henkel, M. Dynamical symmetries and causality in non-equilibrium phase transitions. *Symmetry* **2015**, *7*, 2108–2133.
47. Henkel, M.; Schott, R.; Stoimenov, S.; Unterberger, J. The Poincaré algebra in the context of ageing systems: Lie structure, representations, Appell systems and coherent states. *Conflu. Math.* **2012**, *4*, 1250006.
48. Stoimenov, S.; Henkel, M. From conformal invariance towards dynamical symmetries of the collisionless Boltzmann equation. *Symmetry* **2015**, *7*, 1595–1612.
49. Havas, P.; Plebanski, J. Conformal extensions of the Galilei group and their relation to the Schrödinger group. *J. Math. Phys.* **1978**, *19*, 482–488.
50. Henkel, M. Extended scale-invariance in strongly anisotropic equilibrium critical systems. *Phys. Rev. Lett.* **1997**, *78*, 1940–1943.
51. Negro, J.; del Olmo, M.A.; Rodríguez-Marco, A. Nonrelativistic conformal groups. *J. Math. Phys.* **1997**, *38*, 3786–3809.
52. Negro, J.; del Olmo, M.A.; Rodríguez-Marco, A. Nonrelativistic conformal groups II. *J. Math. Phys.* **1997**, *38*, 3810-3831.
53. Henkel, M.; Unterberger, J. Schrödinger invariance and space-time symmetries. *Nuclear Phys.* **2003**, *B660*, 407–435.
54. Barnich, G.; Compère, G. Classical central extension for asymptotic symmetries at null infinity in three spacetime dimensions. *Class. Quantum Gravity* **2007**, *24*, F15–F23; Corrigendum in **2007**, *24*, 3139.

55. Bagchi, A.; Gopakumar, R.; Mandal, I.; Miwa, A. GCA in 2*D*. *J. High Energy Phys.* **2010**, *8*, 1–40, doi:10.1007/JHEP08(2010)004.
56. Duval, C.; Horváthy, P.A. Non-relativistic conformal symmetries and Newton-Cartan structures. *J. Phys. A Math. Theor.* **2009**, *42*, 465206.
57. Cherniha, R.; Henkel, M. The exotic conformal Galilei algebra and non-linear partial differential equations. *J. Math. Anal. Appl.* **2010**, *369*, 120–132.
58. Hosseiny, A.; Rouhani, S. Affine extension of galilean conformal algebra in $2+1$ dimensions. *J. Math. Phys.* **2010**, *51*, 052307.
59. Zhang, P.-M.; Horváthy, P.A. Non-relativistic conformal symmetries in fluid mechanics. *Eur. Phys. J. C* **2010**, *65*, 607–614.
60. Barnich, G.; Gomberoff, A.; González, H.A. Three-dimensional Bondi-Metzner-Sachs invariant two-dimensional field-theories as the flat limit of Liouville theory. *Phys. Rev.* **2007**, *D87*, 124032.
61. It can be shown [57] that there are no CGA-invariant scalar equations (in the classical Lie sense). However, if one considers the Newton-Hooke extension of the CGA on a curved de Sitter/anti-de Sitter space (whose flat-space limit is not isomorphic to the CGA), non-linear representations have been used to find non-linear invariant equations, related to the Pais-Uhlenbeck oscillator, see [62–67] and refs. therein.
62. Chernyavsky, D. Coest spaces and Einstein manifolds with ℓ-conformal Galilei symmetry. *Nuclear Phys. B* **2016**, *911*, 471–479.
63. Masterov, I. Remark on higher-derivative mechanics with ℓ-conformal Galilei symmetry. *J. Math. Phys.* **2016**, *57*, 092901.
64. Krivonos, S.; Lechtenfeld, O.; Sorin, A. Minimal realization of ℓ-conformal Galilei algebra, Pais-Uhlenbeck oscillators and their deformation. *J. High Energy Phys.* **2016**, *1610*, 073.
65. Chernyasky, D.; Galajinsky, A. Ricci-flat space-times with ℓ-conformal Galilei symmetry. *Phys. Lett.* **2016**, *754*, 249–253.
66. Andrezejewski, K.; Galajinsky, A.; Gonera, J.; Masterov, I. Conformal Newton-Hooke symmetry of Pais-Uhlenbeck oscillator. *Nuclear Phys. B* **2014**, *885*, 150–162.
67. Galajinsky, A.; Masterov, I. Dynamical realisation of ℓ-conformal Newton Hooke group. *Phys. Lett. B* **2013**, *723*, 190–195.
68. Henkel, M.; Stoimenov, S. Physical ageing and Lie algebras of local scale-invariance. In *Lie Theory and Its Applications in Physics*; Dobrev, V., Ed.; Springer Proceedings in Mathematics & Statistics; Springer: Heidelberg, Germany, 2015; Volume 111, pp. 33–50.
69. Bargman, V. Unitary ray representations of continuous groups. *Ann. Math.* **1954**, *59*, 1–46.
70. Picone, A.; Henkel, M. Local scale-invariance and ageing in noisy systems. *Nuclear Phys. B* **2004**, *688*, 217–265.
71. Samko, S.G.; Kilbas, A.A.; Marichev, O.I. *Fractional Integrals and Derivatives*; Gordon and Breach: Amsterdam, The Netherlands, 1993.
72. Di Nezza, E.; Palatucci, G.; Valdinoci, E. Hitchhiker's guide to the fractional Sobolev spaces. *Bull. Sci. Math.* **2012**, *136*, 521–573.
73. Cinti, E.; Ferrari, F. Geometric inequalities for fractional Laplace operators and applications. *Nonlinear Differ. Equ. Appl.* **2015**, *22*, 1699–1714.
74. Gel'fand, I.M.; Shilov, G.E. *Generalized Functions, Volume 1: Properties and Operations*; Academic Press: New York, NY, USA, 1964.
75. Sethuraman, S. On microscopic derivation of a fractional stochastic Burgers equation. *Commun. Math. Phys.* **2016**, *341*, 625–665.
76. Ovsienko, V.; Roger, C. Generalisations of Virasoro group and Virasoro algebras through extensions by modules of tensor-densities on S^1. *Indag. Math.* **1998**, *9*, 277–288.
77. Henkel, M. Schrödinger-invariance and strongly anisotropic critical systems. *J. Stat. Phys.* **1994**, *75*, 1023–1061.
78. Henkel, M. From dynamical scaling to local scale-invariance: A tutorial. *Eur. Phys. J. Spec. Top.* **2017**, to be published.

Chapter 5:

symmetry

MDPI

Article

Classical and Quantum Burgers Fluids: A Challenge for Group Analysis

Philip Broadbridge

Department of Mathematics and Statistics, La Trobe University, Bundoora VIC 3086, Australia;
E-Mail: P.Broadbridge@latrobe.edu.au

Academic Editor: Roman M. Cherniha
Received: 3 July 2015 / Accepted: 24 September 2015 / Published: 9 October 2015

Abstract: The most general second order irrotational vector field evolution equation is constructed, that can be transformed to a single equation for the Cole–Hopf potential. The exact solution to the radial Burgers equation, with constant mass influx through a spherical supply surface, is constructed. The complex linear Schrödinger equation is equivalent to an integrable system of two coupled real vector equations of Burgers type. The first velocity field is the particle current divided by particle probability density. The second vector field gives a complex valued correction to the velocity that results in the correct quantum mechanical correction to the kinetic energy density of the Madelung fluid. It is proposed how to use symmetry analysis to systematically search for other constrained potential systems that generate a closed system of vector component evolution equations with constraints other than irrotationality.

Keywords: Burgers equation; integrability; Schrödinger equation; Madelung fluid

1. Introduction

Generally speaking, integrable equations are related to linear equations either by a classical Darboux transformation (c-integrable) or an inverse scattering transform (s-integrable) [1]. There are two main pathways that use Lie symmetry groups to identify integrable equations. The first is the detection of extended symmetries of order three or higher. Unlike first-order contact symmetries and their equivalent second-order "vertical" symmetries, higher-order symmetry transformations cannot be closed at some finite order [2]. While the very demanding condition of existence of a third-order symmetry is still not a sufficient condition for integrability, it is a useful and practical sieve. Known examples of equations with higher-order symmetries, are often members of a hierarchy of commuting integrable symmetries at successively higher orders, connected by a symmetry recursion operator (e.g., [3,4]). This approach has the advantage that it may reveal equations that are integrable in either sense of being s-integrable or c-integrable.

The second pathway involves detection of a general solution of a linear equation within the Lie point symmetry group or the Lie group of potential symmetries of a Darboux integrable equation. That method has the advantage of an inbuilt algorithm for finding the linearising transformation [5,6].

Classification of integrable scalar evolution equations in one space dimension, is well understood [7]. An inverse scattering transform has been found for some systems of N-waves in two and three dimensions [8]. However, it remains challenging to apply symmetry methods to classify integrable *systems* of parabolic evolution equations in *more than one* space dimension. As a test bed for such methods, in the following sections, some directly integrable vector-valued extensions of Burgers' scalar equation in three spatial dimensions, are considered.

The only source of nonlinearity in the Navier–Stokes momentum transport equation is the deceptively innocuous-looking quadratic inertial term within the convective time derivative

$$\frac{Du^j}{Dt} = \frac{\partial u^j}{\partial t} + \mathbf{u} \cdot \nabla \mathbf{u}^j.$$

Naturally, one seeks to gain insight from simplified transport models that at least retain this nonlinear term. For example, in gas dynamics it is common to assume the inviscid first-order Euler equations [9,10]. In one space dimension, there is the integrable transport model, the Burgers equation

$$u_t + uu_x = \nu u_{xx} \tag{1}$$

This equation resembles the momentum transport equation of incompressible Newtonian fluid but of course one-dimensional incompressible flow is trivial. Therefore solutions are considered with u_x non-zero. In this sense, Equation (1) is often used as a prototype model for compressible gas dynamics, but with the shocks smoothed by the non-zero viscosity [11,12]. The equation has found direct applications also in other areas, such as sedimentation [13] and soil-water transport [14], in which u represents a scalar concentration variable.

The one-dimensional Burgers equation has long been known to be exactly transformable to the classical linear heat diffusion equation by the Cole–Hopf transformation [15,16], previously given as an exercise in the text by Forsyth ([17], p. 102, Ex. 3). However this linearisation applies to the three dimensional prototype transport equation only after an additional constraint is appended. The Cole–Hopf transformation was applied in [18] to the three dimensional Burgers equation but necessarily with the additional constraint of irrotational flow. Matskevich [19] investigated how the Cole–Hopf transformation could simplify the Burgers equation in invariant form adapted to flow on a pseudo-Riemannian manifold. The outcome was that on a manifold with constant non-zero Ricci curvature scalar, Burgers' equation transforms to a reaction-diffusion equation for scalar Cole–Hopf potential ψ, with linear diffusion term but nonlinear reaction term proportional to $\psi \log \psi$. In Section 2 here, the reverse question is easily answered, namely after specifying that the Cole–Hopf potential satisfies a general *linear or semi-linear* second-order reaction-diffusion equation. What is the most general form of the integrable nonlinear vector equation that results from the Cole–Hopf transformation in the reverse direction ?

In fact any system of the following form is integrable:

$$\begin{aligned} u_{k,0} = \alpha^{ij}(\mathbf{r})[u_{k,ij} + c_1 u_{k,i}\, u_j + c_1 u_i u_{k,j}] + \alpha^{ij}_{\ ,k}[c_1 u_i u_j + u_{i,j}] + \\ + b^i_{\ ,k} u_i + b^i u_{k,i} + \frac{1}{c_1}\gamma_{,k}(\mathbf{r});\ k = 1, \cdots 3 \end{aligned} \tag{2}$$

$$\epsilon^{ijk}\partial_j u_k = 0 \ \ (\nabla \times \mathbf{u} = \mathbf{0}) \tag{3}$$

where $\alpha(\mathbf{r})$, $\mathbf{b}(\mathbf{r})$ and $\gamma(\mathbf{r})$ are differentiable symmetric tensor-valued, vector-valued and scalar-valued functions respectively, and c_1 is a non-zero constant. Here, $u_{j,k} = \partial_k u_j = \partial u_j / \partial x^k$, x^0 is the time coordinate, $x^j{:}j = 1 \cdots n$ are the space coordinates, repeated indices are summed and ϵ^{ijk} is the alternating symbol that is $+1(-1)$ for even(odd) permutations (ijk) of (123). The n-dimensional version of Equation (2) ($n > 3$) remains integrable when $\mathbf{u}$ is the gradient of some scalar potential, as shown in the next section. Unlike in one dimension, in higher dimensions it is more convenient to use index notation, especially when the coordinates are allowed to be non-Cartesian.

2. Extension of Cole–Hopf Transformation to n-Dimensions

Suppose that the scalar function $\psi(\mathbf{r}, t)$ is a classical solution of the general linear second-order parabolic equation

$$\psi_{,0} = \alpha^{ij}(\mathbf{r})\psi_{,ij} + b^i(\mathbf{r})\psi_{,i} + \gamma(\mathbf{r})\psi - c_2\gamma(\mathbf{r}); \quad (\mathbf{r}, t) \in \Omega \times \Re^+, \tag{4}$$

with $\psi > c_2$, defined on a closed subset Ω of $\Re^n$. Define a velocity potential ϕ by

$$\psi = e^{c_1\phi} + c_2.$$

$\phi(\mathbf{r}, t)$ is a generalisation of a scalar velocity potential, which as a consequence of Equation (4), satisfies

$$\phi_{,0} = \alpha^{ij}(\mathbf{r}) \left[\phi_{,ij} + c_1\phi_{,i}\phi_{,j}\right] + \beta^i(\mathbf{r})\phi_{,i} + \frac{\gamma(\mathbf{r})}{c_1} \tag{5}$$

Then $u_k = \phi_{,k}$ satisfies the generalised Burgers Equation (2). Note that under this transformation, the c_2-dependent terms cancel, so that this parameter does not appear in Equation (2). For this reason it is usually convenient to assume the homogeneous version of Equation (4) with $c_2 = 0$. Note also that Equation (2) is sufficiently general to allow the diffusion term to be an isotropic kinematic viscosity coefficient multiplied by the Laplace-Beltrami operator for a Riemannian manifold, acting on u^i. That is the right hand side is

$$\frac{\nu}{\sqrt{|g|}}\partial_k \left(\sqrt{|g|}g^{kj}\partial_j u^i\right)$$

plus terms of order 1 and 0. Here, g^{km} is the inverse of the metric tensor, $g^{km}g_{mp} = \delta^k{}_p$. This acts as a raising operator from a covariant vector to a contravariant vector:

$$u^k = g^{km}u_m.$$

Even on a flat Euclidean space, when non-Cartesian coordinates are used, one needs to distinguish between contravariant components of a vector (denoted by superscript indices) and covariant components (denoted by subscript indices). In the usual Einstein summation convention, repeated indices (one superscript and one subscript) are summed when a dyadic tensor product is contracted, contravariant rank being reduced by 1 and covariant rank likewise being reduced by 1.

Similarly Equation (2) is sufficiently general to allow the partial derivative to be extended to a covariant derivative

$$\nabla_i u_j = \partial_i u_j - \Gamma^k_{ij}(\mathbf{r})u_k,$$

where Γ^k_{ij} is the Christoffel symbol for the usual Levi–Civita connection coefficients. This at least allows one to express Burgers' equation on flat space, in terms of a general coordinate system. For example, in plane polar coordinates, the connection coefficients account for the centripetal acceleration component of radial fluid acceleration, that is proportional to the square of the circumferential component of fluid velocity.

As can be seen from [19], if a nonlinear source term of the form $c_3(\psi - c_2)\log(\psi - c_2)$ is added to Equation (4), the Cole–Hopf transformation results in an additional linear component $c_3 u_k$ in the source term of Burgers' equation. In the case of three spatial dimensions, a source term of this type in a constant-coefficient reaction-diffusion equation for ψ, results in an 11-dimensional Lie point symmetry algebra [20], spanned by the generators of common translations in space and time, the common rotations in space, plus four other independent special symmetry generators. Coordinates x^i and t may be rescaled so that without loss of generality, the free parameter c_3 may be assumed to be either $+1$ or -1. Then the ψ equation may be taken to be

$$\psi_{,t} = \nabla^2\psi + c_3\psi\log\psi; \quad c_3 = \pm 1, \tag{6}$$

while the four independent generators may be taken to be

$$\Gamma_8 = e^{c_3 t}\frac{\partial}{\partial \psi}; \; \Gamma_{8+j} = e^{c_3 t}\frac{\partial}{\partial x^j} - \frac{c_3}{2}x^j e^{c_3 t} u\frac{\partial}{\partial u}; \; j = 1\cdots 3. \tag{7}$$

It is a natural question to ask what is the image of other types of nonlinear source terms in the ψ equation, under the reverse Cole–Hopf transformation. It is a fact that only a source of the above form will lead to a closed system of equations for the vector components u_i. When any other nonlinear source term of the form $\Lambda(\mathbf{r})R(\psi)$ is assumed, the additional potential variable ϕ will appear in the system of equations for u_k, with an additional forcing term of the form

$$\frac{d}{d\phi}\left[e^{-c_1\phi}R(e^{c_1\phi} + c_2)\right] u_k + e^{-c_1\phi}R(e^{c_1\phi} + c_2)\Lambda_{,k}. \tag{8}$$

Quadratic $u_i u_j$ terms necessarily appear in Equation (2) whenever α^{ij} depends on x^k. This dependence may originate intrinsically from a curvilinear coordinate system or extrinsically from spatial dependence of viscosity. That variation could be induced for example, by controlling the spatially variable temperature.

The three-dimensional Cole–Hopf transformation has been applied in [21] to a quadratically forced Burgers equation representing transport in a solid medium. In the following two sections, the limited application to gas dynamics will be briefly revisited.

3. Prototype Vector Transport Equations

Consider the prototype vector transport equation with an additional external conservative force:

$$u^j_{,0} + u^k u^j_{,k} = \nu\, u^{j,k}_{\;\;k} + \Xi^{,j}\; ; \; j,k = 1,\cdots,n \tag{9}$$

which follows from the choice,

$$\alpha_i^{\;j} = \nu\delta_i^{\;j}; \; c_1 = \frac{-1}{2\nu}; \; b^i = 0; \; \gamma = c_1\Xi.$$

With $n = 3$, Equation (9) is the same as the Navier-Stokes momentum equation for an incompressible Newtonian fluid after we identify

$$\Xi^{,j} = -\frac{1}{\rho}p^{,j} - V^{,j},$$

that is a pressure gradient plus an external conservative force. However instead of appending the usual incompressibility condition, by analogy with the one-dimensional Burgers equation, we allow the divergence of $\mathbf{u}$ to be non-zero.

It has been well known since the origins of fluid mechanics that the theory of incompressible irrotational flow is linear since the velocity potential satisfies Laplace's equation. In fact, the prototype vector transport Equation (9) remains linearisable when it is supplemented by the potential condition Equation (11) for all *gradient solutions* with *compressible* flow vectors u^j. The prototype Equation (9) is significantly different from the Navier-Stokes momentum equation for a *compressible* fluid

$$u^j_{,0} + u^k\, u^j_{,k} = \frac{\mu}{\rho}u^{j,k}_{\;\;k} + \frac{1}{3}\frac{\mu}{\rho}\partial^j(u^k_{\;,k}) - \frac{1}{\rho}p^{,j} - V^{,j} \tag{10}$$

which combined with

$$u_i = \partial_i\Phi, \tag{11}$$

gives

$$\Phi^{,j}_{\;\;0} + \Phi^{,k}\Phi^{,j}_{\;\;k} = \nu\Phi^{,jk}_{\;\;\;k} - \frac{1}{\rho}p^{,j} - V^{,j}. \tag{12}$$

where $\nu = \frac{4}{3}\frac{\mu}{\rho}$ (e.g., [22]). For compressible Newtonian fluid flow, ρ varies in space and time, and it satisfies the equation of continuity

$$\rho_{,0} + \left(\rho u^j\right)_{,j} = 0. \tag{13}$$

The system consisting of Equation (11) combined with prototype vector transport Equation (9) with ν constant, implies by integration,

$$\Phi_{,0} + \frac{1}{2}(\Phi^{,k}\Phi_{,k}) = \nu\Phi^{,k}_{\ \ k} + \Xi, \tag{14}$$

with Ξ determined up to an additive function of x^0. This integrable multi-dimensional integrable scalar equation generalises Bernoulli's law to the case of non-zero viscosity. By the change of variable

$$\Psi = e^{-\Phi/2\nu} \ ; \ \Phi = -2\nu \log \Psi, \tag{15}$$

Equation (14) is equivalent to the linear heat equation with linear source,

$$\Psi_{,0} = \nu\Psi^{,k}_{\ \ k} - \frac{1}{2\nu}\Xi\Psi. \tag{16}$$

The fact that the one-dimensional Burgers equation can still be linearised when the external forcing term $\Xi'(x)$ is included, has been discovered and re-discovered since the 1970s in various contexts [23–25]. It has been used to investigate the effect of a random external force [26–28] and it has been used to directly model flow in unsaturated soil with extraction by plant roots [29].

3.1. Radial Burgers Equation and Approach to a Spherical Shock

One standard type of irrotational solution is the radial solution of the form $\mathbf{u} = U(r)\hat{\mathbf{e}}_r$ where $r = ||\mathbf{r}||$ is the Euclidean norm and $\hat{\mathbf{e}}_r = \mathbf{r}/r$. Radial solutions of Equation (9) must satisfy the n-dimensional radial forced Burgers equation

$$\begin{aligned} U_{,t} + UU_{,r} &= \nu\partial_r\left[r^{1-n}\partial_r\left(r^{n-1}U\right)\right] + \Xi_{,r} \\ &= \nu U_{,rr} + (n-1)\nu r^{-1}U_{,r} + (1-n)\nu r^{-2}U + \Xi_{,r}. \end{aligned} \tag{17}$$

This is evidently an integrable equation. From any solution to the radial form of the linear reaction-diffusion Equation (16), namely $\Psi = R(r,t)$, with

$$R_{,t} = \nu r^{1-n}\partial_r\left(r^{n-1}\partial_r R\right) - \frac{1}{2\nu}\Xi(r,t)R, \tag{18}$$

$U = \frac{-2\nu R_{,r}}{R}$ is a solution to Equation (17).

One very important three-dimensional solution in gas dynamics is that of a gas at higher density and higher radial velocity exploding radially outwards at $t = 0$ through a small two-dimensional spherical surface $r = a$, displacing initially stationary fluid downstream [30]. As in the well-known travelling wave solution to the one-dimensional Burgers equation, such a solution would introduce some viscous smoothing to the shock front of gas dynamics. Assume that the velocity of the gas at $r = a$ is $U(a) = Q/4\pi a^2$, where Q is the source strength. From the methods of Chapter 9 of [31], one such solution for the radial Cole-Hopf potential is

$$R = 1 + \left[\frac{1}{hr} - \frac{a}{r}\right]\left[erfc\left(\frac{r-a}{2\sqrt{\nu t}}\right) - e^{h(r-a)+h^2\nu t}erfc\left(\frac{r-a}{2\sqrt{\nu t}} + h\sqrt{\nu t}\right)\right], \tag{19}$$

where $h = \frac{1}{a} - \frac{Q}{8\pi\nu a^2}$. As ν is taken to be small, this solution approaches a sharp shock with constant radial speed. The radial solution, depicted in Figure 1, is in dimensionless units after rescaling by length scale a and time scale $t_s = a/U(a)$ by which the supply surface has radius $r/a = 1$, the fluid speed at $(r/a = 1)$ is $ut_s/a = 1$, the asymptotic travelling wave speed is $ct_s/a = 1/2$, in agreement with the Rankine-Hugoniot relations for a shock, and the Reynolds number is $Re = aU(a)/\nu$.

This solution, like others that have been produced, is not consistent with the physics of gas dynamics [32], in comparison with approximate analytical solutions of the full physical gas dynamics system when pressure, density and entropy are properly taken into account [30]. However, the exact radial Burgers solution does have some appealing features, such as a realistic inertial term in the momentum equation, and its approach to a viscous shock, that make this exact solution a useful bench test for computational fluids software packages.

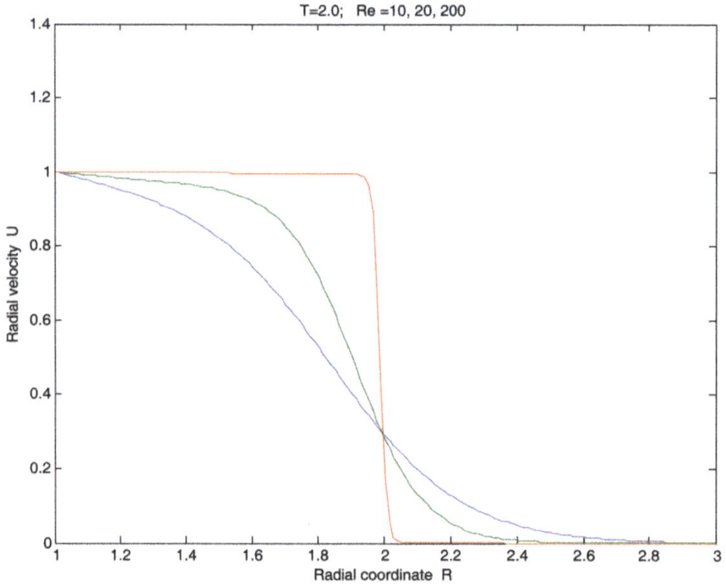

Figure 1. Solution to radial Burgers equation with constant mass supply.

4. Application of Cole–Hopf to the Schrödinger Equation

The single-particle Schrödinger wave function obeys a linear evolution equation that is analogous to Equation (4) except that it is necessarily complex valued and wave-like after the viscosity coefficient is replaced by a pure imaginary number:

$$\partial_t \Psi = \frac{i}{2}\nabla^2\Psi - iV(\mathbf{r})\Psi \tag{20}$$

For convenience, Equation (20) has been rescaled so that the quantum of action is 1 and the mass is 1. Then the real non-negative particle density satisfies the Liouville conservation equation

$$\partial_t \rho + \nabla \cdot \mathbf{J} = 0 \tag{21}$$

where $\rho = \Psi^*\Psi = |\Psi|^2$ and the particle current density is

$$J_k = \frac{i}{2}[\Psi^*\Psi_{,k} - \Psi\Psi^*_{,k}] = \mathcal{I}m\{\Psi\Psi^*_{,k}\}. \tag{22}$$

Now the Cole–Hopf transformation is simply

$$u_k = \partial_k \Phi; \quad \Psi = e^{i\Phi},$$

which results in the complex valued vector transport equation

$$\partial_t u_j + \mathbf{u} \cdot \nabla u_j = \frac{i}{2} \nabla^2 u_j - V_{,j} \,. \tag{23}$$

Note that the force per unit mass correctly emerges as $-\nabla V$. From the real valued particle density and real valued current density, one may construct a real valued fluid velocity

$$\mathbf{v} = \mathbf{J}/\rho = -\mathcal{I}m\{\nabla \log \Psi\} = \mathcal{R}e\{\nabla \Phi\} = \mathcal{R}e\{\mathbf{u}\}.$$

Note that $\mathcal{R}e\{\Phi\}$ is the quantum mechanical phase, whereas $\mathcal{I}m\{\Phi\} = -\frac{1}{2}\log \rho$. Then the imaginary part of $\mathbf{u}$ is

$$\mathbf{w} = -\frac{1}{2} \nabla \log \rho. \tag{24}$$

The velocity components v^j do not satisfy a closed system of transport equations. Instead, they are coupled to the components w^j :

$$\partial_t v_j + \mathbf{v} \cdot \nabla v_j = \frac{-1}{2} \nabla^2 w_j + \mathbf{w} \cdot \nabla w_j - V_{,j} \tag{25}$$

$$\partial_t w_j + \mathbf{v} \cdot \nabla w_j = \frac{1}{2} \nabla^2 v_j - \mathbf{w} \cdot \nabla v_j \tag{26}$$

When the supplementary conditions $\nabla \times \mathbf{v} = \mathbf{0}$ and $\nabla \times \mathbf{w} = \mathbf{0}$ are appended, this is a coupled system of nonlinear transport equations. Presumably, similar integrable multi-component systems could be constructed not only from a complex potential but from a quaternion potential or an octonian potential.

The Cole–Hopf transformation links the vector fields $\mathbf{v}$ and $\mathbf{w}$ as real and imaginary components of a gradient vector $\nabla(-i \log \Psi)$. Without further modifications to the model, $\mathbf{w}$ is uniquely determined by ρ, as in Equation (24) so that ρ and $\mathbf{v}$ form a closed system. That system is Madelung's original hydrodynamic analogue [33] that followed soon after Schrödinger's publication of the complex wave equation. It has been pointed out [34] that although the quantum vector field $\mathbf{w}$ is not independent, there is a use of it in classical fluid mechanics to refocus on the volume-weighted velocity rather than the mass-weighted velocity [35]. The two velocity fields $\mathbf{v}$ and $\mathbf{w}$ that are linked by the Cole–Hopf transformation, are also linked in the measurement of physical quantities. Using the Dirac formalism [36], the expectation of the energy of a particle in a conservative force field, in a pure state $|\Psi>$ is

$$< E > = < \Psi | (\frac{\hat{\mathbf{p}} \cdot \hat{\mathbf{p}}}{2} + V) | \Psi >,$$

where $\hat{p} \equiv -i\nabla$ in the Schrödinger representation. This can be shown to be equal to

$$\int_{\mathcal{R}^3} \rho(\mathbf{r}) [\frac{1}{2}(|\mathbf{v}|^2 + |\mathbf{w}|^2) + V(\mathbf{r})] \, d\mathbf{r} \tag{27}$$

Hence, $\frac{1}{2}\rho|\mathbf{u}|^2$ is the quantum mechanical correction to the classical kinetic energy density of the Madelung fluid. It may be interesting to extend the irrotational vector fields $\mathbf{v}$ and $\mathbf{w}$ to be independent fields by adding independent solenoidal contributions:

$$
\begin{aligned}
\mathbf{v} &= \nabla \mathcal{I}m\{\log \Psi\} + \nabla \times \mathbf{B}, \\
\mathbf{w} &= -\frac{1}{2}\nabla \log \rho + \nabla \times \mathbf{C}, \\
\mathbf{u} &= \mathbf{v} + i\mathbf{w} = \nabla(-i \log \Psi) + \nabla \times \mathbf{A};\ \mathbf{A} = \mathbf{B} + i\mathbf{C}.
\end{aligned}
$$

After breaking the condition $\nabla \times \mathbf{u} = \mathbf{0}$, the system of equations for u_i would no longer be equivalent to a linear equation for Ψ but to a system of nonlinear equations for Ψ and A^i. The nonlinear interactions would become negligible asymptotically if there were shear viscosity to dissipate the vorticity, so that A^i could be neglected after some time.

5. Some Relevant Questions in Symmetry Analysis

It is well known (e.g., [37]) that after neglecting linear superpositions, the generators of Lie point symmetries of a *linear* PDE for $\psi(\mathbf{r}, t)$ take the restricted infinitesimal form

$$a(\mathbf{r}, t)\partial_t + \mathbf{b}(\mathbf{r}, t) \cdot \nabla + c(\mathbf{r}, t)\Psi\partial_\Psi \tag{28}$$

which has an equivalent vertical Lie contact symmetry (e.g., [37]), in infinitesimal form

$$\bar{\Psi} = \Psi + \epsilon[c(\mathbf{r}, t)\Psi - a(\mathbf{r}, t)\partial_t\Psi - \mathbf{b}(\mathbf{r}, t) \cdot \nabla\Psi] + \mathcal{O}(\epsilon^2) \tag{29}$$

The Lie point symmetry classification of the Schrödinger equation with a general potential energy function in two and three dimensions, was completed by Boyer [38]. After applying such an invariance transformation, the gradient of $-i \log \bar{\Psi}$ is still an irrotational vector that satisfies the complex valued vector transport Equation (23). Conversely, if a point transformation leaves the system of complex vector transport equation plus condition of irrationality invariant, then the new solution $\bar{\mathbf{u}}(\mathbf{r}, t)$ may be integrated to construct a potential $\bar{\Phi}(\mathbf{r}, t)$ that is unique up to an additive complex function of t, equivalent to multiplying Ψ by an arbitrary spatially uniform gauge function $\Psi_g(t)$ that has no effect on calculating physical expectation values (e.g., [36]).

From the class of nonlinear scalar parabolic equations, integrability of the one-dimensional Burgers equation hierarchy can be detected by an extended higher-order Lie symmetry analysis (e.g., [2]) or by a potential Lie symmetry analysis associated with local conservation laws (e.g., [6]). The system (2) and (3) of six partial differential equations for three functions $u_i(\mathbf{r}, t)$ of four independent variables x^j and t, is integrable. That fact was found by extending the Cole–Hopf transformation that was known from the one-dimensional version. It was not found from symmetry analysis. Using the Cole–Hopf transformation, one may reconstruct Lie–Bäcklund symmetries of the integrable vector system from those of the associated linear scalar equation. From the transformed solution $\bar{\Psi}(\mathbf{r}, t)$ of the scalar equation, the gradient operation $\mathbf{u}(\mathbf{r}, t) = -2\nu\nabla \log \bar{\Psi}$, preserves the irrotational condition of the vector system. It also must preserve the governing transport equation of the vector transport equation for $\mathbf{u}$. For example, the linear heat equation $\partial_t\Psi(\mathbf{r}; t) = \nabla^2\Psi(\mathbf{r}, t)$ is invariant under differentiation in any fixed direction. An elementary third-order Lie–Bäcklund symmetry is

$$\bar{\Psi} = \Psi + \zeta^{ijk}\Psi_{,ijk} \tag{30}$$

with ζ^{ijk} the fixed components of a totally symmetric tensor. Then by the substitution $\Psi_{,i} = -\frac{1}{2}\Psi u_i$, the corresponding transformation for $\mathbf{u}$ is

$$\bar{u}_\ell = u_\ell + \frac{1}{4}\zeta^{ijk}[u_{i,\ell}u_j u_k + 2u_{\ell,i}u_j u_k \tag{31}$$

$$-2u_{i,j}u_{k,\ell} - 4u_{\ell,i}u_{j,k} - 4u_{i,j\ell}u_k - 2u_{\ell,ij}u_k + 4u_{i,\ell jk}] \tag{32}$$

In the one-dimensional case, all indices are 1 and this reduces to a known Lie–Bäcklund symmetry of the standard Burgers equation (e.g., [7]):

$$\bar{u} = u + \xi^{111}[u_{xxx} - \frac{3}{2}uu_{xx} - \frac{3}{2}u_x^2 + \frac{3}{4}u^2u_x]. \tag{33}$$

In the case of the complex Burgers fluid representation of quantum particle dynamics, incorporation of a solenoidal component of the complex fluid velocity **u** would extend wave mechanics to have not only a scalar wave function, which is a function of the fluid velocity potential, but also a vector potential A^j that satisfies a system of nonlinear PDE coupled to Ψ. Neither of the two principal founders of wave mechanics, de Broglie and Schrödinger, accepted the Copenhagen interpretation of the probability of outcomes of measurement [39,40]. Perhaps they would have found a supplementary classical vector potential more palatable. Since the linear Schrödinger equation correctly describes an evolving particle except when the wave function collapses to a single eigenstate during the decoherence effect of observation by filtering, it can only possibly be the act of measurement that introduces vorticity to the quantum fluid. The coupled nonlinear interaction between A^j and Ψ may have a complicated ergodic dynamics. The probability density of an energy eigenstate may be the measure of a region in state space that becomes the basin of attraction for a particular eigenstate when the filtering observation is carried out. After the imposed vorticity decays, the state will again evolve according to the Schrödinger equation.

Symmetry classification of some Burgers type systems is carried out in [41] (higher-order symmetries) and [42] (Lie and conditional symmetries).

A point symmetry classification of the potential system, with side constraints other than the irrotational condition, looks to be within the capability of symbolic packages, perhaps after making some reasonable ansätze on the functions B_i and C_i. However the complexity of the calculation grows rapidly with the number of variables, for example when one proceeds to tensor transport equations.

In one space dimension, there are other integrable nonlinear diffusion equations that are obtainable from a linear equation for the potential variable, by a change of variable. Under the group of contact transformations, the equivalence classes of these integrable equations are represented by canonical forms [7] that include the linear equations, the Burgers class,

$$u_t = \partial_x[u^{-2}\partial_x u] \text{ and} \tag{34}$$

$$u_t = \partial_x[u^{-2}\partial_x u] + 1. \tag{35}$$

For example, if $u = \phi_x$, then the linear equation $x_t = x_{\phi\phi}$ is sufficient for Equation (34). In one dimension, the hodograph transformation is crucial [43] but in three dimensions it has no simple analogue.

6. Conclusions

From any exact solution of the linearly forced linear heat equation in n spatial dimensions, one may construct an exact compressible solution to the prototype vector transport equation via a generalisation Equation (14) of the potential Burgers equation to higher dimensions. The linearisation procedure applies in three dimensions to compressible irrotational flows but not to rotational incompressible flows. For example, there is a simple integrable radial Burgers equation, for which the radial Cole–Hopf potential $R(r,t)$ obeys the radial heat diffusion equation. In three dimensions, $rR(r,t)$ must satisfy the classical one-dimensional diffusion equation (Chapter 9 of [31]). This fact has been used to construct an exact radial viscous gas flow with a shock. The simple examples provided above have zero forcing term. However, exact solutions may be constructed similarly for simple conservative force fields such as uniform gravity. Conceivably, hydrodynamic statistical distributions could be calculated from the randomly forced 3D vector transport equation just as for the one-dimensional Burgers equation [44].

Just like the one-dimensional Burgers equation, this has limited relevance for the interesting physical phenomena of fluid mechanics that involve evolution of vorticity.

By way of contrast, in wave mechanics it is a linear second-order evolution equation, the Schrödinger equation, that is physically relevant. Just as for the case of the linear heat equation, one may apply the reverse Cole–Hopf transformation to the Schrödinger equation, leading to a complex Burgers-type equation that is physically relevant. This is an integrable system of nonlinear transport equations for two real velocity-like vectors. Both the real and imaginary parts of the Burgers velocity have direct physical interpretations, while the squared modulus of the complex velocity is the quantum mechanical correction to kinetic energy density. However, the Cole–Hopf transformation suggests that the integrable gradient flows that are well understood, be extended by adding a rotational component to the velocity-like fields, then carrying out a symmetry classification of the equivalent systems of PDE for the potentials, with additional side conditions other than that of zero vorticity.

Acknowledgments: Acknowledgments

The author thanks Dimetre Triadis and Richard Kleeman for helpful discussions, the reviewers and guest editor who made good suggestions, and the Courant Institute for providing facilities as the second draft was written.

Conflicts of Interest: The author declares no conflict of interest.

References

1. Calogero, F. Why are certain nonlinear PDEs both widely applicable and integrable? In *What is Integrability*; Zakharov, V.E., Ed.; Springer: Berlin, Germany, 1990; pp. 1–61.
2. Anderson, R.L.; Ibragimov, N.H. *Lie Bäcklund Transformations in Applications*; SIAM: Philadelphia, PA, USA, 1979.
3. Olver, P.J. Evolution equations possessing infinitely many symmetries. *J. Math. Phys.* **1977**, *18*, 1212–1215.
4. Fuchssteiner, B. The tangent bundle for multisolitons: Ideal structure for completely integrable systems. In *Nonlinear Evolution Equations and Dynamical Systems*; Carillo, S., Ragnisco, O., Eds.; Springer: Berlin, Germany, 1990; pp. 114–122.
5. Bluman, G.W.; Kumei, S. Symmetry-based algorithms to relate partial differential equations. I. Local symmetries. *Eur. J. Appl. Math.* **1990**, *1*, 189–216.
6. Bluman, G.W.; Kumei, S. Symmetry-based algorithms to relate partial differential equations. II. Linearization by nonlocal symmetries. *Eur. J. Appl. Math.* **1990**, *1*, 217–223.
7. Svinolupov, S.I. Second-order evolution equations with symmetries. *Uspekhi Mat. Nauk* **1985**, *40*, 263–264.
8. Gerdjikov, V.S.; Ivanov, R.I.; Kyuldjiev, A.V. On the N-wave equations and soliton interactions in two and three dimensions. *Wave Motion* **2011**, *8*, 791–804.
9. Liu, T.-P. Multi-dimensional gas flow: Some historical perspectives. *Bull. Inst. Math. Acad. Sin.* **2011**, *6*, 269–291.
10. Andreev, V.K.; Kaptsov, O.V.; Pukhnachev, V.V. *Applications of Group-Theoretical Methods in Hydrodynamics*; Kluwer: Dordrecht, The Netherlands, 1998.
11. Burgers, J.M. A mathematical model illustrating the theory of turbulence. *Adv. Appl. Mech.* **1948**, *1*, 171–199.
12. Tsai, L.C. Viscous shock propagation with boundary effect. *Bull. Inst. Math. Acad. Sin.* **2011**, *6*, 1–25.
13. Blake, J.R.; Colombera, P.M. Sedimentation: A comparison between theory and experiment. *Chem. Eng. Sci.* **1977**, *32*, 221–228.
14. Broadbridge, P.; Srivastava, R.; Yeh, T.-C. Burgers' equation and layered media: Exact solutions and applications to soil-water flow. *Math. Comput. Mod.* **1992**, *16*, 157–162.
15. Cole, J.D. On a quasilinear parabolic equation occurring in aerodynamics. *Q. Appl. Math.* **1951**, *9*, 225–236.
16. Hopf, E. The partial differential equation $u_t + uu_x = \mu u_{xx}$. *Commun. Pure Appl. Math.* **1950**, *3*, 201–230.
17. Forsyth, A.R. *Theory of Differential Equations*; Cambridge University Press: Cambridge, UK, 1906; Volume VI.
18. Nerney, S.; Schmahl, E.D.; Musielak, Z.E. Analytic solutions of the vector Burgers equation. *Q. Appl. Math.* **1996**, *56*, 63–71.
19. Matskevich, S.E. Burgers equation and Kolmogorov–Petrovsky–Piscunov equation on manifolds. *Infin. Dimens. Anal. Quantum Probab. Relat. Top.* **2011**, *14*, 199–208.

20. Dorodnitsyn, V.A.; Knyazeva, I.V.; Svirshchevskii, S.R. Group properties of the nonlinear heat equation with source in the two- and three-dimensional cases. *Differential'niye Uravneniya (Differ. Equ.)* **1983**, *19*, 1215–1223. (In Russian)
21. Leonenko, N.N.; Ruiz-Medina, M.D. Scaling laws for the multidimensional Burgers equation with quadratic external potential. *J. Stat. Phys.* **2006**, *124*, 191–206.
22. Thompson, P.A. *Compressible-Fluid Dynamics*; McGraw-Hill: New York, NY, USA, 1972.
23. Moulana, M.; Nariboli, G.A. On Bäcklund transformations of some second-order nonlinear equations. *J. Math. Phys. Sci.* **1975**, *11*, 491–504.
24. Nimmo, J.J.C.; Crighton, D.G. Bäcklund transformations for nonlinear parabolic equations: The general results. *Proc. R. Soc. Lond. A* **1982**, *384*, 381–401.
25. Pasmanter, R.A. Stability and Bäcklund transform of the forced Burgers equation. *J. Math. Phys.* **1986**, *29*, 2744–2746.
26. Jeng, D.-T. Forced model equation for turbulence. *Phys. Fluids* **1969**, *12*, 2006–2010.
27. Bertini, L.; Cancrini, N.; Jona-Lasinio, G. The stochastic Burgers equation. *Commun. Math. Phys.* **1994**, *165*, 211–232.
28. Da Prato, G.; Debussche, A.; Temam, R. Stochastic Burgers' equation. *Nonlinear Differ. Equ. Appl. NoDEA* **1994**, *1*, 389–402.
29. Broadbridge, P. The forced Burgers equation, plant roots and Schrödinger's eigenfunctions. *J. Eng. Math.* **1999**, *36*, 25–39.
30. Haque, E.; Broadbridge, P.; Sachdev, P.L. Expansion of high pressure gas into air—A more realistic blast-wave model. *Math. Comput. Model.* **2009**, *50*, 1606–1621.
31. Carslaw, H.S.; Jaeger, J.C. *Conduction of Heat in Solids*, 2nd ed.; Clarendon Press: Oxford, UK, 1986.
32. Nerney, S.; Schmahl, E.J.; Musielak, Z.E. Limits to extensions of Burgers' equation. *Q. Appl. Math.* **1996**, *54*, 385–393.
33. Madelung, E. Quantentheorie in hydrodynamischer Form. *Z. Phys.* **1927**, *40*, 322–326.
34. Jüngel, A. Global weak solutions to compressible Navier-Stokes equations for quantum fluids. *SIAM J. Math. Anal.* **2010**, *42*, 1025–1045.
35. Brenner, H. Navier-Stokes revisited. *Physica A* **2005**, *349*, 60–132.
36. Dirac, P.A.M. *Principles of Quantum Mechanics*; Oxford University Press: London, UK, 1930.
37. Bluman, G.W.; Kumei, S. *Symmetries and Differential Equations*; Springer; Berlin, Germany, 1985.
38. Boyer, C.P. The maximal kinematic invariance group for an arbitrary potential. *Helv. Phys. Acta* **1975**, *47*, 589–626.
39. De Broglie, L. *The Current Interpretation of Wave Mechanics: A Critical Study*; Elsevier: Amsterdam, The Netherlands, 1964.
40. Schrödinger, E. The present situation in quantum mechanics: A translation of Schrödinger's "Cat Paradox" paper (transl. Trimmer J. D.). *Proc. Am. Philos. Soc.* **1980**, *124*, 323–338.
41. Svinolupov, S.I. On the analogues of the Burgers equation. *Phys. Lett. A* **1989**, *135*, 32–36.
42. Cherniha, R.; Serov, M. Nonlinear systems of the Burgers-type equations: Lie and Q-conditional symmetries, Ansätze and solutions. *J. Math. Anal. Appl.* **2003**, *282*, 305–328.
43. Clarkson, P.A.; Fokas, A.S.; Ablowitz, M.J. Hodograph transformations of linearizable partial differential equations. *SIAM J. Appl. Math.* **1989**, *49*, 1188–1209.
44. Weinan, E.; Vanden-Eijnden, E. Statistical theory for the stochastic Burgers equation in the inviscid limit. *Commun. Pure Appl. Math.* **2000**, *53*, 852–901.

symmetry

MDPI

Article

New Nonlocal Symmetries of Diffusion-Convection Equations and Their Connection with Generalized Hodograph Transformation

Valentyn Tychynin

Prydniprovs'ka State Academy of Civil Engineering and Architecture, 24a Chernyshevsky Street, Dnipropetrovsk 49005, Ukraine; E-Mail: tychynin@ukr.net

Academic Editor: Roman M. Cherniha
Received: 23 June 2015 / Accepted: 18 September 2015 / Published: 29 September 2015

Abstract: Additional nonlocal symmetries of diffusion-convection equations and the Burgers equation are obtained. It is shown that these equations are connected via a generalized hodograph transformation and appropriate nonlocal symmetries arise from additional Lie symmetries of intermediate equations. Two entirely different techniques are used to search nonlocal symmetry of a given equation: the first is based on usage of the characteristic equations generated by additional operators, another technique assumes the reconstruction of a parametrical Lie group transformation from such operator. Some of them are based on the nonlocal transformations that contain new independent variable determined by an auxiliary differential equation and allow the interpretation as a nonlocal transformation with additional variables. The formulae derived for construction of exact solutions are used.

Keywords: nonlocal symmetries; potential symmetries; formulae of nonlocal superposition; formulae for generation of solutions; generalized hodograph transformation

---

1. Introduction

We continue the nonlocal symmetries search of the diffusion-convection equations, connected by a generalized hodograph transformation (GHT) [1]. For this purpose we use the additional Lie symmetries (that are additional operators of invariance Lie algebras of higher dimension) of the intermediate equations arising at steps of this transformation, which consists of potential substitution and usual hodograph transformation. One can find many references and the extensive bibliography in researches devoted to studying the potential symmetries of nonlinear partial differential equations and systems [2–8]. The notion of potential symmetries of differential equations was introduced by Bluman *et al.* [2,3]. Later, Lisle proposed in [9,10] the concept of potential equivalence transformations. It was successfully applied in [11] for deriving complete list of potential symmetries for wide classes of diffusion-convection equations.

On the other hand a number of interesting results for nonlinear equations are obtained for present day within the nonlocal transformations approach.

Analogously to continuous groups of usual transformations, the theory of groups of Lie–Bäcklund transformations, *i.e.*, continuous groups of transformations involving derivatives of dependent variables, was developed by Anderson and Ibragimov [12]. An effective tool for this method application is the notion of recursion operator proposed by Olver [13].

A method for constructing nonlocally related PDE systems was introduced in [14,15]. Potential and nonlocal symmetries has been investigated for wide classes of nonlinear PDE systems there. It was shown that each point symmetry of a PDE system systematically yields a nonlocally related PDE system. Appropriate nonlocal symmetries were presented in the form including a potential variable.

The use of finite nonlocal transformations allowed to construct the formulae of generation of solutions and nonlocal nonlinear superposition for wide variety of nonlinear partial differential equations. This approach is based on the nonlocal transformations technic [16–21], which we regularly use for study of symmetries of the given nonlinear equations in [1] and in the current paper.

To start with, we remind the main concepts and notions which are necessary for understanding of the subsequent material. Let's assume that finite nonlocal transformation of variables

$$\mathcal{T}: \quad x^i = h^i(y, v_{(k)}), \quad i = 1, \ldots, n, \quad u^K = H^K(y, v_{(k)}), \quad K = 1, \ldots, m$$

exists and maps the given equation

$$F_1(x, u_{(n)}) = 0 \tag{1}$$

into the equation $\Phi(y, v_{(q)}) = 0$ of order $q = n + k$. Suppose, that this equation admits a factorization to another equation

$$F_2(y, v_{(s)}) = 0 \tag{2}$$

i.e., $\Phi(y, v_{(q)}) = \lambda\ F_2(y, v_{(s)})$. Here λ is a differential operator of order $n + k - s$. Then we say that Equations (1) and (2) are connected by the nonlocal transformation $\mathcal{T}$

$$F_1(x, u_{(n)}) = \lambda\ F_2(y, v_{(s)})$$

Here the symbol $u_{(r)}$ denotes the tuple of derivatives of the function u from order zero up to order r. In the case of two independent variables we use the special notation of the variables: $x_1 = x$, $x_2 = t$ and thus $u_t = \partial u/\partial t = \partial_t u$, $u_x = \partial u/\partial x = \partial_x u$.

Nonlocal transformations can be effectively used for construction of the formulae generating solutions in both cases: for nonlocal invariance of equation

$$F_2(y, v_{(n)}) = F_1(y, v_{(n)})$$

and when the connected equations are different. In a special case, when equation F_2 is linear, one can construct the formulae of nonlinear nonlocal superposition of solutions for nonlinear equation F_1 [1,17,22]. We consider a case when the intermediate equations, connected by components of the generalized hodograph transformation exist, and these equations possess invariance Lie algebras of higher dimension. These, the last, we use for construction of corresponding nonlocal symmetries of the given equations and for deriving the formulae generating solutions.

The paper is organized as follows: In Section 2 we introduce a chain of the equations connected by steps of the GHT and compare the Lie invariance algebras obtained. Then (in Section 3) we construct nonlocal symmetries of these equations, which are generated by additional operators of invariance Lie algebras of the intermediate equations. Section 4 is devoted to the construction of finite nonlocal invariance transformations for a given equation and appropriate formulae generating their solutions. Namely, we combine the Lie symmetry transformation, generated by an additional operator X, which has been admitted by the intermediate equation, with a transformation, mapping this intermediate equation into the given one.

2. Lie Symmetries of Given and Intermediate Equations

We aim to consider two diffusion-convection equations from the class

$$u_t - \partial_x(C(u) + K(u) \cdot u_x) = 0 \tag{3}$$

which possess Lie invariance algebras of different dimensions. It is easy to check the mapping of the given equation

$$u_t - \partial_x\left(\frac{1}{2}u^{-1} + u^{-2}u_x\right) = 0 \tag{4}$$

into the Burgers equation

$$z_\eta - \partial_\xi \left(-\frac{1}{2}z^2 + z_\xi \right) = 0 \tag{5}$$

via all steps of the GHT. This transformation was found first by Storm in [23] for linearization of nonlinear heat equations. Later this transformation was re-discovered in [24] and applied in [22,25] for investigation of various nonlinear heat equations. The GHT includes the potential substitution

$$u = v_x(x,t) \tag{6}$$

which is being applied to Equation (4) transforms it into the potential equation

$$v_t - \frac{1}{2}v_x^{-1} - v_x^{-2}v_{xx} = 0 \tag{7}$$

Then the hodograph transformation

$$v(x,t) = \xi, \;\; x = w(\xi,\eta), \;\; t = \eta \tag{8}$$

maps Equation (7) into the equivalent form

$$w_\eta + \frac{1}{2}w_\xi^2 - w_{\xi\xi} = 0 \tag{9}$$

We will refer to Equation (9), Equation (7) obtained as "intermediate" equations.

Having applied one more potential substitution

$$w_\xi = z \tag{10}$$

to Equations (9) we get the Burgers Equation (5). The composition of transformations Equations (6), (8) and (10) introduced above we mark further as $\mathcal{T}$.

Note that the relation of Equation (4) with the equation

$$u_t - \partial_x \left(u^{-2}u_x \right) = 0 \tag{11}$$

is a particular case of the more general result established in [26]. The results concerning linearization of Equation (11) were obtained by Rosen [27]. An example with Equation (11), particularly, was considered in [14,15]. Nonlocal symmetries were described using a potential variable. Here we apply a traditional approach [1] for searching nonlocal symmetries and aim to construct appropriate formulae of generation of solutions to equations considered. Then we use them for construction of exact solutions.

Consider some results of the classical Lie symmetry analysis for the equations connected by all components of the transformation $\mathcal{T}$. The maximal Lie algebra admitted by Equation (4) is four-dimensional (see for example [28])

$$\begin{aligned} &X_1 = \partial_t, \;\; X_2 = \partial_x, \;\; X_3 = 2t\partial_t + u\partial_u, \\ &X_4 = -2e^{\frac{1}{2}x}\partial_x + e^{\frac{1}{2}x}u\partial_u \end{aligned}$$

while invariance algebra of the Burgers Equation (5) is spanned by five operators

$$\begin{aligned} &X_1 = \partial_\eta, \;\; X_2 = \partial_\xi, \;\; X_3 = \eta\partial_\xi + \partial_z, \\ &X_4 = \xi\partial_\xi + 2\eta\partial_\eta - z\partial_z, \\ &X_5 = \eta\xi\partial_\xi + \eta^2\partial_\eta + (\xi - z\eta)\,\partial_z \end{aligned} \tag{12}$$

The operator X_5 does not correspond to Lie symmetry of Equation (4). The appropriate nonlocal symmetry of Equation (4) corresponding to the Lie symmetry X_5 of the Burgers equation has been investigated in details in [1]. Here we will take into account symmetries of the intermediate equations for construction of the appropriate nonlocal symmetries of two given Equations (4) and (5).

The maximal Lie invariance algebra of Equation (7) is spanned by the generators

$$\begin{aligned} &X_1 = \partial_t,\ X_2 = \partial_x,\ X_3 = \partial_v,\ X_4 = v\partial_x + t\partial_v, \\ &X_5 = 2t\partial_t + v\partial_v,\ X_6 = \left(t + \tfrac{1}{2}v^2\right)\partial_x + t^2\partial_t + tv\partial_v \\ &X_7 = b(v,t)e^{\frac{1}{2}x}\partial_x,\ b_t - b_{vv} = 0 \end{aligned} \tag{13}$$

Since Equation (9) is equivalent to Equation (7) it admits the Lie invariance algebra of the same structure

$$\begin{aligned} &X_1 = \partial_\eta,\ X_2 = \partial_\xi,\ X_3 = \partial_w,\ X_4 = \eta\partial_\xi + \xi\partial_w, \\ &X_5 = \xi\partial_\xi + 2\eta\partial_\eta,\ X_6 = \eta\xi\partial_\xi + \eta^2\partial_\eta + \left(\tfrac{1}{2}\xi^2 + \eta\right)\partial_w, \\ &X_7 = b(\xi,\eta)e^{\frac{1}{2}w}\partial_w,\ b_\eta - b_{\xi\xi} = 0 \end{aligned} \tag{14}$$

Note that both equations have two Lie symmetry generators more (X_3 and X_7) than Equation (5) has.

It will be interesting to compare the symmetries obtained above with those ones, which are admitted by a potential system, constructed for Equation (4)

$$v_x - u = 0 \tag{15}$$

$$2v_t - u^{-1} - 2u^{-2}\,u_x = 0 \tag{16}$$

One can easily calculate the Lie symmetry of this system, which is spanned by generators

$$\begin{aligned} &X_1 = \partial_t,\ X_2 = \partial_x,\ X_3 = \partial_v,\ X_4 = v\partial_x - u^2\partial_u + t\partial_v, \\ &X_5 = 2t\partial_t + ut\partial_u + v\partial_v, \\ &X_6 = \left(t + \tfrac{1}{2}v^2\right)\partial_x + t^2\partial_t + u\,(t - vu)\,\partial_u + tv\partial_v, \\ &X_7 = 2b(v,t)e^{\frac{x}{2}}\partial_x - \left(2u^2\partial_v b(v,t) + u\right)e^{\frac{x}{2}}b(v,t)\partial_u, \\ &b_t(v,t) - b_{vv}(v,t) = 0 \end{aligned} \tag{17}$$

Notice that to obtain an operator X_7 we chose it in the form

$$X_7 = 2b(v,t)e^{\frac{x}{2}}\partial_x - \left(2u^2 a(x,t) - u\right)e^{\frac{x}{2}}b(v,t)\partial_u$$

where $a(x,t)$ and $b(v(x,t),t)$ are unknown differentiable functions.

Applying this operator to the system Equations (15) and (16), we get such conditions for these functions:

$$a(x,t)b(v(x,t),t) + D_1 b(v(x,t),t) = 0 \tag{18}$$

$$a(x,t)D_1 b(v(x,t),t)u(x,t) + b(v(x,t),t)a_x(x,t) + D_2 b(v(x,t),t)u(x,t) = 0 \tag{19}$$

Here $D_1 f(\alpha,\beta)$ means the derivative of a function $f(\alpha,\beta)$ with respect to its first argument. Substituting a solution of the first equation

$$a(x,t) = -\frac{D_1 b(v(x,t),t)}{b(v(x,t),t)}$$

and an equality

$$v_x = u \tag{20}$$

into the second equation, we receive the equation for $b(v,t)$

$$b_t(v,t) - b_{vv}(v,t) = 0 \tag{21}$$

In what follows we will use two entirely different techniques to search for nonlocal symmetry of a given equation. The first is based on usage of the characteristic equation generated by an additional operator. Another technique assumes reconstruction of a parametrical Lie group transformation from such operator.

3. Nonlocal Symmetries Generated by Additional Lie Symmetries of the Intermediate Equations

Here we apply additional operators of Lie algebras obtained for intermediate equations to search appropriate nonlocal symmetry of given equations and construct the formulae of generation their solutions.

3.1. *The Operator X 7th Case*

Comparison of Lie algebras Equations (12) and (14) shows that Equation (9) admits an additional (with respect to Equation (5)) generator X_7. The characteristic equation corresponding to sum $X_7 + X_3$ of generators of algebra Equation (14) has the form

$$w + 2\ln|b| = 0$$

One can easily identify this formula with the known substitution linearizing Equation (9). To present this expression via the variable $z(\xi,\eta)$ we will differentiate it with respect to ξ

$$w_\xi + 2\, b_\xi b^{-1} = 0 \tag{22}$$

Then, substituting $w_\xi = z$ into above equation, we find the characteristic equation which determines appropriate nonlocal symmetry of Equation (5)

$$z + 2\, b_\xi b^{-1} = 0 \tag{23}$$

Therefore the "lacking" operator for the Burgers equation should be written in the form

$$X_7^\star = (z + 2\,\partial_\xi \ln b)\,\partial_z \tag{24}$$

Note that an expression Equation (23) is well known *Cole–Hopf substitution*.

Conversely, applying the transformation Equation (8) with its differential consequence $w_\xi = 1/v_x$ to the Equation (22), we find nonlocal symmetry for Equation (7)

$$v_x + b/2b_v = 0$$

Taking into account substitution Equation (20), this equation takes the form

$$u + b/2b_v = 0 \tag{25}$$

Here $b(v,t)$ is an arbitrary solution of the linear heat Equation (21).

An application of the nonlocal transformation Equation (25) to Equation (4) generates an expression which after simplification vanishes identically on the manifold defined by Equation (7) and expression $v_x + b/2b_v = 0$ with its differential consequences.

Usage of a classical linear superposition principle in Equation (25) allows one to get the formula generating solutions of Equation (4)

$$u^{\mathrm{I}}+b^{\mathrm{I}}\left(2b_v^{\mathrm{I}}\right)^{-1}=0,\ u^{\mathrm{II}}+b^{\mathrm{II}}\left(2b_v^{\mathrm{II}}\right)^{-1}=0,\quad\rightarrow\quad u^{\mathrm{III}}=-\left(b^{\mathrm{I}}+b^{\mathrm{II}}\right)\left(2b_v^{\mathrm{I}}+2b_v^{\mathrm{II}}\right)^{-1}\tag{26}$$

On the other hand the generator X_7 of the Lie invariance algebra of the system Equations (15) and (16) determines a potential symmetry of Equation (4) and admits presentation by the characteristic equation

$$u_x+u^2\frac{b_v(v,t)}{b(v,t)}+\frac{1}{2}\,u=0\tag{27}$$

where $u(x;t)$ and $b(v;t)$ satisfy Equations (21) and (20). Thus, the appropriate "lacking" operator for Equation (4) should be written in the form, depending on $b(v,t)$

$$X_7^{\star}=b(v,t)\,\mathrm{e}^{\frac{1}{2}x}\partial_x-\left(u^2\frac{b_v(v,t)}{b(v,t)}+\frac{1}{2}u\right)b(v,t)\,\mathrm{e}^{\frac{1}{2}x}\partial_u\tag{28}$$

Theorem 1. *The characteristic Equation (27) determines nonlocal transformation with additional variables, which connects Equations (4) and (7) taking into account Equation (21) and condition Equation (20). Additional independent variable $v(x,t)$ is determined by Equation (7) and additional dependent variable $b(v,t)$ is an arbitrary solution of the Equation (21).*

To prove the statement formulated above we will first rewrite Equation (27), using equality Equation (20) in the form

$$\partial_x\ln|u|+\partial_x\ln|b(v,t)|+\frac{1}{2}=0\tag{29}$$

and then integrate the result with respect to x

$$u(x,t)=\frac{s(t)\,\mathrm{e}^{-1/2x}}{b(v,t)}\tag{30}$$

Having applied this transformation to Equation (4) and use the conditions

$$b_t(v,t)-b_{vv}(v,t)=0,\quad v_{xx}=v_tv_x^2-1/2v_x,\quad v_x\,(x,t)=\frac{s\,(t)\,\mathrm{e}^{-1/2x}}{b(v,t)}$$

we obtain a result, which vanishes identically after simplification.

The above assertion admits another form.

Theorem 2. *The Equation (4) is invariant under action of the prolonged operator $X_7^{\star}$ on the manifold, determined by Equation (7), condition Equation (20) and the Equation (21) with their differential consequences.*

To deduce the characteristic equation, which corresponds to Equation (29) and determines nonlocal symmetry of the Equation (4) in terms of variables $x,t,u(x,t)$, we need to exclude a variable $v(x,t)$ from expression Equation (30), where $b\,(v,t)$ is a function belonging to the set of solutions of the linear heat Equation (21).

New designations will be necessary for us in what follows. Suppose that we can solve the functional equation $b(v,t)=a$ with respect to the first argument of b. Then a function $\beta(a,t):\ b(v,t)=a\ \rightarrow\ v=\beta(a,t)$, will denote a solution of this equation. Therefore

$$v\,(x,t)=\beta\left(\frac{s\,(t)\,\mathrm{e}^{-1/2x}}{u},t\right)$$

is a solution of Equation (30) with respect to v. Using potential substitution Equation (20) we obtain equality

$$u(x,t) - \partial_x \beta \left(\frac{s(t)\,\mathrm{e}^{-1/2x}}{u(x,t)}, t \right) = 0$$

Differentiating and simplifying this expression, we get the presentation of the onetime integrated characteristic equation which determines nonlocal symmetry of Equation (4)

$$\begin{array}{l} 2u^3 + D_1\beta \left(\frac{s(t)\mathrm{e}^{-1/2x}}{u}, t \right) s(t)\,\mathrm{e}^{-1/2x} u \\ +2D_1\beta \left(\frac{s(t)\mathrm{e}^{-1/2x}}{u}, t \right) s(t)\,\mathrm{e}^{-1/2x} u_x = 0 \end{array} \tag{31}$$

$D_1 f(\alpha, \beta)$ means the derivative of a function $f(\alpha, \beta)$ with respect to its first argument.

For instance, a given simple solution of the linear heat Equation (21)

$$b(v,t) = \left(c_1 \mathrm{e}^{\sqrt{\alpha} v} + c_2 \mathrm{e}^{-\sqrt{\alpha} v} \right) c_3 \mathrm{e}^{\alpha t}$$

allows one to obtain the expression Equation (30) in such particular form:

$$\left(c_1 \mathrm{e}^{\sqrt{\alpha} v} + c_2 \mathrm{e}^{-\sqrt{\alpha} v} \right) c_3 \mathrm{e}^{\alpha t} - \frac{s(t)\,\mathrm{e}^{-1/2x}}{u(x,t)} = 0$$

Here $\alpha,\ c_i, (i = 1,2,3)$ are arbitrary constants. This equation has two solutions for $v(x,t)$

$$v = \ln \left| \frac{s(t)\mathrm{e}^{-1/2x} \pm \sqrt{s^2(t)\mathrm{e}^{-x} - 4c_3^2 \mathrm{e}^{2\alpha t} u^2 c_1 c_2}}{2c_1 c_3 \mathrm{e}^{\alpha t} u} \right| \left(\sqrt{\alpha} \right)^{-1}$$

Excluding the variable v from Equation (31), we can rewrite it as follows

$$u - \partial_x \left(\ln \left| \frac{s(t)\mathrm{e}^{-1/2x} \pm \sqrt{s^2(t)\mathrm{e}^{-x} - 4c_3^2 \mathrm{e}^{2\alpha t} u^2 c_1 c_2}}{2c_1 c_3 \mathrm{e}^{\alpha t} u} \right| \left(\sqrt{\alpha} \right)^{-1} \right) = 0 \tag{32}$$

Finally we obtain two the onetime integrated characteristic equations which determine nonlocal symmetries of Equation (4)

$$2u_x + 2\sqrt{\alpha} u^2 + u \pm \frac{8\sqrt{\alpha} c_1 c_2 c_3^2 \mathrm{e}^{2\alpha t} u^4}{\sqrt{s^2(t)\mathrm{e}^{-x} - 4c_3^2 \mathrm{e}^{2\alpha t} u^2 c_1 c_2} + s(t)\mathrm{e}^{-1/2x}} = 0$$

Choosing in Equation (32), in particular, $c_2 = 0,\ c_3 = 1$, and integrating this equation, we obtain two ansatzes

$$u = \pm \frac{1}{2\sqrt{\alpha} - g(t)\,\mathrm{e}^{1/2x}}$$

which contain arbitrary functions of t, which must be specialized by Equation (4). So, finally we obtain two solutions of this equation

$$u = \pm \frac{1}{2\sqrt{\alpha} - k\,\mathrm{e}^{1/2x}}$$

Note that such functions can be obtained by means of the Lie ansatz.

Theorem 3. *The operator Equation (28) allows one to construct the associated transformation determined by the formulas*

$$2(\mathrm{e}^{\frac{x^{\mathrm{I}}}{2}} + \mathrm{e}^{\frac{x^{\mathrm{II}}}{2}}) + \varepsilon b^{\mathrm{I}}(v^{\mathrm{I}}(x^{\mathrm{I}}, t), t)\mathrm{e}^{\frac{x^{\mathrm{I}} + x^{\mathrm{II}}}{2}} = 0 \tag{33}$$

$$u^{\mathrm{II}}(x^{\mathrm{II}},t) = -\frac{u^{\mathrm{I}}(x^{\mathrm{I}},t)\left(-2+b^{\mathrm{I}}(v^{\mathrm{I}}(x^{\mathrm{I}},t),t)\varepsilon e^{\frac{x^{\mathrm{I}}}{2}}\right)}{2+2u^{\mathrm{I}}(x^{\mathrm{I}},t)D_1b^{\mathrm{I}}(v^{\mathrm{I}}(x^{\mathrm{I}},t),t)\varepsilon e^{\frac{x^{\mathrm{I}}}{2}}} \tag{34}$$

where $\partial_{x^{\mathrm{I}}}v^{\mathrm{I}}(x^{\mathrm{I}},t) = u^{\mathrm{I}}(x^{\mathrm{I}},t)$ *and* $b^{\mathrm{I}}(v^{\mathrm{I}}(x^{\mathrm{I}},t),t)$ *is any arbitrary solution of the Equation (21).*

To prove this theorem we restore a Lie group transformation using the operator Equation (28). So, we obtain

$$2(e^{\frac{x}{2}}+e^{\frac{r}{2}})+\varepsilon b(v(x,t),t)e^{\frac{x+r}{2}}=0, \quad s=t,$$

$$p(r,s) = -\frac{u(x,t)\left(-2+b(v(x,t),t)\varepsilon e^{\frac{x}{2}}\right)}{2+2u(x,t)D_1b(v(x,t),t)\varepsilon e^{\frac{x}{2}}}$$

Making here change of notations for new variables, we find the Equations (33) and (34).

The algorithm of generation of solutions for Equation (4) based on these formulae works, as follows. Let $u^{\mathrm{I}}(x^{\mathrm{I}},t)$ be any known solution of Equation (4), then $v^{\mathrm{I}}(x^{\mathrm{I}},t) = \int u^{\mathrm{I}}(x^{\mathrm{I}},t)dx^{\mathrm{I}} + s(t)$ can be found. Here $s(t)$ is arbitrary function which can be specialized by Equation (29). Substituting obtained v^{I} and *arbitrary* solution of the linear heat equation $b^{\mathrm{I}}(v^{\mathrm{I}}(x^{\mathrm{I}},t),t)$ into Equation (33) and solving the result with respect to x^{I}, we get $x^{\mathrm{I}} = \chi(x^{\mathrm{II}},t)$. Finally, inserting this x^{I} into Equation (34) we find a new solution of Equation (4).

Example 1. Choosing $u^{\mathrm{I}} = e^{-\frac{x^{\mathrm{I}}}{2}}$ and integrating it with respect to x, we obtain $v^{\mathrm{I}} = -2\, e^{-\frac{x^{\mathrm{I}}}{2}} + s(t)$. Specializing arbitrary function by Equation (29), we get $v^{\mathrm{I}} = -2e^{-\frac{x^{\mathrm{I}}}{2}} + c_1$. Set arbitrary solution of the linear heat equation $b^{\mathrm{I}} = e^{c_2t+\sqrt{c_2}v^{\mathrm{I}}}$, then $D_1b^{\mathrm{I}} = \sqrt{c_2}\, e^{c_2t+\sqrt{c_2}v^{\mathrm{I}}}$. Rewriting these formulae and taking into account $v^{\mathrm{I}} = -2e^{\frac{x^{\mathrm{I}}}{2}} + c_1$, we find

$$b^{\mathrm{I}} = e^{c_2t+\sqrt{c_2}(-2e^{\frac{x^{\mathrm{I}}}{2}}+c_1)}$$

and

$$D_1b^{\mathrm{I}} = \sqrt{c_2}\, e^{c_2t+\sqrt{c_2}(-2e^{\frac{x^{\mathrm{I}}}{2}}+c_1)}$$

Substituting a found above b^{I} into Equation (33) and solving this result with respect to x^{I}, we get $x^{\mathrm{I}} = \chi(x^{\mathrm{II}},t)$

$$x^{\mathrm{I}} = x^{\mathrm{II}} - 2\ln\left|\frac{1}{2}\mathrm{LambertW}\left(\sqrt{c_2}\varepsilon e^{\left(c_1\sqrt{c_2}+tc_2-2\sqrt{c_2}\, e^{-\frac{x^{\mathrm{II}}}{2}}\right)} e^{\frac{x^{\mathrm{II}}}{2}} + \sqrt{c_2}\right)\right| + \ln|c_2|$$

After simplification Equation (34) with obtained $x^{\mathrm{I}} = \chi(x^{\mathrm{II}},t)$, b^{I} and D_1b^{I} we find a new solution of Equation (4)

$$u^{\mathrm{II}}(x^{\mathrm{II}},t) = e^{c_1\sqrt{c_2}+tc_2-2\sqrt{c_2}\, e^{-\frac{x^{\mathrm{II}}}{2}}-\frac{x^{\mathrm{II}}}{2}}G^{-1},$$

$$G = e^{c_1\sqrt{c_2}+tc_2-2\sqrt{c_2}\, e^{-\frac{x^{\mathrm{II}}}{2}}}\left(\mathrm{LambertW}\left(\sqrt{c_2}\varepsilon e^{\left(c_1\sqrt{c_2}+tc_2-2\sqrt{c_2}\, e^{-\frac{x^{\mathrm{II}}}{2}}\right)}\right)+1\right)$$

3.2. *The Operator X 4th Case*

The Lie invariance algebra Equation (13) of Equation (7) includes one operator $X_4 = v\cdot\partial_x + t\cdot\partial_v$ more, than invariance algebra of Equation (4) has. This operator yields the characteristic equation

$$v\cdot v_x - t = 0 \tag{35}$$

Differentiating Equation (35) with respect to x and using substitution Equation (20) we obtain

$$v\,u_x + u^2 = 0 \tag{36}$$

Excluding $v(x,t)$ from this equation we get the second order differential equation for u which determines nonlocal symmetry of Equation (4)

$$-3u_x^2 + uu_{xx} = 0 \tag{37}$$

Solving Equation (37), we find non-Lie ansatzes of Equation (4)

$$u = \pm\left(\sqrt{-2F_1(t)\,x - 2F_2(t)}\right)^{-1}$$

Substitution them into Equation (4) and splitting a result with respect to x yields the reduced system of two ordinary differential equations

$$\dot{F_1}(t) - (F_1(t))^2 = 0, \quad \dot{F_2}(t) - F_1(t)\,F_2(t) - (F_1(t))^2 = 0$$

Solution of this system

$$F_1(t) = -(t - c_2)^{-1}, \; F_2(t) = \frac{\ln|t - c_2| - c_1}{t - c_2}$$

allows us to construct appropriate solutions of Equation (4)

$$u = \pm\frac{\sqrt{2}}{2}\sqrt{\frac{t - c_2}{x - \ln|t - c_2| + c_1}} \tag{38}$$

The result obtained is closely connected with potential symmetry of Equation (4) determined by the generator X_4 of Lie algebra Equation (17). The characteristic equations corresponding to this operator, are

$$v \cdot u_x + u^2 = 0 \tag{39}$$

$$v \cdot v_x - t = 0 \tag{40}$$

They, obviously, coincide with Equations (36) and (35) accordingly. The solution of this system

$$u = \frac{t}{\sqrt{2 \cdot x \cdot t + F_1(t) + F_2(t) \cdot t}}, \quad v = \sqrt{2xt + F_1(t)}$$

yields the reduced system of ordinary differential equations

$$F_2(t) = 0, \quad t \cdot \dot{F_1} - F_1(t) + 2t = 0$$

and, consequently, yields appropriate solution to the system Equations (39) and (40)

$$u = \frac{\sqrt{t}}{\sqrt{2x - 2\ln|t| + c_1}}, \quad v = \sqrt{t}\sqrt{2x - 2\ln|t| + c_1}$$

If we set $c_2 = 0$ in Equation (38), we will obtain the previous expression for the function u.

4. Nonlocal Invariance Transformations and Generation of Solutions

Here we construct the one-parameter group of Lie symmetries generated by operator, admitted by the potential system Equations (15) and (16), and it is used for construction of the corresponding formula for generation of solutions of Equation (4). This involves a successive implementation of the transformations $\mathcal{T}$ and the one-parameter group of point transformations associated with the corresponding infinitesimal operator which is admitted by the potential system.

Assume that the partial differential equation

$$F(x,t,u_{(k)}) = 0 \tag{41}$$

should admit at least one conservation law

$$D_t\phi^t(x,t,u_{(r)}) + D_x\phi^x(x,t,u_{(r)}) = 0 \tag{42}$$

where D_t and D_x are total derivatives with respect to the variables t and x, ϕ^t and ϕ^x are conserved density and flux, respectively. When Equation (41) admits representation Equation (42) there exists the potential function v determined by the auxiliary system [2]

$$v_x = \phi^t(x,t,u_{(r)}), \quad v_t = -\phi^x(x,t,u_{(r)}) \tag{43}$$

Potential symmetry of the Equation (41) is determined by the Lie symmetry generator

$$X = \xi_1(x,t,u,v)\partial_x + \xi_2(x,t,u,v)\partial_t + \eta_1(x,t,u,v)\partial_u + \eta_2(x,t,u,v)\partial_v$$

admitted by system Equation (43) with at least one nonzero partial derivative

$$\partial_v\xi_1(x,t,u,v), \quad \partial_v\xi_1(x,t,u,v), \quad \partial_v\eta_1(x,t,u,v)$$

In this case the one-parameter Lie point symmetry group associated with infinitesimal generator X exists

$$\begin{array}{ll} r = x' = f_1(x,t,u,v;\varepsilon), & s = t' = f_2(x,t,u,v;\varepsilon), \\ p(r,s) = u' = g_1(x,t,u,v;\varepsilon), & q(r,s) = v' = g_2(x,t,u,v;\varepsilon) \end{array} \tag{44}$$

Here ε is a group parameter.

As Equation (4) is connected with the potential system Equations (15) and (16), one can construct the projection of the corresponding transformation Equation (44) onto the space of variables $(x,t,u(x,t))$ using substitution Equation (20)

$$\begin{array}{l} r = x' = f_1(x,t,u,\int u\,(x,t)\,\mathrm{d}x;\varepsilon), \\ s = t' = f_2(x,t,u,\int u\,(x,t)\,\mathrm{d}x;\varepsilon), \\ p(r,s) = u' = g_1(x,t,u,\int u\,(x,t)\,\mathrm{d}x;\varepsilon) \end{array} \tag{45}$$

This transformation is the finite Lie–Bäcklund transformation which leaves Equation (4) nonlocal-invariant. Stated above allows formulating the following statement.

Theorem 4. *Equation (4) is nonlocal-invariant under the transformations*

$$\begin{array}{l} r = \ln\left|\frac{2}{2-\varepsilon t}\right| + \frac{(\int u\,\mathrm{d}x)^2\varepsilon}{4-2\varepsilon t} + x, \\ s = \frac{2t}{2-\varepsilon t}, \; p\,(r,s) = \frac{-2u}{\varepsilon t - u\,\varepsilon\int u\,\mathrm{d}x - 2}; \end{array} \tag{46}$$

$$
\begin{aligned}
r &= 2\ln\left|-2\left(-2+b\left(\int u\,\mathrm{d}x,t\right)\varepsilon e^{\frac{1}{2}x}\right)^{-1}\right| + x,\ s = t,\\
p(r,s) &= -\frac{u\left(-2+b(\int u\,\mathrm{d}x,t)\varepsilon e^{\frac{1}{2}x}\right)}{2+2u\ D_1(b)(\int u\,\mathrm{d}x,t)\varepsilon e^{\frac{1}{2}x}}
\end{aligned} \tag{47}
$$

and

$$
r = \frac{1}{2}\varepsilon^2 t + \varepsilon\int u\,\mathrm{d}x + x,\ s = t,\ p(r,s) = \frac{u}{u\,\varepsilon+1} \tag{48}
$$

The proof of this statement for transformation Equation (46). Let us choose, for instance, the operator

$$
X_6 = (t+\frac{1}{2}v^2)\partial_x + t^2\partial_t + (-vu+t)\,u\partial_u + tv\partial_v
$$

This operator belongs to the Lie invariance algebra Equation (17) of the system Equations (15) and (16). The corresponding Lie group transformation has the form

$$
\begin{aligned}
r &= \ln\left|\tfrac{2}{2-\varepsilon t}\right| + \tfrac{v^2\varepsilon}{4-2\varepsilon t} + x, s = \tfrac{2t}{2-\varepsilon t},\\
p(r,s) &= -\tfrac{2u}{\varepsilon t - vu\varepsilon - 2}, q(r,s) = \tfrac{2v}{2-\varepsilon t}
\end{aligned} \tag{49}
$$

Excluding in these formulae v with the help of expression $v = \int u(x,t)\,\mathrm{d}x$, we get a nonlocal transformation in space of variables r, s, p Equation (46). Then we apply the transformation Equation (46) to Equation (4), rewritten in new designations

$$
\partial_s p(r,s) - \partial_r\left(\frac{1}{2}p(r,s)^{-1} + p(r,s)^{-2}\partial_r p(r,s)\right) = 0 \tag{50}
$$

and substitute the integro-differential consequences of Equation (4)

$$
u_{xx} = u^2 u_t + \frac{1}{2}u_x + 2u^{-1}u_x^2 \tag{51}
$$

and

$$
\int u_t \mathrm{d}x = \frac{1}{2}u^{-1} + u^{-2}u_x \tag{52}
$$

into the obtained result. After simplification the last vanishes identically.

To prove this statement for transformation Equation (47) we consider the generator of Lie algebra Equation (17)

$$
X_7 = e^{\frac{1}{2}x} b(v,t)\,\partial_x - u\left(\frac{u b_v}{b(v,t)} + \frac{1}{2}\right) b(v,t)\, e^{\frac{1}{2}x}\partial_u
$$

where $D_2 b(v,t) - D_{11} b(v,t) = 0,\ v = v(x,t)$, which yields the group-invariant transformation of the potential system Equations (15) and (16)

$$
\begin{aligned}
r &= 2\ln\left|-2\left(-2+b(v,t)\,\varepsilon e^{1/2x}\right)^{-1}\right| + x, \qquad s = t,\\
p(r,s) &= -\tfrac{u\left(-2+b(v,t)\varepsilon e^{1/2x}\right)}{2+2u\ D_1(b)(v,t)\varepsilon e^{1/2x}},\ q(r,s) = v
\end{aligned}
$$

Hereinafter $D_i(f)(\alpha,\beta)$ and $D_{i,j}(f)(\alpha,\beta), (i,j = 1,2)$ denote total derivatives of a function $f(\alpha,\beta)$ with respect to i-th and j-th variables of the first and second order accordingly, ε is a group parameter associated with the operator X_7. Having substituted into previous formulae $v(x,t) = \int u\,\mathrm{d}x$ we get a nonlocal transformation Equation (47).

To verify invariance of Equation (50) we apply to it Equation (47) and substitute Equations (51) and (52) and expressions

$$
\begin{aligned}
D_{11} b\left(\int u\,\mathrm{d}x,t\right) &= D_2 b\left(\int u\,\mathrm{d}x,t\right),\\
D_{111} b\left(\int u\,\mathrm{d}x,t\right) &= D_{12} b\left(\int u\,\mathrm{d}x,t\right),
\end{aligned}
$$

$$D_{11}u(x,t) = \frac{2D_2u(x,t)(u(x,t))^3 + u(x,t)D_1u(x,t) + 4(D_1u(x,t))^2}{2\,u(x,t)}$$

into the obtained result. The expression obtained vanishes identically after simplification.

The proof of Theorem 4 for transformation Equation (48). Another operator of the Lie invariance algebra Equation (17)

$$X_4 = v\partial_x - u^2\partial_u + t\partial_v$$

generates the finite Lie-group transformation

$$r = \frac{1}{2}\varepsilon^2 t + \varepsilon v + x, \;\; s = t, \;\; p(r,s) = \frac{u}{u\varepsilon + 1}, \;\; q(r,s) = v + \varepsilon t \tag{53}$$

Projection of this transformation onto the space of variables $(x, t, u(x,t))$ admits the form Equation (48). This transformation leaves the Equation (4) nonlocal-invariant. The statement can be proved like in the previous cases.

Let us consider some applications of the Theorem 4. If $p(r,s)$ is a known solution of the Equation (50), then the new solution of Equation (4) can be constructed by means of solution Equation (46) after subsequent specialization of the arbitrary function appearing as a result of integration in it.

Example 2. Inserting the solution of Equation (4) $u(x,t) = -\frac{1}{2}e^{-\frac{1}{2}x}$ into Equation (46), we obtain the formulae of transformation, which contain an arbitrary function $F(t)$

$$\begin{array}{l} r = \ln\left|\frac{2}{2-t\varepsilon}\right| + x + \frac{\varepsilon(e^{-\frac{1}{2}x}+F(t))^2}{4-2t\varepsilon}, \\ s = \frac{-2t}{-2+\varepsilon t}, \;\; p(r,s) = \frac{2(-\frac{1}{2}e^{-\frac{1}{2}x})}{2\varepsilon t - \frac{1}{2}e^{-\frac{1}{2}x}(e^{-\frac{1}{2}x}+F(t))\varepsilon - 4} \end{array} \tag{54}$$

Specializing $F(t)$ by Equation (4) and supposing $F(t) = 0$ for simplicity of evaluations, we obtain

$$\begin{array}{l} r = \ln\left|\frac{2}{2-t\varepsilon}\right| + x + \frac{\varepsilon e^{-x}}{4-2t\varepsilon}, \\ s = \frac{-2t}{-2+\varepsilon t}, \;\; p(r,s) = \frac{2e^{-1/2x}}{2\varepsilon t + e^{-x}\varepsilon - 4} \end{array} \tag{55}$$

Solving two first equations of the Equation (55) with respect to x and t, we get

$$x = r + \mathrm{LambertW}\left(-\frac{1}{16}\frac{\varepsilon(\varepsilon s+2)^2}{e^r}\right) + \ln\left|2(\varepsilon s+2)^{-1}\right|, \;\; t = \frac{2s}{\varepsilon s + 2}$$

Substituting these x, t into the third equation of system Equation (55) we receive after simplification the solution of Equation (50)

$$p(r,s) = -\frac{\frac{\sqrt{\varepsilon s+2}}{\sqrt{2}}\,e^{-1/2r}\left(1 + \mathrm{LambertW}\left(-\frac{1}{16}\varepsilon(\varepsilon s+2)^2 e^{-r}\right)\right)^{-1}}{\sqrt{-\varepsilon^{-1}e^r\,\mathrm{LambertW}\left(-\frac{1}{16}\varepsilon(\varepsilon s+2)^2 e^{-r}\right)}}$$

Example 3. Let us apply Equation (46) to the solution $u(x,t) = \frac{t}{\sqrt{2t(x-\ln|t|)}}$. We choose for simplicity the arbitrary function appearing as a result of integration, equal to zero. The change of the independent variables has such a form:

$$\begin{array}{l} x = \ln\left|\frac{2s}{\varepsilon s+2}\right| + \frac{2r-2\ln|s|}{\varepsilon s+2}, \\ t = 2\frac{s}{\varepsilon s+2} \end{array}$$

After substitution of new independent variables into the expression

$$p(x,t) = \sqrt{\frac{t}{2(x - \ln|t|)}}$$

we finally get the same solution of the Equation (4)

$$p(r,s) = \sqrt{\frac{s}{2(r - \ln|s|)}}$$

Hence, this solution is an invariant solution with respect to the nonlocal transformation considered.

Equations (47) and (48) too allow generating new solutions of the Equation (4) if $p(r;s)$ is its known solution.

The operator X_4 generates two corresponding characteristic equations

$$v\, v_x - t = 0, \tag{56}$$

$$v\, u_x + u^2 = 0 \tag{57}$$

The first equation determines usual Lie point symmetry of Equation (7). Equation (57) determines potential symmetry of Equation (4). Both these equations are connected by the potential substitution Equation (20).

Solving Equation (57) with respect to v

$$v = -u^2 {u_x}^{-1}$$

and substituting this into Equation (56), we obtain the second order differential equation, determining nonlocal symmetry of Equation (4)

$$u_{xx}u^4 - 2\, {u_x}^2 u^3 + t\, {u_x}^3 = 0$$

and, consequently, appropriate nonlocal ansatzes for a given equation

$$u = \frac{-f_1(t) \pm \sqrt{f_1(t)^2 + 2tf_2(t) + 2tx}}{2(f_2(t) + x)}$$

Here $f_i, (i = 1,2,3)$ are the arbitrary functions of variable t. This ansatz can be used for construction of solutions to Equation (4).

Let us describe symmetry of Equation (4) corresponding to Lie symmetry Equation (56) of Equation (7). First we substitute $v_x = u$ into Equation (56)

$$v\, v_x - t = v\, u - t = 0$$

and solve it for $v(x,t)$. Differentiating the result with respect to x, we get

$$v_x + t \cdot u^{-2} u_x = 0$$

Note that if we wrote obtained above expression using Equation (20) in a form $u^3 + tu_x = 0$, we would make a mistake, because such symmetry is not admitted by Equation (4).

So, using substitution $vv_x - t = 0$ in the previous expression, we get

$$u^2 + v\, u_x = 0$$

(the second characteristic equation for X_4). Solving the last equation for $v(x,t)$ and differentiating it with respect to x, we obtain

$$v_x = -2u + u^2 u_x^{-2} \cdot u_{xx}$$

Using here potential substitution Equation (20) and simplifying result, we get corresponding nonlocal symmetry of the Equation (4)

$$u_{xx}\, u - 3u_x^2 = 0$$

Integrating this equation we find two ansatzes for Equation (4)

$$u = \pm \frac{1}{\sqrt{-2f_1(t)\, x - 2f_2(t)}}$$

They depend on arbitrary functions $f_1(t)$ and $f_2(t)$. Specializing them by Equation (4), we find two solutions

$$u = \pm \frac{\sqrt{t - c_2}}{\sqrt{2c_1 e^3 - 2\ln|-t + c_2| + 2(x + c_3)}}$$

Since Equation (4) is invariant with respect to the time and space translations generated by the operators X_1 and X_2 the last solution, Equation (38) and u-part of solution to the system Equations (39) and (40) represent the same exact solution.

5. Conclusions

Two given diffusion-convection equations, which allow connection by the generalized hodograph transformation are considered. It is shown in this paper that the intermediate equations possess the additional (with respect to given equations) operators of invariance Lie algebras. The appropriate additional operators allow to construct potential and nonlocal symmetries of the given differential equations. We have constructed the one-parameter groups of Lie symmetries generated by operators, admitted by the potential system Equations (15) and (16), and use them for construction of the corresponding formulae for generation of solutions of Equation (4). This involves a successive implementation of appropriate transformations $\mathcal{T}$ and the one-parameter group of point transformations associated with the corresponding infinitesimal operator admitted by the potential system. New formulae for generation of solutions have been constructed. Some of them are based on the nonlocal transformations that contain new independent variable determined by an auxiliary differential equation. This type of transformations allows interpretation as an example of the nonlocal transformation with additional variables. The obtained formulae allow transformation of a known solution of the given equation into another its solution. They have been used for construction of exact solutions, some of which are new. They are either obtained in explicit form or presented in a new parametrical representations, where a functional parameter is given implicitly. All found solutions can be extended to parametric families by means of the Lie symmetry transformations or by using other formulae for generation of solutions.

Acknowledgments: The author thanks the referees for a lot of helpful remarks and suggestions.

Conflicts of Interest: The author declares no conflict of interest.

References

1. Tychynin, V.A.; Petrova, O.V. Nonlocal symmetries and formulae for generation of solutions for a class of diffusion–convection equations. *J. Math. Anal. Appl.* **2011**, *382*, 20–33.
2. Bluman, G.W.; Reid, G.J.; Kumei, S. New classes of symmetries for partial differential equations. *J. Math. Phys.* **1988**, *29*, 806–811.
3. Bluman, G.W.; Kumei, S. *Symmetries and Differential Equations*; Springer: New York, NY, USA, 1989.
4. Anderson, I.M.; Kamran, N.; Olver, P.J. Internal, external and generalized symmetries. *Adv. Math.* **1993**, *100*, 53–100.

5. Reyes, E.G. Nonlocal symmetries and the Kaup–Kupershmidt equation. *J. Phys. A Math. Gen.* **2005**, *46*, 073507:1–073507:19.
6. Bluman, G.W.; Doran-Wu, P. The use of factors to discover potential systems or linearizations. *Acta Appl. Math.* **1995**, *41*, 21–43.
7. Serov, M.; Omelyan, O.; Cherniha, R. Linearization of some systems of nonlinear diffusion equations by means of non-local transformations. *Dop. Nat. Acad. Ukr.* **2004**, *10*, 39–45.
8. Sophocleous, C. Potential symmetries of nonlinear diffusion-convection equations. *J. Phys. A Math. Gen.* **1996**, *29*, 6951–6959.
9. Lisle, I.G. Equivalence Transformations for Classes of Differential Equations. Ph.D. Thesis, University of British Columbia, Vancouver, BC, Canada, 1992.
10. Lisle I.G.; Reid G.J. *Symmetry Classification Using Invariant Moving Frames;* ORCCA Technical Report TR-00-08; University of Western Ontario: Ontario, Canada, 1992.
11. Popovych, R.O.; Ivanova, N.M. Potential equivalence transformations for nonlinear diffusion—Convection equations. *J. Phys. A Math. Gen.* **2005**, *38*, 3145–3155.
12. Anderson, R.L.; Ibragimov, N.H. *Lie-BÄcklund Transformations in Applications;* SIAM: Philadelphia, PA, USA, 1979.
13. Olver, P.J. *Applications of Lie Groups to Differential Equations;* Springer-Verlag: New York, NY, USA, 1993.
14. Bluman, G.W.; Cheviakov A.; Anco, S. *Applications of Symmetry Methods to Partial Differential Equations;* Applied Mathematical Sciences; Springer: New York, NY, USA, 2010; Volume 168, p. 367.
15. Bluman, G.W.; Yang, Z.Z. A symmetry-based method for constructing nonlocally related PDE systems. *J. Math. Phys.* **2013**, *54*, 093504:1–093504:22.
16. Fuschych, W.I.; Tychynin, V.A. *Preprint No 82.33;* Institut of Mathematics: Kiev, Ukraine, 1982.
17. Fuschych, W.I.; Tychynin, V.A. Exact solutions and superposition principle for nonlinear wave equation. *Proc. Acad. Sci. Ukr.* **1990**, *5*, 32–36.
18. Tychynin, V.A. Nonlocal Symmetries and Solutions for Some Classes of Nonlinear Equations of Mathematical Physics. Ph.D. Thesis, Prydniprovsk State Academy of Civil Engineering and Architecture, Dnipropetrovsk, Ukraine, 1994.
19. Tychynin, V.A. Non-local symmetry and generating solutions for Harry–Dym type equations. *J. Phys. A Math. Gen.* **1994**, *14*, 2787–2797.
20. Tychynin, V.A.; Petrova, O.V.; Tertyshnyk, O.M. Symmetries and Generation of Solutions for Partial Differential Equations. *SIGMA* **2007**, *3*, 019:1–019:14.
21. Rzeszut, W.; Vladimirov, V.; Tertyshnyk, O.M.; Tychynin, V.A. Linearizability and nonlocal superposition for nonlinear transport equation with memory. *Rep. Math. Phys.* **2013**, *72*, 235–252.
22. Fushchych, W.I.; Serov, M.I.; Tychynin, V.A.; Amerov, T.K. On non-local symmetry of nonlinear heat equation. *Proc. Acad. Sci. Ukr.* **1992**, *11*, 27–33.
23. Storm, M.L. Heat conduction in simple metals. *J. Appl. Phys.* **1951**, *22*, 940–951.
24. Bluman, G.W.; Kumei, S. On the remarkable nonlinear diffusion equation $\frac{\partial}{\partial x}(a(u+b)-2\frac{\partial u}{\partial x})-\frac{\partial u}{\partial t}=0$. *J. Math. Phys.* **1980**, *21*, 1019–1023.
25. King, J.R. Some non-local transformations between nonlinear diffusion equations. *J. Phys. A Math. Gen.* **1990**, *23*, 5441–5464.
26. Cherniha, R.; Serov, M. Symmetries, Ansätze and Exact Solutions of Nonlinear Second-order Evolution Equations with Convection Terms, II. *Eur. J. Appl. Math.* **2006**, *17*, 597–605.
27. Rosen, G. Nonlinear heat conduction in solid. *Phys. Rev. B* **1979**, *19*, 2398–2399.
28. Cherniha, R.; Serov, M. Symmetries, Ansätze and Exact Solutions of Nonlinear Second-order Evolution Equations with Convection Terms. *Eur. J. Appl. Math.* **1998**, *9*, 527–542.

symmetry

MDPI

Article

New Nonlocal Symmetries of Diffusion-Convection Equations and Their Connection with Generalized Hodograph Transformation

Valentyn Tychynin

Prydniprovs'ka State Academy of Civil Engineering and Architecture, 24a Chernyshevsky Street, Dnipropetrovsk 49005, Ukraine; tychynin@ukr.net

Academic Editor: Roman M. Cherniha
Received: 23 June 2015; Accepted: 18 September 2015; Published: 29 September 2015

Abstract: Additional nonlocal symmetries of diffusion-convection equations and the Burgers equation are obtained. It is shown that these equations are connected via a generalized hodograph transformation and appropriate nonlocal symmetries arise from additional Lie symmetries of intermediate equations. Two entirely different techniques are used to search nonlocal symmetry of a given equation: the first is based on usage of the characteristic equations generated by additional operators, another technique assumes the reconstruction of a parametrical Lie group transformation from such operator. Some of them are based on the nonlocal transformations that contain new independent variable determined by an auxiliary differential equation and allow the interpretation as a nonlocal transformation with additional variables. The formulae derived for construction of exact solutions are used.

Keywords: nonlocal symmetries; potential symmetries; formulae of nonlocal superposition; formulae for generation of solutions; generalized hodograph transformation

---

1. Introduction

We continue the nonlocal symmetries search of the diffusion-convection equations, connected by a generalized hodograph transformation (GHT) [1]. For this purpose we use the additional Lie symmetries (that are additional operators of invariance Lie algebras of higher dimension) of the intermediate equations arising at steps of this transformation, which consists of potential substitution and usual hodograph transformation. One can find many references and the extensive bibliography in researches devoted to studying the potential symmetries of nonlinear partial differential equations and systems [2–8]. The notion of potential symmetries of differential equations was introduced by Bluman *et al.* [2,3]. Later, Lisle proposed in [9,10] the concept of potential equivalence transformations. It was successfully applied in [11] for deriving complete list of potential symmetries for wide classes of diffusion-convection equations.

On the other hand a number of interesting results for nonlinear equations are obtained for present day within the nonlocal transformations approach.

Analogously to continuous groups of usual transformations, the theory of groups of Lie–Bäcklund transformations, *i.e.*, continuous groups of transformations involving derivatives of dependent variables, was developed by Anderson and Ibragimov [12]. An effective tool for this method application is the notion of recursion operator proposed by Olver [13].

A method for constructing nonlocally related PDE systems was introduced in [14,15]. Potential and nonlocal symmetries has been investigated for wide classes of nonlinear PDE systems there. It was shown that each point symmetry of a PDE system systematically yields a nonlocally related PDE system. Appropriate nonlocal symmetries were presented in the form including a potential variable.

The use of finite nonlocal transformations allowed to construct the formulae of generation of solutions and nonlocal nonlinear superposition for wide variety of nonlinear partial differential equations. This approach is based on the nonlocal transformations technic [16–21], which we regularly use for study of symmetries of the given nonlinear equations in [1] and in the current paper.

To start with, we remind the main concepts and notions which are necessary for understanding of the subsequent material. Let's assume that finite nonlocal transformation of variables

$$\mathcal{T}: x^i = h^i\left(y, v_{(k)}\right), i = 1, \cdots, n, u^K = H^K\left(y, v_{(k)}\right), K = 1, \cdots, m$$

exists and maps the given equation

$$F_1\left(x, u_{(n)}\right) = 0 \tag{1}$$

into the equation $\Phi\left(y, v_{(q)}\right) = 0$ of order $q = n + k$. Suppose, that this equation admits a factorization to another equation

$$F_2\left(y, v_{(s)}\right) = 0 \tag{2}$$

i.e., $\Phi\left(y, v_{(q)}\right) = \lambda F_2\left(y, v_{(s)}\right)$. Here λ is a differential operator of order $n + k - s$. Then we say that Equations (1) and (2) are connected by the nonlocal transformation $\mathcal{T}$

$$F_1\left(x, u_{(n)}\right) = \lambda F_2\left(y, v_{(s)}\right)$$

Here the symbol $u_{(r)}$ denotes the tuple of derivatives of the function u from order zero up to order r. In the case of two independent variables we use the special notation of the variables: $x_1 = x$, $x_2 = t$ and thus $u_t = \partial u/\partial t = \partial_t u$, $u_x = \partial u/\partial x = \partial_x u$.

Nonlocal transformations can be effectively used for construction of the formulae generating solutions in both cases: for nonlocal invariance of equation

$$F_2\left(y, v_{(n)}\right) = F_1\left(y, v_{(n)}\right)$$

and when the connected equations are different. In a special case, when equation F_2 is linear, one can construct the formulae of nonlinear nonlocal superposition of solutions for nonlinear equation F_1 [1,17,22]. We consider a case when the intermediate equations, connected by components of the generalized hodograph transformation exist, and these equations possess invariance Lie algebras of higher dimension. These, the last, we use for construction of corresponding nonlocal symmetries of the given equations and for deriving the formulae generating solutions.

The paper is organized as follows: In Section 2 we introduce a chain of the equations connected by steps of the GHT and compare the Lie invariance algebras obtained. Then (in Section 3) we construct nonlocal symmetries of these equations, which are generated by additional operators of invariance Lie algebras of the intermediate equations. Section 4 is devoted to the construction of finite nonlocal invariance transformations for a given equation and appropriate formulae generating their solutions. Namely, we combine the Lie symmetry transformation, generated by an additional operator X, which has been admitted by the intermediate equation, with a transformation, mapping this intermediate equation into the given one.

2. Lie Symmetries of Given and Intermediate Equations

We aim to consider two diffusion-convection equations from the class

$$u_t - \partial_x(C(u) + K(u) \cdot u_x) = 0 \tag{3}$$

which possess Lie invariance algebras of different dimensions. It is easy to check the mapping of the given equation

$$u_t - \partial_x\left(\frac{1}{2}u^{-1} + u^{-2}u_x\right) = 0 \tag{4}$$

into the Burgers equation

$$z_\eta - \partial_\xi\left(-\frac{1}{2}z^2 + z_\xi\right) = 0 \tag{5}$$

via all steps of the GHT. This transformation was found first by Storm in [23] for linearization of nonlinear heat equations. Later this transformation was re-discovered in [24] and applied in [22,25] for investigation of various nonlinear heat equations. The GHT includes the potential substitution

$$u = v_x(x,t) \tag{6}$$

which is being applied to Equation (4) transforms it into the potential equation

$$v_t - \frac{1}{2}v_x^{-1} - v_x^{-2}v_{xx} = 0 \tag{7}$$

Then the hodograph transformation

$$v(x,t) = \xi, x = w(\xi,\eta), t = \eta \tag{8}$$

maps Equation (7) into the equivalent form

$$w_\eta + \frac{1}{2}w_\xi^2 - w_{\xi\xi} = 0 \tag{9}$$

We will refer to Equation (9), Equation (7) obtained as "intermediate" equations.

Having applied one more potential substitution

$$w_\xi = z \tag{10}$$

to Equations (9) we get the Burgers Equation (5). The composition of transformations Equations (6), (8) and (10) introduced above we mark further as $\mathcal{T}$.

Note that the relation of Equation (4) with the equation

$$u_t - \partial_x\left(u^{-2}u_x\right) = 0 \tag{11}$$

is a particular case of the more general result established in [26]. The results concerning linearization of Equation (11) were obtained by Rosen [27]. An example with Equation (11), particularly, was considered in [14,15]. Nonlocal symmetries were described using a potential variable. Here we apply a traditional approach [1] for searching nonlocal symmetries and aim to construct appropriate formulae of generation of solutions to equations considered. Then we use them for construction of exact solutions.

Consider some results of the classical Lie symmetry analysis for the equations connected by all components of the transformation $\mathcal{T}$. The maximal Lie algebra admitted by Equation (4) is four-dimensional (see for example [28])

$$X_1 = \partial_t, X_2 = \partial_x, X_3 = 2t\partial_t + u\partial_u,$$
$$X_4 = -2e^{\frac{1}{2}x}\partial_x + e^{\frac{1}{2}x}u\partial_u$$

while invariance algebra of the Burgers Equation (5) is spanned by five operators

$$\begin{gathered} X_1 = \partial_\eta, X_2 = \partial_\xi, X_3 = \eta\partial_\xi + \partial_z, \\ X_4 = \xi\partial_\xi + 2\eta\partial_\eta - z\partial_z, \\ X_5 = \eta\xi\partial_\xi + \eta^2\partial_\eta + (\xi - z\eta)\partial_z \end{gathered} \tag{12}$$

The operator X_5 does not correspond to Lie symmetry of Equation (4). The appropriate nonlocal symmetry of Equation (4) corresponding to the Lie symmetry X_5 of the Burgers equation has been investigated in details in [1]. Here we will take into account symmetries of the intermediate equations for construction of the appropriate nonlocal symmetries of two given Equations (4) and (5).

The maximal Lie invariance algebra of Equation (7) is spanned by the generators

$$\begin{gathered} X_1 = \partial_t, X_2 = \partial_x, X_3 = \partial_v, X_4 = v\partial_x + t\partial_v, \\ X_5 = 2t\partial_t + v\partial_v, X_6 = \left(t + \tfrac{1}{2}v^2\right)\partial_x + t^2\partial_t + tv\partial_v \\ X_7 = b(v,t)e^{\frac{1}{2}x}\partial_x, b_t - b_{vv} = 0 \end{gathered} \tag{13}$$

Since Equation (9) is equivalent to Equation (7) it admits the Lie invariance algebra of the same structure

$$\begin{gathered} X_1 = \partial_\eta, X_2 = \partial_\xi, X_3 = \partial_w, X_4 = \eta\partial_\xi + \xi\partial_w, \\ X_5 = \xi\partial_\xi + 2\eta\partial_\eta, X_6 = \eta\xi\partial_\xi + \eta^2\partial_\eta + \left(\tfrac{1}{2}\xi^2 + \eta\right)\partial_w, \\ X_7 = b(\xi,\eta)e^{\frac{1}{2}w}\partial_w, b_\eta - b_{\xi\xi} = 0 \end{gathered} \tag{14}$$

Note that both equations have two Lie symmetry generators more (X_3 and X_7) than Equation (5) has. It will be interesting to compare the symmetries obtained above with those ones, which are admitted by a potential system, constructed for Equation (4)

$$v_x - u = 0 \tag{15}$$

$$2v_t - u^{-1} - 2u^{-2}u_x = 0 \tag{16}$$

One can easily calculate the Lie symmetry of this system, which is spanned by generators

$$\begin{gathered} X_1 = \partial_t, X_2 = \partial_x, X_3 = \partial_v, X_4 = v\partial_x - u^2\partial_u + t\partial_v, \\ X_5 = 2t\partial_t + ut\partial_u + v\partial_v, \\ X_6 = \left(t + \tfrac{1}{2}v^2\right)\partial_x + t^2\partial_t + u(t - vu)\partial_u + tv\partial_v, \\ X_7 = 2b(v,t)e^{\frac{x}{2}}\partial_x - (2u^2\partial_v b(v,t) + u)e^{\frac{x}{2}}b(v,t)\partial_u, \\ b_t(v,t) - b_{vv}(v,t) = 0 \end{gathered} \tag{17}$$

Notice that to obtain an operator X_7 we chose it in the form

$$X_7 = 2b(v,t)e^{\frac{x}{2}}\partial_x - \left(2u^2 a(x,t) - u\right)e^{\frac{x}{2}}b(v,t)\partial_u$$

where $a(x,t)$ and $b(v(x,t),t)$ are unknown differentiable functions.

Applying this operator to the system Equations (15) and (16), we get such conditions for these functions:

$$a(x,t)b(v(x,t),t) + D_1 b(v(x,t),t) = 0 \tag{18}$$

$$a(x,t)D_1 b(v(x,t),t)u(x,t) + b(v(x,t),t)a_x(x,t) + D_2 b(v(x,t),t)u(x,t) = 0 \tag{19}$$

Here $D_1 f(\alpha,\beta)$ means the derivative of a function $f(\alpha,\beta)$ with respect to its first argument. Substituting a solution of the first equation

$$a(x,t) = -\frac{D_1 b(v(x,t),t)}{b(v(x,t),t)}$$

and an equality

$$v_x = u \tag{20}$$

into the second equation, we receive the equation for $b(v,t)$

$$b_t(v,t) - b_{vv}(v,t) = 0 \tag{21}$$

In what follows we will use two entirely different techniques to search for nonlocal symmetry of a given equation. The first is based on usage of the characteristic equation generated by an additional operator. Another technique assumes reconstruction of a parametrical Lie group transformation from such operator.

3. Nonlocal Symmetries Generated by Additional Lie Symmetries of the Intermediate Equations

Here we apply additional operators of Lie algebras obtained for intermediate equations to search appropriate nonlocal symmetry of given equations and construct the formulae of generation their solutions.

3.1. The Operator X 7th Case

Comparison of Lie algebras Equations (12) and (14) shows that Equation (9) admits an additional (with respect to Equation (5)) generator X_7. The characteristic equation corresponding to sum $X_7 + X_3$ of generators of algebra Equation (14) has the form

$$w + 2\ln|b| = 0$$

One can easily identify this formula with the known substitution linearizing Equation (9). To present this expression via the variable $z(\xi,\eta)$ we will differentiate it with respect to ξ

$$w_\xi + 2b_\xi b^{-1} = 0 \tag{22}$$

Then, substituting $w_\xi = z$ into above equation, we find the characteristic equation which determines appropriate nonlocal symmetry of Equation (5)

$$z + 2b_\xi b^{-1} = 0 \tag{23}$$

Therefore the "lacking" operator for the Burgers equation should be written in the form

$$X_7^\star = \left(z + 2\partial_\xi \ln b\right)\partial_z \tag{24}$$

Note that an expression Equation (23) is well known *Cole–Hopf substitution*.

Conversely, applying the transformation Equation (8) with its differential consequence $w_\xi = 1/v_x$ to the Equation (22), we find nonlocal symmetry for Equation (7)

$$v_x + b/2b_v = 0$$

Taking into account substitution Equation (20), this equation takes the form

$$u + b/2b_v = 0 \tag{25}$$

Here $b(v,t)$ is an arbitrary solution of the linear heat Equation (21).

An application of the nonlocal transformation Equation (25) to Equation (4) generates an expression which after simplification vanishes identically on the manifold defined by Equation (7) and expression $v_x + b/2b_v = 0$ with its differential consequences.

Usage of a classical linear superposition principle in Equation (25) allows one to get the formula generating solutions of Equation (4)

$$u^{\mathrm{I}} + b^{\mathrm{I}}\left(2b^{\mathrm{I}}_v\right)^{-1} = 0, u^{\mathrm{II}} + b^{\mathrm{II}}\left(2b^{\mathrm{II}}_v\right)^{-1} = 0, \rightarrow u^{\mathrm{III}} = -\left(b^{\mathrm{I}} + b^{\mathrm{II}}\right)\left(2b^{\mathrm{I}}_v + 2b^{\mathrm{II}}_v\right)^{-1} \quad (26)$$

On the other hand the generator X_7 of the Lie invariance algebra of the system Equations (15) and (16) determines a potential symmetry of Equation (4) and admits presentation by the characteristic equation

$$u_x + u^2\frac{b_v(v,t)}{b(v,t)} + \frac{1}{2}u = 0 \quad (27)$$

where $u(x;t)$ and $b(v;t)$ satisfy Equations (21) and (20). Thus, the appropriate "lacking" operator for Equation (4) should be written in the form, depending on $b(v,t)$

$$X_7^\star = b(v,t)e^{\frac{1}{2}x}\partial_x - \left(u^2\frac{b_v(v,t)}{b(v,t)} + \frac{1}{2}u\right)b(v,t)e^{\frac{1}{2}x}\partial_u \quad (28)$$

Theorem 1. *The characteristic Equation (27) determines nonlocal transformation with additional variables, which connects Equations (4) and (7) taking into account Equation (21) and condition Equation (20). Additional independent variable $v(x,t)$ is determined by Equation (7) and additional dependent variable $b(v,t)$ is an arbitrary solution of the Equation (21).*

To prove the statement formulated above we will first rewrite Equation (27), using equality Equation (20) in the form

$$\partial_x \ln|u| + \partial_x \ln|b(v,t)| + \frac{1}{2} = 0 \quad (29)$$

and then integrate the result with respect to x

$$u(x,t) = \frac{s(t)e^{-1/2x}}{b(v,t)} \quad (30)$$

Having applied this transformation to Equation (4) and use the conditions

$$b_t(v,t) - b_{vv}(v,t) = 0, v_{xx} = v_t v_x^2 - 1/2v_x, v_x(x,t) = \frac{s(t)e^{-1/2x}}{b(v,t)}$$

we obtain a result, which vanishes identically after simplification.

The above assertion admits another form.

Theorem 2. *The Equation (4) is invariant under action of the prolonged operator $X_7^\star$ on the manifold, determined by Equation (7), condition Equation (20) and the Equation (21) with their differential consequences.*

To deduce the characteristic equation, which corresponds to Equation (29) and determines nonlocal symmetry of the Equation (4) in terms of variables $x, t, u(x,t)$, we need to exclude a variable $v(x,t)$ from expression Equation (30), where $b(v,t)$ is a function belonging to the set of solutions of the linear heat Equation (21).

New designations will be necessary for us in what follows. Suppose that we can solve the functional equation $b(v,t) = a$ with respect to the first argument of b. Then a function $\beta(a,t) : b(v,t) = a \rightarrow v = \beta(a,t)$, will denote a solution of this equation. Therefore

$$v(x,t) = \beta\left(\frac{s(t)e^{-1/2x}}{u}, t\right)$$

is a solution of Equation (30) with respect to v. Using potential substitution Equation (20) we obtain equality

$$u(x,t) - \partial_x \beta\left(\frac{s(t)e^{-1/2x}}{u(x,t)}, t\right) = 0$$

Differentiating and simplifying this expression, we get the presentation of the onetime integrated characteristic equation which determines nonlocal symmetry of Equation (4)

$$\begin{gathered} 2u^3 + D_1\beta\left(\tfrac{s(t)e^{-1/2x}}{u}, t\right)s(t)e^{-1/2x}u \\ +2D_1\beta\left(\tfrac{s(t)e^{-1/2x}}{u}, t\right)s(t)e^{-1/2x}u_x = 0 \end{gathered} \quad (31)$$

$D_1 f(\alpha, \beta)$ means the derivative of a function $f(\alpha, \beta)$ with respect to its first argument.

For instance, a given simple solution of the linear heat Equation (21)

$$b(v,t) = \left(c_1 e^{\sqrt{\alpha}v} + c_2 e^{-\sqrt{\alpha}v}\right)c_3 e^{\alpha t}$$

allows one to obtain the expression Equation (30) in such particular form:

$$\left(c_1 e^{\sqrt{\alpha}v} + c_2 e^{-\sqrt{\alpha}v}\right)c_3 e^{\alpha t} - \frac{s(t)e^{-1/2x}}{u(x,t)} = 0$$

Here $\alpha, c_i, (i = 1,2,3)$ are arbitrary constants. This equation has two solutions for $v(x,t)$

$$v = \ln\left|\frac{s(t)e^{-1/2x} \pm \sqrt{s^2(t)e^{-x} - 4c_3^2 e^{2\alpha t}u^2 c_1 c_2}}{2c_1 c_3 e^{\alpha t}u}\right|\left(\sqrt{\alpha}\right)^{-1}$$

Excluding the variable v from Equation (31), we can rewrite it as follows

$$u - \partial_x\left(\ln\left|\frac{s(t)e^{-1/2x} \pm \sqrt{s^2(t)e^{-x} - 4c_3^2 e^{2\alpha t}u^2 c_1 c_2}}{2c_1 c_3 e^{\alpha t}u}\right|\left(\sqrt{\alpha}\right)^{-1}\right) = 0 \quad (32)$$

Finally we obtain two the onetime integrated characteristic equations which determine nonlocal symmetries of Equation (4)

$$2u_x + 2\sqrt{\alpha}u^2 + u \pm \frac{8\sqrt{\alpha}c_1 c_2 c_3^2 e^{2\alpha t}u^4}{\sqrt{s^2(t)e^{-x} - 4c_3^2 e^{2\alpha t}u^2 c_1 c_2} + s(t)e^{-1/2x}} = 0$$

Choosing in Equation (32), in particular, $c_2 = 0, c_3 = 1$, and integrating this equation, we obtain two ansatzes

$$u = \pm\frac{1}{2\sqrt{\alpha} - g(t)e^{1/2x}}$$

which contain arbitrary functions of t, which must be specialized by Equation (4). So, finally we obtain two solutions of this equation

$$u = \pm\frac{1}{2\sqrt{\alpha} - ke^{1/2x}}$$

Note that such functions can be obtained by means of the Lie ansatz.

Theorem 3. *The operator Equation (28) allows one to construct the associated transformation determined by the formulas*

$$2\left(e^{\frac{x^{\mathrm{I}}}{2}} + e^{\frac{x^{\mathrm{II}}}{2}}\right) + \varepsilon b^{\mathrm{I}}\left(v^{\mathrm{I}}\left(x^{\mathrm{I}}, t\right), t\right)e^{\frac{x^{\mathrm{I}} + x^{\mathrm{II}}}{2}} = 0 \quad (33)$$

$$u^{\mathrm{II}}\left(x^{\mathrm{II}},t\right) = -\frac{u^{\mathrm{I}}(x^{\mathrm{I}},t)\left(-2+b^{\mathrm{I}}(v^{\mathrm{I}}(x^{\mathrm{I}},t),t)\varepsilon e^{\frac{x^{\mathrm{I}}}{2}}\right)}{2+2u^{\mathrm{I}}(x^{\mathrm{I}},t)D_1 b^{\mathrm{I}}(v^{\mathrm{I}}(x^{\mathrm{I}},t),t)\varepsilon e^{\frac{x^{\mathrm{I}}}{2}}} \tag{34}$$

where $\partial_{x^{\mathrm{I}}} v^{\mathrm{I}}(x^{\mathrm{I}},t) = u^{\mathrm{I}}(x^{\mathrm{I}},t)$ *and* $b^{\mathrm{I}}(v^{\mathrm{I}}(x^{\mathrm{I}},t),t)$ *s any arbitrary solution of the Equation (21).*

To prove this theorem we restore a Lie group transformation using the operator Equation (28). So, we obtain

$$2\left(e^{\frac{x}{2}}+e^{\frac{r}{2}}\right)+\varepsilon b(v(x,t),t)e^{\frac{x+r}{2}}=0, s=t,$$

$$p(r,s) = -\frac{u(x,t)\left(-2+b(v(x,t),t)\varepsilon e^{\frac{x}{2}}\right)}{2+2u(x,t)D_1 b(v(x,t),t)\varepsilon e^{\frac{x}{2}}}$$

Making here change of notations for new variables, we find the Equations (33) and (34).

The algorithm of generation of solutions for Equation (4) based on these formulae works, as follows. Let $u^{\mathrm{I}}(x^{\mathrm{I}},t)$ be any known solution of Equation (4), then $v^{\mathrm{I}}(x^{\mathrm{I}},t) = \int u^{\mathrm{I}}(x^{\mathrm{I}},t)dx^{\mathrm{I}} + s(t)$ can be found. Here $s(t)$ is arbitrary function which can be specialized by Equation (29). Substituting obtained v^{I} and *arbitrary* solution of the linear heat equation $b^{\mathrm{I}}(v^{\mathrm{I}}(x^{\mathrm{I}},t),t)$ into Equation (33) and solving the result with respect to x^{I}, we get $x^{\mathrm{I}} = \chi(x^{\mathrm{II}},t)$. Finally, inserting this x^{I} into Equation (34) we find a new solution of Equation (4).

Example 1. Choosing $u^{\mathrm{I}} = e^{-\frac{x^{\mathrm{I}}}{2}}$ and integrating it with respect to x, we obtain $v^{\mathrm{I}} = -2e^{-\frac{x^{\mathrm{I}}}{2}} + s(t)$. Specializing arbitrary function by Equation (29), we get $v^{\mathrm{I}} = -2e^{-\frac{x^{\mathrm{I}}}{2}} + c_1$. Set arbitrary solution of the linear heat equation $b^{\mathrm{I}} = e^{c_2 t+\sqrt{c_2}v^{\mathrm{I}}}$, then $D_1 b^{\mathrm{I}} = \sqrt{c_2}e^{c_2 t+\sqrt{c_2}v^{\mathrm{I}}}$. Rewriting these formulae and taking into account $v^{\mathrm{I}} = -2e^{\frac{x^{\mathrm{I}}}{2}} + c_1$, we find

$$b^{\mathrm{I}} = e^{c_2 t+\sqrt{c_2}(-2e^{\frac{x^{\mathrm{I}}}{2}}+c_1)}$$

and

$$D_1 b^{\mathrm{I}} = \sqrt{c_2}e^{c_2 t+\sqrt{c_2}(-2e^{\frac{x^{\mathrm{I}}}{2}}+c_1)}$$

Substituting a found above b^{I} into Equation (33) and solving this result with respect to x^{I}, we get $x^{\mathrm{I}} = \chi(x^{\mathrm{II}},t)$

$$x^{\mathrm{I}} = x^{\mathrm{II}} - 2\ln\left|\frac{1}{2}\mathrm{LambertW}\left(\sqrt{c_2}\varepsilon e^{(c_1\sqrt{c_2}+tc_2-2\sqrt{c_2}e^{-\frac{x^{\mathrm{II}}}{2}})}e^{\frac{x^{\mathrm{II}}}{2}}+\sqrt{c_2}\right)\right| + \ln|c_2|$$

After simplification Equation (34) with obtained $x^{\mathrm{I}} = \chi(x^{\mathrm{II}},t)$, b^{I} and $D_1 b^{\mathrm{I}}$ we find a new solution of Equation (4)

$$u^{\mathrm{II}}\left(x^{\mathrm{II}},t\right) = e^{c_1\sqrt{c_2}+tc_2-2\sqrt{c_2}e^{-\frac{x^{\mathrm{II}}}{2}}-\frac{x^{\mathrm{II}}}{2}}G^{-1},$$

$$G = e^{c_1\sqrt{c_2}+tc_2-2\sqrt{c_2}e^{-\frac{x^{\mathrm{II}}}{2}}}\left(\mathrm{LambertW}\left(\sqrt{c_2}\varepsilon e^{(c_1\sqrt{c_2}+tc_2-2\sqrt{c_2}e^{-\frac{x^{\mathrm{II}}}{2}})}\right)+1\right)$$

3.2. The Operator X 4th Case

The Lie invariance algebra Equation (13) of Equation (7) includes one operator $X_4 = v\cdot\partial_x + t\cdot\partial_v$ more, than invariance algebra of Equation (4) has. This operator yields the characteristic equation

$$v\cdot v_x - t = 0 \tag{35}$$

Differentiating Equation (35) with respect to x and using substitution Equation (20) we obtain

$$vu_x + u^2 = 0 \tag{36}$$

Excluding $v(x,t)$ from this equation we get the second order differential equation for u which determines nonlocal symmetry of Equation (4)

$$-3u_x^2 + uu_{xx} = 0 \tag{37}$$

Solving Equation (37), we find non-Lie ansatzes of Equation (4)

$$u = \pm\left(\sqrt{-2F_1(t)x - 2F_2(t)}\right)^{-1}$$

Substitution them into Equation (4) and splitting a result with respect to x yields the reduced system of two ordinary differential equations

$$\dot{F_1}(t) - (F_1(t))^2 = 0, \dot{F_2}(t) - F_1(t)F_2(t) - (F_1(t))^2 = 0$$

Solution of this system

$$F_1(t) = -(t - c_2)^{-1}, F_2(t) = \frac{\ln|t - c_2| - c_1}{t - c_2}$$

allows us to construct appropriate solutions of Equation (4)

$$u = \pm\frac{\sqrt{2}}{2}\sqrt{\frac{t - c_2}{x - \ln|t - c_2| + c_1}} \tag{38}$$

The result obtained is closely connected with potential symmetry of Equation (4) determined by the generator X_4 of Lie algebra Equation (17). The characteristic equations corresponding to this operator, are

$$v \cdot u_x + u^2 = 0 \tag{39}$$

$$v \cdot v_x - t = 0 \tag{40}$$

They, obviously, coincide with Equations (36) and (35) accordingly. The solution of this system

$$u = \frac{t}{\sqrt{2 \cdot x \cdot t + F_1(t) + F_2(t) \cdot t}}, v = \sqrt{2xt + F_1(t)}$$

yields the reduced system of ordinary differential equations

$$F_2(t) = 0, t \cdot \dot{F_1} - F_1(t) + 2t = 0$$

and, consequently, yields appropriate solution to the system Equations (39) and (40)

$$u = \frac{\sqrt{t}}{\sqrt{2x - 2\ln|t| + c_1}}, v = \sqrt{t}\sqrt{2x - 2\ln|t| + c_1}$$

If we set $c_2 = 0$ in Equation (38), we will obtain the previous expression for the function u.

4. Nonlocal Invariance Transformations and Generation of Solutions

Here we construct the one-parameter group of Lie symmetries generated by operator, admitted by the potential system Equations (15) and (16), and it is used for construction of the corresponding formula for generation of solutions of Equation (4). This involves a successive implementation of the transformations $\mathcal{T}$ and the one-parameter group of point transformations associated with the corresponding infinitesimal operator which is admitted by the potential system.

Assume that the partial differential equation

$$F\left(x,t,u_{(k)}\right) = 0 \tag{41}$$

should admit at least one conservation law

$$D_t\phi^t\left(x,t,u_{(r)}\right) + D_x\phi^x\left(x,t,u_{(r)}\right) = 0 \tag{42}$$

where D_t and D_x are total derivatives with respect to the variables t and x, ϕ^t and ϕ^x are conserved density and flux, respectively. When Equation (41) admits representation Equation (42) there exists the potential function v determined by the auxiliary system [2]

$$v_x = \phi^t\left(x,t,u_{(r)}\right), v_t = -\phi^x\left(x,t,u_{(r)}\right) \tag{43}$$

Potential symmetry of the Equation (41) is determined by the Lie symmetry generator

$$X = \xi_1(x,t,u,v)\partial_x + \xi_2(x,t,u,v)\partial_t + \eta_1(x,t,u,v)\partial_u + \eta_2(x,t,u,v)\partial_v$$

admitted by system Equation (43) with at least one nonzero partial derivative

$$\partial_v\xi_1(x,t,u,v), \partial_v\xi_1(x,t,u,v), \partial_v\eta_1(x,t,u,v)$$

In this case the one-parameter Lie point symmetry group associated with infinitesimal generator X exists

$$\begin{aligned} r = x' = f_1(x,t,u,v;\varepsilon), s = t' = f_2(x,t,u,v;\varepsilon), \\ p(r,s) = u' = g_1(x,t,u,v;\varepsilon), q(r,s) = v' = g_2(x,t,u,v;\varepsilon) \end{aligned} \tag{44}$$

Here ε is a group parameter.

As Equation (4) is connected with the potential system Equations (15) and (16), one can construct the projection of the corresponding transformation Equation (44) onto the space of variables $(x,t,u(x,t))$ using substitution Equation (20)

$$\begin{aligned} r = x' = f_1(x,t,u,\textstyle\int u(x,t)\mathrm{d}x;\varepsilon), \\ s = t' = f_2(x,t,u,\textstyle\int u(x,t)\mathrm{d}x;\varepsilon), \\ p(r,s) = u' = g_1(x,t,u,\textstyle\int u(x,t)\mathrm{d}x;\varepsilon) \end{aligned} \tag{45}$$

This transformation is the finite Lie–Bäcklund transformation which leaves Equation (4) nonlocal-invariant. Stated above allows formulating the following statement.

Theorem 4. *Equation (4) is nonlocal-invariant under the transformations*

$$\begin{aligned} r = \ln\left|\tfrac{2}{2-\varepsilon t}\right| + \tfrac{(\int u\mathrm{d}x)^2\varepsilon}{4-2\varepsilon t} + x, \\ s = \tfrac{2t}{2-\varepsilon t}, p(r,s) = \tfrac{-2u}{\varepsilon t - u\varepsilon\int u\mathrm{d}x - 2}; \end{aligned} \tag{46}$$

$$\begin{array}{c} r = 2\ln\left|-2\left(-2+b(\int u\mathrm{d}x,t)\varepsilon e^{\frac{1}{2}x}\right)^{-1}\right| + x, s = t, \\ p(r,s) = -\frac{u\left(-2+b(\int u\mathrm{d}x,t)\varepsilon e^{\frac{1}{2}x}\right)}{2+2uD_1(b)(\int u\mathrm{d}x,t)\varepsilon e^{\frac{1}{2}x}} \end{array} \tag{47}$$

and

$$r = \frac{1}{2}\varepsilon^2 t + \varepsilon\int u\mathrm{d}x + x, s = t, p(r,s) = \frac{u}{u\varepsilon+1} \tag{48}$$

The proof of this statement for transformation Equation (46). Let us choose, for instance, the operator

$$X_6 = \left(t + \frac{1}{2}v^2\right)\partial_x + t^2\partial_t + (-vu+t)u\partial_u + tv\partial_v$$

This operator belongs to the Lie invariance algebra Equation (17) of the system Equations (15) and (16). The corresponding Lie group transformation has the form

$$\begin{array}{c} r = \ln\left|\frac{2}{2-\varepsilon t}\right| + \frac{v^2\varepsilon}{4-2\varepsilon t} + x, s = \frac{2t}{2-\varepsilon t}, \\ p(r,s) = -\frac{2u}{\varepsilon t - vu\varepsilon - 2}, q(r,s) = \frac{2v}{2-\varepsilon t} \end{array} \tag{49}$$

Excluding in these formulae v with the help of expression $v = \int u(x,t)\mathrm{d}x$, we get a nonlocal transformation in space of variables r, s, p Equation (46). Then we apply the transformation Equation (46) to Equation (4), rewritten in new designations

$$\partial_s p(r,s) - \partial_r\left(\frac{1}{2}p(r,s)^{-1} + p(r,s)^{-2}\partial_r p(r,s)\right) = 0 \tag{50}$$

and substitute the integro-differential consequences of Equation (4)

$$u_{xx} = u^2 u_t + \frac{1}{2}u_x + 2u^{-1}u_x^2 \tag{51}$$

and

$$\int u_t \mathrm{d}x = \frac{1}{2}u^{-1} + u^{-2}u_x \tag{52}$$

into the obtained result. After simplification the last vanishes identically.

To prove this statement for transformation Equation (47) we consider the generator of Lie algebra Equation (17)

$$X_7 = e^{\frac{1}{2}x}b(v,t)\partial_x - u\left(\frac{ub_v}{b(v,t)} + \frac{1}{2}\right)b(v,t)e^{\frac{1}{2}x}\partial_u$$

where $D_2 b(v,t) - D_{11}b(v,t) = 0, v = v(x,t)$, which yields the group-invariant transformation of the potential system Equations (15) and (16)

$$\begin{array}{c} r = 2\ln\left|-2\left(-2+b(v,t)\varepsilon e^{1/2x}\right)^{-1}\right| + x, s = t, \\ p(r,s) = -\frac{u\left(-2+b(v,t)\varepsilon e^{1/2x}\right)}{2+2uD_1(b)(v,t)\varepsilon e^{1/2x}}, q(r,s) = v \end{array}$$

Hereinafter $D_i(f)(\alpha,\beta)$ and $D_{i,j}(f)(\alpha,\beta), (i,j = 1,2)$ denote total derivatives of a function $f(\alpha,\beta)$ with respect to i-th and j-th variables of the first and second order accordingly, ε is a group parameter associated with the operator X_7. Having substituted into previous formulae $v(x,t) = \int u\mathrm{d}x$ we get a nonlocal transformation Equation (47).

To verify invariance of Equation (50) we apply to it Equation (47) and substitute Equations (51) and (52) and expressions

$$\begin{array}{c} D_{11}b(\int u\mathrm{d}x,t) = D_2 b(\int u\mathrm{d}x,t), \\ D_{111}b(\int u\mathrm{d}x,t) = D_{12}b(\int u\mathrm{d}x,t), \end{array}$$

$$D_{11}u(x,t) = \frac{2D_2u(x,t)(u(x,t))^3 + u(x,t)D_1u(x,t) + 4(D_1u(x,t))^2}{2u(x,t)}$$

into the obtained result. The expression obtained vanishes identically after simplification.

The proof of Theorem 4 for transformation Equation (48). Another operator of the Lie invariance algebra Equation (17)

$$X_4 = v\partial_x - u^2\partial_u + t\partial_v$$

generates the finite Lie-group transformation

$$r = \frac{1}{2}\varepsilon^2 t + \varepsilon v + x, s = t, p(r,s) = \frac{u}{u\varepsilon + 1}, q(r,s) = v + \varepsilon t \tag{53}$$

Projection of this transformation onto the space of variables $(x, t, u(x,t))$ admits the form Equation (48). This transformation leaves the Equation (4) nonlocal-invariant. The statement can be proved like in the previous cases.

Let us consider some applications of the Theorem 4. If $p(r,s)$ is a known solution of the Equation (50), then the new solution of Equation (4) can be constructed by means of solution Equation (46) after subsequent specialization of the arbitrary function appearing as a result of integration in it.

Example 2. Inserting the solution of Equation (4) $u(x,t) = -\frac{1}{2}e^{-\frac{1}{2}x}$ into Equation (46), we obtain the formulae of transformation, which contain an arbitrary function $F(t)$

$$\begin{array}{c} r = \ln\left|\frac{2}{2-t\varepsilon}\right| + x + \frac{\varepsilon(e^{-\frac{1}{2}x}+F(t))^2}{4-2t\varepsilon}, \\ s = \frac{-2t}{-2+\varepsilon t}, p(r,s) = \frac{2(-\frac{1}{2}e^{-\frac{1}{2}x})}{2\varepsilon t - \frac{1}{2}e^{-\frac{1}{2}x}(e^{-\frac{1}{2}x}+F(t))\varepsilon - 4} \end{array} \tag{54}$$

Specializing $F(t)$ by Equation (4) and supposing $F(t) = 0$ for simplicity of evaluations, we obtain

$$\begin{array}{c} r = \ln\left|\frac{2}{2-t\varepsilon}\right| + x + \frac{\varepsilon e^{-x}}{4-2t\varepsilon}, \\ s = \frac{-2t}{-2+\varepsilon t}, p(r,s) = \frac{2e^{-1/2x}}{2\varepsilon t + e^{-x}\varepsilon - 4} \end{array} \tag{55}$$

Solving two first equations of the Equation (55) with respect to x and t, we get

$$x = r + \mathrm{LambertW}\left(-\frac{1}{16}\frac{\varepsilon(\varepsilon s+2)^2}{e^r}\right) + \ln\left|2(\varepsilon s+2)^{-1}\right|, t = \frac{2s}{\varepsilon s+2}$$

Substituting these x, t into the third equation of system Equation (55) we receive after simplification the solution of Equation (50)

$$p(r,s) = -\frac{\frac{\sqrt{\varepsilon s+2}}{\sqrt{2}}e^{-1/2r}\left(1+\mathrm{LambertW}\left(-\frac{1}{16}\varepsilon(\varepsilon s+2)^2 e^{-r}\right)\right)^{-1}}{\sqrt{-\varepsilon^{-1}e^r \mathrm{LambertW}\left(-\frac{1}{16}\varepsilon(\varepsilon s+2)^2 e^{-r}\right)}}$$

Example 3. Let us apply Equation (46) to the solution $u(x,t) = \frac{t}{\sqrt{2t(x-\ln|t|)}}$. We choose for simplicity the arbitrary function appearing as a result of integration, equal to zero. The change of the independent variables has such a form:

$$\begin{array}{c} x = \ln\left|\frac{2s}{\varepsilon s+2}\right| + \frac{2r-2\ln|s|}{\varepsilon s+2}, \\ t = 2\frac{s}{\varepsilon s+2} \end{array}$$

After substitution of new independent variables into the expression

$$p(x,t) = \sqrt{\frac{t}{2(x - \ln|t|)}}$$

we finally get the same solution of the Equation (4)

$$p(r,s) = \sqrt{\frac{s}{2(r - \ln|s|)}}$$

Hence, this solution is an invariant solution with respect to the nonlocal transformation considered.

Equations (47) and (48) too allow generating new solutions of the Equation (4) if $p(r;s)$ is its known solution.

The operator X_4 generates two corresponding characteristic equations

$$vv_x - t = 0, \tag{56}$$

$$vu_x + u^2 = 0 \tag{57}$$

The first equation determines usual Lie point symmetry of Equation (7). Equation (57) determines potential symmetry of Equation (4). Both these equations are connected by the potential substitution Equation (20).

Solving Equation (57) with respect to v

$$v = -u^2 {u_x}^{-1}$$

and substituting this into Equation (56), we obtain the second order differential equation, determining nonlocal symmetry of Equation (4)

$$u_{xx}u^4 - 2{u_x}^2 u^3 + t{u_x}^3 = 0$$

and, consequently, appropriate nonlocal ansatzes for a given equation

$$u = \frac{-f_1(t) \pm \sqrt{f_1(t)^2 + 2tf_2(t) + 2tx}}{2(f_2(t) + x)}$$

Here $f_i, (i = 1, 2, 3)$ are the arbitrary functions of variable t. This ansatz can be used for construction of solutions to Equation (4).

Let us describe symmetry of Equation (4) corresponding to Lie symmetry Equation (56) of Equation (7). First we substitute $v_x = u$ into Equation (56)

$$vv_x - t = vu - t = 0$$

and solve it for $v(x,t)$. Differentiating the result with respect to x, we get

$$v_x + t \cdot u^{-2} u_x = 0$$

Note that if we wrote obtained above expression using Equation (20) in a form $u^3 + tu_x = 0$, we would make a mistake, because such symmetry is not admitted by Equation (4).

So, using substitution $vv_x - t = 0$ in the previous expression, we get

$$u^2 + vu_x = 0$$

(the second characteristic equation for X_4). Solving the last equation for $v(x,t)$ and differentiating it with respect to x, we obtain

$$v_x = -2u + u^2 u_x^{-2} \cdot u_{xx}$$

Using here potential substitution Equation (20) and simplifying result, we get corresponding nonlocal symmetry of the Equation (4)

$$u_{xx}u - 3u_x^2 = 0$$

Integrating this equation we find two ansatzes for Equation (4)

$$u = \pm \frac{1}{\sqrt{-2f_1(t)x - 2f_2(t)}}$$

They depend on arbitrary functions $f_1(t)$ and $f_2(t)$. Specializing them by Equation (4), we find two solutions

$$u = \pm \frac{\sqrt{t - c_2}}{\sqrt{2c_1 e^3 - 2\ln|-t + c_2| + 2(x + c_3)}}$$

Since Equation (4) is invariant with respect to the time and space translations generated by the operators X_1 and X_2 the last solution, Equation (38) and u-part of solution to the system Equations (39) and (40) represent the same exact solution.

5. Conclusions

Two given diffusion-convection equations, which allow connection by the generalized hodograph transformation are considered. It is shown in this paper that the intermediate equations possess the additional (with respect to given equations) operators of invariance Lie algebras. The appropriate additional operators allow to construct potential and nonlocal symmetries of the given differential equations. We have constructed the one-parameter groups of Lie symmetries generated by operators, admitted by the potential system Equations (15) and (16), and use them for construction of the corresponding formulae for generation of solutions of Equation (4). This involves a successive implementation of appropriate transformations $\mathcal{T}$ and the one-parameter group of point transformations associated with the corresponding infinitesimal operator admitted by the potential system. New formulae for generation of solutions have been constructed. Some of them are based on the nonlocal transformations that contain new independent variable determined by an auxiliary differential equation. This type of transformations allows interpretation as an example of the nonlocal transformation with additional variables. The obtained formulae allow transformation of a known solution of the given equation into another its solution. They have been used for construction of exact solutions, some of which are new. They are either obtained in explicit form or presented in a new parametrical representations, where a functional parameter is given implicitly. All found solutions can be extended to parametric families by means of the Lie symmetry transformations or by using other formulae for generation of solutions.

Acknowledgments: The author thanks the referees for a lot of helpful remarks and suggestions.

Conflicts of Interest: The author declares no conflict of interest.

References

1. Tychynin, V.A.; Petrova, O.V. Nonlocal symmetries and formulae for generation of solutions for a class of diffusion–convection equations. *J. Math. Anal. Appl.* **2011**, *382*, 20–33. [CrossRef]
2. Bluman, G.W.; Reid, G.J.; Kumei, S. New classes of symmetries for partial differential equations. *J. Math. Phys.* **1988**, *29*, 806–811. [CrossRef]
3. Bluman, G.W.; Kumei, S. *Symmetries and Differential Equations*; Springer: New York, NY, USA, 1989.
4. Anderson, I.M.; Kamran, N.; Olver, P.J. Internal, external and generalized symmetries. *Adv. Math.* **1993**, *100*, 53–100. [CrossRef]

5. Reyes, E.G. Nonlocal symmetries and the Kaup–Kupershmidt equation. *J. Phys. A Math. Gen.* **2005**, *46*, 073507:1–073507:19. [CrossRef]
6. Bluman, G.W.; Doran-Wu, P. The use of factors to discover potential systems or linearizations. *Acta Appl. Math.* **1995**, *41*, 21–43. [CrossRef]
7. Serov, M.; Omelyan, O.; Cherniha, R. Linearization of some systems of nonlinear diffusion equations by means of non-local transformations. *Dop. Nat. Acad. Ukr.* **2004**, *10*, 39–45.
8. Sophocleous, C. Potential symmetries of nonlinear diffusion-convection equations. *J. Phys. A Math. Gen.* **1996**, *29*, 6951–6959. [CrossRef]
9. Lisle, I.G. Equivalence Transformations for Classes of Differential Equations. Ph.D. Thesis, University of British Columbia, Vancouver, BC, Canada, 1992.
10. Lisle, I.G.; Reid, G.J. *Symmetry Classification Using Invariant Moving Frames*; ORCCA Technical Report TR-00-08; University of Western Ontario: Ontario, Canada, 1992.
11. Popovych, R.O.; Ivanova, N.M. Potential equivalence transformations for nonlinear diffusion—Convection equations. *J. Phys. A Math. Gen.* **2005**, *38*, 3145–3155. [CrossRef]
12. Anderson, R.L.; Ibragimov, N.H. *Lie-BÄcklund Transformations in Applications*; SIAM: Philadelphia, PA, USA, 1979.
13. Olver, P.J. *Applications of Lie Groups to Differential Equations*; Springer-Verlag: New York, NY, USA, 1993.
14. Bluman, G.W.; Cheviakov, A.; Anco, S. *Applications of Symmetry Methods to Partial Differential Equations*; Applied Mathematical Sciences; Springer: New York, NY, USA, 2010; Volume 168, p. 367.
15. Bluman, G.W.; Yang, Z.Z. A symmetry-based method for constructing nonlocally related PDE systems. *J. Math. Phys.* **2013**, *54*, 093504:1–093504:22. [CrossRef]
16. Fuschych, W.I.; Tychynin, V.A. *Preprint No 82.33*; Institut of Mathematics: Kiev, Ukraine, 1982.
17. Fuschych, W.I.; Tychynin, V.A. Exact solutions and superposition principle for nonlinear wave equation. *Proc. Acad. Sci. Ukr.* **1990**, *5*, 32–36.
18. Tychynin, V.A. Nonlocal Symmetries and Solutions for Some Classes of Nonlinear Equations of Mathematical Physics. Ph.D. Thesis, Prydniprovsk State Academy of Civil Engineering and Architecture, Dnipropetrovsk, Ukraine, 1994.
19. Tychynin, V.A. Non-local symmetry and generating solutions for Harry–Dym type equations. *J. Phys. A Math. Gen.* **1994**, *14*, 2787–2797. [CrossRef]
20. Tychynin, V.A.; Petrova, O.V.; Tertyshnyk, O.M. Symmetries and Generation of Solutions for Partial Differential Equations. *SIGMA* **2007**, *3*, 019:1–019:14. [CrossRef]
21. Rzeszut, W.; Vladimirov, V.; Tertyshnyk, O.M.; Tychynin, V.A. Linearizability and nonlocal superposition for nonlinear transport equation with memory. *Rep. Math. Phys.* **2013**, *72*, 235–252. [CrossRef]
22. Fushchych, W.I.; Serov, M.I.; Tychynin, V.A.; Amerov, T.K. On non-local symmetry of nonlinear heat equation. *Proc. Acad. Sci. Ukr.* **1992**, *11*, 27–33.
23. Storm, M.L. Heat conduction in simple metals. *J. Appl. Phys.* **1951**, *22*, 940–951. [CrossRef]
24. Bluman, G.W.; Kumei, S. On the remarkable nonlinear diffusion equation $\frac{\partial}{\partial x}\left(a(u+b)-2\frac{\partial u}{\partial x}\right)-\frac{\partial u}{\partial x}=0$. *J. Math. Phys.* **1980**, *21*, 1019–1023. [CrossRef]
25. King, J.R. Some non-local transformations between nonlinear diffusion equations. *J. Phys. A Math. Gen.* **1990**, *23*, 5441–5464. [CrossRef]
26. Cherniha, R.; Serov, M. Symmetries, Ansätze and Exact Solutions of Nonlinear Second-order Evolution Equations with Convection Terms, II. *Eur. J. Appl. Math.* **2006**, *17*, 597–605. [CrossRef]
27. Rosen, G. Nonlinear heat conduction in solid. *Phys. Rev. B* **1979**, *19*, 2398–2399. [CrossRef]
28. Cherniha, R.; Serov, M. Symmetries, Ansätze and Exact Solutions of Nonlinear Second-order Evolution Equations with Convection Terms. *Eur. J. Appl. Math.* **1998**, *9*, 527–542. [CrossRef]

symmetry

MDPI

Article

Bäcklund Transformations for Integrable Geometric Curve Flows

Changzheng Qu [1,*], Jingwei Han [2] and Jing Kang [3]

1 Department of Mathematics, Ningbo University, Ningbo 315211, China
2 School of Information Engineering, Hangzhou Dianzi University, Hangzhou 310018, China; jingweih@hdu.edu.cn
3 Department of Mathematics, Northwest University, Xi'an 710069, China; jingkang@nwu.edu.cn
* Author to whom correspondence should be addressed; quchangzheng@nbu.edu.cn; Tel.: +86-574-8760-9976.

Academic Editor: Roman M. Cherniha
Received: 1 May 2015; Accepted: 30 July 2015; Published: 3 August 2015

Abstract: We study the Bäcklund transformations of integrable geometric curve flows in certain geometries. These curve flows include the KdV and Camassa-Holm flows in the two-dimensional centro-equiaffine geometry, the mKdV and modified Camassa-Holm flows in the two-dimensional Euclidean geometry, the Schrödinger and extended Harry-Dym flows in the three-dimensional Euclidean geometry and the Sawada-Kotera flow in the affine geometry, *etc*. Using the fact that two different curves in a given geometry are governed by the same integrable equation, we obtain Bäcklund transformations relating to these two integrable geometric flows. Some special solutions of the integrable systems are used to obtain the explicit Bäcklund transformations.

Keywords: invariant geometric flow; Bäcklund transformation; integrable system; differential invariant

MSC: 37K35, 37K25, 53A55

1. Introduction

Bäcklund transformations are a powerful tool to explore various properties of integrable nonlinear partial differential equations [1,2]. They can be used to obtain more exact solutions of integrable systems from a particular solution. The classical Bäcklund transformations are local geometric transformations, which are used to construct surfaces of constant negative Gaussian curvature [1]. This provides a geometric construction of new pseudospherical surfaces from a particular solution of an integrable partial differential equation. Indeed, solutions of the sine-Gordon equation describe pseudospherical surfaces. Applying Bäcklund transformations n times to a particular solution of sine-Gordon equation, one can obtain a family of solutions of sine-Gordon equation. These solutions can be obtained using the Bianchi's permutability formula through purely algebraic means [2]. In [3], Chern and Tenenblat performed a complete classification to a class of nonlinear evolution equations which describe pseudospherical surfaces. It is noted that a nonlinear PDE describes pseudospherical surface if it admits $sl(2)$ prolongation structure. More generally, a Bäcklund transformation is typically a system of first-order partial differential equations relating two equations, and usually depending on an additional parameter. In particular, a Bäcklund transformation which relates solutions of the same equations is called an auto-Bäcklund transformation. In [4], Wahlquist and Estabrook [4] provides a systematic method to construct Bäcklund transformations of integrable systems by using the prolongation structure approach. Other effective methods to construct Bäcaklund transformations of integrable systems were also proposed in a number of literatures, see for example [2,5–12] and many more references.

A particular nice feature of integrable systems is their relationship with invariant geometric flows of curves and surfaces in certain geometries. Those flows are invariant with respect to the symmetry groups of the geometries [13]. A number of integrable equations have been shown to be related to motions of curves in Euclidean geometry, centro-equiaffine geometry, affine geometry, homogeneous manifolds and other geometries *etc.*, and many interesting results have been obtained [14–41]. Such relationship is helpful to explore geometric realization of several properties of integrable systems, for example, bi-Hamiltonian structure, recursion operator, Miura transformation and Bäcklund transformation *etc.* On the other hand, the topological properties of closed curves are shown to be related to the infinite number of symmetries and the associated sequence of invariants [11]. The relationship between integrable systems and geometric curve flows in $\mathbb{R}^3$ was studied in 1970s by Hasimoto [14], who showed that the integrable cubic Schrödinger equation is equivalent to the binormal motion flow of space curves in $\mathbb{R}^3$ (called vortex-filament flow or localized induction equation) by using a transformation relating the wave function of the Schrödinger equation to the curvature and torsion of curves (so-called Hasimoto transformation). Furthermore, using the Hasimoto transformation, Lamb [16] verified that the mKdV equation and the sine-Gordon equations arise from the invariant curve flows in $\mathbb{R}^3$. Marí-Beffa, Sanders and Wang [25] noticed that the Hasimoto transformation is a gauge transformation relating the Frenet frame and parallel frame. The well-known integrable equations including the KdV equation, the modified KdV equation, the Sawada-Kotera equation,the Kaup-Kuperschmidt equation and Boussinesq equation were also shown to arise from the invariant plane or space curve flows respectively in centro-equiaffine geometry [18,21,35,40], Euclidean geometry [15,17,21], two-dimensional affine geometry [21,40], projective geometry [37,39] and three-dimensional affine geometry [23].

In this paper, we are mainly concerned with Bäcklund transformations for integrable geometric curve flows in certain geometries. Our work is inspired by the following result.

Proposition 1.1. *[26] Let $\gamma(s)$ be a smooth curve of constant torsion τ in $\mathbb{R}^3$, parametrized by arclength s. Let* $\mathbf{T}$, $\mathbf{N}$ *and* $\mathbf{B}$ *be a Frenet frame, and $k(s)$ the curvature of γ. For any constant C, suppose $\beta = \beta(s, k(s), C)$ is a solution of the differential equation*

$$\tfrac{d\beta}{ds} = C \sin\beta - k. \tag{1}$$

then

$$\widetilde{\gamma}(s) = \gamma(s) + \frac{2C}{C^2 + \tau^2}(\cos\beta\mathbf{T} + \sin\beta\mathbf{N})$$

is a curve of constant torsion τ, also parametrized by arclength s.

Note that this transformation can be obtained by restricting the classical Bäcklund transformation for pseudospherical surfaces to the asymptotic lines of the surfaces with constant torsion.

We will restrict our attention to the geometric plane curve flows

$$\gamma_t = f\mathbf{N} + g\mathbf{T} \tag{2}$$

and space curve flows

$$\gamma_t = f\mathbf{T} + g\mathbf{N} + h\mathbf{B} \tag{3}$$

in Euclidean, centro-equiaffine and affine geometries, where $\mathbf{T}$ and $\mathbf{N}$ in Equation (2) denote frame vectors of planar curves, and $\mathbf{T}$, $\mathbf{N}$ and $\mathbf{B}$ in Equation (3) are frame vectors of spacial curves,f, g and h depend on the curvatures of the curves γ and their derivatives with respect to the arclength parameter, namely, these geometric flows are invariant with respect to the symmetry groups of the geometries.

For a planar or a spacial curve $\gamma(t,s)$ in a given geometry, let $\widetilde{\gamma}(t,s)$ be another curve related to γ through the following Bäcklund transformation

$$\widetilde{\gamma}(t,s) = \gamma(t,s) + \alpha(t,s)\mathbf{N} + \beta(t,s)\mathbf{T} \tag{4}$$

or

$$\widetilde{\gamma}(t,s) = \gamma + \alpha(t,s)\mathbf{T} + \beta(t,s)\mathbf{N} + \chi(t,s)\mathbf{B}. \tag{5}$$

Throughout the paper, we assume that both curve flows for γ and $\widetilde{\gamma}$ are governed by the same integrable system, that means the curvatures of the curves $\widetilde{\gamma}$ determined by the flows (4) or (5) satisfy the integrable systems as for the curves γ. It turns out that the functions $\alpha(t,s)$, $\beta(t,s)$ and $\chi(t,s)$ for space case and $\alpha(t,s)$ and $\beta(t,s)$ for planar case satisfy systems of nonlinear evolution equations. Solving these systems then yields Bäcklund transformations between the two flows for γ and $\widetilde{\gamma}$.

The outline of this paper is as follows. In Section 2, we first study the Bäcklund transformations of planar curve flows in $\mathbb{R}^2$, which include the modified KdV flow and the modified Camassa-Holm flow. Bäcklund transformations of integrable space curve flows in $\mathbb{R}^3$ including the Schrödinger flow and the extended Harry-Dym flow will be discussed in Section 3. In Section 4, we consider the Bäcklund transformations of the KdV and Camassa-Holm flows for planar curves in centro-equiaffine geometry. Finally in Section 5, we discuss the Bäcklund transformations of the Sawada-Kotera flow in two-dimensional affine geometry.

2. Bäcklund Transformations of Integrable Curve Flows in $\mathbb{R}^2$

The invariant geometric curve flows in $\mathbb{R}^2$ were discussed extensively from many points of view in the last three decades. A number of interesting results have been obtained. It was shown that the non-stretching plane curve flows in $\mathbb{R}^2$ are related closely to the integrable systems including the modified KdV equation [15,17,21] and the modified Camassa-Holm equation [42]. In this section, we consider the Bäcklund transformations of those integrable flows.

Let us consider the flows for planar curves in $\mathbb{R}^2$, governed by

$$\gamma_t = f\mathbf{n} + h\mathbf{t}, \tag{6}$$

where $\mathbf{t}$ and $\mathbf{n}$ denote the unit tangent and normal vectors of the curves, respectively, which satisfy the Serret-Frenet formulae

$$\mathbf{t}_s = k\mathbf{n}, \mathbf{n}_s = -k\mathbf{t}, \tag{7}$$

where k is the curvature of the curve γ, s is the arclength of the curve and $ds = g dp$, p is a free parameter. The velocities f and h in Equation (6) depend on k and it's derivatives with respect to the arclength parameter s. Let θ be the angle between the tangent and a fixed direction. Then $\mathbf{t} = (\cos\theta, \sin\theta)$, $\mathbf{n} = (-\sin\theta, \cos\theta)$, and $d\theta = k ds$. Based on the flow (6), it is easy to show that the time evolutions of those geometric invariants are given by [17]

$$\begin{aligned} \mathbf{t}_t &= (f_s + kh)\mathbf{n}, \\ \mathbf{n}_t &= -(f_s + kh)\mathbf{t}, \\ g_t &= g(h_s - kf), \end{aligned} \tag{8}$$

and

$$\begin{aligned} \theta_t &= (f_s + kh), \\ k_t &= \left(\frac{d\theta}{ds}\right)_t = f_{ss} + k_s h + k^2 f. \end{aligned} \tag{9}$$

Assume that the flow is intrinsic, namely the arclength does not depend on time. Then equation

$$h_s = kf \tag{10}$$

follows from Equation (8).

2.1. The Modified KdV Flow in $\mathbb{R}^2$

In [17], Goldstein and Petrich proved that the modified KdV equation arises from an non-stretching curve flow in Equation $\mathbb{R}^2$. Indeed, let $f = k_s$, $h = \frac{1}{2}k^2$ in Equation (6), then k satisfies the modified KdV equation

$$k_t = k_{sss} + \frac{3}{2}k^2 k_s. \tag{11}$$

The corresponding curve flow is

$$\gamma_t = k_s \mathbf{n} + \frac{1}{2}k^2 \mathbf{t}, \tag{12}$$

which is the so-called modified KdV flow [17].

Let $\widetilde{\gamma}$ be another curve in $\mathbb{R}^2$ related to γ by

$$\widetilde{\gamma}(t,s) = \gamma(t,s) + \alpha(t,s)\mathbf{n} + \beta(t,s)\mathbf{t}. \tag{13}$$

Assume that $\widetilde{\gamma}(t,s)$ is also governed by the modified KdV flow, namely it satisfies

$$\widetilde{\gamma}_t = \widetilde{k}_{\widetilde{s}}\widetilde{\mathbf{n}} + \frac{1}{2}\widetilde{k}^2\widetilde{\mathbf{t}}, \tag{14}$$

where $\widetilde{s}$ is the arclength parameter of $\widetilde{\gamma}$; $\widetilde{\mathbf{t}}$ and $\widetilde{\mathbf{n}}$ denote the unit tangent and normal vector of $\widetilde{\gamma}$, respectively. A direct computation shows that $\widetilde{\mathbf{t}}$ and $\widetilde{\mathbf{n}}$ are related to $\mathbf{t}$ and $\mathbf{n}$ by

$$\widetilde{\mathbf{t}} = \frac{F_0\mathbf{t} + G_0\mathbf{n}}{\sqrt{F_0^2 + G_0^2}}, \quad \widetilde{\mathbf{n}} = -\frac{G_0\mathbf{t} + F_0\mathbf{n}}{\sqrt{F_0^2 + G_0^2}}, \tag{15}$$

where

$$F_0 = 1 + \beta_s - k\alpha, G_0 = \alpha_s + k\beta. \tag{16}$$

It is inferred from Equation (13) that

$$\widetilde{\gamma}_t = \left[\alpha_t + k_s + \left(k_{ss} + \tfrac{1}{2}k^3\right)\beta\right]\mathbf{n} + \left[\beta_t + \tfrac{1}{2}k^2 - \left(k_{ss} + \tfrac{1}{2}k^3\right)\alpha\right]\mathbf{t}. \tag{17}$$

Differentiating Equation (13) with respect to s, after using the Serret-Frenet formulae (7), yields

$$\widetilde{\gamma}_s = ((1 + \beta_s - k\alpha)\mathbf{t} + \alpha_s + k\beta)\mathbf{n} := F_0\mathbf{t} + G_0\mathbf{n}. \tag{18}$$

It follows that the arclength $\widetilde{s}$ of $\widetilde{\gamma}$ is related to s of γ by

$$d\widetilde{s} = \sqrt{F_0^2 + G_0^2}ds. \tag{19}$$

Furthermore, differentiating Equation (18) with respect to $\widetilde{s}$, and using Equation (19) yields

$$\widetilde{k}\widetilde{\mathbf{n}} = \widetilde{\gamma}_{\widetilde{s}\widetilde{s}} = \left(F_0^2 + G_0^2\right)^{-2}(F_1\mathbf{t} + G_1\mathbf{n}) := F_2\mathbf{t} + G_2\mathbf{n}, \tag{20}$$

where

$$F_2 = \frac{F_1}{(F_0^2 + G_0^2)^2}, G_2 = \frac{G_1}{(F_0^2 + G_0^2)^2}.$$

and

$$\begin{aligned} F_1 = \; & -k(1 + \beta_s - k\alpha)^2(\alpha_s + k\beta) + (\beta_{ss} - k_s\alpha - 2k\alpha_s - k^2\beta)(\alpha_s + k\beta)^2 \\ & -(1 + \beta_s - k\alpha)(\alpha_s + k\beta)(\alpha_{ss} + k_s\alpha + k\alpha_s), \\ G_1 = \; & k(1 + \beta_s - k\alpha)^3 + k(1 + \beta_s - k\alpha)(\alpha_s + k\beta)^2(\alpha_{ss}k_s\alpha + k\alpha_s) \\ & -(1 + \beta_s - k\alpha)(\alpha_s + k\beta)(\beta_{ss} - k_s\alpha - k\alpha_s). \end{aligned}$$

From Equation (20), we also have

$$\begin{aligned} \widetilde{k}_{\widetilde{s}} &= \frac{F_0(G_{2s}+kF_2)-G_0(F_{2s}-kG_2)}{F_0^2+G_0^2} := F_3, \\ \widetilde{k}^2 &= -\frac{G_0(G_{2s}+kF_2)+F_0(F_{2s}-kG_2)}{F_0^2+G_0^2} := G_3. \end{aligned} \tag{21}$$

Substituting Equations (15), (17) and (21) into Equation (14), we see that the modified KdV flow is invariant with respect to the Bäcklund transformation (13) if and only if α and β satisfy the following system

$$\begin{aligned} \alpha_t + k_s + \beta\left(k_{ss}+\frac{1}{2}k^3\right) &= \frac{1}{\sqrt{F_0^2+G_0^2}}\left(\frac{1}{2}G_0G_3+F_0F_3\right), \\ \beta_t + \frac{1}{2}k^3 - \alpha\left(k_{ss}+\frac{1}{2}k^3\right) &= \frac{1}{\sqrt{F_0^2+G_0^2}}\left(\frac{1}{2}F_0G_3-G_0F_3\right). \end{aligned} \tag{22}$$

Theorem 2.1. *The modified KdV flow* (12) *is invariant with respect to the Bäcklund transformation* (13) *if $\alpha(t,s)$ and $\beta(t,s)$ satisfy the system* (22)*, where G_0, F_0, F_3 and G_3 are given in Equations* (16) *and* (21).

It is noticed that a class of Bäcklund transformations for smooth and discrete plane curves in Euclidean space governed by the modified KdV equation were discussed in [12], which are derived by using the Bäcklund transformations of the potential modified KdV equation.

2.2. The Modified Camassa-Holm Flow

The modified Camassa-Holm equation

$$m_t + \left((u^2-u_s^2)m\right)_s + au_{sss} = 0, m = u - u_{ss} \tag{23}$$

can be derived using the general approach of the tri-Hamiltonian duality from the modified KdV equation [42]. A direct consequence of such approach shows us that the modified Camassa-Holm equation is an integrable equation with bi-Hamiltonian structure. Interestingly, it has peaked solutions and can describe wave breaking phenomena [43]. It was also shown in [43] that the modified Camassa-Holm equation arises from a non-stretching planar curve flow in $\mathbb{R}^2$. Indeed, let $f = u_s$, $h = \frac{1}{2}(u^2 - u_s^2)$ in Equation (6), then the corresponding modified Camassa-Holm flow is

$$\gamma_t = u_s\mathbf{n} + \tfrac{1}{2}(u^2-u_s^2)\mathbf{t}, \tag{24}$$

where u satisfies the modified Camassa-Holm Equation (23) with $a = 1$, where $m = k = u - u_{ss}$ is the curvature of the curve γ. Denote $\Lambda = 1 - \partial_s^2$, then $u = \Lambda^{-1}k$. Assume that $\widetilde{\gamma}$ is another curve related to γ by Equation (13), a direct computation shows

$$\widetilde{\gamma}_t = (u_s + \alpha_t + \beta(f_s + kh))\mathbf{n} + \left(\tfrac{1}{2}(u^2-u_s^2) + \beta_t - \alpha(f_s+kh)\right)\mathbf{t}. \tag{25}$$

Using Equation (19), the corresponding geometric invariants of $\widetilde{\gamma}$ can be expressed in

$$\begin{aligned} \widetilde{\Lambda} &= 1 - \partial_{\widetilde{s}}^2 = 1 - \left(\frac{ds}{d\widetilde{s}}\partial_s\right)\left(\frac{ds}{d\widetilde{s}}\partial_s\right), \\ \widetilde{u} &= \left(1-\partial_{\widetilde{s}}^2\right)^{-1}\widetilde{k}, \\ \widetilde{u}_{\widetilde{s}} &= 1-\partial_{\widetilde{s}}^2)^{-1}\widetilde{k}_{\widetilde{s}}, \\ \widetilde{u}^2 - \widetilde{u}_{\widetilde{s}}^2 &= \left[\left(1-\partial_{\widetilde{s}}^2\right)^{-1}\left(\widetilde{k}+\widetilde{k}_{\widetilde{s}}\right)\right]\left[\left(1-\partial_{\widetilde{s}}^2\right)^{-1}\left(\widetilde{k}-\widetilde{k}_{\widetilde{s}}\right)\right], \end{aligned} \tag{26}$$

where $\widetilde{k} = \sqrt{F_1^2 + G_1^2}/(F_0^2 + G_0^2)^2$. Assume that $\widetilde{\gamma}$ is also governed by the modified Camassa-Holm flow (24), namely, it satisfies

$$\widetilde{\gamma}_t = \widetilde{u}_{\widetilde{s}}\widetilde{\mathbf{n}} + \tfrac{1}{2}(\widetilde{u}^2 - \widetilde{u}_{\widetilde{s}}^2)\widetilde{\mathbf{t}}. \tag{27}$$

Substituting Equations (15), (25) and (26) into Equation (27) and comparing the coefficients of **t** and **n** in the resulting equation, we arrive at the following system for $\alpha(t,s)$ and $\beta(t,s)$

$$\begin{aligned} \alpha_t + u_s + \beta\Big[u_{ss} + \tfrac{1}{2}(u - u_{ss})(u^2 - u_s^2)\Big] &= \frac{\widetilde{u}_{\widetilde{s}}F_0}{\sqrt{F_0^2+G_0^2}} + \frac{1}{2}\frac{(\widetilde{u}^2-\widetilde{u}_{\widetilde{s}}^2)}{\sqrt{F_0^2+G_0^2}}, \\ \beta_t + \tfrac{1}{2}(u^2 - u_s^2) - \alpha\Big[u_{ss} + \tfrac{1}{2}(u - u_{ss})(u^2 - u_s^2)\Big] &= -\frac{\widetilde{u}_{\widetilde{s}}G_0}{\sqrt{F_0^2+G_0^2}} + \frac{1}{2}\frac{(\widetilde{u}^2-\widetilde{u}_{\widetilde{s}}^2)}{\sqrt{F_0^2+F_0^2}}, \end{aligned} \tag{28}$$

where u satisfies the modified Camassa-Holm Equation (23) with $a = 1$. Consequently, we have the following result.

Theorem 2.2. *The modified Camassa-Holm flow* (24) *is invariant with respect to the Bäcklund transformation* (13) *if $\alpha(t,s)$ and $\beta(t,s)$ satisfy the system* (28)*, where G_0 and F_0 are given in Equation* (16).

3. Bäcklund Transformations for Space Curve Flows in $\mathbb{R}^3$

In this section, we consider the integrable flows for space curves in $\mathbb{R}^3$

$$\gamma_t = U\mathbf{n} + V\mathbf{b} + W\mathbf{t}, \tag{29}$$

where **t**, **n** and **b** are the tangent, normal and binormal vectors of the space curve γ, respectively. The velocities U, V and W depend on the curvature and torsion as well as their derivatives with respect to arclength s. It is well know that the vectors **t**, **n** and **b** satisfy the Serret-Frenet formulae

$$\begin{aligned} \mathbf{t}_s &= k\mathbf{n}, \\ \mathbf{n}_s &= -k\mathbf{t} + \tau\mathbf{b}, \\ \mathbf{b}_s &= -\tau\mathbf{n}, \end{aligned} \tag{30}$$

where k and τ are curvature and torsion of γ. Governed by the flow (29), the time evolutions of these geometric invariants fulfill [14,15]

$$\begin{aligned} \mathbf{t}_t &= \left(\frac{\partial U}{\partial s} - \tau V + kW\right)\mathbf{n} + \left(\frac{\partial V}{\partial s} + \tau U\right)\mathbf{b}, \\ \mathbf{n}_t &= -\left(\frac{\partial U}{\partial s} - \tau V + kW\right)\mathbf{t} + \left[\frac{1}{k}\frac{\partial}{\partial s}\left(\frac{\partial V}{\partial s} + \tau U\right) + \frac{\tau}{k}\left(\frac{\partial U}{\partial s} - \tau V + kW\right)\right]\mathbf{b}, \\ \mathbf{b}_t &= -\left(\frac{\partial V}{\partial s} + \tau U\right)\mathbf{t} - \left[\frac{1}{k}\frac{\partial}{\partial s}\left(\frac{\partial V}{\partial s} + \tau U\right) + \frac{\tau}{k}\left(\frac{\partial U}{\partial s} - \tau V + kW\right)\right]\mathbf{n}, \\ g_t &= g\left(\frac{\partial W}{\partial s} - kU\right), \end{aligned} \tag{31}$$

where $g = |\gamma_p|$ denotes the metric of the curve γ. A direct computation leads to the equations for the curvature k and torsion τ:

$$\begin{aligned} \frac{\partial \tau}{\partial t} &= \frac{\partial}{\partial s}\left[\frac{1}{k}\frac{\partial}{\partial s}\left(\frac{\partial V}{\partial s} + \tau U\right) + \frac{\tau}{k}\left(\frac{\partial U}{\partial s} - \tau V\right) + \tau\int^s kU ds'\right] + k\tau U + k\frac{\partial V}{\partial s}, \\ \frac{\partial k}{\partial t} &= \frac{\partial^2 U}{\partial s^2} + \left(k^2 - \tau^2\right)U + \frac{\partial k}{\partial s}\int^s kU ds' - 2\tau\frac{\partial V}{\partial s} - k\frac{\partial \tau}{\partial s}V. \end{aligned} \tag{32}$$

Assume that the flow is intrinsic, namely the arclength does not depend on time, it implies from Equation (31) that

$$\frac{\partial W}{\partial s} = kU. \tag{33}$$

From Equation (32), using the following Hasimoto transformation

$$\phi = k\eta, \eta = \exp\left[i \int^{s} \tau(t,s')ds'\right], \tag{34}$$

we get the equation for ϕ

$$\begin{aligned} \frac{\partial \phi}{\partial t} = \quad & \left(\frac{\partial^2}{\partial s^2} + |\phi|^2 + i\phi \int^s ds'\tau\phi^* + \frac{\partial \phi}{\partial s} \int^s ds'\phi^*\right)(U\eta), \\ & + \left(i\frac{\partial^2}{\partial s^2} + i|\phi|^2 + \phi \int^s ds'\tau\phi^* - i\phi \int^s ds'\frac{\partial \phi^*}{\partial s'}\right)(V\eta), \end{aligned} \tag{35}$$

where ϕ^* denotes the complex conjugate of ϕ.

Let $U = 0$, $V = k$ and $W = 0$. Then we derive from Equation (32) the Schrödinger equation

$$i\phi_t + \phi_{ss} + \frac{1}{2}\Big|\phi\Big|2\phi = 0. \tag{36}$$

Let $U = -k_s$, $V = -k\tau$. Then $W = -\frac{1}{2}k^2$, and ϕ satisfies the mKdV system [15]

$$\phi_t + \phi_{sss} + \frac{3}{2}\Big|\phi\Big|2\phi_s = 0.$$

We now consider the case of $U = W = 0$. Denote $\theta(t,s) = \int^s \tau(s',t)ds'$, $G = V\eta$. It follows from Equation (35) that ϕ satisfies the equation

$$i\phi_t + G_{ss} + \Big|\phi\Big|2G - \phi \int^s G(\cos\theta - i\sin\theta)k_{s'}ds' = 0. \tag{37}$$

Let $\widetilde{u} = k\cos\theta$, $\widetilde{v} = k\sin\theta$ and $G = G_1 + iG_2$. Then Equation (33) is separated into the two equations

$$\begin{aligned} \widetilde{u}_t &= -G_{2,ss} - \widetilde{v}\partial_s^{-1}[k(G_1\cos\theta + G_2\sin\theta)_s], \\ \widetilde{v}_t &= G_{1,ss} + \widetilde{u}\partial_s^{-1}[k(G_1\cos\theta + G_2\sin\theta)_s]. \end{aligned} \tag{38}$$

Furthermore, letting $\widetilde{u} = u + v_s$, $\widetilde{v} = v - u_s$ and choosing $V = \partial_s^{-1}\left[(u^2 + v_s^2)/k\right]$, we find that u and v satisfy the following system [41]

$$\begin{aligned} (u + v_s)_t &= -G_{2,ss} - (v - u_s)\left(u^2 + v^2\right), \\ (v - u_s)_t &= G_{1,ss} + (u + v_s)\left(u^2 + v^2\right), \end{aligned} \tag{39}$$

where $G_1 = 2\cos\theta\partial_s^{-1}(v\cos\theta - u\sin\theta)$, $G_2 = 2\sin\theta\partial_s^{-1}(v\cos\theta - u\sin\theta)$, which is related to the dual system of the Schrödinger equation [42].

3.1. The Schrödinger Flow

Corresponding to the Schrödinger Equation (36), the Schrödinger flow is given by [14]

$$\gamma_t = k\mathbf{b}. \tag{40}$$

In this case, the time evolution of frame vectors is governed by

$$\begin{aligned}\mathbf{t}_t &= -\tau k\mathbf{n} + k_s\mathbf{b},\\ \mathbf{n}_t &= -\left(\frac{k_{ss}}{k} - \tau^2\right)\mathbf{b} + \tau k\mathbf{t},\\ \mathbf{b}_t &= -k_s\mathbf{t} - \left(\frac{k_{ss}}{k} - \tau^2\right)\mathbf{n}.\end{aligned} \tag{41}$$

We now construct Bäcklund transformation of the Schrödinger flow (40)

$$\widetilde{\gamma} = \gamma + \alpha(t,s)\mathbf{t} + \beta(t,s)\mathbf{n} + \chi(t,s)\mathbf{b}, \tag{42}$$

where α, β and χ are the functions of t and s, to be determined. Using Equation (30), (40) and (41), a direct computation leads to

$$\widetilde{\gamma}_s = (1 + \alpha_s - \beta k)\mathbf{t} + (\beta_s + \alpha k - \chi\tau)\mathbf{n} + (\chi_s + \beta\tau)\mathbf{b}, \tag{43}$$

and

$$\begin{aligned}\widetilde{\gamma}_t = & (\alpha_t + \beta\tau k - \chi k_s)\mathbf{t} + \left[\beta_t - \chi\left(\tfrac{k_{ss}}{k} - \tau^2\right) - \alpha\tau k\right]\mathbf{n}\\ & + \left[\chi_t + k + \alpha k_s + \beta\left(\tfrac{k_{ss}}{k} - \tau^2\right)\right]\mathbf{b}.\end{aligned} \tag{44}$$

Then the arclength parameter $\widetilde{s}$ of curve $\widetilde{\gamma}$ is related to s by

$$d\widetilde{s} = \left|\widetilde{\gamma}_s\right| ds = \sqrt{(1 + \alpha_s - \beta k)^2 + (\beta_s + \alpha k - \chi\tau)^2 + (\chi_s + \beta\tau)^2} ds := F ds.$$

The tangent vector of the curve $\widetilde{\gamma}$ is determined by

$$\widetilde{\mathbf{t}} = \widetilde{\gamma}_s \frac{ds}{d\widetilde{s}} = A_1\mathbf{t} + A_2\mathbf{n} + A_3\mathbf{b},$$

where $A_1 = F^{-1}(1 + \alpha_s - \beta k)$, $A_2 = F^{-1}(\beta_s + \alpha k - \chi\tau)$, $A_3 = F^{-1}(\chi_s + \beta\tau)$. Further computation from Equation (43) yields

$$\widetilde{\gamma}_{\widetilde{s}\widetilde{s}} = \widetilde{\gamma}_{\widetilde{s}s}\frac{ds}{d\widetilde{s}} = \frac{A_{1s} - kA_2}{F}\mathbf{t} + \frac{A_{2s} + kA_1 - \tau A_3}{F}\mathbf{n} + \frac{A_{3s} + \tau A_2}{F}\mathbf{b},$$

which gives the curvature of $\widetilde{\gamma}$:

$$\widetilde{k} = \frac{\sqrt{(A_{1s} - kA_2)^2 + (A_{2s} + kA_1 - \tau A_3)^2 + (A_{3s} + \tau A_2)^2}}{F} := \frac{H}{F}. \tag{45}$$

Using the Serret-Frenet formulae, we obtain the normal and binormal vectors of $\widetilde{\gamma}$ given by

$$\begin{aligned}\widetilde{\mathbf{n}} &= \frac{A_{1s} - kA_2}{H}\mathbf{t} + \frac{A_{2s} + kA_1 - \tau A_3}{H}\mathbf{n} + \frac{A_{3s} + \tau A_2}{H}\mathbf{b} := B_1\mathbf{t} + B_2\mathbf{n} + B_3\mathbf{b},\\ \widetilde{\mathbf{b}} &= \frac{C_1\mathbf{t} + C_2\mathbf{n} + C_3\mathbf{b}}{\sqrt{C_1^2 + C_2^2 + C_3^2}},\end{aligned} \tag{46}$$

where $C_1 = F^{-1}(B_{1s} - kB_2 + HA_1)$, $C_2 = F^{-1}(kB_1 + B_{2s} - \tau B_3 + HA_2)$ and $C_3 = F^{-1}(\tau B_2 + B_{3s} + HA_3)$.

Assume that the curve $\widetilde{\gamma}$ also fulfills the Schrödinger flow, that is

$$\widetilde{\gamma}_t = \widetilde{k}\widetilde{\mathbf{b}}. \tag{47}$$

Plugging Equations (44), (45) and (46) into Equation (47), we arrive at the following result.

Theorem 3.1. *The Schrödinger flow (40) is invariant with respect to the Bäcklund transformation (42) if α, β and χ satisfy the system*

$$\begin{aligned} \alpha_t + \beta\tau k - \chi k_s &= \tfrac{H}{F}\frac{C_1}{\sqrt{C_1^2+C_2^2+C_3^2}}, \\ \beta_t - \chi\left(\tfrac{k_{ss}}{k} - \tau^2\right) - \alpha\tau k &= \tfrac{H}{F}\frac{C_2}{\sqrt{C_1^2+C_2^2+C_3^2}}, \\ \chi_t + k + \alpha k_s + \beta\left(\tfrac{k_{ss}}{k} - \tau^2\right) &= \tfrac{H}{F}\frac{C_3}{\sqrt{C_1^2+C_2^2+C_3^2}}. \end{aligned}$$

3.2. The Extended Harry-Dym Flow

The extended Harry-Dym flow [19]

$$\gamma_t = \tau^{-\frac{1}{2}}\mathbf{b}, \tag{48}$$

is obtained by setting $U = M = 0$, and $V = \tau^{-\frac{1}{2}}$ in the space curve flow (29). Here we consider the curve flow with constant curvature k. Let $k = 1$, it follows from Equation (32) that the torsion of γ satisfies the extended Harry-Dym equation [19]

$$\tau_t = \left[\left(\tau^{-\frac{1}{2}}\right)_{ss} - \tau^{\frac{3}{2}} + \tau^{-\frac{1}{2}}\right]_s, \tag{49}$$

which is equivalent to the flow (48). Making use of the transformation $v = \tau^{-1/2}$, we get the equation

$$\left(v^{-1}\right)_t = \tfrac{1}{2}\left(vv_{ss} - \tfrac{1}{2}v_s^2 + \tfrac{1}{2}v^2 - \tfrac{3}{2}v^{-2}\right)_s.$$

In terms of the change of variables $dx = \sqrt{2}v^{-1}ds + \frac{1}{\sqrt{2}}\left(vv_{ss} - \frac{1}{2}v_s^2 + \frac{1}{2}v^2 - \frac{3}{2}v^{-2}\right)dt$, it is deduced that

$$\frac{\partial v}{\partial t} = \frac{\partial}{\partial x}\left[v\left(\frac{\partial}{\partial x}\left(\frac{v_x}{v}\right) - \frac{1}{2}\left(\frac{v_x}{v}\right)^2\right) + \frac{1}{4}v^3 - \frac{3}{4}v^{-1}\right]. \tag{50}$$

Again we set $v = e^{\varphi}$, then it is inferred from Equation (50) that φ satisfies the Calogero's modified KdV equation

$$\varphi_t = \varphi_{xxx} - \frac{1}{2}\varphi_x^3 + \frac{3}{2}\varphi_x \cosh 2\varphi.$$

We now construct Bäcklund transformations to the extended Harry-Dym flow (48). In this case, the corresponding time evolution of frame vectors $\mathbf{t}$, $\mathbf{n}$ and $\mathbf{b}$ are given by

$$\begin{aligned} \mathbf{t}_t &= -\tau^{\frac{1}{2}}\mathbf{n} - \frac{1}{2}\tau^{-\frac{3}{2}}\tau_s\mathbf{b}, \\ \mathbf{n}_t &= \tau^{\frac{1}{2}}\mathbf{t} + \left(\left(\tau^{-\frac{1}{2}}\right)ss - \tau^{\frac{3}{2}}\right)\mathbf{b}, \\ \mathbf{b}_t &= \frac{1}{2}\tau^{-\frac{3}{2}}\tau_s\mathbf{t} - \left(\left(\tau^{-\frac{1}{2}}\right)_{ss} - \tau^{\frac{3}{2}}\right)\mathbf{n}. \end{aligned} \tag{51}$$

In terms of Equation (51), a direct computation gives

$$\begin{aligned} \widetilde{\gamma}_t = \quad & \left[\tau^{-\frac{1}{2}} + \alpha\left(\tau^{-\frac{1}{2}}\right)_s + \beta\left(\left(\tau^{-\frac{1}{2}}\right)_{ss} - \tau^{\frac{3}{2}}\right) + \chi_t\right]\mathbf{b}, \\ & + \left[\beta_t - \alpha\tau^{\frac{1}{2}} - \chi\left(\left(\tau^{-\frac{1}{2}}\right)_{ss} - \tau^{\frac{3}{2}}\right)\right]\mathbf{n} + \left[\alpha_t + \beta\tau^{\frac{1}{2}} + \chi\left(\tau^{-\frac{1}{2}}\right)_s\right]\mathbf{t}. \end{aligned} \tag{52}$$

Assume that a new curve $\widetilde{\gamma}(t,s)$ is governed by the extended Harry-Dym flow, that means $\widetilde{\gamma}$ satisfies

$$\widetilde{\gamma}_t = \widetilde{\tau}^{-\frac{1}{2}}\widetilde{\mathbf{b}}, \tag{53}$$

where $\widetilde{\tau}$ and $\widetilde{b}$ are the torsion and binormal vector of $\widetilde{\gamma}$, respectively, which is related to the geometric invariants of γ through

$$\begin{aligned} \widetilde{\tau} &= \frac{C_1(B_{1,s} - kB_2) + C_2(B_1 k + B_{2,s} - \tau B_3 + HA_2) + C_3(\tau B_2 + B_{3s} + HA_3)}{F\sqrt{C_1^2 + C_2^2 + C_3^2}}, \\ \widetilde{\mathbf{b}} &= \frac{C_1\mathbf{t} + C_2\mathbf{n} + C_3\mathbf{b}}{\sqrt{C_1^2 + C_2^2 + C_3^2}}. \end{aligned} \tag{54}$$

Plugging Equations (52) and (54) into Equation (53) implies that the extended Harry-Dym equation is invariant with respect to the Bäcklund transformation (42) if α, β and χ satisfy the following system

$$\begin{aligned} \alpha_t + \beta\tau^{\frac{1}{2}} + \chi\left(\tau^{-\frac{1}{2}}\right)_s &= \widetilde{\tau}^{-\frac{1}{2}}\frac{C_1}{\sqrt{C_1^2+C_2^2+C_3^2}}, \\ \beta_t - \alpha\tau^{\frac{1}{2}} - \chi\left(\left(\tau^{-\frac{1}{2}}\right)_{ss} - \tau^{\frac{3}{2}}\right) &= \widetilde{\tau}^{-\frac{1}{2}}\frac{C_2}{\sqrt{C_1^2+C_2^2+C_3^2}}, \\ \chi_t + \tau^{-\frac{1}{2}} + \alpha\left(\tau^{-\frac{1}{2}}\right)_s + \beta\left(\left(\tau^{-\frac{1}{2}}\right)_{ss} - \tau^{\frac{3}{2}}\right) &= \widetilde{\tau}^{-\frac{1}{2}}\frac{C_3}{\sqrt{C_1^2+C_2^2+C_3^2}}. \end{aligned}$$

4. Bäcklund Transformations of the KdV and Camassa-Holm Flows

Integrable curve flows in the centro-equiaffine geometry were discussed extensively in [21,24, 33,35,40]. It turns out that the KdV equation arises naturally from a non-stretching curve flow in centro-equiaffine geometry.

For a planar curve $\gamma(p)$ in the centro-equiaffine geometry, which satisfies $[\gamma, \gamma_p] \neq 0$, one can reparametrize it by the special parameter s satisfying $[\gamma, \gamma_s] = 1$, where the parameter s is said to be centro-equiaffine arclength. It follows that in terms of the free parameter p, the centro-equiaffine arclength is represented by

$$ds = [\gamma, \gamma_p]dp.$$

Furthermore, the centro-equiaffine curvature of the curve $\gamma(s)$ is defined to be

$$\phi = [\gamma_s, \gamma_{ss}].$$

Consider the planar curve flow in the centro-equiaffine geometry, specified by

$$\gamma_t = f\mathbf{N} + h\mathbf{T}, \tag{55}$$

where $\mathbf{N}$ and $\mathbf{T}$ are normal and tangent vectors of γ. One can compute the time evolution of $\mathbf{N}$ and $\mathbf{T}$ to get

$$\begin{pmatrix} \mathbf{T} \\ \mathbf{N} \end{pmatrix}_t = \begin{pmatrix} h_s - f & f_s + \phi h \\ -h & -f \end{pmatrix}\begin{pmatrix} \mathbf{T} \\ \mathbf{N} \end{pmatrix}. \tag{56}$$

The Serret-Frenet formulae for curves in centro-equiaffine geometry reads

$$\mathbf{T}_s = \phi\mathbf{N}, \mathbf{N}_s = -\mathbf{T}. \tag{57}$$

Assume that the flow is intrinsic, a direct computation shows that the curvature ϕ satisfies

$$\phi_t = (D_s^2 + 4\phi + 2\phi_s\partial^{-1})f. \tag{58}$$

Letting $f = \phi_s$ in Equation (58), we get the KdV equation

$$\phi_t = \phi_{sss} + 6\phi\phi_s. \tag{59}$$

The corresponding KdV flow is

$$\gamma_t = \phi_s \mathbf{N} + 2\phi \mathbf{T}, \tag{60}$$

which was introduced firstly by Pinkall [18]. Now we consider the Bäcklund transformation of the KdV flow (60)

$$\widetilde{\gamma}(t,s) = \gamma(t,s) + \alpha \mathbf{N} + \beta \mathbf{T}, \tag{61}$$

where α and β are functions of t and s.

We now construct the Bäcklund transformations of the KdV flow. Differentiating Equation (61) with respect to t and using Equation (60), we get

$$\widetilde{\gamma}_t = \left[\alpha_t + (1-\alpha)\phi_s + \beta\left(\phi_{ss} + 2\phi^2\right)\right]\mathbf{N} + [2(1-\alpha)\phi + \beta_t + \beta\phi_s]\mathbf{T}. \tag{62}$$

Assume that the curve $\widetilde{\gamma}$ is also governed by the KdV flow, namely it satisfies

$$\widetilde{\gamma}_t = \widetilde{\phi}_{\widetilde{s}} \widetilde{\mathbf{N}} + 2\widetilde{\phi}\widetilde{\mathbf{T}}, \tag{63}$$

where $\widetilde{s}$ is the arclength of $\widetilde{\gamma}$, which satisfies $d\widetilde{s} = (1-\alpha+\beta_s)ds$. In Equation (63), $\widetilde{\mathbf{T}}$ and $\widetilde{\mathbf{N}}$ are tangent and normal vectors of $\widetilde{\gamma}$, which are related to $\mathbf{T}$ and $\mathbf{N}$ through

$$\widetilde{\mathbf{T}} = \widetilde{\gamma}_{\widetilde{s}} = \widetilde{\gamma}_s \frac{ds}{d\widetilde{s}} = \mathbf{T} + \frac{\alpha_s + \beta\phi}{1-\alpha+\beta_s}\mathbf{N},$$

$$\widetilde{\mathbf{N}} = -\widetilde{\gamma} = -\beta\mathbf{T} + (1-\alpha)\mathbf{N}. \tag{64}$$

Further computation using Equation (62) leads to

$$\widetilde{\gamma}_{\widetilde{s}\widetilde{s}} = \widetilde{\gamma}_{\widetilde{s}s}\frac{ds}{d\widetilde{s}} = \frac{\phi + \left(\frac{\alpha_s+\beta\phi}{1-\alpha+\beta_s}\right)_s}{1-\alpha+\beta_s}\mathbf{N} - \frac{\alpha_s+\beta\phi}{(1-\alpha+\beta_s)^2}\mathbf{T}. \tag{65}$$

It follows from Equations (64) and (65) that the centro-equiaffine curvature of $\widetilde{\gamma}$ is given by

$$\widetilde{\phi} = [\widetilde{\gamma}_{\widetilde{s}}, \widetilde{\gamma}_{\widetilde{s}\widetilde{s}}] = \frac{\phi + \left(\frac{\alpha_s+\beta\phi}{1-\alpha+\beta_s}\right)_s}{1-\alpha+\beta_s} + \frac{\alpha_s+\beta\phi}{(1-\alpha+\beta_s)^3}. \tag{66}$$

Plugging Equations (64) and (66) into the right hand side of Equation (63), and comparing the coefficients of $\mathbf{T}$ and $\mathbf{N}$ with Equation (62), we deduce the following result.

Theorem 4.1. *The KdV flow is invariant with respect to the Bäcklund transformation* (61) *if α and β satisfy the system*

$$\begin{aligned} \alpha_t + (1-\alpha)\phi_s + \beta(\phi_{ss} + 2\phi^2) &= (1-\alpha)\widetilde{\phi}_{\widetilde{s}} + 2\tfrac{\alpha_s+\beta\phi}{1-\alpha+\beta_s}\widetilde{\phi}, \\ \beta_t - 2(1-\alpha)\phi + \beta\phi_s &= 2\widetilde{\phi} - \beta\widetilde{\phi}_{\widetilde{s}}, \end{aligned} \tag{67}$$

where $\widetilde{\phi}$ is determined by Equation (66).

Example 4.1. It is easy to see that $\phi = 0$ is a trivial solution of the KdV equation. Let $\phi = 0$, then

$$\widetilde{\phi} = \frac{\left(\frac{\alpha_s}{1-\alpha+\beta_s}\right)_s}{1-\alpha+\beta_s} + \frac{\alpha_s}{(1-\alpha+\beta_s)^3},$$

and system (67) becomes

$$\left\{ \begin{aligned} \alpha_t &= (1-\alpha)\left(\alpha_{\widetilde{s}\widetilde{s}} + \frac{\alpha_{\widetilde{s}}}{(1-\alpha+\beta_s)^2}\right)_{\widetilde{s}} + 2\alpha_{\widetilde{s}}\left(\alpha_{\widetilde{s}\widetilde{s}} + \frac{\alpha_{\widetilde{s}}}{(1-\alpha+\beta_s)^2}\right), \\ \beta_t &= 2\left(\alpha_{\widetilde{s}\widetilde{s}} + \frac{\alpha_{\widetilde{s}}}{(1-\alpha+\beta_s)^2}\right) - \beta\left(\alpha_{\widetilde{s}\widetilde{s}} + \frac{\alpha_{\widetilde{s}}}{(1-\alpha+\beta_s)^2}\right)_{\widetilde{s}}. \end{aligned} \right. \tag{68}$$

This is a third-order quasi-linear system, it is difficult to solve it. For simplicity, we seek its time-independent solutions: $\alpha = \alpha(\tilde{s})$, $\beta = \beta(\tilde{s})$. Denote

$$H = \alpha_{\tilde{s}\tilde{s}} + \frac{\alpha_{\tilde{s}}}{(1-\alpha+\beta_s)^2}.$$

Then system (68) reduces to

$$(1-\alpha)H_{\tilde{s}} + 2\alpha_{\tilde{s}}H = 0,$$
$$\beta H_{\tilde{s}} - 2H = 0.$$

Integrating it, we arrive at

$$H = c_0(1-\alpha)^2, \beta = \tfrac{\alpha-1}{\alpha_{\tilde{s}}}, \tag{69}$$

where $c_0 \neq 0$ is an integration constant. Employing the chain rule and $d\tilde{s} = (1-\alpha+\beta_s)ds.$, we have

$$\beta_s = (1-\alpha+\beta_s)\beta_{\tilde{s}}.$$

Solving it for β_s, we obtain

$$\beta_s = \tfrac{(1-\alpha)\beta_{\tilde{s}}}{1-\beta_{\tilde{s}}}.$$

A direct computation using Equation (69) yields

$$\beta_{\tilde{s}} = 1 + \tfrac{(1-\alpha)\alpha_{\tilde{s}\tilde{s}}}{\alpha_{\tilde{s}}^2}.$$

It follows from the above two equations that

$$1-\alpha+\beta_s = \tfrac{\alpha_{\tilde{s}}^2}{\alpha_{\tilde{s}\tilde{s}}}.$$

In terms of $\alpha_{\tilde{s}}$, H can be denoted as

$$H = \alpha_{\tilde{s}\tilde{s}} + \tfrac{\alpha_{\tilde{s}\tilde{s}}^2}{\alpha_{\tilde{s}}^3}.$$

Hence the first equation in Equation (69) becomes

$$\alpha_{\tilde{s}\tilde{s}} + \tfrac{\alpha_{\tilde{s}\tilde{s}}^2}{\alpha_{\tilde{s}}^3} - c_0(1-\alpha)^2 = 0. \tag{70}$$

Using the hodograph transformation

$$y = 1-\alpha(\tilde{s}), \tilde{s} = w(y),$$

we get the equation for $w(y)$

$$w_y^{-3}\left(w_{yy}^2 - w_{yy}\right) - c_0y^2 = 0.$$

This equation is reduced to the first-order ordinary differential equation

$$h^{-3}\left(h_y^2 - h_y\right) - c_0y^2 = 0$$

by setting $h = w_y$. Consequently, we derive a Bäcklund transformation (61) of the KdV flow (60), where $\alpha(\tilde{s})$ satisfies Equation (70) and $\beta(\tilde{s})= (\alpha\,(\tilde{s}-1)/\alpha_{\tilde{s}}(\tilde{s})$.

Next we consider the Bäcklund transformation of the Camassa-Holm flow. Let $f = v_s(t,s)$ and $g = 2v(t,s)$, $v = \left(1-\partial_s^2\right)^{-1}\phi$, then flow (55) becomes

$$\gamma_t = v_s\mathbf{N} + 2v\mathbf{T}, \tag{71}$$

which gives the Camassa-Holm equation [6,44]

$$v_t - v_{sst} + v_{sss} + 6vv_s - 4v_sv_{ss} - 2vv_{sss} = 0. \tag{72}$$

Therefore, Equation (71) is called the Camassa-Holm flow. Similar to the discussion for the modified Camassa-Holm equation, we have the following result.

Theorem 4.2. *The Camassa-Holm flow* (71) *admits the the Bäcklund transformation* (61) *if* $\alpha(t,s)$ *and* $\beta(t,s)$ *satisfy the system*

$$\begin{aligned} \alpha_t + (1-\alpha)v_s + \beta\left((1-2v)v_{ss} + 2v^2\right) &= (1-\alpha)\widetilde{v}_{\widetilde{s}} + 2\widetilde{v}G_1, \\ \beta_t + 2v(1-\alpha) + \beta v_s &= 2\widetilde{v}F_1 - \beta\widetilde{v}_{\widetilde{s}}, \end{aligned}$$

where $\widetilde{s}$ *is the arclength of* $\widetilde{\gamma}$*, determined by* $d\widetilde{s} = [(1-\alpha)(1-\alpha+\beta_s) + \beta(\alpha_s + \phi\beta)]ds$, $\widetilde{v} = \left(1-\partial_{\widetilde{s}}^2\right)^{-1}(F_1G_2 - G_1F_2)$, *with*

$$\begin{aligned} F_1 &= \frac{1-\alpha+\beta_s}{H}, G_1 = \frac{\alpha_s + \beta\phi}{H}, \\ F_2 &= \frac{F_{1,s} - G_1}{H}, G_2 = \frac{G_{1,s} + \phi F_1}{H}, \\ H &= (1-\alpha)(1-\alpha+\beta_s) + \beta(\alpha_s + \beta\phi). \end{aligned}$$

5. Bäcklund Transformations of the Sawada-Kotera Flow

Motions of curves in the affine geometry were discussed in [13,21,23,33,40]. It is well-known that the Sawada-Kotera equation arises from a non-stretching curve flow in affine geometry.

For a planar curve $\gamma(p)$ satisfying $\left[\gamma_p, \gamma_{pp}\right] \neq 0$ in affine geometry, we can reparametrize it by the special parameter s satisfying $[\gamma_s, \gamma_{ss}] = 1$, where the parameter s is said to be the arclength. So the affine arclength can be expressed by

$$ds = \left[\gamma_p, \gamma_{pp}\right]^{\frac{1}{3}}dp.$$

Consider the planar curve flow in affine geometry, governed by

$$\gamma_t = f\mathbf{N} + h\mathbf{T}, \tag{73}$$

where $\mathbf{N}$ and $\mathbf{T}$ are affine normal and tangent vectors of γ. The Serret-Frenet formulae for curves in affine geometry reads

$$\mathbf{T}_s = \mathbf{N}, \mathbf{N}_s = -\mu\mathbf{T}, \tag{74}$$

where μ is the curvature of the curve γ, defined by

$$\mu = \left[\gamma_{ss}, \gamma_{sss}\right]. \tag{75}$$

One can compute the time evolution of $\mathbf{N}$ and $\mathbf{T}$, to get

$$\begin{pmatrix} \mathbf{T} \\ \mathbf{N} \end{pmatrix}_t = \begin{pmatrix} h_s - \mu f & f_s + \mu h \\ H_1 & H_2 \end{pmatrix}\begin{pmatrix} \mathbf{T} \\ \mathbf{N} \end{pmatrix}, \tag{76}$$

where $H_1 = h_{ss} - 2\mu f_s - \mu h$, $H_2 = f_{ss} + 2h_s - \mu f$. Assume that the flow is intrinsic, that means the arclength does not depend on time. It is inferred from $\left[\frac{\partial}{\partial t}, \frac{\partial}{\partial s}\right] = 0$ that

$$h = -\tfrac{1}{3} f_s + \tfrac{2}{3}\partial_s^{-1}(\mu f).$$

A direct computation gives the equation for the curvature [21]

$$\mu_t = \tfrac{1}{3}\left(D_s^4 + 5\mu D_s^2 + 4\mu_s D_s + \mu_{ss} + 4\mu^2 + 2\mu_s \partial^{-1}\mu\right) f. \tag{77}$$

Letting $f = -3\mu_s$ in Equation (77), we obtain the Sawada-Kotera equation [45]

$$\mu_t + \mu_5 + 5\mu\mu_3 + 5\mu_1\mu_2 + 5\mu^2\mu_1 = 0. \tag{78}$$

The corresponding Sawada-Kotera flow is [21]

$$\gamma_t = -3\mu_s \mathbf{N} + (\mu_{ss} - \mu^2)\mathbf{T}. \tag{79}$$

We now consider the Bäcklund transformation of the Sawada-Kotera flow (79), determined by Equation (61), where $\mathbf{N}$ and $\mathbf{T}$ are respectively the affine normal and tangent of γ, $\alpha(t,s)$ and $\beta(t,s)$ depend on t and s.

Using the Serret-Frenet formulae (74) and the Sawada-Kotera flow (79), we first have

$$\begin{aligned} \widetilde{\gamma}_t = & \left[\alpha_t - 3\mu_s - (\mu_3 + \mu\mu_s)\alpha - (2\mu_{ss} + \mu^2)\beta\right]\mathbf{N} \\ & + \left[\beta_t + \mu_{ss} - \mu^2 + (\mu_4 + 3\mu\mu_2 + \mu_s^2 + \mu^2)\alpha + (\mu_3 + \mu\mu_s)\beta\right]\mathbf{T}. \end{aligned} \tag{80}$$

On the other hand, assume that the new curve $\widetilde{\gamma}$ is also governed by the Sawada-Kotera flow, which satisfies

$$\widetilde{\gamma}_t = -3\widetilde{\mu}_{\widetilde{s}}\widetilde{\mathbf{N}} + (\widetilde{\mu}_{\widetilde{s}\widetilde{s}} - \widetilde{\mu}^2)\widetilde{\mathbf{T}}, \tag{81}$$

where $\widetilde{s}$ is the arclength of $\widetilde{\gamma}$, defined by $d\widetilde{s} = [\widetilde{\gamma}_s, \widetilde{\gamma}_{ss}]^{\frac{1}{3}} ds$. In terms of the Sawada-Kotera flow, a direct computation yields

$$\begin{aligned} \widetilde{\gamma}_s &= (1 - \alpha\mu + \beta_s)\mathbf{T} + (\alpha_s + \beta)\mathbf{N} := F_1\mathbf{T} + F_2\mathbf{N}, \\ \widetilde{\gamma}_{ss} &= (F_{1,s} - \mu F_2)\mathbf{T} + (F_1 + F_{2,s})\mathbf{N} := F_3\mathbf{T} + F_4\mathbf{N}. \end{aligned} \tag{82}$$

Thus the arclength parameter of $\widetilde{\gamma}$ can be determined by

$$d\widetilde{s} = (F_1^2 + F_1F_{2,s} - F_2F_{1,s} + \mu F_2^2)^{\frac{1}{3}} ds := H_1 ds. \tag{83}$$

Using this and the flow (61), one can determine the tangent and normal vectors of $\widetilde{\gamma}$ by

$$\begin{aligned} \widetilde{\mathbf{T}} = & \tfrac{1}{H_1}(F_1\mathbf{T} + F_2\mathbf{N}), \\ \widetilde{\mathbf{N}} = & \tfrac{1}{H_1}\left[\left(\left(\tfrac{F_1}{H_1}\right)_s - \mu\tfrac{F_2}{H_1}\right)\mathbf{T} + \left(\tfrac{F_1}{H_1} + \left(\tfrac{F_2}{H_1}\right)_s\right)\mathbf{N}\right] \\ & := H_2\mathbf{T} + H_3\mathbf{N}. \end{aligned} \tag{84}$$

Thus the affine curvature of $\widetilde{\gamma}$ is

$$\widetilde{\mu} = H_2(H_2 + H_{3,s}) - H_3(H_{2,s} - \mu H_3) := H_4. \tag{85}$$

Further computation gives

$$\widetilde{\mu}_{\widetilde{s}} = \tfrac{H_{4,s}}{H_1} := H_5, \widetilde{\mu}_{\widetilde{s}\widetilde{s}} = \tfrac{1}{H_1}\left(\tfrac{H_{4,s}}{H_1}\right)_s := H_6.$$

It follows that

$$
\begin{aligned}
&-3\widetilde{\mu}_{\widetilde{s}}\widetilde{\mathbf{N}} + \left(\widetilde{\mu}_{\widetilde{s}\widetilde{s}} - \widetilde{\mu}^2\right)\widetilde{\mathbf{T}} \\
&= \left[-3H_2H_5 + \frac{F_1}{H_1}\left(H_6 - H_4^2\right)\right]\mathbf{T} + \left[-3H_3H_5 + \frac{F_2}{H_1}\left(H_6 - H_4^2\right)\right]\mathbf{N}.
\end{aligned}
$$

Hence we have proved the following result.

Theorem 5.1. *The Sawada-Kotera flow* (79) *is invariant with respect to the Bäcklund transformation* (61) *if α and β satisfy the system*

$$
\begin{aligned}
\alpha_t - (\mu_3 + \mu\mu_s)\alpha - 3\mu_s - (2\mu_{ss} + \mu^2)\beta &= -3H_3H_5 + \tfrac{F_2}{H_1}\left(H_6 - H_4^2\right), \\
\beta_t + (\mu_3 + \mu\mu_s)\beta + (\mu_4 + 3\mu\mu_{ss} + \mu_s^2 + \mu^2)\alpha + \mu_{ss} - \mu^2 &= -3H_2H_5 + \tfrac{F_1}{H_1}\left(H_6 - H_4^2\right).
\end{aligned}
$$

Acknowledgments: The work of Changzheng Qu is supported by the NSF of China (Grant No. 11471174) and NSF of Ningbo (Grant No. 2014A610018). The work of Han is supported by Zhejiang Provincial NSF of China (Grant No. LQ12A01002) and Development Project for Visitors at Universities at Zhejiang Province (Grant No. FX2012013). The work of Kang is supported partially by the NSF of China (Grant No. 11471260).

Author Contributions: Changzheng Qu proposed the idea to study Bäcklund transformations of geometric curve flows. He discussed Bäcklund transformations of geometric curve flows in $\mathbb{R}^2$ and centro-equiaffine geometries, and provided detailed proofs to Theorem 2.1, 2.2, 4.1 and 4.2. Jingwei Han studied Bäcklund transformations of geometric curve flows in $\mathbb{R}^3$, and provided a detailed proof to Theorem 3.1, and carried out computation on the extended Harry-Dym flow. Jing Kang studied Bäcklund transformations of geometric curve flows in affine geometry, and provided a detailed proof to Theorem 5.1. Example 4.1 is given by Changzheng Qu and Jing Kang. Introduction is prepared by Changzheng Qu and Jingwei Han.

Conflicts of Interest: The authors declare no conflict of interest.

References

1. Bäcklund, A.V. *Concerning Surfaces with Constant Negative Curvature*; Coddington, E.M., Translator; New Era Printing Co.: Lancaster, PA, USA, 1905.
2. Rogers, C.; Schief, W.K. *Bäcklund and Darboux Transformations Geometry and Modern Applications in Soliton Theory*; Cambridge University Press: Cambridge, UK, 2002.
3. Chern, S.S.; Tenenblat, K. Pseudospherical surfaces and evolution equations. *Stud. Appl. Math.* **1986**, *74*, 55–83.
4. Wahlquist, H.D.; Estabrook, F.B. Prolongation structures of nonlinear evolution equations. *J. Math. Phys.* **1975**, *16*, 1–7. [CrossRef]
5. Hirota, R. *The Direct Method in Soliton Theory*; Translated from the 1992 Japanese original and edited by Nagal, A.; Nimmo, J.; Gilson, C. With foreword by Hietarinta, J. and Nimmo, J.; Cambridge University Press: Cambridge, UK, 2004.
6. Fuchssteiner, B.; Fokas, A.S. Symplectic structures, their Bäcklund transformations and hereditary symmetries. *Phys. D* **1981/1982**, *4*, 47–66. [CrossRef]
7. Chern, S.S.; Terng, C.L. An analogue of Bäcklund theorem in affine geometry. *Rocky Montain Math. J.* **1980**, *10*, 105–124. [CrossRef]
8. Tenenblat, K.; Terng, C.L. Bäcklund theorem for n-dimensional submanifolds of $\mathbb{R}^{2n-1}$. *Ann. Math.* **1980**, *111*, 477–490. [CrossRef]
9. Terng, C.L.; Uhlenbeck, K. Bäcklund transformations and loop group actions. *Commun. Pur. Appl. Math.* **2000**, *53*, 1–75. [CrossRef]
10. Hirota, R.; Hu, X.B.; Tang, X.Y. A vector potential KdV equation and vector Ito equation: Soliton solutions, bilinear Bäcklund transformations and Lax-pairs. *J. Math. Anal. Appl.* **2003**, *288*, 326–348. [CrossRef]
11. Calini, A.; Ivey, T. Bäcklund transformations and knots of constant torsion. *J. Knot Theor. Ramf.* **1998**, *7*, 719–746. [CrossRef]
12. Inoguchi, J.; Kajiwara, K.; Matsuura, N.; Ohta, Y. Motion and Backlund transformations of discrete plane curves. *Kyushu J. Math.* **2012**, *66*, 303–324. [CrossRef]

13. Olver, P.J. *Equivalence, Invariants, and Symmetry*; Cambridge University Press: Cambridge, UK, 1995.
14. Hasimoto, H. A soliton on a vortext filament. *J. Fluid Mech.* **1972**, *51*, 477–485. [CrossRef]
15. Nakayama, K.; Segur, H.; Wadati, M. Integrability and the motion of curves. *Phys. Rev. Lett.* **1992**, *16*, 2603–2606. [CrossRef]
16. Lamb, G.L. Soliton on moving space curves. *J. Math. Phys.* **1977**, *18*, 1654–1661. [CrossRef]
17. Goldstein, R.E.; Petrich, D.M. The Korteweg-de Vries hierarchy as dynamics of closed curves in the plane. *Phys. Rev. Lett.* **1991**, *67*, 3203–3206. [CrossRef] [PubMed]
18. Pinkall, U. Hamiltonian flows on the space of star-shaped curves. *Result. Math.* **1995**, *27*, 328–332. [CrossRef]
19. Schief, W.K.; Rogers, C. Binormal motion of curves of constant curvature and torsion. Generation of soliton surfaces. *R. Soc. Lond. Proc. Ser. A Math. Phys. Eng. Sci.* **1999**, *455*, 3163–3188. [CrossRef]
20. Ivey, T. Integrable geometric evolution equations for curves. *Contemp. Math.* **2001**, *285*, 71–84.
21. Chou, K.S.; Qu, C.Z. Integrable equations arising from motions of plane curves. *Phys. D* **2002**, *162*, 9–33. [CrossRef]
22. Chou, K.S.; Qu, C.Z. Integrable equations arising from motions of plane curves II. *J. Nonlinear Sci.* **2003**, *13*, 487–517. [CrossRef]
23. Chou, K.S.; Qu, C.Z. Integrable motions of space curves in affine geometry. *Chaos Solitons Fractals* **2002**, *14*, 29–44. [CrossRef]
24. Chou, K.S.; Qu, C.Z. The KdV equation and motion of plane curves. *J. Phys. Soc. Jan.* **2001**, *70*, 1912–1916. [CrossRef]
25. Marí Beffa, G.; Sanders, J.A.; Wang, J.P. Integrable systems in three-dimensional Riemannian geometry. *J. Nonlinear Sci.* **2002**, *12*, 143–167. [CrossRef]
26. Calini, A.; Ivey, T. Finite-gap solutions of the vortex filament equation genus one solutions and symmetric solutions. *J. Nonlinear Sci.* **2005**, *15*, 321–361. [CrossRef]
27. Marí Beffa, G. Hamiltonian evolution of curves in classical affine geometries. *Phys. D* **2009**, *238*, 100–115. [CrossRef]
28. Marí Beffa, G. Bi-Hamiltonian flows and their realizations as curves in real semisimple homogeneous manifolds. *Pac. J. Math.* **2010**, *247*, 163–188. [CrossRef]
29. Marí Beffa, G.; Olver, P.J. Poisson structure for geometric curve flows in semi-simple homogeneous spaces. *Regul. Chaotic. Dyn.* **2010**, *15*, 532–550. [CrossRef]
30. Anco, S.C. Bi-Hamitonian operators, integrable flows of curves using moving frames and geometric map equations. *J. Phys. A Math. Gen.* **2006**, *39*, 2043–2072. [CrossRef]
31. Anco, S.C. Hamitonian flows of curves in G/SO(N) and hyperbolic/evolutionary vector soliton equations. *SIGMA* **2006**, *2*, 1–17.
32. Anco, S.C. Group-invariant solution equations and bi-Hamitonian geometric curve flows in Riemannian symplectic spaces. *J. Geom. Phys.* **2008**, *58*, 1–27. [CrossRef]
33. Marí Beffa, G. Hamiltonian evolutions of curves in classical affine geometries. *Phys. D* **2009**, *238*, 100–115. [CrossRef]
34. Olver, P.J. Invariant submanifld flows. *J. Phys. A* **2008**, *41*. [CrossRef]
35. Calini, A.; Ivey, T.; Marí Beffa, G. Remarks on KdV-type flows on star-shaped curves. *Phys. D* **2009**, *238*, 788–797. [CrossRef]
36. Wo, W.F.; Qu, C.Z. Integrable motions of curves in $\mathbb{S}^1 \times \mathbb{R}$. *J. Geom. Phys.* **2007**, *57*, 1733–1755. [CrossRef]
37. Li, Y.Y.; Qu, C.Z.; Shu, S.C. Integrable motions of curves in projective geometries. *J. Geom. Phys.* **2010**, *60*, 972–985. [CrossRef]
38. Song, J.F.; Qu, C.Z. Integrable systems and invariant curve flows in centro-equiaffine symplectic geometry. *Phys. D* **2012**, *241*, 393–402. [CrossRef]
39. Musso, E. Motion of curves in the projective plane introducing the Kaup-Kuperschmidt hierarchy. *SIGMA* **2012**, *8*. [CrossRef]
40. Calini, A.; Ivey, T.; Marí Beffa, G. Integrable flows for starlike curves in centroaffine space. *SIGMA* **2013**, *9*. [CrossRef]
41. Qu, C.Z.; Song, J.F.; Yao, R.X. Multi-Component Intergrable Systems and Invariant Curve Flow in Certain Geometries. *SIGMA* **2013**, *9*. [CrossRef]
42. Olver, P.J.; Rosenau, P. Tri-Hamiltonian duality between solitons and solitary-wave solutions having compact support. *Phys. Rev. E* **1996**, *53*, 1900–1906. [CrossRef]

43. Gui, G.L.; Liu, Y.; Olver, P.J.; Qu, C.Z. Wave-breaking and peakons for a modified Camassa-Holm equation. *Commun. Math. Phys.* **2013**, *319*, 731–759. [CrossRef]
44. Camassa, R.; Holm, D.D. An integrable shallow water equation with peaked solitons. *Phys. Rev. Lett.* **1993**, *71*, 1661–1664. [CrossRef] [PubMed]
45. Sawada, K.; Kotera, T. A method for finding N-soliton solutions of the KdV equation and KdV-like equation. *Prog. Theor. Phys.* **1974**, *51*, 1335–1367. [CrossRef]

symmetry

MDPI

Article

On the Incompleteness of Ibragimov's Conservation Law Theorem and Its Equivalence to a Standard Formula Using Symmetries and Adjoint-Symmetries

Stephen C. Anco

Department of Mathematics and Statistics, Brock University, St. Catharines, ON L2S3A1, Canada; sanco@brocku.ca

Academic Editor: Roman M. Cherniha
Received: 1 November 2016; Accepted: 17 February 2017; Published: 27 February 2017

Abstract: A conservation law theorem stated by N. Ibragimov along with its subsequent extensions are shown to be a special case of a standard formula that uses a pair consisting of a symmetry and an adjoint-symmetry to produce a conservation law through a well-known Fréchet derivative identity. Furthermore, the connection of this formula (and of Ibragimov's theorem) to the standard action of symmetries on conservation laws is explained, which accounts for a number of major drawbacks that have appeared in recent work using the formula to generate conservation laws. In particular, the formula can generate trivial conservation laws and does not always yield all non-trivial conservation laws unless the symmetry action on the set of these conservation laws is transitive. It is emphasized that all local conservation laws for any given system of differential equations can be found instead by a general method using adjoint-symmetries. This general method is a kind of adjoint version of the standard Lie method to find all local symmetries and is completely algorithmic. The relationship between this method, Noether's theorem and the symmetry/adjoint-symmetry formula is discussed.

Keywords: conservation law; symmetry; adjoint-symmetry; Fréchet derivative identity; Ibragimov's theorem

---

1. Introduction

The most well-known method for finding conservation laws of differential equations (DEs) is Noether's theorem [1], which is applicable to any system of one or more DEs admitting a variational formulation in terms of a Lagrangian. Noether's theorem shows that every local symmetry preserving the variational principle of a given variational system yields a non-trivial local conservation law. Moreover, for variational systems that do not possess any differential identities, every non-trivial local conservation law arises from some local symmetry that preserves the variational principle.

However, there are many physically and mathematically interesting DEs that are not variational systems, and this situation has motivated much work in the past few decades to look for some generalization of Noether's theorem which could be applied to non-variational DEs. One direction of work has been to replace the need for a variational principle by introducing some other structure, but still making use of the local symmetries of a given DE system to produce local conservation laws. In fact, a general formula is available that yields local conservation laws from local symmetries combined with solutions of the adjoint of the symmetry determining equations. This formula first appears (to the knowledge of the author) in a 1986 paper by Caviglia [2] and was later derived independently in a 1990 Russian paper by Lunev [3], as well as in a 1997 paper by the author and Bluman [4]. In the latter paper, solutions of the adjoint of the symmetry determining equations were called adjoint-symmetries; these solutions are also known as cosymmetries in the literature on integrable systems [5]. Essentially the same formula appears in a more abstract form in the cohomological framework for finding conservation laws, summarized in References [6–8].

In recent years, a similar conservation law formula has been popularized by Ibragimov [9–13] and subsequently extended by others [14–17], where a "nonlinear self-adjointness" condition is required to hold for the given DE system. However, in several papers [17–19], this formula sometimes is seen to produce only trivial conservation laws, and sometimes, the formula does not produce all admitted conservation laws. Furthermore, in a number of papers [14–17,20–22], the use of translation symmetries is mysteriously avoided, and other more complicated symmetries are used instead, without explanation.

The purpose of the present paper is to make several relevant remarks:

(1) Ibragimov's conservation law formula is a simple re-writing of a special case of the earlier formula using symmetries and adjoint-symmetries;
(2) Ibragimov's "nonlinear self-adjointness" condition in its most general form is equivalent to the existence of an adjoint-symmetry for a general DE system and reduces to the existence of a symmetry in the case of a variational DE system;
(3) this formula does not always yield all admitted local conservation laws, and it produces trivial conservation laws whenever the symmetry is a translation and the adjoint-symmetry is translation-invariant;
(4) the computation to find adjoint-symmetries (and, hence, to apply the formula) is just as algorithmic as the computation of local symmetries;
(5) most importantly, if all adjoint-symmetries are known for a given DE system (whether or not it has a variational formulation), then they can be used directly to obtain all local conservation laws, providing a kind of generalization of Noether's theorem to general DE systems.

All of these remarks have been pointed out briefly in Reference [23], and Remark (2) has been discussed in References [16,17], but it seems worthwhile to give a comprehensive discussion for all of the remarks (1)–(5), with examples, as the formula continues to be used in recent papers when a complete, general method for finding all local conservation laws is available instead. In particular, for any given DE system, a full generalization of the content of Noether's theorem is provided by a direct method using adjoint-symmetries, based on the framework shown in References [24,25] and presented in an algorithmic fashion in References [4,26–28]. In the case when a DE system is variational, adjoint-symmetries reduce to symmetries, and the direct method reproduces the relationship between symmetries and conservation laws in Noether's theorem, but without the need for a Lagrangian. A detailed review and further development of this general method appears in Reference [29]. Consequently, there is no need for any kind of special methods to find local conservation laws, just as there is no need to use special methods to find local symmetries, because the relevant determining equations can be solved in a direct algorithmic manner.

The remainder of the present paper is organized as follows. Remarks (1) and (2) will be demonstrated in Section 2. Remark (3), along with some further consequences and properties related to the action of symmetries, will be explained in Section 3. Remarks (4) and (5) will be briefly discussed in Section 4. Throughout, the class of DEs $u_{tt} - u_{xx} + a(u)(u_t^2 - u_x^2) + b(u)u_t + c(u)u_x + m(u) = 0$ will be used as a running example to illustrate the main points, and the notation in Ibragimov's work will be used to allow the simplest possible comparison of the results. Some concluding remarks are made in Section 5.

Many examples of conservation laws of wave equations and other evolution equations can be found in References [24,30,31] and the references therein.

2. Symmetries, Adjoint-Symmetries and "Nonlinear Self-Adjointness"

As preliminaries, a few basic tools from variational calculus will be reviewed. Let $x = (x^1, \ldots, x^n)$ be $n \geq 1$ independent variables and $u = (u^1, \ldots, u^m)$ be $m \geq 1$ dependent variables, and let $\partial^k u$

denote all k-th order partial derivatives of u with respect to x. Introduce an index notation for the components of x and u: x^i, $i = 1, \ldots, n$; and u^α, $\alpha = 1, \ldots, m$. In this notation, the components of $\partial^k u$ are given by $u^\alpha_{i_1 \cdots i_k}$, $\alpha = 1, \ldots, m$, $i_q = 1, \ldots, n$, with $q = 1, \ldots, k$. Summation is assumed over each pair of repeated indices in any expression. The coordinate space $J = (x, u, \partial u, \partial^2 u, \ldots)$ is called the jet space associated with the variables x, u. A *differential function* is a locally smooth function of finitely many variables in J. Total derivatives with respect to x applied to differential functions are denoted $D_i = \frac{\partial}{\partial x^i} + u^\alpha_i \frac{\partial}{\partial u^\alpha} + u^\alpha_{ij} \frac{\partial}{\partial u^\alpha_j} + \cdots$.

The necessary tools that will now be introduced are the Fréchet derivative and its adjoint derivative, the Euler operator and its product rule, and the Helmholtz conditions.

Given a set of $M \geq 1$ differential functions $f_a(x, u, \partial u, \ldots, \partial^N u)$, $a = 1, \ldots, M$, with differential order $N \geq 1$, the *Fréchet derivative* is the linearization of the functions as defined by

$$
\begin{aligned}
(\delta_w f)_a &= \Big(\frac{\partial}{\partial \epsilon} f_a(x, u + \epsilon w, \partial(u + \epsilon w), \ldots, \partial^N(u + \epsilon w))\Big)\Big|_{\epsilon=0} \\
&= w^\alpha \frac{\partial f_a}{\partial u^\alpha} + w^\alpha_i \frac{\partial f_a}{\partial u^\alpha_i} + \cdots + w^\alpha_{i_1 \cdots i_N} \frac{\partial f_a}{\partial u^\alpha_{i_1 \cdots i_N}}.
\end{aligned} \tag{1}
$$

This linearization can be viewed as a local directional derivative in jet space, corresponding to the action of a generator $\hat{\mathbf{X}} = w^\alpha \partial_{u^\alpha}$ in characteristic form, $\hat{\mathbf{X}}(f) = \delta_w f$, where $w = (w^1(x, u, \partial u, \ldots, \partial^k u), \ldots, w^m(x, u, \partial u, \ldots, \partial^k u))$ is a set of m arbitrary differential functions.

It is useful also to view the Fréchet derivative as a linear differential operator acting on w. Then, integration by parts defines the *Fréchet adjoint derivative*

$$
(\delta^*_v f)_\alpha = v^a \frac{\partial f_a}{\partial u^\alpha} - D_i\Big(v^a \frac{\partial f_a}{\partial u^\alpha_i}\Big) + \cdots + (-1)^N D_{i_1} \cdots D_{i_N}\Big(v^a \frac{\partial f_a}{\partial u^\alpha_{i_1 \cdots i_N}}\Big) \tag{2}
$$

which is a linear differential operator acting on a set of $M \geq 1$ arbitrary differential functions $v = (v^1(x, u, \partial u, \ldots, \partial^k u), \ldots, v^M(x, u, \partial u, \ldots, \partial^k u))$.

These two derivatives (1) and (2) are related by

$$
v^a(\delta_w f)_a - w^\alpha(\delta^*_v f)_\alpha = D_i \Psi^i(w, v; f) \tag{3}
$$

where the associated vector $\Psi^i(v, w; f)$ is given by the explicit formula

$$
\begin{aligned}
\Psi^i(w, v; f) = {}& w^\alpha v^a \frac{\partial f_a}{\partial u^\alpha_i} + (D_j w^\alpha) v^a \frac{\partial f_a}{\partial u^\alpha_{ji}} - w^\alpha D_j\Big(v^a \frac{\partial f_a}{\partial u^\alpha_{ji}}\Big) + \cdots \\
&+ \sum_{q=1}^{N} (-1)^{q-1} (D_{j_1} \cdots D_{j_{N-q}} w^\alpha) D_{i_1} \cdots D_{i_{q-1}}\Big(v^a \frac{\partial f_a}{\partial u^\alpha_{j_1 \cdots j_{N-q} i_1 \cdots i_{q-1} i}}\Big).
\end{aligned} \tag{4}
$$

The *Euler operator* E_{u^α}, or variational derivative, is defined in terms of the Fréchet derivative through the variational relation

$$
\delta_w f = w^\alpha E_{u^\alpha}(f) + D_i \Phi^i(w; f) \tag{5}
$$

which is obtained from integration by parts, yielding

$$
E_{u^\alpha}(f) = \frac{\partial f}{\partial u^\alpha} - D_i\Big(\frac{\partial f}{\partial u^\alpha_i}\Big) + \cdots + (-1)^N D_{i_1} \cdots D_{i_N}\Big(\frac{\partial f}{\partial u^\alpha_{i_1 \cdots i_N}}\Big) = \frac{\delta f}{\delta u^\alpha} \tag{6}
$$

where

$$
\Phi^i(w; f) v = \Psi^i(w, v; f). \tag{7}
$$

Here, for simplicity, $f(x, u, \partial u, \ldots, \partial^N u)$ is a single differential function. In particular, an explicit formula for $\Phi^i(w; f)$ is given by

$$\Phi^i(w;f) = w^\alpha \frac{\partial f}{\partial u^\alpha_i} + (D_j w^\alpha)\frac{\partial f}{\partial u^\alpha_{ji}} - w^\alpha D_j \frac{\partial f}{\partial u^\alpha_{ji}} + \cdots + \sum_{q=1}^{N} (-1)^{q-1} (D_{j_1} \cdots D_{j_{N-q}} w^\alpha) D_{i_1} \cdots D_{i_{q-1}} \frac{\partial f}{\partial u^\alpha_{j_1 \cdots j_{N-q} i_1 \cdots i_{q-1} i}} \tag{8}$$

from expression (4).

The Euler operator (6) has the following three important properties: First, it obeys the product rule

$$E_{u^\alpha}(fg) = (\delta^*_g f)_\alpha + (\delta^*_f g)_\alpha. \tag{9}$$

Second, its kernel

$$E_{u^\alpha}(f) = 0 \tag{10}$$

is given by total divergences

$$f = D_i F^i \tag{11}$$

holding for some differential vector function F^i. Third, its image consists of differential functions

$$E_{u^\alpha}(f) = g_\alpha \tag{12}$$

characterized by the Helmholtz conditions

$$(\delta_w g)_\alpha = (\delta^*_w g)_\alpha \tag{13}$$

where w^α is a set of arbitrary differential functions.

There are several common alternative notations for the Fréchet derivative and its adjoint: $\delta_w f = f'(w)$ and $\delta^*_v f = f'^*(v)$ appear in the literature on integrable systems and in Reference [29]; $\delta_w f = D_w f$ and $\delta^*_v f = D^*_v f$ are used in Olver's book [24]; $\delta_w f = L[u]w$ is used in the early work of Anco and Bluman [4,26,27] and in the book [31]. In contrast, Ibragimov [9,12] uses $\delta^*_v f = f^*[u, v]$.

2.1. Conservation Laws and Symmetries

Consider an N-th-order system of $M \geq 1$ DEs

$$F = (F_1(x, u, \partial u, \ldots, \partial^N u), \ldots, F_M(x, u, \partial u, \ldots, \partial^N u)) = 0. \tag{14}$$

The space of solutions $u(x)$ of the system will be denoted $\mathcal{E}$. When the number of independent variables x is $n = 1$, each DE is an ordinary differential equation (ODE), whereas when the number of independent variables x is $n \geq 2$, each DE is a partial differential equation (PDE). The number, m, of dependent variables u need not be the same as the number, M, of DEs in the system.

A *local infinitesimal symmetry* [24,31,32] of a given DE system (14) is a generator

$$\mathbf{X} = \xi^i(x, u, \partial u, \ldots, \partial^r u)\partial/\partial x^i + \eta^\alpha(x, u, \partial u, \ldots, \partial^r u)\partial/\partial u^\alpha \tag{15}$$

whose prolongation leaves invariant the DE system

$$\mathrm{pr}\mathbf{X}(F)|_{\mathcal{E}} = 0 \tag{16}$$

which holds on the whole solution space $\mathcal{E}$ of the system. (In this determining equation, the notation $\mathcal{E}$ means that the given DE system, as well as its differential consequences, are to be used). The differential functions ξ^i and η^α in the symmetry generator are called the *symmetry characteristic functions*. When acting on the solution space $\mathcal{E}$, an infinitesimal symmetry generator can be formally exponentiated to produce a one-parameter group of transformations $\exp(\epsilon \text{pr}\mathbf{X})$, with parameter ϵ, where the infinitesimal transformation is given by

$$u^\alpha(x) \to u^\alpha(x) + \epsilon\big(\eta^\alpha(x, u(x), \partial u(x), \dots, \partial^r u(x)) - u^\alpha_i(x)\xi^i(x, u(x), \partial u(x), \dots, \partial^r u(x))\big) + O(\epsilon^2) \tag{17}$$

for all solutions $u(x)$ of the DE system.

Two infinitesimal symmetries are equivalent if they have the same action (17) on the solution space $\mathcal{E}$ of a given DE system. An infinitesimal symmetry is thereby called *trivial* if it leaves all solutions $u(x)$ unchanged. This occurs iff its characteristic functions satisfy the relation

$$\eta^\alpha|_\mathcal{E} = (u^\alpha_i \xi^i)|_\mathcal{E}. \tag{18}$$

The corresponding generator (15) of a trivial symmetry is thus given by

$$\mathbf{X}_{\text{triv.}}|_\mathcal{E} = \xi^i \partial/\partial x^i + \xi^i u^\alpha_i \partial/\partial u^\alpha \tag{19}$$

which has the prolongation $\text{pr}\mathbf{X}_{\text{triv.}}|_\mathcal{E} = \xi^i D_i$. Conversely, any generator of this form on the solution space $\mathcal{E}$ represents a trivial symmetry. Thus, any two generators that differ by a trivial symmetry are equivalent. The differential order of an infinitesimal symmetry is defined to be the smallest differential order among all equivalent generators.

Any symmetry generator is equivalent to a generator given by

$$\hat{\mathbf{X}} = \mathbf{X} - \mathbf{X}_{\text{triv.}} = P^\alpha \partial/\partial u^\alpha, \quad P^\alpha = \eta^\alpha - \xi^i u^\alpha_i, \tag{20}$$

under which u is infinitesimally transformed while x is invariant, due to the relation

$$\text{pr}\mathbf{X} - \text{pr}\hat{\mathbf{X}} = \xi^i D_i. \tag{21}$$

This generator (20) defines the *characteristic form* for the infinitesimal symmetry. The symmetry invariance (16) of the DE system can then be expressed by

$$\text{pr}\hat{\mathbf{X}}(F)|_\mathcal{E} = 0 \tag{22}$$

holding on the whole solution space $\mathcal{E}$ of the given system. Note that the action of $\text{pr}\hat{\mathbf{X}}$ is the same as a Fréchet derivative (1), and hence, an equivalent, modern formulation [24,29,31] of this invariance (22) is given by the *symmetry determining equation*

$$(\delta_P F)_a|_\mathcal{E} = 0. \tag{23}$$

(Recall, the notation $\mathcal{E}$ means that the given DE system, as well as its differential consequences, are to be used in these determining equations.)

In jet space J, a group of transformations $\exp(\epsilon \text{pr}\mathbf{X})$ with a non-trivial generator $\mathbf{X}$ in general will not act in a closed form on x, u and derivatives $\partial^k u$ up to a finite order, except [24,31] for point transformations acting on (x, u) and contact transformations acting on $(x, u, \partial u)$. Moreover, a contact transformation is a prolonged point transformation when the number of dependent variables is

$m > 1$ [24,31]. A *point symmetry* is defined as a symmetry transformation group on (x, u), whose generator is given by characteristic functions of the form

$$\mathbf{X} = \xi(x,u)^i \partial/\partial x^i + \eta^\alpha(x,u)\partial/\partial u^\alpha \tag{24}$$

corresponding to the infinitesimal point transformation

$$x^i \to x^i + \epsilon\, \xi^i(x,u) + O(\epsilon^2), \quad u^\alpha \to u^\alpha + \epsilon\, \eta^\alpha(x,u) + O(\epsilon^2). \tag{25}$$

Likewise, a *contact symmetry* is defined as a symmetry transformation group on $(x, u, \partial u)$ whose generator corresponds to an infinitesimal transformation that preserves the contact relations $u^\alpha_i = D_i u^\alpha$. The set of all admitted point symmetries and contact symmetries for a given DE system comprises its group of *Lie symmetries*. The corresponding generators of this group comprise a Lie algebra [24,31,32].

A *local conservation law* of a given DE system (14) is a divergence equation

$$D_i C^i|_{\mathcal{E}} = 0 \tag{26}$$

which holds on the whole solution space $\mathcal{E}$ of the system, where

$$C = (C^1(x,u,\partial u,\ldots,\partial^r u),\ldots,C^n(x,u,\partial u,\ldots,\partial^r u)) \tag{27}$$

is the *conserved current vector*. In the case when one of the independent variables represents a time coordinate and the remaining $n-1$ independent variables represent space coordinates, namely $x = (t, x^1, \ldots, x^{n-1})$, then $C^1 = T$ is a *conserved density* and $(C^2, \ldots, C^n) = \vec{X}$ is a *spatial flux vector*, while the conservation law has the form of a local continuity equation $(D_t T + \mathrm{Div}\vec{X})|_{\mathcal{E}} = 0$. (Similarly to the symmetry determining equation, the notation $\mathcal{E}$ here means that the given DE system, as well as its differential consequences, are to be used).

A conservation law (26) is *locally trivial* if

$$C^i|_{\mathcal{E}} = D_j \Theta^{ij} \tag{28}$$

holds for some differential antisymmetric tensor function $\Theta^{ij}(x,u,\partial u,\ldots,\partial^{r-1}u)$ on $\mathcal{E}$, since any total curl is identically divergence free, $D_i(D_j\Theta^{ij}) = D_i D_j \Theta^{ij} = 0$ due to the commutativity of total derivatives. Two conservation laws are said to be *locally equivalent* if, on the solution space $\mathcal{E}$, their conserved currents differ by a locally trivial current (28). The *differential order of a conservation law* is defined to be the smallest differential order among all locally equivalent conserved currents. (Sometimes a local conservation law is itself defined as the equivalence class of locally equivalent conserved currents).

For a given DE system (14), the set of all non-trivial local conservation laws (up to local equivalence) forms a vector space on which the local symmetries of the system have a natural action [24,31,33]. In particular, the infinitesimal action of a symmetry (15) on a conserved current (27) is given by [24]

$$C^i_{\mathbf{X}} = \mathrm{pr}\mathbf{X}(C^i) + C^i D_i \xi^i - C^j D_j \xi^i. \tag{29}$$

When the symmetry is expressed in characteristic form (20), its action has the simple form

$$C^i_{\hat{\mathbf{X}}} = \mathrm{pr}\hat{\mathbf{X}}(C^i) = \delta_P C^i. \tag{30}$$

The conserved currents $C^i_{\mathbf{X}}$ and $C^i_{\hat{\mathbf{X}}}$ are locally equivalent,

$$(C^i_{\hat{\mathbf{X}}} - C^i_{\mathbf{X}})|_{\mathcal{E}} = D_j \Theta^{ij} \tag{31}$$

with

$$\Theta^{ij} = \xi^i C^j - \xi^j C^i \tag{32}$$

which follows from the relation (21).

A DE system is *variational* if it arises as the Euler–Lagrange equations of a local Lagrangian. This requires that the number of equations in the system is the same as the number of dependent variables, $M = m$, and that the differential order N of the system is even, in which case the system is given by

$$F_\alpha = E_{u^\alpha}(L), \quad \alpha = 1, \ldots, M = m \tag{33}$$

where the Lagrangian is a differential function

$$L(x, u, \partial u, \ldots, \partial^{N/2} u). \tag{34}$$

The necessary and sufficient conditions [24,29,31,32] for a given DE system (14) to be variational consist of the Helmholtz conditions (13), which are given by

$$(\delta_w F)_\alpha = (\delta^*_w F)_\alpha \tag{35}$$

where w^α is a set of arbitrary differential functions. Note that these conditions (35) are required to hold identically in jet space J (and not just on the solution space $\mathcal{E}$ of the DE system).

2.2. Ibragimov's Conservation Law Formula

The starting point is the well-known observation [24] that any N-th-order system of $M \geq 1$ DEs (14) can be embedded into a larger system by appending an "adjoint variable" for each DE in the system, where this set of $M \geq 1$ variables $v = (v^1, \ldots, v^M)$ is taken to satisfy the adjoint of the linearization of the original DE system. Specifically, the enlarged DE system is given by

$$F_a(x, u, \partial u, \ldots, \partial^N u) = 0, \quad a = 1, \ldots, M \tag{36}$$

$$(\delta^*_v F)_\alpha = F^*_\alpha(x, u, v, \partial u, \partial v, \ldots, \partial^N u, \partial^N v) = 0, \quad \alpha = 1, \ldots, m \tag{37}$$

for $u^\alpha(x)$ and $v^a(x)$, in Ibragimov's notation. This system (36)–(37) comprises the Euler–Lagrange equations of the Lagrangian function

$$L = v^a F_a(x, u, \partial u, \ldots, \partial^N u) \tag{38}$$

since, clearly,

$$E_{v^a}(L) = F_a, \quad E_{u^\alpha}(L) = (\delta^*_v F)_\alpha \tag{39}$$

through the product rule (9).

All solutions $u(x)$ of the original DE system (36) give rise to solutions of the Euler–Lagrange system (39) by letting $v(x)$ be any solution (for instance $v = 0$) of the DEs (37). Conversely, all solutions $(u(x), v(x))$ of the Euler–Lagrange system (39) yield solutions of the original DE system (36) by projecting out $v(x)$.

This embedding relationship can be used to show that every symmetry of the original DE system (36) can be extended to a variational symmetry of the Euler–Lagrange system (39). The proof is simplest when the symmetries are formulated in characteristic form (20).

Let

$$\hat{\mathbf{X}} = P^\alpha(x, u, \partial u, \ldots, \partial^r u)\partial/\partial u^\alpha \tag{40}$$

be any local symmetry generator (in characteristic form) admitted by the DE system (36). Under some mild regularity conditions [29] on the form of these DEs, the symmetry determining Equation (23) implies that the characteristic functions P^α satisfy

$$(\delta_P F)_a = R_P(F)_a \tag{41}$$

where

$$R_P = R_P{}^b_a + R_P{}^{bi}_a D_i + R_P{}^{bij}_a D_i D_j + \cdots + R_P{}^{bi_1\cdots i_r}_a D_{i_1} \cdots D_{i_r} \tag{42}$$

is some linear differential operator whose coefficients $R_P{}^b_a, R_P{}^{bi}_a, \ldots, R_P{}^{bi_1\cdots i_r}_a$ are differential functions that are non-singular on solution space $\mathcal{E}$ of the DE system (14). Now, consider the action of this symmetry generator (40) on the Lagrangian (38). From the operator relation (41) followed by integration by parts, the symmetry action is given by

$$\mathrm{pr}\hat{\mathbf{X}}(L) = v^a R_P(F)_a = F_a R^*_P(v)^a + D_i \hat{\Theta}^i \tag{43}$$

where

$$R^*_P = R^{*b}_{Pa} - R^{*bi}_{Pa} D_i + R^{*bij}_{Pa} D_i D_j + \cdots + (-1)^r R^{*bi_1\cdots i_r}_{Pa} D_{i_1} \cdots D_{i_r} \tag{44}$$

is the adjoint of the operator (42), with the non-singular coefficients

$$\begin{aligned}
R^{*b}_{Pa} &= R_P{}^b_a - (D_j R_P{}^{bj}_a) + \cdots + (-1)^r (D_{j_1} \cdots D_{j_r} R_P{}^{bj_1\cdots j_r}_a),\\
R^{*bi}_{Pa} &= R_P{}^{bi}_a - \tbinom{2}{1}(D_j R_P{}^{bji}_a) + \cdots + (-1)^{r-1}\tbinom{r}{r-1}(D_{j_1} \cdots D_{j_{r-1}} R_P{}^{bj_1\cdots j_{r-1}i}_a),\\
R^{*bij}_{Pa} &= R_P{}^{bij}_a - \tbinom{3}{1}(D_k R_P{}^{bkij}_a) + \cdots + (-1)^{r-2}\tbinom{r}{r-2}(D_{j_1} \cdots D_{j_{r-2}} R_P{}^{bj_1\cdots j_{r-2}ij}_a),\\
&\vdots\\
R^{*bi_1\cdots i_r}_{Pa} &= R_P{}^{bi_1\cdots i_r}_a D_{i_1} \cdots D_{i_r}.
\end{aligned} \tag{45}$$

Although the Lagrangian is not preserved, the expression (43) for the symmetry action shows that if the symmetry is extended to act on v via

$$\hat{\mathbf{X}}^{\text{ext.}} = P^\alpha \partial/\partial u^\alpha - R^*_P(v)^a \partial/\partial v^a, \tag{46}$$

then, under this extended symmetry, the Lagrangian will be invariant up to a total divergence,

$$\mathrm{pr}\hat{\mathbf{X}}^{\text{ext.}}(L) = v^a R_P(F)_a - F_a R^*_P(v)^a = D_i \hat{\Theta}^i. \tag{47}$$

This completes the proof. A useful remark is that the vector $\hat{\Theta}^i$ in the total divergence (47) is a linear expression in terms of F_a (and total derivatives of F_a), and hence, this vector vanishes whenever $u(x)$ is a solution of the DE system (36). Consequently, $\hat{\Theta}^i$ is a trivial current for the Euler–Lagrange system (39).

Some minor remarks are that the proof given by Ibragimov [9] does not take advantage of the simplicity of working with symmetries in characteristic form and also glosses over the need for some regularity conditions on the DE system so that the symmetry operator relation (41) will hold. Moreover, that proof is stated only for DE systems in which the number of equations is the same as the number of dependent variables, $M = m$.

Now, since the extended symmetry (46) is variational, Noether's theorem can be applied to obtain a corresponding conservation law for the Euler–Lagrange system (39), without the need for any additional conditions. The formula in Noether's theorem comes from applying the variational identity (5) to the Lagrangian (38), which yields

$$\mathrm{pr}\hat{\mathbf{X}}(L) = \phi^a F_a + v^a(\delta_P F)_a = \hat{\mathbf{X}}(v^a)E_{v^a}(L) + \hat{\mathbf{X}}(u^\alpha)E_{u^\alpha}(L) + D_i\Phi^i(P;L) \tag{48}$$

for any generator

$$\hat{\mathbf{X}} = P^\alpha \partial/\partial u^\alpha + \phi^a \partial/\partial v^a. \tag{49}$$

The total divergence term $D_i\Phi^i(P;L)$ is given by the formula (8) derived using the Euler operator (6). This yields

$$\begin{aligned}\Phi^i(P;L) = {} & P^\alpha v^a \frac{\partial F_a}{\partial u^\alpha_i} + (D_j P^\alpha) v^a \frac{\partial F_a}{\partial u^\alpha_{ji}} - P^\alpha D_j\Big(v^a \frac{\partial F_a}{\partial u^\alpha_{ji}}\Big) + \cdots \\ & + \sum_{q=1}^{N} (-1)^{q-1} (D_{j_1}\cdots D_{j_{N-q}} P^\alpha) D_{i_1}\cdots D_{i_{q-1}}\Big(v^a \frac{\partial F_a}{\partial u^\alpha_{j_1\cdots j_{N-q} i_1 \cdots i_{q-1} i}}\Big).\end{aligned} \tag{50}$$

When this variational identity (48) is combined with the action (47) of the variational symmetry (46) on the Lagrangian, the following Noether relation is obtained:

$$D_i(\hat{\Psi}^i - \Phi^i(P;L) = \phi^a F_a + P^\alpha F^*_\alpha, \quad \phi^a = -R^*_P(v)^a \tag{51}$$

where F^*_α is expression (37). Since F_a, F^*_α and $\hat{\Psi}^i$ vanish when $(u(x), v(x))$ is any solution of the Euler–Lagrange system (39), the Noether relation (51) yields a local conservation law

$$D_i \hat{C}^i|_{\mathcal{E}(u,v)} = 0, \quad \hat{C}^i = \Phi^i(P;L) \tag{52}$$

where $\mathcal{E}(u,v)$ denotes the solution space of the system (39) (including its differential consequences). This conservation law is locally equivalent to the conservation law formula underlying Ibragimov's work [9,12], which is given by

$$D_i C^i|_{\mathcal{E}(u,v)} = 0, \quad C^i = \hat{C}^i - \xi^i L \tag{53}$$

where $C^i|_{\mathcal{E}(u)} = \hat{C}^i|_{\mathcal{E}(u)}$ since $L|_{\mathcal{E}(u)} = 0$. Strangely, nowhere does Ibragimov (or subsequent authors) point out that the term $\xi^i L$ in the conserved current trivially vanishes on all solutions $(u(x), v(x))$ of the Euler–Lagrange system!

Hence, the following result has been established.

Proposition 1. *Any DE system* (36) *can be embedded into a larger Euler–Lagrange system* (39) *such that every symmetry* (40) *of the original system can be extended to a variational symmetry* (46) *of the Euler–Lagrange system. Noether's theorem then yields a conservation law* (52) *for all solutions* $(u(x), v(x))$ *of the Euler–Lagrange system* (39).

A side remark is that the locally equivalent conservation law (53) also can be derived from Noether's theorem if the extended symmetry (46) is expressed in canonical form

$$\mathbf{X}^{\text{ext.}} = \xi^i \partial/\partial x^i + \eta^\alpha \partial/\partial u^\alpha + (\xi^i v^a_i - R^*_P(v)^a)\partial/\partial v^a \tag{54}$$

as obtained from relations (20)–(21). In particular, the corresponding form of the variational identity (48) becomes

$$\mathrm{pr}\mathbf{X}^{\text{ext.}}(L) + (D_i\xi^i)L = \hat{\mathbf{X}}(v^a)E_{v^a}(L) + \hat{\mathbf{X}}(u^\alpha)E_{u^\alpha}(L) + D_i(\xi^i L + \Phi^i(P;L)) \tag{55}$$

where $P^\alpha = \eta^\alpha - \xi^i u^\alpha_i$ and $Q^a = -R^*_P(v)^a$, while the action of the symmetry (54) on the Lagrangian is given by

$$\begin{aligned}\mathrm{pr}\mathbf{X}^{\text{ext.}}(L) + (D_i\xi^i)L &= (D_i(\xi^i v^a) - R^*_P(v)^a)F_a + v^a \mathrm{pr}\mathbf{X}^{\text{ext.}}(F^a) \\ &= (D_i(\xi^i v^a) - R^*_P(v)^a)F_a + v^a(R_P(F)^a + \xi^i D_i F^a) = D_i\Theta^i\end{aligned} \tag{56}$$

where Θ^i vanishes whenever $u(x)$ is a solution of the DE system (36). Hence, the Noether relation obtained from combining Equations (55) and (56) yields the conserved current $C^i = \Phi^i(P, v; F) - \xi^i L$ modulo the locally trivial current Θ^i. If the original symmetry (40) being used is a point symmetry, then this trivial current Θ^i can be shown to vanish identically, which is the situation considered in Ibragimov's papers [9,12] and in nearly all subsequent applications in the literature.

2.3. "Nonlinear Self-Adjointness"

The conservation law (52) holds for all solutions $(u(x), v(x))$ of the Euler–Lagrange system (39). It seems natural to restrict this to solutions of the original DE system (36) for $u(x)$ by putting $v = 0$. However, the resulting conserved current is trivial, $\Phi^i(P; L)|_{v=0} = \Phi^i(P; 0) = 0$, because L is a linear expression in terms of v. Consequently, some other way must be sought to project the solution space $\mathcal{E}(u, v)$ of the Euler–Lagrange system onto the solution space $\mathcal{E}$ of the original DE system (36).

Ibragimov's first paper [9] proposes to put $v = u$, which is clearly a significant restriction on the form of the original DE system (36). In particular, this requires that $F^*_\alpha|_{v=u} = F_a$ hold identically, where the DE system is assumed to have the same number of equations as the number of dependent variables, $M = m$, which allows the indices $a = \alpha$ to be identified. He calls such a DE system $F_a = 0$ "strictly self-adjoint". This definition is motivated by the case of a linear DE system, since linearity implies that $(\delta_u F)_a = F_a$ and $(\delta^*_u F)_\alpha = F^*_\alpha|_{v=u}$ are identities, whereby a linear DE system with $M = m$ is "strictly self-adjoint" iff it satisfies $(\delta F)_\alpha = (\delta^* F)_\alpha$, which is the condition for the self-adjointness of a linear system. However, for nonlinear DE systems, the definition of "strictly self-adjoint" conflicts with the standard of definition [6,24] in variational calculus that a general DE system $F_a = 0$ is self-adjoint iff its associated Fréchet derivative operator is self-adjoint, $(\delta F)_a = (\delta^* F)_\alpha$, which requires $M = m$.

Ibragimov subsequently [10] proposed to have $v = \phi(u)$, which he called "quasi-self-adjointness". A more general proposal $v = \phi(x, u)$ was then introduced first in Reference [14] and shortly later appears in Ibragimov's next paper [12], with the condition that $F^*_\alpha|_{v=\phi(x,u)} = \lambda_\alpha{}^\beta F_\beta$ must hold for some coefficients $\lambda_\alpha{}^\beta$, again with $M = m$. This condition is called "weak self-adjointness" in Reference [14] and "nonlinear self-adjointness" in Reference [12]. Ibragimov also mentions an extension of this definition to $v = \phi(x, u, \partial u, \ldots, \partial^s u)$, but does not pursue it. Later, he applies this definition in Reference [13] to a specific PDE, where $\lambda_\alpha{}^\beta$ is extended to be a linear differential operator. However, unlike in the previous papers, no conservation laws are found from using this extension. A subsequent paper [15] then uses this extension, which is called "nonlinear self-adjointness through a differential substitution", to obtain conservation laws for several similar PDEs. Finally, the same definition is stated more generally in Reference [17] for DE systems with $M = m$:

$$F^*_\alpha|_{v=\phi(x,u,\partial u,\ldots,\partial^s u)} = \lambda_\alpha{}^\beta F_\beta + \lambda_\alpha{}^{\beta i} D_i F_\beta + \cdots + \lambda_\alpha{}^{\beta i_1 \cdots i_p} D_{i_1} \cdots D_{i_p} F_\beta \tag{57}$$

where the coefficients $\lambda_\alpha{}^\beta, \lambda_\alpha{}^{\beta i}, \ldots, \lambda_\alpha{}^{\beta i_1 \cdots i_p}$ are differential functions.

These developments lead to the following conservation law theorem, which is a generalization of Ibragimov's main theorem [9,12] to arbitrary DE systems (not restricted by $M = m$), combined with the use of a differential substitution [12,15,17].

Theorem 1. *Suppose a system of DEs* (14) *satisfies*

$$F^*_\alpha|_{v=\phi} = \lambda_\alpha{}^a F_a + \lambda_\alpha{}^{ai} D_i F_a + \cdots + \lambda_\alpha{}^{a i_1 \cdots i_p} D_{i_1} \cdots D_{i_p} F_a \tag{58}$$

for some differential functions $\phi^a(x, u, \partial u, \ldots, \partial^s u)$ *and* $\lambda_\alpha{}^a(x, u, \partial u, \ldots, \partial^s u)$, $\lambda_\alpha{}^{ai}(x, u, \partial u, \ldots, \partial^s u), \ldots,$ $\lambda_\alpha{}^{a i_1 \cdots i_p}(x, u, \partial u, \ldots, \partial^s u)$ *that are non-singular on the solution space* $\mathcal{E}$ *of the DE system, where* F^*_α *is the adjoint linearization* (2) *of the system. Then, any local symmetry*

$$\mathbf{X} = \xi^i(x, u, \partial u, \ldots, \partial^r u)\partial/\partial x^i + \eta^\alpha(x, u, \partial u, \ldots, \partial^r u)\partial/\partial u^\alpha \tag{59}$$

admitted by the DE system yields a local conservation law (26) *given in an explicit form by the conserved current* (50) *with* $v^a = \phi^a$ *and* $P^\alpha = \eta^\alpha - \xi^i u^\alpha_i$.

An important remark is that all of the functions $\phi^a, \lambda_\alpha{}^a, \lambda_\alpha{}^{ai}, \ldots, \lambda_\alpha{}^{ai_1\cdots i_p}$ must be non-singular on $\mathcal{E}$, as otherwise, the condition (58) can be satisfied in a trivial way. This point is not mentioned in any of the previous work [9,12,14,15,17].

The "nonlinear self-adjointness" condition (58) turns out to have a simple connection to the determining equations for symmetries. This connection is somewhat obscured by the unfortunate use of non-standard definitions and non-standard notation in References [9,12]. Nevertheless, it is straightforward to show that Equation (58) is precisely the adjoint of the determining Equation (23) for symmetries formulated as an operator Equation (41).

2.4. Adjoint-Symmetries and a Formula for Generating Conservation Laws

For any given DE system (14), the adjoint of the symmetry determining Equation (23) is given by

$$(\delta^*_Q F)_\alpha|_\mathcal{E} = 0 \tag{60}$$

for a set of differential functions $Q^a(x, u, \partial u, \ldots, \partial^r u)$. (Similarly to the symmetry determining equation, the notation $\mathcal{E}$ here means that the given DE system, as well as its differential consequences, are to be used). These differential functions are called an *adjoint-symmetry* [4], in analogy to the characteristic functions of a symmetry (40), and so, Equation (60) is called the *adjoint-symmetry determining equation*. As shown in Reference [29], this analogy has a concrete geometrical meaning in the case when a DE system is an evolutionary system $F_\alpha = u^\alpha_t - f_\alpha(x, u, \partial_x u, \ldots, \partial^N_x u) = 0$ with $M = m$ and $x = (t, x^1, \ldots, x^{n-1})$, where t is a time coordinate and x^i, $i = 1, \ldots, n-1$, are space coordinates. In this case, Q^α can be viewed as the coefficients of a one-form or a covector $Q^\alpha \mathbf{d}u^\alpha$, in analogy to P^α being the coefficients of a vector $P^\alpha\partial/\partial u^\alpha$. The condition for $P^\alpha\partial/\partial u^\alpha$ to be a symmetry can be formulated as $(\mathcal{L}_f P^\alpha\partial/\partial u^\alpha)|_\mathcal{E} = 0$ where $\mathcal{L}_f$ denotes the Lie derivative [24,29] with respect to the time evolution vector $\hat{\mathbf{X}} = f_\alpha\partial/\partial u^\alpha$. Then, the condition for $Q^\alpha \mathbf{d}u^\alpha$ to be an adjoint-symmetry is equivalent to $(\mathcal{L}_f Q^\alpha \mathbf{d}u^\alpha)|_\mathcal{E} = 0$. (Note the awkwardness in the index positions here comes from Ibragimov's choice of index placement F_α for a DE system with $M = m$. A better notation would be F^α and F^a when $M \neq m$, which is used in References [4,26,27,31].)

In the case when a DE system is variational (33), the symmetry determining equation is self-adjoint, since $(\delta^*_Q F)_\alpha = (\delta_Q F)_\alpha$. Then, the adjoint-symmetry determining Equation (60) reduces to the symmetry determining Equation (23), with $Q^a(x, u, \partial u, \ldots, \partial^r u) = P^\alpha(x, u, \partial u, \ldots, \partial^r u)$, where the indices $a = \alpha$ can be identified, due to $M = m$. Consequently, adjoint-symmetries of any variational DE system are the same as symmetries.

Other aspects of adjoint-symmetries and their connection to symmetries are discussed in Reference [34].

Now, under some mild regularity conditions [29] on the form of a general DE system (14), the adjoint-symmetry determining Equation (60) implies that the functions Q^a satisfy

$$(\delta^*_Q F)_\alpha = R_Q(F)_\alpha \tag{61}$$

where

$$R_Q = R_Q{}^b_\alpha + R_Q{}^{bi}_\alpha D_i + R_Q{}^{bij}_\alpha D_i D_j + \cdots + R_Q{}^{bi_1\cdots i_r}_\alpha D_{i_1}\cdots D_{i_r} \tag{62}$$

is some linear differential operator whose coefficients $R_Q{}^b_\alpha, R_Q{}^{bi}_\alpha, \ldots, R_Q{}^{bi_1\cdots i_r}_\alpha$ are differential functions that are non-singular on the solution space $\mathcal{E}$ of the DE system (14). In Ibragimov's notation $F^*_\alpha(x, u, v, \partial u, \partial v, \ldots, \partial^N u, \partial^N v) = (\delta^*_v F)_\alpha$, the adjoint-symmetry Equation (61) coincides with the "nonlinear self-adjointness" condition (58) in Theorem 1, where the operator on the right-hand side of Equation (58) is precisely the adjoint-symmetry operator (62).

Therefore, the following equivalence has been established.

Proposition 2. *For a general DE system* (14), *the condition* (58) *of "nonlinear self-adjointness" coincides with the condition of existence of an adjoint-symmetry* (60). *When a DE system is variational* (33), *these conditions reduce to the condition of the existence of a symmetry.*

One remark is that the formulation of "nonlinear self-adjointness" given here is more general than what appears in References [12,15,17] since those formulations assume that the DE system has the same number of equations as the number of dependent variables, $M = m$. Another remark is that the meaning of "nonlinear self-adjointness" shown here in the case of variational DE systems has not previously appeared in the literature.

Example: Consider the class of semilinear wave equations $u_{tt} - u_{xx} + a(u)(u_t^2 - u_x^2) + b(u)u_t + c(u)u_x + m(u) = 0$ for $u(t,x)$, with a nonlinearity coefficient $a(u)$, damping coefficients $b(u), c(u)$ and a mass-type coefficient $m(u)$. In Reference [20], the conditions under which a slightly more general family of wave equations is "nonlinearly self-adjoint" (58) are stated for $v = \phi(u)$. These results will be generalized here by considering $v = \phi(t,x,u)$. A first observation is that this class of wave equations admits an equivalence transformation $u \to \tilde{u} = f(u)$, with $f' \neq 0$, which can be used to put $a = 0$ by $f(u) = \int \exp(A(u))du$ where $A' = a$. (Equivalence transformations were not considered in Reference [20], and so, their results are considerably more complicated than is necessary). This transformation gives

$$u_{tt} - u_{xx} + b(u)u_t + c(u)u_x + m(u) = 0. \tag{63}$$

In Ibragimov's notation, the condition of "nonlinear self-adjointness" with $v = \phi(t,x,u)$ is given by $0 = E_u(vF)|_{v=\phi}$ where

$$F = u_{tt} - u_{xx} + b(u)u_t + c(u)u_x + m(u). \tag{64}$$

This yields

$$(D_t^2\phi - D_x^2\phi - bD_t\phi - cD_x\phi + m'\phi)\big|_{F=0} = 0. \tag{65}$$

For comparison, the determining Equation (23) for local symmetries $\hat{\mathbf{X}} = P(t,x,u,u_t,u_x,\ldots)\partial/\partial u$ (in characteristic form) is given by

$$(D_t^2P - D_x^2P + bD_tP + cD_xP + (u_tb' + u_xc' + m')P = 0)\big|_{F=0} = 0. \tag{66}$$

Its adjoint is obtained by multiplying by $Q(t,x,u,u_t,u_x,\ldots)$ and integrating by parts, which yields $(D_t^2Q - D_x^2Q - D_t(bQ) - D_x(cQ) + (u_tb' + u_xc' + m')Q = 0)\big|_{F=0} = 0$. After the D_x terms are expanded out, this gives the determining Equation (60) for local adjoint-symmetries

$$(D_t^2Q - D_x^2Q - bD_tQ - cD_xQ + (b'u_t + c'u_x + m')Q)\big|_{F=0} = 0 \tag{67}$$

which coincides with the "nonlinear self-adjointness" condition (65) extended to differential substitutions [12,14,16] given by $v = Q(t,x,u,u_t,u_x,\ldots)$. All adjoint-symmetries of lowest-order form $Q(t,x,u)$ can be found in a straightforward way. After $Q(t,x,u)$ is substituted into the determining Equation (67) and u_{tt} is eliminated through the wave Equation (63), the determining equation splits with respect to the variables u_t and u_x, yielding a linear overdetermined system of four equations (after some simplifications):

$$Q_{tt} - Q_{xx} - bQ_t - cQ_x - mQ_u + m'Q = 0, \tag{68}$$

$$Q_{tu} - bQ_u = 0, \quad Q_{xu} + cQ_u = 0, \tag{69}$$

$$Q_{uu} = 0. \tag{70}$$

It is straightforward to derive and solve this determining system by Maple. Hereafter, the conditions

$$b' \neq 0, \quad c' \neq 0, \quad m'' \neq 0, \quad m(0) = 0 \tag{71}$$

will be imposed, which corresponds to studying wave equation (63) whose lower-order terms are nonlinear and homogeneous. The general solution of the determining system (68)–(70) then comprises three distinct cases (as obtained using the Maple package 'rifsimp'), after merging. This leads to the following complete classification of solution cases shown in Table 1. The table is organized by listing each solution Q and the conditions on b, c, m for which it exists. (From these conditions, a classification of maximal linear spaces of multipliers can be easily derived). Note that if the transformation $u \to \tilde{u} = \int \exp(A(u))du$ is inverted, then Q transforms to $\tilde{Q} = \exp(A(u))Q$. (Also note that, under the restriction $Q = \phi(u)$ considered in Reference [20], the classification reduces to just the first case with $m =$ const. and $Q = 1$).

Table 1. Adjoint-symmetries ("nonlinear self-adjointness").

$Q(t,x,u)$	$b(u)$	$c(u)$	$m(u)$	Conditions
$e^{m_2 t + m_3 x}$	arb.	arb.	$m_1 u + \int(m_2 b + m_3 c)du$	$m_1 = m_3^2 - m_2^2$
$e^{\alpha x + \beta t}$	$b_0 + b_1 m'$	$c_0 + c_1 m'$	arb.	$b_1\beta + c_1\alpha = 1$ $\beta(\beta - b_0) = \alpha(\alpha + c_0)$
$e^{\gamma x} q(x \mp t)$	$b_0 + b_1 m'$	$c_0 + c_1 m'$	arb.	$\gamma = \pm b_0 = -c_0$, $b_1 = 1/b_0, c_1 = -1/c_0$, $q(\xi) =$ arb.

The Fréchet derivative operator in the symmetry determining Equation (23) and the adjoint of this operator in the adjoint-symmetry determining Equation (60) are related by the integration-by-parts formula (3). For a general DE system (14), this formula is given by

$$Q^a(\delta_P F)_a - P^\alpha(\delta_Q^* F)_\alpha = D_i \Psi^i(P, Q; F) \tag{72}$$

where the vector $\Psi^i(P, Q; F)$ is given by the explicit expression (4) with $v = Q$, $w = P$, and $f = F$. As shown in References [2–4], this vector $\Psi^i(P, Q; F)$ will be a conserved current

$$D_i \Psi^i(P, Q; F)|_{\mathcal{E}} = 0 \tag{73}$$

whenever the differential functions P^α and Q^a respectively satisfy the symmetry and adjoint-symmetry determining equations. Moreover, it is straightforward to see

$$\Psi^i(P, Q; F) = \Phi^i(P; L)|_{v=Q}, \tag{74}$$

which follows from relation (7), where $\Phi^i(P; L)$ is the Noether conserved current (50) and L is the Lagrangian (38). Alternatively, the equality (74) can be derived indirectly by applying formula (72) to the variational identity (48) with $v = Q$, giving

$$\begin{aligned} \mathrm{pr}\hat{\mathbf{X}}(L) &= \phi^a F_a + v^a(\delta_P F)_a = \phi^a F_a + P^\alpha(\delta_v^* F)_\alpha + D_i \Psi^i(P, v; F) \\ &= \hat{\mathbf{X}}(v^a) E_{v^a}(L) + \hat{\mathbf{X}}(u^\alpha) E_{u^\alpha}(L) + D_i \Psi^i(P, v; F) \end{aligned} \tag{75}$$

which implies $\Psi^i(P, v; F) = \Phi^i(P; L)$ holds (up to the possible addition of a total curl).

When the relation (74) is combined with Propositions 1 and 2, the following main result is obtained.

Theorem 2. *For any DE system* (14) *admitting an adjoint-symmetry* (60) *(namely, a "nonlinearly self-adjoint system" in the general sense), the conserved current* (50) *derived from applying Noether's theorem to the extended Euler–Lagrange system* (39) *using any given symmetry* (46) *is equivalent to the conserved current obtained using the adjoint-symmetry/symmetry formula* (72).

This theorem shows that the "nonlinear self-adjointness" method based on Ibragimov's theorem as developed in papers [9,12,14,15,17] for DE systems with $M = m$ is just a special case of the adjoint-symmetry/symmetry formula (72) introduced for general DE systems in prior papers References [2–4], which were never cited. Moreover, the adjoint-symmetry/symmetry formula (72) has the advantage that there is no need to extend the given DE system by artificially adjoining variables to get a Euler–Lagrange system.

Another major advantage of the adjoint-symmetry/symmetry formula is that it can be used to show how the resulting local conservation laws are, in general, not necessarily non-trivial and comprise only a subset of all of the non-trivial local conservation laws admitted by a given DE system. In particular, in many applications of Theorem 1, it is found that some non-trivial symmetries, particularly translation symmetries, only yield trivial conservation laws [17–19], and that some local conservation laws are not produced even when all admitted symmetries are used. These observations turn out to have a simple explanation through the equivalence of Theorem 1 and the adjoint-symmetry/symmetry formula (72), as explained in the next section.

Example: For the semilinear wave Equation (63), the extended Euler–Lagrange system in Ibragimov's notation consists of

$$F = u_{tt} - u_{xx} + b(u)u_t + c(u)u_x + m(u) = \frac{\delta L}{\delta v} = 0, \tag{76}$$

$$F^* = v_{tt} - v_{xx} - bv_t - cv_x + (b'u_t + c'u_x + m')v = \frac{\delta L}{\delta u} = 0, \tag{77}$$

where F^* is defined by the adjoint-symmetry Equation (67) with $Q = v$, and where the Lagrangian (38) is simply $L = vF = v(u_{tt} - u_{xx} + b(u)u_t + c(u)u_x + m(u))$ in terms of the variables u and v. Consider any point symmetry of the wave Equation (76) for u, given by a generator

$$\mathbf{X} = \tau(t,x,u)\partial/\partial t + \xi(t,x,u)\partial/\partial x + \eta(t,x,u)\partial/\partial u. \tag{78}$$

Its equivalent characteristic form is $\hat{\mathbf{X}} = P\partial/\partial u$, with $P = \eta - \tau u_t - \xi u_x$ satisfying the symmetry determining Equation (66) on the space of solutions $u(x)$ of the wave Equation (76). Every point symmetry can be extended to a variational symmetry (54) admitted by the Euler–Lagrange system, which is given by the generator $\mathbf{X}^{\text{ext.}} = \mathbf{X} + (\tau v_t + \xi v_x - R_P^*(v))\partial/\partial v$ where R_P^* is the adjoint of the operator R_P defined by relation (41) for the point symmetry holding off of the solution space of the wave Equation (76). In particular, R_P can be obtained by a straightforward computation of $\delta_P F = R_P(F)$, where the terms in $\delta_P F$ are simplified by using the equations $\tau_u = \xi_u = 0$, $\tau_t = \xi_x$ and $\tau_x = \xi_t$ that arise from splitting the determining Equation (66). This yields

$$R_P = -\tau D_t - \xi D_x + \eta_u - (\tau_t + \xi_x), \tag{79}$$

and thus

$$R_P^* = \tau D_t + \xi D_x + \eta_u. \tag{80}$$

Hence, the variational symmetry is simply

$$\mathbf{X}^{\text{ext.}} = \tau\partial/\partial t + \xi\partial/\partial x + \eta\partial/\partial u - \eta_u v\partial/\partial v \tag{81}$$

which is a point symmetry.

The action of this variational symmetry on the Lagrangian $L = vF$ is given by

$$\text{pr}\mathbf{X}^{\text{ext.}}(L) = -\eta_u vF + v\text{pr}\mathbf{X}(F) = -(\tau_t + \xi_x)vF = -(D_t\tau + D_x\xi)L \tag{82}$$

since $\text{pr}\mathbf{X}(F) = \tau D_t F + \xi D_x F + R_P(F) = (\eta_u - (\tau_t + \xi_x))F$. This symmetry action then can be combined with the variational identity (55) to get the Noether relation

$$D_t(\tau L + \Phi^t(P;L)) + D_x(\tau L + \Phi^x(P;L)) = -\hat{\mathbf{X}}^{\text{ext.}}(v)F - \hat{\mathbf{X}}^{\text{ext.}}(u)F^* \tag{83}$$

using $F = E_v(L)$ and $F^* = E_u(L)$, where

$$\Phi^t(P;L) = vD_tP(b(u)v - v_t)P, \quad \Phi^x(P;L) = -vD_xP + (c(u)v + v_x)P \tag{84}$$

are obtained from formula (50). This yields a conservation law

$$(D_tC^t + D_xC^x)|_{\mathcal{E}(u,v)} = 0, \quad C^t = \Phi^t(P;L) - \tau L, \quad C^x = \Phi^x(P;L) - \xi L \tag{85}$$

on the solution space $\mathcal{E}(u,v)$ of the Euler–Lagrange system $F = 0$, $F^* = 0$. Since $L|_{\mathcal{E}(u,v)} = 0$, this conservation law is locally equivalent to the conservation law (52) which is given by

$$(D_t\hat{C}^t + D_x\hat{C}^x)|_{\mathcal{E}(u,v)} = 0, \quad \hat{C}^t = \Phi^t(P;L), \quad \hat{C}^x = \Phi^x(P;L). \tag{86}$$

Moreover, from the identity (72) relating the symmetry Equation (66) and the adjoint-symmetry Equation (67), the conserved current $(\hat{C}^t, \hat{C}^x)$ in the conservation law (86) is the same as the conserved current (Ψ^t, Ψ^x) in the adjoint-symmetry/symmetry formula

$$\Psi^t(P,Q;F)|_{Q=v} = \Phi^t(P;L), \quad \Psi^x(P,Q;F)|_{Q=v} = \Phi^x(P;L) \tag{87}$$

where

$$\begin{aligned} \Psi^t(P,Q;F) &= QD_tP + (b(u)Q - D_tQ)P, \\ \Psi^x(P,Q;F) &= -QD_xP + (c(u)Q + D_xQ)P. \end{aligned} \tag{88}$$

In Reference [20], the conservation law formula (85) is used to obtain a single local conservation law for a special case of the wave Equation (63) given by $b = -c = -\ln(u)$ and $d = 0$, corresponding to $\tilde{u}_{tt} - \tilde{u}_{xx} - (\tilde{u}_t^2 - \tilde{u}_x^2) + \tilde{u}(\tilde{u}_t - \tilde{u}_x) = 0$ after an equivalence transformation $u \to \tilde{u} = e^{-u}$ is made. The formula is applied to the adjoint-symmetry $\tilde{Q} = e^{-\tilde{u}}$ and the point symmetry $\tilde{\mathbf{X}} = e^{(t+x)/2}\partial/\partial\tilde{u}$ with characteristic $\tilde{P} = e^{(t+x)/2}$, which respectively correspond to $Q = 1$ and $\mathbf{X} = e^{(t+x)/2}u\partial/\partial u$ with $P = e^{(t+x)/2}u$. The likely reason why the obvious translation symmetries $\tilde{\mathbf{X}} = \partial/\partial t$ and $\tilde{\mathbf{X}} = \partial/\partial x$ were not considered in Reference [20] is that these symmetries lead to locally trivial conservation laws when $\tilde{Q} = e^{-\tilde{u}}$ is used.

To illustrate the situation, consider the translation symmetries

$$\mathbf{X}_1 = \partial/\partial t, \quad \mathbf{X}_2 = \partial/\partial x \tag{89}$$

admitted by the wave Equation (63) for arbitrary $b(u)$, $c(u)$, $m(u)$. The characteristic functions of these two symmetries are, respectively, $P = -u_t$ and $P = -u_x$. Local conservation laws can be obtained by applying formula (85), or its simpler equivalent version (86), with $v = Q(t,x,u)$ being the adjoint-symmetries classified in Table 1. The resulting conserved currents (Ψ^t, Ψ^x), modulo locally trivial currents, are shown in Table 2.

Notice that for $Q = \text{const.}$ the conserved currents (Ψ^t, Ψ^x) obtained from the two translation symmetries vanish. This implies that Ibragimov's theorem (85) yields just trivial conserved currents (Ψ^t, Ψ^x) for some cases of the wave Equation (63) when a non-trivial conserved current exists. A full explanation of why this occurs will be given in the next section.

Table 2. Conserved currents from the adjoint-symmetry/symmetry formula.

		$X = \partial/\partial t$	$X = \partial/\partial x$
Conditions	Q	Ψ^t, Ψ^x	Ψ^t, Ψ^x
$m = m_1 u + m_2 \int b\,du + m_3 \int c\,du$ $m_1 = m_3^2 - m_2^2$	$e^{m_3 x + m_2 t}$	$m_2 Q(u_t - m_2 u + \int b\,du)$, $m_2 Q(m_3 u - u_x + \int c\,du)$	$m_3 Q(u_t - m_2 u + \int b\,du)$, $m_3 Q(m_3 u - u_x + \int c\,du)$
$b = b_0 + b_1 m'$ $c = c_0 + c_1 m'$ $b_1\beta + c_1\alpha = 1$ $\beta(\beta - b_0) = \alpha(\alpha + c_0)$	$e^{\alpha x + \beta t}$	$\beta Q(u_t - \beta u + \int b\,du)$, $\beta Q(\alpha u - u_x + \int c\,du)$	$\alpha Q(u_t - \beta u + \int b\,du)$, $\alpha Q(\alpha u - u_x + \int c\,du)$
$b = \pm(\gamma + \frac{1}{\gamma} m')$ $c = -\gamma + \frac{1}{\gamma} m'$	$e^{\gamma x} q(x \mp t)$	$-e^{\gamma x}(q'' u \pm q'(u_t + \int b\,du))$, $\pm e^{\gamma x}((\gamma q' - q'')u + q(u_x \mp \int b\,du))$	$e^{\gamma x}(\pm(q'' + \gamma q')u + (q' + \gamma q)(u_t + \int b\,du))$, $e^{\gamma x}((q'' - \gamma^2 q)u - (q' + \gamma q)(u_x \mp \int b\,du))$

3. Properties of Conservation Laws Generated by the Adjoint-Symmetry/Symmetry Formula and Ibragimov's Theorem

To determine when a conserved current is locally trivial or when two conserved currents are locally equivalent, it is useful to have a characteristic (canonical) form for local conservation laws, in analogy to the characteristic form for local symmetries.

Any local conservation law (26) can be expressed as a divergence identity [24]

$$D_i C^i = R_C{}^a F_a + R_C{}^{ai} D_i F_a + \cdots + R_C{}^{ai_1\cdots i_r} D_{i_1} \cdots D_{i_r} F_a \tag{90}$$

by moving off of the solution space $\mathcal{E}$ of the system, where $R_C{}^a, R_C{}^{ai}, \ldots, R_{C}{}_a^{ai_1\cdots i_r}$ are some differential functions that are non-singular on $\mathcal{E}$, under some mild regularity conditions [29] on the form of the DEs (14). Integration by parts on the terms on the right-hand side in this identity (90) then yields

$$D_i \tilde{C}^i = Q_C^a F_a \tag{91}$$

with

$$Q_C^a = R_C{}^a - D_i R_C{}^{ai} + \cdots + (-1)^r D_{i_1} \cdots D_{i_r} R_C{}^{ai_1\cdots i_r}, \tag{92}$$

where

$$\tilde{C}^i|_{\mathcal{E}} = C^i|_{\mathcal{E}} \tag{93}$$

reduces to the conserved vector in the given conservation law (26). Hence,

$$(D_i \tilde{C}^i)|_{\mathcal{E}} = 0 \tag{94}$$

is a locally equivalent conservation law. The identity (91) is called the *characteristic equation* [24] for the conservation law (26), and the set of differential functions (92) is called the *conservation law multiplier* [24]. In general, a set of functions $f^a(t, x, u, \partial u, \partial^2 u, \ldots \partial^s u)$ will be a multiplier if it is non-singular on $\mathcal{E}$ and its summed product with the DEs F_a in the system has the form of a total divergence.

For a given local conservation law, the multiplier arising from the integration by parts formula (92) will be unique iff the coefficient functions in the characteristic Equation (90) are uniquely determined by the conserved vector C^i. This uniqueness holds straightforwardly for any DE system consisting of a single equation that can be expressed in a solved form for a leading derivative [29]. For DE systems containing more than one equation, some additional technical requirements are necessary [24].

In particular, it is necessary that a DE system have no differential identities [24], and it is sufficient that a DE system have a generalized Cauchy–Kovalevskaya form [24–27]. A concrete necessary and sufficient condition, which leads to the following uniqueness result, is stated in Reference [29].

Proposition 3. *For any closed DE system* (14) *having a solved form in terms of leading derivatives and having no differential identities, a conserved current is locally trivial* (28) *iff its corresponding multiplier* (92) *vanishes when evaluated on the solution space of the system.*

This class of DE systems includes nearly all systems of physical interest, apart from systems such as the Maxwell equations and the incompressible fluid equations, which possess differential identities. Often the distinction between systems with and without differential identities is overlooked in the literature on conservation law multipliers.

The importance of Proposition 3 is that, in a wide class of DE systems, it establishes that a unique characteristic form for locally equivalent conservation laws is provided by multipliers. From this result, it is now straightforward to derive a simple condition to detect when a local conservation law given by the adjoint-symmetry/symmetry formula (72) is locally trivial (28).

Let $P^\alpha(x,u,\partial u,\ldots,\partial^r u)$ be the characteristic functions defining a symmetry (23), and let $Q^a(x,u,\partial u,\ldots,\partial^s u)$ be a set of differential functions defining an adjoint-symmetry (60). Then, the adjoint-symmetry/symmetry formula (72) yields a local conservation law (73). The characteristic equation of this conservation law is given by substituting the symmetry identity (41) and the adjoint-symmetry identity (61) into the formula (72) to get

$$D_i\Psi^i(P,Q;F) = Q^a R_P(F)_a - P^\alpha R_Q(F)_\alpha. \tag{95}$$

Integration by parts gives

$$D_i\tilde{\Psi}^i(P,Q;F) = (R_P^*(Q)^a - R_Q^*(P)^a)F_a \tag{96}$$

where

$$\tilde{\Psi}^i(P,Q;F)|_{\mathcal{E}} = \Psi^i(P,Q;F)|_{\mathcal{E}}. \tag{97}$$

Hence, the conservation law multiplier is given by [35]

$$Q^a_\Psi = R_P^*(Q)^a - R_Q^*(P)^a. \tag{98}$$

This yields the following result.

Proposition 4. *The adjoint-symmetry/symmetry formula* (72) *for a given DE system* (14) *produces a locally trivial conservation law if the condition*

$$(R_P^*(Q)^a - R_Q^*(P)^a)|_{\mathcal{E}} = 0 \tag{99}$$

holds for the given symmetry and adjoint-symmetry pair, where $P^\alpha(x,u,\partial u,\ldots,\partial^r u)$ *is the set of characteristic functions of the symmetry* (23) *and* $Q^a(x,u,\partial u,\ldots,\partial^s u)$ *is the set of functions defining the adjoint-symmetry* (60)*. This condition* (99) *is also sufficient whenever the DE system* (14) *belongs to the class stated in Proposition 3.*

Through the equivalence stated in Theorem 2, which relates the adjoint-symmetry/symmetry formula (72) and the generalized version of Ibragimov's conservation law formula in Theorem 1, it follows that "nonlinear self-adjointness" through a differential substitution with $v = Q$ produces a conservation law (50) that is locally trivial when $Q^a(x,u,\partial u,\ldots,\partial^s u)$ and $P^\alpha(x,u,\partial u,\ldots,\partial^r u)$ satisfy condition (99).

A useful remark is that the triviality condition (99) can be checked directly, without the need to derive the local conservation law itself.

Example: For the semilinear wave Equation (63), consider the conserved currents obtained in Table 2, which are generated from the three adjoint-symmetries

$$Q_1 = e^{m_2 t + m_3 x}, \quad Q_2 = e^{\alpha x + \beta t}, \quad Q_3 = e^{\gamma x} q(x \mp t), \tag{100}$$

and the two translation symmetries (89). The operators $\delta_P F = R_P(F)$ associated with the characteristics

$$P_1 = -u_t, \quad P_2 = -u_x \tag{101}$$

of these two symmetries (89) are given by the formula (79), which yields

$$R_{P_1} = -D_t, \quad R_{P_2} = -D_x. \tag{102}$$

For adjoint-symmetries of the form $Q(t, x, u)$, the operator $\delta^*_Q F = R_Q(F)$ is easily found to be

$$R_Q = Q_u. \tag{103}$$

Hence, the operators associated with the three adjoint-symmetries (100) are simply

$$R_{Q_1} = R_{Q_2} = R_{Q_3} = 0. \tag{104}$$

The triviality condition (99) is then given by

$$R^*_{P_1}(Q_l) - R^*_{Q_l}(P_1) = D_t Q_l = 0, \quad R^*_{P_2}(Q_l) - R^*_{Q_l}(P_2) = D_x Q_l = 0, \quad l = 1,2,3. \tag{105}$$

This shows that the two conserved currents obtained from Q_1 will be trivial when $m_2 = 0$ and $m_3 = 0$ hold, respectively, and that likewise, the two conserved currents obtained from Q_2 will be trivial when $\beta = 0$ and $\alpha = 0$ hold, respectively. Similarly, for Q_3, the first conserved current will be trivial when $q' = 0$ holds, while the second conserved current will be trivial when $q' + \gamma q = 0$ holds, corresponding to $q = e^{-\gamma(x \pm t)}$. These trivial cases can be seen to occur directly from the explicit expressions for the conserved currents (Ψ^t, Ψ^x) in Table 2.

In general, while the adjoint-symmetry/symmetry formula (72) (and, hence, Ibragimov's theorem) looks very appealing, it has major drawbacks that in many examples [14–17,20–22] the selection of a symmetry must be fitted to the form of the adjoint-symmetry to produce a non-trivial conservation law, and that in other examples [17–19] no non-trivial conservation laws are produced when only translation symmetries are available. More importantly, it is *not* (as is sometimes claimed) a generalization of Noether's theorem to non-variational DE systems.

As a reinforcement of these statements, consider the situation of variational DE systems, where adjoint-symmetries coincide with symmetries. Then, the adjoint-symmetry/symmetry formula (72) produces a conserved current directly from any pair of symmetries admitted by a given variational DE system. However, from Noether's theorem, this conserved current must also arise directly from some variational symmetry of the system. Moreover, if the pair of symmetries being used are variational symmetries that happen to commute with each other, then the resulting conserved current turns out to be trivial, as shown in Reference [36].

To understand these aspects and other properties of the formula, the determining equations for multipliers are needed.

3.1. Multiplier Determining equations

All conservation law multipliers for any given DE system can be determined from the property (10) and (11) that a differential function is a total divergence iff it is annihilated by the Euler operator (6).

Specifically, when this property is applied directly to the characteristic Equation (91) for local conservation laws, it yields the determining equations

$$E_{u^\alpha}(Q^a_C F_a) = 0, \quad \alpha = 1, \ldots, m \tag{106}$$

which are necessary and sufficient [24] for a set of differential functions $Q^a_C(x, u, \partial u, \ldots, \partial^s u)$ to be a multiplier for a local conservation law (26). Note the Equation (106) must hold identically in jet space (and not just on the solution space $\mathcal{E}$ of the DE system).

The multiplier determining the Equation (106) have a close connection to the determining Equation (60) for adjoint-symmetries. This can be immediately seen from the product rule (9) obeyed by the Euler operator, which gives

$$0 = E_{u^\alpha}(Q^a_C F_a) = (\delta^*_{Q_C} F)_\alpha + (\delta^*_F Q_C)_\alpha \tag{107}$$

holding identically in jet space $J(x, u, \partial u, \partial^2 u, \ldots)$. Notice that if this Equation (107) is restricted to the solution space $\mathcal{E} \subset J$ of the given DE system (14), then it coincides with the adjoint-symmetry determining Equation (60). Hence, every conservation law multiplier is an adjoint-symmetry. This is a well-known result [4,24,26,27,29,31]. What is not so well known are the other conditions [4,29] that an adjoint-symmetry must satisfy to be a conservation law multiplier. These conditions arise from splitting the determining Equation (107) with respect to F_a and its total derivatives. As shown in Reference [29], the splitting can be derived by using the adjoint-symmetry identity (61) combined with the expression

$$(\delta^*_F Q)_\alpha = F_a \frac{\partial Q^a}{\partial u^\alpha} - D_i\Big(F_a \frac{\partial Q^a}{\partial u^\alpha_i}\Big) + \cdots + (-1)^s D_{i_1} \cdots D_{i_s}\Big(F_a \frac{\partial Q^a}{\partial u^\alpha_{i_1 \cdots i_s}}\Big). \tag{108}$$

Then, in the determining Equation (107), the coefficients of F_a, $D_i F_a$, and so on yield the system of equations [29]:

$$(\delta^*_{Q_C} F)_\alpha|_{\mathcal{E}} = 0 \tag{109}$$

and

$$R_{Q\alpha}^{\ a} + E_{u^\alpha}(Q^a_C) = 0 \tag{110}$$

$$R_{Q\alpha}^{\ a i_1 \cdots i_q} + (-1)^q E^{(i_1 \cdots i_q)}_{u^\alpha}(Q^a_C) = 0, \quad q = 1, \ldots, s \tag{111}$$

where $R_{Q\alpha}^{\ a}$ and $R_{Q\alpha}^{\ a i_1 \cdots i_q}$ are the coefficient functions of the linear differential operator (62) determined by Equation (109), and where E_{u^α} is the Euler operator (6) and $E^{(i_1 \cdots i_q)}_{u^\alpha}$ is a higher-order Euler operator defined by [24,29]

$$E^{(i_1 \cdots i_q)}_{u^\alpha}(f) = \frac{\partial f}{\partial u^\alpha_{i_1 \cdots i_q}} - \binom{q+1}{1} D_j\Big(\frac{\partial f}{\partial u^\alpha_{i_1 \cdots i_q j}}\Big) + \cdots + (-1)^r \binom{q+r}{r} D_{j_1} \cdots D_{j_r}\Big(\frac{\partial f}{\partial u^\alpha_{i_1 \cdots i_q j_1 \cdots j_r}}\Big), \quad q = 1, 2, \ldots \tag{112}$$

for an arbitrary differential function $f(x, u, \partial u, \ldots, \partial^s u)$. This system (109)–(111) constitutes a determining system for conservation law multipliers. Its derivation requires the same technical conditions on the form of the DE system (14) as stated in Proposition 3.

Theorem 3. *The determining Equation* (107) *for conservation law multipliers of a general DE system* (14) *is equivalent to the linear system of equations* (109)–(111). *In particular, multipliers are adjoint-symmetries* (109) *satisfying Helmholtz-type conditions* (110)–(111) *which are necessary and sufficient for an adjoint-symmetry to have the variational form* (92) *derived from a conserved current.*

This well known result [4,26,27,29] gives a precise relationship between adjoint-symmetries and multipliers, or equivalently between "nonlinear self-adjointness" and multipliers. In particular, it provides necessary and sufficient conditions for an adjoint-symmetry to be a multiplier. The simplest situation is when adjoint-symmetries of the lowest-order form $Q^{\alpha}(x,u)$ are considered, which corresponds to "nonlinear self-adjointness" without differential substitutions. In this case, the only condition is Equation (110), which reduces to $R_{Q^a_\alpha} + \dfrac{\partial Q^a}{\partial u^\alpha} = 0$. This condition is, in general, non-trivial. (Unfortunately, some recent work [16] incorrectly asserts that, for any DE system, every adjoint-symmetry of the form $Q^{\alpha}(x,u)$ is a multiplier).

When Theorem 3 is applied to variational DE systems, it yields the following well-known connection [4,26,27,29] with Noether's theorem.

Corollary 1. *For a variational DE system* (33)*, the multiplier determining system* (109)–(111) *reduces to a determining system for variational symmetries. In particular, the determining equation for adjoint-symmetries* (109) *coincides with the determining equation for symmetries* (23)*, and the Helmholtz-type conditions* (110)–(111) *coincide with the necessary and sufficient conditions for a symmetry to be variational (namely, that* $\mathrm{pr}\hat{\mathbf{X}}(L) = D_i\Gamma^i$ *holds for some differential vector function* Γ^i*, where L is the Lagrangian* (34)*)).*

Note that, in this modern formulation of Noether's theorem, the use of a Lagrangian is completely by-passed through the Helmholtz-type conditions (110) and (111).

Example: For the semilinear wave Equation (63), the determining equation for multipliers of lowest-order form $Q_C(t,x,u)$ is given by

$$0 = E_u(Q_C F) = \delta^*_{Q_C} F + \delta^*_F Q_C. \tag{113}$$

Since $Q_C(t,x,u)$ does not depend on derivatives of u, this determining equation splits with respect to the variables u_t, u_x, u_{tt}, u_{xx}, giving an overdetermined linear system which can be derived and solved directly by Maple. This provides the simplest computational route to finding all multipliers of lowest-order form. The connection between multipliers and adjoint-symmetries arises when the determining Equation (113) is instead split into the two terms $\delta^*_{Q_C} F$ and $\delta^*_F Q_C$, which are given by

$$\delta^*_{Q_C} F = D_t^2 Q_C - D_x^2 Q_C - bD_t Q_C - cD_x Q_C + (b' u_t + c' u_x + m') Q_C = R_{Q_C}(F) \tag{114}$$

$$\delta^*_F Q_C = \frac{\partial Q_C}{\partial u} F = E_u(Q_C) F \tag{115}$$

where the operator R_{Q_C} is obtained from expression (103). Hence, on the solution space $\mathcal{E}$ of the wave Equation (63), the multiplier determining equation reduces to the adjoint-symmetry Equation (67). Off of the solution space $\mathcal{E}$, the multiplier determining equation then becomes

$$0 = R_{Q_C}(F) + E_u(Q_C) F = 2\frac{\partial Q_C}{\partial u} F \tag{116}$$

which splits with respect to F, yielding

$$\frac{\partial Q_C}{\partial u} = 0. \tag{117}$$

This Helmholtz-type Equation (117) together with the adjoint-symmetry Equation (67) constitutes the determining system (109)–(111) for finding all lowest-order multipliers $Q_C(t,x,u)$ admitted by the wave Equation (63).

The Helmholtz-type Equation (117) directly shows that all adjoint-symmetries of the form $Q(t,x)$ are conservation laws multipliers $Q_C(t,x)$, and so, the three adjoint-symmetries (100) each determine a non-trivial conserved current through the characteristic equation

$$Q_C F = D_t \hat{C}^t + D_x \hat{C}^x. \tag{118}$$

These conserved currents $(\hat{C}^t, \hat{C}^x)$ can be derived in terms of the multipliers $Q_C(t,x,u)$ in several different ways. One simple way is by applying integration by parts to the terms in $Q_C F$ to get a total time derivative $D_t\hat{C}^t$ plus a total space derivative $D_x\hat{C}^x$, which yields $(\hat{C}^t, \hat{C}^x)$. Another way is by taking $\hat{C}^t(t,x,u,u_t,u_x)$ and $\hat{C}^x(t,x,u,u_t,u_x)$ as unknowns and splitting the characteristic equation with respect to u_{tt}, u_{tx}, u_{xx} to get a linear system of determining equations that can be integrated. The resulting conserved currents are shown in Table 3. The specific relationship between these conserved currents and the conserved currents derived in Table 2 will be explained in the next subsection.

Table 3. Conserved currents.

Conditions	Q_C	$\hat{C}^t$	$\hat{C}^x$
$m = m_1 u + m_2 \int b\, du$ $+ m_3 \int c\, du$ $m_1 = m_3^2 - m_2^2$	$e^{m_3 x + m_2 t}$	$e^{m_3 x + m_2 t}(u_t - m_2 u + \int b\, du)$	$e^{m_3 x + m_2 t}(m_3 u - u_x + \int c\, du)$
$b = b_0 + b_1 m'$ $c = c_0 + c_1 m'$ $b_1\beta + c_1\alpha = 1$ $\beta(\beta - b_0) = \alpha(\alpha + c_0)$	$e^{\alpha x + \beta t}$	$e^{\alpha x + \beta t}(u_t - \beta u + \int b\, du)$	$e^{\alpha x + \beta t}(\alpha u - u_x + \int c\, du)$
$b = \pm(\gamma + \frac{1}{\gamma} m')$ $c = -\gamma + \frac{1}{\gamma} m'$	$e^{\gamma x} q(x \mp t)$	$e^{\gamma x}(q(u_t + \int b\, du) \pm q' u)$	$e^{\gamma x}((q' - \gamma q)u - (u_x \mp \int b\, du))$

3.2. Conservation Laws Produced by a Multiplier/Symmetry Pair

From Theorem 3 and Proposition 3, every multiplier admitted by a given DE system determines, up to local equivalence, a conserved current for the system. Since multipliers are adjoint-symmetries, the adjoint-symmetry/symmetry formula (72) can be applied by using any multiplier (92) together with any symmetry (40). The resulting conserved current produced this way is given by

$$Q_C^a(\delta_P F)_a - P^\alpha(\delta^*_{Q_C} F)_\alpha = D_i \Psi^i(P, Q_C; F) \tag{119}$$

where $P^\alpha(x,u,\partial u,\ldots,\partial^r u)$ is a given symmetry characteristic and $Q_C^a(x,u,\partial u,\ldots,\partial^s u)$ is a given multiplier. The following result characterizing these conserved currents will now be established for DE systems in the class stated in Proposition 3. The case of DE systems consisting of a single DE has appeared previously in Reference [23].

Theorem 4. *Let $\Psi^i(P, Q_C; F)$ be the conserved current produced from the adjoint-symmetry/symmetry formula* (72) *by using any multiplier $Q_C^a(x,u,\partial u,\ldots,\partial^s u)$ together with any symmetry characteristic $P^\alpha(x,u,\partial u,\ldots,\partial^r u)$. This conserved current $\Psi^i(P, Q_C; F)$ is locally equivalent to a conserved current* (30) *that is given by the infinitesimal action of the symmetry $\hat{\mathbf{X}}_P = P^\alpha \partial/\partial u^\alpha$ applied to the conserved current C^i determined by the multiplier Q_C^a. In particular, $\Psi^i(P, Q_C; F)$ and C^i are related by*

$$(\Psi^i(P, Q_C; F) - \mathrm{pr}\hat{\mathbf{X}}_P(C^i))|_{\mathcal{E}} = D_j \Theta^{ij} \tag{120}$$

for some differential antisymmetric tensor function $\Theta^{ij}(x,u,\partial u,\ldots,\partial^k u)$.

The proof consists of showing that both conserved currents $\Psi^i(P, Q_C; F)$ and $\mathrm{pr}\hat{\mathbf{X}}_P(C^i)$ have the same multiplier. Consider the local conservation law determined by the multiplier Q_C^a. The symmetry $\hat{\mathbf{X}}_P$ applied to the characteristic Equation (91) of this conservation law yields

$$\mathrm{pr}\hat{\mathbf{X}}_P(D_i\tilde{C}^i) = \mathrm{pr}\hat{\mathbf{X}}(Q_C^a F_a) = \delta_P(Q_C^a F_a) = (\delta_P Q_C)^a F_a + Q_C^a(\delta_P F_a). \tag{121}$$

The second term in this equation can be expressed as

$$Q_C^a(\delta_P F_a) = Q_C^a R_P(F)_a = F_a R_P^*(Q_C)^a + D_i\Gamma^i(Q_C, F; P) \tag{122}$$

using the symmetry identity (41) combined with integration by parts, where $\Gamma^i(Q_C, F; P)|_\mathcal{E} = 0$. Next, the first term in Equation (121) can be expressed as

$$\begin{aligned}(\delta_P Q_C)^a F_a &= P^\alpha(\delta_F^* Q_C)_\alpha + D_i\Psi^i(P, Q_C; F) = -P^\alpha(\delta_{Q_C}^* F)_\alpha + D_i\Psi^i(P, Q_C; F)\\ &= -P^\alpha R_{Q_C}(F)_\alpha + D_i\Psi^i(P, Q_c; F)\end{aligned} \tag{123}$$

through the Fréchet derivative identity (3) combined with the multiplier determining Equation (107) and the adjoint-symmetry identity (61). Integration by parts then yields

$$(\delta_P Q_C)^a F_a = -F_a R_{Q_C}^*(P)^a + D_i(\Psi^i - \Gamma^i(P, F; Q_C)) \tag{124}$$

where $\Gamma^i(P, F; Q_C)|_\mathcal{E} = 0$. Substitution of expressions (124) and (122) into Equation (121) gives

$$\text{pr}\hat{\mathbf{X}}_P(D_i\tilde{C}^i) = (R_P^*(Q_C)^a - R_{Q_C}^*(P)^a)F_a + D_i(\Psi^i(P, Q_c; F) + \Gamma^i(Q_C, F; P) - \Gamma^i(P, F; Q_x)). \tag{125}$$

Finally, since $\text{pr}\hat{\mathbf{X}}_P$ commutes with total derivatives [24,29], this yields

$$D_i(\text{pr}\hat{\mathbf{X}}_P(C^i) + \tilde{\Gamma}^i) = Q_\Psi^a F_a \tag{126}$$

where $\tilde{\Gamma}^i|_\mathcal{E} = 0$ is a locally trivial conserved current, and where

$$Q_\Psi^a = R_P^*(Q_C)^a - R_Q^*(P)^a \tag{127}$$

is the multiplier (98) of the local conservation law (96) from the adjoint-symmetry/symmetry formula (72) with $Q^a = Q_C^a$. This completes the proof.

Theorem 4 is a generalization of a similar result [2,3,36] for variational DE systems, where the adjoint-symmetry/symmetry formula (72) reduces to a formula using any pair of symmetries.

Corollary 2. *For a variational DE system, let $\Psi^i(P, Q_C; F)$ be the conserved current produced from the adjoint-symmetry/symmetry formula (72) by using any symmetry characteristic $P^\alpha(x, u, \partial u, \dots, \partial^r u)$ together with any multiplier $Q_C^\alpha(x, u, \partial u, \dots, \partial^s u)$ given by a variational symmetry characteristic. The conserved current $\Psi^i(P, Q_C; F)$ is locally equivalent to a conserved current (30) that is given by the infinitesimal action of the symmetry $\hat{\mathbf{X}}_P = P^\alpha\partial/\partial u^\alpha$ applied to the conserved current C^i determined by the multiplier Q_C^α. Moreover, through Noether's theorem, the multiplier of this conserved current $\Psi^i(P, Q_C; F)$ is the characteristic of a variational symmetry given by the commutator of the symmetries $\hat{\mathbf{X}}_P = P^\alpha\partial/\partial u^\alpha$ and $\hat{\mathbf{X}}_{Q_C} = Q_C^\alpha\partial/\partial u^\alpha$.*

Several basic properties of the adjoint-symmetry/symmetry formula (72) can be deduced from Theorem 4, as first shown in Reference [23] for DE systems consisting of a single DE.

Theorem 5. *(i) For a given DE system (14), let Q_C^a be the multiplier for a local conservation law in which the components of the conserved current C^i have no explicit dependence on x. Then, using any translation symmetry $\mathbf{X} = a^i\partial/\partial x^i$, with characteristic $P^\alpha = -a^i u_i^\alpha$ where a^i is a constant vector, the conserved current $\Psi^i(P, Q_C; F)$ is locally trivial. (ii) For a given DE system (14) that possesses a scaling symmetry $\mathbf{X} = a_{(i)}x^i\partial/\partial x^i + b_{(\alpha)}u^\alpha\partial/\partial u^\alpha$, where $a_{(i)}, b_{(\alpha)}$ are constants, let Q_C^a be the multiplier for a local conservation law in which the components of the conserved current C^i are scaling homogeneous. Then, using the characteristic $P^\alpha = b_{(\alpha)}u^\alpha - a_{(i)}x^i u_i^\alpha$ of the scaling symmetry, the conserved current $\Psi^i(P, Q_C; F)$ is locally equivalent to a multiple w of the conserved current C^i determined by Q_C^a. This multiple, $w =$* const.*, is the scaling weight of the conserved integral given by $\int_{\partial\Omega} C^i dS_i$ where Ω is any closed domain in $\mathbb{R}^n$ and $\partial\Omega$ is its boundary surface.*

The proof is a straightforward extension of the proof in Reference [23] and will be omitted.

Part (i) of this theorem explains the observations made in many recent papers in which Ibragimov's theorem gave only trivial local conservation laws. This will happen whenever the only local symmetries admitted by a DE system are translations and the only admitted adjoint-symmetries have no dependence on x.

Part (ii) of the theorem first appeared in Reference [28]. It shows that the local conservation laws admitted by any DE system with a scaling symmetry can be obtained from an algebraic formula using the conservation law multipliers. This explains why in many recent papers, the use of scaling symmetries in Ibragimov's theorem has produced non-trivial local conservation laws.

A more important point comes from putting together Theorems 3 and 4. Together, these two theorems show that the adjoint-symmetry/symmetry formula (72) cannot produce any "new" local conservation laws, since any local conservation law admitted by a given DE system must already arise directly from a multiplier. Moreover, for this formula to generate all of the local conservation laws for a given DE system, it seems plausible that the set of admitted symmetries needs to act transitively on a set of admitted local conservation laws, so then every multiplier arises from some symmetry applied to some multiplier. The need for a transitive action is especially clear from Corollary 2, since if a pair of commuting variational symmetries is used in the formula, then the resulting local conservation law will have a trivial multiplier, and hence, will be a locally trivial conservation law.

These significant deficiencies should discourage the unnecessary use of the adjoint-symmetry/symmetry formula (72), and consequently the unnecessary use of Ibragimov's theorem, when local conservation laws are being sought for a given DE system. It is much simpler and more direct to find all multipliers and then to derive the conserved currents determined by these multipliers, as will be explained further in the next section.

Example: For the semilinear wave Equation (63), Table 2 shows the conserved currents obtained from the adjoint-symmetry/symmetry formula (72). Each of these conserved currents can be checked to satisfy the characteristic Equation (118) with $\hat{C}^t = \Psi^t$ and $\hat{C}^x = \Psi^x$, where the resulting multipliers Q_Ψ are shown in Table 4. There is a simple relationship (127) between each multiplier Q_Ψ and the adjoint-symmetry/symmetry pair Q, P used to generate the conserved current (Ψ^t, Ψ^x). In particular, from expressions (100)–(104) for the symmetry characteristics, adjoint-symmetries and their associated operators R_P and R_Q, the relationship (127) yields

$$\begin{aligned} Q_{\Psi(P_1,Q_l;F)} &= R^*_{P_1}(Q_l) - R^*_{Q_l}(P_1) = D_t Q_l, \quad l = 1,2,3 \\ Q_{\Psi(P_2,Q_l;F)} &= R^*_{P_2}(Q_l) - R^*_{Q_l}(P_2) = D_x Q_l, \quad l = 1,2,3 \end{aligned} \tag{128}$$

in accordance with Table 4.

In particular, consider the case when $m(u)$ is zero and both $b(u), c(u)$ are arbitrary, so then the only admitted multiplier of lowest-order form $Q(t,x,u)$ is $Q = 1$ (up to a multiplicative constant), as shown by Table 1. In this case, it is straightforward to show that (by solving the relevant determining equations) there are no first-order multipliers and that the only admitted point symmetries are generated by the translations (89). Consequently, when the set of multipliers $Q(t,x,u,u_t,u_x)$ is considered, a single non-trivial conservation law $C^t = u_t + \int b(u)\,du$, $C^x = -u_x + \int c(u)\,du$ is admitted by the wave equation $u_{tt} - u_{xx} + b(u)u_t + c(u)u_x = 0$ with $b(u)$ and $c(u)$ arbitrary, whereas all of the conserved currents (Ψ^t, Ψ^x) obtained from Ibragimov's theorem (85) or from the simpler equivalent adjoint-symmetry/symmetry formula (86), are trivial! Note that, correspondingly, the symmetry action on the set of non-trivial conservation laws given by the set of multipliers $Q(t,x,u,u_t,u_x)$ is not transitive. This example succinctly illustrates the incompleteness of these formulas for generating conservation laws.

Table 4. Multipliers from the adjoint-symmetry/symmetry formula.

P	Q	Ψ^t, Ψ^x	Q_Ψ
$-u_t$	$e^{m_3x+m_2t}$	$m_2e^{m_3x+m_2t}(u_t - m_2u + \int b\,du)$, $m_2e^{m_3x+m_2t}(m_3u - u_x + \int c\,du)$	$m_2e^{m_3x+m_2t}$ $= D_t(e^{m_3x+m_2t})$
$-u_x$	$e^{m_3x+m_2t}$	$m_3e^{m_3x+m_2t}(u_t - m_2u + \int b\,du)$, $m_3e^{m_3x+m_2t}(m_3u - u_x + \int c\,du)$	$m_3e^{m_3x+m_2t}$ $= D_x(e^{m_3x+m_2t})$
$-u_t$	$e^{\alpha x+\beta t}$	$\beta e^{\alpha x+\beta t}(u_t - \beta u + \int b\,du)$, $\beta e^{\alpha x+\beta t}(\alpha u - u_x + \int c\,du)$	$\beta e^{\alpha x+\beta t}$ $= D_t(e^{\alpha x+\beta t})$
$-u_x$	$e^{\alpha x+\beta t}$	$\alpha e^{\alpha x+\beta t}(u_t - \beta u + \int b\,du)$, $\alpha e^{\alpha x+\beta t}(\alpha u - u_x + \int c\,du)$	$\alpha e^{\alpha x+\beta t}$ $= D_x(e^{\alpha x+\beta t})$
$-u_t$	$e^{\gamma x}q$	$-e^{\gamma x}(q''u \pm q'(u_t + \int b\,du))$, $\pm e^{\gamma x}((\gamma q' - q'')u + q(u_x \mp \int b\,du))$	$\mp e^{\gamma x}q$ $= D_t(e^{\gamma x}q)$
$-u_x$	$e^{\gamma x}q$	$e^{\gamma x}(\pm(q'' + \gamma q')u + (q' + \gamma q)(u_t + \int b\,du))$, $e^{\gamma x}((q'' - \gamma^2 q)u - (q' + \gamma q)(u_x \mp \int b\,du))$	$e^{\gamma x}(q' + \gamma q)$ $= D_x(e^{\gamma x}q)$

4. A Direct Construction Method to Find All Local Conservation Laws

The results stated in Proposition 3 and Theorems 3, 4 and 5 have been developed in References [4,23,26–28] and extended in References [29,35]. This collective work provides a simple, algorithmic method to find *all* local conservation laws for any given system of DEs. The method is based on the general result that all local conservation laws arise from multipliers as given by the solutions of a linear system of determining equations, where the multipliers are simply adjoint-symmetries subject to certain Helmholtz-type conditions.

Consequently, all multipliers can be found by either of the two following methods [4]: (1) directly solve the full determining system for multipliers; or (2) first, solve the determining equation for adjoint-symmetries, and next, check which of the adjoint-symmetries satisfy the Helmholtz-type conditions. The adjoint-symmetry determining equation is simply the adjoint of the symmetry determining equation, and hence, it can be solved by the standard algorithmic procedure used for solving the symmetry determining equation [24,31,32]. Likewise, the same procedure works equally well for solving the multiplier determining system.

A natural question is, in practice, at which differential orders $s \geq 0$ will multipliers or adjoint-symmetries $Q^a(x, u, \partial u, \ldots, \partial^s u)$ be found?

One answer is that the same situation arises for symmetries. Normally, point symmetries are sought first, since many DE systems admit point symmetries, and since relatively fewer DE systems admit contact symmetries or higher-order symmetries. Indeed, the existence of a sufficiently high-order symmetry is one main definition of an integrable system [37], as this can indicate the existence of an infinite hierarchy of successively higher-order symmetries. For multipliers, the most physically important conserved currents always have a low differential order. Based on numerous examples, a concrete definition of a *low-order multiplier* that seems to characterize these physically important conserved currents, and distinguishes them from higher-order conserved currents arising for integrable systems, has been introduced in recent work [29,35].

Another answer is that it is straightforward just to find all multipliers or adjoint-symmetries with a specified differential order $s = 0, 1, 2, \ldots$, going up to any desired maximum finite order. Moreover, in some situations, a standard descent/induction argument [38–40] can be used to find the multipliers or adjoint-symmetries to all orders $s \geq 0$.

Once a set of multipliers has been found for a given DE system, the corresponding conserved currents are straightforward to find in an explicit form. Several different methods are available.

One algorithmic method is the direct integration of the characteristic equation [31,41] defining the conserved current. Another algorithmic method is the use of a homotopy integral formula. This method has several versions [4,24,26,27,29], all of which involve trade-offs between the simplicity of the integration versus the flexibility of avoiding singularities (if any) in the integrand.

However, purely algebraic methods for the construction of conserved currents from multipliers are known. One algebraic method is the use of a scaling formula [23,28,31], which is given by the adjoint-symmetry/symmetry formula. This applies only to DE systems that admit a scaling symmetry, but it has recently been extended to general DE systems by incorporating a dimensional analysis method as shown in Reference [29]. In particular, with the use of this dimensional-scaling method, the construction of conserved currents becomes completely algebraic.

Therefore, the general method just outlined provides a completely algorithmic computational way to derive all local conservation laws for any given DE system. In particular, there is no need to resort to any special methods or ansatzs (such as the "abc" technique [42], partial Lagrangians [43], "nonlinear self-adjointness" [9,12,14,15,17], undetermined coefficients [44]), which at best just yield a subset of all of the local conservation laws admitted by a DE system or just apply to restricted classes of DE systems.

Example: The semilinear wave Equation (63) can be expected to admit conserved currents that depend nonlinearly on u_t and u_x, in addition to the previous conserved currents in Table 3, all of which have linear dependence on u_t and u_x. The multipliers (100) for the latter conserved currents have the form $Q_C(t,x)$. Conserved currents that depend nonlinearly on u_t and u_x will arise from multipliers $Q_C(t,x,u,u_t,u_x)$ that have explicit dependence on u_t and u_x. It is straightforward to find all such multipliers by using Maple to set up and solve the multiplier Equation (113), which splits with respect to the variables $u_{tt}, u_{tx}, u_{xx}, u_{ttt}, u_{txx}, u_{ttx}, u_{txx}, u_{xxx}$, giving an overdetermined linear system. Alternatively, the multiplier Equation (113) can be split instead into the two terms $\delta^*_{Q_C}F$ and $\delta^*_F Q_C$, which provides a direct connection between multipliers and adjoint-symmetries. In particular, the first term in the multiplier Equation (113) consists of

$$\delta^*_{Q_C}F = D_t^2 Q_C - D_x^2 Q_C - bD_tQ_C - cD_xQ_C + (b'u_t + c'u_x + m')Q_C = R_{Q_C}(F) \tag{129}$$

where the operator R_{Q_C} is found to be given by

$$\begin{aligned} R_Q = {} & \frac{\partial Q_C}{\partial u_t}D_tF + \frac{\partial Q_C}{\partial u_x}D_xF + \frac{\partial^2 Q_C}{\partial u_t \partial u_t}F + \frac{\partial Q_C}{\partial u} - 2b\frac{\partial Q_C}{\partial u_t} + 2u_t\frac{\partial^2 Q_C}{\partial u \partial u_t} \\ & + 2u_{tx}\frac{\partial^2 Q_C}{\partial u_x \partial u_t} + 2(u_{xx} - bu_t - cu_x - d)\frac{\partial^2 Q_C}{\partial u_t \partial u_t} \end{aligned} \tag{130}$$

for multipliers $Q_C(t,x,u,u_t,u_x)$, through $u_{tt} = u_{xx} - b(u)u_t - c(u)u_x - m(u)$. The second term in the multiplier Equation (113) is given by

$$\delta^*_F Q_C = \frac{\partial Q_C}{\partial u}F - D_t\Big(\frac{\partial Q_C}{\partial u_t}F\Big) - D_x\Big(\frac{\partial Q_C}{\partial u_x}F\Big) = -\frac{\partial Q_C}{\partial u_t}D_tF - \frac{\partial Q_C}{\partial u_x}D_xF + E_u(Q)F \tag{131}$$

where

$$\begin{aligned} E_u(Q) = {} & -\frac{\partial^2 Q_C}{\partial u_t \partial u_t}F + \frac{\partial Q_C}{\partial u} - u_t\frac{\partial^2 Q_C}{\partial u \partial u_t} - u_x\frac{\partial^2 Q_C}{\partial u \partial u_x} - 2u_{tx}\frac{\partial^2 Q_C}{\partial u_x \partial u_t} \\ & + (bu_t + cu_x + d - 2u_{xx})\frac{\partial^2 Q_C}{\partial u_t \partial u_t} \end{aligned} \tag{132}$$

for multipliers $Q_C(t,x,u,u_t,u_x)$. On the solution space $\mathcal{E}$ of the wave Equation (63), the terms (131) vanish, while the other terms (129) reduce to the adjoint-symmetry Equation (67). Off of the solution space $\mathcal{E}$, these terms (129) and (131) become a linear combination of F, D_tF, D_xF, whose coefficients must vanish separately. This splitting is found to yield a single Helmholtz-type equation

$$2\frac{\partial Q_C}{\partial u} - b\frac{\partial Q_C}{\partial u_t} + u_t\frac{\partial^2 Q_C}{\partial u \partial u_t} - u_x\frac{\partial^2 Q_C}{\partial u \partial u_x} - (bu_t + cu_x + d)\frac{\partial^2 Q_C}{\partial u_t \partial u_t} = 0. \tag{133}$$

Taken together, this Helmholtz-type Equation (133) and the adjoint-symmetry Equation (67) constitute the determining system (109)–(111) for finding all first-order multipliers $Q_C(t, x, u, u_t, u_x)$ admitted by the wave Equation (63).

The most computationally effective way to solve Equations (133) and (67) in the determining system is by changing variables from t, x, u, u_t, u_x to $\mu = \frac{1}{2}(t + x)$, $\nu = \frac{1}{2}(t - x)$, u, $u_\mu = u_t + u_x$, $u_\nu = u_t - u_x$, based on null coordinates for the wave Equation (63). In these new variables, the general solution of the determining system consists of three distinct cases (as obtained using the Maple package 'rifsimp'), after the nonlinearity and homogeneity conditions (71) are imposed on $b(u), c(u), d(u)$. The resulting multipliers, after merging cases, are shown in Table 5. Each multiplier determines a non-trivial conserved current through the characteristic Equation (118). These conserved currents $(\hat{C}^t, \hat{C}^x)$ can be derived in terms of the multipliers $Q_C(t, x, u, u_t, u_x)$ in the same way discussed previously for lowest-order multipliers. The results are shown in Table 6.

Table 5. First-order multipliers.

Conditions	Q_C
$\frac{2m_1m_2}{m - m_1} = \frac{4m_1}{b \pm c} = \int (b \mp c)\, du$	$\frac{2m_1 + (b \pm c)(u_t \pm u_x)}{2m + (b \pm c)(u_t \pm u_x)}$
$m = (m_1 + \frac{1}{4}\int (b - c)\, du)(b + c)$, $(1 - \gamma)b = (1 + \gamma)c$	$\frac{((1-\gamma)u_t + (1+\gamma)u_x)(b^2 - c^2)}{((b+c)(u_t + u_x) + 2m)((b-c)(u_t - u_x)2m)}$

Table 6. First-order conserved currents.

Conditions	$\hat{C}^t$, $\hat{C}^x$
$\frac{2m_1m_2}{m - m_1} = \frac{4m_1}{b \pm c} = \int (b \mp c)\, du$	$\gamma \ln\left(\frac{b \pm c}{2m_1 + (b \pm c)(\gamma + u_t \pm u_x)}\right) + u_t + \frac{1}{2}\int (b \pm c)\, du$, $\mp \gamma \ln\left(\frac{b \pm c}{2m_1 + (b \pm c)(\gamma + u_t \pm u_x)}\right) - u_x + \frac{1}{2}\int (c \pm b)\, du + \gamma x$
$m = (m_1 + \frac{1}{4}\int (b - c)\, du)(b + c)$, $(1 - \gamma)b = (1 + \gamma)c$	$\ln\left(\frac{(\gamma(\int (b+c)\, du + 2(u_t - u_x)) + m_1)^{\frac{1}{\gamma}}}{\gamma\int (b+c)\, du + 2(u_t + u_x) + m_1}\right)$, $\ln\left((\gamma(\int (b+c)\, du + 2(u_t - u_x)) + m_1)^{\frac{1}{\gamma}} \times (\gamma\int (b+c)\, du + 2(u_t + u_x) + m_1)\right)$

5. Concluding Remarks

The conservation law theorem stated by Ibragimov in References [9,12] for "nonlinear self-adjoint" DEs and subsequent extensions of this theorem in References [14,15,17] are not new. In its most general form, this theorem is simply a re-writing of a standard formula [2–4] that uses a pair consisting of a symmetry and an adjoint-symmetry to produce a conservation law through a well-known Fréchet derivative identity [2,3,24,29,31]. Unfortunately, no references to prior literature are provided in Ibragimov's papers, which may give the impression that the results are original. One aspect that is novel is the derivation of the formula by using an auxiliary Lagrangian, although it does not in any way simplify either the formula or its content. Moreover, the condition of "nonlinear self-adjointness" is nothing but a re-writing of the condition that a DE system admits an adjoint-symmetry [4,29], and this condition automatically holds for any DE system that admits a local conservation law.

The present paper shows how the symmetry/adjoint-symmetry formula is directly connected to the action of symmetries on conservation laws, which explains a number of major drawbacks in trying to use the formula and, hence, in applying Ibragimov's theorem, as a method to generate conservation laws. In particular, the formula can generate trivial conservation laws and does not always yield all non-trivial conservation laws unless the symmetry action on the set of these conservation laws is transitive, which cannot be known until all conservation laws have been found.

A broader point, which is more important, is that there is a completely general method [29,31] using adjoint-symmetries [2–4,26,27] to find all local conservation laws for any given DE system. This method is a kind of adjoint version of the standard Lie method to find all local symmetries. The method is algorithmic [29], and the required computations are no more difficult than the computations used to find local symmetries.

Acknowledgments: The author is supported by an NSERC Discovery grant. M. Gandarias is thanked for helpful discussions.

Conflicts of Interest: The author declares no conflict of interest. The funding sponsor had no role in the design of the study; in the collection, analyses, or interpretation of data; in the writing of the manuscript, and in the decision to publish the results.

References

1. Noether, E. Invariante variations probleme. *Transp. Theory Stat. Phys* **1971**, *1*, 186–207.
2. Caviglia, G. Symmetry transformations, isovectors, and conservation laws. *J. Math. Phys.* **1986**, *27*, 972–978.
3. Lunev, F.A. An analogue of the Noether theorem for non-Noether and nonlocal symmetries. *Theor. Math. Phys.* **1991**, *84*, 816–820.
4. Anco, S.C.; Bluman, G. Direct construction of conservation laws from field equations. *Phys. Rev. Lett.* **1997**, *78*, 2869–2873.
5. Blaszak, M. *Multi-Hamiltonian Theory of Dynamical Systems*; Springer: Heildelberg, Germany, 1998.
6. Krasil'shchik, I.S.; Vinogradov, A.M. (Eds.) *Symmetries and Conservation Laws for Differential Equations of Mathematical Physics*; Translations of Mathematical Monographs 182; American Mathematical Society: Providence, RI, USA, 1999.
7. Verbotevsky, A. Notes on the horizontal cohomology. In *Secondary Calculus and Cohomological Physics*; Contemporary Mathematics 219; American Mathematical Society: Providence, RI, USA, 1997.
8. Zharinov, V.V. *Lecture Notes on Geometrical Aspects of Partial Differential Equations*; Series on Soviet and East European Mathematics 9; World Scientific: River Edge, NJ, USA, 1992.
9. Ibragimov, N.H. A new conservation theorem. *J. Math. Anal. Appl.* **2007**, *333*, 311–328.
10. Ibragimov, N.H. Quasi self-adjoint differential equations. *Arch. ALGA* **2007**, *4*, 55–60.
11. Ibragimov, N.H. Nonlinear self-adjointness in constructing conservation laws. *Arch. ALGA* **2010**, *7/8*, 1–86.
12. Ibragimov, N.H. Nonlinear self-adjointness and conservation laws. *J. Phys. A Math. Theor.* **2011**, *44*, 432002–432009.
13. Galiakberova, L.R.; Ibragimov, N.H. Nonlinear self-adjointness of the KricheverNovikov equation. *Commun. Nonlinear Sci. Numer. Simul.* **2014**, *19*, 361–363.
14. Gandarias, M.L. Weak self-adjoint differential equations. *J. Phys. A Math. Theor.* **2011**, *44*, 262001–262006.
15. Gandarias, M.L. Nonlinear self-adjointness through differential substitutions. *Commun. Nonlinear Sci. Numer. Simul.* **2014**, *19*, 3523–3528.
16. Zhang, Z.-Y. On the existence of conservation law multiplier for partial differential equations. *Commun. Nonlinear Sci. Numer. Simul.* **2015**, *20*, 338–351.
17. Zhang, Z.-Y.; Xie, L. Adjoint symmetry and conservation law of nonlinear diffusion equations with convection and source terms. *Nonlinear Anal. Real World Appl.* **2016**, *32*, 301–313.
18. Freire, I.L. New classes of nonlinearly self-adjoint evolution equations of third- and fifth-order. *Commun. Nonlin. Sci. Numer. Simul.* **2013**, *18*, 493–499.
19. Gandarias, M.L. Conservation laws for some equations that admit compacton solutions induced by a non-convex convection. *J. Math. Anal. Appl.* **2015**, *420*, 695–702.

20. Ibragimov, N.H.; Torrisi, M.; Tracina, R. Quasi self-adjoint nonlinear wave equations. *J. Phys. A Math. Theor.* **2010**, *43*, doi:10.1088/1751-8113/43/44/442001.
21. Ibragimov, N.H.; Torrisi, M.; Tracina, R. Self-adjointness and conservation laws of a generalized Burgers equation. *J. Phys. A Math. Theor.* **2011, 44,** 145201–145205.
22. Freire, I.L.; Sampaio, J.C.S. Nonlinear self-adjointness of a generalized fifth-order KdV equation. *J. Phys. A Math. Theor.* **2012,** 44, 032001–032007.
23. Anco, S.C. Symmetry properties of conservation laws. *Int. J. Mod. Phys. B* **2016,** *30,* doi:10.1142/S0217979216400038.
24. Olver, P. *Applications of Lie Groups to Differential Equations*; Springer: New York, NY, USA, 1986.
25. Martinez Alonso, L. On the Noether map. *Lett. Math. Phys.* **1979,** *3,* 419–424.
26. Anco, S.C.; Bluman, G. Direct construction method for conservation laws of partial differential equations. I. Examples of conservation law classifications, *Eur. J. Appl. Math.* **2002,** *13,* 545–566.
27. Anco, S.C.; Bluman, G. Direct construction method for conservation laws of partial differential equations. II. General treatment, *Eur. J. Appl. Math.* **2002,** *13,* 567–585.
28. Anco, S.C. Conservation laws of scaling-invariant field equations. *J. Phys. A Math. Gen.* **2003,** *36,* 8623–8638.
29. Anco, S.C. Generalization of Noether's theorem in modern form to non-variational partial differential equations. In *Recent progress and Modern Challenges in Applied Mathematics, Modeling and Computational Science;* Fields Institute Communications 79; Springer: New York, NY, USA, 2017.
30. *CRC Handbook of Lie Group Analysis of Differential Equations, Volume I: Symmetries, Exact Solutions, and Conservation Laws;* Ibragimov, N.H., Ed.; CRC Press: Boca Raton, FL, USA, 1994.
31. Bluman, G.; Cheviakov, A.; Anco, S.C. *Applications of Symmetry Methods to Partial Differential Equations;* Springer Applied Mathematics Series 168; Springer: New York, NY, USA, 2010.
32. Bluman, G.; Anco, S.C. *Symmetry and Integration Methods for Differential Equations;* Springer Applied Mathematics Series 154; Springer: New York, NY, USA, 2002.
33. Khamitova, R.S. The structure of a group and the basis of conservation laws. *Theoret. Math. Phys.* **1982,** *52,* 244–251.
34. Anco, S.C. Brock University, St. Catharines, ON, Canada. Unpublished work, 2017.
35. Anco, S.C.; Kara, A.H. Symmetry invariance of conservation laws of partial differential equations. *Eur. J. Appl. Math.* **2017,** in press.
36. Anco, S.C.; Bluman, G. Derivation of conservation laws from nonlocal symmetries of differential equations. *J. Math. Phys.* **1996,** *37,* 2361–2375.
37. Mikhailov, A.V.; Shabat, A.B.; Sokolov, V.V. Symmetry approach to classification of integrable equations. In *What Is Integrability;* Springer: Berlin, Germany, 1999.
38. Duzhin, S.V.; Tsujishita, T. Conservation laws of the BBM equation. *J. Phys. A* **1984,** *17,* 3267–3276.
39. Anco, S.C.; Pohjanpelto, J. Classification of local conservation laws of Maxwell's equations. *Acta Appl. Math.* **2001,** *69,* 285–327.
40. Anco, S.C.; Pohjanpelto, J. Conserved currents of massless spin s fields. *Proc. R. Soc.* **2003,** *459,* 1215–1239.
41. Wolf, T. A comparison of four approaches to the calculation of conservation laws. *Eur. J. Appl. Math.* **2002,** *13,* 129–152.
42. Morawetz, C. Variations on conservation laws for the wave equation. *Bull. Am. Math. Soc.* **2000,** *37,* 141–154.
43. Kara, A.H.; Mahomed, F.M. Noether-type symmetries and conservation laws via partial Lagrangians. *Nonlinear Dyn.* **2006,** *45,* 367–383.
44. Poole, D.; Hereman, W. Symbolic computation of conservation laws for nonlinear partial differential equations in multiple space dimensions. *J. Symb. Comput.* **2011,** *46,* 1355–1377.

symmetry

MDPI

Article

Conservation Laws of Discrete Evolution Equations by Symmetries and Adjoint Symmetries

Wen-Xiu Ma [1,2]

[1] College of Mathematics and Physics, Shanghai University of Electric Power, Shanghai 200090, China
[2] Department of Mathematics and Statistics, University of South Florida, Tampa, FL 33620, USA; E-Mail: mawx@cas.usf.edu; Tel.: +1-813-9749563; Fax: +1-813-9742700

Academic Editor: Roman M. Cherniha
Received: 24 April 2015 / Accepted: 20 May 2015 / Published: 22 May 2015

Abstract: A direct approach is proposed for constructing conservation laws of discrete evolution equations, regardless of the existence of a Lagrangian. The approach utilizes pairs of symmetries and adjoint symmetries, in which adjoint symmetries make up for the disadvantage of non-Lagrangian structures in presenting a correspondence between symmetries and conservation laws. Applications are made for the construction of conservation laws of the Volterra lattice equation.

Keywords: symmetry; adjoint symmetry; conservation law

---

1. Introduction

Noether's theorem tells us that a symmetry of a differential equation leads to a conservation law of the same equation, if the equation is derived from a Lagrangian, namely, it has a Lagrangian formulation [1,2]. The Lagrangian formulation of the equation is essential for presenting conservation laws from symmetries, and many physically important examples can be found in [1–4]. A natural question arises whether there is any correspondence between symmetries and conservation laws for differential equations not derivable from any Lagrangian. We would, in this paper, like to show that it is possible to give a positive answer to the above question if we adopt adjoint symmetries. More precisely, we want to exhibit that using symmetries and adjoint symmetries together can lead to conservation laws for both Lagrangian and non-Lagrangian equations.

A good attempt to use adjoint symmetries in computing conservation laws of differential equations was made in [5], and the approach utilizes adjoint symmetries, in which an adjoint invariance condition equivalently requires the existence of an adjoint symmetry. More generally, nonlinear self-adjointness was introduced on the basis of adjoint systems and successfully applied to construction of conservation laws of differential equations [6,7]. In that theory, the nonlinear self-adjointness means that the second set of dependent variables in an adjoint system stands for an adjoint symmetry.

We would like to consider regular differential-difference equations, and so, they can be written as evolution equations. Equations of this type contain difference equations, since any difference equation can be considered as a stationary equation of discrete evolution equations. We will utilize pairs of symmetries and adjoint symmetries to present a direct formula for constructing conservation laws, and thus conserved densities, for evolution equations. Our approach will be used to compute conservation laws for the Volterra lattice equation. The general theory justifies that symmetries really reflect conservation laws. However, the adoption of adjoint symmetries has not attracted much attention within the mathematical physics community. It is expected that our findings could stimulate to expose more mathematical properties of adjoint symmetries as well as develop efficient algorithms for computing adjoint symmetries.

The paper is structured as follows. In Section 2, a general theory will be formulated for constructing conservation laws and thus conserved densities from symmetries and adjoint symmetries for discrete evolution equations. In Section 3, an example will be analyzed, along with new conserved densities. Finally in Section 4, a few of concluding remarks will be given with some discussion.

2. General Theory

We will use a plain language to formulate the framework of our results on the correspondence between conservation laws and pairs of symmetries and adjoint symmetries.

Let the potential vector u be defined by

$$u = (u^1, \cdots, u^q)^T,\ u^i = u^i(n,t),\ 1 \le i \le q,\ n = (n_1, \cdots, n_p) \in \mathbb{Z}^p,\ t \in \mathbb{R}, \tag{1}$$

and E_i, $1 \le i \le p$, denote the shift operators for the variables n_i, $1 \le i \le p$, *i.e.*,

$$(E_i u)(n) = u(n_1, \cdots, n_{i-1}, n_i + 1, n_{i+1}, \cdots, n_p),\ n = (n_1, \cdots, n_p) \in \mathbb{Z}^p,\ 1 \le i \le p. \tag{2}$$

we introduce

$$E^\alpha = E_1^{\alpha_1} \cdots E_p^{\alpha_p},\ E^\alpha u = (E^\alpha u^1, \cdots, E^\alpha u^q) = (u_\alpha^1, \cdots, u_\alpha^q),\ (E^\alpha u^i)(n) = u^i(n+\alpha), \tag{3}$$

where $\alpha = (\alpha_1, \cdots, \alpha_p), n = (n_1, \cdots, n_p) \in \mathbb{Z}^p$, and $n + \alpha = (n_1 + \alpha_1, \cdots, n_p + \alpha_p)$. We denote by $\mathcal{A}$ and $\mathcal{B}$ the space of all local C^∞ functions in n, t, u and $E^\alpha u$ to some finite order α, and the space of all C^∞ functions in n, t, u and $E^\alpha u$ to some finite order α. Moreover, we assume that $\mathcal{A}^r$ and $\mathcal{B}^r$ denote the r-th order tensor products of $\mathcal{A}$ and $\mathcal{B}$, respectively, *i.e.*,

$$\mathcal{A}^r = \underbrace{\mathcal{A} \otimes \cdots \otimes \mathcal{A}}_{r},\ \mathcal{B}^r = \underbrace{\mathcal{B} \otimes \cdots \otimes \mathcal{B}}_{r}. \tag{4}$$

The locality here means that for a function $f(u)$ in $\mathcal{A}$, any value $(f(u))(n)$ is completely determined by the values of u at finitely many points $n \in \mathbb{Z}^p$. The space $\mathcal{A}$ contains functions of polynomial type, $P(u, Eu, \cdots, E^\alpha u)$, where P is a polynomial in its variables, and the space $\mathcal{B}$ contains non-local functions:

$$(f(u))(n) = \sum_{|\alpha| \le m} \sum_{k \ge n} (E^\alpha u)(k),\ m \ge 1,\ |\alpha| = |\alpha_1| + \cdots + |\alpha_p|.$$

For a local vector function $X = X(u) = (X_1, \cdots, X_r)^T \in \mathcal{A}^r$, we can compute its Gateaux derivative operator as follows:

$$X' = X'(u) = (V_j(X_i))_{r \times q} = \begin{bmatrix} V_1(X_1) & V_2(X_1) & \cdots & V_q(X_1) \\ V_1(X_2) & V_2(X_2) & \cdots & V_q(X_2) \\ \vdots & \vdots & \ddots & \vdots \\ V_1(X_r) & V_2(X_r) & \cdots & V_q(X_r) \end{bmatrix},\ V_i(X_j) = \sum_{\alpha \in \mathbb{Z}^p} \frac{\partial X_j}{\partial u_\alpha^i} E^\alpha, \tag{5}$$

and its (formal) adjoint operator

$$X'^\dagger = X'^\dagger(u) = (V_i^\dagger(X_j))_{q \times r},\ V_i^\dagger(X_j) = \sum_{\alpha \in \mathbb{Z}^p} E^{-\alpha} \frac{\partial X_j}{\partial u_\alpha^i}, \tag{6}$$

where the action of the operator $E^{-\alpha} f$ on g is given by $(E^{-\alpha} f) g = E^{-\alpha}(fg),\ f, g \in \mathcal{B}$. The operator $X'^\dagger$ can be explained as an adjoint operator of X', if the inner products

$$\langle Y, Z \rangle = \sum_{n \in \mathbb{Z}^p} \sum_{i=1}^{s} Y_i(n) Z_i(n),\ Y = (Y_1, \cdots, Y_s)^T, Z = (Z_1 \cdots, Z_s)^T \in \mathcal{B}^s,\ s \ge 1, \tag{7}$$

are well defined over some selected space of u. For example, for the adjoint operator $E_i^\dagger = E_i^{-1}$ of each shift operator E_i $(1 \le i \le p)$, we have

$$\langle E_i^\dagger Y, Z\rangle = \langle E_i^{-1} Y, Z\rangle = \langle Y, E_i Z\rangle,\ Y, Z \in \mathcal{B}.$$

Let us now consider an discrete evolution equation

$$u_t = K(n,t,u),\ K \in \mathcal{A}^q. \tag{8}$$

its linearized equation and adjoint linearized equation are defined by

$$(\sigma(n,t,u))_t = K'(n,t,u)\sigma(n,t,u),\ \sigma \in \mathcal{B}^q, \tag{9}$$
$$(\rho(n,t,u))_t = -K^{'\dagger}(n,t,u)\rho(n,t,u),\ \rho \in \mathcal{B}^q, \tag{10}$$

respectively. Here K' and $K^{'\dagger}$ denotes the Gateaux derivative operator and the adjoint Gateaux derivative operator of K with respect to u, respectively; and f_t denotes the total derivative of f with respect to t.

Definition 1. *A vector field $\sigma \in \mathcal{B}^q$ is called a symmetry of the discrete evolution Equation (8), if it satisfies the linearized Equation (9) when u solves Equation (8). A vector field $\rho \in \mathcal{B}^q$ is called an adjoint symmetry of the discrete evolution equation Equation (8), if it satisfies the adjoint linearized Equation (10) when u solves (8).*

It is easy to see that two local vector fields $\sigma \in \mathcal{A}^q$ and $\rho \in \mathcal{A}^q$ are a symmetry and an adjoint symmetry of the discrete evolution Equation (8), if and only if they satisfy

$$\frac{\partial \sigma(n,t,u)}{\partial t} = K'(n,t,u)\sigma(n,t,u) - \sigma'(n,t,u)K(n,t,u), \tag{11}$$
$$\frac{\partial \rho(n,t,u)}{\partial t} = -K^{'\dagger}(n,t,u)\rho(n,t,u) - \rho'(n,t,u)K(n,t,u), \tag{12}$$

respectively, when u solves Equation (8). Here σ' and ρ' are the Gateaux derivative operators of σ and ρ, and $\frac{\partial}{\partial t}$ denotes the partial derivative with respect to t.

Definition 2. *If a relation*

$$h_t = \sum_{i=1}^{p}(E_i - 1)f_i,\ h, f_i \in \mathcal{B},\ 1 \le i \le p, \tag{13}$$

holds when u solves the discrete evolution Equation (8), then Equation (13) is called a conservation law of Equation (8), h a conserved density of Equation (8), and $f = (f_1, \cdots, f_p)^T$ a conversed flux of Equation (8) corresponding to h. A conserved density $h \in \mathcal{B}$ is called trivial, if there exist $g_i \in \mathcal{B}$, $1 \le i \le p$, such that $h = \sum_{i=1}^{p}(E_i - 1)g_i$ holds on the solution set.

A conserved quantity means a quantity which does not vary with respect to time t on the solution set. A quantity defined by $\mathcal{I} = \sum_{n\in Z^p} I(n)$, $I = I(n,t,u) \in \mathcal{B}$, is called a functional. We would first like to show a relation between conserved quantities and adjoint symmetries.

Proposition 1. *Let $\mathcal{I} = \mathcal{I}(n,u)$ be a functional which does not depend explicitly on time t. Then, $\mathcal{I}$ is a conserved quantity of a discrete evolution equation $u_t = K$, $K = K(n,t,u) \in \mathcal{B}^q$, if and only if its variational derivative $\frac{\delta \mathcal{I}}{\delta u}$ is an adjoint symmetry of the same equation.*

Proof. Let $\mathcal{S}$ be the Schwartz space and the Gateaux derivative of an object $P = P(u)$ with respect to Y be defined by

$$P'[Y] = P'(u)[Y] = \frac{\partial P(n,t,u+\varepsilon Y)}{\partial \varepsilon}\Big|_{\varepsilon=0},\ Y \in \mathcal{B}^q.$$

We denote the variational derivative $\frac{\delta \mathcal{I}}{\delta u}$ by G, and so, the definition of variational derivatives tells that

$$\mathcal{I}'[Y] = \langle G, Y\rangle,\ Y \in \mathcal{S}^q,$$

where the inner product is defined as in Equation (7). Then the differentiability of functions of C^∞ class guarantees that on the solution set, we have

$$\mathcal{I}_t = \frac{\partial \mathcal{I}}{\partial t} + \frac{\partial \mathcal{I}(n, u+\varepsilon u_t)}{\partial \varepsilon}\Big|_{\varepsilon=0} = \langle G, u_t\rangle = \langle G, K\rangle.$$

It now follows that

$$\begin{aligned}(\mathcal{I}_t)'[Y] &= \langle G'[Y], K\rangle + \langle G, K'[Y]\rangle \\ &= \langle G'^{\dagger}K, Y\rangle + \langle K'^{\dagger}G, Y\rangle \\ &= \langle G'K, Y\rangle + \langle K'^{\dagger}G, Y\rangle \\ &= \langle \frac{\partial G}{\partial t} + G'K + K'^{\dagger}G, Y\rangle,\ Y \in \mathcal{S}^q,\end{aligned}$$

where $G'^{\dagger} = G'$ and $\frac{\partial G}{\partial t} = 0$ were used. This last equality implies that $\mathcal{I}$ is conseved if and only if G is an adjoint symmetry. □

We use a conventional assumption below: an empty product of shift operators E^{α_i}, $\alpha_i \in \mathbb{Z}$, $1 \le i \le p$, is understood to be the identity operator. For example, the operator $\Pi_{i=k}^{l} E_i^2$ implies the identity operator, when $k > l$. We would now like to prove the following lemma, in order to give a direct formula for constructing conservation laws of the discrete evolution Equation (8).

Lemma 1. *Let f and g be two C^∞ functions in variables $n_1, \cdots, n_p$. Then for any $\alpha = (\alpha_1, \cdots \alpha_p) \in \mathbb{Z}^p$, we have*

$$\begin{aligned} & fE^{\alpha}g - (E^{-\alpha}f)g = f(E_1^{\alpha_1}\cdots E_p^{\alpha_p}g) - (E_1^{-\alpha_1}\cdots E_p^{-\alpha_p}f)g \\ = & \sum_{i=1}^{p}(E_i - 1)\Big[\sum_{\beta_i=1}^{\alpha_i}(E_1^{-\alpha_1}\cdots E_{i-1}^{-\alpha_{i-1}}E_i^{-\beta_i}f)(E_i^{\alpha_i-\beta_i}E_{i+1}^{\alpha_{i+1}}\cdots E_p^{\alpha_p}g) \\ & - \sum_{\beta_i=1}^{-\alpha_i}(E_1^{-\alpha_1}\cdots E_{i-1}^{-\alpha_{i-1}}E_i^{-\alpha_i-\beta_i}f)(E_i^{-\beta_i}E_{i+1}^{\alpha_{i+1}}\cdots E_p^{\alpha_p}g)\Big], \end{aligned} \tag{14}$$

where the value of an empty sum is conventionally zero.

Proof. First note that we have

$$\begin{aligned} & aE_i^k b - (E_i^{-k}a)b \\ = & (E_i - 1)\Big[\sum_{l=1}^{k}(E_i^{-l}a)(E_i^{k-l}b) - \sum_{l=1}^{-k}(E_i^{-k-l}a)(E_i^{-l}b)\Big] \\ = & \begin{cases} (E_i - 1)\sum_{l=1}^{k}(E_i^{-l}a)(E_i^{k-l}b), & k \ge 0, \\ (E_i - 1)\sum_{l=1}^{-k}(E_i^{-k-l}a)(E_i^{-l}b), & k < 0, \end{cases} \end{aligned} \tag{15}$$

for any $k \in \mathbb{Z}$ and any two C^∞ functions a and b in variables $n_1, \cdots, n_p$. Then we decompose that

$$\begin{aligned} & fE^\alpha g - (E^{-\alpha} f)g = f(E_1^{\alpha_1} \cdots E_p^{\alpha_p} g) - (E_1^{-\alpha_1} \cdots E_p^{-\alpha_p} f)g \\ = & \sum_{i=1}^{p} \big[(E_1^{-\alpha_1} \cdots E_{i-1}^{-\alpha_{i-1}} f)(E_i^{\alpha_i} \cdots E_p^{\alpha_p} g) - (E_1^{-\alpha_1} \cdots E_i^{-\alpha_i} f)(E_{i+1}^{\alpha_{i+1}} \cdots E_p^{\alpha_p} g)\big]. \end{aligned}$$

It now follows from Equation (15) that each term in the above sum can be computed as follows:

$$\begin{aligned} & (E_1^{-\alpha_1} \cdots E_{i-1}^{-\alpha_{i-1}} f)(E_i^{\alpha_i} \cdots E_p^{\alpha_p} g) - (E_1^{-\alpha_1} \cdots E_i^{-\alpha_i} f)(E_{i+1}^{\alpha_{i+1}} \cdots E_p^{\alpha_p} g) \\ = & (E_1^{-\alpha_1} \cdots E_{i-1}^{-\alpha_{i-1}} f)[E_i^{\alpha_i}(E_{i+1}^{\alpha_{i+1}} \cdots E_p^{\alpha_p} g)] \\ & -[E_i^{-\alpha_i}(E_1^{-\alpha_1} \cdots E_{i-1}^{-\alpha_{i-1}} f)](E_{i+1}^{\alpha_{i+1}} \cdots E_p^{\alpha_p} g) \\ = & (E_i - 1)\Big[\sum_{\beta_i=1}^{\alpha_i} (E_1^{-\alpha_1} \cdots E_{i-1}^{-\alpha_{i-1}} E_i^{-\beta_i} f)(E_i^{\alpha_i-\beta_i} E_{i+1}^{\alpha_{i+1}} \cdots E_p^{\alpha_p} g) \\ & \quad - \sum_{\beta_i=1}^{-\alpha_i} (E_1^{-\alpha_1} \cdots E_{i-1}^{-\alpha_{i-1}} E_i^{-\alpha_i-\beta_i} f)(E_i^{-\beta_i} E_{i+1}^{\alpha_{i+1}} \cdots E_p^{\alpha_p} g)\Big], \ 1 \le i \le p, \end{aligned}$$

where the value of an empty sum is conventionally zero. This allows us to conclude that the equality Equation (14) holds for any $\alpha = (\alpha_1, \cdots, \alpha_p) \in \mathbb{Z}^p$. The proof is finished. □

Theorem 1. *Let $\sigma = (\sigma_1, \cdots, \sigma_q)^T \in \mathcal{B}^q$ and $\rho = (\rho_1, \cdots, \rho_q)^T \in \mathcal{B}^q$ be a symmetry and an adjoint symmetry of the discrete evolution Equation (8), respectively. Then we have a conservation law of the discrete evolution Equation (8):*

$$\begin{aligned} & (\sigma^T \rho)_t = (\sum_{i=1}^{q} \sigma_i \rho_i)_t \\ = & \sum_{k=1}^{p} (E_k - 1) \sum_{i,j=1}^{q} \sum_{\alpha \in \mathbb{Z}^p} \Big[\sum_{\beta_k=1}^{\alpha_k} (E_1^{-\alpha_1} \cdots E_{k-1}^{-\alpha_{k-1}} E_k^{-\beta_k} \rho_i \frac{\partial K_i}{\partial u_\alpha^j})(E_k^{\alpha_k-\beta_k} E_{k+1}^{\alpha_{k+1}} \cdots E_p^{\alpha_p} \sigma_j) \\ & - \sum_{\beta_k=1}^{-\alpha_k} (E_1^{-\alpha_1} \cdots E_{k-1}^{-\alpha_{k-1}} E_k^{-\alpha_k-\beta_k} \rho_i \frac{\partial K_i}{\partial u_\alpha^j})(E_k^{-\beta_k} E_{k+1}^{\alpha_{k+1}} \cdots E_p^{\alpha_p} \sigma_j)\Big], \end{aligned} \tag{16}$$

where $\alpha = (\alpha_1, \cdots, \alpha_p)$, and the value of an empty sum is conventionally zero. Therefore, $\sigma^T \rho$ is a conserved density of the discrete evolution Equation (8).

Proof. Let us compute that

$$\begin{aligned} & (\sigma^T \rho)_t = \sigma_t^T \rho + \sigma^T \rho_t = \rho^T \sigma_t + \sigma^T \rho_t \\ = & \rho^T K' \sigma - \sigma^T K'^{\dagger} \rho = \sum_{i,j=1}^{q} (\rho_i V_j(K_i) \sigma_j - \sigma_j V_j^{\dagger}(K_i) \rho_i) \\ = & \sum_{i,j=1}^{q} \sum_{\alpha \in \mathbb{Z}^p} (\rho_i \frac{\partial K_i}{\partial u_\alpha^j} E^\alpha \sigma_j - \sigma_j E^{-\alpha} \rho_i \frac{\partial K_i}{\partial u_\alpha^j}). \end{aligned}$$

By using Lemma 1, for all $1 \le i, j \le q$, $\alpha \in \mathbb{Z}^p$, we have

$$\rho_i \frac{\partial K_i}{\partial u^j_\alpha} E^\alpha \sigma_j - \sigma_j E^{-\alpha} \rho_i \frac{\partial K_i}{\partial u^j_\alpha}$$
$$= \sum_{k=1}^{p} (E_k - 1) \Big[\sum_{\beta_k=1}^{\alpha_k} (E_1^{-\alpha_1} \cdots E_{k-1}^{-\alpha_{k-1}} E_k^{-\beta_k} \rho_i \frac{\partial K_i}{\partial u^j_\alpha})(E_k^{\alpha_k - \beta_k} E_{k+1}^{\alpha_{k+1}} \cdots E_p^{\alpha_p} \sigma_j)$$
$$- \sum_{\beta_k=1}^{-\alpha_k} (E_1^{-\alpha_1} \cdots E_{k-1}^{-\alpha_{k-1}} E_k^{-\alpha_k - \beta_k} \rho_i \frac{\partial K_i}{\partial u^j_\alpha})(E_k^{-\beta_k} E_{k+1}^{\alpha_{k+1}} \cdots E_p^{\alpha_p} \sigma_j) \Big],$$

where the value of an empty sum is conventionally zero. If then follows that the equality Equation (16) is true, and thus, $\sigma^T \rho$ is a conserved density of the discrete evolution Equation (8). The proof is finished. □

This theorem tells us that a pair of a symmetry and an adjoint symmetry naturally yields a conservation law. Moreover, the expression of the conserved density is only dependent on the pair of a symmetry and an adjoint symmetry, but the expression of the conserved flux is dependent on the pair of a symmetry and an adjoint symmetry as well as the underlying equation.

If for two functions $h_1, h_2 \in \mathcal{B}$, there exist functions $g_i \in \mathcal{B}, 1 \le i \le p$, such that

$$h_1 - h_2 = \sum_{i=1}^{p} (E_i - 1) g_i, \tag{17}$$

then we say that h_1 is equivalent to h_2, denoted by $h_1 \sim h_2$. Obviously, this is an equivalence relation, and can be used in classifying conserved densities. A trivial conserved density is equivalent to zero.

3. Applications to the Volterra Lattice Equation

Now we go on to illustrate by an example rich structures of the conservation laws resulted from symmetries and adjoint symmetries.

Let us consider the Volterra lattice equation [8]:

$$u_t = K(u) = u(E^{-1}u - Eu),\ u = u(n,t),\ n \in \mathbb{Z},\ t \in \mathbb{R}. \tag{18}$$

Its linearized equation and adjoint linearized equation read

$$\sigma_t = (E^{-1}u - Eu)\sigma + uE^{-1}\sigma - uE\sigma, \tag{19}$$
$$\rho_t = (Eu - E^{-1}u)\rho - E(u\rho) + E^{-1}(u\rho), \tag{20}$$

respectively. There exist infinitely many symmetries [9]:

$$K_i = \Phi^i(u(E^{-1}u - Eu)),\ i \ge 0, \tag{21}$$
$$\tau_i = \Phi^i(t[K_0, u] + u) = \Phi^i(tK_0 + u) = \Phi^i(tu(E^{-1}u - Eu) + u),\ i \ge 0, \tag{22}$$

where the hereditary recursion operator Φ is defined by

$$\Phi = u(1 + E^{-1})(-(Eu)E^2 + u)(E-1)^{-1}u^{-1}.$$

Moreover, these symmetries constitute a Lie algebra [9]:

$$[K_i, K_i] = 0,\ [K_i, \tau_j] = (i+1)K_{i+j},\ [\tau_i, \tau_j] = (i-j)\tau_{i+j},\ i, j \ge 0, \tag{23}$$

where $[K, S]$ is defined by $[K, S] = K'S - S'K$, K' and S' being the corresponding Gateaux derivative operators.

By an inspection, the function defined by

$$S_0 = u^{-1} \tag{24}$$

is an adjoint symmetry of the Volterra lattice Equation (18). This means that S_0 satisfies the adjoint linearized Equation (20) while u solves Equation (18). The proof just needs a simple and direct computation:

$$(S_0)_t = -u^{-2}u_t,\ (Eu - E^{-1}u)S_0 - E(uS_0) + E^{-1}(uS_0) = (Eu - E^{-1}u)u^{-1},$$

from which it directly follows that if u solves Equation (18), S_0 satisfies Equation (20). Now, using the principle in Theorem 1, we have infinitely many conserved densities, $u^{-1}K_i$ and $u^{-1}\tau_i$, $i \geq 0$. In particular, since we have

$$\begin{aligned} &S_0\tau_1 = S_0(tK_1 + \Phi u) \sim u^{-1}\Phi u, \\ &(\Phi u)(n) = u(n)[-(n+2)u(n+1) - u(n) + (n-1)u(n-1)], \end{aligned}$$

we obtain a nontrivial local conserved density of the Volterra lattice Equation (18):

$$(h_1)(n) = -(n+2)u(n+1) - u(n) + (n-1)u(n-1). \tag{25}$$

This generates a conservation law

$$h_{1t} = (E-1)[([n]+1)u(Eu) - ([n]-1)(E^{-2}u)(E^{-1}u)], \tag{26}$$

where $[n]$ is the operator

$$([n]f)(m) = mf(m),\ f \in \mathcal{B},\ m \in \mathbb{Z}.$$

Except $S_0\tau_1 = S_0(tK_0 + u) \sim 1$, all other products of the same type, $u^{-1}\tau_i$, $i \geq 2$, give us nontrivial nonlocal conserved densities of the Volterra lattice Equation (18). But all $u^{-1}K_i$, $i \geq 0$, are trivial conserved densities, which can be seen directly from the recursion structure of symmetries [10].

Based on Proposition 1, the Hamiltonian formulation of the Volterra soliton hierarchy [9] guarantees that the Volterra lattice Equation (18) has a hierarchy of adjoint symmetries:

$$\rho_i = \Psi^i 1 = \frac{\delta \mathcal{H}_i}{\delta u},\ \Psi = \Phi^\dagger = u^{-1}(E-1)^{-1}(-(Eu)E^2 + u)(1 + E^{-1})u,\ i \geq 0, \tag{27}$$

which generate, by Theorem 1, the conserved densities: $\rho_i K_j$, $\rho_i\tau_j$, $i,j \geq 0$. The nontrivial conserved density generated from $\rho_0\tau_1$ is, due to $K_1 \sim 0$, equivalent to

$$(h_2)(n) = (\Phi u)(n) = u(n)[-(n+2)u(n+1) - u(n) + (n-1)u(n-1)], \tag{28}$$

whose associated conservation law reads

$$h_{2t} = (E-1)\{([n]+1)[(E^{-1}u)u^2 + (E^{-1}u)u(Eu)] - ([n]-1)[(E^{-2}u)^2u + (E^{-2}u)(E^{-1}u)u]\}. \tag{29}$$

Moreover, we see that all symmetries, $K_i = \rho_0 K_i$ and $\tau_i = \rho_0\tau_i$, $i \geq 0$, are also conserved densities, and that the conserved densities $\rho_i\tau_0 \sim \rho_i u$, $i \geq 0$, present all standard ones associated with the Hamiltonian functionals $\{\mathcal{H}_i | i \geq 0\}$ [9], upon noting

$$u\frac{\delta \mathcal{H}}{\delta u} = u\sum_{\alpha \in \mathcal{Z}} E^{-\alpha}\frac{\partial H}{\partial(E^\alpha u)} \sim \sum_{\alpha \in \mathcal{Z}} (E^\alpha u)\frac{\partial H}{\partial(E^\alpha u)} = H,\ \mathcal{H} = \sum_{n \in \mathcal{Z}} H(n),\ H \in \mathcal{A}.$$

The Volterra lattice Equation (18) also has the following Lax pair [11]:

$$U(u,\lambda) = \begin{bmatrix} 0 & u \\ -1 & \lambda \end{bmatrix},\ V(u,\lambda) = \begin{bmatrix} \lambda^2 - u & -\lambda E^{-1}u \\ \lambda & -E^{-1}u \end{bmatrix}, \tag{30}$$

which means that the Volterra lattice Equation (18) is equivalent to the discrete zero curvature equation $U_t = (EV)U - UV$. Let λ_s, $1 \leq s \leq N$, be arbitrary constants, and introduce N replicas of the spectral problems:

$$E\phi^{(s)} = U(u,\lambda_s)\phi^{(s)},\ \phi_t^{(s)} = V(u,\lambda_s)\phi^{(s)},\ \phi^{(s)} = (\phi_{1s},\phi_{2s})^T,\ 1 \leq s \leq N, \tag{31}$$

and N replicas of the adjoint spectral problems:

$$E\psi^{(s)} = (U^T)^{-1}(u,\lambda_s)\psi^{(s)},\ \psi_t^{(s)} = -V^T(u,\lambda_s)\psi^{(s)},\ \psi^{(s)} = (\psi_{1s},\psi_{2s})^T,\ 1 \leq s \leq N. \tag{32}$$

Then, we have a class of adjoint symmetries represented in terms of eigenfunctions and adjoint eigenfunctions [11]:

$$T_0 = \frac{1}{u}(P_2^T A Q_1 + P_2^T Q_2), \tag{33}$$

where the matrix A, and the two N dimensional vector functions P_i and Q_i are defined by

$$A = \text{diag}(\lambda_1, \cdots, \lambda_N),\ P_i = (\phi_{i1}, \cdots, \phi_{iN})^T,\ Q_i = (\psi_{i1}, \cdots, \psi_{iN})^T,\ 1 \leq i \leq 2. \tag{34}$$

Now, Theorem 1 guarantees that we have the conserved densities: $T_0 K_i$ and $T_0 \tau_i$, $i \geq 0$. This particularly gives the following two conserved densities:

$$T_0 K_0 = (E^{-1}u - Eu)(P_2^T A Q_1 + P_2^T Q_2), \tag{35}$$
$$T_0 \tau_0 = t(E^{-1}u - Eu)(P_2^T A Q_1 + P_2^T Q_2) + (P_2^T A Q_1 + P_2^T Q_2). \tag{36}$$

Again by Theorem 1, their corresponding fluxes read

$$-(P_2^T A Q_1 + P_2^T Q_2)E^{-1}K_0 - (E(P_2^T A Q_1 + P_2^T Q_2))K_0, \tag{37}$$
$$-(P_2^T A Q_1 + P_2^T Q_2)E^{-1}\tau_0 - (E(P_2^T A Q_1 + P_2^T Q_2))\tau_0, \tag{38}$$

respectively. The involvement of eigenfunctions and adjoint eigenfunctions exhibits strong nonlocality of this kind of conservation laws.

4. Concluding Remarks

We have established a direct approach for constructing conservation laws and thus conserved densities of discrete evolution equations, whether the evolution equations are derivable from a Lagrangian or not. Our approach utilizes pairs of symmetries and adjoint symmetries, in which adjoint symmetries make up for the disadvantage of non-Lagrangian structures in establishing a correspondence between symmetries and conservation laws. The approach has been applied to the generation of conserved densities of the Volterra lattice equation.

We remark that on one hand, all evolution equations become the first-order ordinary differential equations (ODEs), when there is no spatial shift appeared in the equations. On the other hand, if we remove the time derivative, *i.e.*, there is no time involved in evolution equations, then we immediately obtain conservation laws for difference equations. Moreover, our results pave a way to construct conservation laws from symmetries for self-adjoint discrete evolution equations or difference equations, since symmetries are also adjoint symmetries of self-adjoint discrete evolution equations or difference equations. We also point out that our idea of constructing conservation laws by symmetries and adjoint symmetries is quite similar to that of carrying out binary nonlinearization under symmetry constraints [12,13]. In the theory of binary nonlinearization [12,13], adjoint symmetries are used to establish a balance with non-Lie adjoint symmetries generated from both spectral problems and adjoint spectral problems.

There are many other approaches or theories on conservation laws of differential equations and differential-difference equations. In the case of integrable equations, Hamiltonian formulations [14,15] generate usual conservation laws associated with Hamiltonian functionals, and a bi-Hamiltonian formulation presents a recurrence relation of conserved densities in the resulting conservation laws [16]. Moreover, Lax pairs are used to make Riccati type equations that ratios of eigenfunctions need to satisfy, and series expansions of solutions of the resulting Riccati type equations around the spectral parameter yield conservation laws (see, e.g., [17,18]). For continuous evolution equations, adjoint symmetries are also called conserved covariants [19], and it is recognized that the product of a symmetry and an adjoint symmetry presents a conserved density (see, e.g., [19–21]) and that a functional $\mathcal{I}$ is conserved if and only if its variational derivative $\frac{\delta \mathcal{I}}{\delta u}$ is an adjoint symmetry (see, e.g., [13,20,21]).

There exists a geometrical theory to deal with adjoint symmetries of the second-order ODEs [22]. Adjoint symmetries are also used to show separability of finite-dimensional Hamiltonian systems (see, e.g., [23]) and links to integrating factors of the second-order ODEs (see, e.g., [24]). An important character for symmetries is the existence of Lie algebraic structures, and such Lie algebraic structures can be resulted from a kind of Lie algebras of Lax operators corresponding to symmetries (see [25–27] for the continuous case and [9,28] for the discrete case). Therefore, we are curious about whether there exist any Lie algebraic structures for adjoint symmetries and what kind of Lie algebraic structures we can have if they exist. We finally make a remark about this question. A natural binary operation for adjoint symmetries could be taken as

$$[\![\rho_1, \rho_2]\!] = (\rho_1')^\dagger \rho_2 - (\rho_2')^\dagger \rho_1,$$

where $(\rho_1')^\dagger$ and $(\rho_2')^\dagger$ denote their adjoint Gateaux derivative operators. However, this doesn't keep the space of adjoint symmetries closed. For example, for the Volterra lattice Equation (18), we have two adjoint symmetries $\rho_1 = 1$ and $\rho_2 = u^{-1}$, which generate

$$[\![\rho_1, \rho_2]\!] = [\![1, u^{-1}]\!] = u^{-2},$$

but this function u^{-2} is not an adjoint symmetry of the Volterra lattice equation.

Acknowledgments: The work was supported in part by NNSFC under the grants 11371326, 11271008 and 1371361, the Fundamental Research Funds for the Central Universities (2013XK03), the Natural Science Foundation of Shandong Province (Grant No. ZR2013AL016), Zhejiang Innovation Project of China (Grant No. T200905), the First-class Discipline of Universities in Shanghai and the Shanghai University Leading Academic Discipline Project (No. A.13-0101-12-004), and Shanghai University of Electric Power.

Conflicts of Interest: The author declares no conflict of interest.

References

1. Bluman, G.; Kumei, S. *Symmetries and Differential Equations*; Springer-Verlag: New York, NY, USA, 1989.
2. Olver, P.J. *Applications of Lie Groups to Differential Equations*; Graduate Texts in Mathematics 107; Springer-Verlag: New York, NY, USA, 1993.
3. Ibragimov, N.H. *Transformation Groups Applied to Mathematical Physics*; Mathematics and its Applications (Soviet Series); D. Reidel Publishing Co.: Dordrecht, The Netherlands, 1985.
4. Morawetz, C.S. Variations on Conservation Laws for the Wave Equation. *Bull. Amer. Math. Soc.* **2000**, *37*, 141–154.
5. Anco, S.C.; Bluman, G. Direct Computation of Conservation Laws from Field Equations. *Phys. Rev. Lett.* **1997**, *78*, 2869–2873.
6. Ibragimov, N.H. A New Conservation Theorem. *J. Math. Anal. Appl.* **2007**, *333*, 311–328.

7. Ibragimov, N.H. Nonlinear Self-Adjointness and Conservation Laws. *J. Phys. A Math. Theor.* **2011**, *44*, doi:10.1088/1751-8113/44/43/432002.
8. Volterra, V. *Leçons sur la Théorie Mathématique de la Lutte pour la vie*; Gauthier-Villars: Paris, France, 1931. (In French)
9. Ma, W.X.; Fuchssteiner, B. Algebraic Structure of Discrete Zero Curvature Equations and Master Symmetries of Discrete Evolution Equations. *J. Math. Phys.* **1999**, *40*, 2400–2418.
10. Fuchssteiner, B.; Ma, W.X. An Approach to Master Symmetries of Lattice Equations. In *Symmetries and Integrability of Difference Equations*; Clarkson, P.A., Nijhoff, F.W., Eds.; London Mathematical Society Lecture Note Series 255; Cambridge University Press: Cambridge, UK, 1999; pp. 247–260.
11. Ma, W.X.; Geng, X.G. Bäcklund Transformations of Soliton Systems from Symmetry Constraints. In *Bäcklund and Darboux Transformations—The Geometry of Solitons*; CRM Proceedings and Lecture Notes. 29; American Mathematical Society: Providence, RI, USA, 2001; pp. 313–323.
12. Ma, W.X.; Strampp, W. An Explicit Symmetry Constraint for the Lax Pairs and the Adjoint Lax Pairs of AKNS Systems. *Phys. Lett. A* **1994**, *185*, 277–286.
13. Ma, W.X.; Zhou, R.G. Adjoint Symmetry Constraints Leading to Binary Nonlinearization. *J. Nonlinear Math. Phys.* **2002**, *9*, 106–126.
14. Tu, G.Z. On Liouville Integrability of Zero-Curvature Equations and the Yang Hierarchy. *J. Phys. A Math. Gen.* **1989**, *22*, 2375–2392.
15. Ma, W.X.; Chen, M. Hamiltonian and Quasi-Hamiltonian Structures Associated with Semi-Direct Sums of Lie Algebras. *J. Phys. A Math. Gen.* **2006**, *39*, 10787–10801.
16. Magri, F. A Simple Model of the Integrable Hamiltonian Equation. *J. Math. Phys.* **1978**, *19*, 1156–1162.
17. Alberty, J.M.; Koikawa, T.; Sasaki, R. Canonical Structure of Soliton Equations I. *Phys. D Nonlinear Phenom.* **1982**, *5*, 43–65.
18. Zhang, D.J.; Chen, D.Y. The Conservation Laws of Some Discrete Soliton Systems. *Chaos Solitons Fractals* **2002**, *14*, 573–579.
19. Fuchssteiner, B.; Fokas, A.C. Symplectic Structures, Their Bäcklund Transformations and Hereditary Symmetries. *Phys. D Nonlinear Phenom.* **1981**, *4*, 47–66.
20. Tu, G.Z.; Qin, M.Z. Relationship Between Symmetries and Conservation Laws of Nonlinear Evolution Equations. *Chin. Sci. Bull.* **1979**, *24*, 913–917. (In Chinese)
21. Fu, W.; Huang, L.; Tamizhmani, K.M.; Zhang, D.J. Integrability Properties of the Differential-Difference Kadomtsev-Petviashvili Hierarchy and Continuum Limits. *Nonlinearity* **2013**, *26*, 3197–3229.
22. Morando, P.; Pasquero, S. The Symmetry in the Structure of Dynamical and Adjoint Symmetries of Second-Order Differential Equations. *J. Phys. A Math. Gen.* **1995**, *28*, 1943–1955.
23. Sarlet, W.; Ramos, A. Adjoint Symmetries, Separability, and Volume Forms. *J. Math. Phys.* **2000**, *41*, 2877–2888.
24. Mohanasubha, R.; Chandrasekar, V.K.; Senthilvelan, M.; Lakshmanan, M. Interplay of Symmetries, Null Forms, Darboux Polynomials, Integrating Factors and Jacobi Multipliers in Integrable Second-Order Differential Equations. *Proc. R. Soc. Lond. Ser. A Math. Phys. Eng. Sci.* **2014**, *470*, doi:10.1098/rspa.2013.0656.
25. Ma, W.X. The Algebraic Structures of Isospectral Lax Operators and Applications to Integrable Equations. *J. Phys. A Math. Gen.* **1992**, *25*, 5329–5343.
26. Ma, W.X. Lax Representations and Lax Operator Algebras of Isospectral and Nonisospectral Hierarchies of Evolution Equations. *J. Math. Phys.* **1992**, *33*, 2464–2476.
27. Ma, W.X. Lie Algebra Structures Associated with Zero Curvature Equations and Generalized Zero Curvature Equations. *Br. J. Appl. Sci. Tech.* **2013**, *3*, 1336–1344.
28. Ma, W.X. A Method of Zero Curvature Representation for Constructing Symmetry Algebras of Integrable Systems. In *Proceedings of the 21st International Conference on the Differential Geometry Methods in Theoretical Physics*; Ge, M.L., Ed.; World Scientific: Singapore, 1993; pp. 535–538.

Chapter 6:

symmetry

MDPI

Article

Symbolic and Iterative Computation of Quasi-Filiform Nilpotent Lie Algebras of Dimension Nine

Mercedes Pérez [1], Francisco Pérez [2] and Emilio Jiménez [1,*]

[1] Department of Mechanical Engineering, University of La Rioja, C/Luis de Ulloa 26004, Spain; mercedes.perez@unirioja.es

[2] Department of Applied Mathematics I, University of Sevilla, Av. Reina Mercedes 41012, Spain; mpparte@gmail.com

* Author to whom correspondence should be addressed; emilio.jimenez@unirioja.es; Tel.: +34-941-299502.

Academic Editor: Roman M. Cherniha

Received: 30 June 2015; Accepted: 16 September 2015; Published: 1 October 2015

Abstract: This paper addresses the problem of computing the family of two-filiform Lie algebra laws of dimension nine using three Lie algebra properties converted into matrix form properties: Jacobi identity, nilpotence and quasi-filiform property. The interest in this family is broad, both within the academic community and the industrial engineering community, since nilpotent Lie algebras are applied in traditional mechanical dynamic problems and current scientific disciplines. The conditions of being quasi-filiform and nilpotent are applied carefully and in several stages, and appropriate changes of the basis are achieved in an iterative and interactive process of simplification. This has been implemented by means of the development of more than thirty Maple modules. The process has led from the first family formulation, with 64 parameters and 215 constraints, to a family of 16 parameters and 17 constraints. This structure theorem permits the exhaustive classification of the quasi-filiform nilpotent Lie algebras of dimension nine with current computational methodologies.

Keywords: Lie algebra; nilpotence; quasi-filiform algebra; Maple

---

1. Introduction

1.1. State of the Art

Traditionally, Lie algebras have been used in physics in the context of symmetry groups of dynamical systems, as a powerful tool to study the underlying conservation laws [1,2]. At present, space-time symmetries and symmetries related to degrees of freedom are considered. For instance, non-trivial Heidelberg algebra arises right in the base of the Hamiltonian mechanics. Hamiltonian mechanics describes the state of a dynamic system with $2n$ variables (n coordinates and n momenta), and the other interesting observable physics quantities are functions of them. Thus, the observables commute with the Hamiltonian respecting the Poisson bracket, and they constitute a Lie algebra of infinite dimension. Furthermore, a description in quantum mechanics is obtained by an algebra of Hermitian operators in a Hilbert space with the bracket product as the commutator. In such a case, the Heisenberg algebra arises if n is one, and the generalized Heisenberg algebra results for other values of n, since the traditional canonical variables preserve the Poisson bracket. In general, a transformation is said to be symplectic if it preserves the Poisson bracket. Therefore, the study of symplectic structures of nilpotent Lie algebras is worthwhile as a wide generalization of the Heisenberg algebra. These symplectic Lie algebras appear in the study of traditional dynamic problems, like the problem of the two bodies or the problem of the three bodies, as well as in current studies in solid state physics [3], modern geometry [4] or particle physics [5]. Furthermore, Lie theory is closely connected to control theory in the controllability and optimization of the tracking without drift of complex dynamical systems as a rolling sphere. Some other applications can be consulted in [6–8]. Hence, it is convenient to classify the families of Lie algebras as large as possible.

The matter of Lie algebra classification comes down to classifying the solvable semisimple algebras [9–11], since Levi's decomposition theorem [12] permits one to state that any Lie algebra can be decomposed into a semidirect sum of its radical, *i.e.*, its maximal solvable ideal, and a semisimple part called Levi's subalgebra. The classification of semisimple Lie algebras in **C** is presently associated with Dynkin diagrams (1945). However, the solvable Lie algebra classification problem comes down in a sense (Goze and Khakimdjanov [12]) to the nilpotent Lie algebra classification.

Numerous researchers have tackled the problem of nilpotent Lie algebra classification. However, their studies were restricted to the filiform case, due to the difficulties arising from a nilpotence index higher than the dimension, providing a great number of parameters without restrictions among them. The first lists of algebras were obtained by K. Umlauf [13] in 1891 in his PhD thesis, providing the lists of all of the laws of dimensions less than or equal to six and all of dimensions 7, 8 and 9 that allow a basis $\{X_0, X_1, ..., X_{n-1}\}$, such that it satisfies $[X_0, X_i] = X_{i+1}$ $(1 \leqslant i \leqslant n-2)$, in **R** or **C**. Nilpotent Lie algebra classification had important progress thanks to Goze and Ancochea [14] with the definition of a more powerful invariant than the known invariants up to that moment: the characteristic series or Goze's invariant. These authors achieved the classification of the complex nilpotent Lie algebras of dimension seven and of the complex filiform Lie algebras of dimension eight [15]. Gómez and Echarte [16] classified the complex filiform Lie algebras of dimension nine using Goze's invariant. Gómez *et al.* [17] classified the symplectic filiform Lie algebras that are not two-to-two symplectic-isomorphic of dimensions less than or equal to 10 in 2001. Higher dimensions were tackled by Boza *et al.* [18] and Echarte *et al.* [19,20] in the last ten years. The more the dimension increases, the more and more complex is the determination of exhaustive lists of Lie algebras, so new computation methodologies are a present field of research [21–23].

Cabezas *et al.* (1998) [24] study a family of Lie algebras that they call p-filiform with dimension n and Goze's invariant $(n-p, 1, ..., 1)$. Since filiform algebras have Goze's invariant $(n-1, 1)$, they are included in the p-filiform family as one-filiform Lie algebras; analogously, the quasi-filiform algebras are the two-filiform algebras [25], and the abelian algebras are the $(n-1)$-filiform algebras. In a sense, the study of the quasi-filiform Lie algebras appears natural, since they are only known until dimension eight. On the other hand, in 1999, Camacho [26] studied the $(n-5)$-filiform and $(n-6)$-filiform Lie algebras, closing the classification of the p-filiform Lie algebras up to dimension eight. Another classification of the $(n-5)$-filiform Lie algebras is provided by Ancochea and Campoamor [27]. Their research line is used as the context for our piece of research, leaning on the current availability of symbolic manipulation programs, such as Maple, which allow the user to perform the tedious algebra and routine computations [28–30]. The present paper tackles the proof of the structure theorem of quasi-filiform Lie algebras of dimension nine. The classification and a complete casuistry of that family of Lie algebras was published in [31], based on the results of [32]. We strongly recommend the reading of [33–37] to become familiar with Lie algebra terminology and concepts.

After this state of the art, Subsection 1.2. is included to declare the terminology that has been developed from the 1990s to the present. Section 2 is devoted to the symbolic and iterative computational proof of the structure theorem of the laws of every complex quasi-filiform Lie algebra of dimension nine, which is the original contribution of the present paper. Finally, Section 3 summarizes the computational work developed for the appropriate changes of the basis to demonstrate the general theorem.

1.2. *Terminology*

The abelian algebra of dimension n is the only one with Goze's invariant $(1, ..., 1)$; in metabelian algebras, the characteristic series is $(2, ...2, 1, ...1)$; in Heisenberg algebras, it is $(2, 1, ..., 1)$; in filiform algebras, it is $(n-1, 1)$; and in quasi-filiform algebras, it is $(n-2, 1, 1)$. From now on, let us use the term $\mathbf{Jac}(x,y,z)$ for the Jacobi identity: $\mathbf{Jac}(x,y,z) \Leftrightarrow \mu(x,\mu(y,z)) + \mu(y,\mu(z,x)) + \mu(z,\mu(x,y)) = 0$. Additionally, let us use $\mathcal{B}^2(\mathbf{C^n})$ for the space of the bilinear applications of $\mathbf{C^n} \times \mathbf{C^n}$ in $\mathbf{C^n}$, and let us choose a basis $\{e_0, e_2, ..., e_{n-1}\}$ of $\mathbf{C^n}$. An element α of $\mathcal{B}^2$ can be determined from a set of scalars C_{ij}^k, called structure constants, defined by $\alpha(e_i, e_j) = \sum_{k=0}^{n-1} C_{ij}^k e_k$; thus, $\mathcal{B}^2$ can have a structure of affine space. Then, a Lie algebra $\mathfrak{g}$ can be considered as an element of $\mathcal{B}^2$; the set $\mathcal{L}_n$ of Lie algebras in $\mathbf{C^n}$ is the affine algebraic set that is defined by the following polynomial expressions:

$$C_{ii}^k = 0, \forall i, k 0 \leqslant i, k \leqslant n-1 \tag{1a}$$

$$C_{ij}^k = -C_{ji}^k \forall i, j, k 0 \leqslant i, j, k \leqslant n-1 \tag{1b}$$

$$\sum_{k=0}^{n-1} \left(C_{ij}^k . C_{kl}^s + C_{jl}^k . C_{ki}^s + C_{li}^k . C_{kj}^s \right) = 0, 0 \leqslant i, j, l, s \leqslant n-1 \tag{1c}$$

and it is parametrized by the $\frac{n^3 - n^2}{2}$ structure constants C_{ij}^k.

If $\mathfrak{g}$ is a Lie algebra, the series of ideals defined by:

$$\mathcal{D}^0(\mathfrak{g}) = \mathfrak{g} \tag{2a}$$

$$\mathcal{D}^{k+1}(\mathfrak{g}) = \left[\mathcal{D}^k(\mathfrak{g}), \mathcal{D}^k(\mathfrak{g})\right], k \in \mathbf{N} \cup \{0\} \tag{2b}$$

is called the derived series of $\mathfrak{g}$, which satisfies $\mathfrak{g} = \mathcal{D}^0(\mathfrak{g}) \supseteq \mathcal{D}^1(\mathfrak{g}) \supseteq ... \supseteq \mathcal{D}^i(\mathfrak{g})$... If there exists an integer k, such that $\mathcal{D}^k(\mathfrak{g}) = \{0\}$, the algebra is said to be solvable; in such a case, the smaller integer that satisfies the previous condition is called the solvability index of $\mathfrak{g}$.

Levi's theorem [33] states that every Lie algebra $\mathfrak{g}$ can be decomposed in a semidirect sum of its radical (the maximal solvable ideal) and semisimple subalgebras (Levi's subalgebra). This result reduces in a sense the Lie algebra classification problem to the classification of the solvable algebras [[10], since semisimple algebra classification is known.

If $\mathfrak{g}$ is a Lie algebra, the series of ideals defined by:

$$\mathcal{C}^0(\mathfrak{g}) = \mathfrak{g} \tag{3a}$$

$$\mathcal{C}^{k+1}(\mathfrak{g}) = \left[\mathcal{C}^k(\mathfrak{g}), \mathfrak{g}\right], k \in \mathbf{N} \cup \{0\} \tag{3b}$$

is called the lower central series of $\mathfrak{g}$. It satisfies that $\mathfrak{g} = \mathcal{C}^0(\mathfrak{g}) \supseteq \mathcal{C}^1(\mathfrak{g}) \supseteq ... \supseteq \mathcal{C}^i(\mathfrak{g})$... If there exists an integer k, such that $\mathcal{C}^k(\mathfrak{g}) = \{0\}$, the algebra is said to be nilpotent; in such a case, the smaller integer that satisfies the previous condition is called the nilpotence index or nilindex of $\mathfrak{g}$. If the dimension is n and the nilindex is $n-1$, the algebras obtained are called filiform; they are said to be quasi-filiform if their nilindex is $n-2$. The abelian Lie algebras are the algebras with nilindex one.

The characteristic series or Goze's invariant is defined as the maximum of the Segre symbols of the nilpotent linear applications $ad(X)$, where X is an element of the derived subalgebra complementary. In other words, if $\mathfrak{g}$ is a complex nilpotent Lie algebra of finite dimension n, for every $X \in \mathfrak{g} - [\mathfrak{g}, \mathfrak{g}]$, the series of the characteristic subspace dimensions of the nilpotent operator $ad(X)$ in decreasing order is denoted by $c(X) = (c_1(X), c_2(X),, ..., 1)$. Reordering the set of series in lexicographical order, the characteristic series is defined by $c(\mathfrak{g}) = \sup\{c(X) : X \in \mathfrak{g} - [\mathfrak{g}, \mathfrak{g}]\}$. This invariant has been used to classify the nilpotent Lie algebras of dimension seven. Obviously, $c(\mathfrak{g})$ is an invariant for the isomorphisms, and by construction, there exists at least a vector $X \in \mathfrak{g} - [\mathfrak{g}, \mathfrak{g}]$ that satisfies $c(\mathfrak{g}) = c(X)$; every vector that satisfies the previous condition is called the characteristic vector of the algebra.

If $\mathfrak{g}$ is a p-filiform Lie algebra of dimension n (*i.e.*, nilpotent with characteristic series $(n-p,1,...,1)$, then there exists a basis, which will be denoted as $\{X_0, X_1, ..., X_p, Y_1, ..., Y_{n-p-1}\}$, that satisfies:

$$[X_0, X_i] = X_{i+1} 1 \leqslant i \leqslant p-1 \tag{4a}$$

$$[X_0, X_{n-p}] = 0 \tag{4b}$$

$$[X_0, Y_j] = 0 1 \leqslant j \leqslant n-p-1 \tag{4c}$$

This basis is called the adapted basis of the algebra, where X_0 is a characteristic vector.

2. Structure Theorem

This section presents the development of the structure theorem of the family of laws of complex quasi-filiform Lie algebras (QFLA) of dimension nine. Our objective was to find the simplest expression of the family of laws. Every QFLA of dimension nine can have an adapted basis $\{x_0, x_1, ..., x_8\}$, such that:

$$[x_0, x_i] = x_{i+1}, 1 \leqslant i \leqslant 6; [x_0, x_i] = 0, 7 \leqslant i \leqslant 8 \tag{5}$$

A first approximation of the family can be obtained just with the application of the Jacobi identity to the three-tuple (x_0, x_i, x_j), where x_i, x_j are basis vectors different from x_0 vector [31].

A condition that sometimes is more difficult to apply is the nilpotence. The Engel theorem puts nilpotence on a level with ad-nilpotence for Lie algebras. Therefore, a Lie algebra $\mathfrak{g}$ is nilpotent if and only if the characteristic polynomial of the matrix $Adj(x)$ is λ^9, for every vector x of $\mathfrak{g}$. Anyway, this condition is often difficult to apply, so the moment in the process when the nilpotence condition is applied or, much better, when the condition is applied for each vector has to be chosen carefully.

The condition of being quasi-filiform can be also interpreted in terms of matrices. Thus, the vector candidate of characteristic vectors, *i.e.*, the vectors in $\mathfrak{g} - [\mathfrak{g}, \mathfrak{g}]$, has to satisfy that the respective adjoint matrices do not have non-null minors of order $\leqslant 7$. As in the case of the nilpotence, this condition has to be applied with caution and probably in several stages.

Theorem 1. *The laws of every complex quasi-filiform Lie algebra of dimension nine can be described by the following family with 16 parameters and 17 polynomial restriction equations:*

$$[x_0, x_i] = x_{i+1}, 1 \leqslant i \leqslant 6 \tag{6a}$$

$$[x_1, x_2] = \alpha_1 x_4 + \alpha_2 x_5 + \alpha_3 x_6 + \alpha_4 x_7 + \alpha_5 x_8 \tag{6b}$$

$$[x_1, x_3] = \alpha_1 x_5 + \alpha_2 x_6 + \alpha_3 x_7 \tag{6c}$$

$$[x_1, x_4] = \alpha_6 x_5 + \alpha_7 x_6 + \alpha_8 x_7 + \alpha_9 x_8 \tag{6d}$$

$$[x_1, x_5] = 2\alpha_6 x_6 + (2\alpha_7 - \alpha_1) x_7 \tag{6e}$$

$$[x_1, x_6] = \alpha_{10} x_7 + \alpha_{11} x_8 \tag{6f}$$

$$[x_1, x_8] = \alpha_{12} x_3 + \alpha_{13} x_4 + \alpha_{14} x_5 + \alpha_{15} x_6 + \alpha_{16} x_7 \tag{6g}$$

$$[x_2, x_3] = -\alpha_6 x_5 + (\alpha_1 - \alpha_7) x_6 + (\alpha_2 - \alpha_8) x_7 - \alpha_9 x_8 \tag{6h}$$

$$[x_2, x_4] = -\alpha_6 x_6 + (\alpha_1 - \alpha_7) x_7 \tag{6i}$$

$$[x_2, x_5] = (2\alpha_6 - \alpha_{10}) x_7 - \alpha_{11} x_8 \tag{6j}$$

$$[x_2, x_8] = \alpha_{12} x_4 + \alpha_{13} x_5 + \alpha_{14} x_6 + \alpha_{15} x_7 \tag{6k}$$

$$[x_3, x_4] = (-3\alpha_6 + \alpha_{10}) x_7 + \alpha_{11} x_8 \tag{6l}$$

$$[x_3, x_8] = \alpha_{12} x_5 + \alpha_{13} x_6 + \alpha_{14} x_7 \tag{6m}$$

$$[x_4, x_8] = \alpha_{12}x_6 + \alpha_{13}x_7 \tag{6n}$$

$$[x_5, x_8] = \alpha_{12}x_7 \tag{6o}$$

subject to:

$$\alpha_5\alpha_{12} = 0 \tag{7a}$$

$$\alpha_6\alpha_{12} = 0 \tag{7b}$$

$$\alpha_6\alpha_{13} = 0 \tag{7c}$$

$$\alpha_9\alpha_{12} = 0 \tag{7d}$$

$$\alpha_9\alpha_{13} = 0 \tag{7e}$$

$$\alpha_9\alpha_{14} = 0 \tag{7f}$$

$$\alpha_{10}\alpha_{12} = 0 \tag{7g}$$

$$\alpha_{11}\alpha_{12} = 0 \tag{7h}$$

$$\alpha_{11}\alpha_{13} = 0 \tag{7i}$$

$$\alpha_{11}\alpha_{14} = 0 \tag{7j}$$

$$\alpha_{11}\alpha_{15} = 0 \tag{7k}$$

$$\alpha_{11}\alpha_{16} = 0 \tag{7l}$$

$$\alpha_{11}(3\alpha_1 - \alpha_7) = 0 \tag{7m}$$

$$\alpha_{12}(\alpha_1 - \alpha_7) = 0 \tag{7n}$$

$$\alpha_5\alpha_{13} - 2\alpha_6^2 - \alpha_9\alpha_{15} \tag{7o}$$

$$2(\alpha_2 - \alpha_8)\alpha_{12} + 3(\alpha_1 - \alpha_7)\alpha_{13} + 2(\alpha_6 - \alpha_{10})\alpha_{14} = 0 \tag{7p}$$

$$\alpha_5\alpha_{14} - 2(2\alpha_1 + \alpha_7)\alpha_6 - \alpha_9\alpha_{16} + (3\alpha_1 - \alpha_7)\alpha_{10} = 0 \tag{7q}$$

Proof of Theorem 1. Let $\mathfrak{g}$ be a nilpotent Lie algebra of dimension n and the characteristic series $(n-2, 1, 1)$. Let $x_0 \in \mathfrak{g} - [\mathfrak{g}, \mathfrak{g}]$ be a characteristic vector of $\mathfrak{g}$. Then, there is a basis of $\mathfrak{g}$, $\{x_i : 0 \leqslant i \leqslant n-1\}$, such that $[x_0, x_i] = x_{i+1}$, $1 \leqslant i \leqslant n-3$, and the other bracket products of x_0 are null. On the whole, all of the bracket products can be described by:

$$[x_i, x_j] = \sum_{k=0}^{n-1} C_{ij}^k . x_k, 0 \leqslant i, j \leqslant n-1 \tag{8}$$

where C_{ij}^k are the algebra structure constants. It is simple to prove that for a nilpotent Lie algebra of dimension n and characteristic series $(n-2, 1, 1.., 1)$, like $\mathcal{C}^m\mathfrak{g} = \langle x_{n-2} \rangle$, it is true that $x_{n-2} \in Z(\mathfrak{g})$. Then, in our case:

$$[x_{n-2}, x_j] = 0, \forall j 0 \leqslant j \leqslant n-1 \rightarrow C_{(n-2)j}^k = 0, \forall j, k 0 \leqslant j, k \leqslant n-1 \tag{9}$$

It is well known that the application of the anticommutativity to Jacobi identity will provide $\mathbf{Jac}(y, x, z) \equiv [x, [y, z]] = [y, [x, z]] - [z, [x, y]]$. In order to maintain this identity, the coefficients of x_i, $0 \leqslant i \leqslant n-1$, must be the same at both sides of the equation. Our objective is to study the case $n = 9$; therefore, the coefficients' identification is tackled in an iterative and interactive way. A Maple module called EcuJac has been developed to obtain all of the equations resulting from the application of the aforementioned conditions. Figure 1 illustrates how the Maple module for applying the Jacobi identity conditions has been written. EcuJac is executed iteratively, and each time, it prints a number of equations, so that the number of equations can be reduced in the next iteration. The other proprietary modules executed in EcuJac (Leyes, ObtEcus, OrdApIIG and SimplEcu) are not provided, but their functionalities are commented on in Figure 1.

```
EcuJac:=proc(VV,list,pr)

# list=1 print equations
# pr=1 print all Jacobis

local i,j,i1,i2,i3,i4:
global Nig,nec,nja,Nid,E,IG,ID,Ecus,IEcus,F:

Nig:=1:
for i from 1 by 1 while Nig<>0 do
    # Module Leyes: to print the equations of the laws of an algebra in different formats
    if pr=1 then Leyes(C,1,"x","x",EKus,neK,1): end if:
    nec:=0: nja:=0: unassign('Nig','Nid'):
    # Module ObtEcus: to obtain the restriction equations from the application of Jacobi conditions
    ObtEcus(VV,E,IEE,IG,Nig,ID,Nid,pr):
    print(nja,"Number of Jacobis",nec,"Number of equations",Nig,"Number of identities between coefficient
        functions",Nid,"Number of coefficient identities");
    # Module OrdApllG: to sort out the restriction equations
    OrdApllG(VV,IG,Nig,pr):
    Ecus:={seq(E[i],i=1..nec)}:
    IEcus:={seq(IEE[i],i=1..nec)}:
    if list=1 then
        # Module SimplEcu: to simplify the restriction equations
        SimplEcu(pr):
        for j from 1 to nec do
            i4:=IFF[j] mod 100:
            i3:=(IFF[j] mod 10000-i4)/100:
            i2:=(IFF[j] mod 1000000-i3*100-i4)/10000:
            i1:=(IFF[j] mod 100000000-i2*10000-i3*100-i4)/1000000:
            print(" *** Eq. ",j," : From coeff. of x[",i4,"] in Jac(",i1,i2,i3,") *** ",F[j]):
        end do:
    end if:
end do:
Ecus:={seq(E[i],i=1..nec)}:
IEcus:={seq(IEE[i],i=1..nec)}:
SimplEcu(pr):
if list=1 then
    for j from 1 to nec do
        i4:=IFF[j] mod 100:
        i3:=(IFF[j] mod 10000-i4)/100:
        i2:=(IFF[j] mod 1000000-i3*100-i4)/10000:
        i1:=(IFF[j] mod 100000000-i2*10000-i3*100-i4)/1000000:
        print(" *** Eq. ",j," : From coeff. of x[",i4,"] in Jac(",i1,i2,i3,") *** ",F[j]):
    end do:
end if:
end proc:
```

Figure 1. Module for the application of the Jacobi identity conditions.

Before each new iteration of EcuJac, the simplest conditions are applied, and the process is repeated until there are no restrictions of simple application. Thus, after a first iteration of module EcuJac with substitutions like:

$$\text{From} Jac(x_0, x_6, x_8) \to C_{68}^1 = C_{68}^2 = C_{68}^3 = C_{68}^4 = C_{68}^5 = C_{68}^6 = 0 \tag{10a}$$

$$\text{From} Jac(x_0, x_5, x_8) \to$$
$$C_{58}^1 = C_{58}^2 = C_{58}^3 = C_{58}^4 = C_{58}^5 = C_{68}^8 = 0, C_{68}^7 = C_{58}^6 \tag{10b}$$

$$\text{From} Jac(x_0, x_4, x_8) \to$$
$$C_{48}^1 = C_{48}^2 = C_{48}^3 = C_{48}^4 = C_{58}^8 = 0, C_{58}^6 = C_{48}^5, C_{58}^7 = C_{48}^6 \tag{10c}$$

215 equations are obtained; some of them are repeated, and others are identities. Selecting one of the simplest equations, like the one that corresponds to the coefficient of x_6 in $\mathbf{Jac}(x_0, x_3, x_5) \to C_{35}^5 - C_{45}^6 - C_{36}^6 = 0$, it is possible to achieve the subsequent substitutions, like

$C_{45}^5 = C_{16}^3, C_{45}^6 = C_{16}^4$, and to compute again the Jacobi equations (EcuJac), obtaining 151 equations. Subsequent iterations provide 124 equations, 78 equations and 55 equations. Then, the first description of the laws is:

$$[x_0, x_i] = x_{i+1}, 1 \leqslant i \leqslant 6 \tag{11a}$$

$$\begin{aligned}[x_1, x_2] = {} & C_{12}^1 x_1 + C_{12}^2 x_2 + C_{12}^3 x_3 + C_{12}^4 x_4 + C_{12}^5 x_5 \\ & + C_{12}^6 x_6 + C_{12}^7 x_7 + C_{12}^8 x_8\end{aligned} \tag{11b}$$

$$[x_1, x_3] = C_{12}^1 x_2 + C_{12}^2 x_3 + C_{12}^3 x_4 + C_{12}^4 x_5 + C_{12}^5 x_6 + C_{12}^6 x_7 \tag{11c}$$

$$[x_1, x_4] = C_{14}^3 x_3 + C_{14}^4 x_4 + C_{14}^5 x_5 + C_{14}^6 x_6 + C_{14}^7 x_7 + C_{14}^8 x_8 \tag{11d}$$

$$\begin{aligned}[x_1, x_5] = {} & (2C_{14}^3 - C_{12}^1) x_4 + (2C_{14}^4 - C_{12}^2) x_5 \\ & + (2C_{14}^5 - C_{12}^3) x_6 + (2C_{14}^6 - C_{12}^4) x_7\end{aligned} \tag{11e}$$

$$[x_1, x_6] = (\tfrac{5}{3} C_{14}^3 - C_{12}^1) x_5 + (\tfrac{5}{3} C_{14}^4 - C_{12}^2) x_6 + C_{16}^7 x_7 + C_{16}^8 x_8 \tag{11f}$$

$$\begin{aligned}[x_1, x_8] = {} & C_{18}^2 x_2 + C_{18}^3 x_3 + C_{18}^4 x_4 + C_{18}^5 x_5 \\ & + C_{18}^6 x_6 + C_{18}^7 x_7 + C_{18}^8 x_8\end{aligned} \tag{11g}$$

$$\begin{aligned}[x_2, x_3] = {} & (C_{12}^1 - C_{14}^3) x_3 + (C_{12}^2 - C_{14}^4) x_4 + (C_{12}^3 - C_{14}^5) x_5 \\ & + (C_{12}^4 - C_{14}^6) x_6 + (C_{12}^5 - C_{14}^7) - C_{14}^8 x_8\end{aligned} \tag{11h}$$

$$[x_2, x_4] = (C_{12}^1 - C_{14}^3) x_4 + (C_{12}^2 - C_{14}^4) x_5 + (C_{12}^3 - C_{14}^5) x_6 + (C_{12}^4 - C_{14}^6) x_7 \tag{11i}$$

$$[x_2, x_5] = \tfrac{1}{3} C_{14}^3 x_5 + \tfrac{1}{3} C_{14}^4 x_6 + (2C_{14}^5 - C_{12}^3 - C_{16}^7) x_7 - C_{16}^8 x_8 \tag{11j}$$

$$[x_2, x_6] = (\tfrac{5}{3} C_{14}^3 - C_{12}^1) x_6 + (\tfrac{5}{3} C_{14}^4 - C_{12}^2) x_7 \tag{11k}$$

$$[x_2, x_8] = C_{18}^2 x_3 + C_{18}^3 x_4 + C_{18}^4 x_5 + C_{18}^5 x_6 + C_{18}^6 x_7 \tag{11l}$$

$$[x_3, x_4] = \left(C_{12}^1 - \tfrac{4}{3} C_{14}^3\right) x_5 + \left(C_{12}^2 - \tfrac{4}{3} C_{14}^4\right) x_6 + (2C_{12}^3 - 3C_{14}^5 + C_{16}^7) x_7 + C_{16}^8 x_8 \tag{11m}$$

$$[x_3, x_5] = \left(C_{12}^1 - \tfrac{4}{3} C_{14}^3\right) x_6 + \left(C_{12}^2 - \tfrac{4}{3} C_{14}^4\right) x_7 \tag{11n}$$

$$[x_3, x_6] = (\tfrac{5}{3} C_{14}^3 - C_{12}^1) x_7 \tag{11o}$$

$$[x_3, x_8] = C_{18}^2 x_4 + C_{18}^3 x_5 + C_{18}^4 x_6 + C_{18}^5 x_7 \tag{11p}$$

$$[x_4, x_5] = (2C_{12}^1 - 3C_{14}^3) x_7 \tag{11q}$$

$$[x_4, x_8] = C_{18}^2 x_5 + C_{18}^3 x_6 + C_{18}^4 x_7 \tag{11r}$$

$$[x_5, x_8] = C_{18}^2 x_6 + C_{18}^3 x_7 \tag{11s}$$

$$[x_6, x_8] = C_{18}^2 x_7 \tag{11t}$$

Subject to the restrictions detailed in Tables 1 and 2.

With the basic change of basis:

$$\begin{cases} y_i = x_i, i \neq 8 \\ y_8 = C_{18}^2 . x_0 + x_8 \end{cases} \tag{12}$$

It can be supposed that $C_{18}^2 = 0$, Equations (11f), (11l), (11p) and (11r) are simplified and Equation (11t) disappears.

Table 1. First group of constraints.

(1)	$C^2_{18}C^8_{18} = 0$
(2)	$-\frac{10}{3}C^3_{14}C^3_{18} + 2C^3_{18}C^1_{12} - C^2_{18}C^8_{18} = 0$
(3)	$-5C^4_{18}C^3_{14} + 3C^4_{18}C^1_{12} - \frac{10}{3}C^3_{18}C^4_{14} + 2C^3_{18}C^2_{12} - C^3_{18}C^8_{18} = 0$
(4)	$-\frac{2}{3}C^3_{18}C^3_{14} - C^2_{18}C^8_{18} = 0$
(5)	$-\frac{2}{3}C^3_{18}C^4_{14} - C^4_{18}C^3_{14} - C^3_{18}C^8_{18} = 0$
(6)	$2C^7_{16}C^3_{18} + 2C^5_{18}C^3_{14} - C^4_{18}C^4_{14} - 4C^3_{18}C^5_{14} + 2C^3_{18}C^3_{12} - 2C^5_{18}C^1_{12} - C^4_{18}C^8_{18} = 0$
(7)	$2C^8_{16}C^3_{18} = 0$
(8)	$2C^3_{18}C^3_{14} - 2C^3_{18}C^1_{12} - C^2_{18}C^8_{18} = 0$
(9)	$2C^3_{18}C^4_{14} - 2C^3_{18}C^2_{12} - 3C^4_{18}C^1_{12} + 3C^4_{18}C^3_{14} - C^3_{18}C^8_{18} = 0$
(10)	$2C^3_{18}C^5_{14} - 2C^3_{18}C^3_{12} + 3C^4_{18}C^4_{14} - 3C^4_{18}C^2_{12} - 2C^5_{18}C^1_{12} + \frac{4}{3}C^5_{18}C^3_{14} - C^4_{18}C^8_{18} = 0$
(11)	$2C^6_{14}C^3_{18} - 2C^3_{18}C^4_{12} - 2C^5_{18}C^2_{12} - 3C^4_{18}C^3_{12} + 3C^4_{18}C^5_{14} + \frac{4}{3}C^5_{18}C^4_{14} - \frac{5}{3}C^6_{18}C^3_{14} - C^5_{18}C^8_{18} = 0$
(12)	$2C^7_{14}C^3_{18} + 3C^4_{18}C^6_{14} - 3C^4_{18}C^4_{12} + 2C^7_{16}C^5_{18} - C^7_{18}C^1_{12} - 2C^3_{18}C^5_{12} - 2C^5_{18}C^5_{14} - \frac{5}{3}C^6_{18}C^4_{14} - C^6_{18}C^8_{18} = 0$
(13)	$2C^8_{14}C^3_{18} + 2C^8_{16}C^5_{18} - C^8_{18}C^1_{12} = 0$
(14)	$C^8_{16}C^2_{18} = 0$
(15)	$C^8_{16}C^3_{18} = 0$
(16)	$-\frac{10}{3}(C^3_{14})^2 + \frac{16}{3}C^3_{14}C^1_{12} - C^8_{14}C^2_{18} - 2(C^1_{12})^2 + C^8_{16}C^4_{18} = 0$
(17)	$\frac{14}{3}C^3_{14}C^1_{12} - 2(C^1_{12})^2 - \frac{20}{9}(C^3_{14})^2 + C^8_{16}C^4_{18} = 0$
(18)	$\frac{4}{3}C^4_{14}C^1_{12} - 2C^2_{12}C^1_{12} + \frac{5}{3}C^3_{14}C^2_{12} + \frac{5}{9}C^3_{14}C^4_{14} + C^8_{16}C^5_{18} = 0$
(19)	$\frac{20}{9}(C^3_{14})^2 - \frac{14}{3}C^3_{14}C^1_{12} + 2(C^1_{12})^2 - C^8_{16}C^4_{18} = 0$
(20)	$-\frac{4}{3}C^4_{14}C^1_{12} + 2C^2_{12}C^1_{12} - \frac{5}{3}C^3_{14}C^2_{12} - \frac{5}{9}C^3_{14}C^4_{14} - C^8_{16}C^5_{18} = 0$
(21)	$\frac{10}{3}C^7_{16}C^3_{14} - 3C^7_{16}C^1_{12} + \frac{5}{3}C^2_{12}C^4_{14} + C^2_{18}C^8_{12} - \frac{10}{3}C^5_{14}C^3_{14} + 2C^5_{14}C^1_{12} - \frac{25}{9}(C^4_{14})^2 - C^6_{18}C^8_{16} = 0$
(22)	$C^3_{14}C^1_{12} - \frac{4}{3}(C^3_{14})^2 - C^8_{14}C^2_{18} = 0$
(23)	$4C^3_{14}C^2_{12} - \frac{31}{3}C^3_{14}C^4_{14} + 6C^4_{14}C^1_{12} - C^8_{14}C^3_{18} - 2C^2_{12}C^1_{12} = 0$
(24)	$-\frac{2}{3}C^3_{14}C^1_{12} + \frac{4}{9}(C^3_{14})^2 = 0$
(25)	$-2C^4_{14}C^1_{12} + \frac{28}{9}C^3_{14}C^4_{14} - \frac{2}{3}C^3_{14}C^2_{12} = 0$
(26)	$3C^7_{16}C^1_{12} - \frac{10}{3}C^7_{16}C^3_{14} - 3C^5_{14}C^1_{12} - 2C^2_{12}C^4_{14} + \frac{2}{3}C^3_{12}C^3_{14} + \frac{8}{3}C^5_{14}C^3_{14} + \frac{8}{3}(C^4_{14})^2 = 0$
(27)	$3C^8_{16}C^1_{12} - \frac{10}{3}C^8_{16}C^3_{14} = 0$

Table 2. Second group of constraints.

(28)	$\frac{8}{3}(C^3_{14})^2 - \frac{16}{3}C^3_{14}C^1_{12} - C^8_{16}C^4_{18} + 2(C^1_{12})^2 = 0$
(29)	$\frac{23}{9}C^3_{14}C^4_{14} - \frac{10}{3}C^4_{14}C^1_{12} - C^8_{16}C^5_{18} + 2C^2_{12}C^1_{12} - \frac{7}{3}C^3_{14}C^2_{12} = 0$
(30)	$-\frac{2}{3}C^5_{14}C^3_{14} + \frac{2}{3}C^3_{12}C^3_{14} - \frac{1}{9}(C^4_{14})^2 - \frac{1}{3}C^2_{12}C^4_{14} - C^6_{18}C^8_{16} - C^5_{14}C^1_{12} + C^2_{18}C^8_{12} = 0$
(31)	$\frac{8}{3}C^6_{14}C^3_{14} - 3C^6_{14}C^1_{12} + 2C^2_{12}C^5_{14} + \frac{2}{3}C^4_{12}C^3_{14} - C^7_{18}C^8_{16} - 2C^3_{12}C^2_{12} + 5C^3_{12}C^4_{14} + \frac{7}{3}C^7_{16}C^4_{14} + C^3_{18}C^8_{12} - \frac{22}{3}C^4_{14}C^5_{14} = 0$
(32)	$\frac{7}{3}C^8_{16}C^4_{14} - C^8_{18}C^8_{16} = 0$
(33)	$\frac{13}{3}C^3_{14}C^1_{12} - 2(C^1_{12})^2 - \frac{8}{3}C^{32}_{14} + C^8_{16}C^4_{18} - C^8_{14}C^2_{18} = 0$
(34)	$\frac{10}{3}C^4_{14}C^1_{12} - 4C^2_{12}C^1_{12} - \frac{32}{9}C^3_{14}C^4_{14} + \frac{13}{3}C^3_{14}C^2_{12} + C^8_{16}C^5_{18} - C^8_{14}C^3_{18} = 0$
(35)	$2C^5_{14}C^1_{12} - \frac{4}{3}C^5_{14}C^3_{14} - 2C^3_{12}C^1_{12} + \frac{4}{3}C^3_{12}C^3_{14} + \frac{10}{3}C^2_{12}C^4_{14} - \frac{8}{9}(C^4_{14})^2 - 2(C^2_{12})^2 + C^6_{18}C^8_{16} - C^4_{18}C^8_{14} = 0$
(36)	$4C^6_{14}C^1_{12} - \frac{13}{3}C^6_{14}C^3_{14} - \frac{5}{3}C^4_{12}C^3_{14} - \frac{7}{3}C^7_{16}C^4_{14} + C^7_{18}C^8_{16} - 2C^3_{12}C^2_{12} + 2C^2_{12}C^5_{14} - 2C^3_{12}C^4_{14} + \frac{13}{3}C^4_{14}C^5_{14} - C^5_{18}C^8_{14} = 0$
(37)	$-\frac{7}{3}C^8_{16}C^4_{14} + C^8_{18}C^8_{16} = 0$
(38)	$-C^3_{14}C^1_{12} - C^8_{14}C^2_{18} = 0$
(39)	$2C^3_{14}C^2_{12} - 2C^2_{12}C^1_{12} - C^3_{14}C^4_{14} - C^8_{14}C^3_{18} = 0$
(40)	$C^5_{14}C^1_{12} - 2C^5_{14}C^3_{14} + 3C^2_{12}C^4_{14} - (C^4_{14})^2 - 2(C^2_{12})^2 - 2C^3_{12}C^1_{12} + 2C^3_{12}C^3_{14} + C^2_{18}C^8_{12} - C^4_{18}C^8_{14} = 0$
(41)	$C^6_{14}C^1_{12} - \frac{5}{3}C^6_{14}C^3_{14} + 4C^2_{12}C^5_{14} - 3C^4_{14}C^5_{14} - 4C^3_{12}C^2_{12} + 3C^3_{12}C^4_{14} + C^3_{18}C^8_{12} - C^4_{12}C^3_{14} - C^5_{18}C^8_{14} = 0$
(42)	$-C^6_{18}C^8_{14} + 2C^5_{12}C^1_{12} - 2C^2_{12}C^4_{12} + 4C^2_{12}C^6_{14} - \frac{8}{3}C^4_{14}C^6_{14} - 4C^5_{12}C^3_{14} + 4C^3_{12}C^5_{14} - 2(C^3_{12})^2 + C^4_{18}C^8_{12} - 2(C^5_{14})^2 = 0$
(43)	$3C^7_{14}C^2_{12} + 3C^3_{12}C^6_{14} - 3C^5_{12}C^4_{14} - 4C^4_{12}C^5_{14} - C^7_{14}C^4_{14} + 3C^7_{16}C^4_{12} - C^7_{16}C^6_{14} + C^5_{18}C^8_{12} - C^6_{12}C^1_{12} - C^7_{18}C^8_{14} - 2C^6_{14}C^5_{14} + C^4_{12}C^3_{12} + \frac{2}{3}C^6_{12}C^3_{14} = 0$
(44)	$3C^8_{14}C^2_{12} - C^8_{14}C^4_{14} + 3C^8_{16}C^4_{12} - C^8_{16}C^6_{14} - C^8_{18}C^8_{14} = 0$
(45)	$-\frac{10}{3}C^3_{18}C^3_{14} + 2C^3_{18}C^1_{12} = 0$
(46)	$\frac{8}{3}C^3_{18}C^3_{14} - 2C^3_{18}C^1_{12} = 0$
(47)	$\frac{8}{3}C^3_{18}C^4_{14} - 2C^3_{18}C^2_{12} - 3C^4_{18}C^1_{12} + 4C^4_{18}C^3_{14} = 0$
(48)	$6C^3_{18}C^5_{14} - 4C^3_{18}C^3_{12} - 2C^7_{16}C^3_{18} + 4C^4_{18}C^4_{14} - 3C^4_{18}C^2_{12} - \frac{2}{3}C^5_{18}C^3_{14} = 0$
(49)	$-\frac{2}{3}C^3_{14}C^1_{12} + \frac{10}{9}(C^3_{14})^2 + C^8_{14}C^2_{18} = 0$
(50)	$2(C^1_{12})^2 - \frac{13}{3}C^3_{14}C^1_{12} + 2(C^3_{14})^2 - C^8_{16}C^4_{18} = 0$
(51)	$5C^3_{14}C^1_{12} - 2(C^1_{12})^2 - \frac{28}{9}(C^3_{14})^2 - C^8_{14}C^2_{18} + C^8_{16}C^4_{18} = 0$
(52)	$\frac{16}{3}C^4_{14}C^1_{12} - \frac{20}{3}C^3_{14}C^4_{14} - 4C^2_{12}C^1_{12} + 5C^3_{14}C^2_{12} - C^8_{14}C^3_{18} + C^8_{16}C^5_{18} = 0$
(53)	$C^6_{18}C^8_{16} - C^4_{18}C^8_{14} + 5C^5_{14}C^1_{12} - 2(C^2_{12})^2 + \frac{16}{3}C^2_{12}C^4_{14} - 2C^3_{12}C^1_{12} + \frac{10}{3}C^7_{16}C^3_{14} + \frac{2}{3}C^3_{12}C^3_{14} - 3C^7_{16}C^1_{12} - \frac{32}{9}(C^4_{14})^2 - 4C^5_{14}C^3_{14} = 0$
(54)	$\frac{10}{3}C^8_{16}C^3_{14} - 3C^8_{16}C^1_{12} = 0$
(55)	$-4C^3_{18}C^1_{12} + 6C^3_{18}C^3_{14} = 0$

A new computation of the Jacobi equations provides 52 restrictions and selecting:

$$\text{fromcoeff.of} x_6 \text{in} \mathbf{Jac}(x_1, x_4, x_5) \quad \rightarrow \quad \frac{1}{3}(3C^1_{12} - 4C^3_{14})C^3_{14} = 0 \tag{13a}$$

$$\text{fromcoeff.of}x_2\text{in}\mathbf{Jac}(x_1, x_2, x_3) \quad \rightarrow \quad C_{12}^1 C_{14}^3 = 0 \tag{13b}$$

It is deduced that $C_{14}^3 = 0$, and:

$$\text{fromcoeff.of}x_8\text{in}\mathbf{Jac}(x_1, x_3, x_5) \quad \rightarrow \quad -3C_{12}^1 C_{16}^8 = 0 \tag{14a}$$

$$\text{fromcoeff.of}x_4\text{in}\mathbf{Jac}(x_1, x_3, x_4) \quad \rightarrow \quad -2\left(C_{12}^1\right)^2 + C_{16}^8 C_{18}^4 = 0 \tag{14b}$$

It is deduced that $C_{12}^1 = 0$ and Equation (11) are simplified and Equations (11o) and (11q) disappear.

A new computation of the Jacobi equations provides 31 restrictions and selecting:

$$\begin{gathered}\text{fromcoeff.of}x_7\text{in}\mathbf{Jac}(x_1, x_4, x_6) \\ C_{16}^8 C_{18}^4 = 0\end{gathered} \tag{15a}$$

$$\begin{gathered}\text{fromcoeff.of}x_6\text{in}\mathbf{Jac}(x_1, x_3, x_4) \\ \tfrac{10}{3} C_{12}^2 C_{14}^4 - \tfrac{8}{9}\left(C_{14}^4\right)^2 - 2\left(C_{12}^2\right)^2 + C_{18}^6 C_{16}^8 - C_{18}^4 C_{14}^8 = 0\end{gathered} \tag{15b}$$

$$\begin{gathered}\text{fromcoeff.of}x_4\text{in}\mathbf{Jac}(x_1, x_2, x_3) \\ 3C_{12}^2 C_{14}^4 - \left(C_{14}^4\right)^2 - 2\left(C_{12}^2\right)^2 - C_{18}^4 C_{14}^8 = 0\end{gathered} \tag{15c}$$

$$\begin{gathered}\text{fromcoeff.of}x_7\text{in}\mathbf{Jac}(x_2, x_3, x_4) \\ \tfrac{16}{3} C_{12}^2 C_{14}^4 - \tfrac{32}{9}\left(C_{14}^4\right)^2 - 2\left(C_{12}^2\right)^2 + C_{18}^6 C_{16}^8 - C_{18}^4 C_{14}^8 = 0\end{gathered} \tag{15d}$$

It is deduced that $C_{14}^4 = 0$, and another simplification of the laws in (11) is applied subject to 18 restrictions.

In accordance to Engel theorem, if a Lie algebra is nilpotent, then it is ad-nilpotent, *i.e.*, the matrices associated with the adjoints of all of the Lie algebra elements have all of their eigenvalues null. In this moment, a subprogram in Maple is used to calculate the characteristic polynomials of the adjoints of all of the vectors in the basis in an iterative and interactive way. From:

$$Adj(x_1) = \lambda^3\left(\lambda - C_{12}^2\right)^2\left(\lambda + C_{12}^2\right)^2\left(\lambda^2 - C_{18}^8\lambda - C_{14}^8 C_{18}^4\right) = 0 \tag{16}$$

it is deduced that $C_{12}^2 = 0$ and $C_{18}^8 = 0$. Applying the resulting substitutions and with a new computation of the Jacobi equations, the restrictions are reduced to 15, and the laws for QFLA of dimension nine are described by:

$$[x_0, x_i] = x_{i+1}, 1 \leqslant i \leqslant 6 \tag{17a}$$

$$[x_1, x_2] = C_{12}^3 x_3 + C_{12}^4 x_4 + C_{12}^5 x_5 + C_{12}^6 x_6 + C_{12}^7 x_7 + C_{12}^8 x_8 \tag{17b}$$

$$[x_1, x_3] = C_{12}^3 x_4 + C_{12}^4 x_5 + C_{12}^5 x6 + C_{12}^6 x_7 \tag{17c}$$

$$[x_1, x_4] = C_{14}^5 x_5 + C_{14}^6 x_6 + C_{14}^7 x_7 + C_{14}^8 x_8 \tag{17d}$$

$$[x_1, x_5] = (2C_{14}^5 - C_{12}^3)x_6 + (2C_{14}^6 - C_{12}^4)x_7 \tag{17e}$$

$$[x_1, x_6] = C_{16}^7 x_7 + C_{16}^8 x_8 \tag{17f}$$

$$[x_1, x_8] = C_{18}^3 x_3 + C_{18}^4 x_4 + C_{18}^5 x_5 + C_{18}^6 x_6 + C_{18}^7 x_7 + C_{18}^8 x_8 \tag{17g}$$

$$[x_2, x_3] = (C_{12}^3 - C_{14}^5)x_5 + (C_{12}^4 - C_{14}^6)x_6 + (C_{12}^5 - C_{14}^7) - C_{14}^8 x_8 \tag{17h}$$

$$[x_2, x_4] = (C_{12}^3 - C_{14}^5)x_6 + (C_{12}^4 - C_{14}^6)x_7 \tag{17i}$$

$$[x_2, x_5] = (2C_{14}^5 - C_{12}^3 - C_{16}^7)x_7 - C_{16}^8 x_8 \tag{17j}$$

$$[x_2, x_8] = C_{18}^3 x_4 + C_{18}^4 x_5 + C_{18}^5 x_6 + C_{18}^6 x_7 \tag{17k}$$

$$[x_3, x_4] = (2C_{12}^3 - 3C_{14}^5 + C_{16}^7) x_7 + C_{16}^8 x_8 \tag{17l}$$

$$[x_3, x_8] = C_{18}^3 x_5 + C_{18}^4 x_6 + C_{18}^5 x_7 \tag{17m}$$

$$[x_4, x_8] = C_{18}^3 x_6 + C_{18}^4 x_7 \tag{17n}$$

$$[x_5, x_8] = C_{18}^3 x_7 \tag{17o}$$

With the change of basis:

$$\begin{cases} y_i = x_i, i \neq 1 \\ y_1 = -C_{12}^3 x_0 + x_1 \end{cases} \tag{18}$$

It can be supposed that $C_{12}^3 = 0$. Let us consider the adoption of the simplified notation shown in Table 3. Then, the laws for complex QFLA of dimension nine coincide with Equation (6).

Table 3. Notation for the quasi-filiform Lie algebra (QFLA) parameters.

$\alpha_1 = C_{12}^4$	$\alpha_2 = C_{12}^5$	$\alpha_3 = C_{12}^6$	$\alpha_4 = C_{12}^7$
$\alpha_5 = C_{12}^8$	$\alpha_6 = C_{14}^5$	$\alpha_7 = C_{14}^6$	$\alpha_8 = C_{14}^7$
$\alpha_9 = C_{14}^8$	$\alpha_{10} = C_{16}^7$	$\alpha_{11} = C_{16}^8$	$\alpha_{12} = C_{18}^3$
$\alpha_{12} = C_{18}^4$	$\alpha_{14} = C_{18}^5$	$\alpha_{15} = C_{18}^6$	$\alpha_{16} = C_{18}^7$

Finally, the conditions to consider with the notation in Table 3 are:

$$\text{fromcoeff.of} x_7 \text{in} \mathbf{Jac}(x_1, x_4, x_8) \rightarrow 2\alpha_{10}\alpha_{12} = 0 \tag{19a}$$

$$\text{fromcoeff.of} x_5 \text{in} \mathbf{Jac}(x_1, x_2, x_8) \rightarrow 2\alpha_6\alpha_{12} = 0 \tag{19b}$$

$$\text{fromcoeff.of} x_6 \text{in} \mathbf{Jac}(x_1, x_2, x_8) \rightarrow 2\alpha_{12}(-\alpha_1 + \alpha_7) + 3\alpha_6\alpha_{13} = 0 \tag{19c}$$

$$\text{fromcoeff.of} x_7 \text{in} \mathbf{Jac}(x_1, x_2, x_8) \rightarrow$$

$$\rightarrow 2\alpha_{12}(-\alpha_2 + \alpha_8) + 3\alpha_{13}(-\alpha_1 + \alpha_7) + 2\alpha_{14}(-\alpha_6 + \alpha_{10}) = 0 \tag{19d}$$

$$\text{fromcoeff.of} x_7 \text{in} \mathbf{Jac}(x_1, x_5, x_6) \rightarrow \alpha_{11}\alpha_{12} = 0 \tag{19e}$$

$$\text{fromcoeff.of} x_7 \text{in} \mathbf{Jac}(x_1, x_4, x_6) \rightarrow \alpha_{11}\alpha_{13} = 0 \tag{19f}$$

$$\text{fromcoeff.of} x_7 \text{in} \mathbf{Jac}(x_1, x_3, x_6) \rightarrow \alpha_{11}\alpha_{14} = 0 \tag{19g}$$

$$\text{fromcoeff.of} x_7 \text{in} \mathbf{Jac}(x_1, x_2, x_6) \rightarrow \alpha_{11}\alpha_{15} = 0 \tag{19h}$$

$$\text{fromcoeff.of} x_7 \text{in} \mathbf{Jac}(x_1, x_4, x_5) \rightarrow \alpha_9\alpha_{12} = 0 \tag{19i}$$

$$\text{fromcoeff.of} x_7 \text{in} \mathbf{Jac}(x_1, x_3, x_4) \rightarrow \alpha_9\alpha_{14} - \alpha_{11}\alpha_{16} = 0 \tag{19j}$$

$$\text{fromcoeff.of} x_5 \text{in} \mathbf{Jac}(x_1, x_2, x_4) \rightarrow \alpha_9\alpha_{13} = 0 \tag{19k}$$

$$\text{fromcoeff.of} x_6 \text{in} \mathbf{Jac}(x_1, x_2, x_4) \rightarrow \alpha_5\alpha_{12} - \alpha_{11}\alpha_{16} = 0 \tag{19l}$$

$$\text{fromcoeff.of} x_6 \text{in} \mathbf{Jac}(x_1, x_2, x_3) \rightarrow -2\alpha_6^2 + \alpha_5\alpha_{13} - \alpha_9\alpha_{15} = 0 \tag{19m}$$

$$\text{fromcoeff.of} x_7 \text{in} \mathbf{Jac}(x_1, x_2, x_3) \rightarrow$$

$$\rightarrow -2\alpha_6(2\alpha_1 + \alpha_7) + \alpha_{10}(3\alpha_1 - \alpha_7) - \alpha_9\alpha_{16} + \alpha_5\alpha_{14} = 0 \tag{19n}$$

$$\text{fromcoeff.of} x_8 \text{in} \mathbf{Jac}(x_1, x_2, x_3) \rightarrow \alpha_{11}(3\alpha_1 - \alpha_7) = 0 \tag{19o}$$

Thus, the restrictions simplified and rewritten coincide with Equation (7). Q.E.D.

3. Concluding Remarks

In this paper, the proof of the theorem of the structure of the laws of every complex quasi-filiform Lie algebra of dimension nine has been presented. Symbolic and iterative computation has been indispensable in this piece of research. A PC Pentium 4 of 2.4 Ghz and the programming language Maple 6®have been used in the process. It has been necessary to program modules to tackle, among others, the following functions for the general treatment of processes on: the storage and recovery of intermediate data of hypermatrices and restriction equations; the storage and recovery of matrices of the change of the basis; the search of special substitutions; the print of laws in different formats; the print and checking of hypermatrices. Furthermore, modules have been developed for specific treatment on: the development of hypermatrices and general variables; the application of anticommutativity; the application of conditions from Jacobi equations; simplification; the application of ad-nilpotence; calculation of the lower central series. The library modules developed represent approximately 3000 lines of code. The massive application of the changes of the basis and the characteristic vector has permitted obtaining the general family of QFLA laws of dimension nine.

Acknowledgments: The authors would like to thank J. R. Gómez and J. J. López from University of Sevilla for their support and orientation on the knowledge of Lie Algebras.

Author Contributions: M.P. and F.P. conceived main proof and analyzed the data; E.J. contributed analysis tools and wrote the paper.

Conflicts of Interest: The authors declare no conflict of interest.

References

1. Gilmore, R. *Lie Groups, Lie Algebras, and Some of Their Applications*; Dover Publications: Mineola, NY, USA, 2005.
2. Sattinger, D.H.; Weaver, O.L. *Lie Groups and Algebras with Applications to Physics, Geometry and Mechanics*; Springer-Verlag New York Inc.: New York, NY, USA, 1986.
3. Yao, Y.; Ji, J.; Chen, D.; Zeng, Y. The quadratic-form identity for constructing the Hamiltonian structures of the discrete integrable systems. *Comput. Math. Appl.* **2008**, *56*, 2874–2882. [CrossRef]
4. Benjumea, J.C.; Echarte, F.J.; Núñez, J.; Tenorio, A.F. A Method to Obtain the Lie Group Associated With a Nilpotent Lie Algebra. *Comput. Math. Appl.* **2006**, *51*, 1493–1506. [CrossRef]
5. Georgi, H. *Lie Algebras in Particle Physics: From Isospin to Unified Theories (Frontiers in Physics)*; Westview Press: Boulder, CO, USA, 1999.
6. Brockett, R.W. Lie algebras and Lie groups in control theory. In *Geometric Methods in System Theory, ser. NATO Advanced Study Institutes Series*; Mayne, D., Brockett, R., Eds.; Springer: Amsterdam, The Netherlands, 1973; Volume 3, pp. 43–82.
7. Sachkov, Y. Control Theory on Lie Groups. *J. Math. Sci.* **2009**, *156*, 381–439. [CrossRef]
8. Zimmerman, J. Optimal control of the Sphere S^n Rolling on E^n. *Math. Control Signals Syst.* **2005**, *17*, 14–37. [CrossRef]
9. Malcev, A.I. On semi-simple subgroups of Lie groups. *Izv. Akad. Nauk SSSR Ser. Mat.* **1944**, *8*, 143–174.
10. Malcev, A.I. On solvable Lie algebras. *Izv. Akad. Nauk SSSR Ser. Mat.* **1945**, *9*, 329–356.
11. Onishchik, A.L.; Vinberg, E.B. *Lie Groups and Algebraic Groups III, Structure of Lie Groups and Lie Algebras*; Springer-Verlag: Berlin, Germany; Heidelberg, Germany, 1994.
12. Goze, M.; Khakimdjanov, Y. *Nilpotent Lie Algebras*; Kluwer Academic Publishers: Dordrecht, The Netherlands, 1996.
13. Umlauf, K.A. Über die Zusammensetzung der Endlichen Continuenlichen Transformationsgruppen, Insbesondere der Gruppen vom Range Null. Ph.D. Thesis, University of Leipzig, Leipzig, Germany, 1891.
14. Ancochea, J.M.; Goze, M. Sur la classification des algèbres de Lie nilpotentes de dimension 7. *C. R. Acad. Sci. Paris* **1986**, *302*, 611–613.
15. Ancochea, J.M.; Goze, M. On the varieties of nilpotent Lie algebras of dimension 7 and 8. *J. Pure Appl. Algebra* **1992**, *77*, 131–140.

16. Gómez, J.R.; Echarte, F.J. Classification of complex filiform Lie algebras of dimension 9. *Rend. Sem. Fac. Sc. Univ. Cagliari* **1991**, *61*, 21–29.
17. Gómez, J.R.; Jiménez-Merchán, A.; Khakimdjanov, Y. Symplectic Structures on the Filiform Lie Algebras. *J. Pure Appl. Algebra* **2001**, *156*, 15–31. [CrossRef]
18. Boza, L.; Fedriani, E.M.; Nuñez, J. Complex Filiform Lie Algebras of Dimension 11. *Appl. Math. Comput.* **2003**, *141*, 611–630. [CrossRef]
19. Echarte, F.J.; Núñez, J.; Ramírez, F. Relations among invariants of complex filiform Lie algebras. *Appl. Math. Comput.* **2004**, *147*, 365–376. [CrossRef]
20. Echarte, F.J.; Nuñez, J.; Ramirez, F. Description of Some Families of Filiform Lie Algebras. *Houst. J. Math.* **2008**, *34*, 19–32.
21. Benjumea, J.C.; Nuñez, J.; Tenorio, A.F. Computing the Law of a Family of Solvable Lie Algebras. *Int. J. Algebra Comput.* **2009**, *19*, 337–345. [CrossRef]
22. Burde, D.; Eick, B.; de Graaf, W. Computing faithful representations for nilpotent Lie algebras. *J. Algebra* **2009**, *322*, 602–612. [CrossRef]
23. Ceballos, M.; Nuñez, J.; Tenorio, A.F. The Computation of Abelian Subalgebras in Low-Dimensional Solvable Lie Algebras. *WSEAS Trans. Math.* **2010**, *9*, 22–31.
24. Cabezas, J.M.; Gómez, J.R.; Jiménez-Merchán, A. Family of p-filiform Lie algebras. In *Algebra and Operator Theory: Proceedings of the Colloquium in Taskent (Uzbekistan)*; Kluwer Academic Publishers: Dordrecht, The Netherlands, 1997; pp. 93–102.
25. Camacho, L.M.; Gómez, J.R.; González, A.; Omirov, B.A. Naturally Graded Quasi-Filiform Leibniz Algebras. *J. Symb. Comput.* **2009**, *44*, 27–539. [CrossRef]
26. Camacho, L.M. Álgebras de Lie P-Filiformes. Ph.D. Thesis, Universidad de Sevilla, Seville, Spain, 1999.
27. Ancochea, J.M.; Campoamor, O.R. Classification of (*n*-5)-filiform Lie Algebras. *Linear Algebra Appl.* **2001**, *336*, 167–180. [CrossRef]
28. Eick, B. Some new simple Lie algebras in characteristic 2. *J. Symb. Comput.* **2010**, *45*, 943–951. [CrossRef]
29. Schneider, C. A Computer-Based Approach to the Classification of Nilpotent Lie Algebras. *Exp. Math.* **2005**, *14*, 153–160. [CrossRef]
30. Sendra, J.R.; Perez-Diaz, S.; Sendra, J.; Villarino, C. *Introducción a la Computación Simbólica y Facilidades Maple*; Addlink Media: Madrid, Spain, 2009.
31. Pérez, M.; Pérez, F.; Jiménez, E. Classification of the Quasifiliform Nilpotent Lie Algebras of Dimension 9. *J. Appl. Math.* **2014**, *2014*, 1–12. [CrossRef]
32. Pérez, F. Clasificación de las Álgebras de Lie Cuasifiliformes de Dimensión 9. Ph.D. Thesis, Universidad de Sevilla, Seville, Spain, 2007.
33. Bäuerle, G.G.A.; De Kerf, E.A. *Lie Algebras Part 1, Studies in Mathematical Physics 1*; Elsevier: Amsterdam, The Netherlands, 1990.
34. Benjumea, J.C.; Fernandez, D.; Márquez, M.C.; Nuñez, J.; Vilches, J.A. *Matemáticas Avanzadas y Estadística para Ciencias e Ingenierías*; Secretariado de Publicaciones de la Universidad de Sevilla: Sevilla, Spain, 2006.
35. Erdmann, K.; Wildon, M.J. *Introduction to Lie Algebras*; Springer: Berlin, Germany; Heidelberg, Germany, 2006.
36. Jacobson, N. *Lie Algebras*; Dover Publications, Inc.: Mineola, NY, USA, 1979.
37. Onishchik, A.L.; Arkadij, L.; Vinberg, E.B. *Lie Groups and Algebraic Groups*; Springer-Verlag: Berlin, Germany; Heidelberg, Germany, 1990.
38. Agrachev, A.; Sachkov, Y. *Control Theory from the Geometric Viewpoint*; Springer: Berlin, Germany; Heidelberg, Germany, 2004.

symmetry

MDPI

Article

An Elementary Derivation of the Matrix Elements of Real Irreducible Representations of $\mathfrak{so}(3)$

Rutwig Campoamor-Stursberg

Instituto de Matemática Interdisciplinar and Department of Geometry and Topology, Universidad Complutense de Madrid, Plaza de Ciencias 3, E-28040 Madrid, Spain; rutwig@ucm.es

Academic Editor: Roman M. Cherniha
Received: 30 June 2015; Accepted: 10 September 2015; Published: 14 September 2015

Abstract: Using elementary techniques, an algorithmic procedure to construct skew-symmetric matrices realizing the real irreducible representations of $\mathfrak{so}(3)$ is developed. We further give a simple criterion that enables one to deduce the decomposition of an arbitrary real representation R of $\mathfrak{so}(3)$ into real irreducible components from the characteristic polynomial of an arbitrary representation matrix.

Keywords: real representation; matrix element; tensor product

1. Introduction

Albeit the fact that the representation theory of semisimple Lie algebras in general, and the orthogonal algebras $\mathfrak{so}(n)$ and their various reals forms in particular, is well known and constitutes nowadays a standard tool in (physical) applications (see, e.g., [1,2] and the references therein), specific results in the literature concerning the explicit matrix construction of the matrices corresponding to real irreducible representations of $\mathfrak{so}(n)$ are rather scarce. Even if the structural properties of such representations can be derived from the complex case [3], the inherent technical difficulties arising in the analysis of irreducible representations over the real field make it cumbersome to determine an algorithmic procedure that provides the specific real representation matrices explicitly.

Even for the lowest dimensional case, that of $\mathfrak{so}(3)$, the description of real irreducible representations is generally restricted to multiplets of low dimension appearing in specific problems [4]. One interesting work devoted exclusively to the real irreducible representations from the perspective of harmonic analysis is given in [5]. Most of the applications of $\mathfrak{so}(3)$ make use of the angular momentum operators or the Gel'fand–Zetlin formalism, hence describing the states by means of eigenvalues of a complete set of diagonalizable commuting operators. However, for real irreducible representations of $\mathfrak{so}(3)$, corresponding to rotations in the representation space, no such bases of states of this type are possible, as no inner labeling diagonalizable operator over the real numbers can exist, the external being the Casimir operator [6]. In spite of this fact, real representations are of considerable practical importance, as they provide information on the embedding of $\mathfrak{so}(3)$ into other simple algebras and, thus, constitute interesting tools to determine the stability of semidirect sums of Lie algebras [7]. The hierarchy of real irreducible representations of simple Lie algebras is therefore deeply connected to the embedding problem and the branching rules. In this context, it is desirable to develop a simple algorithmic method for the construction of real irreducible representations R of $\mathfrak{so}(3)$ in terms of skew-symmetric matrices, as these correspond naturally to the embedding of $\mathfrak{so}(3)$ as a subalgebra of $\mathfrak{so}(\dim R)$.

In this work, we propose such a procedure, based on the elementary properties of rotation matrices. It is shown that the class of a real irreducible representation R is completely determined by the characteristic polynomial of a matrix in R. This further enables one to deduce the decomposition of an arbitrary real representation of $\mathfrak{so}(3)$ into real irreducible factors from the properties of the characteristic polynomial of a matrix within the representation.

1.1. Real Representations of $\mathfrak{so}(3)$

Recall that for $\mathfrak{sl}(2,\mathbb{C})$, the standard basis is given by $\{h,e,f\}$ with commutators:

$$[h,e]=2e,\ [h,f]=-2f,\ [e,f]=h. \tag{1}$$

Let D_J denote the irreducible representation of $\mathfrak{sl}(2,\mathbb{C})$ of dimension $(J+1)$, where $J=0,\frac{1}{2},1,\frac{3}{2},\cdots$. For the basis $\{\mathbf{e}_1,\cdots,\mathbf{e}_{2J+1}\}$ of the representation space, the matrices D_J for the generators h,e,f are easily recovered from the matrix elements:

$$\langle \mathbf{e}^i|D_J(h)|\,\mathbf{e}_j\rangle=\delta_i^j(2J+1-2i);\quad \langle \mathbf{e}^i|D_J(e)|\,\mathbf{e}_j\rangle=\delta_{i+1}^j(2J+1-i)$$

$$\langle \mathbf{e}^i|D_J(f)|\,\mathbf{e}_j\rangle=\delta_i^{j+1}(i-1). \tag{2}$$

As is well known, the Lie algebra $\mathfrak{sl}(2,\mathbb{C})$ admits two real forms, the normal real form $\mathfrak{sl}(2,\mathbb{R})$ obtained by restriction of scalars, as well as the compact real form $\mathfrak{so}(3)$ obtained from the Cartan map:

$$X_1=\frac{\mathrm{i}}{2}h,\ X_2=\frac{1}{2}(e-f),\ X_3=\frac{\mathrm{i}}{2}(e+f) \tag{3}$$

and satisfying the brackets:

$$[X_i,X_j]=\varepsilon_{ijk}\,X_k,\ 1\leq i,j,k\leq 3. \tag{4}$$

While the matrices of the representation D_J define a real representation of $\mathfrak{sl}(2,\mathbb{R})$ for the compact real form $\mathfrak{so}(3)$, the matrices of D_J are complex, given by:

$$D_J(X_1)=\frac{\mathrm{i}}{2}D_J(h),\ D_J(X_2)=\frac{1}{2}(D_J(e)-D_J(f)),\ D_J(X_3)=\frac{\mathrm{i}}{2}(D_J(e)+D_J(f)). \tag{5}$$

In many applications, the representation space of D_J is best described by states of the type:

$$|\,\mu,\ J(J+1)\rangle,\ \mu=-J,\cdots,J \tag{6}$$

on an appropriate basis, as, e.g., that commonly used in the theory of angular momentum [8]. It must be observed, however, that such bases are not suitable for real representations, as geometric rotation matrices are not diagonalizable over the real field $\mathbb{R}$.

The problem of classifying the real irreducible representations of the compact real forms of semisimple Lie algebras was systematically considered by Cartan and Karpelevich, being later expanded for arbitrary real Lie algebras by Iwahori [9]. According to these works, real representations are distinguished by the decomposition of their complexification. More precisely, if Γ is a real representation of the (real) Lie algebra $\mathfrak{g}$, then:

1. Γ is called of first class, denoted by Γ^I, if $\Gamma\otimes_{\mathbb{R}}\mathbb{C}$ is a complex irreducible representation of $\mathfrak{g}$.
2. Γ is called of second class, denoted by Γ^{II}, if $\Gamma\otimes_{\mathbb{R}}\mathbb{C}$ is a complex reducible representation of $\mathfrak{g}$.

Following this distinction, the representations D_J of $\mathfrak{so}(3)$ with $J\in\mathbb{N}$ belong to the first class. This in particular implies the existence of an invertible matrix $U\in GL(2J+1,\mathbb{C})$, such that for $1\leq k\leq 3$:

$$R_J^I(X_k)=U\,D_J(X_k)\,U^{-1} \tag{7}$$

is a real matrix [9]. For half-integer values $J \in \frac{1}{2}\mathbb{N}$, no such transition matrices U can exist, and in order to obtain a real representation, the dimension of the representation space must be doubled:

$$D_J(X_k) \mapsto D_J^{II}(X_k) = \begin{pmatrix} \mathrm{Re}D_J(a_k) & -\mathrm{Im}D_J(a_k) \\ \mathrm{Im}D_J(a_k) & \mathrm{Re}D_J(a_k) \end{pmatrix}. \tag{8}$$

As a consequence, even dimensional irreducible real representations of $\mathfrak{so}(3)$ only exist for $n = 4q$ with $q \geq 1$ (details on the double-covering $SU(2) \to SO(3)$ can be found, e.g., in [10]).

Albeit not usually referred to in the literature, the class of a real representation of a (simple) Lie algebra is deeply connected to the embedding problem of (complex) semisimple Lie algebras [11]. In particular, it determines whether an algebra is irreducibly embedded into another. Recall that an embedding $j : \mathfrak{s}' \to \mathfrak{s}$ of semisimple Lie algebras is called irreducible if the lowest dimensional irreducible representation Γ of $\mathfrak{s}$ remains irreducible when restricted to $\mathfrak{s}'$ [11]. Irreducible embeddings play an important role in applications, as they allow one to construct bases of a Lie algebra $\mathfrak{s}$ in terms of a basis of irreducibly-embedded subalgebras and irreducible tensor operators [12].

From the analysis of $\mathfrak{so}(3)$ representations, it is straightforward to establish the following embeddings:

1. For $J = 2$, $\mathfrak{so}(3)$ is a maximal subalgebra irreducibly embedded into $\mathfrak{sp}(4) \simeq \mathfrak{so}(5)$.
2. For $J = 3$, $\mathfrak{so}(3)$ is irreducibly embedded into $\mathfrak{so}(7)$ through the chain:

$$\mathfrak{so}(3) \subset G_{2,-14} \subset \mathfrak{so}(7).$$

3. For any integer $J \geq 4$, $\mathfrak{so}(3)$ is a maximal subalgebra irreducibly embedded into $\mathfrak{so}(2J+1)$.
4. For $J = \frac{3}{2}$, $\mathfrak{so}(3)$ is embedded into $\mathfrak{so}(4)$ through the chain:

$$\mathfrak{so}(3) \subset \mathfrak{sp}(4) \subset \mathfrak{su}(4) \subset \mathfrak{so}(7) \subset \mathfrak{so}(8)$$

5. For half-integers $J \geq \frac{5}{2}$, $\mathfrak{so}(3)$ is embedded into $\mathfrak{so}(4J+2)$ through the chain:

$$\mathfrak{so}(3) \subset \mathfrak{sp}(2J+1) \subset \mathfrak{su}(2J+1) \subset \mathfrak{so}(4J+2)$$

In this context, a natural construction of real irreducible representations of $\mathfrak{so}(3)$ should be by means of skew-symmetric matrices that realize these embeddings.

2. Construction of the Matrices $R_J^I(X_k)$

As already observed, for integer J, the representation R_J^I given by (7) is of first class. Therefore, $\mathfrak{so}(3)$ can be represented as a subalgebra of the compact Lie algebra $\mathfrak{so}(2J+1)$. In particular, we can find a transition matrix $U \in GL(2J+1, \mathbb{C})$, such that the matrices:

$$R_J^I(X_k) = U\, D_J(X_k)\, U^{-1} \tag{9}$$

are skew-symmetric for $k = 1, 2, 3$, thus describe the embedding.

The construction of skew-symmetric real matrices $R_J^I(X_k)$ satisfying the similarity Condition (9) is essentially based on the following two properties of the (complex) representation matrices $D_J(X_k)$, the proof of which is straightforward using Equation (5):

Lemma 1. *Let J be a positive integer. The following conditions hold:*

1. *The characteristic and minimal polynomials $p_J(z)$ and $q_J(z)$ of the matrices $D_J(X_k)$ in (5) coincide and are given by:*

$$p_J(z) = q_J(z) = -z\left(z^2+1\right)\left(z^2+4\right)\cdots\left(z^2+J^2\right) \tag{10}$$

for $k = 1,2,3$.

2. *In the representation* D_J, *the Casimir operator* C_2 *of* $\mathfrak{so}(3)$ *is given by:*

$$C_2 = D_J(X_1)^2 + D_J(X_2)^2 + D_J(X_3)^2 = -J(J+1)\,\mathrm{Id}_{2J+1}. \tag{11}$$

We show that, up to multiplicative factors, these properties are sufficient to construct skew-symmetric matrices $R_J^I(X_k)$, such that:

$$\left[R_J^I(X_i), R_J^I(X_j)\right] = \varepsilon_{ijk} R_J^I(X_k) \tag{12}$$

holds and Equation (7) is satisfied. In particular, there is no need to consider the transition matrix U explicitly. As a starting point, for any $1 \leq \alpha \leq J$, we define the 2×2 matrices:

$$M_\alpha = \begin{pmatrix} 0 & -\alpha \\ \alpha & 0 \end{pmatrix}. \tag{13}$$

We further define the $(2J+1) \times (2J+1)$-block matrix:

$$R_J^I(X_3) = \begin{pmatrix} M_J & & & \\ & \ddots & & \\ & & M_1 & \\ & & & 0 \end{pmatrix}. \tag{14}$$

It is obvious that $R_J^I(X_3)$ belongs to $\mathfrak{so}(2J+1)$ and that the minimal and characteristic polynomials of $R_J^I(X_3)$ coincide. These polynomials are given by (10). It follows at once that $R_J^I(X_3)$ is similar to the matrices $D_J(X_k)$ for any $k = 1,2,3$. Now, to construct skew-symmetric matrices $R_J^I(X_1)$ and $R_J^I(X_2)$ satisfying (12), we consider block matrices of the type:

$$S = \begin{pmatrix} 0 & A_1 & & & & & \\ B_1 & 0 & A_2 & & & & \\ & B_2 & \ddots & & & & \\ & & & 0 & A_{J-1} & & \\ & & & B_{J-1} & 0 & -v^T \\ & & & & v & 0 \end{pmatrix}, \tag{15}$$

where A_l, B_l are 2×2 real matrices for $1 \leq l \leq J-1$ and $v = (v_1, v_2)$ is a vector. As S is assumed to be a skew-symmetric matrix, for any index l, we have:

$$B_l + A_l^T = 0. \tag{16}$$

The choice of the matrix form is motivated by the fact that each block M_l of $R_J^I(X_3)$ describes a rotation in the two-plane generated by the vectors $\{\mathbf{e}_l, \mathbf{e}_{l+1}\}$. With this block structure, it is straightforward to verify that the commutator of A_3 and S has the following structure:

$$\left[R_J^I(X_3), S\right] = \begin{pmatrix} 0 & C_1 & & & & & \\ D_1 & 0 & C_2 & & & & \\ & D_2 & \ddots & & & & \\ & & & 0 & C_{J-1} & & \\ & & & D_{J-1} & 0 & -w^T \\ & & & & w & 0 \end{pmatrix}, \tag{17}$$

where $w = (-v_2, v_1)$ and for $1 \leq l \leq J-1$ the identities:

$$C_l = M_{J+1-l}A_l - A_l M_{J-l};\ D_l = M_{J-l}B_l - B_l M_{J+1-l}. \tag{18}$$

hold. The matrix $[A_3, S]$ is still skew-symmetric, as can be easily shown using (16) and the skew-symmetry of the (2×2)-matrices M_α. For each l, we have:

$$\begin{aligned} C_l^T + D_l &= A_l^T M_{J+1-l}^T - M_{J-l}^T A_l^T + M_{J-l}B_l - B_l M_{J+1-l} \\ &= B_l M_{J+1-l} - M_{J-l}B_l + M_{J-l}B_l - B_l M_{J+1-l} = 0. \end{aligned} \tag{19}$$

As the matrix S is composed of 2×2-blocks (with the exception of the vector v), the A_l can be essentially of two types: either A_l is a diagonal matrix or it is skew-symmetric. A generic S-matrix will thus depend at most on $3J-1$ parameters. In order to facilitate the computation of representatives to describe the real representation R_J^I, we consider all blocks A_l being of the same type (by a change of basis, an equivalent matrix representative with 2×2-blocks of a different type can be obtained). Without loss of generality, we make the choice:

$$A_l = \begin{pmatrix} 0 & a_l \\ -a_l & 0 \end{pmatrix},\ 1 \leq l \leq J-1. \tag{20}$$

By Equation (16), we have $B_l = A_l$; hence, the matrix S depends on $(J+1)$ parameters. For the commutator matrix $\left[R_J^I(X_3), S\right]$, it now follows at once from (18) that:

$$C_l = \begin{pmatrix} a_l & 0 \\ 0 & a_l \end{pmatrix} = -D_l \tag{21}$$

for any $1 \leq l \leq J-1$. The blocks C_l correspond to the second possible type (diagonal) for the blocks A_l, showing that the result does not depend on the particular form chosen initially for the blocks.

If we now compute the iterated commutator $\left[S, \left[R_J^I(X_3), S\right]\right]$, we obtain a matrix having the same block structure as $R_J^I(X_3)$ and given explicitly by:

$$\left[S, \left[R_J^I(X_3), S\right]\right] = \begin{pmatrix} E_1 & & & \\ & \ddots & & \\ & & E_J & \\ & & & 0 \end{pmatrix}, \tag{22}$$

where

$$E_1 = \begin{pmatrix} 0 & -2a_1^2 \\ 2a_1^2 & 0 \end{pmatrix};\ E_k = \begin{pmatrix} 0 & 2a_{k-1}^2 - 2a_k^2 \\ -2a_{k-1}^2 + 2a_k^2 & 0 \end{pmatrix},\ 2 \leq k \leq J-1 \tag{23}$$

and

$$E_J = \begin{pmatrix} 0 & 2a_k^2 - v_1^2 - v_2^2 \\ -2a_k^2 + v_1^2 + v_2^2 & 0 \end{pmatrix}. \tag{24}$$

Assuming that the blocks A_l are given by (16), we define $R_J^I(X_1) = S$. Following Equation (12):

$$R_J^I(X_2) = \left[R_J^I(X_3), R_J^I(X_1)\right]. \tag{25}$$

As a consequence, the matrix on the right hand side of the commutator (22) must coincide with $R_J^I(X_3)$. Comparing the entries leads to the quadratic system:

$$
\begin{aligned}
J &= 2a_1^2, \\
J - l &= 2\left(a_l^2 - a_{l-1}^2\right), \quad 2 \le l \le J-2 \\
v_1^2 + v_2^2 - 2a_{J-1}^2 &= 1.
\end{aligned} \tag{26}
$$

Up to the sign, the solution to this system is given by:

$$
a_l = \pm\sqrt{\frac{2l\,J - l(l-1)}{4}},\ 1 \le l \le J-1;\ v_1 = \pm\sqrt{\frac{J(J+1)}{2} - v_2^2}, \tag{27}
$$

where $v_2 \le \sqrt{\frac{J(J+1)}{2}}$ is free. This shows that the matrices $R_J^I(X_k)$ transform like the $\mathfrak{so}(3)$ generators (4). As these matrices must satisfy the similarity Condition (7) with the matrices (5), the Casimir operator must have the form (11). In particular, this implies that the following matrix identity must be fulfilled:

$$
R_J^I(X_1)^2 + R_J^I(X_2)^2 = \begin{pmatrix} \lambda_1 & & \\ & \ddots & \\ & & \lambda_{2J+1} \end{pmatrix}, \tag{28}
$$

where

$$
\begin{aligned}
\lambda_{2q-1} = \lambda_{2q} &= (q-1)^2 - J(2q-1), \quad 1 \le q \le J \\
\lambda_{2J+1} &= -J(J+1).
\end{aligned} \tag{29}
$$

A routine computation shows that the preceding system is satisfied identically for the values obtained in (27). Therefore, the three matrices $R_J^I(X_k)$ have (10) as their characteristic and minimal polynomial, and thus, there exists a complex matrix U transforming the matrices (5) onto the real matrices $R_J^I(X_k)$. We observe that the value of v_2 is not determined by either the commutator (12) or the Condition (28). This parameter is however inessential, as it merely indicates the possibility of considering linear combinations of the matrices $R_J^I(X_1)$ and $R_J^I(X_2)$. In fact, taking the case $J = 1$, the realization above gives the matrices:

$$
R_J^I(X_2) = \begin{pmatrix} & & -\sqrt{1-v_2^2} \\ & & v_2 \\ \sqrt{1-v_2^2} & -v_2 & \end{pmatrix},\ R_J^I(X_1) = \begin{pmatrix} & & v_2 \\ & & \sqrt{1-v_2^2} \\ -v_2 & -\sqrt{1-v_2^2} & \end{pmatrix}. \tag{30}
$$

For $v_2 = 0$, these matrices reduce to the standard rotation matrices in $\mathbb{R}^3$ corresponding to the adjoint representation of $\mathfrak{so}(3)$. For this reason, in the following, we set $v_2 = 0$ without loss of generality. As the signs in (27) can further be chosen freely, we make the following choice:

$$
a_l = \sqrt{\frac{2l\,J - l(l-1)}{4}},\ 1 \le l \le J-1;\ v_1 = \sqrt{\frac{J(J+1)}{2}}. \tag{31}
$$

The matrices $R_J^I(X_k)$ constructed with these values satisfy Equation (7) and clearly belong to $\mathfrak{so}(2J+1)$, showing that the linear map:

$$
\varphi_J : \mathfrak{so}(3) \to \mathfrak{so}(2J+1); \quad X_k \mapsto R_J^I(X_k) \tag{32}
$$

defines a Lie algebra homomorphism and an irreducible embedding. We observe that choosing different signs for the parameters a_l gives rise to an embedding belonging to the same conjugation class in $\mathfrak{so}(2J+1)$.

Let $\{\mathbf{e}_1, \cdots, \mathbf{e}_{2J+1}\}$ denote a basis of the representation space of the real representation R_J^I. Further, let $\left[\frac{n}{2}\right]$ denote the integer part of $\frac{n}{2}$. Then, the matrix elements are easily described in terms of the coefficients in (31) as:

$$\begin{aligned}\left\langle \mathbf{e}^k \middle| R_J^I(X_1) \middle| \mathbf{e}_l \right\rangle &= \left(\tfrac{1+(-1)^{k-1}}{2}\right)\left(\delta_{k+3}^l\, a_{\left(\left[\frac{k+1}{2}\right]\right)} + \delta_k^{l+1} a_{\left(\left[\frac{k-1}{2}\right]\right)}\right) - \left(a_J + \sqrt{\tfrac{J^2+J}{2}}\right)\times \\ &\quad \left(\delta_{2J+1}^l \delta_k^{2J} - \delta_{2J}^l \delta_k^{2J+1}\right) - \left(\tfrac{1+(-1)^k}{2}\right)\left(\delta_{k+1}^l a_{\left(\left[\frac{k}{2}\right]\right)} + \delta_k^{l+3} a_{\left(\left[\frac{k-2}{2}\right]\right)}\right).\end{aligned} \tag{33}$$

$$\left\langle \mathbf{e}^k \middle| R_J^I(X_2) \middle| \mathbf{e}_l \right\rangle = \delta_{k+2}^l\, a_{\left(\left[\frac{k+1}{2}\right]\right)} - \delta_k^{l+2} a_{\left(\left[\frac{k-1}{2}\right]\right)} - \left(a_J + \sqrt{\frac{J^2+J}{2}}\right)\left(\delta_{2J+1}^l \delta_k^{2J-1} - \delta_{2J-1}^l \delta_k^{2J+1}\right). \tag{34}$$

$$\left\langle \mathbf{e}^k \middle| R_J^I(X_3) \middle| \mathbf{e}_l \right\rangle = \frac{\left(1+(-1)^k\right)\delta_k^{l+1}(2J+2-k) + \left((-1)^k - 1\right)\delta_l^{k+1}(2J+1-k)}{4}, \tag{35}$$

where $1 \leq k, l \leq 2J+1$.

The first non-trivial case for which the method applies is $J = 2$ in dimension five. According to (5), the complex matrices of the irreducible representation D_2 are given by the diagonal matrix $D_2(X_1) = \Delta(2\mathrm{i}, \mathrm{i}, 0, -\mathrm{i}, -2\mathrm{i})$ and:

$$D_2(X_2) = \begin{pmatrix} 0 & 2 & 0 & 0 & 0 \\ -\frac{1}{2} & 0 & \frac{3}{2} & 0 & 0 \\ 0 & -1 & 0 & 1 & 0 \\ 0 & 0 & -\frac{3}{2} & 0 & \frac{1}{2} \\ 0 & 0 & 0 & -2 & 0 \end{pmatrix}, \quad D_2(X_3) = \begin{pmatrix} 0 & 2\mathrm{i} & 0 & 0 & 0 \\ \frac{\mathrm{i}}{2} & 0 & \frac{3\mathrm{i}}{2} & 0 & 0 \\ 0 & \mathrm{i} & 0 & \mathrm{i} & 0 \\ 0 & 0 & \frac{3\mathrm{i}}{2} & 0 & \frac{\mathrm{i}}{2} \\ 0 & 0 & 0 & 2\mathrm{i} & 0 \end{pmatrix}.$$

In this form, however, the matrices are not skew-symmetric, and hence, the properties of the representation are not easily recognized. Using the matrix elements deduced in (33)–(35), we can easily construct the corresponding real matrices $R_2^I(X_k)$. Their explicit expression is:

$$R_2^I(X_1) = \begin{pmatrix} 0 & 0 & 0 & 1 & 0 \\ 0 & 0 & -1 & 0 & 0 \\ 0 & 1 & 0 & 0 & 0 \\ -1 & 0 & 0 & 0 & \sqrt{3} \\ 0 & 0 & 0 & -\sqrt{3} & 0 \end{pmatrix}, \quad R_2^I(X_2) = \begin{pmatrix} 0 & 0 & 1 & 0 & 0 \\ 0 & 0 & 0 & 1 & 0 \\ -1 & 0 & 0 & 0 & -\sqrt{3} \\ 0 & -1 & 0 & 0 & 0 \\ 0 & 0 & \sqrt{3} & 0 & 0 \end{pmatrix}, \quad R_2^I(X_3) = \begin{pmatrix} 0 & -2 & 0 & 0 & 0 \\ 2 & 0 & 0 & 0 & 0 \\ 0 & 0 & 0 & -1 & 0 \\ 0 & 0 & 1 & 0 & 0 \\ 0 & 0 & 0 & 0 & 0 \end{pmatrix}. \tag{36}$$

These matrices are linear combinations of the basis elements of the compact orthogonal Lie algebra $\mathfrak{so}(5)$, hence defining an embedding $\mathfrak{so}(3) \subset \mathfrak{so}(5)$. If, moreover, $\{\mathbf{e}_1, \cdots, \mathbf{e}_5\}$ denotes the canonical basis of the representation space, we can easily check that:

$$\begin{gathered} R_2^I(X_1)\mathbf{e}_1 = -\mathbf{e}_4,\, R_2^I(X_2)\mathbf{e}_1 = -\mathbf{e}_3,\, R_3^I(X_1)\mathbf{e}_1 = 2\mathbf{e}_2, \\ \left(R_2^I(X_1)\right)^2\mathbf{e}_1 = -\mathbf{e}_1 + \sqrt{3}\mathbf{e}_5,\ \left(R_2^I(X_2)\right)^2\mathbf{e}_1 = -\mathbf{e}_1 - \sqrt{3}\mathbf{e}_5, \end{gathered}$$

showing that the action of $\mathfrak{so}(3)$ is actually irreducible. It is routine to check that for $1 \leq j \leq 3$, the similarity relation $R_2^I(X_j) = U\, D_2(X_j)\, U^{-1}$ is satisfied for the transition matrix:

$$U = \begin{pmatrix} \frac{1}{2\sqrt{3}} & 0 & \sqrt{3} & 0 & \frac{\sqrt{3}}{2} \\ 0 & \frac{-2\mathrm{i}}{\sqrt{3}} & 0 & \frac{-2\mathrm{i}}{\sqrt{3}} & 0 \\ 0 & \frac{-2}{\sqrt{3}} & 0 & \frac{2}{\sqrt{3}} & 0 \\ \frac{\mathrm{i}}{\sqrt{3}} & 0 & 0 & 0 & \frac{-\mathrm{i}}{\sqrt{3}} \\ -\frac{1}{2} & 0 & 1 & 0 & -\frac{1}{2} \end{pmatrix}.$$

3. Construction of the Matrices $R_J^{II}(X_k)$

In contrast to the case of integer J, the matrices $D_J^{II}(X_k)$ are already given over the reals, as a consequence of the dimension doubling in the representation space. It is straightforward to see that the matrices $D_J^{II}(X_k)$ can be written in terms of tensor products as:

$$D_J^{II}(X_k) = \begin{pmatrix} 1 & 0 \\ 0 & 1 \end{pmatrix} \otimes \mathrm{Re}D_J(X_k) + \begin{pmatrix} 0 & -1 \\ 1 & 0 \end{pmatrix} \otimes \mathrm{Im}D_J(X_k). \tag{37}$$

We observe that $D_J^{II}(X_1)$ is skew-symmetric by construction, as $D_J(X_1)$ is diagonal with purely imaginary entries. In general, however, $D_J^{II}(X_2)$ and $D_J^{II}(X_3)$ are not skew-symmetric, and therefore, the representation is not given in terms of elements belonging to the (compact) Lie algebra $\mathfrak{so}(4J+2)$. The two properties required to construct the skew-symmetric matrices realizing the representation $R_{\frac{J}{2}}^{II}$ are again the characteristic polynomial and the eigenvalue of the Casimir operator. The procedure to find such matrices is formally very similar to the previous case, up to the necessary modifications due to the tensor product (37). For this reason, we merely indicate the mains steps, skipping the detailed computations.

For any $\frac{J}{2} \in \frac{1}{2}\mathbb{N}$, the characteristic and minimal polynomials of $D_J^{II}(X_k)$ are respectively given by:

$$p_J(z) = \frac{1}{2^{2J+2}}\left(1+4z^2\right)^2\left(9+4z^2\right)^2\cdots\left(J^2+4z^2\right)^2,\ q_J(z) = \sqrt{p_J(x)}. \tag{38}$$

The eigenvalue of the Casimir operator on such a representation is given by:

$$C_2\left(D_J^{II}\right) = -\frac{J(J+1)}{4}\,\mathrm{Id}_{2J+1}. \tag{39}$$

In this case, the 2×2-matrices to start from are of the type:

$$N_\beta = \begin{pmatrix} 0 & -\frac{\beta}{2} \\ \frac{\beta}{2} & 0 \end{pmatrix} \tag{40}$$

where $1 \leq \beta \leq J$ is an odd integer. With these blocks, we define the $(4J+2)\times(4J+2)$-block matrix:

$$R_J^{II}(X_3) = \begin{pmatrix} N_J & & & & & \\ & \ddots & & & & \\ & & N_1 & & & \\ & & & -N_1 & & \\ & & & & \ddots & \\ & & & & & -N_J \end{pmatrix}. \tag{41}$$

For this rotation matrix, it is easy to verify that the characteristic and minimal polynomials satisfy Equation (38). Next, we consider matrices of the type:

$$S = \begin{pmatrix} 0 & A_1 & & & \\ -A_1^T & 0 & A_2 & & \\ & -A_2^T & \ddots & & \\ & & & 0 & A_J \\ & & & -A_J^T & 0 \end{pmatrix}, \tag{42}$$

where the A_l are 2×2-matrices. We observe that, without loss of generality, these can be taken as in (16). Repeating the same argument as for the integer case, the commutator $\left[R^{II}_{\frac{J}{2}}(X_3), S\right]$ is a skew-symmetric matrix having the same block structure as (42). We thus define the matrix $R^{II}_{\frac{J}{2}}(X_1) = S$ and also $R^{II}_{\frac{J}{2}}(X_2) = \left[R^{II}_{\frac{J}{2}}(X_3), S\right]$. Developing explicitly the commutators of these matrices, it can be proven easily that the A_l-blocks satisfy the constraint:

$$A_l + A_{J-l} = 0, 1 \leq l \leq \left[\frac{J}{2}\right]. \tag{43}$$

Hence, the number of parameters for a generic matrix S is bounded by $3\left(\left[\frac{J}{2}\right] + 1\right)$. Now, imposing the condition $\left[S, \left[R^{II}_{\frac{J}{2}}(X_3), S\right]\right] = R^{II}_{\frac{J}{2}}(X_3)$, we are again led to a quadratic system in the coefficients of N_β and A_l. In this case, however, the solution can be computed up to the sign, and no free parameters appear (this is a consequence of the constraint (43)).

Making, e.g., the choice of skew-symmetric blocks A_l and fixing the positive sign for the solution of the quadratic system, the matrix elements of $R^{II}_{\frac{J}{2}}(X_k)$ for $k = 1, 2, 3$ are given by the formulae:

$$\begin{aligned}\left\langle \mathbf{e}^k \middle| R^{II}_{\frac{J}{2}}(X_1) \middle| \mathbf{e}_l \right\rangle &= \left(\tfrac{1+(-1)^{k-1}}{2}\right)\left(\delta^l_{k+3}\, a_{([\frac{k+1}{2}])} + \delta^{l+1}_k a_{([\frac{k-1}{2}])}\right) - \left(\tfrac{1+(-1)^k}{2}\right) \times \\ &\quad \left(\delta^l_{k+1} a_{([\frac{k}{2}])} + \delta^{l+3}_k a_{([\frac{k-2}{2}])}\right)\end{aligned} \tag{44}$$

$$\left\langle \mathbf{e}^k \middle| R^{II}_{\frac{J}{2}}(X_2) \middle| \mathbf{e}_l \right\rangle = \delta^l_{k+2}\, a_{([\frac{k+1}{2}])} - \delta^{l+2}_k a_{([\frac{k-1}{2}])}. \tag{45}$$

$$\left\langle \mathbf{e}^k \middle| R^{II}_{\frac{J}{2}}(X_3) \middle| \mathbf{e}_l \right\rangle = \frac{\left(1 + (-1)^k\right)\delta^{l+1}_k (2J + 2 - k) + \left((-1)^k - 1\right)\delta^{k+1}_l (2J + 1 - k)}{4} \tag{46}$$

As a byproduct of the method, we remark that the matrix elements (33)–(35), as well as those in (44)–(46) provide a prescription to realize the Lie algebra $\mathfrak{so}(3)$ in terms of vectors fields in $\mathbb{R}^{2J+1}$ and $\mathbb{R}^{4J+2}$, respectively. More specifically, if M is the representation matrix of an element $Y \in \mathfrak{so}(3)$, the associated vector field $\hat{Y}$ is given by:

$$\hat{Y} := \left\langle \mathbf{e}^k \middle| M \middle| \mathbf{e}_l \right\rangle x_k \frac{\partial}{\partial x_l}. \tag{47}$$

4. Tensor Products of Real Irreducible Representations

While the tensor products of complex representations of $\mathfrak{so}(3)$ are well known and easily found by means of the formula:

$$D_J \otimes D_{J'} = D_{J+J'} \oplus \cdots \oplus D_{|J-J'|}, \tag{48}$$

for the tensor products of the real irreducible representations, the preceding formula is generally no longer valid, due to the division into the first and second class [8]. As a consequence, in general, such a tensor product will not be always multiplicity free, *i.e.*, the irreducible real representations appearing in the decomposition may have multiplicity greater than one. This is easily seen using the corresponding complexification, to which Formula (48) applies. A simple computation shows that for the tensor products of real irreducible representations R^I_J and $R^{II}_{\frac{J'}{2}}$ of $\mathfrak{so}(3)$, three possibilities are given:

1. $J, J' \in \mathbb{N}$ and $J \geq J'$:

$$R^I_J \otimes R^I_{J'} = \sum_{\alpha=0}^{2J'} R^I_{J+J'-\alpha}. \tag{49}$$

The tensor product is multiplicity free, and the irreducible factors are all of Class I. This actually corresponds exactly to the tensor product of the complex representations D_J.

2. $J \in \mathbb{N}, J' \equiv 1 \pmod 2$:

$$R_J^I \otimes R_{\frac{J'}{2}}^{II} = \sum_{\alpha=0}^{2J'} R_{\frac{|2J+J'-\alpha|}{2}}^{II}. \tag{50}$$

The irreducible factors are all of Class II and have multiplicity one; hence, the product is also multiplicity free.

3. $J, J' \equiv 1 \pmod 2$:

$$R_{\frac{J}{2}}^{II} \otimes R_{\frac{J'}{2}}^{II} = \sum_{\alpha=0}^{2J'} 4\, R_{\frac{|J+J'-\alpha|}{2}}^{I}. \tag{51}$$

As expected, in this case, the irreducible factors are all of Class I, and the tensor product is not multiplicity free. All factors have the same multiplicity $\lambda = 4$.

As follows from (38) when compared to (10), given an arbitrary matrix of a real irreducible representation of $\mathfrak{so}(3)$, its class can be immediately deduced from the characteristic polynomial. Actually, a stronger assertion can be obtained using this property. The main fact in this context is that the representation matrices of the three generators X_1, X_2, X_3 of $\mathfrak{so}(3)$ have the same characteristic and minimal polynomials. This enables us to determine easily the characteristic polynomial for any linear combination $X = \sum_{k=1}^{3} \lambda_k X_k$ and any real irreducible representation:

1. If R_J^I is a representation of first class, then $R_J^I(X)$ has characteristic polynomial:

$$p_J(z) = -z \prod_{\alpha=1}^{J} \left(z^2 + \xi\, \alpha^2\right), \tag{52}$$

where $\xi = \lambda_1^2 + \lambda_2^2 + \lambda_3^2$. Moreover, the minimal polynomial satisfies $q_J(z) = p_J(z)$.

2. If $R_{\frac{J}{2}}^{II}$ is a representation of the second class, then $R_{\frac{J}{2}}^{II}(X)$ has characteristic polynomial:

$$p_{\frac{J}{2}}(z) = \frac{1}{2^{2J+2}} \prod_{\beta=0}^{\frac{J-1}{2}} \left(4z^2 + \xi\,(2\beta+1)^2\right)^2 \tag{53}$$

where $\xi = \lambda_1^2 + \lambda_2^2 + \lambda_3^2$. In this case, $q_{\frac{J}{2}}(z) = \sqrt{p_{\frac{J}{2}}(z)}$.

It is worthy to be observed that the quadratic factor (z^2+1) must appear in any representation with integer J, while $(4z^2+1)^2$ appears for any half-integer. This implies that the common factor ξ can be easily found from the corresponding characteristic polynomial when the latter is rewritten taking into account (10) and (38). This fact further enables us to deduce the decomposition of an arbitrary real representation of $\mathfrak{so}(3)$ by simply analyzing the characteristic polynomial of a matrix within this representation. Let us inspect this fact more closely.

Let

$$R = \mu_0 R_0^I \oplus \mu_1 R_{J_1}^I \oplus \cdots \oplus \mu_r R_{J_r}^I \oplus \nu_1 R_{\frac{J_1'}{2}}^{II} \oplus \cdots \oplus \nu_s R_{\frac{J_s'}{2}}^{II} \tag{54}$$

be the decomposition of R into real irreducible factors, where μ_k, ν_l are positive integers, such that:

$$\dim R = \sum_{k=0}^{r} \mu_k (2J_k + 1) + \sum_{l=1}^{s} \nu_l (2J_l' + 2) \tag{55}$$

holds and $J_k, J'_l \neq 0$ for $k, l \neq 0$. Without loss of generality, we can suppose that $J_1 < J_2 < \cdots < J_r$ and $J'_1 < J'_2 < \cdots < J'_s$. The polynomial $p(z)$ of $R(X)$ thus factorizes as the product:

$$p(z) = p_0^{\mu_0}(z) p_{J_1}^{\mu_1}(z) \cdots p_{J_r}^{\mu_r}(z)\, p_{\frac{J'_1}{2}}(z)^{\nu_1} \cdots p_{\frac{J'_s}{2}}(z)^{\nu_s}. \quad (56)$$

As follows from (52) and (53), there exists a common factor ξ in all quadratic factors of $p(z)$. For $0 \leq \sigma \leq r$ and $1 \leq \tau \leq s$, define further:

$$m_\sigma = \sum_{k=\sigma}^{r} \mu_k;\ n_\tau = 2\sum_{l=\tau}^{s} \nu_l. \quad (57)$$

Expanding the polynomial $p(z)$, we obtain the expression:

$$\begin{aligned} p(z) &= -z^{m_0}\left(\prod_{\alpha=1}^{J_1}(z^2+\xi\,\alpha^2)\right)^{m_1}\left(\prod_{\alpha=1+J_1}^{J_2}(z^2+\xi\,\alpha^2)\right)^{m_2}\cdots\left(\prod_{\alpha=J_{r-1}+1}^{J_r}(z^2+\xi\,\alpha^2)\right)^{m_r}\times \\ &\prod_{l=1}^{s} 2^{-(2J+2)\nu_l}\left(\prod_{\beta=0}^{\frac{J'_1-1}{2}}\left(4z^2+\xi\,(2\beta+1)^2\right)\right)^{n_1}\cdots\left(\prod_{\beta=\frac{J'_{s-1}-1}{2}}^{\frac{J'_s-1}{2}}\left(4z^2+\xi\,(2\beta+1)^2\right)\right)^{n_s}. \end{aligned} \quad (58)$$

Starting from the polynomial (58), we can go backwards and deduce the precise decomposition (54) of R by merely inspecting the multiplicities of the different quadratic factors. In practice, the coefficients of the polynomial simplify, so that the factor ξ must be first deduced from the quadratic real irreducible factors, having in mind that for irreducible representations of the first class and second class, they are of the form given in (52) and (53). On the other hand, the values $J_1, \cdots, J_r$ and $\frac{J'_1}{2}, \cdots, \frac{J'_s}{2}$ of the irreducible factors are uniquely determined as the highest values in the quadratic factors $(z^2+\rho^2)$ and $(4z^2+\omega^2)$ preceding a variation in the multiplicity. Therefore, the number of irreducible factors in the decomposition of R is given by the number of different multiplicities of the quadratic factors and that of z. The corresponding multiplicity of each irreducible factor of R is easily obtained by the following prescription:

1. The multiplicity of z, given by m_0, indicates the number of irreducible factors of Class I.
2. The multiplicity of $R^I_{J_r}$ is given by m_r, whereas the multiplicity of $R^I_{J_k}$ is given by $m_k - m_{k+1}$ for $r-1 \geq k \geq 1$.
3. The multiplicity of the trivial representation R^I_0 is given by $m_0 - m_1$.
4. The multiplicity of $R^{II}_{\frac{J'_s}{2}}$ is given by $\frac{1}{2}n_s$, whereas the multiplicity of $R^{II}_{\frac{J'_l}{2}}$ is given by $\frac{n_l - m_{l+1}}{2}$ for $s-1 \geq l \geq 1$.

This proves that the essential information concerning the real irreducible factors of a real representation is codified in the factorization of the characteristic polynomial of an arbitrary matrix. This proves the following criterion:

Theorem 2. *Let R be an arbitrary real representation of $\mathfrak{so}(3)$ and $X \in \mathfrak{so}(3)$. Then, the decomposition of R as the sum of real irreducible representations is completely determined by the characteristic polynomial $p(z)$ of the matrix $R(X)$.*

As an example that illustrates the method, suppose that the matrix X belonging to a real representation R of $\mathfrak{so}(3)$ has characteristic polynomial:

$$p(z) = \lambda\left(25+2z^2\right)^8\left(225+2z^2\right)^6\left(625+2z^2\right)^4\left(1225+2z^2\right)^2, \quad (59)$$

where $\lambda \neq 0$. The exponents are $n_1 = 8$, $n_2 = 6$, $n_3 = 4$ and $n_4 = 2$; thus, it follows at once that R must be a sum of four irreducible factors of Class II, as z does not appear in the factorization of $p(z)$ into real irreducible factors. Taking into account Expression (38), the polynomial can be rewritten as:

$$p(z) = 2^{-20}\lambda\left(50+4z^2\right)^8\left(450+4z^2\right)^6\left(1250+4z^2\right)^4\left(2450+4z^2\right)^2, \tag{60}$$

Hence, we can extract the common factor $\xi = 50$. The values of J for the irreducible components are:

$$J_1^2 = 1,\, J_2^2 = \frac{450}{50} = 9,\, J_3^2 = \frac{1250}{50} = 25,\, J_4^2 = \frac{2450}{50} = 49. \tag{61}$$

On the other hand, $\nu_{J_4} = 1$, $\nu_3 = \frac{n_3-n_4}{2} = 1$, $\nu_2 = \frac{n_2-n_3}{2} = 1$ and $\nu_1 = \frac{n_1-n_2}{2} = 1$, showing that X is a matrix belonging to the representation $R^{II}_{\frac{1}{2}} \oplus R^{II}_{\frac{3}{2}} \oplus R^{II}_{\frac{5}{2}} \oplus R^{II}_{\frac{7}{2}}$.

5. Conclusions

By means of elementary techniques of Lie algebras and matrix theory, explicit formulae to construct real matrices of real irreducible representations of the first and second class of the compact Lie algebra $\mathfrak{so}(3)$ have been obtained. The procedure is based on the important observation that, as a consequence of the Cartan map (3), the representation matrices of the $\mathfrak{so}(3)$-generators in an irreducible representation have the same characteristic and minimal polynomial, a fact that is not true on the usual Cartan–Weyl basis. This enables us to characterize the class of a real representation according to the structure of these polynomials. Using the latter enables one to construct skew-symmetric matrices for any irreducible real representation. The real matrices so constructed actually realize the embedding of $\mathfrak{so}(3)$ into the compact Lie algebras $\mathfrak{so}(2J+1)$ and $\mathfrak{so}(4J+2)$, respectively, depending on whether J is an integer or half-integer and, hence, corresponding to matrices of the representation subduced by the restriction of the defining representation of the orthogonal Lie algebras. As an application of the method, it has been shown that for an arbitrary real representation R of $\mathfrak{so}(3)$, the decomposition of R into irreducible factors can be deduced from the characteristic polynomial of an arbitrary matrix in the representation. This provides in particular a useful practical criterion to determine whether a given matrix belongs to an irreducible real representation.

We finally remark that the realizations in terms of vector fields (47) that are deduced from the matrix elements (33)–(35), as well as those in (44)–(46), are potentially of interest in the context of point symmetries of ordinary differential equations. Systems of ordinary differential equations have been exhaustively studied by means of the Lie method (see, e.g., [13–15] and the references therein), albeit for systems containing arbitrary functions as parameters, there still remains some work to be done. In this context, indirect approaches as that developed in [16] characterizing systems in terms of specific realizations of Lie algebras constitute an alternative procedure that can be useful for applications.

As an elementary application of the real representations of $\mathfrak{so}(3)$ to the Lie symmetry method, consider the representation R_J^I for $J = 2$. Using the prescription given in (47), the vector fields in $\mathbb{R}^5$ associated with the matrices (36) are the following:

$$\begin{aligned} \hat{X}_1 &= -x_4\frac{\partial}{\partial x_1} + x_3\frac{\partial}{\partial x_2} - x_2\frac{\partial}{\partial x_3} + (x_1 - \sqrt{3}x_5)\frac{\partial}{\partial x_4} + \sqrt{3}x_4\frac{\partial}{\partial x_5}, \\ \hat{X}_2 &= -x_3\frac{\partial}{\partial x_1} - x_4\frac{\partial}{\partial x_2} + (x_1 + \sqrt{3}x_5)\frac{\partial}{\partial x_3} + x_2\frac{\partial}{\partial x_4} - \sqrt{3}x_3\frac{\partial}{\partial x_5}, \\ \hat{X}_3 &= 2x_2\frac{\partial}{\partial x_1} - 2x_1\frac{\partial}{\partial x_2} + x_4\frac{\partial}{\partial x_3} - x_3\frac{\partial}{\partial x_4}. \end{aligned} \tag{62}$$

Now, let $\Phi(t) \neq 0$ be an arbitrary function, and consider the equations of motion:

$$\ddot{x}_i = \Phi(t)\frac{\partial V}{\partial x_i},\ 1 \leq i \leq 5 \tag{63}$$

associated with the Lagrangian:

$$L = \frac{1}{2}\left(\dot{x}_1^2 + \cdots + \dot{x}_5^2\right) + \Phi(t)\, V(x_1, \cdots, x_5), \tag{64}$$

where $V(x_1, \cdots, x_5) = \alpha_{i_1 \cdots i_5} x_1^{i_1} \cdots x_5^{i_5}$ is a homogeneous cubic polynomial. After some computation, it can be shown that the preceding vector fields are point symmetries of (63) only if $V(x_1, \cdots, x_5)$ has the following form:

$$V(x_1, \cdots, x_5) = \alpha\left(6\left(x_1^2 + x_2^2\right)x_5 + 3\left(\sqrt{3}x_1 - x_5\right)x_3^2 - 3\left(\sqrt{3}x_1 + x_5\right)x_4^2 - 2x_5^3 + 6\sqrt{3}x_2x_3x_4\right),$$

where $\alpha \in \mathbb{R}$. The realization (62) of $\mathfrak{so}(3)$ obtained from the representation R_2^J further imposes some restrictions on the existence of additional point symmetries. A generic point symmetry $Z = \xi(t, \mathbf{x})\frac{\partial}{\partial t} + \eta^j(t, \mathbf{x})\frac{\partial}{\partial x_j}$ of (63) has components:

$$\begin{array}{rc} \xi(t,\mathbf{x}) = & b_4t^2 + b_5t + b_6, \\ \eta^1(t,\mathbf{x}) = & -b_1x_4 + b_2x_3 + 2b_3x_2 + b_4t\, x_1 + \frac{1}{2}b_5x_1 + b_7x_1, \\ \eta^2(t,\mathbf{x}) = & b_1x_3 + b_2x_4 - 2b_3x_1 + b_4t\, x_2 + \frac{1}{2}b_5x_2 + b_7x_2, \\ \eta^3(t,\mathbf{x}) = & -b_1x_2 - b_2\left(x_1 + \sqrt{3}x_5\right) + b_3x_4 + b_4t\, x_3 + \frac{1}{2}b_5x_3 + b_7x_3, \\ \eta^4(t,\mathbf{x}) = & b_1\left(x_1 - \sqrt{3}x_5\right) - b_2x_2 - b_3x_3 + b_4t\, x_4 + \frac{1}{2}b_5x_4 + b_7x_4, \\ \eta^5(t,\mathbf{x}) = & \sqrt{3}b_1x_4 + \sqrt{3}b_2x_3 + b_4t\, x_5 + \frac{1}{2}b_5x_5 + b_7x_5 \end{array} \tag{65}$$

where the coefficients $b_4, \cdots, b_7$ are subjected to the constraint:

$$(10b_4t + 5b_5 + 2b_7)\Phi(t) + \left(2b_4t^2 + 2b_5t + 2b_6\right)\frac{d\Phi}{dt} = 0. \tag{66}$$

It follows that for non-constant generic functions $\Phi(t)$, the symmetry algebra is isomorphic to $\mathfrak{so}(3)$, whereas if $\Phi(t)$ satisfies the separable ordinary differential Equation (66), at most two additional point symmetries can be found. It is easily verified that if the system possesses five point symmetries (these are determined by the coefficients b_6 and $b_7 = -\frac{5}{2}b_5$, corresponding to the time translation and a scaling symmetry, respectively), then $\Phi(t)$ is necessarily a constant. It may be observed that, in any case, the symmetries generating the $\mathfrak{so}(3)$-subalgebra are also Noether symmetries. We thus conclude that for functions $\Phi(t)$ not satisfying the constraint (66), the algebras of point and Noether symmetries coincide.

For the remaining values of J, a similar ansatz as the previous one can be applied to obtain criteria that ensure that a non-linear system of ordinary differential equations exhibits an exact $\mathfrak{so}(3)$-symmetry. Work in this direction is currently in progress.

Acknowledgments: The author acknowledges the referees for useful suggestions that improved the manuscript. This work was partially supported by the research project MTM2013-43820-P of the Ministerio de Economía y Competitividad (Spain).

Conflicts of Interest: The author declares no conflict of interest.

References

1. Kachurik, I.I.; Klimyk, A.U. Matrix elements for the representations of *SO*(*n*) and *SO*$_0$(*n*,1). *Rep. Math. Phys.* **1984**, *20*, 333–346. [CrossRef]
2. Mladenova, C.D.; Mladenov, I.M. About parametric representations of *SO*(*n*) matrices and plane rotations. *AIP Conf. Proc.* **2012**, *1487*, 280–287.
3. Onishchik, A.L. *Lectures on Real Semisimple Lie Algebras and their Representations*; European Math. Soc.: Zürich, Switzerland, 2003.
4. Turkowski, P. Classification of multidimensional spacetimes. *J. Geom. Phys.* **1987**, *4*, 119–132. [CrossRef]

5. Gordienko, V.M. Matrix elements of real representations of the groups $O(3)$ and $SO(3)$. *Sibirsk. Mat. Zh.* **2002**, *43*, 51–63.
6. Campoamor-Stursberg, R. Internal labelling problem: An algorithmic procedure. *J. Phys. A Math. Theor.* **2011**, *44*, 025234:1–025234:18. [CrossRef]
7. Mendes, R.V. Deformations, stable theories and fundamental constants. *J. Phys. A Math. Gen.* **1994**, *29*, 8091–8104. [CrossRef]
8. Devanathan, V. *Angular Momentum Techniques in Quantum Mechanics*; Kluwer Academic Press: New York, NY, USA, 2002.
9. Iwahori, N. On real irreducible representations of Lie algebras. *Nagoya Math. J.* **1959**, *14*, 59–83.
10. Gel'fand, I.M.; Minlos, R.A. *Representations of the Rotation Group and the Lorentz Group and Their Applications*; Fizmatgiz: Moscow, Russia, 1958.
11. Dynkin, E.G. Semisimple subalgebras of semisimple Lie algebras. *Mat. Sbornik N.S.* **1952**, *30*, 349–462.
12. Pan, F.; Bao, L.; Zhang, Y.Z.; Draayer, J.P. Construction of basis vectors for symmetric irreducible representations of $O(5) \supset \cap O(3)$. *Eur. Phys. J. Plus* **2014**, *129*. [CrossRef]
13. Pilley, T.; Leach, P.G.L. A general approach to the symmetries of differential equations. *Probl. Nonlinear Anal. Engrg. Systems Internat. J.* **1997**, *2*, 33–39.
14. Gray, R.J. The Lie point symmetry generators admitted by systems of linear differential equations. *Proc. Royal Soc. A* **2014**, *470*. [CrossRef]
15. Ibragimov, N.H. *Elementary Lie Group Analysis of Ordinary Differential Equations*; John Wiley & Sons: New York, NY, USA, 1999.
16. Campoamor-Stursberg, R.; Guerón, J. Linearizing systems of second-order ODEs via symmetry generators spanning a simple subalgebra. *Acta Appl. Math.* **2013**, *127*, 105–115.

symmetry

MDPI

Article

Lorentz Transformations from Intrinsic Symmetries

Sheng D. Chao

Institute of Applied Mechanics, National Taiwan University, Taipei 106, Taiwan; sdchao@spring.iam.ntu.edu.tw; Tel.: +886-2-3366-5066

Academic Editor: Roman M. Cherniha
Received: 20 June 2016; Accepted: 1 September 2016; Published: 9 September 2016

Abstract: We reveal the frame-exchange space-inversion (FESI) symmetry and the frame-exchange time-inversion (FETI) symmetry in the Lorentz transformation and propose a symmetry principle stating that the space-time transformation between two inertial frames is invariant under the FESI or the FETI transformation. In combination with the principle of relativity and the presumed nature of Euclidean space and time, the symmetry principle is employed to derive the proper orthochronous Lorentz transformation without assuming the constancy of the speed of light and specific mathematical requirements (such as group property) a priori. We explicitly demonstrate that the constancy of the speed of light in all inertial frames can be derived using the velocity reciprocity property, which is a deductive consequence of the space–time homogeneity and the space isotropy. The FESI or the FETI symmetry remains to be preserved in the Galilean transformation at the non-relativistic limit. Other similar symmetry operations result in either trivial transformations or improper and/or non-orthochronous Lorentz transformations, which do not form groups.

Keywords: symmetry principle; Lorentz transformation; special relativity

PACS: 03.30.+p

1. Introduction

The importance of the Lorentz transformation (LT) in the special theory of relativity can hardly be overemphasized. Physical laws are Lorentz-covariant between two inertial frames; namely, the form of a physical law is invariant under the LT. This is called the Lorentz symmetry. The proper orthochronous LT forms a group and reduces to the Galilean transformation (GT) as the speed of light approaches infinity. The mathematical structure of the LT is simple, while the conceptual change involved in interpreting it properly is profound. This explains the relentless interest in re-deriving and re-deciphering the LT, even a century after the birth of the theory.

Einstein's original derivation of the LT [1] was based on the principle of relativity and the assumed constancy of the speed of light (Einstein's second postulate). It is now known that the second postulate is not a necessary ingredient in the axiomatic development of the theory. It has been shown, as far back as 1910s [2,3], that the LT can be derived using the velocity reciprocity property for the relative velocity of two inertial frames and a mathematical requirement of the transformation to be a one-parameter linear group [4–7]. In fact, the mathematical form of the LT was known before Einstein published his seminal paper. Pauli provided a brief historical background of the theoretical development of the LT before Einstein's 1905 paper [4]. In particular, it was Poincaré who first recognized the group property of the LT and named it after Lorentz [8]. Both the velocity reciprocity property [9,10] and the linearity property [11,12] can be deduced from the presumed space–time homogeneity and the space isotropy, which are the embedded characteristics of Euclidean space and time [13,14]. Therefore, special relativity can be formulated on a weaker base of assumptions than Einstein's, and special relativity becomes purely kinematic with no connection to any specific interactions or dynamical processes. These efforts are more than pedantic pursuits for intellectual satisfaction, but greatly extend the original scope for unifying electrodynamics and mechanics. To the present knowledge it is generally believed that the Lorentz symmetry serves as a universal principle to describe the world manifold in which all fundamental processes take place, except, perhaps, for the quantum gravity phenomena [15]. To put it in a modern context and future perspective, recent interests in reformulating the logical foundation of special relativity have been mainly invoked by the experimental search for the evidence of the Lorentz-violating effects [15–17]. For example, relaxing some presumed postulates can lead to a "Very Special Relativity" proposed by Cohen and Glashow [18–20] or an extension of special relativity by Hill and Cox [21] that is applicable to relative velocities greater than the speed of light. These extensions largely widen our scope of exploring the more fundamental side of Lorentz symmetry and give impetus to further experimental research.

A prevailing theme in the literature is to reformulate special relativity in terms of intrinsic space–time symmetry principles [22–26], where the form of the space–time transformation is invariant under the symmetry operations, and auxiliary mathematical requirements such as group property can be minimized. The practice of replacing the mathematical requirement of group property by a more fundamental symmetry principle is appealing. Not only is it more axiomatically natural from the physical point of view, but it also provides a perspective capable of admitting fundamentally new physical concepts. In this paper, we sort out the possible space–time symmetries which leave the LT invariant under the corresponding symmetry operations. One of the most surprising observations is that the LT is intrinsically related to some *discrete* space–time symmetry, while the LT itself forms the basis to describe the continuous Lorentz symmetry to gauge physical laws. We reveal two symmetry operations under which the LT is invariant; namely, (1) the frame-exchange space-inversion (FESI); $x' \leftrightarrow -x$, $c't' \leftrightarrow ct$; and (2) the frame-exchange time-inversion (FETI); $x' \leftrightarrow x$, $c't' \leftrightarrow -ct$. To best demonstrate the utility of the proposed symmetry principle, we re-derive the LT without assuming the mathematical group property a priori. We show that either the FESI or the FETI can lead to the proper orthochronous LT, and both are preserved for the GT. This is the main contribution of this work. Additionally, as has been known for a long time, we demonstrate that the second postulate can be obtained by using the velocity reciprocity property [9,10]. Moreover, the necessary condition for physical causality is shown to be a deductive consequence of this symmetry principle.

2. Derivation of the Lorentz Transformation

Originally motivated, we start with Einstein's simple derivation of the LT in a popular science exposition of relativity in 1916, in which he employed a symmetrized form of the space–time transformation [27]. Although there have been a number of analyses on the 1905 paper [28–30], this later formulation seems to receive little attention for its implications. In our point of view, the formulation has the great advantage of providing streamlined reasoning and heuristic inspiration, and is therefore suitable for presentation of the intrinsic symmetry hidden in the LT.

Let us now proceed to derive the LT. Consider the two inertial coordinate systems K and K' depicted in Figure 1a. The x- and x'-axes of both systems are assumed to coincide permanently, and the origins of the two systems coincide at $t = 0$. If a light-ray is transmitted along the positive direction of x and x', then the propagation of the light-ray is described by $x - ct = 0$ in K and $x' - c't' = 0$ in K', respectively, where c and c' are the light-speed measurements in K and K'. Our purpose is to find a system of transformation equations connecting x, t in K and x', t' in K'. It is obvious that the simplest equations must be linear in order to account for the presumed homogeneity property of space and time, which can be formally proved [11,12,27]. Following Einstein [27], a symmetrized form of the transformation reads:

$$x' - c't' = \lambda\,(x - ct) \tag{1}$$

where λ is a constant which may depend on the constant velocity v. Similar considerations, when applying to the light-ray being transmitted along the negative direction of x and x', lead to:

$$x' + c't' = \mu (x + ct) \tag{2}$$

where μ is another constant not necessary to be the same as λ [27]. Different from Einstein's derivation, we do not use the second postulate; namely, we do not require $c' = c$ at this point. We now show that the constancy of the speed of light can be obtained from the velocity reciprocity property. For a proper observer who is "at rest" in K, K' is "moving" with a constant velocity v towards the positive x-axis. The origin of K' is specified by $x' = 0$ in Equations (1) and (2); so, we have:

$$\frac{\lambda - \mu}{\lambda + \mu} = \frac{v}{c} \tag{3}$$

Similarly (see Figure 1b), for a proper observer who is "at rest" in K', K is "moving" with a constant velocity v towards the negative x'-axis (the velocity reciprocity property) [9,10]. The origin of K is specified by $x = 0$ in Equations (1) and (2), so we have:

$$\frac{\lambda - \mu}{\lambda + \mu} = \frac{v}{c'} \tag{4}$$

Equations (3) and (4) lead to $c' = c$. As we mentioned, this fact has been known for a long time. Here we explicitly demonstrate it.

To determine the specific form of the transformation, we employ the symmetry principle. If Equations (1) and (2) are invariant under the FESI or the FETI transformation, it is found that λ and μ are mutually in inverse proportion to each other:

$$\lambda\mu = 1 \tag{5}$$

For example, applying the FETI symmetry on Equation (1), we obtain $x + ct = \lambda(x' + c't')$. Using Equation (2), we have $x + ct = \lambda\mu(x + ct)$, thus yielding Equation (5). Together with Equation (3), we obtain:

$$\begin{aligned} \lambda &= \sqrt{\tfrac{c+v}{c-v}} \\ \mu &= \sqrt{\tfrac{c-v}{c+v}} \end{aligned} \tag{6}$$

and c must be greater than v in order to constrain λ and μ being real numbers. Substituting Equation (6) into (1) and (2), we obtain:

$$\begin{aligned} x' &= \tfrac{1}{\sqrt{1-v^2/c^2}} (x - vt) \\ t' &= \tfrac{1}{\sqrt{1-v^2/c^2}} \left(t - \tfrac{v}{c^2}x\right) \end{aligned} \tag{7}$$

which is the proper orthochronous Lorentz transformation.

3. Discussion

There are several interesting points to note from the above derivation of the LT. First, using the symmetrized form of the transformation and the symmetry principle, all of the mathematical formulas are essentially symmetric. This clearly gives some aesthetic satisfaction; second, this formulation demonstrates that the second postulate is not at all necessary to be assumed a priori [31–37]. The velocity reciprocity property alone leads to the constancy of the speed of light. This may justify the numerous studies re-deriving the LT by dispensing with the second postulate. On the other hand, if we did use Einstein's second postulate, Equation (5), together with Equations (1)–(3), it suffices to obtain the LT while the velocity reciprocity property, Equation (4), now becomes a deduction. However, in this way, we could not see that the existence of an invariant quantity with a dimension of speed is related to the space isotropy. As has been constantly criticized by others, the (experimentally found) constancy of light-speed is just an exhibition of the nature of space–time, but not a special property of any specific theory such as electrodynamics. It just happens that light is propagating in vacuum in this specific speed; third, the FESI or the FETI symmetry principle replaces the group assumption of the transformation in determining the functional form of the transformation parameters. The resulting LT is proper and orthochronous, and thus forms a group post priori. Finally, the necessary condition of c being a limiting velocity for the physical requirement of causality can be obtained without resorting to auxiliary postulates. It is simply the result of the requirement of the transformation parameters being real numbers.

Other similar space–time operations result in either trivial transformations or improper and/or non-orthochronous LTs, which do not form groups [38]. For example, the following operations:

$$\begin{gathered} x' \leftrightarrow x, c't' \leftrightarrow ct \\ x' \leftrightarrow -x, c't' \leftrightarrow -ct \\ x' \leftrightarrow ct, c't' \leftrightarrow x \\ x' \leftrightarrow -ct, c't' \leftrightarrow -x \end{gathered} \tag{8}$$

lead to the trivial transformation, $\lambda = \mu = 1$, while the following operations:

$$\begin{gathered} x' \leftrightarrow ct, c't' \leftrightarrow -x \\ x' \leftrightarrow -ct, c't' \leftrightarrow x \\ x' \leftrightarrow c't', x \leftrightarrow -ct \\ x' \leftrightarrow -c't', x \leftrightarrow ct \end{gathered} \tag{9}$$

result in improper and/or non-orthochronous LTs. Field [25] was able to derive the proper orthochronous LT using the space–time exchange (STE); $x' \leftrightarrow ct'$, $x \leftrightarrow ct$ (for completeness, the operation of $x' \leftrightarrow -ct'$, $x \leftrightarrow -ct$, termed STE′, is also a proper choice, although it was not discussed in the paper). Notice that these operations are performed in the *same* frames, respectively. It can be seen that Equations (1) and (2) are also invariant under the STE (or the STE′) operation. However, it has been pointed out [39] that the STE symmetry is not exactly preserved for the GT, although the "broken symmetry" has its own subtleties [40,41]. On the other hand, the FESI or FETI symmetry remains to be valid at the non-relativistic limit as $c \to \infty$, as can be shown easily.

Another compact presentation which is consistent with the FESI or the FETI symmetry utilizes an involutive form of the transformation [42–44]. Starting with a change of sign of the spatial coordinate in K only; $x \to -x$, Equations (1) and (2) read:

$$x' - c't' = -\lambda\,(x + ct) \tag{10}$$

and:

$$x' + c't' = -\mu\,(x - ct) \tag{11}$$

respectively. If one now assumes that the above equations are involutive; namely, they are invariant under the operations $x' \leftrightarrow x$ and $c't' \leftrightarrow ct$, one obtains Equation (5). Unfortunately, the resulting LT is an improper orthochronous LT, and the symmetry is not preserved for the GT (for completeness, if one starts with $t \to -t$ in K only and assumes the transformation equations are involutive, one obtains yet another improper orthochronous LT). To obtain the physically acceptable proper orthochronous LT, one has to reverse the sign of x (in K only) post priori, thus making the whole methodology ad hoc. Although this involutive formulation has the mathematical advantage of utilizing well-established matrix algebra (e.g., the transformation matrix is involutory), it is not suitable to be promoted to a physical principle.

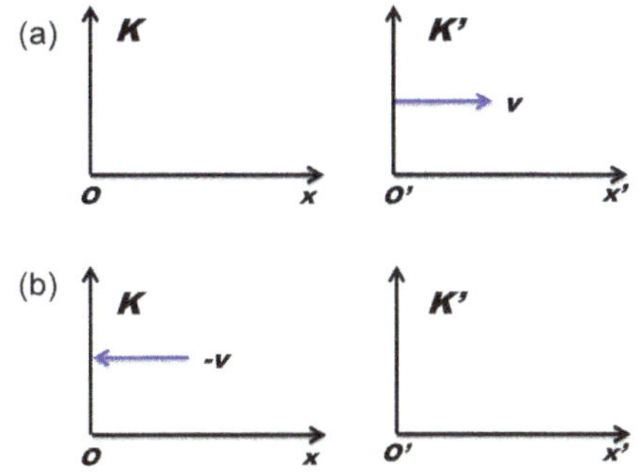

Figure 1. Two inertial frames $K(O)$ and $K'(O')$ with (**a**) K' moving with a relative constant velocity v viewed by a proper observer in K; and (**b**) K moving with a relative constant velocity $-v$ viewed by a proper observer in K'.

4. Conclusions

In concluding this paper, among the possible discrete-type space–time symmetry operations studied herein which leave the coordinate transformation between two inertial frames formally similar to the LT, we have found that only the FESI or the FETI satisfies the following two requirements: (1) the final resulting LT is proper and orthochronous and thus forms a group; and (2) the symmetry remains to be valid at the non-relativistic limit. We demonstrate the utility of the revealed symmetry principle through a derivation of the LT which closely follows the logic of Einstein in 1916. The mathematical requirement of the group property for the space–time transformation is not assumed a priori, but becomes a natural result due to the intrinsic symmetry principle of space–time.

Acknowledgments: This work was partly supported by the National Taiwan University. This work is financially supported by the Ministry of Science and Technology (MOST) of Taiwan through MOST 104-2221-E-002-032-MY3.

Conflicts of Interest: The author declares no conflict of interest.

References

1. Einstein, A. Zur elecktrodynamik bewegter korper. *Ann. Phys. (Leipzig)* **1905**, *17*, 891–921. (In German) [CrossRef]
2. Von Ignatowsky, W.A. Einige allgemeine bemerkungen zum relativitatsprinzip. *Verh. Deutch. Phys. Ges.* **1910**, *12*, 788–796. (In German)
3. Frank, P.; Rothe, H. Uber die transformation der raumzeitkoordinaten von ruhenden auf bewegte systeme. *Ann. Phys. (Leipzig)* **1911**, *34*, 825–855. (In German) [CrossRef]
4. Pauli, W. *Theory of Relativity*; Pergamon: London, UK, 1958.
5. Arzelies, H. *Relativistic Kinematics*; Pergmon: New York, NY, USA, 1966.
6. Lee, A.R.; Kalotas, T.M. Lorentz transformations from the first postulate. *Am. J. Phys.* **1975**, *43*, 434–437. [CrossRef]
7. Lėvy-Leblond, J.-M. One more derivation of the Lorentz transformation. *Am. J. Phys.* **1976**, *44*, 271–277. [CrossRef]
8. Poincaré, H. On the Dynamics of the Electron. *Comptes Rendus* **1905**, *140*, 1504–1508. (In French)
9. Berzi, V.; Gorini, V. Reciprocity principle and the Lorentz transformations. *J. Math. Phys.* **1969**, *10*, 1518–1524. [CrossRef]
10. Bacry, H.; Lėvy-Leblond, J.-M.J. Possible kinematics. *Math. Phys.* **1968**, *9*, 1605–1614. [CrossRef]
11. Eisenberg, L.J. Necessity of the linearity of relativistic transformations between inertial systems. *Am. J. Phys.* **1967**, *35*. [CrossRef]
12. Baird, L.C. Linearity of the Lorentz transformation. *Am. J. Phys.* **1976**, *44*, 167–171. [CrossRef]
13. Einstein, A. *The Principle of Relativity*; Methuen: London, UK, 1923.

14. Einstein, A. *The Meaning of Relativity*; Princeton University Press: Princeton, NJ, USA, 1955.
15. Liberati, S. Tests of Lorentz invariance: A 2013 update. *Class. Quantum Gravity* **2013**, *30*. [CrossRef]
16. Mattingly, D. Modern tests of Lorentz invariance. *Living Rev. Relativ.* **2005**, *8*. [CrossRef]
17. Kostelecky, V.A. Gravity, Lorentz violation, and the standard model. *Phys. Rev. D* **2004**, *69*, 105009. [CrossRef]
18. Cohen, A.G.; Glashow, S.L. Very special relativity. *Phys. Rev. Lett.* **2006**, *97*, 021601. [CrossRef] [PubMed]
19. Gibbons, G.W.; Gomis, J.; Pope, C.N. Generla Very special relativity is Finsler geometry. *Phys. Rev. D* **2007**, *76*, 081701. [CrossRef]
20. Bogoslovsky, G.Y. Lorentz symmetry violation without violation of relativistic symmetry. *Phys. Lett. A* **2006**, *350*, 5–10. [CrossRef]
21. Hill, J.M.; Cox, B.J. Einstein's special relativity beyond the speed of light. *Proc. R. Soc. A* **2012**, *468*. [CrossRef]
22. Brennich, H. Süssmann's deduction of Lorentz group. *Z. Naturforschung* **1969**, *24*, 1853–1854.
23. Rindler, W. *Essential Relativity*; Springer: New York, NY, USA, 1977.
24. Süssmann, G. A purely kinematical derivation of the Lorentz group. *Opt. Commun.* **2000**, *179*, 479–483. [CrossRef]
25. Field, J.H. Space–time exchange invariance: Special relativity as a symmetry principle. *Am. J. Phys.* **2001**, *69*, 569–575. [CrossRef]
26. Friedman, Y.; Gofman, Y. Relativistic linear spacetime transformations based on symmetry. *Found. Phys.* **2002**, *32*, 1717–1736. [CrossRef]
27. Einstein, A. *Relativity: The Special and General Theory*; Lawson, R.W., Translator; Crown: New York, NY, USA, 1961; Appendix I; pp. 115–120.
28. Miller, A.I. *Albert Einstein's Special Relativity: Emergence (1905) and Early Interpretation (1905–1911)*; Addison-Wesley: Reading, MA, USA, 1981; pp. 207–219.
29. Martinez, A.A. Kinematic subtleties in Einstein's first derivation of the Lorentz transformations. *Am. J. Phys.* **2004**, *72*, 790–798. [CrossRef]
30. Rynasiewicz, R. The optics and electrodynamics of 'On the electrodynamics of moving bodies'. *Ann. Phys. (Leipzig)* **2005**, *14* (Suppl. S1), 38–57. [CrossRef]
31. Mermin, N.D. Relativity without light. *Am. J. Phys.* **1984**, *52*, 119–124. [CrossRef]
32. Schwartz, H.M. Deduction of the general Lorentz transformations from a set of necessary assumptions. *Am. J. Phys.* **1984**, *52*, 346–350. [CrossRef]
33. Singh, S. Lorentz transformations in Mermin's relativity without light. *Am. J. Phys.* **1986**, *54*, 183–184. [CrossRef]
34. Schwartz, H.M. A simple new approach to the deduction of the Lorentz transformations. *Am. J. Phys.* **1985**, *53*, 1007–1008. [CrossRef]
35. Lucas, J.R.; Hodgson, P.E. *Space Time and Electromagnetism*; Oxford University Press: Oxford, UK, 1990.
36. Field, J.H. A new kinematic derivation of the Lorentz transformation and the particle description of light. *Helv. Phys. Acta* **1997**, *70*, 542–564.
37. Coleman, B. A dual first-postulate basis for special relativity. *Eur. J. Phys.* **2003**, *24*, 301–313. [CrossRef]
38. Wigner, E.P. On unitary representations of the inhomogeneous Lorentz group. *Ann. Math.* **1939**, *40*, 149–204. [CrossRef]
39. De Lange, O.L. Comment on "Space–time exchange invariance: Special relativity as a symmetry principle," by J.H. Field [Am. J. Phys. 69 (5), 569–575 (2001)]. *Am. J. Phys.* **2002**, *70*, 78–79. [CrossRef]
40. Field, J.H. Space–time symmetry is broken. *Fund. J. Mod. Phys.* **2015**, *8*, 25–34.
41. Field, J.H. Differential equations, Newton's laws of motion and relativity. *Fund. J. Mod. Phys.* **2015**, *8*, 147–162.
42. Süssmann, G. Foundation of Lorentzian groups on symmetry-assumptions and relativity-assumptions alone. *Z. Naturforschung* **1969**, *24*, 495–498.
43. Fowles, G.R. Self-inverse form of the Lorentz transformation. *Am. J. Phys.* **1977**, *45*, 675–676. [CrossRef]
44. Cook, R.J. Comment on "Self-inversion form of the Lorentz transformation". *Am. J. Phys.* **1979**, *47*, 117–118. [CrossRef]

MDPI AG
St. Alban-Anlage 66
4052 Basel, Switzerland
Tel. +41 61 683 77 34
Fax +41 61 302 89 18
http://www.mdpi.com

Symmetry Editorial Office
E-mail: symmetry@mdpi.com
http://www.mdpi.com/journal/symmetry

www.ingramcontent.com/pod-product-compliance
Lightning Source LLC
Chambersburg PA
CBHW051705210326
41597CB00032B/5379
* 9 7 8 3 0 3 8 4 2 5 2 6 7 *